THE IDENTIFICATION OF DARK MATTER

Proceedings of the Fifth International Workshop on

THE IDENTIFICATION OF DARK MATTER

Edinburgh, UK   6–10 September 2004

edited by

Neil J. C. Spooner
Vitaly Kudryavtsev

University of Sheffield

World Scientific

NEW JERSEY • LONDON • SINGAPORE • BEIJING • SHANGHAI • HONG KONG • TAIPEI • CHENNAI

Published by

World Scientific Publishing Co. Pte. Ltd.
5 Toh Tuck Link, Singapore 596224
USA office: 27 Warren Street, Suite 401-402, Hackensack, NJ 07601
UK office: 57 Shelton Street, Covent Garden, London WC2H 9HE

British Library Cataloguing-in-Publication Data
A catalogue record for this book is available from the British Library.

THE IDENTIFICATION OF DARK MATTER
Proceedings of the 5th International Workshop

Copyright © 2005 by World Scientific Publishing Co. Pte. Ltd.

All rights reserved. This book, or parts thereof, may not be reproduced in any form or by any means, electronic or mechanical, including photocopying, recording or any information storage and retrieval system now known or to be invented, without written permission from the Publisher.

For photocopying of material in this volume, please pay a copying fee through the Copyright Clearance Center, Inc., 222 Rosewood Drive, Danvers, MA 01923, USA. In this case permission to photocopy is not required from the publisher.

ISBN 981-256-344-X

Printed in Singapore by B & JO Enterprise

Preface

This book contains written versions of the presentations made at the 5th International Workshop on the Identification of Dark Matter (IDM2004) held in Edinburgh, Scotland on 6th to 10th September 2004. The objective of the *Identification of Dark Matter* workshop series, started in 1996 with a meeting in Sheffield (IDM96), is to assess critically the status of work trying to *identify* what constitutes the dark matter in the Universe. In particular, to consider what techniques, both observational and experimental, are currently being used; how successful these are and what new techniques are likely to improve prospects for identifying possible dark matter candidates in the future. Special emphasis was placed on recent results obtained in searches for baryonic and non-baryonic dark matter. Following the format adopted at IDM96, IDM98, IDM2000 and IDM2002 the meeting included reviews on major topics related to dark matter, but was largely devoted to short contributed talks.

A general aim of the workshop was to bring together in one dedicated meeting astronomers and particle physicists working specifically in the dark matter field. For instance, those working on microlensing searches for MACHOs, direct and indirect searches for WIMPs, searches for axions, searches for dark matter candidates at accelerators, as well as many other areas. People presented original work or informed reviews of the subject. Astrophysics and particle physics sessions interleaved to provide an interdisciplinary atmosphere.

Several social events were held as an antidote to the science. A reception was held at the National Galleries of Scotland. The workshop banquet was held at the Signet Library. A public talk was given by Peter Kalmus in the National Museum.

Many thanks must go to the Department of Physics and Astronomy at the University of Sheffield and the School of Physics at the University of Edinburgh for their strenuous efforts in making the workshop run so smoothly. Particular thanks to Dr. Matt Robinson for organizing the computing facilities. Sponsorship from the Particle Physics and Astronomy Research Council, Institute of Physics and the companies who participated in the industrial exhibition was very welcome. Finally we thank all the delegates, without whom there would have been no workshop or proceedings.

Neil Spooner

International Scientific Committee

F. Avignone (USC)
R. Bernabei (Rome)
B. Carr (QMW)
D. Cline (UCLA)
J. Ellis (CERN)
K. Freese (Michigan)
H. Klapdor-Kleingrothaus (MPI)
L. Krauss (Case Western)
J. Peacock (Edinburgh)
M. Rowan-Robinson (ICL)
P. Sikivie (Florida)
N. Smith (RAL)
T. Sumner (ICL)

L. Bergstrom (Stockholm)
V. Berezinsky (LNGS)
J. Carr (CPPM)
M. Dress (Munich)
E. Fiorini (Milano)
C. Frenk (Durham)
G. Kane (Michigan)
P. Nath (NEU)
J. Quenby (ICL)
L. Roszkowski (Sheffield)
J. Silk (Oxford)
N. Spooner (Sheffield)
K. Van Bibber (LLNL)

Local Organising Committee

M. Carson
J. Davies
J. Kirkpartick
P. Lightfoot
A. Murphy
N. Spooner

S. Cartwright
C. Ghag
V. Kudryavtsev
J. Milner
M. Robinson
L. Thompson

Session Chairmen

K. Olive
R. Schild
R. Gaitskell
E. Baltz
S. Cartwright
P. Sikivie
Y. Ramachers
A. Goobar
G. Gerbier
E. Daw
B. Cabrera

J. Jochum
J-L. Vuilleumier
K. Eitel
V. Kudryavtsev
M. Kachelriess
L. Widrow
L. Roszkowski
J. Edsjö
I. Irastorza
P. Belli

CONTENTS

Preface   v

SESSION A. DARK MATTER IN THE UNIVERSE — THEORY AND OBSERVATION

Session A1: Cosmology, Large Scale Structure and Dark Energy

Six Years of Dark Energy: Present and Prospects   1
    A. Goobar

Studying Dark Matter with CMB + LSS[†]   –
    J. Peacock

Probing the Dark Universe with Weak Gravitational Lensing[†]   –
    A. Taylor

Locked Quintessence and Cold Dark Matter   11
    M. Axenides

The Cluster Soft Excess: New Faces of an Old Enigma   18
    R. Lieu* and J. P. D. Mittaz

XMM Observations of High Redshift X-Ray Clusters and Cosmological Interpretation   25
    A. Blanchard*, S. Vauclair, R. Sadar and the XMM-Newton Ω-Project Collaboration

Dark Energy and Non-Linear Perturbations   31
    C. Van De Bruck* and D. F. Mota

---

*Speaker.
[†]Contribution not received.

Cosmological Singularities and Dark Energy	37
  W. Piechocki

Emergence of Space-Time Localization and Cosmic	43
Decoherence: More on Irreversible Time, Dark Energy,
Anti-Matter and Black-Holes
  A. Magnon

Relic Neutrinos as the Origin of Dark Energy[†]	–
  N. Weiner

Session A2: Halos, Halo Models and Dark Matter

Dark Matter on Galactic Scales (Or the Lack Thereof)	49
  M. R. Merrifield

Galactic Models and the Search for Dark Matter	59
  L. M. Widrow

Weighing the Dark Matter Halo	65
  J. L. Bourjaily

Cold Dark Matter Flows and Caustics	71
  P. Sikivie

Small-Scale Dark Matter Clumps	81
  V. S. Berezinsky, V. I. Dokuchaev* and Y. N. Eroshenko

The Prolate Shape of the Galactic Dark Matter Halo	87
  A. Helmi

A Bird's Eye View of M31 and its Satellite Galaxies	92
  A. McConnachie et al.

Dark Matter Halos: Shapes, the Substructure Crisis, and   98
Indirect Detection
    A. R. Zentner*, S. M. Koushiappas and S. Kazantzidis

The Power Spectrum of CDM on Sub-Galactic Scales   104
    S. Hofmann*, A. M. Green and D. J. Schwarz

Session A3: Particle Physics and SUSY

SUSY Dark Matter   110
    K. A. Olive

SUSY Dark Matter at the LHC[†]   –
    D. Miller

Gravitino Cold Dark Matter and Implications for Leptogenesis   –
and the LHC[†]
    L. Roszkowski

Supersymmetric WIMPs[†]   –
    L. Roszkowski

Searches for Supersymmetry at High Energy Colliders   122
    B. Heinemann

CP Phases, Dark Matter and the b Quark Mass   128
    M. E. Gómez*, T. Ibrahim, P. Nath and S. Skadhauge

Neutralino Dark Matter in Supergravity Theories with   135
Non-Universal Scalar and Gaugino Masses
    D. G. Cerdeño

SUSY and Dark Matter at the Linear Collider[†]   –
    G. Weiglein

Markov Chain Monte Carlo Exploration of SUSY†
   E. Baltz

SESSION B. BARYONIC SEARCHES

RIP: The MACHO Era (1974–2004)   141
   N. W. Evans* and V. Belokurov

Baryonic Dark Matter — An Outsider's View   151
   J. J. Quenby

Evidence for Dark Matter in the Form of Compact Bodies   159
   M. R. S. Hawkins

Identification of Black-Hole Dark Matter   165
   A. Cooray* and N. Seto

News from the Dark Mass at the Center of the Milky Way   171
   A. Eckart*, R. Schödel, C. Straubmeier, N. Mouaward and S. Pfalzner

Microlensing Events Towards LMC and M31   177
   P. Jetzer* and S. C. Novati

Some Consequences of the Baryonic Dark Matter Population   183
   R. E. Schild

SESSION C. NON-BARYONIC SEARCHES

Session C1: WIMP Detectors

DAMA: Results and Perspectives   189
   R. Bernabei et al.

CDMS Soudan First Results and Status   200
   M. R. Dragowsky

From EDELWEISS-I to EDELWEISS-II 206
  V. Sanglard for the EDELWEISS Collaboration

The CRESST Dark Matter Search 212
  B. Majorovits et al.

ZEPLIN I: First Limits on Nuclear Recoil Rate 218
  G. J. Alner et al.

Limits on WIMP Cross-Sections from the NAIAD Experiment 224
at Boulby
  G. J. Alner et al.

Recent Results from the SIMPLE Dark Matter Search 230
  T. A. Girard et al.

Status of KIMS Experiment Searching for WIMP with 236
CsI(Tℓ) Crystals
  J. W. Kwak et al.

DRIFT: Status and Prospects 242
  J. C. Davies and N. J. C. Spooner on behalf of the
  DRIFT Collaboration*

XMASS Experiment 248
  S. Moriyama

Status of ZEPLIN II and ZEPLIN III 254
  H. Wang

Status and Plans for the XENON Dark Matter Experiment 260
  R. J. Gaitskell on behalf of the XENON Collaboration

An Engineering Design Study for a Large-Scale Xenon 277
WIMP Detector
  E. J. Daw

CDMS Future Directions[†]
  B. Cabrera   –

Status of the ANAIS Experiment at Canfranc   283
  M. Martínez et al.

Model Independent Experiment Limits on Spin-Dependent WIMPs   289
  F. Giuliani

WIMP Direct Detection: Halo Modelling and Small Scale Structure   295
  A. M. Green

Some Issues Related to the Direct Detection of SUSY Dark Matter   301
  J. D. Vergados

Can WIMP Spin Dependent Couplings Explain DAMA?   309
  C. Savage*, K. Freese and P. Gondolo

Direct Detection of Neutralino Dark Matter in the NMSSM   315
  A. M. Teixeira

Spherical Statistics for WIMP Direct Detection   321
  B. Morgan* and A. M. Green

Dark Matter Signal and Signal Modulation for a Cold Flow   327
  F.-S. Ling

Exploiting the Materials Signature in Cryogenic WIMP Detectors   333
  H. Kraus et al.

The GENIUS-Test-Facility and the HDMS Detector in Gran Sasso   339
  H. V. Klapdor-Kleingrothaus* and I. V. Krivosheina

WARP: A WIMP Double Phase Argon Detector   348
  R. Brunetti et al.

COUPP, A Heavy-Liquid Bubble Chamber for WIMP Detection   355
    J. Bolte et al.

MIMAC-HE3: A New Detector for Non-Baryonic Dark   360
Matter Search
    D. Santos* and E. Moulin

The ULTIMA Project: Ultra-Low Temperature Instrumentation   366
for Measurements in Astrophysics
    Yu. M. Bunkov*, C. B. Winkelmann and H. Godfrin

R&D Status of the NEWAGE Experiment   372
    K. Miuchi et al.

Dark Matter Search with Direction Sensitive Scintillators   378
    H. Sekiya et al.

Performance of a Scintillating Sapphire Bolometer for the   384
ROSEBUD Experiment
    J. Amaré et al.

Development of Low Background CsI(Tℓ) Crystals   390
    H. S. Lee et al.

Properties of Liquid Rare Gas Scintillation for WIMP Searches   396
    A. Hitachi

Low Temperature Tests of Photomultipliers for Use in Liquid   402
Xenon Experiments
    R. J. Hollingworth* and J. E. McMillan

Detection of Dark Electric Matter Objects Falling Out from   408
Earth-Crossing Orbits
    E. M. Drobyshevski

Session C2: Axions

First Results from the CERN Axion Solar Telescope (CAST)   414
    S. Andriamonje et al.

PVLAS Results on Laser Production of Axion-Like Dark Matter Candidate Particles   420
    E. Zavattini et al.

Status of the Axion Dark-Matter Experiment (ADMX)   426
    D. Kinion

Session C3: Underground Laboratories

Laboratori Nazionali del Gran Sasso and the ILIAS Initiative   432
    N. Ferrari

Laboratoire Souterrain de Modane: Status and Projects at Fréjus Site   440
    G. Gerbier

The Canfranc Underground Laboratory. Present and Future   447
    J. Morales et al.

The Boulby Underground Laboratory   453
    A. S. Murphy for the UK Dark Matter Collaboration

SNOLAB   460
    D. Sinclair

DUSEL: North America Deep Underground Science and Engineering Laboratory[†]   –
    R. J. Gaitskell

Session C4: Background Studies

Neutron Studies at the Canfranc Underground Laboratory   465
G. Luzón et al.

Study of the Muon-Induced Neutron Background with the LVD Detector   471
H. Menghetti

CDMS Backgrounds and Monte Carlo Studies[†]   –
R. J. Gaitskell

Status of Neutron Background Study in CRESST   477
H. Wulandari, W. Rau, F. Von Feilitzsch and J. Jochum*

Neutron Background in a Time Projection Chamber for WIMP Searches   483
M. J. Carson et al.

Simulation of Low Energy Neutron Recoils with GEANT4   489
S. Scholl, M. Bauer and J. Jochum*

Simulations of Muon-Induced Neutron Background with GEANT4   494
M. Bauer, J. Jochum and S. Scholl*

Muon-Induced Neutron Production and Detection with GEANT4 and FLUKA   499
*H. M. Araújo and V. A. Kudryavtsev**

Update on Neutron Studies in EDELWEISS and LSM[†]   –
G. Gerbier

Veto Performance for Large-Scale Xenon Dark Matter Detectors   505
M. J. Carson et al.*

Measurement of Low Level Neutron Fluxes: Status and Prospects   511
J. E. McMillan

CRESST II Background Discrimination: Detection of ^{180}W 517
Natural Decay in a Pure α-Spectrum
  C. Cozzini* et al.

Session C5: Indirect Techniques

Indirect Detection of Neutralinos 523
  J. Edsjö

EGRET Excess of Diffuse Galactic Gamma Rays Interpreted as a 533
Signal of Dark Matter Annihilation
  W. De Boer

Indirect Dark Matter Searches with the AMS-02 Detector in Space 544
  W. De Boer for the AMS Collaboration

Search for Dark Matter with GLAST and PAMELA 551
  A. Morselli*, A. Lionetto and V. Zdravković

Dark Matter Search by High-Energy Gamma Rays and Electrons 557
with CALET
  K. Yoshida for the CALET Collaboration

The ANTARES Neutrino Telescope and its Dark Matter Capabilities 563
  J. Hößl on behalf of the ANTARES Collaboration

WMAP Excess Interpreted as WIMP Annihilation[†] –
  D. Finkbeiner

Indirect Searches for Kaluza-Klein Dark Matter 569
  D. Hooper

Superheavy Dark Matter 575
  M. Kachelriess

Seismic Moon Search for Strange Quark Matter 581
    *W. B. Banerdt, T. Chui, E. T. Herrin, D. Rosenbaum and V. L. Teplitz**

Session C6: Neutrino

Direct Neutrino Mass Experiments 587
    K. Eitel

The Search for θ_{13} with the Double-Chooz Experiment 598
    J. Jochum

NOSTOS: A New Low-Energy Neutrino Experiment 607
    S. Aune et al.

Neutrino, Other Bursts in Cosmology 615
    L. Stodolsky

Present and Future of Neutrinoless Double Beta Decay Experiments 623
    A. Nucciotti

First Evidence for Lepton Number Violation and the Majorana 633
Character of Neutrinos
    H. V. Klapdor-Kleingrothaus

Results from CUORICINO and Prospects for CUORE 647
    S. Cebrian et al.*

The EXO Double Beta Decay Experiment 653
    J.-L. Vuilleumier on behalf of the EXO Collaboration

Summary[†] –
    R. J. Gaitskell

List of Participants 659

SIX YEARS OF DARK ENERGY: PRESENT AND PROSPECTS

A. GOOBAR[*]

Physics Department,
Stockholm University, AlbaNova University Center
S-106 91 Stockholm, SWEDEN
E-mail: ariel@physto.se

Observational cosmology experienced a dramatic paradigm shift about six years ago. The measurements of magnitude vs redshift of very distant type Ia supernovae indicated that the expansion rate of the universe is increasing. The acceleration of the universe requires the existence of a 'dark energy' component that overcomes the gravitational self-attraction of matter, such as the vacuum energy density associated with the *cosmological constant* (Λ). Understanding the nature of dark energy, weather constant or dynamical, is among the most fundamental questions in contemporary science. Some of the most critical systematic effects involved in the use of Type Ia supernovae as distance indicators are discussed.

1. Cosmological parameters from Type Ia supernovae

In the *Standard Model* of cosmology the Universe started with a *Big Bang*. The expansion of an isotropic and homogeneous Universe is described by the Friedmann-Lemâitre-Robertson-Walker model (or FLRW model, for short).

The free parameters of the FLRW model are the energy contributions from radiation, matter, vacuum fluctuations and possible some yet unspecified form of dark energy. At the present epoch, the energy density in the form of radiation ρ_{rad} can be neglected in comparison with the matter density ρ_m. Furthermore, the densities of the cosmic fluids are normally expressed in units of the critical density, i.e:

$$\Omega_M \equiv \frac{8\pi G}{3H_0^2}\rho_m^0, \quad \Omega_\Lambda \equiv \frac{\Lambda}{3H_0^2} \quad \Omega_K \equiv \frac{-k}{a_0^2 H_0^2}$$

There are only two independent contributions to the energy density since in the FLRW model the sum of all energies is unity: $\Omega_M + \Omega_\Lambda + \Omega_K = 1$

---

[*]for the **Supernova Cosmology Project**

In the next sections we will use both Ω_Λ and Ω_X to refer to dark energy, the latter case being the more general, i.e. with arbitrary properties, such as time dependence.

1.1. *Cosmological parameters from "standard candles"*

A source of known strength, a *standard candle* can be used to measure relative distances to provide information on the mass density and vacuum energy density of the universe, see Goobar & Perlmutter (1995)[1]. Ω_M and Ω_X denote the present-day energy density parameters of ordinary matter and dark energy, which is characterized by the equation of state parameter, $w(z)$, where $p_X = w \cdot \rho_X$. For the specific case of the cosmological constant, $w = -1$, i.e. $p_\Lambda = -\rho_\Lambda$.

The measured magnitude of a supernova at redshift z depends on the absolute magnitude, and the luminosity distance, d_L, which in turn is a function of the cosmological parameters:

$$H_0 d_L = \begin{cases} (1+z)\frac{1}{\sqrt{-\Omega_K}} \sin(\sqrt{-\Omega_K}\, I), & \Omega_k < 0 \\ (1+z)\, I, & \Omega_K = 0 \\ (1+z)\frac{1}{\sqrt{\Omega_K}} \sinh(\sqrt{\Omega_K}\, I), & \Omega_K > 0 \end{cases} \quad (1)$$

$$I = \int_0^z \frac{dz'}{H'(z')}, \quad (2)$$

$$H'(z) = H(z)/H_0 =$$
$$\sqrt{(1+z)^3\, \Omega_M + f(z)\, \Omega_X + (1+z)^2\, \Omega_K}, \quad (3)$$

$$f(z) = \exp\left[3 \int_0^z dz' \frac{1+w(z')}{1+z'}\right], \quad (4)$$

As the measurements are performed through broad-band filters one has to K-corrected for the fact that different parts of the supernova spectrum are detected depending on the redshift z of the source, see eg. Kim, Goobar & Perlmutter (1996)[2].

1.2. *Current results*

Since the first discovery of a high-z supernova using subtractions of CCD images in the early nineties[3,4,5], the Supernova Cosmology Project (SCP) and the High-Z team (HZT)[6,7] have been successfully studying high-z supernovae with the purpose to measure the contents of the universe. The most recent SCP results, see Knop et al (2003)[8] are shown on Figure 1. Also shown are results from cross-cutting techniques; estimates of Ω_M from cluster data[9] and on the total energy content from the anisotropies in the cosmic microwave background[10]. Thus, a "cosmic concordance" model emerges suggesting $(\Omega_M, \Omega_\Lambda) \approx (0.3, 0.7)$. The results are in good agreement with the independent supernova results presented in Tonry et al. (2003)[11] and Riess et al (2004)[12].

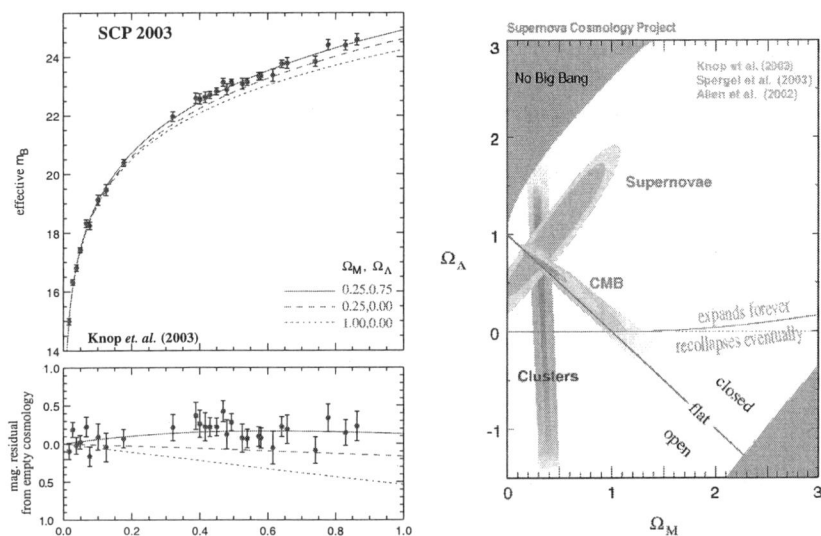

Figure 1. Left: most recent SNIa Hubble diagram from the Supernova Cosmology Project[8]. Right: 68% an 90 % statistical confidence limits from the SNIa data combined with independent measurements from Allen et al.(2002)[9] and Spergel et al.(2003)[10].

1.3. Identified systematic effects

Knop et al (2003)[8] made a very thorough study of the potential systematic effects (not included in the CL regions in Figure 1). To summarize, these include: non-Type Ia contamination, Malmquist bias, uncertainties in K-corrections and SN colors[13] and the related issues with possible extinction by dust in the host galaxy or the intergalactic medium[14,15,16], possible supernova brightness evolution or possible redshift dependence on the shape-brightness relation, gravitational lensing uncertainties[17,18], instrumental corrections, problems associated with the lightcurve fitting technique or host galaxy subtraction. Furthermore, there could be new particle physics effects, eg axion-photon oscillations[19,20,21,22]. For the considered redshift range in Knop et al. (2003)[8] the largest uncertainty stems from host galaxy extinction corrections, especially where knowledge of the intrinsic U-B color is required, as shown in Figure 2(a). Assuming a flat universe, as suggested by the CMB anisotropy measurements, Knop et al. (2003)[8] found: $\Omega_\Lambda = 0.75^{+0.06}_{-0.07} \pm 0.04$ and $w = -1.05^{+0.15}_{-0.20} \pm 0.09$, where the first errors are statistical and last are from identified systematics. Clearly, by only adding supernovae and thereby decreasing the statistical uncertainty as $1/\sqrt{N}$ will not necessarily lead to a drastic improvement in our knowledge of cosmological parameters. A viable way to make cosmological predictions minimizing the uncertainties due to extinction by dust is through observations of supernovae at larger wavelengths. Nobili (2003)[23] has built a Hubble diagram based on restframe I-band SNIa data up to z~0.5. The results are consistent with what has been observed in restframe B-band at the same redshifts, as shown in Figure 2(b). Here, instead the major problem are the low statistics.

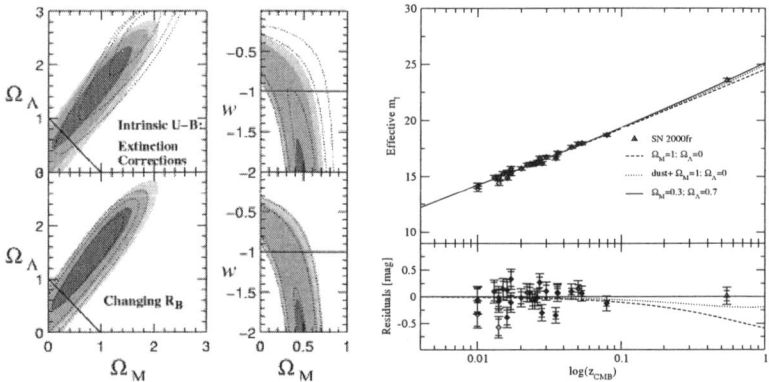

Figure 2. Left: Host galaxy extinction corrections were found to be the largest identified systematic uncertainties in[8]. The dashed lines show the bias introduced by systematic uncertainties on the intrinsic color of SNIa and possible evolution of dust properties. Right: building a Hubble diagram in restframe I-band is a viable way to avoid large corrections for host galaxy extinction[23].

1.3.1. *Brightness evolution of SNIa*

A systematic shift in the intrinsic brightness of SNIa is potentially a problematic source of uncertainty for attempts to make precision estimates of cosmological parameters. Folatelli (2004)[24] and Garavini (2004)[25] have developed techniques to quantify possible redshift dependence in the features of supernova spectra. By defining an equivalent-width like quantity, possible evolution of the spectral features in SNIa can be looked for, as shown in Figure 3. The measured equivalent widths for nine SNe in the redshift range $0.3 < z < 0.9$ were found to be consistent with the average at low redshift. Although very encouraging, larger data-sets and further theoretical studies are required to translate the results in Figure 3 into a firm upper limit of possible magnitude shift vs redshift.

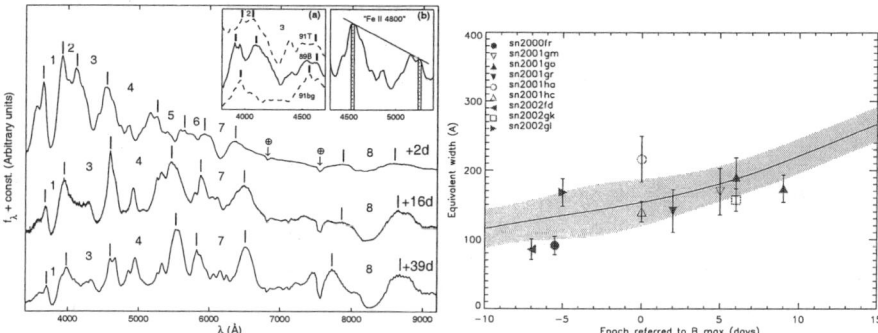

Figure 3. An equivalent-width like measure is used to explore the possibility that the spectral features evolve with redshift. The measured equivalent widths for nine SNe in the redshift range $0.3 < z < 0.9$ was found to be consistent with the average pot low-z, shown by a gray band.

2. New supernova data-sets

The available supernova data has already revolutionized our view of the universe. More supernovae, probing different redshift ranges of minimizing certain classes of systematics are already available. In what follows, some of the newly acquired supernova data by the Supernova Cosmology Project is described.

2.1. Ground based discoveries

One of the essential ingredients to tighten up our knowledge of cosmological parameters is to extend the redshift range covered[1]. For that purpose, the SCP has conducted two search runs at the 8-meter Subaru telescope using the wide field camera (SuprimeCam). Out of thirty discoveries, ten supernovae have redshifts $z > 0.9$ and were subsequently monitored with HST and four also in the NIR with VLT/ISAAC[26]. NIR follow-up provides very important cross-check points to the much wider HST/NICMOS filters, which correspond to several contiguous restframe optical filters.

2.2. Space based discoveries

During the spring of 2004, the SCP took turns with the group lead by A. Riess to follow supernovae discovered with HST/ACS with images in F850LP filter (Z-band), 45 days apart. Typically, a dozen candidates were found in each of the four search periods. However, due to the limited follow-up time for Z, J and H imaging and grism spectroscopy, only three candidates were followed for which photometric redshifts of the host galaxies indicates extremely high z ~ 1.7.

3. Challenges: dust and lensing

Measuring the brightness of SNIa at extreme high redshifts is potentially an extremely powerful way of studying the expansion history. However, weak lensing uncertainties due to the inhomogeneity of the universe rapidly increase with redshift. Amanullah, Mörtssell & Goobar (2002)[27] showed that around $z \sim 1.5$ this effect is comparable to the intrinsic spread in SNIa brightness. Moreover, due to the skewed magnification probability, especially if the fraction of compact objects in the universe is large, a non-negligible bias would result in the $\Omega_M - \Omega_\Lambda$ and $\Omega_M - w$ planes, even for several hundreds of very high redshift supernovae.

Extinction corrections for supernovae at very high-z galaxies may be problematic for (at least) two reasons. First, accurate knowledge of the intrinsic UV properties of SNe are required to use the color to deduce the reddening. Second, the extinction law for very high-z galaxies is essentially unknown. Eg, we note that the extinction corrections for the $z > 0.9$ SNe in Riess et al (2004)[12] are larger than at lower redshifts. A careful assessment of this source of error is still missing.

4. Theoretical systematics

Revealing the true nature of dark energy (DE) has become one of the most important tasks in cosmology. Thus, a model independent reconstruction of DE would be a more appealing alternative than testing all models proposed in the literature separately. In two recent papers [28,29] attempts were made by Alam et al. to reconstruct the dark energy equation of state parameter $w_{DE}(z)$ in a model independent manner using the latest supernova. In these two papers a truncated Taylor series was used to model the dark energy density $\rho_{DE}(z)$. The results indicate an evolution of $w_{DE}(z)$, a behavior they call metamorphosis. From the reported analysis, it would seem that this is a significant effect prompting for "exotic" models for dark energy.

However, Jönsson et al (2004)[30] showed that, even with seemingly 'perfect' data, the parametrization used by Alam et al. could introduce an artificial time dependence in w_{DE}. Thus, the robustness of any particularly fit must always be tested by eg Monte-Carlo techniques. SNOC[31] is public Monte-Carlo simulation package for testing the accuracy of cosmological parameter fitting, but also for investigating potentials biases due to astrophysical effects.

5. The next generation of SN experiments

Figure 4 shows how the accuracy in the magnitude–redshift method increases as supernovae at higher redshifts are added to the sample. In particular, at redshifts above $z \sim 1$ one can study the transition from acceleration to deceleration as the mass density term contribution, enhanced by the the shrinking volume as $(1 + z)^3$, overtakes the effect of Ω_Λ.

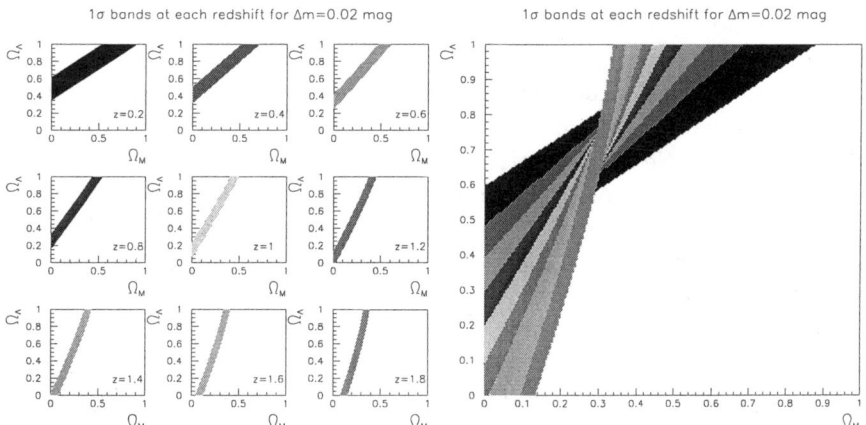

Figure 4. Left:68 % CL-regions in the $\Omega_M - \Omega_\Lambda$ parameter space defined by each redshift bin ($\Delta z = 0.2$) assuming a total uncertainty in the mean brightness of $\Delta m = 0.02$ /bin. Right:The bands are superimposed. The resulting CL region is defined by the common area.

Several projects with the aim to discover thousands of high-z supernovae are being proposed. Currently the ESSENCE and SNLS collaborations are gathering hundreds of supernovae with redshifts up to z∼1. One of the most interesting projects for the future is the SNAP satellite, a 2-m telescope equipped with an optical and NIR mosaic camera with a field of view of ∼ 0.7 square deg.

In addition of having the capability of discovering about 2500 SNe a year up to a redshift $z \sim 2$, the design of the SNAP satellite also includes an integral field spectrograph. This will allow for detailed spectroscopic studies of the supernovae and their host galaxies. Thus, systematic uncertainties on the measured supernova brightnesses are supposed to stay below 0.02 mag in which case one can expect to measure Ω_M and Ω_Λ simultaneously to about 2% and 5% respectively.

An accurate measurement of the possible time evolution of the equation of state parameter is very hard to do. Assuming a linear expansion, $w(z) = w_0 + w_1 \cdot z$, is sufficient for the small redshift range $z < 2$, one additional parameter has to be considered. Figure 5(a) (from[32]) shows the fit of simulated data corresponding to one year of the SNAP satellite. The accuracy on the estimate of the nature of the Dark Energy will depend on independent knowledge, especially, of the Ω_M from e.g. weak lensing measurements. The SNAP satellite, with is large field of view, will also provide extremely accurate measurements of cosmic shear. In addition, dedicated low-z supernova searches will be required in order to bound the intercept of the Hubble diagram, $\mathcal{M}$.

Strongly gravitationally lensed SNe could be detected in large numbers in SNAP, probably on the order of several hundred[33]. Time-delay measurements of lensed SNe are potentially interesting as they provide independent measurements of cosmological parameters, mainly H_0, but also the energy density fractions and the equation of state of dark energy. The results are independent of, and would therefore complement the Type Ia program, as shown in Figure 5(b).

6. The Nature of Dark Matter

With Type Ia supernovae it may be also possible to shed light on the nature of Dark Matter. Gravitational lensing in the inhomogeneous path that the beam of high-z supernovae follow from the source to us, affects the dispersion of the data points in the Hubble diagram. Thus, with a large sample of high-z supernovae, it is possible to measure the fraction of compact objects in the universe from the residuals of the Hubble diagram. While the compact objects are likely to be of astrophysical nature, e.g.

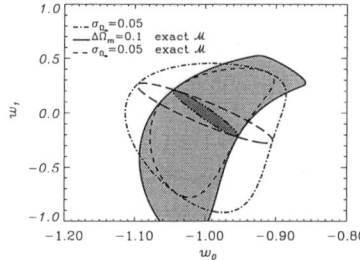 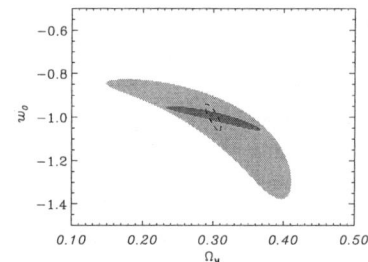

Figure 5. (a) Left: 68.3 % confidence regions for (w_0, w_1) in the one-year SNAP scenario. The elongated ellipses correspond to the assumption of exact knowledge of Ω_m: the dash-dot-dot-dotted line is with exact $\mathcal{M}$ and the long-dashed line corresponds to no knowledge of $\mathcal{M}$. The larger, non-elliptic regions assume prior knowledge of Ω_m: the dash-dotted line assumes that Ω_m is known with a Gaussian prior for which $\sigma_{\Omega_m-\text{prior}} = 0.05$; the short-dashed line assumes the same prior and exact knowledge of $\mathcal{M}$; finally, the solid line is with Ω_m confined to the interval $\Omega_m \pm 0.1$ and exact knowledge of $\mathcal{M}$. (b) Right:68.3 % CL region of $\Omega_M - w_0$ fit from lensed supernovae in the SNAP 3-year data. The dark (green) region shows the smaller confidence region that would result if h would be exactly known from independent measurements. The dashed line shows the expected statistical uncertainty from a 3 year SNAP data sample of Type Ia SNe.

faint stars or black holes, a smooth Dark Matter component would indicate that the missing mass is in the form of particles, such as the lightest stable supersymmetric particles. In [34] we used Monte-Carlo simulations to show that with one year of SNAP data, the fraction of compact objects can be measured with 5% absolute precision.

7. Summary

Type Ia supernovae have been shown to be extremely useful distance indicators. Their use lead to a dramatic paradigm shift about six years ago leading to the recognition of the need of a dominant dark energy component of the universe to explain the ensemble of cosmological data to date. Present and future SN surveys will play a critical role in studies of the nature of dark energy. However, we note that the systematic uncertainties in current estimates are within a factor two from the statistical uncertainties. Thus, future observational programs, producing large number of high-z SNe must also be able to have a significantly better control of systematics than at present.

References

1. A. Goobar & S. Perlmutter, ApJ, **450**, 14 (1995)
2. A. Kim, A. Goobar & S. Perlmutter, S., PASP, **108**, 190 (1996)
3. S. Perlmutter, et al., ApJL, **440**, 41 (1995)
4. S. Perlmutter, et al., ApJ, **483**, 565 (1997)
5. S. Perlmutter, et al., ApJ, **517**, 565 (1999)
6. B. Schmidt et al., ApJ, **507**, 46 (1998)
7. A. G. Riess, et al., ApJ, **117**,707 (1999)
8. R. Knop, et al, ApJ, **598**, 102 (2003)
9. A. W. Allen, R. W. Schmidt & A. C. Fabian, A. C., MNRAS, **334**, L11 (2002)
10. D. N. Spergel, et al., ApJS, **148**, 175S (2003)
11. J. L. Tonry, et al, ApJ, **594**, 1 (2003)
12. A. G. Riess, A. et al. ApJ, **607**, 665 (2004)
13. S. Nobili, A. Goobar, R. Knop, P.,Nugent, A&A, **404**, 901 (2003)
14. A. Aguirre, ApJ, **512**, L19 (1999)
15. A. Aguirre, ApJ, **525**, 583 (1999)
16. A: Goobar, L. Bergström & E. Mörtsell, A&A, **384**, 1 (2002)
17. E. Mörtsell, C. Gunnarsson,& A. Goobar, ApJ, **561**, 106 (2001)
18. N. Benitez, et al., ApJ, **577L**, 1 (2002)
19. C. Csaki,N. Kaloper, N. & J. Terning, Phys.Rev.Lett, **88**, 161302, (2002)
20. E. Mörtsell, L. Bergström & A. Goobar, Phys.Rev.D, **66**, 047702 (2002)
21. E. Mörtsell, & A. Goobar, JCAP, **09**, 009M (2003)
22. L. Östman & E. Mörtsell, E., astro-ph/0410501 (2004)
23. S. Nobili,PhD thesis, Stockholm University (2004)
24. G. Folatelli, PhD thesis, Stockholm University (2004)
25. G. Garavini, PhD thesis, Stockholm University (2004)
26. N. Yasuda, et al, AAS, **203**, 82.11 (2003)
27. R. Amanullah,E. Mörtsell & A. Goobar, A&A, **397**, 819 (2003)
28. U. Alam, V. Sahni, T. D. Saini & A. A. Starobinsky, astro-ph/0311364 (2003)
29. U. Alam, V. Sahni V & A. A. Starobinsky, astro-ph/0403687 (2004)
30. J. Jönsson, A. Goobar, R. Amanullah, & L. Bergström, L, JCAP, **007**, 17 (2004)
31. A. Goobar, E. Mörtsell, R. Amanullah, M. Goliath, L. Bergström, L. & T. Dahlen, A&A, **392**, 757 (2002)
32. G.Goliath, R.Amanullah, P.Astier, A.Goobar and R.Pain, A&A, **380**, 6 (2001)
33. A. Goobar, E. Mörtsell,, R. Amanullah & P. Nugent, A&A, **393**, 25 (2002)
34. E. Mörtsell, A. Goobar & L. Bergström, ApJ, **559**, 53 (2001)

LOCKED QUINTESSENCE AND COLD DARK MATTER

MINOS AXENIDES

*Institute of Nuclear Physics, National Center for Scientific Research
'Demokritos',
Agia Paraskevi Attikis, Athens 153 10, Greece*

A supersymmetric hybrid potential model with low energy supersymmetry breaking scale ($M_S \sim 1 - 10 Tev$) is presented for both dark matter and dark energy. Cold dark matter is associated with a light modulus field ($\sim 10 - 100 Mev$) undergoing coherent oscillations around a saddle point false vacuum with the presently observed energy density ($\rho_0 \sim 10^{-12} eV^4$). The latter is generated by its coupling to a light dark energy scalar field ($\sim 10^{-18} eV$) which is trapped at the origin ("locked quintessence"). Through naturally attained initial conditions the model is consistent with cosmic coincidence reproducing LCDM cosmology. An exit from the cosmic acceleration phase is estimated to occur within some eight Hubble times.

1. Introduction

There is a growing observational evidence to the fact that we live in a spatially flat Universe ($\Omega_{tot} \approx 1$) in a state of cosmic acceleration [1,2,3,4]. Most of its content, by weight ($\Omega_{tot} - \Omega_{bar} \sim 0.96$), cannot be accounted for by the standard model of particle physics. It is believed to be associated with an invisible sector of Matter and Energy of, remarkably, almost equal energy density in a cosmic coincidence. Dark Matter($\Omega_{DM} \sim 0.3$), responsible for the growth of structure in our Universe, is believed to be non-baryonic in nature with small free streaming length behaving as a non-relativistic gas (Cold Dark Matter-CDM). It is typically associated with weakly interacting massive particles (WIMPs) such as axions, axinos, neutralinos, gravitinos, string moduli and others[5].

Dark Energy ($\Omega_{DE} \sim 0.7$), on the other hand, is probably a homogeneous perfect fluid component($p \sim w\rho$) with negative pressure ($w < -\frac{1}{3}$) giving rise to the observed cosmic acceleration(for a review see [6]). In its most popular version it is attributed to the Cosmological Constant ($w = -1$) whose value must be fine tuned to an unprecedent degree to be in accordance with the observational data ($\frac{\Lambda}{8\pi G} \sim 10^{-47}$). The emerging LCDM Cosmology, although economical and succesful is not lacking of theoretical shortcomings. Indeed a constant vacuum energy inevitably leads to eternal accelerated expansion , technically implying the presence of causal horizons and hence non-existence of well defined in and out states in the formulation of the underlying quantum theory such as String theory[7].

Alternative scenarios employ dynamical scalars , such as Quintessence fields ($-1 < w < -1/3$) [8] which possess time varying energy density as they roll down their monotonically decreasing potential energies . They typically predict an exit from the present accelerating phase. Eventhough these models dispense with the theoretical problems of the Cosmological Constant scenario they dont lack unnatural fine tunnings[9] associated typically with both their initial conditions, present value and/or their small mass ($M_Q \sim 10^{-33}$ eV). In the context of supergravity theories such a light field is difficult to be understood because the flatness of its potential is lifted by excessive supergravity corrections or due to the action of non-renormalizable terms, which become important at displacements of order M_P.

Cosmic acceleration in the very early universe has been extensively studied in supersymmetric hybrid models[14]. There the vacuum energy density required in order to generate the necessary number of e-foldings is fed into the slow rolling inflaton through its coupling to a second scalar field the "waterfall" which is kept trapped along the inflaton track. In a fast-roll variation of this scenario, also dubbed "locked inflation"[11,12], the inflaton field undergoes rapid coherent oscillations around its "Saddle point" vacuum before it is displaced away from it, prolonging consequently the inflationary phase.

Interestingly it has been known for quite a while that coherent oscillations of massive (pseudo)scalar weakly interacting particles, such as the axion can mimic Cold Dark Matter ($w = 0$) [15]. We have recently produced an interacting model that realizes LCDM Cosmology by putting these two ingredients together in the very late universe [13]. Other interacting models for Dark Matter and Dark Energy can be found in Ref. [10].

Our model is a given by a standard Supersymmetric Hybrid Potential with only two characteristic energy scales : the Planck Mass ($M_{Pl} \approx 10^{19} GeV$) and a low energy SUSY breaking scale ($M_S \approx M_{3/2} \approx 1 TeV$). We assume that the dark matter particle is a modulus Φ, corresponding to a flat direction of supersymmetry. The modulus field is undergoing coherent oscillations, which are equivalent to a collection of massive Φ-particles ($M_\Phi = \frac{M_S^2}{M_{Pl}} \sim 10 - 100 MeV$), that are the required WIMPs. A second scalar field (Ψ) interacts with (Φ) in a standard way ($\lambda\Phi^2\Psi^2$). This can be thought of as our quintessence field and it corresponds to a flat direction lifted by non-renormalizable terms. Even though the Ψ-field is a light scalar ($M_\Psi \approx \frac{M_S^3}{M_{Pl}^2} \approx 10^{-18}$ eV), it is much more massive than the m_Q mentioned above, so as not to be in danger from supergravity corrections to its potential[16,17]. Our quintessence field is coupled to our dark matter in a hybrid manner, which is quite natural in the context of a supersymmetric theory. Due to this coupling, the oscillating Φ, keeps Ψ 'locked' on top of a potential hill, giving rise to the desired dark energy. When the amplitude of the Φ-oscillations decreases enough, the dark energy dominates the Universe, causing the observed accelerated expansion as dictated by the cosmic coincidence. Within some eight Hubble times, when the oscillating amplitude falls below the width of the saddle, the 'locked' quintessence field is released and rolls down to its global minimum. The system reaches the true vacuum and accelerated expansion ceases.

We assume a spatially flat Universe, according to the WMAP observations [1]. We use natural units such that $\hbar = c = 1$ and Newton's gravitational constant is $8\pi G = M_{Pl}^{-2}$, where $M_{Pl} = 10^{18}$ GeV is the reduced Planck mass.

2. A Supersymmetric Hybrid Model

Consider two real scalar fields Φ and Ψ interacting through a hybrid type of potential of the form[13]

$$V(\Phi, \Psi) = \frac{1}{2}m_\Phi^2 \Phi^2 + \frac{1}{2}\lambda\Phi^2\Psi^2 + \frac{1}{4}\alpha(\Psi^2 - M_s^2)^2, \quad (1)$$

where ($\lambda \leq 1$) and ($\alpha = \frac{M_s^4}{M_{Pl}^4}$). Dark Matter is associated with Φ and Dark Energy with Ψ. All parameters are expressed in terms of two fundamental energy scales: the Planck mass ($M_{Pl} \sim 10^{18}$ GeV) and the Susy breaking scale which is also taken to be the Gravitino mass ($M_s \sim m_{3/2} \sim 1 TeV$). They should be considered in the framework of gauge mediated supersymmetry[18]. Standard features of the potential are:
• Two global minima $(\Phi, \Psi) = (0, M_s)$ with an unstable saddle point at $(\Phi, \Psi) = (0, 0)$
• Ψ possesses a Φ dependent curvature $(m_\Psi^{\text{eff}})^2 = \lambda\Phi^2 - \alpha M_s^2)$ with a width of

$$\Phi_w = \frac{1}{\sqrt{\lambda}} \frac{M_s^3}{M_{Pl}^2} \quad (2)$$

- Cosmic Coincidence ($\Omega_{DE} - \Omega_{DM} \sim O(1)$) at present Hubble time demands small scalar masses : $m_\Phi \sim \frac{M_s^2}{M_{Pl}} \approx 10 - 100 MeV$ and $m_\psi \sim \frac{M_s^3}{M_{Pl}^2} \approx 10^{-18}$eV. The latter is conceivable to be due to accidental cancellations in the Kähler potential or some other accidental symmetry protecting m_Ψ.

Our physical system acts as a two component perfect fluid with energy density ($\rho_{tot} = \rho_\Phi + \rho_\Psi$) which gets diluted as the universe expands. When the system finds itself rolling at $\Phi \geq \Phi w$ it is energetically favorable for Ψ to be trapped in its origin $\Psi \cong 0$. Φ performs coherent oscillations in a quadratic potential

$$V(\Phi, \Psi = 0) = \frac{1}{2} m_\Phi^2 \Phi^2 + V_0. \tag{3}$$

around a saddle point false vaccuum with energy density given by

$$V_0 = \frac{1}{4} \alpha M_s^4 \sim \frac{M_s^8}{M_{Pl}^4} \sim 10^{-120} M_{Pl}^4 \tag{4}$$

the observed present vacuum energy density. It is associated with the DE condensate Ψ acting as an effective cosmological constant, i.e. behaving as a perfect fluid component with an equation of state ($p = -V_0$). In the high temperature phase the energy density is dominated by the kinetic energy of the Φ oscillations which behaves as a pressureles non-relativistic component of a collection of massive particles, hence Cold Dark Matter(CDM), which is given by :

$$\rho_\Phi = \frac{1}{2} \dot{\Phi}^2 + \frac{1}{2} m_\Phi^2 \Phi^2, \tag{5}$$

where the dot denotes derivative with respect to the cosmic time t. The model therefore identifies the following cosmic phases:
- [CDM domination] The overall density is dominated by the coherent oscillations of Φ in Eq. (5), when the oscillation amplitude is larger than

$$\Phi_\Lambda \sim \frac{\sqrt{\alpha} M^2}{m_\Phi} \sim \left(\frac{m_\Psi}{m_\Phi}\right) M_s \sim \frac{M_s^2}{M_{Pl}} \tag{6}$$

They behave as a collection of non-relativistic particles whose energy gets diluted accordingly as ($\rho_\Phi \propto R^{-3}$)
- [Locked Quintessence] The energy density is dominated by the Saddle Point Vacuum of eq.(4) for the range of Φ amplitude oscillations

$$\Phi_w < \Phi_0 < \Phi_\Lambda \tag{7}$$

The characteristic time scale that Φ spends on the saddle ($\Phi_0 < \Phi_w$) is ($\Delta t_w \sim \frac{\Phi_w}{m_\Phi \Phi_0}$). As long as it is smaller than the time scale ($\Delta t_\Psi \approx \frac{1}{m_\Psi}$)

it takes for Ψ to start to roll away from the top of the hill rapid coherent oscillations of Φ persist (Locked Quintessence). The effect is present due to the ratio of masses chosen ($\frac{m_\Phi}{m_\Psi} \approx \frac{M_{Pl}}{M_s} \approx 10^{16}$). A (quasi) de Sitter expansion phase sets in with $a \simeq a_0 \exp(H_0 \Delta t)$, where $\Delta t = t - t_0$ and $H_0 \simeq \sqrt{V_0}/\sqrt{3} M_{Pl} = $ constant. For the oscillating Φ we have $\Phi \propto \sqrt{\rho_\Phi} \propto a^{-3/2}$. We can thus obtain an estimate of the length of the cosmic acceleration phase.

• [Post-Acceleration Phase] Our two fluid system will release its stored vacuum energy when Ψ will start to roll away from the top of the hill away from its present false vacuum state into its future true vacuum of zero energy density when ($\Phi_0 = \Phi_w = \Phi_\Lambda$)

$$\Phi_w \simeq \Phi_\Lambda \exp(-\tfrac{3}{2} H_0 \Delta t_w) \Rightarrow \quad \Delta t_w \simeq \frac{2}{3}\left[\ln\left(\frac{M_{Pl}}{M_S}\right) + \ln\sqrt{\lambda}\right] H_0^{-1}, \quad (8)$$

We see that the period of acceleration may last up to wight Hubble times (e-foldings) depending on the value of λ.

3. Dark Matter and Dark Energy Requirements

• Coherent Oscillations of the modulus Φ field in a quadratic potential behave as a collection of massive non-relativistic particles. In order that we may identify them with a realistic CDM component ($\Omega \approx \tfrac{1}{3}$) they must persist until today with the Φ quanta not having decayed, namely satisfying

$$\Gamma_\Phi < H_0 , \qquad (9)$$

where $H_0 \sim \sqrt{\rho_0}/M_{Pl}$ is the Hubble parameter at present. Using that $\Gamma_\Phi \sim g_\Phi^2 m_\Phi$ we find the bound

$$m_\Phi \leq 10^{-20} M_{Pl} , \qquad (10)$$

where we used that the coupling g_Φ of Φ with its decay products lies in the range $\frac{m_\Phi}{M_{Pl}} \leq g_\Phi \leq 1$, with the lower bound corresponding to the gravitational decay of Φ, for which $\Gamma_\Phi \sim [m_\Phi^3]/M_{Pl}^2$. We may conclude that Φ has to be a rather light field with mass $\lesssim$ 10-100 MeV.

• We must require that our dark matter field Φ should not decay into Ψ-particles, through their mutual coupling, until the present time either perturbatively ($\Phi \to \phi\,\phi$) or non-perturbatively through parametric resonance. The perturbative condition reads

$$\Gamma_{\Phi \to \phi\phi} \simeq \frac{\lambda^2 \Phi_0^{\,2}}{8\pi m_\Phi} < H_0 . \qquad (11)$$

Since $\bar{\Phi} \propto a^{-3/2}$, it becomes obvious that the above constraint is the tightest in the early times after the amplitude of oscillations become ($\Phi_0 \approx \frac{m_\Phi}{\sqrt{\lambda}}$) which takes place in the radiation era. By imposing it we get an upper-bound condition for λ :

$$\lambda < \frac{m_\Phi}{M_{Pl}} \left(\frac{M_{Pl}}{T_{eq}}\right)^{2/5} \sim 10^{-19}. \qquad (12)$$

- The condition that the oscillations of Φ are dominated by V_0 of eq. 4 in the present Hubble era imply that ($\Phi_0 \leq \Phi_\Lambda$) gets to be satisfied when ($\sqrt{\lambda}\left(\frac{m_{Pl}}{M_s}\right) > 1$) which, in turn, gives us a lower bound for the coupling constant

$$\lambda > 10^{-30}. \qquad (13)$$

- The onset of Φ-oscillations must occur in the radiation era ($T > 1\text{eV}$) when ($H_{osc} \sim m_\Phi$) in the aftermath of an early phase of inflation being followed right afterwards by reheating[19]. Their fractional contribution to the energy density is ($\frac{\rho_\Phi}{\rho} \propto a \propto H^{-\frac{1}{2}}$). They eventually dominate the energy density of the Universe. By requiring this to take place at ($T_{eq} = 1\text{eV}$) we find the initial displacement of Φ to be much smaller than the Planck scale namely ($\Phi_{osc} \sim 10^{-6} M_{Pl} \ll M_{Pl}$). However the inclusion of supergravity corrections to the potential ($\Delta m_\Phi^2 \propto H(t)^2$)[20] lift the flatness of the Φ direction so that Φ begins to roll down long before ($H \sim m_\Phi$). Its motion is, however, overdamped by the excessive friction of a large Hubble parameter (compared to its mass) imposing a freeze out to the value of Φ until H is reduced enough for the quadratic oscillations to commence.
- Similar in spirit analysis can be applied to the study of the initial conditions for the Quintessence field Ψ which has to find itself near the origin ($\Psi \leq M_s$) in order to get "locked" when the Φ oscillations begin. The oscillations of Ψ begin immediately after reheating with ($\Psi \propto \sqrt{\rho_\Psi} \propto H^{3/4}$). It can be analytically demonstrated that our original assumption for ($\Psi \approx 0$) is well justified.
- The smallness of the saddle point vacuum energy does not only require a small mass for our tachyonic field Ψ but a small VEV as well ($M_s \sim 1$ TeV). This can be done through higher order non-renormalizable terms or logarithmic loop corrections[16]. Clearly the level of fine tuning implied by ($m_\Phi \sim 10^{15} H_0 \sim 10^9 H_{eq}$) is much less severe than the one required in most quintessence models ($m_Q \sim H_0$). As a consequence and in contrast to quintessence models Sugra corrections in the matter era are negligible.

4. Conclusions

We have presented a unified model of dark matter and dark energy in the context of low-scale gauge-mediated supersymmetry breaking. Our LQCDM model retains the predictions of LCDM Cosmology, while avoiding eternal acceleration and achieving coincidence without significant finetuning. The initial conditions of our model are naturally attained due to the effect of supergravity corrections to the scalar potential in the early Universe, following a period of primordial inflation. Our oscillating Φ-condensate does not have to be the dark matter necessarily. Indeed, it is quite possible that Ψ-remains locked on top of the false vacuum while ρ_Φ is negligible at present. It is easy to see that indeed ($\frac{\rho_\Phi^{min}}{\rho_0} \sim 10^{-30}\lambda^{-1}$). Depending on λ, Φ may contribute only by a small fraction to dark matter at present, while still being able to lock quintessence and cause the observed accelerated expansion at present. This option appears less appealing to us.

References

1. D. N. Spergel et al., Astrophys. J. Suppl. **148**, 175 (2003).
2. M. Tegmark et al. [SDSS Collaboration], Phys. Rev. D **69** (2004) 103501
3. M. Colless, astro-ph/0305051.
4. S. Perlmutter et al. [Supernova Cosmology Project Collaboration], Astrophys. J. **517**, 565 (1999); A. G. Riess et al. [Supernova Search Team Collaboration], Astron. J. **116**, 1009 (1998).
5. G. Jungman, M. Kamionkowski and K. Griest, Phys. Rept. **267** (1996) 195
6. P. J. E. Peebles and B. Ratra, Rev. Mod. Phys. **75**, 559 (2003).
7. W. Fischler, A. Kashani-Poor, R. McNees and S. Paban, JHEP **0107**, 003 (2001);
8. I. Zlatev, L. M. Wang and P. J. Steinhardt, Phys. Rev. Lett. **82**, 896 (1999);
9. C. F. Kolda and D. H. Lyth, Phys. Lett. B **458**, 197 (1999).
10. L. Amendola and D. Tocchini-Valentini, Phys. Rev. D **64** (2001) 043509;
11. R. Easther, J. Khoury and K. Schalm, JCAP **0406** (2004) 006
12. K. Dimopoulos and M. Axenides, arXiv:hep-ph/0310194.
13. M. Axenides and K. Dimopoulos, JCAP **0407** (2004) 010
14. A. D. Linde, Phys. Rev. D **49** (1994) 748
15. M. S. Turner, Phys. Rev. D **28**, 1243 (1983)
16. A. R. Liddle and D. H. Lyth, *Cosmological Inflation and Large-Scale Structure* (Cambridge Univ. Press, Cambridge U.K., 2000).
17. M. Dine, L. Randall and S. Thomas, Nucl. Phys. B **458**, 291 (1996); Phys. Rev. Lett. **75**, 398 (1995).
18. G. F. Giudice and R. Rattazzi, Phys. Rept. **322** (1999) 419
19. L. Kofman, A. D. Linde and A. A. Starobinsky, Phys. Rev. D **56** (1997) 3258.
20. D. H. Lyth and T. Moroi, JHEP **0405** (2004) 004.

THE CLUSTER SOFT EXCESS: NEW FACES OF AN OLD ENIGMA

R. LIEU & J. P. D. MITTAZ

Physics Department, UAH
Huntsville
AL 35899, USA
E-mail: lieur@cspar.uah.edu

Until the advent of XMM-Newton, the cluster soft excess (CSE) was the subject of some controversy due to both data analysis issues and uncertainties with the soft excess emission mechanism. XMM-Newton observations have finally laid to rest any doubts as to the existence of the CSE and have also given tantalising clues as to the nature of its emission mechanism. Here we report on the analysis of XMM-Newton observations of a number of CSE clusters in an attempt to improve the analysis and understanding of the CSE. Included as part of the study is an analysis of the effects of background subtraction, which calls to question the integrity of the claimed O VII line discovery, though not the soft excess itself. We also give details of both thermal and non-thermal fits to the CSE cluster Abell 3112.

1. Introduction

The cluster soft excess is an excess of flux seen above the hot ICM in a number of clusters first discovered by Lieu et al. (1996), but its origin still remains a mystery. Currently there are two prevalent models, one thermal, normally assumed to arise from warm (~ 0.2keV) gas, and one non-thermal. Observations with XMM-Newton have the potential to provide vital new information to resolve this issue and recently there have been claims of Oxygen lines associated with the CSE (Kaastra et al. 2003, Finoguenov, Briel & Henry 2003). If true this is very important to our understanding of the CSE since it implies that its origin must be thermal rather than non-thermal. However, there are several issues that first need to be addressed before we can be sure that the emission is indeed thermal. The most important is that of background subtraction - to date most CSE studies have used the background dataset of Lumb et al. (2002) exclusively and therefore have not taken into account Galactic background variations across the sky. As we will show, this can cause an overestimate of the CSE as well as the spurious detection of O VII.

2. Using in-situ backgrounds

To investigate if variations in the Galactic background are important to the analysis of XMM-Newton data on the CSE we must determine in-situ backgrounds i.e. we must use observations which contain uncontaminated background signal. This will either be when the clusters fit inside the XMM-Newton field of view, or offset pointing which contain both cluster and background flux, Of all the CSE clusters currently studied with XMM-Newton to data, only four fit approximately within the XMM-Newton field of view and we discuss two of these, Abell S1101 and Abell 3112. These two clusters are amongst the most important CSE clusters: Abell S1101 has the largest reported ROSAT PSPC soft excess (Bonamente et al. 2001) and Abell 3112 has one of the largest XMM-Newton soft excesses (Nevalainen et al. 2003). For both clusters the estimated contamination of the background by residual cluster emission in the 12'-14' annulus used to estimate the background is small. For these two cluster we can therefore use a double subtraction method (see for example Arnaud et al. 2002) to remove the correct background.

From the point of view of previous analyses of these clusters it is important to compare the different background subtraction techniques. We have therefore compared the Lumb et al. (2002) background used in, for example, Kaastra et al. (2003,2004) (lowest spectrum) with the double subtraction background method using an in-situ background (middle spectrum). Figure 1 shows that there is considerable difference between the Lumb background and the double subtraction background particularly in the case of Abell S1101. Using the Lumb et al. background alone can give rise to a spurious soft X-ray signal which could be confused with a soft excess. However, the model fit to the double background subtracted spectrum (shown as a solid line on the top spectra in Figure 1 together with the fit residuals show that there is still evidence for a soft excess. Therefore, even taking into account the correct Galactic background subtraction, the soft excess still persists.

3. The O VII/O VII line region

One of the key findings of the Kaastra et al. (2003,2004) works was the apparent detection of O VII in the outer parts of a number of clusters including Abell S1101. The line detection was seen at the outer regions of

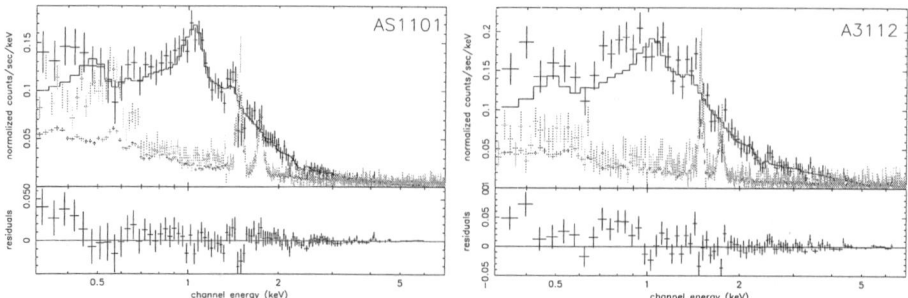

Figure 1. The top spectrum shows the background subtracted spectra from EPIC MOS1 for the 4'-6' arcminute annulus for AS1101 and A3112. The other spectra show the Lumb background (lowest spectrum) and the double subtracted background (middle spectrum). All backgrounds have been normalised to the 10-12 keV source flux. Note the strong O VIII line seen in the AS1101 background spectra. The solid line is a model for the hot ICM fitted to the 1 - 7 keV band of the background subtracted spectrum.

clusters and Kaastra et al. directly associated the emission with the CSE. They also related this emission with large extended soft X-ray emission seen in ROSAT all sky survey maps, directly linking the emission with the supercluster environment and hence to the Warm Hot Intergalactic Medium (WHIM). However, in our analysis we find no such line, although the *background* spectrum of the Abell S1101 field (12'-14') shows strong O VII emission. At first sight, the detection of this line seems to verify the claims of Kaastra et al. but this can be directly tested as the O VII is strong enough to be resolved and we can therefore determine the redshift of the line directly.

The left panel of Figure 2 shows the O VII/O VIII line region together with a model including a continuum model with two gaussians placed at the redshifted energy of the O VII/O VIII lines. It is clear that putting the Oxygen lines at the redshift of the cluster does not match the observed location of the lines. The right hand panel of Figure 2 shows the case where the lines have been put at zero (Galactic) redshift and it is clear that the lines are much more consistent with a Galactic redshift rather than the cluster redshift. Indeed, if we fit the observed line energies we get 0.569 ± 0.003 keV for O VII and 0.651 ± 0.007 keV for O VIII, completely consistent with zero (Galactic) emission and completely inconsistent with the redshifted line energies 0.538 keV and 0.618 keV. We must therefore conclude that the line emission and hence the majority of the emission at the 12'-14' annulus is Galactic in origin and therefore not associated with the cluster. This then not only verifies our use of the double subtraction method

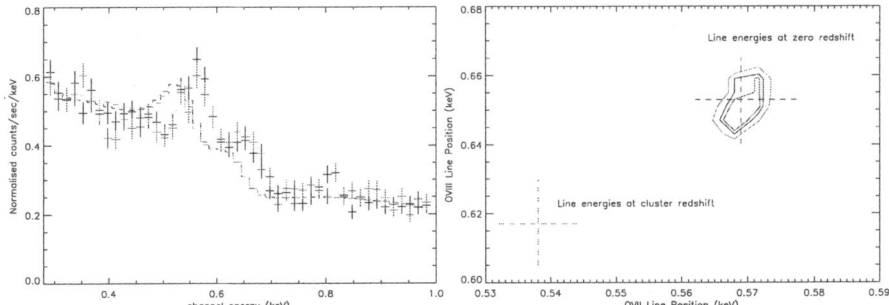

Figure 2. The left panel shows the O VII/O VIIIline region of Abell S1101 showing the two MOS instruments (black = MOS1, red = MOS2), together with models including gaussians placed at the redshift of Abell S1101 (z = 0.058). It is obvious that the redshifed Oxygen lines do not represent the observed Oxygen lines. The right panel shows the uncertainties to the energies of the O VII and O VIII lines based on fits to the XMM-Newton data. The lines are clearly at the Galactic (zero) redshift rather than the redshift of the cluster

since it shows that the background annulus is dominated by background emission, but calls doubt on the claims of Kaastra et al. (2003,2004) of the association of the CSE with an extended supercluster WHIM emission component.

4. A case study: Abell 3112

In order to investigate the impact of our background subtraction method on the observed properties of the soft excess, we have first looked at the cluster with the strongest XMM-Newton CSE, A3112. This cluster was previously studied by Nevalainen et al. (2003) using the Lumb data as the sole background estimator. Figure 1 shows that the Lumb background is not too dissimilar to double subtraction background with the double subtraction background being 30-40% higher.

Figure 3 shows a broad band (0.3-7 keV) single temperature fit to all three EPIC instruments to the 1'-2' arcminute annulus. Note that we have only fitted the 0.5 - 7 keV band for the PN due to calibration problems below 0.5 keV (Kirsch et al. 2004). This single temperature fit is very poor with strong residuals at low energies and a bad $\chi^2 = 1.52$. These are all characteristics of the presence of a strong soft excess in this cluster. We have therefore fitted extra models to the spectrum, on representing a thermal soft excess, one representing a non-thermal model. The fits are shown in Figure 4 with the χ^2 values for these fits are $\chi^2_\nu = 1.03$

(null hypothesis = 0.28) for the thermal soft excess and $\chi_\nu^2 = 1.05$ (null hypothesis = 0.16) for the non-thermal model respectively. Both models have improved significantly over a single temperature fit. However, it is not possible to distinguish between a thermal or non-thermal model. Hopefully, a more detailed analysis of this and other CSE clusters with XMM-Newton will be able to resolve the true emission mechanism.

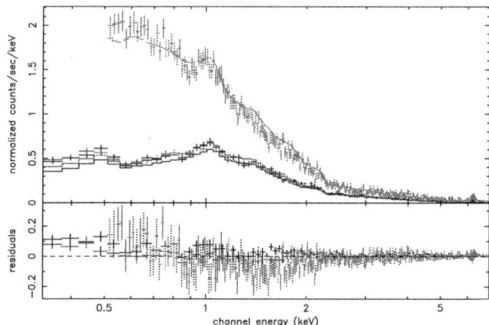

Figure 3. Single temperature broad band fit (0.3-7 keV) to the 1'-2' arcminute annulus for all three EPIC cameras for Abell 3112. Clear low energy residuals and a very poor fit ($\chi_\nu^2 = 1.52$ null hypothesis = 10^{-16}) show the presence of a strong soft excess.

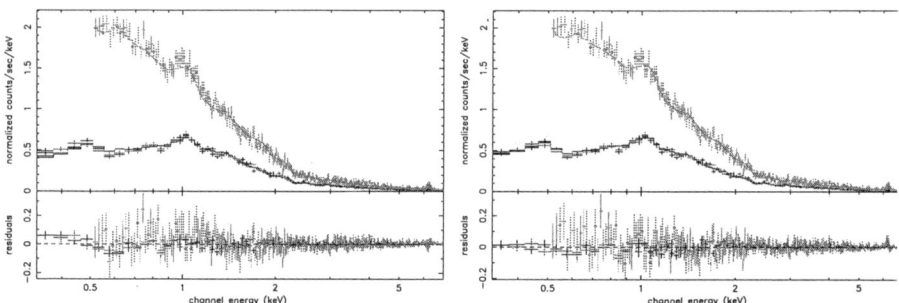

Figure 4. Left panel shows a two temperature model fit to the data (a thermal soft excess) and the right panel shows a hot ICM plus power-law (non-thermal) fit to the same data. The χ^2 values for these fits are $\chi_\nu^2 = 1.03$ (null hypothesis = 0.28) for the thermal soft excess and $\chi_\nu^2 = 1.05$ (null hypothesis = 0.16) for the non-thermal model. Both models have improved significantly over a single temperature fit.

5. Discussion

We have re-assessed the XMM-Newton data analysis of two clusters with reported soft excesses. We find that using a separate background dataset without taking into account the variation of the soft X-ray Galactic background by using in-situ background measurements can lead to an underestimate of the true background and therefore an overestimate of the cluster soft excess. It can also lead to the spurious detection of Galactic O VII lines which in the past have been ascribed to the clusters themselves. This calls into question many of the conclusions reported in Kaastra et al. (2003,2004). We do find, however, that with the correct background subtraction the cluster soft excesses still exists even though no O VII lines can be seen. Fundamentally, this demonstrates that if we are to correctly understand and parameterise the CSE using XMM-Newton data, the correct background subtraction is absolutely vital. By using the correct background subtraction we then find that instead of having a clear cut emission model as has been recently proposed, we are back to the situation where it is not clear what mechanism gives rise to the CSE.

We have also assessed the physicallity of the different models. With regard to the status of thermal emission from the CSE we cannot completely discount it, but there is no definitive evidence supporting it. On the non-thermal side, emission at the centre may require a non-thermal explanation, but on the outskirts it is not clear if there can be enough energy inputted into the relativistic CR population to give rise to the strength of the CSE. Hopefully, new observations and new instruments such as Astro-E2 will be able to finally resolve this issue.

References

1. Arnaud, M., Neumann, D., Aghanim, N., Gastaud, R., Majerowicz, S., Hughes, J., 2001, A&A, 365, L80
2. Bonamente, M., Lieu, R. & Mittaz, J., 2001, ApJ, 561, L63
3. Kaastra, J.S., Lieu, R., Tamura, T., Paerels, F.B.S., den Herder, J.W., 2003, A & A, 397, 445
4. Kaastra, J.S., Lieu, R., Tamura, T., Paerels, F.B.S., den Herder, J.W., 2004, in 'Soft X-ray emission from clusters of galaxies and related phenomena', eds. R. Lieu & J, Mittaz (Kluwer Dordrecht), p37
5. Kirsch, M.G.F., Altieri, B., Chen, B., Haberl, F., Metcalfe, L., Pollock, A.M.T., Read, A.M., Saxton, R.D., Sembay, S., Smith, M.J.S., 2004, astro-ph/0407257
6. Lieu, R., Mittaz, J.P.D. et al., 1996, Ap,J., 458, L5
7. Lumb, D.H., Warwick, R.S., Page, M., DeLuca, A., 2002, A&A, 389, 93

8. Nevalainen, J., Lieu, R., Bonamente, M., & Lumb, D. 2003, ApJ, 584, 716

XMM OBSERVATIONS OF HIGH REDSHIFT X-RAY CLUSTERS AND COSMOLOGICAL INTERPRETATION

A. BLANCHARD, S. VAUCLAIR, R.SADAT AND THE XMM-NEWTON
Ω–PROJECT COLLABORATION.
*LATT, UPS, CNRS, UMR 5572, 14 Av Ed.Belin
31 400 Toulouse, France
E-mail: alain.blanchard@ast.obs-mip.fr*

We present the first results of the Ω project, a large XMM program devoted to observe distant SHARC clusters. For the first time a measurement of the $L - T$ evolution with XMM has been obtained. We found clear evidence for a positive evolution of the $L-T$ relation, in agreement with previous analysis based on ASCA and Chandra observations. Its cosmological implication is also discussed based on a new analysis of the modeling of different X-ray surveys : EMSS, RDCS, MACS, SHARC, 160 deg². It is found that a high matter density model fits remarkably well all these surveys while concordance models produce far more faint clusters than observed counts, independently of the local amplitude σ_8, provide that the local abundance of clusters is actually matched, *in agreement with independent previous analyzes following the same strategy*. This apparent severe failure of the concordance model could be the indication of a deviation from the expected scaling of the $M - T$ relation with redshift. However, no signature of such possibility is found in existing data. We conclude that a self consistent represen- tation of clusters abundance evolution can be obtained only in an Einstein-de Sitter universe.

1. Introduction

The XMM-Ω project (Bartlett et al. 2001) was conducted in order to provide an accurate estimation of the possible evolution of the luminosity–temperature relation at high redshift for clusters of medium luminosity which constitute the bulk of X-ray selected samples, allowing to remove a major source of degeneracy in the determination of Ω_M from cluster number counts in flux limited number counts

The question of the value of matter density in the present day universe is generally considered as a problem essentially solved. However, although the flatness of the Universe is established beyond reasonable doubts since almost ten years, the only direct for a cosmological constant comes from the Hubble diagram of distant supernovae (and the possible detection of the cross correlation of CMB and surveys of the local matter content: Boughn & Crittenden, 2004). The WMAP signal for instance, as well as other LSS properties of the universe can be well reproduced in an Einstein de Sitter model (Blanchard et al., 2003). The first author of the present paper therefore firmly believes that further evidences in favor of the actual existence of a cosmological constant are needed before it can be regarded as an established scientific fact.

2. Observed evolution of the $L - T$ relation of X-ray clusters

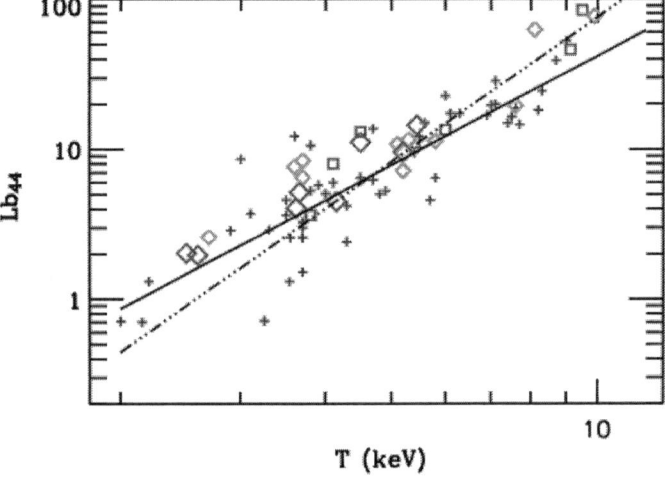

Figure 1. Temperature–luminosity of X-ray clusters: crosses are local clusters from a flux selected sample (Blanchard et al, 2000), grey diamonds are distant clusters from Chandra (Vikhlinin et al., 2002) in the redshift range $0.4 \leq z \leq 0.625$, large dark diamonds are clusters from the XMM Ω project, squares are other XMM clusters within the same redshift range.

D.Lumb et al. (2004) present the results of the X-ray measurements of 8 distant clusters with redshifts between 0.45 and 0.62. By comparing to various local $L - T$ relations, clear evidence for evolution in the $L - T$ relation has been found. The possible evolution has been modeled in the following way:

$$L_x = L_6(0) \left(\frac{T}{6keV}\right)^\alpha (1+z)^\beta \qquad (1)$$

where $L_6(0) \left(\frac{T}{6keV}\right)^\alpha$ is the local $L-T$ relation. β is found to be of the order of 0.6 ± 0.3 in an Einstein-de Sitter cosmology (Lumb et al., 2004; Vauclair et al., 2003). This result is entirely consistent with previous analyzes (Sadat et al., 1998; Vikhlinin et al, 2002) and other XMM data (see figure 1).

3. Cosmological interpretation

The evolution of the abundance of X-rays clusters is known to be a powerful cosmological test (Oukbir & Blanchard, 1992). Indeed the evolution of the number of clusters of a given mass is a sensitive function of the cosmological density of the Universe, very weakly depending on other quantities when properly normalized (Blanchard & Bartlett, 1998). Attempts to apply this test have been performed but still from a very limited number of clusters (typically 10 at redshift 0.35) (Henry, 1997; Viana and Liddle, 1998; Eke et al., 1998; Blanchard et al., 2000). In Blanchard et al., it was found that $\Omega = 0.86 \pm 0.25(1\sigma)$, so that a concordance model is away at only a 2-σ level, while systematics differences explain the values obtained from the various authors. On the other hand, number counts allow one to use samples comprising much more clusters. Indeed using simultaneously different existing surveys: EMSS, SHARC, RDCS, MACs NEP and 160 deg^2 one can use information provided by more than 300 clusters with $z > 0.3$ (not necessarily independent). In order to model clusters number counts, for which temperatures are not known, it is necessary to have a good knowledge of the $L-T$ relation over the redshift range which is investigated, which information has been provided by XMM and Chandra. Number counts can then be computed:

$$\begin{aligned} N(>f_x, z, 2\Delta z) &= \Omega \int_{z-\Delta z}^{z+\Delta z} \frac{\partial N}{\partial z}(L_x > 4\pi D_l^2 f_x) dz \\ &= \Omega \int_{z-\Delta z}^{z+\Delta z} N(>T(z)) dV(z) \\ &= \Omega \int_{z-\Delta z}^{z+\Delta z} \int_{M(z)}^{+\infty} N(M,z) dM dV(z) \end{aligned} \qquad (2)$$

where $T(z)$ is the temperature threshold corresponding to the flux f_x as given by the observations, being therefore independent of the cosmological model. For most surveys the above formula has to be adapted to the fact that the area varies with the flux limit, and eventually with redshift. Several ingredients are needed: the local abundance of clusters as given by the

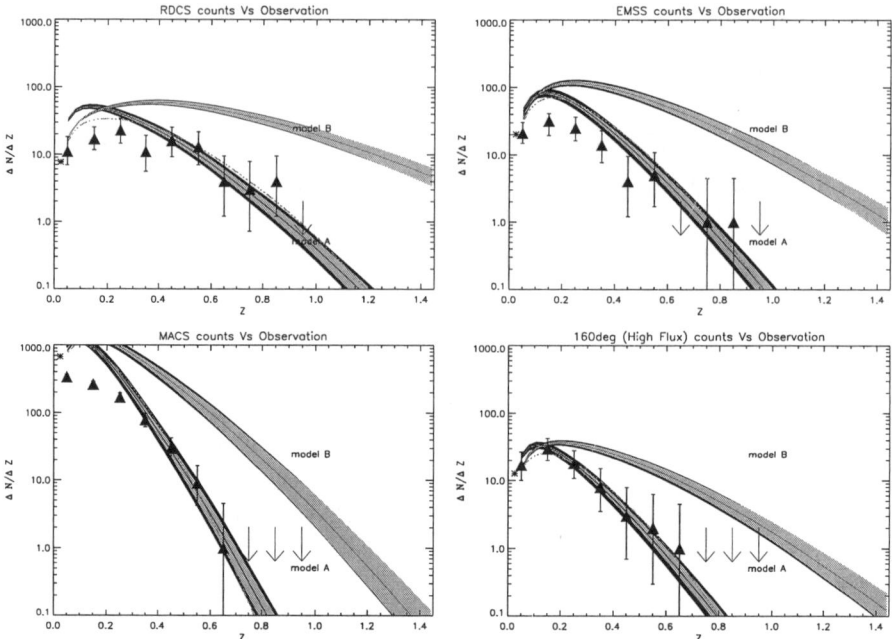

Figure 2. Theoretical number counts in bins of redshift ($\Delta z = 0.1$) for the different surveys: RDCS, EMSS, MACS and 160deg^2 (high flux, corresponding to fluxes $f_x > 2\,10^{-13}$ erg/s/cm^2). Observed numbers are triangles with 95% confidence interval on the density assuming poissonian statistics (arrows are 95% upper limits). The upper curves are the predictions in the concordance model (model B). The lower curves are for critical universe (model A). Uncertainties on σ_8 and on $L-T$ evolution lead to the grey area(see Vauclair et al., 2003).

temperature distribution function ($N(T)$), the mass-temperature relation and its evolution, the mass function and the knowledge of the dispersion. Uncertainties in these quantities result in -systematics- uncertainties in the modeling which have been found to be comparable to statistical uncertainties. Figure 2 illustrates the counts obtained with a standard mass temperature relation: $T = 4\text{keV} M_{15}^{2/3}(1+z)$, the SMT mass function (Sheth et al., 2001), and the $L-T$ relation observed by XMM with its uncertainty. These counts were computed for different existing surveys to which they can be compared. A likelihood analysis on independent samples lead to $< \Omega_M <$ (Vauclair et al., 2005). During this analysis numerous possible source of systematics were investigated with great detail (local samples, normalization of the $M-T$ relation, local $L-T$ relation, dispersion in the various relations). We have also check that the local luminosity in our models is in rough agreement with local surveys (without requesting it explicitly).

4. Possible loopholes

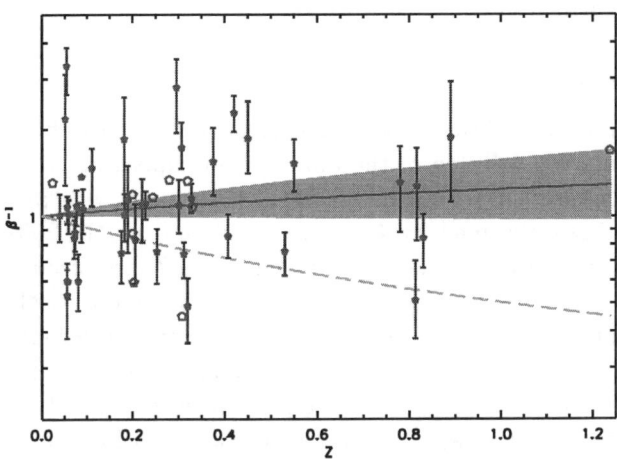

Figure 3. The ratio between thermal energy of the gas measured by T_x and the kinetic energy of galaxies measured by their velocity dispersion for a sample of clusters with $T_x \geq 6$ keV with redshift spanning from 0 to 1.2. No sign of evolution is found. The best fit is the continuous line, grey area is the formal one σ region, dashed line is the level necessary to make the concordance in agreement with the x-ray clusters counts.

Special attention has been paid to selection functions. For instance if flux limit, or identically flux calibration in faint surveys, is erroneous by a factor of 2–3 the concordance would be much closer to existing surveys. However typical uncertainty is considered to be of the order of 20%. We have identified only one possible realistic way to reproduce number counts in a concordance model, that is by assuming that the redshift evolution of the $M - T$ relation is not standard: $T \propto M_{15}^{2/3}$ (i.e. removing the $1 + Z$ factor). This is conceivably possible if a large fraction of the thermal energy of the gas originates from other processes than the the gravitational collapse (although it remain to be shown that this is actually possible in a realistic way). It is possible to test observationally this latter possibility. For this aim we have collected some existing measurements of velocity dispersion σ for massive clusters (selected to be with temperature greater than 6 keV).

The quantity:

$$\beta^{-1} \propto \frac{T_x}{\sigma^2}$$

should evolved with redshift accordingly to $(1+z)^{-1}$ if the $M-T$ relation evolved accordingly to the above non-standard scheme (and should remains constant in the standard case). We found no sign of such a non-standard behavior which is in principle ruled out at nearly the 3–σ level.

5. Conclusion

The major results obtained with the Ω project are the first XMM measurement of the evolution of the luminosity-temperature with redshift. A positive evolution has been detected, in agreement with previous results including those obtained by Chandra (Vikhlinin et al., 2002). The second important result is that this evolving $L-T$ produced counts in the concordance model which are inconsistent with the observed counts in all existing published surveys. This could be the sign of a high density universe or a deviation from the expected scaling of the $M-T$ relation with redshift. Our investigation of the ratio $\frac{T_x}{\sigma^2}$ shows no sign of such deviation. The distribution of x-ray selected clusters favors a high density universe, alleviating the need for a cosmological constant.

References

1. Bartlett, J. et al., proceedings of the XXI rencontre de Moriond, astro-ph/0106098 (2001).
2. Blanchard, A. and Bartlett, J. A&A **332**, 49L (1998).
3. A. Blanchard, R. Sadat, J. Bartlett. and M. Le Dour, A&A **362**, 809 (2000).
4. Blanchard, A., Douspis, M., M. Rowan-Robinson and S. Sarkar, A&A **412**, 35 (2003).
5. S. Boughn and R. Crittenden 2004, Nature **427**, 45
6. V. R. Eke, S. Cole, C. Frenk, and J. P. Henry, MNRAS **298**, 1145 (1998).
7. J. P. Henry, ApJ **489**, L1 (1997).
8. D. H. Lumb et al., A&A **420**, 853 (2004).
9. A. Markevitch, ApJ **504**, 27 (1998).
10. J. Oukbir and A. Blanchard, A&A **262**, L21 (1992).
11. R. Sadat, A. Blanchard. and J. Oukbir, A&A **329**, 21 (1998).
12. R. K. Sheth, H. J. Mo and G. Tormen, MNRAS **323**, 1 (2001).
13. T. P. Viana and A. R. Liddle, MNRAS **303**, 535 (1999).
14. S. C. Vauclair et al, A&AL **412**, L37 (2003).
15. S. C. Vauclair et al, in preparation (2005).
16. A. Vikhlinin et al., ApJL **578**, 107 (2002).

DARK ENERGY AND NON–LINEAR PERTURBATIONS

C. VAN DE BRUCK
Department of Applied Mathematics
The University of Sheffield
Hounsfield Road
Sheffield S3 2RH
United Kingdom

D. F. MOTA
Astrophysics Department
University of Oxford
Keble Road
OX1 3RH
United Kingdom

Dark energy might have an influence on the formation of non–linear structures during the cosmic history. For example, in models in which dark energy couples to dark matter, it will be non–homogeneous and might have some influence on the collapse of a dark matter overdensity. We use the spherical collapse model to estimate how much influence dark energy might have.

1. Introduction

One of the most important goals of contemporary cosmology is to unreveal the properties of dark energy. This energy form is thought to be responsible for the observed accelerating expansion of the present day universe. There are several methods used to study the properties of dark energy. The important ones make use of the anisotropies in the Cosmic Microwave Background Radiation (CMB), the evolution of large scale structure formation (LSS) and/or the distances of high redshift supernovae (see e.g. the overview by Peebles and Ratra[1]).

In this contribution we address the question whether dark energy can have some impact on the formation of non–linear structures in the universe, such as clusters of galaxies or galaxies itself. In doing so, we assume that dark energy is a scalar field, pervading the universe. It will obviously depend on the properties of dark energy if this scalar field has any influence on non–linear structure formation. For example, if dark energy couples to dark matter, there is an extra force between the dark matter particles, mediated by dark energy. This possibility was first discussed in detail by Wetterich[2] and afterwards in particular by Amendola[3]. However, even if dark energy does not couple to dark matter, backreaction effects of the gravitational field might influence the evolution of dark energy inside a non–linear overdensity.

Here, we use the spherical collapse model to study the influence of dark energy on non–linear structure formation.

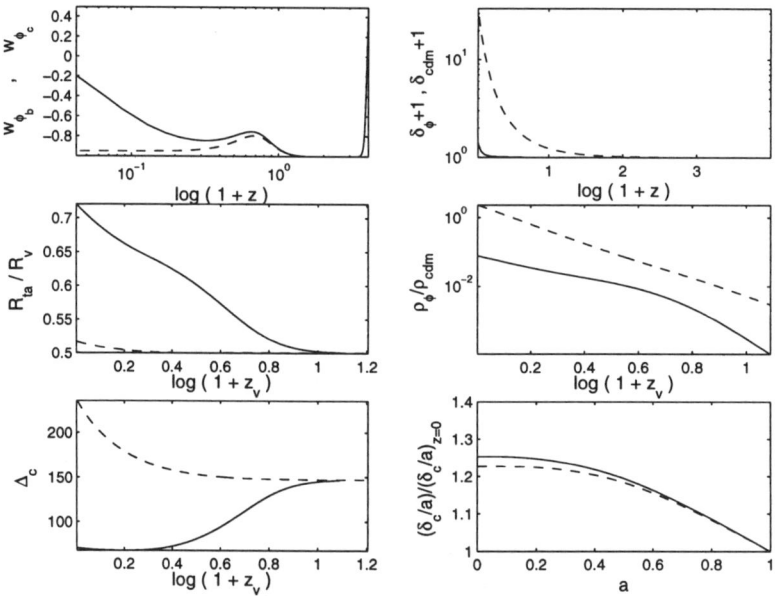

Figure 1. Double exponential potential (see text). Top left panel: Full collapse of dark energy: evolution of w_ϕ in the background (dashed line) and inside the overdensity (solid line) as a function of redshift. Top right panel: evolution of $\rho_{\phi,\text{overdensity}}/\rho_{\phi,\text{outside}}$ (solid line) and $\rho_{\text{cdm,overdensity}}/\rho_{\text{cdm,outside}}$ as a function of redshift in the case of clustering of dark energy. Middle left panel: R_v/R_t as a function of virialisation redshift in the case of homogeneous dark energy (dashed line) and collapsing dark energy (solid line). Middle right panel: The ratio $\rho_\phi/\rho_{darkmatter}$ inside the overdensity as a function of virialisation redshift. Bottom left: Δ_c as a function of virialisation redshift in the case of homogeneous (dashed line) and inhomogeneous (solid line) dark energy. Bottom right: the linear density contrast as a function of the scale factor a.

2. The spherical collapse model

The spherical collapse model is based on the assumption that an overdensity can be treated as a homogeneous and isotropic, closed "sub-universe", embedded in our universe. For a cold dark matter universe, this assumption is justified by Birkhoff's theorem. However, as soon there is a second fluid, such as radiation, dark energy, etc., this (idealised) over-density can, and will, exchange energy with its surroundings. In the case of dark energy studied here, however, the exchange will only be a small fraction of the total energy, since dark energy is subdominant for most parts of the cosmic history. Therefore, it should be a not too bad description once the energy out- or inflow is specified. However, the spherical collapse model does not specify the energy outflow (which we denote by Γ) of dark energy into the surroundings of the overdensity. Therefore, we have to make assumptions about Γ. In our work[4] we considered two extreme cases. In the first case, we assumed that dark energy does not cluster at all but is homogeneous throughout space. In the second case we assumed that it fully collapses along with dark matter. Clearly, the reality might be somewhere between these two possibilities. However, we will get an idea about the difference to be expected.

The typical time-evolution of a spherical overdensity is as follows: Initially, the overdensity expands with almost the same rate as the universe. However, since the density is higher, the expansion of the overdensity will eventually slow down until it starts to contract. The point at which the expansion turns into contraction is called *turnaround*. Without dissipation, the overdensity would collapse to a singularity. However, in reality, energies inside the overdensities viralise and the sphere ends up at some final radius (virialisation). In our calculations, we follow the evolution of the sphere until virialisation is reached. We have studied several different potentials and refer to our paper[4] for the details of the calculation. Here, we consider only three different potentials as models for dark energy. The first one is a double exponential potential[5]. The form is $V(\phi) = M(\exp(\alpha\phi) + \exp(\beta\phi))$. The second is the well known supergravity potential[6], which is $V(\phi) = M\exp(\phi^2)/\phi^\gamma$. The third is a exponential potential with power–law modifications[7], i.e. $V(\phi) = M(A + (\phi - B)^2)\exp(-\gamma\phi)$.

The results of our calculations can be found in Figs.1 – 3. Apart from the equation of state of dark energy inside and outside the overdensity and the density contrast of dark energy and dark matter, we calculate the ratio of

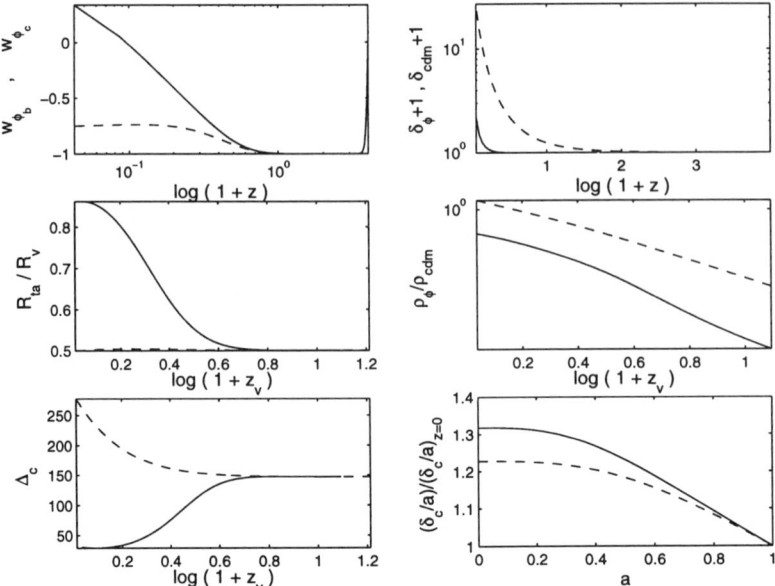

Figure 2. The same as Figure 1, but for the supergravity potential (see text).

the radius at turnaround R_t to the radius at virialisation R_v as a function of redshift at which the overdensity virialises (z_v). The latter is equal to 0.5 in the case of the standard cold dark matter model and has a slight dependence on the cosmological constant in a ΛCDM model [8]. It can be seen from Figs. 1, 2 and 3 that in the case of a homogeneous dark energy component the ratio R_t/R_v depends on the model of dark energy, but is still of order 0.5. If dark energy collapses together with dark matter, this quantity depends strongly on the virialisation redshift. As a result, the density contrast at the time of virialisation $\Delta_c = \rho_{\text{cdm,inside}}(z_v)/\rho_{\text{cdm,outside}}(z_v)$ becomes strongly dependent on z_v as well.

On the other hand, the ratio of the energy densities of dark matter and dark energy depends on the clustering properties of dark energy, but is small at the typical redshift of cluster formation, even if dark energy clusters.

We have also checked the dependence on our assumptions. For example, a more conservative assumption would be that dark energy only clusters *after* turnaround, i.e. once the overdensity is decoupled from the rest of the universe. Before that, the field is homogeneous. In this case, there is still a

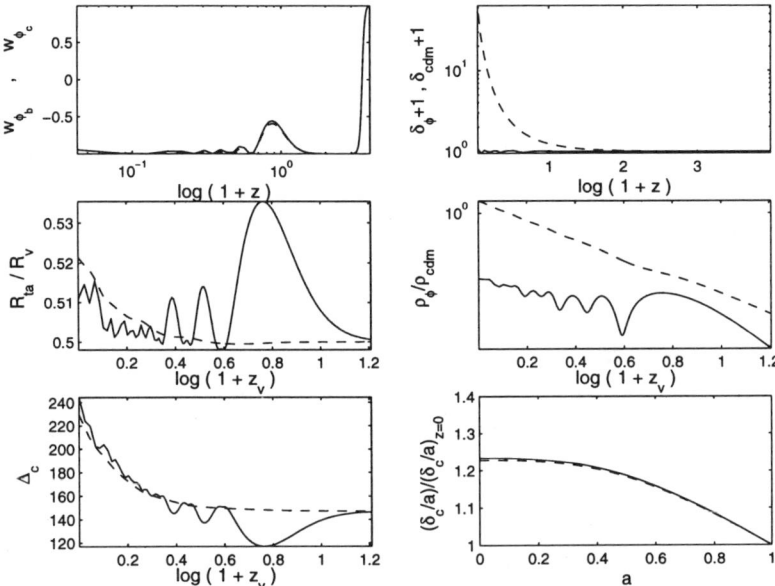

Figure 3. The same as Figure 1 but for the modified exponential potential (see text).

big difference from the homogeneous case (see our work[4] for more details).

3. Outlook

The spherical collapse model susggests, that dark energy can have an important impact on non–linear structure formation, even if it is dynamically unimportant for most of the time during the cosmic history. We have considered two extreme cases, namely that dark energy either fully collapses together with dark matter or that dark energy is homogeneous throughout space. Its interesting to note that even *if* dark energy collapses with dark matter, it will stay in the linear regime (or sometimes in the quasi non–linear regime, in which the density contrast is of order 1), whereas dark matter is in the highly non–linear regime. In fact, although dark matter enters a highly non–linear regime, the dark energy density contrast deviates from unity only slightly.

The aim of our approach was not to make predictions for structure formation, but rather to investigate if dark energy can at all have a significant impact on the details of structure formation. Clearly, our work shows that it can have, but the answer depends strongly on the details of the theory.

Our work is rather limited, since the spherical collapse model can not predict how much dark energy will flow out of the overdensity. For this, a fully relativistic approach has to be taken, in order to calculate the amount exactly. However, our work clearly indicates that the spherical collapse model has to be used with care, when comparing models of dark energy with data.

On the other hand, our work raises some questions for future work: even if dark energy clusters strongly, the differences to a theory where it does not cluster should not be too large, since at the redshift of structure formation (i.e. $z \geq 3$) the differences between the theories are small. How will we be able to differentiate between the theories using only structure formation data? Note that the case of a dark energy which collapses with dark matter is rather extreme, so its likely that any signal of dark energy clustering will be even smaller, once the exact value of Γ is known.

Acknowledgements

C.v.d.B. was supported by PPARC. D.F.M. is supported by Funda cao Ciencia e a Tecnologia.

References

1. P. J. E. Peebles and B. Rathra, Rev. Mod. Phys. **75**, 559 (2003)
2. C. Wetterich, Astronomy and Astrophysics **301**, 321(1995)
3. L. Amendola, Phys.Rev. D **62**, 043511, (2000)
4. D. F. Mota and C. van de Bruck, Astronomy and Astrophysics **421**, 71 (2004)
5. T. Barreiro, E. Copeland and N. J. Nunes, Phys. Rev. D **61**, 127301 (2000)
6. P. Brax and J. Martin, Phys. Lett. B **468** 40 (1999)
7. A. Albrecht and C. Skordis, Phys. Rev. Lett. **84**, 2076 (2000)
8. O. Lahav, P. B. Lilje, J. R. Primach and M. J. Rees, Month. Not. Roy. Astron. Soc. **251**, 128 (1991)

COSMOLOGICAL SINGULARITIES AND DARK ENERGY

W. PIECHOCKI

Sołtan Institute for Nuclear Studies,
Hoża 69,
00-681 Warszawa, Poland
E-mail: piech@fuw.edu.pl

We consider classical and quantum dynamics of a free particle in de Sitter's spacetimes with different topologies to see what happens to space-time singularities of removable type in quantum theory. Our results indicate that taking account of global properties of space-time enables quantization of particle dynamics in all considered cases. We expect that understanding of the nature of curvature singularities may bring some new ideas concerning the nature of the dark energy.

1. Introduction

The struggle for quantum gravity has lasted about 70 years. It seems that understanding of the nature of spacetime singularities in a quantum context may be the core of the problem. The understanding may mean changing some of the principles underlying quantum mechanics or general relativity, or the necessity of introducing a new unknown so far source of gravitational field that we may call dark energy. An inside into the problem may be achieved by studying some suitable toy models which include both spacetime singularities and quantum rules. In what follows we present results concerning one of such models that is a quantization of dynamics of a test particle in singular spacetime. More specifically, we consider classical and quantum dynamics of a test particle in de Sitter's spacetimes with different topologies to see what happens to spacetime singularities of removable type in quantum theory. For more details concerning this lecture we recommend Refs. 1 and 2.

2. Classical dynamics

The considered spacetimes, V_p and V_h, are de Sitter's type. They are defined to be

$$V_p = (\mathbf{R} \times \mathbf{R}, \hat{\mathbf{g}}) \quad \text{and} \quad \mathbf{V_h} = (\mathbf{R} \times \mathbf{S}, \hat{\mathbf{g}}). \tag{1}$$

In both cases the metric $g_{\mu\nu} := (\hat{g})_{\mu\nu}$ ($\mu,\nu = 0,1$) may be defined by the line-element

$$ds^2 = dt^2 - exp(2t/r_0)\, dx^2, \tag{2}$$

where r_0 is a positive real constant. It is clear that (1) includes all possible topologies of de Sitter's type spacetimes in two dimensions. V_p is a plane with global $(t,x) \in \mathbf{R}^2$ coordinates. V_h is defined to be a one-sheet hyperboloid embedded in 3d Minkowski space. There exists an isometric immersion map of V_p into V_h

$$V_p \ni (t,x) \longrightarrow (y^0, y^1, y^2) \in V_h, \tag{3}$$

where

$$y^0 := r_0 \sinh(t/r_0) + \frac{x^2}{2r_0} \exp(t/r_0),$$

$$y^1 := -r_0 \cosh(t/r_0) + \frac{x^2}{2r_0} \exp(t/r_0), \quad y^2 := -x \exp(t/r_0),$$

and where

$$(y^2)^2 + (y^1)^2 - (y^0)^2 = r_0^2. \tag{4}$$

Equation (3) defines a map of V_p onto a non-compact half of V_h.

It is known that V_p is geodesically incomplete. However, all incomplete geodesics in V_p can be extended to complete ones in V_h, i.e. V_p has removable type singularities.

An action integral, $\mathcal{A}$, describing a free relativistic particle of mass m_0 in gravitational field $g_{\mu\nu}$ reads

$$\mathcal{A} = \int_{\tau_1}^{\tau_2} L(\tau)\, d\tau, \quad L(\tau) := -m_0 \sqrt{g_{\mu\nu}(\tau)\, \dot{x}^\mu(\tau) \dot{x}^\nu(\tau)}, \tag{5}$$

where τ is an evolution parameter, x^μ are space-time coordinates and $\dot{x}^\mu := dx^\mu/d\tau$.

The Lagrangian (5) is invariant under the reparametrization $\tau \to f(\tau)$. This gauge symmetry leads to the constraint

$$G := g^{\mu\nu} p_\mu p_\nu - m_0^2 = 0, \tag{6}$$

where $g^{\mu\nu}$ is the inverse of $g_{\mu\nu}$ and $p_\mu := \partial L/\partial \dot{x}^\mu$ are canonical momenta.

Since we assume that a free particle does not modify the geometry of spacetime, the local symmetry of the system is defined by the set of all Killing vector fields of space-time. The corresponding dynamical integrals have the form

$$D = p_\mu X^\mu, \quad \mu = 0, 1, \tag{7}$$

where X^μ is a Killing vector field. The physical phase space Γ is defined to be the space of all particle trajectories consistent with the dynamics of a particle and with the constraint (6).

2.1. Particle on hyperboloid

One can show[1] that the dynamical integrals J_a ($a = 0, 1, 2$) of a particle on hyperboloid satisfy the commutation relations of $sl(2, \mathbf{R})$ algebra

$$\{J_0, J_1\} = -J_2, \quad \{J_0, J_2\} = J_1, \quad \{J_1, J_2\} = J_0, \tag{8}$$

and that they are related by (6) as follows

$$J_1^2 + J_2^2 - J_0^2 = \kappa^2, \quad \kappa := m_0 r_0. \tag{9}$$

It can be shown[1] that the equations for a particle trajectory read

$$J_a y^a = 0, \quad J_2 y^1 - J_1 y^2 = r_0^2 p_\rho, \tag{10}$$

where $p_\rho := \partial L/\partial \dot{\rho} < 0$, since we consider timelike trajectories.

Each point (J_0, J_1, J_2) of the hyperboloid (9) defines uniquely a particle trajectory (10) on (4) admissible by the dynamics and consistent with the constraint (6). Thus, the one-sheet hyperboloid (9) defines the physical phase space Γ_h. It is clear that the proper orthochronous Lorentz group $SO_0(1, 2)$ is the symmetry group of Γ_h.

2.2. Particle on plane

One may verify[1] that the dynamical integrals P, K and M satisfy the commutation relations of $sl(2, \mathbf{R})$ algebra in the form

$$\{P, K\} = P, \quad \{K, M\} = M, \quad \{P, M\} = 2K. \tag{11}$$

The constraint (6) reads

$$K^2 - PM = \kappa^2, \quad \kappa = m_0 r_0. \tag{12}$$

By analogy to V_h case one may expect that each triple (P, K, M) satisfying (12) determines a trajectory of a particle. However, not all such trajectories are consistent with particle dynamics: For $P = 0$ there are two lines $K = \pm\kappa$ on the hyperboloid (12). Since by assumption $\dot{t} > 0$, we have $p_t = \partial L/\partial \dot{t} = -m\dot{t}\,(\dot{t} - \dot{x}\exp(2t/r_0))^{-1/2} < 0$. Since $K - xP = -r_0 p_t$, we get $K - xP > 0$, i.e. $K > 0$ for $P = 0$. Therefore, the line $(P = 0, K = -\kappa)$ is not available for the dynamics. The hyperboloid (12) without this line defines the physical phase space Γ_p (topologically equivalent to $\mathbf{R}^2$).

The explicit formulae for particle trajectories

$$x(t) = M/2K, \quad \text{for} \quad P = 0 \tag{13}$$

and

$$x(t) = \left[K - \sqrt{\kappa^2 + (r_0 P)^2 \exp(-2t/r_0)}\right]/P, \quad \text{for} \quad P \neq 0, \tag{14}$$

where (14) takes into account that $K - xP > 0$. The space of trajectories defined by (13) and (14) represents the phase space Γ_p.

2.3. Observables

In case of V_h system we choose J_0, J_1 and J_2 as the observables so they satisfy $sl(2,\mathbf{R})$ algebra. We parametrize the hyperboloid (9) as follows

$$J_0 = J, \quad J_1 = J\cos\beta - \kappa\sin\beta, \quad J_2 = J\sin\beta + \kappa\cos\beta, \tag{15}$$

where $J \in \mathbf{R}$ and $0 \leq \beta < 2\pi$.

The phase space Γ_p cannot be invariant under the action of $SO_0(1,2)$ group. In fact, the dynamical integral M is not well defined globally. Thus, the set of observables of V_p system consists only of the integrals P and K satisfying the subalgebra of (11)

$$\{P, K\} = P. \tag{16}$$

The algebra (16) is isomorphic to the algebra $\mathbf{aff}(1,\mathbf{R})$ of the affine group $\mathbf{Aff}(1,\mathbf{R})$. Thus the algebras of observables of V_p and V_h systems are quite different.

3. Quantization

3.1. Quantum particle on hyperboloid

Making use of the Schrödinger representation for the canonical coordinates J and β (we set $\hbar = 1$ through the paper) in (15) leads to

$$\hat{J}_0 \psi(\beta) = \left[-i\frac{d}{d\beta}\right]\psi(\beta), \tag{17}$$

$$\hat{J}_1\psi(\beta) = \left[\cos\beta \ \hat{J}_0 - (\kappa - \frac{i}{2})\sin\beta\right]\psi(\beta), \tag{18}$$

$$\hat{J}_2\psi(\beta) = \left[\sin\beta \ \hat{J}_0 + (\kappa - \frac{i}{2})\cos\beta\right]\psi(\beta), \tag{19}$$

where $\psi \in \Omega_\theta \subset L^2(\mathbf{S})$, $0 \leq \theta < 2\pi$, and where $L^2(\mathbf{S})$ is the space of square-integrable complex functions on a unit circle $\mathbf{S}$. The subspace Ω_θ is defined to be

$$\Omega_\theta := \{\psi \in L^2(\mathbf{S}) \mid \psi \in \mathbf{C}^\infty[\mathbf{0}, 2\pi], \ \psi^{(\mathbf{n})}(\mathbf{0}) = \mathbf{e}^{\mathbf{i}\theta}\psi^{(\mathbf{n})}(2\pi), \ \mathbf{n} = \mathbf{0}, \mathbf{1}, \mathbf{2}, ...\}. \tag{20}$$

The unbounded operators $\hat{J}_a$ ($a = 0, 1, 2$) are well defined because Ω_θ is a dense subspace of the Hilbert space $L^2(\mathbf{S})$. It is clear that Ω_θ is a common invariant domain for all $\hat{J}_a$ and their products. One can verify that

$$[\hat{J}_a, \hat{J}_b]\psi = -i\widehat{\{J_a, J_b\}}\psi, \quad \psi \in \Omega_\theta, \tag{21}$$

and that the representation (17)-(20) is essentially self-adjoint on Ω_θ.

The representation (17)-(20) is parametrized by θ which labels unitarily non-equivalent representations of $sl(2, \mathbf{R})$ algebra corresponding to the unitary representations of various covering groups of $SO_0(1, 2)$, so it is connected with the ambiguity in the choice of the symmetry group of phase space. However, taking account of time-reversal invariance of the system[2] fixes the parameter to its two values: $\theta = 0$ and $\theta = \pi$. Considering the Casimir operator

$$\hat{C}\psi = [\hat{J}_1^2 + \hat{J}_2^2 - \hat{J}_0^2]\psi = (\kappa^2 + 1/4)\psi, \quad \psi \in \Omega_\theta, \tag{22}$$

we can see that our representation coincides with the principal series representation of $SO_0(1, 2)$ group. The case $\theta = 0$ ($\theta = \pi$) corresponds to the single (double) valued representation.

3.2. *Quantum particle on plane*

In this case the global and local symmetries are incompatible: the algebra of all observables is $sl(2, \mathbf{R})$, whereas the Lie algebra corresponding to the global symmetry is $aff(1, \mathbf{R})$. In such cases we propose modification of the quantization method by assuming that the algebra of observables should be consistent with the *global symmetry* of the phase space.

To quantize the algebra (16) we use the fact that all unitary irreducible representations of $\mathbf{Aff}(1, \mathbf{R})$ group are known. There exist only two (non-trivial) unitarily non-equivalent representations

$$U_s(g) : \mathcal{H}_\Lambda \longrightarrow \mathcal{H}_\Lambda, \quad s = -, + \tag{23}$$

where $g \in \mathbf{Aff}(1,\mathbf{R})$, and where $\mathcal{H}_\Lambda$ is a suitable Hilbert space[1]. The operators $U_s(g)$ are defined to be

$$U_s[g(a,b)]\psi(x) := \exp(-isbx)\,\psi(ax), \quad \psi \in \mathcal{H}_\Lambda, \quad s = -, + \qquad (24)$$

where $g(a,b) \in \mathbf{Aff}(1,\mathbf{R})$ and $(a,b) \in \mathbf{R}_+ \times \mathbf{R}$ parametrize the group elements. The application of Stone's theorem to (24) defines two sets of operators $\hat{A}_s$ and $\hat{B}_s$

$$\hat{A}_s\varphi(x) = -ix\frac{d}{dx}\varphi(x), \quad \hat{B}_s\varphi(x) = -sx\varphi(x), \quad \varphi \in \Lambda, \quad s = -, + \qquad (25)$$

where the space Λ is defined as

$$\Lambda := C_0^\infty(\mathbf{R}_+), \quad \mathbf{R}_+ := \{\mathbf{x} \in \mathbf{R} \mid \mathbf{x} > \mathbf{0}\}. \qquad (26)$$

One may show that these operators are essentially self-adjoint on Λ. One can also verify that

$$[\hat{A}_s, \hat{B}_s]\varphi = -i\hat{B}_s\varphi, \quad \varphi \in \Lambda, \quad s = -, + \qquad (27)$$

which demonstrates that (25) is the representation of the algebra $\mathbf{aff}(1,\mathbf{R})$.

Therefore, there are possible only two (up to unitary equivalence) quantum systems corresponding to a single classical dynamics of V_p system.

4. Conclusions

Our main result is that taking account both local and global properties of spacetime makes possible the imposition of quantum rules into the dynamics of a particle. Local properties of a given spacetime described by the metric tensor and Lie algebra of the Killing vector fields do not specify the system uniquely because spacetimes with different global properties may have isomorphic Lie algebras. Quantization of a gravitational system should go beyond the Einstein equations, since they describe spacetime mainly locally.

Our results concern removable type singularities of spacetime. It is a challenge to extend our analysis to spacetimes with essential type singularities, i.e. including not only incomplete geodesics, but also blowing up curvature invariants. We expect that consolidation of quantum rules and curvature singularities may bring some new ideas concerning the nature of the dark energy.

References

1. W. Piechocki, *Class. Quantum Grav.* **20**, 2491 (2003).
2. W. Piechocki, *Class. Quantum Grav.* **21**, A331 (2004).

EMERGENCE OF SPACE-TIME LOCALIZATION AND COSMIC DECOHERENCE : MORE ON IRREVERSIBLE TIME, DARK ENERGY, ANTI-MATTER AND BLACK-HOLES

ANNE MAGNON

Blaise Pascal University, Clermont-Ferrand, France
matiman@wanadoo.fr

A non geometric cosmology is presented, based on logic of observability, where logical categories of our perception set frontiers to comprehensibility. The Big-Bang singularity finds here a substitute (comparable to a "quantum jump") : a logical process (tied to self-referent and divisible totality) by which information emerges, focalizes on events and recycles, providing a transition from incoherence to causal coherence. This jump manufactures causal order and space-time localization, as exact solutions to Einstein's equation, where the last step of the process disentangles complex Riemann spheres into real null-cones (a geometric overturning imposed by self-reference, reminding us of our ability to project the cosmos within our mental sphere). Concepts such as antimatter and dark energy (dual entities tied to bifurcations or broken symmetries, and their compensation), are presented as hidden in the virtual potentialities, while irreversible time appears with the recycling of information and related flow. Logical bifurcations (such as the "part-totality" category, a quantum of information which owes its recycling to non localizable logical separations, as anticipated by unstability or horizon dependence of the quantum vacuum) induce broken symmetries, at the (complex or real) geometric level [eg. the antiselfdual complex non linear graviton solutions, which break duality symmetry, provide a model for (hidden) anti-matter, itself compensated with dark-energy, and providing, with space-time localization, the radiative gravitational energy (Bondi flux and related bifurcations of the peeling off type), as well as mass of isolated bodies]. These bifurcations are compensated by inertial effects (non geometric precursors of the Coriolis forces) able to explain (on logical grounds) the cosmic expansion (a repulsion?) and critical equilibrium of the cosmic tissue. Space-time environment, itself, emerges through the jump, as a censor to totality, a screen to incoherence (as anticipated by black-hole event horizons, cosmic censors able to shelter causal geometry). In analogy with black-hole singularities, the Big-Bang can be viewed as a geometric hint that a transition from incoherence to (causal space-time) localization and related coherence (comprehensibility), is taking place (space-time demolition, a reverse process towards incoherence or information recycling, is expected in the vicinity of singularities, as hinted by black-holes and related "time-machines"). A theory of the emergence of perception (and life?), in connection with observability and the function of partition (able to screen totality), is on its way [interface incoherence-coherence, sleeping and awaking states of localization, horizons of perception etc, are anticipated by black-hole event horizons, beyond which a non causal, dimensionless incoherent regime or memorization process, presents itself with the loss of localization, suggesting a unifying regime (ultimate energies?) hidden in cosmic potentialities]. The decoherence process presented here, suggests an ultimate interaction, expression of the logical relation of subsystems to totality, and to be identified to the flow of information or its recycling through cosmic jump (this is anticipated by the dissipation of distance or hierarchies on null-cones, themselves recycled with information and events). The geometric projection of this unified irreversible dynamics is expressed by unified Yang-Mills field equations (coupled to Einsteinian gravity). An ultimate form of action ("set "-volumes of information) presents itself, whose extrema can be achieved through extremal transfer of information and related partition of cells of information (thus anticipating the mitosis of living cells, possibly triggered at the non localizable level, as imposed by the logical regime of cosmic decoherence: participating subsystems ?). The matching of the objective and subjective facets of (information and) decoherences is perceived as contact with a reality.

1. Introduction

Both Relativity and Quantum Mechanics are attempts to understand our relationship to the observable world, based on a process of localization of events. On one hand, Relativity postulates space-time localization, reducing the event to a geometric point. On the other hand, Quantum Mechanics is more concerned with the mechanism of shaping of the event: if a coin falls on head, instead of tail, a choice seems to be made among two contradictory possibilities, supposed to coexist (principle of superposition) within a virtual entity: the probability wave-function. In spite of one century of controversies, the process of reduction of the wave-function (the choice), has not been well understood. Our aim, here, is to provide a new theory, called "cosmic decoherence", based on logical foundations to observability, the emergence of information, its focalization on events (the Einsteinian space-time localization). This process, which unifies known approaches, finds its source within non localizable logical categories of our perception. The key turns out to be self-reference ([1]-a, b) of the observable and closed world: the necessary invention of subsystems in view of identification (localization), imposes partition of inaccessible totality. This new theory (of decoherence), brings a solution to the everlasting conflict between the complex formalism of quantum mechanics and the real formalism of relativity theory. Self-reference is part of our daily experience, since our thinking activity can project the observable universe (which surrounds us: the outside?) onto our mental sphere of perception, (the inside?). This will lead us to the objective and subjective facets of the concept of information, the adjustment of the two facets (and related decoherences) explaining our perception of a *contact with reality*, along irreversible time-flow.

2. The jump towards space-time localization, causal coherence and observability

Our innovation is to show that space-time environment (as solution to Einstein's equation) which enables localization, in space or time, is shaped by the logical regime in question. The main participation comes from the function of self-reference, implementing the idea that our observable universe invents itself from within. This logical context, though universal, is also present in our ability to imagine the cosmos within our mental sphere of perception. Decoherence, or reduction of (complex) potentialities into (causally related and real) events, will be shown to emerge from this context.

Problems raised by the concept of totality can be approached with the empty (pre-) set $\emptyset$, a non divisible entity familiar to mathematicians: $\emptyset$ is identical to its parts, and in particular to its self-complement ($C_\emptyset \emptyset = \emptyset$). If observability is to set in, parts must be invented from within the totality (in view of identification). Dissociating $\emptyset$ from its self-complement creates the "part-totality" logical binarity [$\emptyset$, $C_\emptyset \emptyset = \emptyset^*$] (category of our thinking activity, primordial distinction, and expression of a partition), by which distinction of one entity wrt. another one, can take place. In brief, the context of observability stands on the duplication of an ultimate atom: a logical barrier by which spontaneous division sets in [thus raising the question of the emergence of life (recycling spheres of perception) in connection with the concept of observable world].The [$\emptyset, \emptyset^*$] self-referent binarity provides a quantum of information, as well as the objective-subjective self-referent (non localizable) separation.

If $\emptyset^*$ is to be identified, irrespective of $\emptyset$, the process must be iterated, leading to the irreversible following sequence:

$$\emptyset \rightarrow \emptyset^* \rightarrow \emptyset^{**} \rightarrow \emptyset^{***} \rightarrow ... \rightarrow \emptyset^*_1..._n^*$$

This sequence also produces $\mathbb{N}$, the sequence of natural integers (1,2,...n, etc), and is associated to irreversible flow of recycling information (by which $\mathbb{N}$ becomes a recycled concept!). With this set up for emergence and flow of information, various mathematical (and recycled) entities become available: real numbers, complex numbers, spinors, bispinors, quaternions or twistors, and finally twistor-spaces. We stop here for reasons of dimension, and also because the incidence of these two dual spaces provides **R,** the Riemann sphere (compactified complex plane, an entangled, and recycled site enabling coexistence of opposite orientations). We shall use **R** as a substitute to the complex probability wave function Ψ : indeed each space-time event emerging at the incidence of two dual twistors ([2-a]), **R** is a site of focalization of information on pre-events. Space-time points, themselves, emerge with causal organization. Their perceived evanescence is due to information recycling. Hence a reduction process (guided by the logical context sourcing observability), interpretable as a decoherence (shaping of transparence) of the primordial cosmos. The jump to causal level corresponds to a disentanglement of **R** into S^+_2 and S^-_2, cross sections of retarded and advanced null cones, respectively. This disentanglement was obtained independently of the logical context, by the Pittsburgh school (Newman et al , in [1-c], p 266, Ref. 14), using analytic methods. The recycling of information (to be identified to irreversible time, and possible model for a unified interaction), is to be traced back to the unstability or non localizability of the logical partition, as suggested by the quantum vacuum unstability (evanescent pair-creation).

3. Decoherence, unified interaction, and its decoupling

The dissipation of distance on null rays suggests a dissipation of localization, hierarchies, and onset of a unifying regime. Besides, black-hole singularities suggest a demolition of space-time, its localization and related causal coherence, its dimensions. A geometric formulation (of the reversed process) can anticipate the situation: we briefly summarize. The disentanglement of R can be viewed as the emergence of a smooth asymptotic 3-sphere that will overturn into null boundary (cone). Asymptotic simplicity being assumed in its vicinity, including for the available gauge fields, one can identify coupling constants (see in [1-c], chap. 3, and related article). This is a hint that hierarchies are disappearing. Here the existence of loops near the (3-sphere) null boundary, is crucially needed (elimination of gauge-field singularities, hinting that R can erase black-hole type singularities, as causality dissipates). The reversed process of decoupling of interactions, consequently appears with the emergence of space-time dimensions (scales). Key ingredients of our geometric formulation of relative motion (space, time-separation, velocity, volume, etc), find here a justification. Additional attributes of the unification process, can be formulated at the geometric level, eg. in terms of unified Yang-Mills field-equations ([1-c], p235). As a result, Lagrangian or Hamiltonian dynamics appear as t-reversible projection of the new irreversible dynamics controled by extremality of information transfer. The emergence of R and its Lorentz group of invariance, also suggests that of other gauge Lie-groups. In absence of geometry, extremality of the transfer of information is achieved with extremality of the distinction, by which entities $\emptyset$ and $\emptyset^*$ separate : $d(\emptyset,\emptyset^*) \equiv (\emptyset \cap \emptyset^*) = \emptyset$. We are meeting logical foundations to the mitosis ([2-b]): duplication and separation of living cells, implements an extremal spreading of information, as well as a non geometric modelization of the concept of action (set-volume of information tied to the set-distance **d**).

The above process of *decoherence* can be associated to irreversible-time (a new "dimension") , while more familiar dimensions must be restricted to the (t-symmetric) geometric environment, including the temporal lapse $t^+ - t^-$: a t-symmetric projection (near the event) of the irreversible process.

4. Antimatter and dark energy as dual entities: ultimate energies

The (extreme form of) energy associated to (irreversible) time-flow can be viewed as **dark-energy**. It emerges from potentialities, as 2-sphere integrals on (suitably rescaled components of) the (real) Weyl curvature tensor, which compensates for broken duality symmetry, or more broadly for bifurcations, hence an inertial connotation. An example is available with non linear (anti-) gravitons: these antiself-dual complex space-times (and related Weyl spinor), if complemented, lead to the real Weyl curvature, a jump which anticipates the emergence of antiparticles, as geometric entities (emerging with events at the "past-future" bifurcation: "peeling off" bifurcations of the real radiative regime, anticipate mass of isolated bodies). [A neutrino analogue of the Sagnac interferometer (optical analogue of the Foucault pendulum) was proposed in ([1-d]) for detection of the neutrino massive connotation, as a participation to compensation (the emergence of time-lapse t^+-t^- through space-time localization, being registered as interference pattern)]. This *enlarged energy* suggests a possible *substitute to the Higgs mechanism*. Event horizons (of the **black-hole** type) find an interpretation here, as boundaries to observable zones (manifestations of a logical context by which causal coherence sets in, black-holes can be compared to the background noise of a crowd, where censorship of the totality enables coherent conversation to set in, eg. between two interlocutors: the possibility of a multitude of worlds shaped by decoherence, eventually intertwined with ours, yet inaccessible to our perception, cannot be excluded). The well-known Schwarzschild time-machine (and black-hole) anticipates information recycling, also suggesting that space-time itself could emerge, not as a geometric box but as a censor invented from within, an implementation of divisible totality. Along these lines, c (the velocity of light), suggests limits to causal geometry and related concepts: space, time-lapse, velocity, etc: our perception of motion being a geometric perspective on information recycling (at a scaleless regime). This suggests a unified interaction tied to our (logical) correlation to totality (as subsystems), and perceived as irreversible time-flow. Beyond black-hole horizons, incoherence preventing focalization or organisation of events, inobservability sets in, which is better interpreted as a regime of potentialities, or incoherent memory. [Our scenario of cosmic decoherence is determined by self-reference (deterministic as opposed to probabilistic), does not require geometry, uncertainty being replaced by the (non metric set-) volume of information focalized on the event]. This clarifies the meaning of space-time singularities: the tearing of causal geometry on zones of incoherence (where ultimate expressions of energy are expected). Entities such as the antiself-dual "nlg-solutions" (bifurcations within complex geometry), and more broadly similar entities contributing to decoherence, can serve as models for **anti-matter**. Hence, both anti-matter or dark-energy seem to be hidden in the cosmic potentialities (non observable level). Their contribution to the objective reduction is of dual nature : on one hand, implementation of a bifurcation or logical choice (eliminating an alternative), which can turn into broken symmetry at the geometric level (eg. broken duality), and on the other hand, compensation leading to energy release (symmetry reinstalled). Anti-matter and dark-energy both hidden in the foundations, contribute to the critical equilibrium (of the logical context or cosmic tissue [1-d, e]) which accompanies decoherence.The related compensation for bifurcations, can be interpreted as a cosmic dilation (in analogy with inertial repulsion) to be traced back to the "part-totality" primordial bifurcation (a non localizable correlation).

5. Our contact with a reality

A principle of structuration (confrontation "symmetry-broken symmetry") being intertwined with the emergence of space-time itself, matter-sources of space-time geometry (the RHS of Einstein's equation) find their origin in the "subjective-objective" connotation of (immaterial) information. These (unavoidable) concepts are inherited from universal logics of observability, and our perception (relying on localization). We thus appear as participants to the cosmic decoherence: the *matching of subjective and objective decoherences* (via mental or environmental censorships) is perceived (by us) as contact with a reality [2-b] (a shared coherence, leading to awareness).

References:

1. A. Magnon :
(a) *"La Conscience Cosmique : la vie est-elle la clé du cosmos observable ?"* (monograph pp250 in press).
(b) *"Beyond Black-Hole Physics: towards a theory of perception "* (monograph pp220 in press).
(c) *"Arrow of Time and Reality : in search of a conciliation"* (monograph, World Scientific 1997, pp300, and references therein).
(d) *"Critical Cosmic State and Dark Energy : a logical pre-geometric approach"*, Conference at the Dark 2002 workshop, Cape-Town University, Edited by H.V.Klapdor-Kleingrothaus, Springer 2002.
(e) *" Les Théories Physiques de Newton à nos Jours "*, Conference « Journées Sciences en Fête 2003», Archives LPC, UBP.

2. R.Penrose :
(a) (with Rindler) *"Spinors and Space-Time"*, Vol.1 & 2, Cambridge University Press (1984).
(b) *"Shadows of the Mind : a search for the missing science of consciousness "*, monograph, Oxford University Press (1994).

DARK MATTER ON GALACTIC SCALES
(OR THE LACK THEREOF)

MICHAEL R. MERRIFIELD

School of Physics & Astronomy
University of Nottingham
University Park, Nottingham
NG7 2RD, United Kingdom
E-mail: michael.merrifield@nottingham.ac.uk

This paper presents a brief review of the evidence for dark matter in the Universe on the scales of galaxies. In the interests of critically and objectively testing the dark matter paradigm on these scales, this evidence is weighed against that from the only other game in town, modified Newtonian dynamics. The verdict is not as clear cut as one might have hoped.

1. Introduction

In the study of dark matter, its properties on the galactic scale are of abiding interest. It is, after all, the signature of a galaxy's dark halo that almost all of the current dark matter experiments are seeking to detect — they are trying to interact with the rather small part of the Milky Way's massive halo that happens to be passing through the experimenters' laboratories at the moment. Any predictions as to detection rates, etc, are therefore dependent on our understanding the properties of galactic-scale dark matter halos.

Galaxies have also played a pivotal role in the entire development of the dark matter paradigm. In one of the founding papers on the subject,[1] Zwicky used the galaxies in the Coma Cluster as test particles to which he applied the virial theorem. This calculation led to an unexpectedly high mass for the cluster, and Zwicky pointed out that if this total were divided amongst the constituent galaxies, then they would each have an average mass of $5 \times 10^{10} M_\odot$ (where $M_\odot$ is the mass of the Sun), whereas their average luminosity was estimated to be only $9 \times 10^7 L_\odot$ (where $L_\odot$ is the Sun's luminosity). Clearly, such galaxies could not be made up of stars like the Sun, and some extra contributor to the mass is required. In this prescient paper, Zwicky also pointed out that gravitational lensing should provide another approach to determining the masses of clusters, and that the rotation curves of individual galaxies — their speed of rotation as a function of radius, $v_c(r)$ — should allow one to probe the distribution of mass within these systems by equating the centripetal acceleration to the gravitational acceleration due to the mass.

Early studies of rotation curves were hampered by the fact that the optical emission lines used to measure Doppler shifts, and hence rotation speeds, could only be detected in the inner parts of galaxies. In general, at these small radii the inferred mass is quite consistent with the rather uncertain census of luminous components.[2] However, the discovery that spiral galaxies are surrounded by huge disks of atomic hydrogen completely changed the picture.[3] Using the Doppler shifts in this gas' 21cm radio emission, it became clear that the rotation speeds of such galaxies remain roughly constant out to the largest radii measured, $v_c \sim$ constant. This is certainly not what would have been predicted on the basis of the known luminous mass, as the measurements were made at radii well beyond most of these components, so one would have expected the velocities to have entered a Keplerian decline, with $v_c \propto r^{-1/2}$. These incontrovertible observations were fundamental ingredients in a huge shift of paradigm, taking astronomers from a position in which evidence for dark matter could be quietly swept under the cosmic rug to one in which it was recognised that dark matter seems to dominate the Universe on almost all scales.

Over the last twenty-five years, this recognition has spawned an entire industry of simulators trying to model the formation of structure in the Universe through the gravitational collapse of dark-matter-dominated material, concentrating on the idea that this mass comprises gravitationally-interacting non-relativistic "cold dark matter" (CDM). In the early days, these simulations lacked the resolution to study the properties of individual galaxies, but more recently very large computer simulations have achieved the dynamic range necessary to study the formation and properties of individual galaxies in CDM cosmology.[4] These simulations now make some fairly definite and robust predictions as to the distribution of dark matter on galactic scales. In particular, they predict that the dark matter should be distributed such that its density distribution has a central power-law cusp, $\rho \propto r^{-\gamma}$. Authors differ somewhat as to the steepness of this cusp,[4] but it seems to have a power-law index of between $\gamma = 1$ and $\gamma = 1.5$. However, all seem to agree that at large radii the density profile steepens to a power-law index of $\gamma = 3$. This all fits reasonably well with the observed flat rotation curves, which would be produced by a density profile with $\gamma = 2$, as the measured rotation curves typically probe intermediate radii between these two extreme regimes.

The simulations also predict the over-all shapes of the dark halos, which are typically oblate with a range of shortest-to-longest axis ratios of $\sim 0.5 \pm 0.2$.[5] This quantity is hard to determine observationally — the rotation curve of a spiral galaxy, for example, provides a useful measure of the radial distribution of mass in a galaxy, but not its distribution perpendicular to the plane of the galaxy's disk. However, probes of this third dimension do exist in the form of the orbits of material in merging satellites, the thickness of the gas layer, and the shapes of hot X-ray emitting gas halos around elliptical galaxies. As far as can be ascertained from these limited data, the observed shapes of halos are consistent with the CDM predictions.[6]

Not everything in the CDM simulations fits so well with the observations, though. In particular, the repeated merging of sub-halos that produces galaxies in these simulations does not produce a simple smooth density distribution. Instead, the simulations predict that one should see large amounts of substructure in the form of small satellite systems around every large galaxy. Although big objects like the Milky Way are accompanied by a retinue of smaller satellites, there seems to be a problem in that CDM simulations predict far more companions than are observed.[7] This galaxy formation picture is also beginning to run into trouble with observations at high redshift: in this hierarchical scenario, the largest most massive galaxies should form last, yet observations indicate that some fraction of very luminous galaxies are already in place quite early in the Universe's history.[8]

2. The Alternative to Dark Matter

Although the dark matter paradigm is almost universally accepted in the astronomical community, it does seem to have its limitations, so it still pays to step back from it occasionally and try to weigh it up against competing theories. In this regard, the only real alternative is the Modified Newtonian Dynamics or MOND hypothesis.[9,10] In this theory, the Newtonian acceleration due to gravity, a_n, is replaced by an acceleration a that obeys the equation $a_n = a \times \mu(|a|/a_0)$. The function μ is chosen so that a varies smoothly from a_n when $a \gg a_0$ to $\sqrt{a_n \times a_0}$ when $a \ll a_0$. The only free parameter in the theory is the characteristic acceleration constant a_0 at which the transition occurs.

This modification to Newton's laws neatly explains the flatness of the outer parts of galaxy rotation curves. At these large radii, Newtonian acceleration due to gravity is simply $a_n \approx \frac{GM}{r^2}$, where M is the galaxy's total mass. Centripetal acceleration here is sufficiently low that we are in the MOND regime, so we can write

$$\frac{v_c^2}{r} = a \approx \sqrt{\frac{GM}{r^2} \times a_0}. \qquad (1)$$

The r dependence cancels from the equation, so we find that

$$v_c \approx (GMa_0)^{1/4}, \qquad (2)$$

independent of radius.

This amendment to Newton's laws is somewhat *ad hoc*, and was clearly reverse-engineered to deal with the issue of flat rotation curves. However, it is quite legitimate to ask whether this solution is any more arbitrary than postulating the existence of invisible mass of an unspecified nature to explain away the perceived problem of galaxies' rotation curves. It is also worth noting that MOND provides explanations for various other astronomical phenomena at no extra charge. For example, there is observed to be a tight correlation between asymptotic rotation speeds and luminosities in galaxies, such that $L \propto v_c^4$. This proportionality, known as the Tully–Fisher relation,[11] has no particularly fundamental explanation in standard Newtonian gravity. However, if all that a galaxy contains is its luminous stars, so that $L \propto M$, then the Tully–Fisher relation would follow trivially from Eq. (2) if MOND were correct.

A further criticism arising from the *ad hoc* nature of MOND is that it has no satisfactory relativistic generalisation. However, this situation changed dramatically earlier this year when a fully relativistic field theory was developed that reduces to Newtonian dynamics in the low-velocity regime and to MOND at low accelerations.[12] This theory for the first time makes a robust prediction as to what gravitational lensing signature one might expect from MOND — another shortcoming often pointed out in the theory — and shows that it will be of the same amplitude as that predicted on the basis of dark matter models.

With this new relativistic formulation, it is also now possible to carry out fully self-consistent tests of MOND such as performing simulations of structure formation in a cosmological context, to see if the theory fits with other astronomical observations as closely as the CDM simulations. Although no-one has yet carried out such simulations, there are already some interesting insights that one can obtain into the subject. The longer range nature of gravity in MOND means that the two-body relaxation time and dynamical friction timescale are dramatically shorter than in Newtonian gravity.[13] One might therefore expect the over-abundance of substructure found in the CDM simulations (see Sec. 1) to be effectively wiped out in a MOND universe, as such structure should relax into a smooth distribution on a relatively short timescale. However, whether this simple heuristic argument works in detail will only become clear once some full cosmological MOND simulations are performed.

3. Case Studies

To further test the dark matter paradigm on galactic scales, let us now turn to a few specific types of galaxy to see how well the data fit the theory. There is not much point in using normal spiral galaxies for such tests, as these systems played a key role in the development of the dark matter paradigm (and, indeed, of MOND as well); it would be very surprising if the theory failed to reproduce the properties that so strongly motivated it in the first place. Instead, we look to other types of galaxy for some independent confirmation or refutation of the theory.

3.1. *Low Surface Brightness Galaxies*

Actually, spiral galaxies are rather poor places to test the dark matter paradigm for other reasons, too. One of the clear predictions in CDM is that the dark matter should have a central cusp. However, the large amount of luminous matter at small radii in these systems means that such cusps will have very little impact on the rotation curves of spiral galaxies. Further, the gravitational interaction between luminous and dark matter could well have redistributed the dark matter, so any primordial central cusp may have been completely wiped out.

Fortunately, a class of galaxies exists in which these issues should not arise. These "low surface brightness galaxies" contain a very low density of luminous material even in the inner parts, so that the observed dynamics should be dominated by the gravitational forces of the dark halo at small radii as well as large radii. Further, the small amount of luminous matter should not have had much ability to redistribute the dark matter, so the central cusp should still be there.

Rather disturbingly, as illustrated in Fig. 1, the rotation curves of these systems do not seem to match up to the predictions of the CDM models. Although they flatten off to the constant rotation velocity characteristic of a dark matter halo, they rise significantly more slowly in their central parts

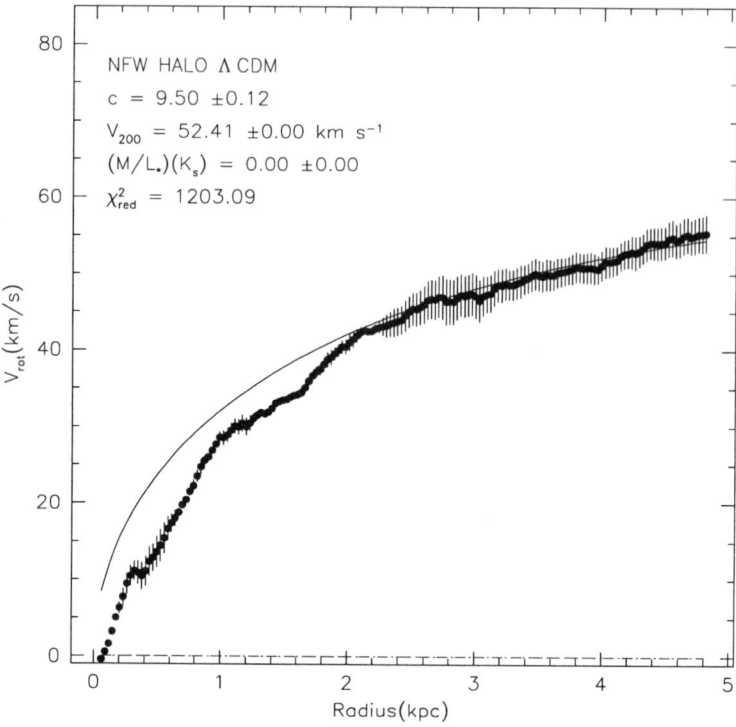

Figure 1. Rotation curve (circular rotation velocity as a function of radius) for the low surface brightness galaxy NGC 6822. The points show data obtained from the 21cm emission of atomic hydrogen, while the line gives the best-fit model assuming that the galaxy's mass is dominated by a centrally-cusped dark matter halo. (Figure kindly provided by W.J.G. de Blok.)

than one would expect for a system with a centrally-concentrated cusped mass distribution.[14]

Once again, MOND seems to do rather better.[15] Not only does it reproduce the observed slowly-rising rotation curves, but in many cases it can also explain the match up between localized features in the rotation curves and small-scale structure in the photometry — if these low surface brightness galaxies were dominated by dark matter halos, then the faint features in the luminous light distribution should have almost no direct impact on the rotation curve. However, one has to keep a critical eye on how illustrative examples are selected: in the case of Ref. [10], the highlighted low surface brightness galaxy in their Fig. 3 shows a beautiful match in the small-scale structure of the rotation curve and the MOND predictions, whereas the broader range of cases in their Figs. 4 and 5 show a number of examples where there is no such correspondence.

3.2. Elliptical Galaxies

The other main class of luminous galaxy are the rather dull-looking elliptical systems. Studies of the dark matter in these objects have been hampered for several reasons. First, they tend to be found in clusters, so it becomes difficult (or even an issue of semantics) to separate any dark matter associated with galaxies from that which should be attributed to the cluster over all. Second, they lack the extensive gas disks that surround spiral galaxies. They therefore do not have a simple kinematic tracer that allows their masses to be measured out to large radii.

For the largest elliptical galaxies, alternative probes such as gravitational lensing[16] and X-ray emitting gas[17] have been used to infer that these systems have extensive massive halos. However, such bright galaxies tend to lie at the centres of clusters, raising the issue of whether the halo belongs to the galaxy or the cluster.

For more "normal" ellipticals, comparable in luminosity to the Milky Way, not a lot of data have been available. However, we have recently developed a new instrument, the Planetary Nebula Spectrograph, specifically to study such objects.[18] This instrument detects and measures the kinematics of planetary nebulae (PNe) in the outer parts of elliptical galaxies. Since PNe are simply stars that have reached the ends of their lives, their kinematics will be representative of the over-all stellar population (but much easier to measure due to the bright emission lines in PNe spectra). The initial results that we have obtained using this instrument were not at all what we expected.[19] As Fig. 2 shows, the random velocities of the PNe do not stay constant out to large radii, as one might expect by analogy with the rotation curve of a comparable spiral galaxy; instead they seem to go into Keplerian decline, as would be expected in the complete absence of dark matter.

There are a number of possible explanations for this result that would not involve making as radical a change as to suggest that these systems are devoid of dark matter. First, there is an extra degree of freedom that we did not have to worry about for spiral galaxies: while the gas in disks around spirals all follows orbits that are close to circular, we have no *a priori* way

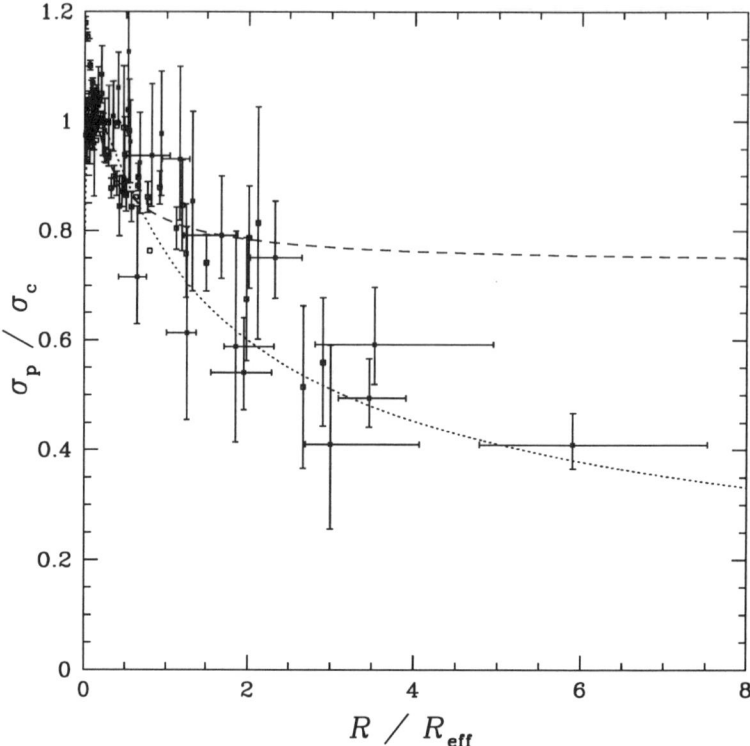

Figure 2. A compilation of the stellar line-of-sight velocity dispersion as a function of projected radius for four intermediate-luminosity elliptical galaxies. The data for different galaxies have been normalized by the central velocity dispersion and the effective radius (the radius within which half of the galaxy's light appears projected) to make them directly comparable. The dashed line shows the predicted trend in this relation if these galaxies were embedded in massive halos comparable to those found around similar luminosity spirals. The dotted line shows the Keplerian decline that would be found in the absence of a dark halo.

of knowing whether the stars in ellipticals (and hence their PNe) follow circular orbits, radial orbits, or something in between. Since objects on radial orbits will move mostly transverse to the line of sight at large radii, one would expect the measured line-of-sight velocity dispersion to drop for a system consisting of such objects. However, it would be a surprising coincidence if such an extreme collection of orbits managed to mimic the behaviour of a Keplerian decline; indeed, detailed orbit modelling indicates that such a solution is not consistent with the data.[19] Alternatively, these galaxies could be surrounded by halos that are so extended that, even at the unprecedentedly large radii to which we are now looking, the dark matter still does not dominate the mass. However, this scenario also seems to conflict with CDM models, which predict that these moderate-luminosity galaxies should have dark halos that are more centrally concentrated than their brighter kin, not less so.[20]

Although this result was not what we were expecting, it did not come as a surprise to the advocates of MOND. Indeed, Ref. [10] had already made the definite prediction that lower-luminosity elliptical galaxies should have velocity dispersion profiles that fall with radius, and subsequent fitting to our data confirmed that the observed approximately Keplerian decline is what would be expected in MOND.[21] Essentially, the absence of any dark matter signature in these systems arises because elliptical galaxies are more centrally concentrated than their spiral cousins, which means that the characteristic accelerations within them stay safely above the MOND a_0 threshold. Thus, their observed kinematics should be consistent with ordinary Newtonian physics, with no non-standard dynamics to be misinterpreted as dark matter.

4. Conclusions

The idea that we live in a Universe dominated by dark matter is so deeply embedded in most astronomers' World views that it is not something that many of us ever question. Nonetheless, all of the evidence that we currently have for dark matter is highly circumstantial, so we should at least compare the hypothesis critically with any alternatives.

In this regard, a comparison between the dark matter hypothesis and modified Newtonian dynamics, at least on the galactic scale, produces a rather equivocal answer. In a number of cases, MOND seems to fit the observations rather better than dark matter. There are also examples where MOND has passed the ultimate test of a scientific theory by making predictions that differ from the dark matter theory, which subsequently turn out to be true. On scales other than the galactic, perhaps MOND does a little less well: even its proponents recognise that there seems to be a problem at the size of clusters, with MOND predicting more mass than can be found within the known luminous components.[22] However, until some serious effort is invested in proper cosmological MOND simulations, we will not be able to establish the context that will determine quite how serious any problems might be with this theory.

As things stand, it is principally a matter of aesthetics as to which idea Occam's Razor favours. Personally, I would still go for the dark matter theory: it is astoundingly arrogant to assume that everything in the Universe should glow in the dark for the benefit of astronomers, so invoking dark matter appeals to me as a way to remind us of our own insignificance. Nonetheless, I recognise the appeal of a simple modification to the law of gravity over the invocation of an entirely unknown form of matter, so am unable to draw any definite conclusions even on this aesthetic issue.

In the context of this particular volume, however, the message is much more clear cut: the direct laboratory detection of massive particles from the Milky Way's halo would provide by far the most convincing confirmation of the whole dark matter paradigm, and would lay this issue to rest once and for all.

References

1. F. Zwicky, *Astrophys. J.* **86**, 217 (1937).
2. S. M. Kent, *Astron. J.* **91**, 1301 (1986).
3. S. M. Kent, *Astron. J.* **93**, 1816 (1987).
4. e.g. J. F. Navarro et al., *M.N.R.A.S.* **349**, 1039 (2004).
5. J. Dubinski, *Astrophys. J.* **431**, 617 (1994).
6. M. R. Merrifield, in *IAU Symposium 220*, 431 (Astronomical Society of the Pacific, San Francisco, 2004).
7. B. Moore et al., *Astrophys. J.* **524**, L19 (1999).
8. R. S. Somerville et al., *Astrophys. J.* **600**, L135 (2004).
9. M. Milgrom, *Astrophys. J.* **270**, 371 (1983).
10. A thorough review can be found in R. H. Sanders and S. S. McGaugh, *Ann. Rev. Astron. Astrophys.* **40**, 263 (2002).
11. R. B. Tully and J. R. Fisher, *Astron. Astrophys.* **54**, 661 (1983).
12. J. D. Bekenstein, *Phys. Rev.* **D70**, 083509 (2004).
13. L. Ciotti and J. Binney, *M.N.R.A.S.* **351**, 285 (2004).
14. W. J. G. de Blok and A. Bosma, *Astron. Astrophys.* **385**, 816 (2002).
15. W. J. G. de Blok and S. S. McGaugh, *Astrophys. J.* **508**, 132 (1998).
16. C. R. Keeton, *Astrophys. J.* **561**, 46 (2001).
17. M. Loewenstein and R. E. White, *Astrophys. J.* **518**, 50 (1999).
18. N. G. Douglas et al., *P.A.S.P.* **114**, 1234 (2002).
19. A. J. Romanowsky et al., *Science* **301**, 1696 (2003).
20. N. R. Napolitano et al., *M.N.R.A.S.*, in press (astro-ph/0411639) (2004).
21. M. Milgrom and R. H. Sanders, *Astrophys. J.* **599**, L28 (2003).
22. R. H. Sanders, *Astrophys. J.* **512**, L23 (1999).

GALACTIC MODELS AND THE SEARCH FOR DARK MATTER

LAWRENCE M. WIDROW

Department of Physics,
Queen's University,
Kingston, K7L 3N6, Canada
E-mail: widrow@astro.queensu.ca

The notion that dark halos are composed of exotic elementary particles grew out of the exchange of ideas between particle theory and cosmology. Direct detection of these particles requires further input from the field of Galactic dynamics. I will explore this connection and also describe a new set of self-consistent, multi-component models for disk galaxies.

1. Introduction

Dark matter detection experiments probe the distribution function (DF) of dark matter in the Galaxy. For example, terrestrial experiments are sensitive to the speed (and in some cases directional) distribution of dark matter in the vicinity of the Earth. At present, our community is focused on finding convincing evidence that Galactic dark matter is indeed in the form of elementary particles. When and if dark matter is discovered, attention will turn to its particle physics properties and to its DF, the latter providing a unique window into the structure and formation history of the Galaxy.

Terrestrial experiments typically quote results in terms of a standard model for the local dark matter distribution in which the density is $\rho_0 = 0.3\,\text{GeV}\,\text{cm}^{-3} \simeq 0.0079\,M_\odot\,\text{pc}^{-3}$ and the velocity distribution is Maxwellian (in Galactocentric coordinates) with rms speed $\langle v^2 \rangle^{1/2} \simeq 270\,\text{km s}^{-1}$ [1]. The Earth is displaced from the center of the Maxwellian by the Sun's motion around the Galactic center and the Earth's motion around the Sun, the latter providing the famous seaonal modulation of the dark matter signal. A standard reference model enables one to compare sensitivities of different experiments and also to make contact with theoretical particle physics models while sidestepping the complications that arise from astrophysical uncertainties.

Numerous authors, many of them at this conference, have explored dark matter detection in non-standard Galactic models. These efforts serve first to quantify astrophysical uncertainties in the dark matter DF and second to alert experimentalists to the potential for unusual "features" in their data that might arise from halo structure. One may consider variations of the standard model that are quantified by a small set of parameters (e.g., velocity space anisotropy or bulk rotation) or models based on analytic DFs. However these approaches explore just a small subset of the space of possible models and may, in fact, veer into unphysical regions of parameter space where models are unstable to bar formation or are incompatible with astronomical observations and/or theories of structure formation.

In this talk I review the observational and theoretical constraints that define the landscape of acceptable Galactic halo models. I then describe new, flexible set of self-consistent Galactic models [2].

2. Constraints on Galactic Model

First and foremost, Galactic models are constrained by astronomical observations of the Galaxy (see Ref.[3,4] and Merrifield's contribution to these proceedings). Astrophysical constraints include measurements of local quantities such as the Oort constants, the local velocity ellipsoid of stars in the solar neighborhood, and the vertical force near the position of the Sun. Equally important are observations of global quantities such as the circular rotation curve, the velocity dispersion profile in the bulge region, and the kinematics of satellites.

Dehnen and Binney have considered a suite of Galactic models that include an exponential disk, bulge, and dark halo [3]. They take the halo density profile to be of the form

$$\rho_{\text{halo}} = \frac{\rho_0}{(r/r_0)^\gamma (1+r/r_0)^{\beta-\gamma}} \quad (1)$$

where γ is allowed to vary from -2 (a "hollow" halo) to 1 (the Navarro, Frenk, and White (NFW) universal halo profile [5]) to 1.8 (nearly that of the singular isothermal sphere). For each of their 22 models, one can calculate the local dark matter density. The results range from $(0.0063 - 0.19)\,M_\odot\,\text{pc}^{-3}$, a spread of a factor of 3 with the "standard" value toward the low end of this range.

Further constraints come from theoretical considerations and N-body simulations. It is well-known that a rotationally supported self-gravitating disk is unstable to the formation of a bar. Conversely, the disk will be stable if much of its gravitational support is provided by a dynamically "hot" system such as a bulge or halo. (This argument was central to the seminal work by Ostriker and Peebles which proposed the existence dark halos [6].) Dynamical studies of the bar instability may allow us to cull model sets such as the one presented [3].

Simulations of structure formation in a dark matter dominated Universe suggest the following:

- Halo centers are cuspy. The exact value of the logarithmic index as $r \to 0$ (γ in Eq. (1)) is a matter of some debate but appears to be between 1 and 1.5. Halo density profiles gradually steepen with radius with $\rho \propto r^{-3}$ providing a good fit to the outer regions of simulated halos.
- Halos are triaxial with axis ratios between 0.6 and 1 with a slight preference for prolate halos over oblate ones [7].
- The velocity distribution of halo particles is biased toward radial orbits. A useful parameter is the velocity anisotry parameter (see, for example, [8]):

$$\beta \equiv 1 - \frac{\langle v_t^2 \rangle}{2\langle v_r^2 \rangle} \ . \qquad (2)$$

$\beta = 0$ for isotropic distribution, $\beta = 1$ for pure radial orbits, and $\beta \to -\infty$ for pure circular orbits. Typical simulated halos have $\beta \simeq 0.6$ [9].
- Halos are clumpy. The amount of substructure in simulated halos appears to be limited by mass resolution of the simulations in the more particles one uses, the more substructure one sees. That said, only a small fraction of the mass in a typical halo appears to be locked up in subclumps (See [10] and references therein) .

These results may have important implications for dark matter detection experiments. For example, velocity-space anisotry has a dramatic effect on the directional signal that would be measured in experiments [11]. Substructure might lead to features in a WIMP energy recoil spectrum or axion energy spectrum [12-14]. A closely related issue concerns the finegrain structure of the dark matter DF. In the cold dark matter scenario, dark matter has negligible velocity dispersion at early times and therefore its DF is a thin, 3-dimensional sheet in 6-dimensional phase space. This sheet is curled up through the process of structure formation but retains its 3-dimensional character. Thus, a perfect detector, namely one with arbitrarily high resolution, would find discrete peaks in velocity space [15-17].

There is one important caveat. The simulations that lead to the results listed above by and large did not include the dissipational matter that forms the disk and bulge.

3. Disk-Bulge-Halo Models for the Milky Way and M31

A new set of self-consistent, multi-component disk galaxy models has been proposed [2]. The models consist of an NFW halo, a Hernquist bulge [18], an exponential disk, and a central supermassive blackhole. The DFs are constructed from chosen functions of the integrals of motion (axisymmetry is assumed) and therefore represent solutions to the coupled collisionless-Poisson equations. Self-consistency of the density and potential are achieved through an iterative scheme as described in Kuijken and Dubinski [19] wherein the predecessors of these models were presented.

The purpose of our models is two-fold. First, they provide high quality initial conditions for N-body experiments. Though our models represent equilibrium axisymmetric systems, they are subject to non-axisymmetric instabilities such as those that lead to the formation of bars and spiral structure. Our models thus allow for the detailed study of these instabilities. In addition, they may be used in investigations of the tidal disruption of satellite systems and mergers of comparably-sized galaxies.

Our models have the 15 free parameters and thus have the flexibility to match a wide range of observations for individual galaxies. For a given set of observations and a given model, one can generate mock data and calculate a χ^2 statistic describing the quality of agreement between model and observations. An efficient parameter search algorithm known as the simplex method (See, Ref. [20] and also [21] for its application to the original Kuijken and Dubinski models) allows one to search quickly the parameter space and hone in on models that describe a particular galaxy. The plural is used here because observational data are not capable of pinpointing a particular model. In particular, the relative contributions of the disk and halo to the rotation curve is poorly constrained because their shapes through the visible parts of most galaxies, are farily similar. The degeneracy is essentially an uncertainty in the mass-to-light ratio of the disk, a quantity which may be constrained by stellar population studies (see, Bell and de Jong [22]). Alternatively, one may use dynamics to constrain the disk-to-halo ratio. Consider a sequence of models described by a single parameter, the mass-to-light ratio of the disk. Each of the models is chosen from the set of disk-bulge-halo models to fit the observational data for a particular galaxy. Other parameters such as the bulge mass and halo scale length vary along the sequence. What are held relatively constant are predictions for observations such as the rotation curve and surface brightness profile. An N-body realization of each of the models evolved using standard techniques and the results are then analysed for the presence of a bar and compared with the actual galaxy. Figure 1 shows two models that fit data for the Galaxy M31. Though the fits are comparable, the models have drastically different stability properties. The model on the left is stable while the model on the right forms a strong bar. A similar analysis of Milky Way models suggests that the disk is lighter than in the standard reference model and therefore the local dark matter density is larger than the oft-quoted $0.3\,\text{GeV}\,\text{cm}^{-3}$.

4. Acknowledgments

This work was supported, in part, by a grant from the Natural Sciences and Engineering Research Council of Canada. I thank my collaborators David Stiff and John Dubinski.

References

1. G. Jungman, M. Kamionkowski, and K. Griest, *Phys. Rep.*, **267**, 195 (1996).
2. L. M. Widrow and J. Dubinski, *in preparation*.
3. W. Dehnen and J. Binney, *Mon. Not. R. Astr. Soc.* **294**, 429 (1998).
4. J. Binney and M. Merrifield Galactic Astronomy, Princeton Univ. Press, Princeton (1998).
5. J. F. Navarro, C. S. Frenk, and S. D. M. White, *Astrophs. J.* **462**, 563 (1996).
6. J. P. Ostriker and P. J. E. Peebles, *Astrophys. J.* **186**, 467 (1973).
7. M. W. Warren, P. J. Quinn, J. K. Salmon, and W. H. Zurek, *Astrophys. J.* **399**, 405, 1992.
8. J. Binney and S. Tremaine Galactic Dynamics, Princeton Univ. Press, Princeton (1987)
9. F. C. van den Bosch, G. F. Lewis, G. Lake, and J. Stadel, *Astrophys. J.* **515**, 50 (1999).
10. A. R. Zentner and J. S. Bullock, *Astrophys. J.* **598**, 49 (2003).
11. B. Morgan, A. M. Green, and N. J. C. Spooner, *astro-ph*/0408047.
12. D. Stiff, L. M. Widrow, and J. Frieman, *Phys. Rev. D* **64**, 083516.
13. K. Freese, P. Gondolo, and L. Stodolsky, *Phys. Rev. D* **64**, 123502.
14. A. M. Green, *Phys. Rev. D* **66**, 083003 (2002).
15. P. Sikivie and J. R. Ipser, *Phys. Lett. B*, **291**, 288 (1992).
16. P. Sikivie, I. I. Tkachev, and Y. Wang, *Phys. Rev. Lett.* **75**, 2911 (1995).

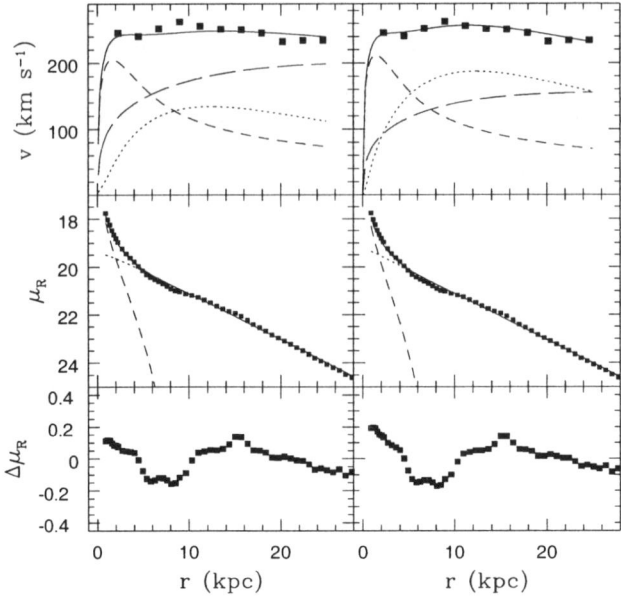

Figure 1. Two models for the galaxy M31. Top: circular rotation curve with bulge (dashed curve), disk (dotted curve), halo (long dashed curve), and total (solid curve). Middle: Surface brightness profile. Bottom: residual between data and model of surface brightness profile. For a discussion of the data (squares) used in the figure, see [21] and references therein. Model on the left has a relatively light disk and is stable against the formation of a bar. The model on the right has a heavy disk and is unstable to bar formation.

17. D. Stiff and L. M. Widrow, *Phys. Rev. Lett.* **90**, 1301 (2003).
18. L. Hernquist, *Astrophys. J.* **356**, 359 (1990).
19. K. Kuijken, K. and J. Dubinski *Mon. Not. R. Astr. Soc.*, **277**, 1341 (1995)
20. W. H. Press, B. P. Flannery, S. A. Teukolksy, and W. T. Vettelring, *Numerical Recipes*, (Cambridge: Cambridge University Press) (1986).
21. L. M. Widrow, K. Perrett, and S. Suyu *Astrophys. J.* **588**, 311 (2003).
22. E. F. Bell and R. S. de Jong, *Astro. Phys. J.* **550**, 212 (2001).

WEIGHING THE DARK MATTER HALO

JACOB L. BOURJAILY*

*Michigan Center for Theoretical Physics,
University of Michigan, Ann Arbor, MI 48109-1120*

The dark matter problem will be solved only when all of the dark matter is accounted for. Although wimps may be discovered in direct detection experiments soon, we will not know what fraction of the dark matter halo they compose until we measure their local density. In this talk, I will offer a novel method to determine the mass of a wimp from direct detection experiments alone using kinematical consistency constraints. I will then describe a general method to estimate the local density of wimps using both dark matter detection and hadron collider data when it becomes available. These results were obtained in collaboration with Gordon Kane at the University of Michigan.

1. Introduction

The direct detection of wimps in the galactic halo would be an enormous triumph of experimental and theoretical particle cosmology, have deep implications for our understanding of the universe, and would explain (at least) some of the dark matter in the universe. However, it is unreasonable to assume that the entire dark matter halo is composed of only the wimp observed.

What fraction of the dark matter halo is represented by a particular wimp is a question that cannot be answered by direct detection experiments alone or colliders alone. Even if a stable, weakly interacting massive particle is discovered at the LHC, no amount of collider data can measure its *actual* local density. Alternatively, even if many direct detection experiments unambiguously observe the scattering of wimps from the halo, these experiments cannot identify the wimp or determine its couplings—crucial to measuring the local density. This point was first raised technically in Ref.[1]. In this talk, I will present a general way to address the cosmological significance of wimps observed in direct detection experiments using data and bounds from colliders. For a more detailed discussion, see Ref.[2].

---
*jbourj@umich.edu

2. The No-Lose Theorem vs. Our Ability to Win

Let us imagine that a weakly interacting massive particle χ has been unambiguously observed in direct detection experiments. There is no reason to suspect that χ is *all* the dark matter. In fact, it is very easy to find models where detectable wimps compose only a fraction of a percent of all the dark matter. Using the framework of the DarkSUSY code[3], we randomly generated some six thousand constrained MSSMs and found that for any signal rate, the relic density fluctuates over at least two orders of magnitude[2a]. It appears as if the detection signal rate and relic density are largely uncorrelated: a wimp with a low relic density may be just as observable as one making up most of the dark matter.

The fact that particles with even small relic densities can have large detection signals has been noted by many authors (see, *e.g.* Ref.[4]) and has sometimes referred to as the 'no-lose theorem:' experimentalists may not lose out on discovering even a very tiny faction of the dark matter halo. However, the no-lose theorem also implies that a wimp discovery could easily represent a negligible fraction of the dark matter. Therefore, although wimps may be discovered in the near future, the dark matter problem will not be solved until the density of wimps has been directly determined, and all the dark matter is accounted for.

3. Dark Matter Direct Detection Rates

In general, the wimp-nucleon elastic scattering scattering rate is a function of the cross section for scattering, nuclear physics describing the detector, and the local velocity profile of the wimp fraction of the dark matter halo. If a detector is composed of nuclei labeled by the index j, each with mass fraction c_j, then the differential rate of wimp scattering at recoil energy q is given by[b],

$$\left.\frac{dR}{dQ}\right|_{Q=q} = \frac{2\rho_\chi}{\pi m_\chi}\sum_j c_j \int_{v_{\min_j}(q)}^{\infty} \frac{f(v,t)}{v}dv \left\{ F_j^2(q)[Z_j f_p + (A_j - Z_j)f_n]^2 \right.$$
$$\left. + \frac{4\pi}{(2J_j+1)}\left[a_0^2 S_{j_{00}}(q) + a_1^2 S_{j_{11}}(q) + a_0 a_1 S_{j_{01}}(q)\right]\right\}, \quad (1)$$

---

[a]Only an upper bound on the relic density, consistent with WMAP, was imposed while generating these models.
[b]A detailed discussion of equation 1 can be found in most modern reviews of dark matter, (see, *e.g.*, Ref.[5]).

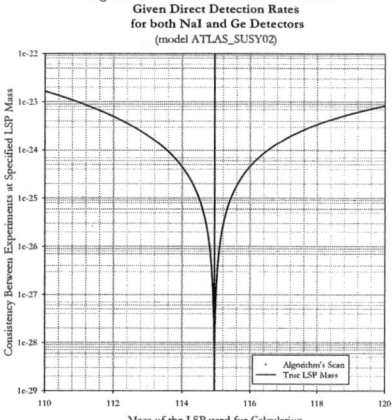

Figure 1. The function $\zeta(m'_\chi)$ where the wimp corresponds to the neutralino in the MSSM specified by ATLAS SUSY point 2. The models and data were generated within the framework of the DarkSUSY package.

where $f(v,t)$ is the halo velocity profile, $F_j^2(q)$ and $S_{j_{mn}}(q)$ are nuclear form factors, $a_0 \equiv a_p + a_n$ and $a_1 \equiv a_p - a_n$, and the constant parameters $f_{p,n}$ and $a_{p,n}$ describe the coherent and incoherent wimp-nucleon scattering cross sections, respectively.

From the equation above, it is clear that to determine the density ρ_χ, one must identify the particle, determine its mass, estimate the halo profile, and 'know' the interaction parameters from the theory describing χ. Each of these require enormous efforts of both dark matter and collider experiments.

4. Determining the Wimp Mass

Initially, perhaps the most important parameter of χ to determine is its mass. Not only does m_χ determine the particle's kinematics, but it may be critical to the identification of the particle[c].

There are two known ways to determine m_χ from direct detection data alone. One method, using the annual modulation crossing energy, was described in a dark matter review article by Primack et. al. in 1988[6][d]. The other method has been developed by the author and is described presently.

If the halo velocity profile and m_χ are known, then direct detection data from different detector materials and different energies can be used to

---

[c] One may hope that accelerators observe a weakly interacting, stable, massive particle with the same mass as χ.

[d] Although it seems unlikely to have originated in a review article, we have been unable to find any earlier reference.

solve for $\sqrt{\rho_\chi} f_{p,n}$ and $\sqrt{\rho_\chi} a_{p,n}$ by inverting equation 1. If the halo velocity profile can be approximated, only m_χ is required to compute $\sqrt{\rho_\chi} f_{p,n}$ and $\sqrt{\rho_\chi} a_{p,n}$ with sufficient data. We can generally expect to have many more measurements than the minimum required to solve the system of equations once wimps are observed.

Because the interaction parameters are constant, all linearly independent combinations of measurements used to solve for the scaled interaction parameters must agree, if the correct mass were used in the derivation. This motivates us to define a 'kinematical consistency' function $\zeta(m'_\chi)$,

$$\zeta(m'_\chi) \equiv \sqrt{\rho_\chi} \sum_{i \neq j} \left\{ (a_p(i) - a_p(j))^2 + (a_n(i) - a_n(j))^2 + \text{similar terms} \right\},$$

which compares the values of $\sqrt{\rho_\chi} f_{p,n}$ or $\sqrt{\rho_\chi} a_{p,n}$ obtained using different independent subsets of the data—indexed by i, j—as a function of m'_χ used to invert the equations. It is obviously necessary that $\zeta(m'_\chi) = 0$ when $m'_\chi = m_\chi$.[e]

To determine the wimp mass, one varies m'_χ until $\zeta(m'_\chi) = 0$. We applied this test to some six thousand random, constrained MSSMs. For every single model tested, the correct mass was determined to near-arbitrary precision. Figure 1 illustrates a typical plot of $\zeta(m'_\chi)$. Notice that it has an extremely sharp minimum within a few GeV of the true wimp mass. It should be noted, however, that experimental uncertainties and resolutions were not considered during these calculations.

5. Neutralino Dark Matter and a Strict Lower Bound on ρ_χ

Before we can compute ρ_χ, the interaction parameters $f_{p,n}$ or $a_{p,n}$ must be 'known.' These interaction parameters depend on very detailed knowledge of the particle physics of χ. Let us consider the specific case in which the discovered particle is the neutralino. The interaction parameters will then depend on many details of the MSSM—these may not be known until well after dark matter particles have been discovered. Nevertheless, even with extremely limited knowledge of the MSSM, we can estimate the parameters using partial data, bounds, and constraints.

For example, we have found that given bounds on $\tan \beta$ and a lower bound on the lightest squark mass, $m_{\tilde{q}}$, there is a strict upper bound for the incoherent χ-quark scattering parameters. In this case, it can be shown,

---

[e]This may not be a sufficient condition, however; although, we have found no example where $\zeta(m'_\chi) = 0$ when $m'_\chi \neq m_\chi$.

that the magnitude of a_u is strictly bounded by

$$a_u \leq \frac{g^2}{16m_W^2}(N_{\tilde{H}_1}^2 - N_{\tilde{H}_2}^2) + \frac{g^2}{8}\frac{1}{(m_{\tilde{q}_\ell}^2 - (m_\chi^2 + m_u)^2}\left\{\frac{17}{18}\tan^2\theta_W N_{\tilde{B}}^2 + \frac{1}{2}N_{\tilde{W}}^2\right.$$
$$+ \frac{m_u^2}{m_W^2 \sin^2\beta_\ell}N_{\tilde{H}_2}^2 + \frac{1}{3}\tan\theta_W |N_{\tilde{B}}||N_{\tilde{W}}|\cos(\alpha_{\tilde{W}})$$
$$\left.+\frac{m_u}{m_W \sin\beta_\ell}|N_{\tilde{W}}||N_{\tilde{H}_2}|\cos(\alpha_{\tilde{H}_2}-\alpha_{\tilde{W}})-\frac{m_u}{m_W \sin\beta_\ell}\tan\theta_W|N_{\tilde{B}}||N_{\tilde{H}_2}|\cos(\alpha_{\tilde{H}_2})\right\},$$

where $\alpha_{\tilde{H}_2}$, and $\alpha_{\tilde{W}}$ are the relative phases between $N_{\tilde{H}_2}, N_{\tilde{W}}$ and $N_{\tilde{B}}$, respectively. This expression has six real unknowns. Notice that by the normalization of the neutralino wave function, the parameter space is compact. Therefore, a_u can be *absolutely* maximized with respect to all six unknowns. It should be emphasized that this analysis is for the most general softly-broken MSSM; no *ad hoc* supersymmetry breaking scenarios—such as mSUGRA—were assumed.

Because we can determine $\sqrt{\rho_\chi}a_{p,n}$, the strong upper bounds on $a_{p,n}$ directly translate into strong lower bounds on ρ_χ. To test this idea, we considered some six thousand randomly generated, constrained MSSMs. For each of these models, upper bounds were calculated for $a_{p,n}$ assuming 10% uncertainty in $\tan\beta$ and a lower bound on the lowest squark mass of either 200 GeV or the actual mass of the lightest squark, whichever is less. The specific gauge content of the neutralino was taken to be known for each model for computational simplicity[f]. Using the upper bounds for $a_{p,n}$, we obtain a lower bound on the local density ρ_χ.

Figure 2 illustrates the results of using this algorithm. Notice that the estimated local density is always strictly less than the true local density. Also, for many models the lower bound is not such a poor estimate.

6. Conclusions

We have seen that, by itself, a discovery of dark matter particles in our galactic halo cannot solve the dark matter problem. However, combined with data from colliders to identify a particle and determine its interaction parameters, we can generally estimate its local density.

We presented a robust, model independent method to determine the mass of a wimp using direct detector data alone. We have shown explicitly

---

[f]If the gauge content of the neutralino was unknown, the interaction parameters could have been maximized with respect to these parameters as described earlier. In general, therefore, the upper bounds plotted are more restrictive than they would be in practice.

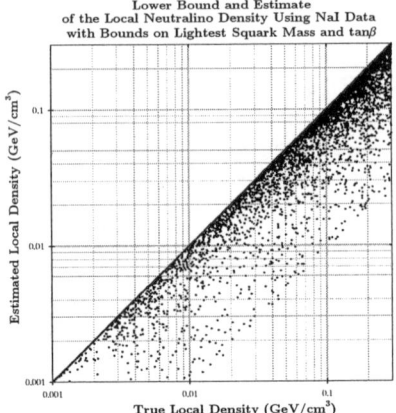

Figure 2. This plot compares the lower bound and estimate of the local denisty computed using the strong upper bound for $a_{p,n}$ to the true local density for each model.

how one can determine or estimate ρ_χ in the case where χ is the neutralino.

Although the dark matter problem may not be solved immediately when wimps are discovered, there are clear and general ways to address their cosmological significance.

Acknowledgements

This research was done in collaboration with Gordon Kane of the University of Michigan and was supported by the Michigan Center for Theoretical Physics and the National Science Foundation's 2004 REU program.

I would like to thank the organizers of the 5th International Workshop on the Identification of Dark Mater for the wonderful program.

References

1. M. Brhlik, D. J. H. Chung, and G. L. Kane, "Weighing the Universe with Accelerators and Detectors," *Int. J. Mod. Phys.*, vol. D10, p. 367, 2001, hep-ph/0005158.
2. J. L. Bourjaily, "Determining the Actual Local Density of Dark Matter Particles," 2004, astro-ph/0410470.
3. P. Gondolo et al., "Darksusy: Computing Supersymmetric Dark Matter Properties Numerically," *JCAP*, vol. 0407, p. 008, 2004, astro-ph/0406204.
4. G. Duda, G. Gelmini, P. Gondolo, J. Edsjo, and J. Silk, "Indirect Detection of a Subdominant Density Component of Cold Dark Matter," *Phys. Rev.*, vol. D67, p. 023505, 2003, hep-ph/0209266.
5. G. Jungman, M. Kamionkowski, and K. Griest, "Supersymmetric Dark Matter," *Phys. Rept.*, vol. 267, pp. 195–373, 1996, hep-ph/9506380.
6. J. R. Primack, D. Seckel, and B. Sadoulet, "Detection of Cosmic Dark Matter," *Ann. Rev. Nucl. Part. Sci.*, vol. 38, p. 751, 1988.

COLD DARK MATTER FLOWS AND CAUSTICS *

P. SIKIVIE

Institute for Fundamental Theory
Department of Physics
University of Florida
Gainesville, FL 32611-8440
E-mail: sikivie@phys.ufl.edu

The late infall of cold dark matter onto an isolated galaxy, such as our own, produces discrete flows and caustics in its halo. The set of caustics includes simple fold catastrophes located on topological spheres surrounding the galaxy, and a series of caustic rings in or near the galactic plane. The caustic rings are closed tubes whose cross-section is an elliptic umbilic catastrophe. The self-similar model of galactic halo formation predicts that the caustic ring radii a_n follow the approximate law $a_n \sim 1/n$. In a study of 32 extended and well-measured external galactic rotation curves evidence was found for this law. Also, the locations of ten sharp rises in the rotation curve of the Milky Way fit the prediction of the self-similar model at the 3% level. Moreover, a triangular feature in the IRAS map of the Galactic plane is consistent with the imprint of a ring caustic upon the baryonic matter. These observations imply that the dark matter in our neighborhood is dominated by a single flow. Estimates of that flow's density and velocity vector are given.

1. Introduction

There are compelling reasons to believe that the dark matter of the universe is constituted in large part of non-baryonic collisionless particles with very small primordial velocity dispersion, such as axions and/or weakly interacting massive particles (WIMPs) [1]. Generically, such particles are called cold dark matter (CDM). Knowledge of the distribution of CDM in galactic halos, and in our own halo in particular, is of paramount importance to understanding galactic structure and predicting signals in experimental searches for dark matter.

One should expect this dark matter to form caustics. A caustic is a place in physical space where the density is very large because the sheet

---

*This work is supported in part by the U.S. Department of Energy under grant DEFG05-86ER-40272.

on which the dark matter particles lie in phase-space has a fold there. Caustics are commonplace in the propagation of light. Examples include the sharp luminous lines at the bottom of a swimming pool on a breezy sunny day, rainbows, the twinkling of stars, and the shimmering of the sea. Caustics occur generically when two conditions are satisfied. First, the flow must be collisionless. Second the flow must have low velocity dispersion. Light propagation is collisionless, and the flow of light from a point source has zero velocity dispersion. Thus caustics are common in light. Caustics in ordinary matter are very unusual because ordinary matter is not normally collisionless. But CDM is collisionless and has very small primordial velocity dispersion. This leads us to expect that caustics are common in the distribution of CDM.

The primordial velocity dispersion of the cold dark matter candidates is indeed very small, of order

$$\delta v_a(t) \sim 3 \cdot 10^{-17} \left(\frac{10^{-5} eV}{m_a}\right)^{\frac{5}{6}} \left(\frac{t_0}{t}\right)^{\frac{2}{3}} \quad (1)$$

for axions, and

$$\delta v_W(t) \sim 10^{-11} \left(\frac{GeV}{m_W}\right)^{\frac{1}{2}} \left(\frac{t_0}{t}\right)^{\frac{2}{3}} \quad (2)$$

for WIMPs. Here t_0 is the present age of the universe and m_a and m_W are respectively the masses of the axion and WIMP. The small velocity dispersion means that the dark matter particles lie on a thin 3-dim. sheet in 6-dim. phase-space. The thickness of the sheet is δv. The sheet cannot break and hence its evolution is constrained by topology.

2. The Phase-Space Structure of Galactic Halos

Where a galaxy forms, the sheet wraps up in phase-space, turning clockwise in any two dimensional cut $(x, \dot{x})$ of that space. x is the physical space coordinate in an arbitrary direction and $\dot{x}$ its associated velocity. The outcome of this process is a discrete set of flows at any physical point in a galactic halo [2]. Two flows are associated with particles falling through the galaxy for the first time ($n = 1$), two other flows are associated with particles falling through the galaxy for the second time ($n = 2$), and so on. Scattering in the gravitational wells of inhomogeneities in the galaxy (e.g. molecular clouds and globular clusters) are ineffective in thermalizing the flows with low values of n. The flows are seen in N-body simulations of galactic halo formation [3] when care is taken to enhance the numerical resolution in the relevant regions of phase-space.

A commonly raised objection to the above picture is that, before the dark matter falls onto a large galaxy such as our own, it has already clustered on smaller scales, making dwarf halos and other types of clumps, in a process called "hierarchical clustering". However, the effect of hierarchical clustering is only to produce an effective velocity dispersion for the infalling dark matter, i.e. a thickening of the phase space sheet. This effective velocity dispersion is at most equal to the velocity dispersion, of order 10 km/s, of dwarf halos and on average should be much less than that. Because the effective velocity dispersion of the infalling dark matter is much less than the 300 km/s velocity dispersion of the Galaxy as a whole, the phase space sheet folds in qualitatively the same way as in the zero velocity dispersion case. The flows and caustics remain.

Caustics appear wherever the projection of the phase-space sheet onto physical space has a fold [4,5]. Generically, caustics are surfaces in physical space. On one side of the caustic surface there are two more flows than on the other. At the surface, the dark matter density is very large. It diverges there in the limit of zero velocity dispersion. There are two types of caustics in the halos of galaxies, inner and outer. The outer caustics are simple fold (A_2) catastrophes located on topological spheres surrounding the galaxy. They occur near where a given outflow reaches its furthest distance from the galactic center before falling back in. The inner caustics are rings [4]. They occur near where the particles with the most angular momentum in a given inflow reach their distance of closest approach to the galactic center before going back out. A caustic ring is a closed tube whose cross-section is an *elliptic umbilic* (D_{-4}) catastrophe [5]. The existence of these caustics and their topological properties are independent of any assumptions of symmetry.

Primordial peculiar velocities are expected to be the same for baryonic and dark matter particles because they are caused by gravitational forces. Later the velocities of baryons and CDM differ because baryons collide with each other whereas CDM is collisionless. However, because angular momentum is conserved, the net angular momenta of the dark matter and baryonic components of a galaxy are aligned. Since the caustic rings are located near where the particles with the most angular momentum in a given infall are at their closest approach to the galactic center, they lie close to the galactic plane.

A specific proposal was made for the radii a_n of caustic rings [4]:

$$\{a_n : n = 1, 2, ...\} \simeq (39, 19.5, 13, 10, 8, ...) \text{kpc} \times \left(\frac{j_{\max}}{0.25}\right) \left(\frac{v_{\text{rot}}}{220 \frac{\text{km}}{\text{s}}}\right) \quad (3)$$

where v_{rot} is the rotation velocity of the galaxy and $j_{\max}$ is a parameter with a specific value for each halo. For large n, $a_n \propto 1/n$. Eq.(3) is predicted by the self-similar infall model [6,7] of galactic halo formation. $j_{\max}$ is then the maximum of the dimensionless angular momentum j-distribution [7]. The self-similar model depends upon a parameter ϵ [6]. In CDM theories of large scale structure formation, ϵ is expected to be in the range 0.2 to 0.35 [7]. Eq.(3) is for $\epsilon = 0.3$. However, in the range $0.2 < \epsilon < 0.35$, the ratios a_n/a_1 are almost independent of ϵ. When $j_{\max}$ values are quoted below, $\epsilon = 0.3$ and $h = 0.7$ will be assumed.

It was pointed out in ref. [7] that including angular momentum in the self-similar infall model results in a depletion of the inner halo and hence

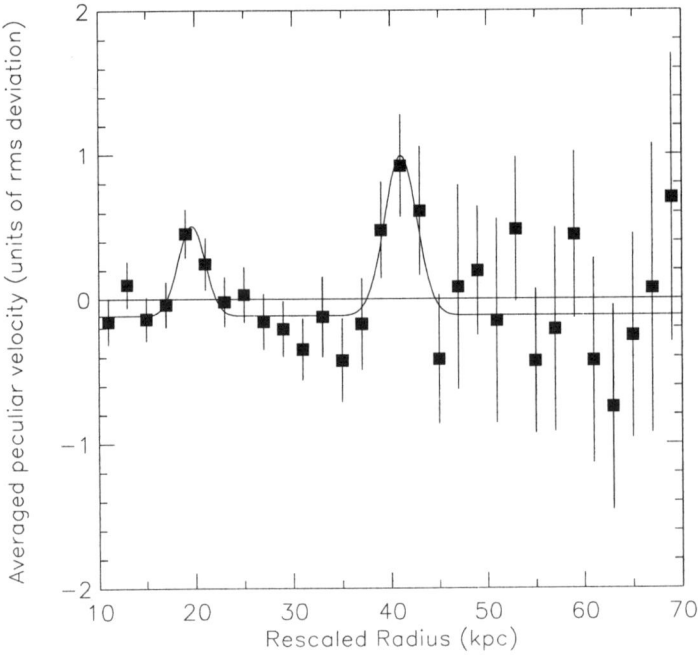

Figure 1. Composite rotation curve constructed in ref. [8]. It combines data on 32 exterior galaxies to test the hypothesis that the caustic ring radii are given by Eq.(3).

an effective core radius. The average amount of angular momentum of the Milky Way halo was estimated [7] by requiring that approximately half of the rotation velocity squared at our location is due to dark matter, the other half being due to ordinary matter. This yields $\bar{j} \sim 0.2$ where $\bar{j}$ is the average of the j-distribution for our halo. $\bar{j}$ and j_{max} are related if some assumption is made about the shape of the j-distribution. For example, if the j-distribution is taken to be that of a rigidly rotating sphere, one has $j_{max} = \frac{4}{\pi}\bar{j}$. Hence $j_{max} \sim 0.25$ for our halo.

Since caustic rings lie close to the galactic plane, they cause bumps in the rotation curve, at the locations of the rings. In ref. [8] a set of 32 extended well-measured rotation curves was analyzed and statistical evidence was found for the $n = 1$ and $n = 2$ caustic rings, distributed according to Eq.(3). In this analysis, each of the 32 individual galactic rotation curves was rescaled according to

$$r \to \tilde{r} = r \left(\frac{220 \text{ km/s}}{v_{rot}} \right) \qquad (4)$$

where v_{rot} is the measured rotation velocity. To isolate the outer halo-dominated portion of the rotation curves, all data with rescaled radii $\tilde{r} < 10$ kpc were removed. Each rotation curve was then fitted to a line or a quadratic polynomial. The residual deviations were normalized to the rms deviation in each fit and then binned together. The result is the composite rotation shown in Fig. 1 for the case where the individual rotation curves were fitted to quadratic polynomials. The composite rotation curve has two peaks, near 20 kpc and 40 kpc, with statistical significance of 3σ and 2.6σ respectively. It implies that the j_{max} distribution is peaked near 0.27. The rotation curve of NGC3198, one of the best measured, by itself shows three faint bumps which are consistent with Eq.(3) and three faint bumps which are consistent with Eq.(3) and $j_{max} = 0.28$ [4]. Also our earlier estimate of j_{max} for the Milky Way halo is close to the peak value of 0.27.

3. Evidence for Ring Caustics in the Milky Way

Eq.(3) with $j_{max} = 0.25$ implies that our halo has caustic rings with radii near 40 kpc/n, where n is an integer. Here we point to evidence [9] in support of this extraordinary claim.

Galactic rotation curves are obtained from HI and CO surveys of the Galactic plane. A list of surveys performed to date is given in ref. [10]. Everything else being equal, CO surveys have far better angular resolution than HI surveys because their wavelength is nearly two orders of magnitude smaller (0.26 cm vs. 21 cm). The most detailed inner Galactic rotation curve appears to be that obtained [11] from the Massachusetts-Stony Brook North Galactic Plane CO survey [12]. It is reproduced in Fig. 2. It shows highly significant rises between 3 and 8.5 kpc. Eq.(3) predicts ten caustic rings between 3 and 8.5 kpc. Allowing for ambiguities in identifying rises, the number of rises in the rotation curve between 3 and 8.5 kpc is in fact approximately ten. Below 3 kpc the predicted rises are so closely spaced that they are unlikely to be resolved in the data. The rises are marked as slanted line segments in Fig. 2.

The effect of a caustic ring in the plane of a galaxy upon its rotation curve was analyzed in ref. [5]. The caustic ring produces a rise in the rotation curve which starts at $r_1 = a_n$, where a_n is the caustic ring radius, and which ends a $r_2 = a_n + p_n$, where p_n is the caustic ring width. The ring widths depend in a complicated way on the velocity distribution of the infalling dark matter at last turnaround [5] and are not predicted by the model. They also need not be constant along the ring.

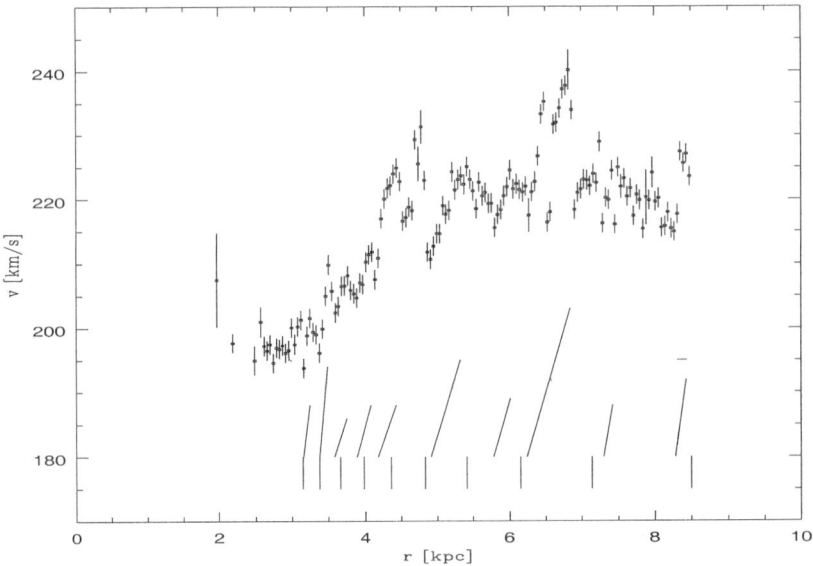

Figure 2. North Galactic rotation curve from ref. [11]. The locations of rises mentioned are indicated by line segments parallel to the rises but shifted downwards. The caustic ring radii for the fit described in the text are shown as vertical line segments. The position of the triangular feature in the IRAS map of the galactic plane near 80° longitude is shown by the short horizontal line segment. It coincides with a rise in the rotation curve.

In the past, rises (or bumps) in galactic rotation curves have been interpreted as due to the presence of spiral arms [13]. Spiral arms may in fact cause some of the rises in rotation curves. However there may be other valid explanations. Two properties of the high resolution rotation curve of Fig. 2 favor the interpretation that its rises are caused by caustic rings of dark matter. First, there are of order ten rises in the range of radii covered (3 to 8.5 kpc). This agrees qualitatively with the predicted number of caustic rings, whereas only three spiral arms are known in that range: Scutum, Sagittarius and Local. Second, the rises are sharp transitions in the rotation curve, both where they start (r_1) and where they end (r_2). Sharp transitions are consistent with caustic rings because the latter have divergent density at $r_1 = a$ and $r_2 = a + p$ in the limit of vanishing velocity dispersion. Finally, there are bumps and rises in rotation curves measured at galactocentric distances much larger than the disk radius, where no spiral arms are seen. In particular, the features found in the composite rotation curve of Fig. 1 are at distances 20 kpc and 40 kpc when scaled to our own galaxy.

The self-similar infall model prediction for the caustic ring radii, Eq.(3), was fitted to the eight rises between 3 and 7 kpc by minimizing $rmsd \equiv [\frac{1}{8}\sum_{n=7}^{14}(1 - \frac{a_n}{r_{1n}})^2]^{\frac{1}{2}}$ with respect to $j_{\max}$, for $\epsilon = 0.30$. The fit yields $j_{\max} = 0.263$ and $rmsd = 3.1\%$. The corresponding caustic ring radii a_n are indicated by short vertical line segments at the bottom of Fig. 2.

Caustic rings of dark matter produce gravitational forces on the baryonic matter. They may reveal themselves in maps of the sky by the gas and dust that they have accreted. Looking tangentially to a ring caustic from a vantage point in the plane of the ring, one may recognize the tricusp [5] shape of the D_{-4} catastrophe. I searched for such features. The IRAS map of the galactic disk in the direction of galactic coordinates $(l, b) = (80°, 0°)$ shows a triangular shape which is strikingly reminiscent of the cross-section of a ring caustic. The relevant IRAS maps are posted at http://www.phys.ufl.edu/~sikivie/triangle/ . They were downloaded from the Skyview Virtual Observatory (http://skyview.gsfc.nasa.gov/). The vertices of the triangle are at $(l, b) = (83.5°, 0.4°), (77.8°, 3.4°)$ and $(77.8°, -2.6°)$. The shape is correctly oriented with respect to the galactic plane and the galactic center. To an extraordinary degree of accuracy it is an isosceles triangle with axis of symmetry parallel to the galactic plane, as is expected for a caustic ring whose transverse dimensions are small compared to its radius. Moreover its position is consistent with the position of a rise in the rotation curve, the one between 8.28 and 8.43 kpc ($n = 5$). The caustic ring radius implied by the image is 8.31 kpc and its dimensions are $p \sim 130$ pc and $q \sim 200$ pc, in the directions parallel and perpendicular to the galactic plane respectively. It therefore predicts a rise which starts at 8.31 kpc and ends at 8.44 kpc, just where a rise is observed. The probability that the coincidence in position of the triangular shape with a rise in the rotation curve is fortuitous is less than 10^{-3}.

In principle, the feature at $(80°, 0°)$ should be matched by another in the opposite tangent direction to the nearby ring caustic, at approximately $(-80°, 0°)$. Although there is a plausible feature there, it is much less compelling than the one in the $(+80°, 0°)$ direction. There are several reasons why it may not appear as strongly. One is that the $(+80°, 0°)$ feature is in the middle of the Local spiral arm, whose stellar activity enhances the local gas and dust emissivity, whereas the $(-80°, 0°)$ feature is not so favorably located. Another is that the ring caustic in the $(+80°, 0°)$ direction has unusually small dimensions. This may make it more visible by increasing its contrast with the background. In the $(-80°, 0°)$ direction, the nearby ring caustic may have larger transverse dimensions.

4. The Big Flow

Our proximity to a caustic ring means that the corresponding flows, i.e. the flows in which the caustic occurs, contribute very importantly to the local dark matter density. Using the results of refs. [4,5,7], we can estimate their densities and velocity vectors. Let us assume, for illustrative purposes, that we are in the plane of the nearby caustic and that its outward cusp is 55 pc away from us, i.e. $a_5 + p_5 = 8.445$ kpc. The densities and velocity vectors on Earth of the $n = 5$ flows are then:

$$d^+ = 1.7\,10^{-24}\,\frac{\text{gr}}{\text{cm}^3}\,,\ d^- = 1.5\,10^{-25}\,\frac{\text{gr}}{\text{cm}^3}\,,\ \vec{v}^{\pm} = (470\,\hat{\phi} \pm 100\,\hat{r})\,\frac{\text{km}}{\text{s}}, \quad (5)$$

where $\hat{r}, \hat{\phi}$ and $\hat{z}$ are the local unit vectors in galactocentric cylindrical coordinates. $\hat{\phi}$ is in the direction of galactic rotation. The velocities are given in the (non-rotating) rest frame of the Galaxy. Because of an ambiguity, it is not presently possible to say whether $d^{\pm}$ are the densities of the flows with velocity $\vec{v}^{\pm}$ or $\vec{v}^{\mp}$. The large size of d^+ is due to our proximity to the outward cusp of the nearby caustic. Its exact value is sensitive to our distance to the cusp. We do not know that distance well enough to estimate d^+ with accuracy. However we can say that d^+ is very large, of order the value given in Eq.(5), perhaps even larger. If we are inside the tube of the nearby caustic, there are two additional flows on Earth, aside from those given in Eq.(5). A list of local densities and velocity vectors for the $n \neq 5$ flows can be found in ref. [14].

Eq.(5) has dramatic implications for dark matter searches. Previous estimates of the local dark matter density, based on isothermal halo profiles, range from 5 to 7.5 10^{-25} $\frac{\text{gr}}{\text{cm}^3}$. The present analysis implies that a single flow (d^+) has of order three times (or more) that much local density.

The sharpness of the rises in the rotation curve and of the triangular feature in the IRAS map implies an upper limit on the velocity dispersion δv_{DM} of the infalling dark matter. Caustic ring singularities are spread over a distance of order $\delta a \simeq \frac{R \, \delta v_{\text{DM}}}{v}$ where v is the velocity of the particles in the caustic, δv_{DM} is their velocity dispersion, and R is their turnaround radius. The sharpness of the IRAS feature implies that its edges are spread over $\delta a \lesssim 20$ pc. Assuming that the feature is due to the $n = 5$ ring caustic, $R \simeq 180$ kpc and $v \simeq 480$ km/s. Therefore $\delta v_{\text{DM}} \lesssim 53$ m/s.

The caustic ring model may explain the puzzling persistence of galactic disk warps [15]. These may be due to outer caustic rings lying somewhat outside the galactic plane and attracting visible matter. The resulting disk warps would not damp away, as is the case in more conventional explanations of the origin of the warps, but would persist on cosmological time scales.

The caustic ring model, and more specifically the prediction Eq.(5) of the locally dominant flow associated with the nearby ring, has important consequences for axion dark matter searches [16], the annual modulation [17,18,19,20,14] and signal anisotropy [21,19] in WIMP searches, the search for γ-rays from dark matter annihilation [22,23], and the search for gravitational lensing by dark matter caustics [24,25]. The model makes predictions for each of these approaches to the dark matter problem.

References

1. E.W. Kolb and M.S. Turner, *The Early Universe*, Addison-Wesley, 1990; M. Srednicki, Editor *Particle Physics and Cosmology: Dark Matter*, Nort-Holland, 1990.
2. P. Sikivie and J.R. Ipser, Phys. Lett. **B291** (1992) 288.
3. D. Stiff and L. Widrow, astro-ph/0301301.
4. P. Sikivie, Phys. Lett. **B432** (1998) 139.
5. P. Sikivie, Phys. Rev. **D60** (1999) 063501.
6. J.A. Filmore and P. Goldreich, Ap.J. **281** (1984) 1; E. Bertschinger, Ap. J. Suppl. **58** (1985) 39.

7. P. Sikivie, I. Tkachev and Y. Wang, Phys. Rev. Lett. **75** (1995) 2911; Phys. Rev. **D56** (1997) 1863.
8. W. Kinney and P. Sikivie, Phys. Rev. **D61** (2000) 087305.
9. P. Sikivie, Phys. Lett. **567** (2003) 1.
10. J. Binney and M. Merrifield, *Galactic Astronomy*, Princeton University Press, 1998, pp 550, 553.
11. D.P. Clemens, Ap.J. **295** (1985) 422.
12. D.B. Sanders et al., Ap.J.S. **60** (1986) 1; D.P. Clemens et al., Ap.J.S. **60** (1986) 297.
13. C. Yuan, Ap. J. **158** (1969) 871; W.B. Burton and W.W. Shane, Proceedings of the 38th IAU Symposium *The Spiral Structure of our Galaxy*, edited by W. Becker and G.I. Kontopoulos, Dordrecht, Reidel, p. 397; W.W. Shane, Astron. and Astroph. **16** (1972) 118.
14. F.-S. Ling, P. Sikivie and S. Wick, astro-ph/0405231, to appear in Phys. Rev. D.
15. R.W. Nelson and S. Tremaine, MNRAS **275** (1995) 897; J. Binney, I.-G. Jiang and S. Dutta, MNRAS **297** (1998) 1237.
16. C. Hagmann et al., Phys. Rev. Lett. **80** (1998) 2043; I. Ogawa, S. Matsuki and K. Yamamoto, Phys. Rev. **D53** (1996) 1740.
17. P. Sikivie, Proceedings of the Second International Workshop on *The Identification of Dark Matter*, edited by N. Spooner and V. Kudryavtsev, World Scientific 1999, p. 68.
18. J. Vergados, Phys. Rev. **D63** (2001) 063511; A. Green, Phys. Rev. **D63** (2001) 103003; G. Gelmini and P. Gondolo, Phys. Rev. **D64** (2001) 023504.
19. D. Stiff, L.M. Widrow and J. Frieman, astro-ph/0106048.
20. P. Sikivie and S. Wick, Phys. Rev. **D66** (2002) 023504;
21. C. Copi, J. Heo and L. Krauss, Phys. Lett. **B461** (1999) 43.
22. L. Bergstrom, J. Edsjo and C. Gunnarsson, Phys. Rev. **D63** (2001) 083515.
23. C. Hogan, Phys. Rev. **D64** (2001) 063515.
24. C. Hogan, Ap. J. **527** (1999) 42.
25. C. Charmousis, V. Onemli, Z. Qiu and P. Sikivie, Phys. Rev. **D67** (2003) 103502.

SMALL-SCALE DARK MATTER CLUMPS

V. S. BEREZINSKY

Laboratori Nazionali del Gran Sasso, INFN,
67010 Assergi (AQ), Italy
E-mail: berezinsky@lngs.infn.it

V. I. DOKUCHAEV AND YU. N. EROSHENKO

Institute for Nuclear Research of the Russian Academy of Sciences,
60th Anniversary of October Prospect 7a, Moscow 117312, Russia
E-mail: dokuchaev@inr.npd.ac.ru; erosh@inr.npd.ac.ru

We study the cosmological origin of small-scale DM clumps with mass $\lesssim 10^3 M_\odot$ in the hierarchical scenario with most conservative assumption of adiabatic gaussian fluctuations. The main included effect (tidal interaction) results in the formation of large core in the center of a clump and in tidal destruction of large fraction of the clumps. The mass distribution of clumps has a cutoff at $M_{\min}$ due to diffusion of DM particles out of a fluctuation and free streaming at later stage. $M_{\min}$ is a model dependent quantity. In the case the neutralino, considered as a pure bino, is a DM particle, $M_{\min} \sim 10^{-8} M_\odot$. The enhancement of annihilation signal due to DM clumpiness in the Galactic halo, valid for arbitrary DM particles, is calculated. For observationally preferable value of index or primeval fluctuation spectrum $n_p \approx 1$, the enhancement of an annihilation signal is described by a factor $2-5$ depending on the density profile in a clump.

1. Introduction

The gravitationally bound structures in the universe are developed from primordial density fluctuations $\delta(\vec{x}, t) = \delta\rho/\rho$. They are produced at inflation from quantum fluctuations. The predicted power spectrum of these fluctuations has a nearly universal form $P(k) \equiv \delta_k^2 \propto k^{n_p}$, with $n_p \simeq 1$. At radiation-dominated epoch the fluctuations grow slowly, $\delta \propto \ln(t/t_i)$. After transition at $t = t_{\rm eq}$ to the matter-dominated epoch, the fluctuations grow as $\delta \propto (t/t_{\rm eq})^{2/3}$. The gravitationally bound objects are formed and detached from cosmological expansion when fluctuations enter the non-linear stage $\delta \geq 1$. The non-linear stage of fluctuation growth has been studied both by analytic calculations [1] and in numerical simulations [2,3,4] for Large

Scale Structure (LSS). The density profile in the inner part of these objects is given by $\rho(r) \propto r^{-\beta}$, with with $\beta \approx 1.7 - 1.9$ in analytic calculations [1], $\beta = 1$ in simulations of NFW [2] and $\beta = 1.5$ in simulations of Moore et al. [3] and Jing and Suto [4]. In this work we apply this approach to the smallest DM objects in the universe, which we shall call *clumps*. The clumps, being the smallest structures, are produced first in the universe, and it makes different our consideration from LSS formation.

The small-scale clumps are formed only if the fluctuation amplitudes in the spectrum are large enough at the corresponding small scales. The inflation models predict the power-law primeval fluctuation spectrum. If the power-law index $n_p \geq 1$, clumps are formed in a wide range of scales. During the universe expansion the small clumps are captured by the larger ones, and the larger clumps consist of the smaller ones and of continuously distributed DM. The convenient analytic formalism, which describes statistically this hierarchical clustering, is the Press-Schechter theory and its extensions, in particular 'excursion set' formalism developed by Bond et al. However, this theory does not include the important process of the tidal destruction of small clumps inside the bigger ones. We take into account this process and obtain the mass function for the small-scale clumps in the Galactic halo.

The theoretical observation of this work is that the role of tidal interaction is crucial: the central core in DM distribution is produced and some fluctuations can be fully disrupted. The LSS are produced much later, when the tidal interactions work in the different regime and their role is less essential. In the case of the power-law spectrum only a small fraction of the captured clumps survives, but even this small fraction is enough to dominate the total annihilation rate in the Galactic halo. We use the standard cosmology with WMAP parameters.

2. Tidal Destruction of Clumps

The destruction of clumps by the tidal interaction occurs at the formation of the hierarchical structures, long time before the galaxy formation. The characteristic epoch is roughly $t \sim t_{\rm eq}$. This interaction arises when two clumps pass near each other and when a clump moves in the external gravitational field of the bigger host to which this clump belongs. In both cases a clump is exited by the external gravitational field, i. e. its constituent particles obtain additional velocities in the center of mass system. The clump is destroyed if its internal energy increase ΔE exceeds the corresponding total energy $|E| \sim GM^2/2R$. In [5] we have calculated the rate of excitation energy production by both aforementioned processes. The dominating process is given by tidal interaction in the gravitational field of the host clumps, with the main contribution from the smallest host clump. We use the Press-Schechter formalism for hierarchical clustering.

The differential fraction of mass, M, in the form of clumps which escape the tidal destruction in the hierarchical objects (survival probability) is found as

$$\xi(n,\nu) \simeq (2\pi)^{-1/2} e^{-\nu^2/2}(n+3) y(\nu), \qquad (1)$$

where $\nu = \delta/\sigma(M)$ is the peak-height of a fluctuation with δ being the amplitude and σ variance (dispersion), $n \approx -3$ is the effective spectral index at $t \sim t_{\rm eq}$, and function $y(\nu)$ is given numerically in [5]. $\xi(n,\nu)$ depends weakly on index of DM distribution in a clump β. Integrating over ν, we obtain

$$\xi_{\rm int} \simeq 0.01(n+3). \qquad (2)$$

Since n is close to -3, only a small fraction of clumps about $0.1 - 0.5\%$ survive the stage of tidal destruction. However, this fraction is enough to dominate the total annihilation rate in the Galactic halo.

3. Core of Dark Matter Clump

The core formation in a fluctuation begins at the linear stage of evolution and continues at the beginning of non-linear stage. The tidal forces diminishes with time as $t^{-4/3}$ (see [5]). Once the core is produced it is not destroyed in the evolution followed. The stage of the core formation continues approximately from $t_{\rm eq}$ to the time of maximal expansion t_s and a little above, when a clump decouples from expansion of universe and contracts in the non-linear regime. Soon after this period, a clump enters the hierarchical stage of evolution, when the tidal forces can destroy it, but surviving clumps retain their cores.

The calculations proceed in the following way (see [5] for details and references). The background gravitational field (including that of the host clumps) is expanded in series in respect to the distance from the point with maximum density in a fluctuation. The motion of a DM particle in this field is studied. The spherically symmetric term of the expansion causes the radial motion of a particle in the oscillation regime. Spherically non-symmetric term describes the tidal interaction. It results in deflection of a particle trajectory from a center (point with maximum density). The average (over statistical ensemble) deflection gives the radius of the core R_c. After statistical averaging, R_c is expressed through the amplitude of the fluctuation δ_{eq} and the variance σ_{eq} (or $\nu = \delta_{eq}/\sigma_{eq}$) as

$$x_c = R_c/R \approx 0.3\nu^{-2} f^2(\delta_{eq}). \tag{3}$$

The fluctuations with $\nu \sim 0.5 - 0.6$ have $x_c \sim 1$, i.e. they are practically destroyed by tidal interactions. Most of galactic clumps are formed from $\nu \sim 1$ peaks, but the main contribution to the annihilation signal is given by the clumps with $\nu \simeq 2.5$ for which $x_c \simeq 0.05$.

4. Clumps of Minimal Mass

The mass spectrum of clumps has a low-mass cutoff at $M = M_{\min}$, which value is determined by a leakage of DM particles from the overdense fluctuations in the early universe. CDM particles at high temperature $T > T_f \sim 0.05 m_\chi$ are in the thermodynamical (chemical) equilibrium with cosmic plasma. After freezing at $t > t_f$ and $T < T_f$, the DM particles remain for some time in *kinetic* equilibrium with plasma, when the temperature of CDM particles T_χ is equal to temperature of plasma T. At this stage the CDM particles are not perfectly coupled to the cosmic plasma. Collisions between a CDM particle and fast particles of ambient plasma result in exchange of momenta and a CDM particle diffuses in the space. Due to diffusion the DM particles leak from the small-scale fluctuations and thus their distribution obtain a cutoff at the minimal mass M_D.

When the energy relaxation time for DM particles $\tau_{\rm rel}$ becomes larger than the Hubble time $H^{-1}(t)$, the DM particles get out of the kinetic equilibrium. This conditions determines the time of kinetic decoupling t_d. At $t \geq t_d$ the CDM matter particles are moving in the free streaming regime and all fluctuations on the scale of free-streaming length λ_{fs} and smaller are washed away. The corresponding minimal mass

$$M_{\rm fs} = (4\pi/3)\rho_\chi(t_0)\lambda_{\rm fs}^3, \tag{4}$$

is much larger than M_D and therefore $M_{\min} = M_{\rm fs}$. In [5] we have performed the calculations using two methods: the transparent physical method, based on the description of diffusion and free streaming, and more formal method based on solution of kinetic equation for DM particles starting from the period of chemical equilibrium. Both methods agree perfectly.

The minimal mass in the mass distribution is given by free-streaming mass and for the case of neutralino (bino) as DM particle it is equal to

$M_{\min} = 1.5 \times 10^{-8} M_\odot$ for the mass of neutralino $m_\chi = 100$ GeV and the mass of selectron and sneutrino $\tilde{M} = 1$ TeV. Our calculations agree reasonably well with that of [6], while M_{min} from [7] coincides with our value for M_D.

5. Annihilation Signal due to Clumps

We calculate the enhancement of annihilation signal due to presence of small clumps in the Galactic halo as

$$\eta = (I_{\rm cl} + I_{\rm hom})/I_{\rm hom}, \qquad (5)$$

where $I_{\rm hom}$ is the flux due to annihilation of unclumpy DM particles homogeneously distributed in halo, and $I_{\rm cl}$ is the flux from the clumps (see Figs. 1). In calculations we used different density profiles in the clumps, the distribution of DM clumps over their masses M and radii R, and the distribution of clumps in the galactic halo. The enhancement depends on the nature of DM particle only through $M_{\min}$. The details of calculations and corresponding plots one can find in [5]. The enhancements η for $n_p = 1$ or less is not large: typically it is not larger than factor 2 -5 for $M_{\min} \sim 10^{-8} M_\odot$. For example, $\eta = 5$ for $n_p = 1.0$ and $M_{\min} = 2 \cdot 10^{-8} M_\odot$. It strongly increases at smaller $M_{\min}$ and larger n_p. For example, for $n_p = 1.1$ and $M_{\min} = 2 \cdot 10^{-8} M_\odot$, enhancement becomes very large, $\eta = 130$ and $\eta = 4 \cdot 10^3$, respectively. Our approach is based on the hierarchical clustering model in which smaller mass objects are formed earlier than the larger ones, i. e. $\sigma_{\rm eq}(M)$ diminishes with the growing of M. This condition is satisfied for objects with mass $M > M_{min} \simeq 2 \cdot 10^{-8} M_\odot$ only if the primordial power spectrum has the value of the power index $n_p > 0.84$. In this case the enhancement of the annihilation signal is absent: $\eta \simeq 1$, for $n_p < 0.9$.

For the Galactic halo we use the NFW density profile [2]:

$$\rho_{\rm DM}(l) = \frac{\rho_0}{(l/L)(1+l/L)^2}, \qquad (6)$$

with $L = 45$ kpc, and ρ_0 fixed by the local density value $\rho_{\rm DM}(r_\odot) = 0.3$ GeV cm^{-3}. With these parameters the halo mass within the virial radius of 100 kpc is $10^{12} M_\odot$.

Finally, we calculated the enhancement due to large clumps with masses in the range $10^8 - 10^{10} M_\odot$ and the number density distribution $\propto dM/M^2$, which are seen in numerical simulations (see e.g. [3]). This distribution of large accidentally coincides with our derived distribution for small clumps

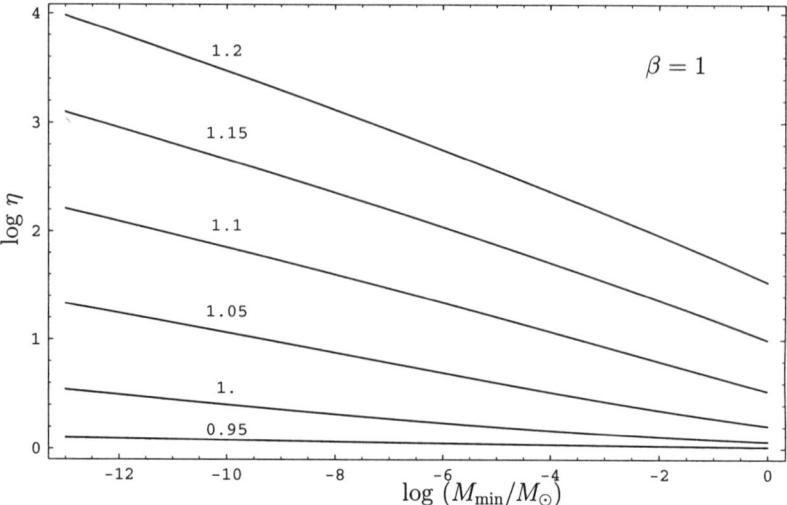

Figure 1. The global enhancement η of the annihilation signal from Eq. (5) as a function of the minimal clump mass M_{min}, for clump density profile with index $\beta = 1$ and for different indices n_p of primeval perturbation spectrum. The curves are marked by the values of n_p.

(2). For NFW profile of DM in halo and mass fraction of large clumps 0.15 the calculated enhancement of annihilation signal from typical ($\nu \simeq 1$) large clumps is rather small, $\eta \simeq 1.07$.

Acknowledgments

This work was supported in part by the Russian Foundation for Basic Research grants 02-02-16762-a, 03-02-16436-a and 04-02-16757-a and the Russian Ministry of Science grants 1782.2003.2 and 2063.2003.2.

References

1. A. V. Gurevich and K. P. Zybin, *Sov. Phys. Usp.* **165**, 723 (1995).
2. J. F. Navarro, C. S. Frenk, and S. D. M. White, *Astrophys. J.* **462**, 563 (1996).
3. B. Moore et al., *Astrophys. J.* **524**, L19 (1999).
4. Y. P. Jing and Y. Suto, *Astrophys. J.* **529**, L69 (2000).
5. V. S. Berezinsky, V. I. Dokuchaev, and Yu. N. Eroshenko, *Phys. Rev.* **D 68**, 103003 (2003); ArXiv:astro-ph/0301551.
6. D. J. Schwarz, S. Hofmann, and H. Stocker, *Phys. Rev.* **D64**, 083507 (2001).
7. K. P. Zybin, M. I. Vysotsky, and A. V. Gurevich, *Phys. Lett.* **A 260**, 262 (1999).

THE PROLATE SHAPE OF THE GALACTIC DARK-MATTER HALO

AMINA HELMI

Kapteyn Astronomical Institute,
P.O. Box 700,
9700 AV Groningen, The Netherlands
E-mail: ahelmi@astro.rug.nl

Knowledge of the distribution of dark-matter in our Galaxy plays a crucial role in the interpretation of dark-matter detection experiments. I will argue here that probably the best way of constraining the properties of the dark-matter halo is through astrophysical observations. These provide constraints on the spatial structure (density profile, shape, local density), on the velocity distribution function and on the presence of substructure (streams or lumps of dark-matter).

1. Introduction

The shapes of dark-matter halos are sensitive probes of the nature of the dark-matter particles themselves. Numerical simulations show that halos made up of cold-dark matter (like e.g. neutralinos) have typical axes ratios of 0.6 - 0.8 (on the galaxies scale), with no preference for an oblate or prolate shape, although often they are triaxial, particularly in the outskirts[1-3] (see Fig.1 for some definitions). Simulations of hot dark-matter halos (constituted by neutrinos) are spherical[4], while if the dark-matter particles are self-interacting they are close to spherical[5-6].

The measurement of the shape of a dark-matter halo is non-trivial, which explains why different techniques using a variety of tracers have found notably disparate values[7]. Measurements of the flattening q_ρ close to the equatorial plane of a galactic system using the evolution of gaseous warps favor spherical or prolate halos[8-9], while the flaring of the HI gas layer constrains the flattening to be between 0.2 and 0.5 (e.g. Ref.[10]) for similar types of disk galaxies. These measurements use tracers located in the inner 10 - 20 kpc of a galaxy, and as such their dynamics is influenced by *all* dominant mass components, not only the dark-matter halo. Better tracers are, for example, polar rings (usually at tens of kpc from the nucleus),

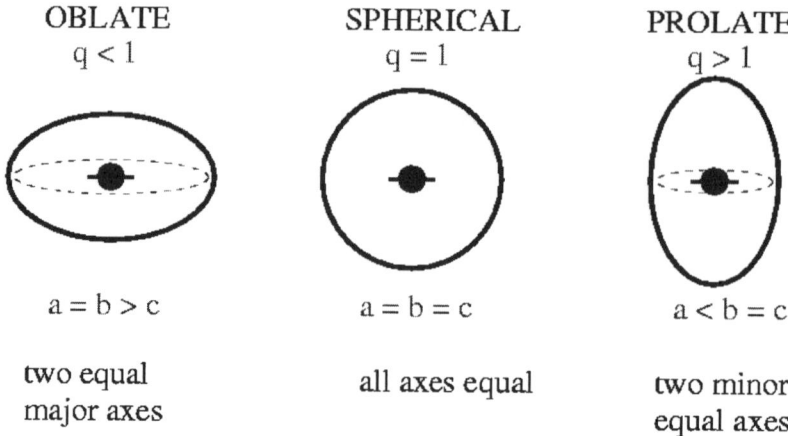

Figure 1. Description of the shape of a halo, whose principal axes are labeled a, b, c. The parameter q represents the flattening of the potential, i.e. $q = c/a$.

X-ray gas in halos, tidal streams, and gravitational lensing.

Attempts to measure the flattening of our Galaxy have not yielded conclusive results. The velocity distribution of population II stars, the flaring of the HI gas layer and the local disk kinematics have provided estimates in the range 0.3 to 1[11-12]. This large variation is partly due to our incomplete knowledge of the Galactic parameters[13].

Undoubtedly, the best tracers are distant halo stars who motions are primarily dictated by the dark halo potential. Tidal streams in the halo are excellent probes, since they represent groups of stars moving on nearly parallel orbits and are hence much more sensitive than a random population of halo stars. For example, if the halo is spherical the orbits of stars in a stream remain in a plane and define a great circle arc on the sky. On the other hand, for a non-spherical mass distribution, the orientation of the plane of motion is not constant and precesses in time. As a consequence the streams become wider in time and are less well-defined on the sky. This simple physical mechanism is therefore very useful to measure the shape of a dark halo. Recently, tidal streams from the Sagittarius dwarf have been detected in the Galactic halo, and have been used by Ref.[14] to argue that the halo is close to spherical.

2. The streams from Sagittarius

We have performed numerical simulations of the disruption of a system like the Sagittarius dwarf orbiting a Galactic potential with 3 components: a bulge, a disk and a dark logarithmic halo

$$\Phi_{\text{halo}} = v_{\text{halo}}^2 \ln(R^2 + z^2/q^2 + d^2), \tag{1}$$

where $d = 12$ kpc and $v_{\text{halo}} = 131.5$ km/s[15]. The parameter q is allowed to vary from 0.8 to 1.25, that is, from an oblate to a prolate configuration. The orbital initial conditions are chosen to satisfy the constraints given by the present position and radial velocity of the main body of Sgr[16]. For each of the q values of the dark halo potentials, we select orbits which have similar (mean) pericenter and apocenter distances as well as comparable L_z (z-component of the angular momentum) satisfying the above mentioned constraints. Our models of the dwarf are also slightly readjusted for each of the dark halo shapes, so as to produce the same remnant system by the present day. This implies that differences in the characteristics of the debris may only be attributed to a change in the flattening of the potential. The other possible free variables, such as the model of the dwarf and the orbital parameters, are (essentially) the same by construction.

Figure 2 shows the sky distribution of the tidal streams formed in the past 4 Gyr for the cases $q = 0.8$, $q = 1$ and $q = 1.25$. The reason for focusing on such relatively young streams is that the debris from Sgr that has been discovered so far is indeed consistent with having been formed over less than this period of time. The differences in the sky distribution of these young tidal streams are almost imperceptible (contrary to the conclusion of Ref.[14]). However, the trends in the distance distribution of the stars (particularly for $\Lambda \sim 200° - 300°$) appear to be better reproduced in the prolate halo simulation[17].

The kinematics of the particles released more than 1.5 Gyr ago do, however, show some unambiguous features that allow direct distinction between the different shapes. This is illustrated in Fig. 3, particularly for $\Lambda \sim 200° - 300°$, where the predicted mean velocity of the stream shows differences larger than 100 km/s which are easily measurable. Recently, the kinematics of several hundred M giant stars located along the tidal streams from Sgr have been measured[19-20]. These are the solid symbols shown in the same figure. It is therefore clear that the data is only consistent with a prolate halo with major-to-minor axis ratio $q_\rho = 5 : 3$ or perhaps even slightly larger.

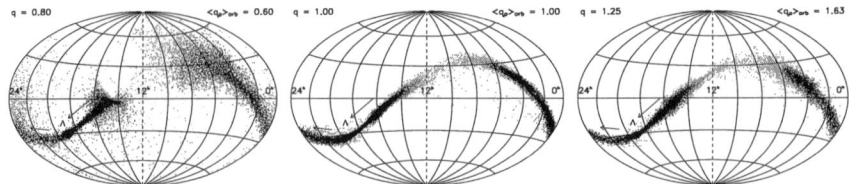

Figure 2. Sky distribution of particles released in the last 4 Gyr in our models of the Sgr dwarf evolution. The black dots correspond to those that became unbound in the past 1.5 Gyr, and are even more narrowly confined to a great circle on the sky. From this figure we define a new angular coordinate along the stream Λ, that increases from $0°$ at the Sgr dwarf to $180°$ along the southern trail, and goes from $360° - 180°$ along the northern trail.

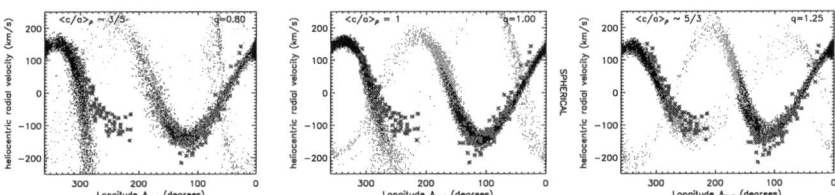

Figure 3. Radial velocity distribution for particles lost in the last 4 Gyr (same color-coding as in Fig.2). The asterisks denote the radial velocities measured by Ref.[19] and presented in Ref.[20]. The latter unambiguously show that a prolate halo provides the best fit[18].

3. Conclusions

This result has several important implications for dark matter detection experiments. Firstly, the local density of dark matter is approximately a factor two lower than expected for a spherical halo, at a fixed circular velocity. Secondly, no streams from the Sagittarius dwarf would be expected to cross the solar neighborhood. Indeed, there is currently no evidence of the presence of such streams near the Sun, even though the predicted density and large velocities should make them rather easy to identify. Moreover, even if there were an observational bias against the detection of such streams, only in the spherical or oblate halo cases the leading stream is predicted to cross the solar neighborhood. The observed trend in distance from the Sun as function of angle Λ for the leading stream (see Law et al. 2004) only appears to be reproduced in the prolate halo case, ruling out the contribution of Sgr streams to the Solar vicinity.

If the Sgr streams do not cross the vicinity of the Sun, could other streams have a large impact on the velocity and density distribution of dark matter particles that go through the earth based detectors? This seems to be highly unlikely. Other streams will have typically much lower densities than Sgr (since this is the most prominent one). Such lower density streams have indeed been found by Ref.[21], but their stellar densities are of the order of 5% of the local stellar halo density. Given the observed seggregation between dark matter and stars in galaxies, one would expect the dark streams to have an even lower density contrast. Such streams would then have a small effect on direct dark-matter detection experiments.

Acknowledgments

I would like to thank NWO, NOVA and LKBF for financial support.

References

1. J. Dubinski and R.G. Carlberg, *ApJ* **378**, 496 (1991).
2. M.S. Warren, P.J. Quinn, J.K. Salmon and W.H. Zurek, *ApJ* **399**, 405 (1992).
3. J. S. Bullock, in Proceedings of the Yale Cosmology Workshop "The Shapes of Galaxies and Their Dark Matter Halos", ed. P. Natarajan (Singapore: World Scientific), p. 109 (2002).
4. L. Mayer, B. Moore, T. Quinn, F. Governato and J. Stadel, *MNRAS* **336** 119 (2002).
5. N. Yoshida, V. Springel, S.D.M. White and G. Tormen, *ApJ* **535**, L103 (2000).
6. R. Davé, D.N. Spergel, P.J. Steinhardt and B.D. Wandelt, *ApJ* **547**, 574 (2002).
7. P.D. Sackett, in ASP Conf. Ser. 182, Galaxy Dynamics, ed. D. Merritt, J. A. Sellwood, & M. Valluri (San Francisco: ASP), p. 393 (1999).
8. P. Hofner and L.S. Sparke, *ApJ* **428**, 466 (1994).
9. M. Ideta, S. Hozumi, T. Tsuchiya and M. Takizawa, *MNRAS* **311**, 733 (2000).
10. R.P. Olling, *AJ* **112**, 481 (1996).
11. R.P. van der Marel *MNRAS* **248**, 515 (1991).
12. P. Amendt and P. Cuddeford *ApJ* **435**, 93 (1994).
13. R. P. Olling and M.R. Merrifield, *MNRAS* **326** 164 (2001).
14. R. Ibata, G. Lewis, M. Irwin, E. Totten and T. Quinn, *ApJ* **551** 294 (2001).
15. K. V. Johnston, L. Hernquist and M. Bolte, *ApJ* **465**, 278 (1996).
16. R. Ibata, R. Wyse, G. Gilmore, M. Irwin and N. Suntzeff, *AJ* **113**, 634 (1997).
17. A. Helmi, *MNRAS* **351**, 643 (2004).
18. A. Helmi, *ApJ* **610**, L97 (2004).
19. S.R. Majewski et al., *AJ* **128**, 245 (2004).
20. D. R. Law, S.R. Majewski, K.V. Johnston and M.F. Skrutskie, in Proc. of "Satellites and tidal streams", eds. F. Prada and D. Martinez-Delgado, in press (astro-ph/0309567).
21. A. Helmi, S.D.M. White, P.T. de Zeeuw and H.S. Zhao, *Nat* **402**, 53 (1999).

A BIRD'S EYE VIEW OF M31 AND ITS SATELLITE GALAXIES

ALAN MCCONNACHIE

Institute of Astronomy,
Madingley Road,
Cambridge, U.K.
E-mail: alan@ast.cam.ac.uk

SCOTT CHAPMAN, ANNETTE FERGUSON, AVON HUXOR, RODRIGO IBATA, MIKE IRWIN, GERAINT LEWIS, NIAL TANVIR

A panoramic survey of M31's outer regions using the Isaac Newton Telescope Wide Field Camera has revealed a substantial and surprising amount of stellar substructure in the halo of this spiral galaxy, some of which can be used to probe the dark matter distribution of the halo. In particular, a giant stellar stream is observed to extend over a radial range of some 120 kpc from the centre of M31. Combining this with radial velocity data, taken with Keck/DEIMOS, allows for numerical modelling of the orbit of the stream and *directly* measures, for the first time, the mass of a giant galaxy's halo out to large galactocentric radius. The dynamical mass of M31 within the volume probed by the stream is $7.5 - 15 \times 10^{11} \, M_\odot$, and a halo of mass $< 5 \times 10^{11} \, M_\odot$ is ruled out at the 99 % confidence level. A complimentary study of M31's satellite galaxies reveals that their distribution is extremely assymetric, and that the gross assymetry correlates strongly with the position of the Milky Way. The causes of such a distribution, and the consequences for the usage of the satellites as tracers of the dynamical mass of M31, are discussed.

The dominant mass component of a large galaxy such as our own is, by far and away, the halo. Although much of this mass is dark, there exists various stellar components - individual stars (including planetary nebulae), globular clusters and satellite galaxies - whose orbital dynamics are dominated by the gravitational potential of the host halo. We are thus able to constrain the size, shape and profile of the dark matter content of nearby large galaxies by using these visible components as tracers.

1. The INT WFC survey of M31

M31, the Andromeda galaxy, is the nearest giant spiral galaxy to the Milky Way and is believed to be similar to our own galaxy. Optical observations show a pristene, highly inclined disk whose major axis is some 2° (~ 30 kpc)

in radius. Two companion dwarf galaxies located at small galactocentric radii - M32 and NGC 205 - are seen superimposed near the disk.

Figure 1 shows a map of the spatial distribution of red giant branch stars over an area of some 40 sq. degrees, centred on M31. This data was taken over a 4 year period with the Isaac Newton Telescope Wide Field Camera (INT WFC) and reveals stellar structures with a surface brightness as low as 32 mags/sq.arcsec [2,11]. It is immediately clear that the stellar halo of M31 is *extremely* inhomogeneous and a wealth of substructure is visible. These 'fossil' structures can be used to attempt to decipher the formation and evolutionary history of M31 [9].

The stream of stars in the bottom left of Figure 1 is 60 kpc in projection [15]. Follow-up observations with the Canada-France-Hawaii 12K Camera has allowed us to study the (very red) stream population as a function of projected position from M31. We find that the stars furthest in projection from M31 are systematically 0.2 magnitudes fainter than the stream component much closer in to M31, and a gradient is observed between these positions. We conclude that the variation in magnitude is due to distance effects and that the stars furthest in projection from M31 are located 100 kpc further away along the line of sight than the brighter stars. This implies that the stellar stream is over 120 kpc long, on a highly radial orbit [3].

This fortuitous alignment allows us to probe the dark matter potential of M31 out to a very large galactocentric radius. By using the DEep Imaging Multi-Object Spectrograph (DEIMOS) on the Keck II telescope, we have been able to measure radial velocities along the stream, and in many other M31 fields, to an accuracy of some 5km/s [16]. By combining this kinematic data with the three dimensional positional data, we can measure the mass of the M31 halo by finding the potential which best helps to reproduce the orbit defined by the stream. We use the multi-component fixed potential model of [1], with a disk of mass $7 \times 10^{10} M_\odot$, a bulge of mass $1.9 \times 10^{10} M_\odot$ and a spherical halo whose mass is to be determined.

Figure 2 shows several projections of our best-fitting orbits. The solid line in each panel represents the orbit which we find matches the available data with the highest accuracy. The dynamical mass in the halo implied by this, and similar matching orbits, is $7.5 - 15 \times 10^{11} M_\odot$. We rule out at the 99 % confidence level a mass in the halo of $< 5 \times 10^{11} M_\odot$. This is the first time that the dynamical mass of a large galactic halo has been able to be measured *directly*. As we accumulate more data, sophisticated N-body realisations will allow us to improve the accuracy of our estimate, and also

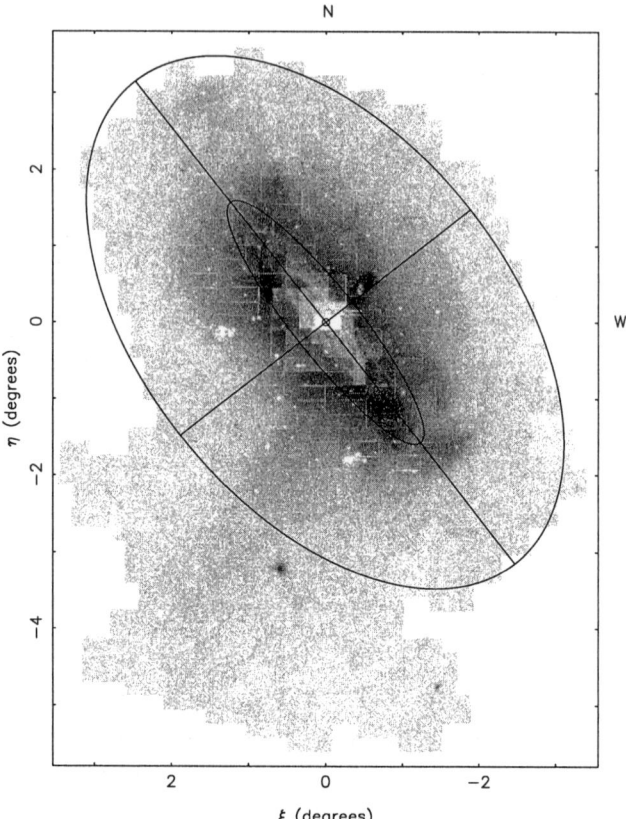

Figure 1. A map showing the spatial distribution of red giant branch stars around M31. The 2° radius ellipse marks the outer boundary of the optical disk of M31, and the 4° radius ellipse corresponds to the original limit of our INT survey. A wealth of substructure is observed. The giant stellar stream is visible in the south-east of the diagram, and the dwarf spheroidal galaxy Andromeda I is also seen along the same line of sight. Andromeda III is visible in the far south of the survey, and Andromeda IX is just visible near to the semi-major axis of M31 in the north-east. Another stream, emanating from the dwarf elliptical galaxy NGC 205, is observed as a loop in the north-west of the diagram.

start estimating the shape of the dark matter potential (see Helmi 2004, *these proceedings*). A second stream in M31, emanating from NGC 205 (upper right of Figure 1) will allow for similar measurement to be made, but over a shorter range in galactocentric distance [4].

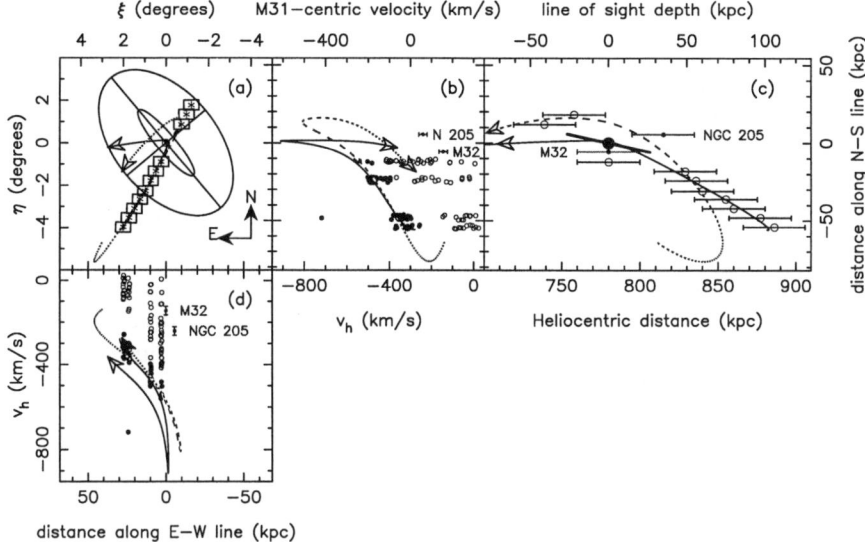

Figure 2. A multi-projection view of the best-fitting orbit to the stream data, represented by the solid line. The position of NGC 205 and M32 is also shown. Their measured radial velocities are inconsistent with being realated to this phase of the stream's orbit.

2. The M31 Satellite Galaxies

A second method of measuring the gravitational potential of large galaxies is through a statistical treatment of the orbital kinematics of their satellites, which are assumed to be unbiased tracers of the underlying potential. The most notable work that has been conducted in this regard in recent years are the studies by M. Wilkinson and N. Evans [10,13,14] for the Galactic and M31 satellite systems.

Following our findings in M31, we extended the INT WFC survey to include the small spiral galaxy M33, and 15 other Local Group dwarf galaxies visible from La Palma, the vast majority of which were members of the M31 subgroup. This has created a large, homogeneous photometric dataset for Local Group galaxies: all the objects have been observed with the same telescope, instrument and filters for similar exposure times. The data were then reduced in the same way using the same pipeline.

Using the tip of the red giant branch, we have derived a set of self-consistent distances to each of the galaxies in our dataset [5,6]. This point in stellar evolution represents the end of the red giant branch phase, immediately prior to the Helium flash. It has been shown observationally and theoretically to be a standard candle for old, metal poor stellar populations [8,12] and is ideal for use in the Local Group. For what follows, it should be emphasised that our distance measurements for the individual galaxies are in good agreement with previous estimates for these bodies.

An Aitoff projection of the satellite system in M31-centric coordinates is shown in Figure 4 and reveals a most intriguing result. The dashed line in this figure represents the projection of the plane, centred on M31, whose pole is *defined* by the position vector of the Milky Way. Thirteen of the fourteen M31 satellites lie on the near side of this plane, towards the Milky Way. If we assume the underlying distribution is isotropic then, taken at face value, such an occurance would happen only 8 times out of 10 000. Clearly, this finding is somewhat strange.

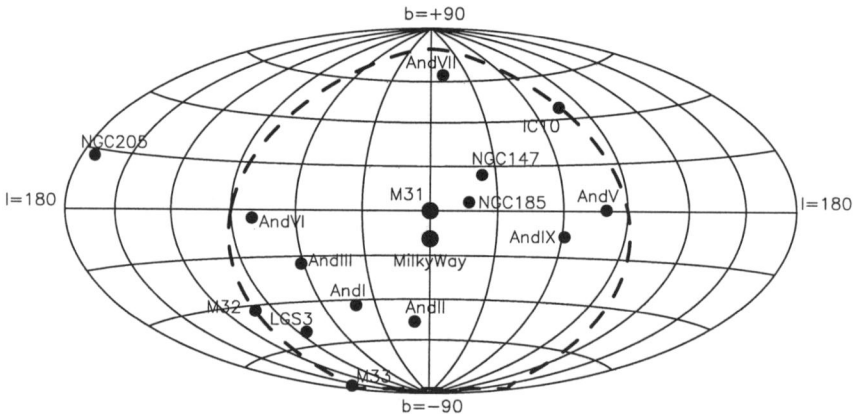

Figure 3. An Aitoff projection of the satellite galaxies of M31 in a native coordinate sytem. l is measured around the disk of M31 (where the Milky Way is defined to lie at $l = 0$) and b is measured from the disk of M31. The satellite distribution is seen to be highly anisotropic. In particular, 13 of the 14 satellites are found on the near side of M31 with respect to the Milky Way (the plane whose pole is defined by the position vector of the Milky Way is shown as a dashed line). Even taking into account the uncertainties in the distances to the satellites, this still suggests that the satellites know about the Milky Way and that they are *not* unbiased tracers of the M31 halo potential.

Such a treatment is of course niave, but even accounting for the uncertainties in the distances, there is a global tendency for the satellites of M31 to be orientated towards the direction of the Milky Way. It transpires that this effect is unlikely to be due to selection. Unknown sources of error in our distances and in previous distance estimates may explain the result, but will have significant repurcussions for the use of the TRGB as a standard candle. We deem it equally likely that this anisotropy has a physical cause, and that the orbital characteristics of the satellites of M31 have been strongly influenced by the Milky Way. In this case, they are clearly *not* unbiased indicators of the M31 potential, unless the M31 dark matter potential is itself extremely lop-sided. It is thus crucial to understand the nature of the anisotropy if we are to be able to confidently use the satellite galaxies as probes of the dark matter in the Local Group [7].

3. Summary

The INT WFC survey of M31 and its environs has revealed a vast amount of faint stellar substructure in M31, giving vital clues to this galaxy's ongoing formation process. The giant stellar stream visible in the south has been used to *directly* estimate the dark matter halo potential of M31 out to 125 kpc, and we find its most likely mass is $7.5 - 15 \times 10^{12} M_\odot$. Analysis of the satellites of M31 has shown that there is a gross anisotropy in their distribution. Until the cause of this unexpected assymetry is known, it is not clear that we will be able to use these objects as tracers of the dark mass in M31 and the Local Group.

References

1. A. Klypin, H.Zhao and R.S.Somerville, *ApJ*, **573**, 597 (2002).
2. A.M.N. Ferguson et al. *AJ*, **124**, 1452 (2002).
3. A.W. McConnachie et al., *MNRAS*, **343**, 1335 (2003).
4. A.W. McConnachie et al., *MNRAS*, **351**, L94 (2004a).
5. A.W. McConnachie et al., *MNRAS*, **350**, 243 (2004b).
6. A.W. McConnachie et al., *MNRAS*, *in press* (2004c).
7. A.W. McConnachie and M.J. Irwin, *MNRAS*, *submitted* (2004d).
8. G.S. Da Costa and T.E. Armandroff, *AJ*, **100**, 162 1990.
9. K.V. Johnston, L. Hernquist and M. Bolte, *ApJ*, **465**, 278 (1996).
10. M.I. Wilkinson and N.W. Evans, *MNRAS*, **310**, 645 (1999).
11. M.J. Irwin et al., *in preparation* (2004).
12. M. Salaris and S. Cassisi, *MNRAS*, **289**, 406 (1997).
13. N.W. Evans and M.I. Wilkinson, it MNRAS, **316**, 929 (2000).
14. N.W. Evans et al., *ApJL*, **540**, 9 (2000)
15. R. Ibata et al., *Nature*, **412**, 49 (2001).
16. R.A. Ibata et al., *MNRAS*, **351**, 117 (2004).

DARK MATTER HALOS: SHAPES, THE SUBSTRUCTURE CRISIS, AND INDIRECT DETECTION

A. R. ZENTNER[1], S. M. KOUSHIAPPAS[2], AND S. KAZANTZIDIS[1,3]

[1] *Kavli Institute for Cosmological Physics & Department of Astronomy and Astrophysics, The University of Chicago, Chicago, IL 60637 USA*
[2] *Department of Physics, Swiss Federal Institute of Technology, ETH Hönggerberg, Zürich, Switzerland*
[3] *Institute for Theoretical Physics, University of Zürich, Zürich, Switzerland*

In this proceeding, we review three recent results. First, we show that halos formed in simulations with gas cooling are significantly rounder than halos formed in dissipationless N-body simulations. The increase in principle axis ratios is $\sim 0.2 - 0.4$ in the inner halo and remains significant at large radii. Second, we discuss the CDM substructure crisis and demonstrate the sensitivity of the crisis to the spectrum of primordial density fluctuations on small scales. Third, we assess the ability of experiments like VERITAS and GLAST to detect γ-rays from neutralino dark matter annihilation in dark subhalos about the MW.

1. Introduction

A preponderance of evidence indicates that galaxies are embedded in massive, extended dark matter (DM) *halos*. Simulations of structure formation in the hierarchical cold dark matter (CDM) paradigm predict that CDM halos are generally triaxial[1,2] that they teem with self-bound *subhalos*[3].

The structure of halos is an important ingredient in modeling the DM direct detection signals[4] and halo shapes have received attention for testing the CDM paradigm as new and improved probes of halo shape have been applied[5,6]. *Dissipationless* simulations predict that Milky Way(MW)-size halos have a mean minor-to-major axis ratio of $c/a \approx 0.6 - 0.7$ with a dispersion of $\sim 0.1^1$, while studies suggest that the coherence of the Sagittarius tidal stream constrains the MW halo to $c/a \gtrsim 0.8^6$. In § 2, we present recent results on the effect of baryonic dissipation on halo shapes in high-resolution, cosmological simulations.

In § 3, we turn to halo substructure. In the MW and M31, there are more than an order of magnitude fewer observed satellites than the predicted number of subhalos of comparable size[3]. Several explanations have

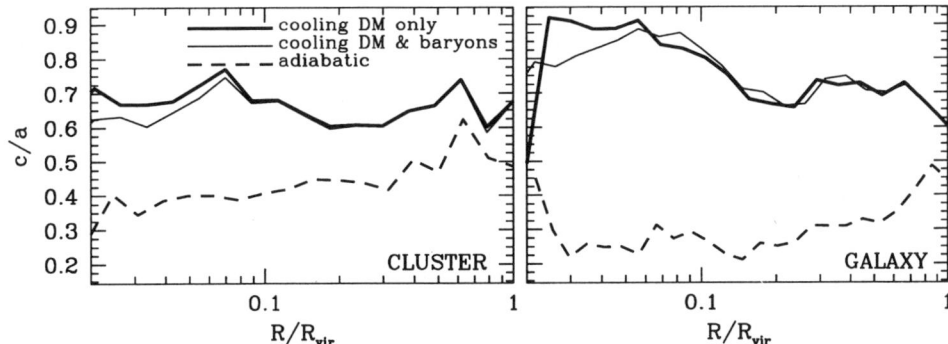

Figure 1. The effect of gas cooling on halo shapes. *Left*: Minor-to-major axis ratio c/a, as a function of major axis length for a cluster-size halo. The *dashed* line shows the shape profile in the adiabatic simulation. The *thick, solid* line shows the shape profile of DM only in the cooling run, while the *thin, solid* line shows the shape profile for DM and baryons in the cooling run. *Right*: Same as the left panel, but for a MW-size galaxy progenitor (see text).

been offered, including alternative DM properties[7] and inefficient galaxy formation in the shallow potentials of small subhalos[8]. We study the sensitivity of the dwarf satellite population to the primordial power spectrum (PPS) of density fluctuations. If the lack of luminous MW satellites is due to inefficient galaxy formation, the MW halo should contain $\gtrsim 10^2$ otherwise dark subhalos. Strong lensing will be one probe of dark subhalos[9]. More speculatively, the annihilation of DM particles in these dense substructures may result in numerous γ-ray sources in the MW halo. We assess the potential for instruments like VERITAS[11] and GLAST[12] to detect such sources in favorable models of supersymmetric (SUSY) DM in § 4.

2. Halo Shapes

We studied the effect of gas cooling on the shapes of DM halos using high-resolution cosmological simulations of cluster and galaxy formation in a concordance ΛCDM cosmology. The simulations were performed with the ART N-body plus Eulerian gasdynamics code[14], the details are in Kazantzidis et al.[13].

Briefly, we analyzed simulations of 8 cluster-size objects of mass 10^{13} $h^{-1}M_\odot$ to 3×10^{14} $h^{-1}M_\odot$. The cluster simulations had a peak force resolution of $\simeq 2.4 h^{-1}$kpc and a DM particle mass of $m_p \simeq 2.7 \times 10^8$ $h^{-1}M_\odot$. We also analyzed a simulation of the early evolution ($z \gtrsim 4$)

of a galaxy that becomes MW-size at $z = 0$ described by Kravtsov[15]. This simulation had $m_\mathrm{p} \simeq 9.2 \times 10^5\ h^{-1}\mathrm{M}_\odot$ and peak resolution $\simeq 183 h^{-1}$kpc. The mass and force resolution are adequate to study the inner regions of halos reliably. For each object, we analyzed two sets of simulations started from the same set of initial conditions. In one set, the gas dynamics were treated adiabatically, without any radiative cooling. The second set of simulations included cooling, and star formation.

We measured halo shapes by diagonalizing the moment of inertia tensor[2]. Our main results are summarized in Figure 2. In the left panel, we show c/a, as a function of major axis length for a representative cluster-size halo. On the right, we show results for the galaxy progenitor. The net effect of baryon dissipation is striking. At small radii, the axis ratios in the cooling simulations are greater by $\Delta(c/a) \gtrsim 0.3$ and the difference persists out to $\sim R_\mathrm{vir}$, where $\Delta(c/a) \sim 0.1$. The baryons in the cluster are mostly in a massive, central, elliptical, while in the galaxy, $\sim 90\%$ of the baryons are in a flattened, gaseous disk. In both cases, the net effect of cooling is weakly dependent upon radius, implying that the effect of baryonic dissipation on halo shapes is not critically sensitive to the detailed morphology of the baryonic component.

3. Halo Substructure

The most accurate technique for studying halo substructure is numerical simulation; however, the computational expense of simulations limits their dynamic range and their applicability in explorations of cosmological parameter space. To overcome this, Zentner and Bullock (ZB)[18] developed an approximate, analytic model for subhalo populations and an updated model has recently been successfully tested against a suite of N-body simulations[19]. The model approximately accounts for the merger statistics of subhalos, dynamical friction, and mass loss and redistribution due to tidal forces. The model allows one to generate hundreds of realizations of MW-like halos and thereby explore the distribution of possible subhalo populations.

In the standard paradigm, structure forms from primordial density fluctuations characterized by a nearly scale-invariant PPS, $P(k) \propto k^n$ with $n \simeq 1$. This basic picture has significant observational support[16]. However, cosmic microwave background anisotropy constrains the PPS on large scales, $k \sim 10^{-2}\ h\mathrm{Mpc}^{-1}$, while halo substructure is sensitive to small scale power, $k \sim 10 - 100\ h\mathrm{Mpc}^{-1}$. ZB studied the effect of variant power

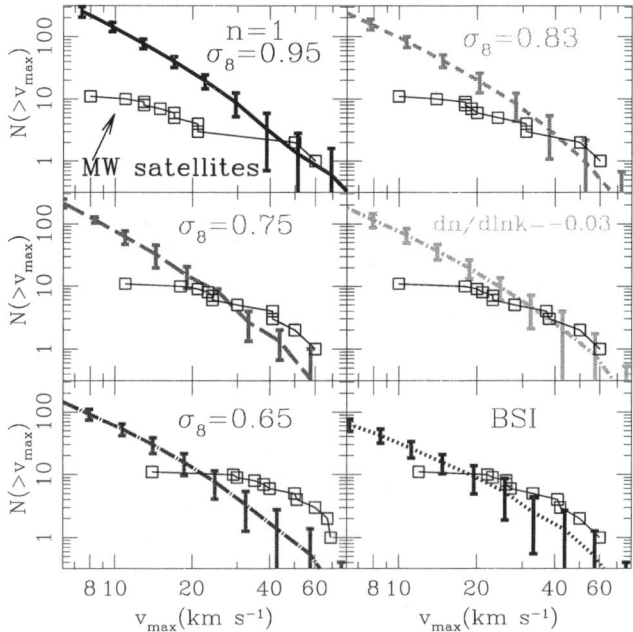

Figure 2. Dwarf satellites and the power spectrum. We show the observed satellite velocity functions (*squares*) and the predicted satellite velocity functions (*thick lines*) for 6 different PS. Clockwise from the top left: standard $n = 1$, $\sigma_8 = 0.95$; $n = 0.94$, $\sigma_8 = 0.83$; WMAP best-fit $n = 1.03$, $dn/d\ln k = -0.03$, $\sigma_8 = 0.84$; BSI; $n = 0.84$, $\sigma_8 = 0.65$; and $n = 0.90$, $\sigma_8 = 0.75$. The models are labeled by σ_8. Lines are the means of 100 model realizations and errorbars represent the 1σ scatter. Observational data are from Mateo[20].

spectra on the MW dwarf satellites. They took several PPS with various motivations, all normalized to COBE: (1) standard $n = 1$, $\sigma_8 = 0.95$; (2) $n = 0.94$, $\sigma_8 = 0.83$; (3) $n = 0.9$, $\sigma_8 = 0.75$; (4) running mass inflation $n = 0.84$, $\sigma_8 = 0.65$; (5) broken scale-invariance (BSI) with a power cut-off at $k_c = 1\ h\mathrm{Mpc}^{-1}$[17]; and (6) the best-fit running spectrum from WMAP $n = 1.03$, $dn/d\ln k = -0.03$, $\sigma_8 = 0.84$. The steps in the calculation are first to generate MW halo substructure realizations for each PPS and to model the velocity dispersions of the embedded stellar components to determine the appropriate subhalo size (labelled by maximum circular velocity $V_{\max}$) in which the observed satellites may be embedded. In this way, one constructs predicted and observed cumulative velocity functions.

Figure 3 summarizes the results. First, one sees that the degree to which the dwarf satellite problem represents a challenge is greatly alleviated

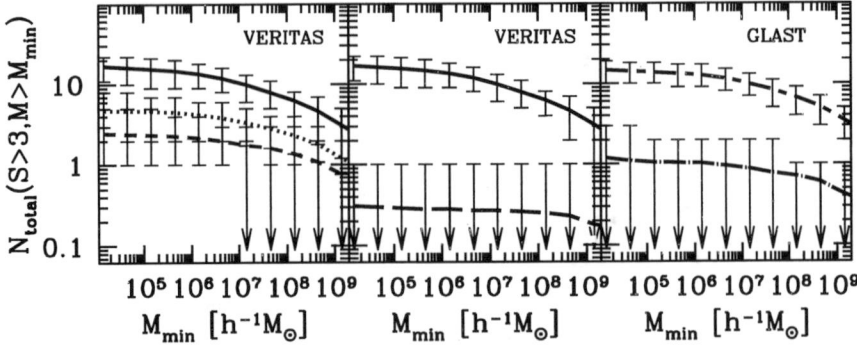

Figure 3. The cumulative number of subhalos of mass $M \geq M_{\min}$ detectable at $S > 3$ on the sky. Results are based on 100 realizations of a MW-size halo. Errorbars indicate the 68% range and down arrows indicate that $> 16\%$ realizations have zero subhalos at that mass. *Left*: The number detectable by VERITAS. The *solid* line shows the highest detection efficiency case of $M_\chi = 500$GeV. For comparison, the *dotted* line shows results for $M_\chi = 200$GeV and the *dashed* line for $M_\chi = 5$TeV. *Middle*: The *solid* line shows our standard result for a ΛCDM cosmology with $n = 1$ and $\sigma_8 = 0.95$. The *dashed* line shows the detectable number of subhalos with the WMAP best-fit running power spectrum, $dn/d\ln k = -0.03$. *Right*: The number detectable with GLAST. The *dashed* line represents the best case of $M_\chi = 50$GeV in a standard ΛCDM cosmology. The *dot-dashed* line shows the potential number of detections for a $M_\chi = 100$GeV neutralino.

in the WMAP best-fit cosmology. The level at which inefficient galaxy formation or a critical mass scale for galaxy formation must be invoked to solve the satellite scarcity problem is degenerate with the PPS on small scales. Second, the MW satellite population by itself provides independent evidence against extreme models, such as the low normalization, $\sigma_8 = 0.65$ model which under-predict substructure.

4. γ-rays from Dark Substructure

Using the subhalo model of § 3, we can assess the ability of experiments like VERITAS and GLAST to detect γ-ray fluxes from neutralino DM (χ) annihilations in the dense, inner regions of subhalos. Koushiappas et al.[21], assumed the most optimistic SUSY parameters consistent with constraints on Ω_M[16] and determined the number of expected detections at a significance $S > 3$, as a function of subhalo mass M. The results (summarized in Figure 4) show that for χ masses $M_\chi \lesssim 100$GeV, the large field of view of glast GLAST and the energy sensitivity of VERITAS will allow them to detect substructure when operated in concert. For $100\text{GeV} \lesssim M_\chi \lesssim 500\text{GeV}$, a detection will rely on serendipity, while for $M_\chi \gtrsim 500\text{GeV}$, substructure detection via the γ-ray signal is unlikely.[21]

Acknowledgments

These results are based on several collaborative works. We thank B. A. Allgood, J. S. Bullock, A. V. Kravtsov, B. Moore, D. Nagai, and T. P. Walker for their invaluable contributions and for allowing us to present our results here. ARZ and SK are funded by the Kavli Institute for Cosmological Physics at The University of Chicago and The National Science Foundation. SMK is funded by the Swiss National Science Foundation.

References

1. e.g. Y. P. Jing and Y. Suto, *ApJ*, **574**, 538 (2002)
2. J. Dubinski and R. G. Carlberg, *ApJ*, **378**, 496 (1991)
3. A. A. Klypin et al., *ApJ*, **522**, 82 (1999); B. Moore et al., *ApJL*, **524**, L19, (1999)
4. For example, see A. Green, *This Proceeding*, and references therein
5. e.g. Kolokotronis et al., *MNRAS*, **320**, 49 (2001); D. A. Buote et al., *ApJ*, **577**, 183 (2002); H. Hoekstra et al., *ApJ*, **606**, 67 (2004)
6. R. Ibata et al., *ApJ*, **551**, 294 (2001); S. Majewski et al., *ApJ*, **599**, 1082 (2003); but see A. Helmi, *MNRAS*, **351**, 643 (2004) and *This Proceeding*
7. e.g. D. N. Spergel and P. J. Steinhardt, *PRL*, **84**, 3760 (2000); Colín et al., *ApJ*, **542** 622 (2000); A. Knebe et al., *MNRAS*, **329**, 813 (2002)
8. e.g. J. S. Bullock et al., *ApJ*, **539**, 517 (2000); R. S. Somerville, 572, L23 (2002); A. V. Kravtsov et al., *ApJ*, **609**, 482 (2004)
9. N. Dalal and C. S. Kochanek, *ApJ*, **572** 25 (2001)
10. L. Bergström et al., *PRD*, **59**, 043506 (1999); C. Calcáneo-Roldán and B. Moore, *PRD*, **62**, 123005 (2000); A. Tasitsiomi and A. Olinto, *PRD*, **66**, 083006 (2002); F. Stoehr et al., *MNRAS*, **345**, 1313 (2003)
11. URL http://veritas.sao.arizona.edu
12. URL http://glast.gsfc.nasa.gov
13. S. Kazantzidis et al., *ApJL*, **611**, L73 (2004)
14. A. V. Kravtsov, A. A. Klypin, and Y. Hoffman, *ApJ*, **571**, 563 (2002)
15. A. V. Kravtsov, *ApJL*, **590**, L1 (2003)
16. D. N. Spergel et al., *ApJS*, **148**, 175 (2003)
17. M. Kamionkowski and A. R. Liddle, *PRL*, **84**, 4525 (2000)
18. A. R. Zentner and J. S. Bullock, *ApJ*, **598**, 49 (2003)
19. A. R. Zentner et al., *ApJ*, **624**, in press (2005)
20. M. Mateo, *ARAA*, **36**, 435 (1998)
21. S. M. Koushiappas et al., *PRD*, **69**, 043501 (2004)

THE POWER SPECTRUM OF CDM ON SUB-GALACTIC SCALES

S. HOFMANN

Physics Department, Stockholm University, AlbaNova University Center,
SE-106 91 Stockholm, Sweden
E-mail: stehof@physto.se

A. M. GREEN

Department of Physics and Astronomy, University of Sheffield,
Sheffield, S3 7RH, UK
E-mail: a.m.green@sheffield.ac.uk

D. J. SCHWARZ

Universität Bielefeld, Fakultät für Physik, 33501 Bielefeld, Germany
E-mail: dschwarz@physik.uni-bielefeld.de

We show that collisional damping and free-streaming of cold dark matter (CDM) induce a sharp cut-off in the CDM power spectrum at about $10^{-6} M_\odot$, which sets the typical scale for the first CDM haloes in the hierarchical picture of structure formation. We present a WMAP normalized primordial power spectrum, which could serve as an input for high resolution CDM simulations. The smallest inhomogeneities typically enter the non-linear regime at a redshift of about 60.

1. Motivation

Cosmological structures are generated by cold dark matter (CDM) density perturbations $\delta\rho$ that eventually collapse to purely gravitationally bound structures. CDM density perturbations are characterised by their mass density contrast $\Delta \equiv \delta\rho/\rho$ with respect to the average CDM density ρ. The mass density contrast as function of redshift z and comoving wavenumber k can be written as

$$\Delta(k,z) = \Delta(k,z_\text{i}) \, T_\Delta^{1/2}(k,z) \, D(k) \,, \tag{1}$$

where $\Delta(k, z_\text{i})$ is the primordial density contrast, $T_\Delta(k, z)$ is the transfer function and $D(k)$ is a damping term.

On large scales $D(k) = 1$, so $\Delta(k, z)$ is the same for different CDM particle candidates. Thus structure formation on large scales is solely governed by gravitational interaction. However this statement can not be extended down to scales, on which particle interactions dominate. In general $\Delta(k, z)$ does contain information on specific CDM particle interactions and it is encoded in $D(k)$.

Below we explain the physics encoded in $D(k)$. We show that $D(k)$ gives rise to characteristic features in the power spectrum of matter density perturbations[1]: a sharp cut-off and a maximum close to this cut-off.

Hierarchical models of structure formation (like the CDM family of models) have increasing amounts of power at smaller scales. It is a demanding challenge to implement this in the initial conditions for CDM simulations. A physical cut-off allows to take into account consistently the history of substructure formation on all scales that contribute to structure formation. The maximum close to the cut-off sets the typical scale for the first CDM haloes. The clumpiness of CDM at the smallest scales is of utmost importance for the interpretation of indirect CDM searches[2,3].

2. Physics behind $D(k)$

Nonequilibrium processes in CDM determine deviations of $D(k)$ from one. At temperatures far above the mass of a CDM particle, $T \gg m$, any deviation of local thermal equilibrium (LTE) in CDM is transferred to the radiation by elastic scatterings. There is a hierarchy of time scales relevant to keep CDM in LTE: the smallest is the interaction time $\tau_0 \equiv r_0/\langle v \rangle$, with r_0 denoting the interaction range (provided by the underlying field theory) and $\langle v \rangle$ is the average CDM velocity. The next scale is given by the time in between to collisions, $\tau_f \equiv l_f/\langle v \rangle$ with l_f denoting the mean free path of CDM particles. The third time scale is the relaxation time $\tau_r \equiv N\tau_f$, with N denoting the number of collisions needed in order to establish or keep LTE in CDM. The hierarchy is

$$\tau_0 \ll \tau_f \ll \tau_r \,, \tag{2}$$

in particular for a WIMP $N \gg 1$, once the temperature drops below the WIMP mass[4].

Kinetic decoupling happens at a temperature $T_{\rm kd}$, when $\tau_r \approx 1/H$. Around $T_{\rm kd}$ the elastic scatterings are not frequent enough in order to maintain LTE. The residual interactions between CDM and radiation give rise to viscous processes in CDM[5]: a local volume expansion around the adiabatic CDM current (bulk viscosity) and a bending of CDM velocities away from it (shear viscosity). In principal there is a third viscosity, heat conduction, but it is subdominant in CDM. Neglecting the heat current, the entropy current in CDM is proportional to the adiabatic current density. However, sources for entropy production in CDM are provided by bulk and shear viscosity.

The viscous processes give rise to modifications of the CDM dispersion relation[5]: since Δ evolves in a medium (radiation), Δ oscillates with complex frequencies ω. The imaginary part $\text{Im}(\omega) \propto -\tau_r\, k^2$ implies a decay rate for Δ. This decay rate is effective from z_i down to the redshift of kinetic decoupling, resulting in exponential damping of Δ characterised by a damping scale k_d. Δ is a solution of the linearised Navier-Stokes equation, which reduce to an oscillator with time dependent friction.

After kinetic decoupling, CDM enters the free streaming regime. Free streaming refers to geodesic motion of CDM particles, which allows them to propagate from overdense to underdense regions, thereby further damping Δ. This effect is taken into account by solving the collisionless Boltzmann equation for Δ in the matter-radiation mixture with the appropriate initial conditions at kinetic decoupling. It turns out that free streaming leads to further damping of Δ. The damping is again proportional to a damping exponential[1], but with a different damping scale $k_{fs} < k_d$.

3. Results

The physical processes described in the last section lead to the following damping term[1]:

$$D(k) = \left[1 - \frac{2}{3}\left(\frac{k}{k_{fs}}\right)^2\right] \exp\left[-\left(\frac{k}{k_{fs}}\right)^2 - \left(\frac{k}{k_d}\right)^2\right]. \quad (3)$$

In the case of a WIMP, we typically find for the characteristic damping scale at kinetic decoupling:

$$k_d \approx \frac{3.8 \times 10^7}{\text{Mpc}} \left(\frac{m}{100\ \text{GeV}}\right)^{1/2} \left(\frac{T_{kd}}{30\ \text{MeV}}\right)^{1/2}. \quad (4)$$

For bino-like CDM, this corresponds to a length scale of $\sim 10^{-2}/H$ at kinetic decoupling. The total CDM mass contained in a sphere with radius π/k_d is[5,6], $M_d \sim 10^{-10} M_\odot$.

The characteristic damping scale given by free streaming is then

$$k_{fs} \approx \frac{1.7 \times 10^6}{\text{Mpc}} \frac{(m/100\ \text{GeV})^{1/2}\, (T_{kd}/30\ \text{MeV})^{1/2}}{1 + \ln(T_{kd}/30\ \text{MeV})/19.2}. \quad (5)$$

The damping scale from free streaming depends on $\omega_m \equiv \Omega_m h^2$ only via the logarithm, we therefore set it equal to WMAP's best fit value[7], $\omega_m = 0.14$. The corresponding length scale at matter-radiation equality $\sim 10^{-8}/H$ and the total mass contained in a sphere with radius π/k_{fs} is[5,6], $M_{fs} \sim 10^{-6} M_\odot$ for bino-like CDM.

We are now in the position to present the dimensionless power spectrum for matter density perturbations (defined as $P_\Delta(k,z) = (k^3/2\pi^2)|\Delta(k,z)|^2$) normalized to the WMAP measurements. For simplicity we assume that gravitational waves have a negligible contribution to the CMB anisotropies and that the density perturbations have a Harrison-Zel'dovich primordial power spectrum ($n = 1$). We find for $k > k_b \sim 10^3 \mathrm{Mpc}^{-1}$ and $z_{eq} \gg z > z_b \sim 150$:

$$\frac{P_\Delta(k,z)}{10^{-7} A} = 1.06 \, c^2 \left[\ln \frac{k}{k_{eq}} + b \right]^2 D^2(k) \left(\frac{1 + z_{eq}}{1 + z} \right)^\nu , \qquad (6)$$

where $k_{eq} \equiv 1/(aH)_{eq}$ and $A = 0.9 \pm 0.1$ according to [xx]. The scale of equality is $k_{eq} = (0.01/\mathrm{Mpc})(\omega_m/0.14)$ and $1 + z_{eq} = 3371(\omega_m/0.14)$. The parameters c, b and ν depend on the baryon to matter fraction $f_b \equiv \Omega_b/\Omega_m$. For $f_b = 0.16(0)$, $c = 1.37(3/2)$, $\nu = 1.8(2)$ and $b = -1.57(-1.74)$. The calculation of $T_\Delta(k,z)$ including baryons, as well as a discussion of the constraints k_b and z_b can be found in Green et al.[1].

Fig. 1 shows $P_\Delta(k,z)$ for bino-like CDM at a redshift of 500, close to the end of the linear regime of structure formation. It can be observed that the cut-off induced by $D(k)$ is indeed very sharp and that the power spectrum has a maximum close to the cut-off.

The cut-off in the power spectrum sets the typical scale for the first haloes in the hierarchical picture of structure formation. Non-linear structure formation starts at a redshift $z_{nl}^{max} \approx 60$, see Fig. 2. Also shown is the redshift z_{nl} at which typical density perturbations on comoving scale R go nonlinear. The plateau at $R < 1$ pc is due to the sharp cut-off in P_Δ.

The size and mass of the first generation of subhaloes that form at z_{nl}^{max} can be estimated within the spherical collapse model. The mean CDM mass within a sphere of comoving radius R is $M(R) = 1.6 \times 10^{-7} M_\odot (\omega_m/0.14)(R/\mathrm{pc})^3$. CDM overdensities that go non-linear have mass twice this value i.e. roughly equal to the mass of Mars. These CDM haloes are however much less compact than Mars. The physical size of the first haloes at turn-around (when their evolution decouples from the cosmic expansion) is $r = 1.05 R/(1 + z_{nl}^{max}) \sim 0.02$ pc. The first haloes then undergo violent relaxation, decreasing in radius by a factor of two so

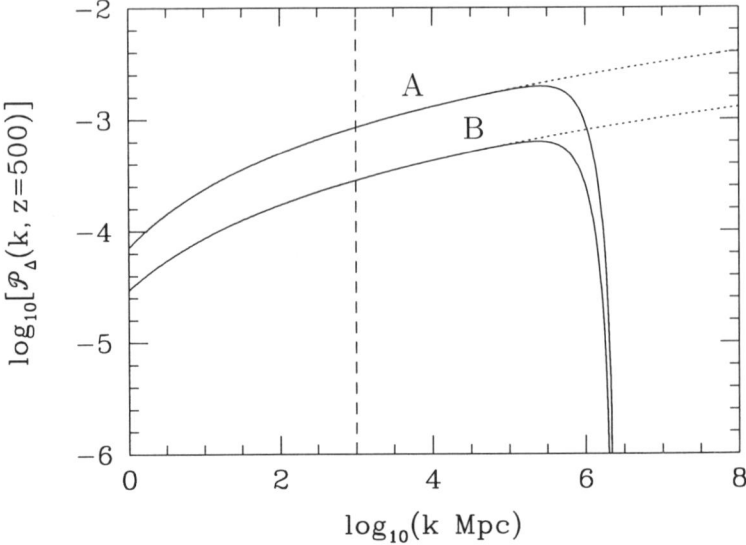

Figure 1. The dimensionless power spectrum of the CDM density contrast at $z = 500$ for two fiducial sets of parameters (full lines): model A (B) has $m = 100(150)$ GeV and a universal sfermion mass $m_{\tilde{f}} = 230(190)$ GeV. Then $\Omega_{\rm bino} = 0.31(0.17)$, which is the WMAP allowed range of values, and $T_{\rm kd} = 33(21)$ MeV, $M_{\rm d} = 9(6) \times 10^{-11} M_\odot$ and $M_{\rm fs} = 9(6) \times 10^{-7} M_\odot$. Without the effects of collisional damping and free streaming, the power spectra would be given by the dotted lines.

that their present day radius is of order tens of milli-pc (comparable to the size of the solar system). Their present day overdensity would be of order $\Delta_{\rm halo} = 7(1 + z_{\rm nl}^{\rm max})^3 \sim 10^6$, several orders of magnitude larger than that of galaxies, which makes them very attractive for indirect CDM searches, if a significant number of them survive.

Therefore, it is a major challenge to estimate the survival probability of the first CDM halos during the hierarchical process of CDM halo formation, but also in consideration of encounters with stars.

Acknowledgments

It is a pleasure to thank the organisers and participants of IDM2004 for a great conference. AMG was supported by the Particle Physics and Astronomy Research Council (UK). SH was supported by the Wenner-Gren Foundation.

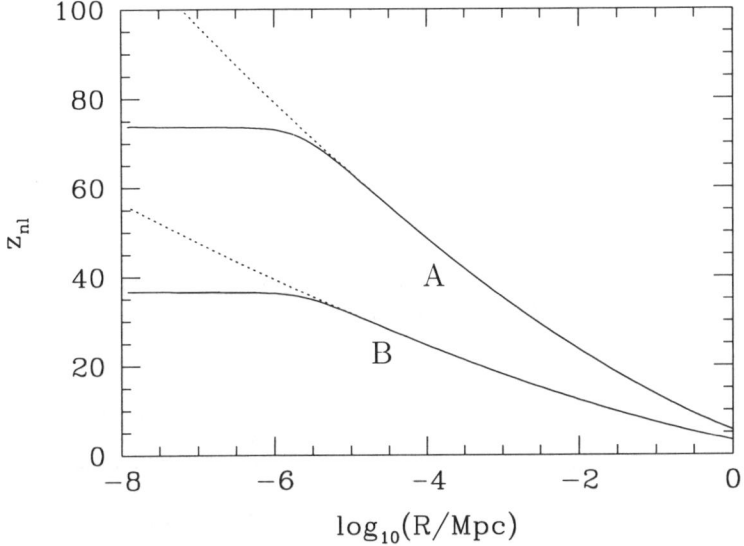

Figure 2. The redshift at which typical fluctuations of comoving scale R become non-linear, for the two models from Fig. 1. The full lines take into account the effects of collisional damping and free streaming, whereas the dashed lines show the behaviour without a cut-off in the power spectrum.

References

1. A. M. Green, S. Hofmann and D. J. Schwarz, *Mon. Not. Roy. Astron. Soc.* **353**, L23 (2004).
2. L. Bergstöm, J. Edsjö, P. Gondolo and P. Ullio, *Phys. Rev.* **D59**, 043506 (1999).
3. L. Bergström, *Rept. Prog. Phys.* **63**, 793 (2000).
4. C. Schmid, D. J. Schwarz and P. Widerin, *Phys. Rev.* **D59**, 043517 (1999).
5. S. Hofmann, D. J. Schwarz and H. Stöcker, *Phys. Rev.* **D 64**, 083507 (2001).
6. V. Berezinsky, V. Dokuchaev and Y. Eroshenko, Y., *Phys. Rev.* **D68**, 103003 (2003).
7. D. N. Spergel et al., *Astrophys. J. Suppl.* **148**, 175 (2003).

SUSY DARK MATTER*

KEITH A. OLIVE

William I. Fine Theoretical Physics Institute,
University of Minnesota, Minneapolis, MN 55455, USA
E-mail: olive@umn.edu

The status of the constrained minimal supersymmetric standard model (CMSSM) will be discussed in light of our current understanding of the relic density after WMAP. A global likelihood analysis of the model is performed. Also considered are models which relax and further constrain the CMSSM. Prospects for dark matter detection in colliders and cryogenic detectors will be briefly discussed.

1. Introduction

Supersymmetric models with conserved R-parity contain one new stable particle which is a candidate for cold dark matter (CDM) [1]. Here, I will assume that the lightest supersymmetric particle (LSP) is the neutralino, however, another possibility is the gravitino which has been discussed recently in the CMSSM context [2].

In general, the neutralino mass eigenstates can be expressed as a linear combination

$$\chi = \alpha \tilde{B} + \beta \tilde{W}^3 + \gamma \tilde{H}_1 + \delta \tilde{H}_2 \qquad (1)$$

The solution for the coefficients α, β, γ and δ for neutralinos that make up the LSP can be found by diagonalizing the mass matrix which depends on $M_1(M_2)$ which are the soft supersymmetry breaking U(1) (SU(2)) gaugino mass terms, μ, the supersymmetric Higgs mixing mass parameter and the two Higgs vacuum expectation values, v_1 and v_2. One combination of these is related to the Z mass, and therefore is not a free parameter, while the other combination, the ratio of the two vevs, $\tan \beta$, is free.

The most general version of the MSSM, despite its minimality in particles and interactions contains well over a hundred new parameters. It is

---

*This work was partially supported by DOE grant DE-FG02-94ER-40823.

often assumed that, at some unification scale, all of the gaugino masses receive a common mass, $m_{1/2}$. The gaugino masses at the weak scale are determined by running a set of renormalization group equations. Similarly, one often assumes that all scalars receive a common mass, m_0, at the GUT scale. These too are run down to the weak scale. The remaining supersymmetry breaking parameters are the trilinear mass terms, A_0, which I will also assume are unified at the GUT scale, and the bilinear mass term B.

The natural boundary conditions at the GUT scale for the MSSM would include μ and B in addition to $m_{1/2}$, m_0, and A_0. In this case, upon running the RGEs down to a low energy scale and minimizing the Higgs potential, one would predict the values of M_Z, $\tan\beta$ (in addition to all of the sparticle masses). Since M_Z is known, it is more useful to analyze supersymmetric models where M_Z is input rather than output. It is also common to treat $\tan\beta$ as an input parameter. This can be done at the expense of shifting μ (up to a sign) and B from inputs to outputs. This model is often referred to as the constrained MSSM or CMSSM. Once these parameters are set, the entire spectrum of sparticle masses at the weak scale can be calculated. In the CMSSM, the solutions for μ generally lead to a neutralino which which very nearly a pure $\tilde{B}$.

2. The CMSSM after WMAP

For a given value of $\tan\beta$, A_0, and $sgn(\mu)$, the resulting regions of acceptable relic density and which satisfy the phenomenological constraints can be displayed on the $m_{1/2} - m_0$ plane. In Fig. 1a, the light shaded region corresponds to that portion of the CMSSM plane with $\tan\beta = 10$, $A_0 = 0$, and $\mu > 0$ such that the computed relic density yields $0.1 < \Omega_\chi h^2 < 0.3$. At relatively low values of $m_{1/2}$ and m_0, there is a large 'bulk' region which tapers off as $m_{1/2}$ is increased. At higher values of m_0, annihilation cross sections are too small to maintain an acceptable relic density and $\Omega_\chi h^2 > 0.3$. Although sfermion masses are also enhanced at large $m_{1/2}$ (due to RGE running), co-annihilation processes between the LSP and the next lightest sparticle (in this case the $\tilde{\tau}_1$) enhance the annihilation cross section and reduce the relic density [3]. This occurs when the LSP and NLSP are nearly degenerate in mass. The dark shaded region has $m_{\tilde{\tau}_1} < m_\chi$ and is excluded.

Also shown in Fig. 1a are the relevant phenomenological constraints. The preferred range of the relic LSP density has been altered significantly by the recent improved determination of the allowable range of the cold dark

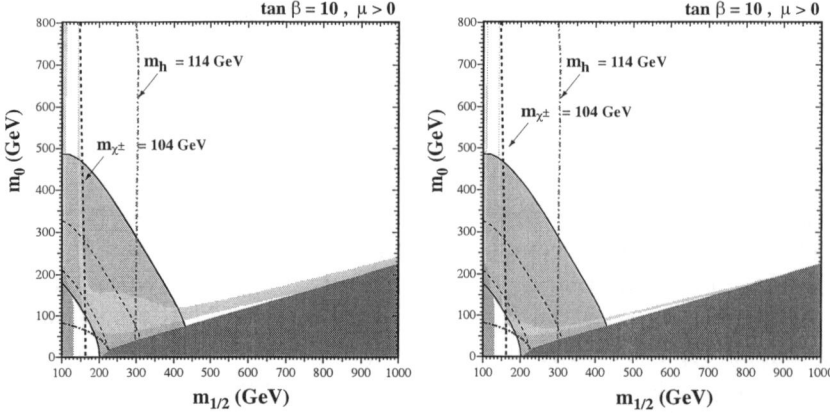

Figure 1. The $(m_{1/2}, m_0)$ planes for (a) $\tan\beta = 10$ and $\mu > 0$, assuming $A_0 = 0$, $m_t = 175$ GeV and $m_b(m_b)^{\overline{MS}}_{SM} = 4.25$ GeV. The near-vertical (red) dot-dashed lines are the contours $m_h = 114$ GeV, and the near-vertical (black) dashed line is the contour $m_{\chi^\pm} = 104$ GeV. Also shown by the dot-dashed curve in the lower left is the corner excluded by the LEP bound of $m_{\tilde{e}} > 99$ GeV. The medium (dark green) shaded region is excluded by $b \to s\gamma$, and the light (turquoise) shaded area is the cosmologically preferred regions with $0.1 \leq \Omega_\chi h^2 \leq 0.3$. In the dark (brick red) shaded region, the LSP is the charged $\tilde{\tau}_1$. The region allowed by the E821 measurement of a_μ at the 2-σ level, is shaded (pink) and bounded by solid black lines, with dashed lines indicating the 1-σ ranges. In (b), the relic density is restricted to the range $0.094 < \Omega_\chi h^2 < 0.129$.

matter density obtained by combining WMAP and other cosmological data: $0.094 < \Omega_{CDM} < 0.129$ at the 2-σ level [4]. In the second panel of Fig. 1, we see the effect of imposing the WMAP range on the neutralino density [5,6,7]. We see immediately that (i) the cosmological regions are generally much narrower, and (ii) the 'bulk' regions at small $m_{1/2}$ and m_0 have almost disappeared, in particular when the laboratory constraints are imposed. Looking more closely at the coannihilation regions, we see that (iii) they are significantly truncated as well as becoming much narrower, since the reduced upper bound on $\Omega_\chi h^2$ moves the tip where $m_\chi = m_{\tilde{\tau}}$ to smaller $m_{1/2}$ so that the upper limit is now $m_{1/2} \lesssim 950$ GeV or $m_\chi \lesssim 400$ GeV.

3. A Likelihood analysis of the CMSSM

In displaying acceptable regions of cosmological density in the $m_0, m_{1/2}$ plane, it has been assumed that the input parameters are known with perfect accuracy so that the relic density can be calculated precisely. While all of the beyond the standard model parameters are completely unknown and therefore carry no formal uncertainties, standard model parameters such as the top and bottom Yukawa couplings are known but do carry significant uncertainties.

The optimal way to combine the various constraints (both phenomenological and cosmological) is via a likelihood analysis. When performing such an analysis, in addition to the formal experimental errors, it is also essential to take into account theoretical errors, which introduce systematic uncertainties that are frequently non-negligible. Recently, we have preformed an extensive likelihood analysis of the CMSSM [8].

The interpretation of the combined Higgs likelihood, $\mathcal{L}_{exp}$, in the $(m_{1/2}, m_0)$ plane depends on uncertainties in the theoretical calculation of m_h. These include the experimental error in m_t and (particularly at large $\tan\beta$) m_b, and theoretical uncertainties associated with higher-order corrections to m_h. Our default assumptions are that $m_t = 175 \pm 5$ GeV for the pole mass, and $m_b = 4.25 \pm 0.25$ GeV for the running $\overline{MS}$ mass evaluated at m_b itself. The theoretical uncertainty in m_h, σ_{th}, is dominated by the experimental uncertainties in $m_{t,b}$, which are treated as uncorrelated Gaussian errors:

$$\sigma_{th}^2 = \left(\frac{\partial m_h}{\partial m_t}\right)^2 \Delta m_t^2 + \left(\frac{\partial m_h}{\partial m_b}\right)^2 \Delta m_b^2. \quad (2)$$

Typically, we find that $(\partial m_h / \partial m_t) \sim 0.5$, so that σ_{th} is roughly 2-3 GeV.

The combined experimental likelihood, $\mathcal{L}_{exp}$, from direct searches at LEP 2 and a global electroweak fit is then convolved with a theoretical likelihood (taken as a Gaussian) with uncertainty given by σ_{th} from (2) above. Thus, we define the total Higgs likelihood function, $\mathcal{L}_h$, as

$$\mathcal{L}_h(m_h) = \frac{\mathcal{N}}{\sqrt{2\pi}\,\sigma_{th}} \int dm'_h \, \mathcal{L}_{exp}(m'_h) \, e^{-(m'_h - m_h)^2 / 2\sigma_{th}^2}, \quad (3)$$

where $\mathcal{N}$ is a factor that normalizes the experimental likelihood distribution. In addition to the Higgs likelihood function, we have included the likelihood function based on $b \to s\gamma$. While the likelihood function based on the measurements of the anomalous magnetic moment of the muon was considered in [8], it will not be discussed here.

Finally, in calculating the likelihood of the CDM density, we take into account the contribution of the uncertainties in $m_{t,b}$. We will see that the theoretical uncertainty plays a very significant role in this analysis. The likelihood for Ωh^2 is therefore,

$$\mathcal{L}_{\Omega h^2} = \frac{1}{\sqrt{2\pi}\sigma} e^{-(\Omega h^{2\,th} - \Omega h^{2\,exp})^2 / 2\sigma^2}, \quad (4)$$

where $\sigma^2 = \sigma_{exp}^2 + \sigma_{th}^2$, with σ_{exp} taken from the WMAP [4] result and σ_{th}^2 from (2), replacing m_h by Ωh^2.

The total likelihood function is computed by combining all the components described above:

$$\mathcal{L}_{tot} = \mathcal{L}_h \times \mathcal{L}_{bs\gamma} \times \mathcal{L}_{\Omega_\chi h^2}(\times \mathcal{L}_{a_\mu}) \tag{5}$$

The likelihood function in the CMSSM can be considered a function of two variables, $\mathcal{L}_{tot}(m_{1/2}, m_0)$, where $m_{1/2}$ and m_0 are the unified GUT-scale gaugino and scalar masses respectively. We adopt a flat prior distribution for $m_{1/2}$, and normalize the volume integral:

$$\int \mathcal{L}_{tot} \, dm_0 \, dm_{1/2} = 1 \tag{6}$$

for each value of $\tan\beta$, combining where appropriate both signs of μ.

Using the fully normalized likelihood function $\mathcal{L}_{tot}$ obtained by combining both signs of μ for each value of $\tan\beta$, we can determine the regions in the $(m_{1/2}, m_0)$ planes which correspond to specific CLs as shown in Fig. 2. The darkest (blue), intermediate (red) and lightest (green) shaded regions are, respectively, those where the likelihood is above 68%, above 90%, and above 95%.

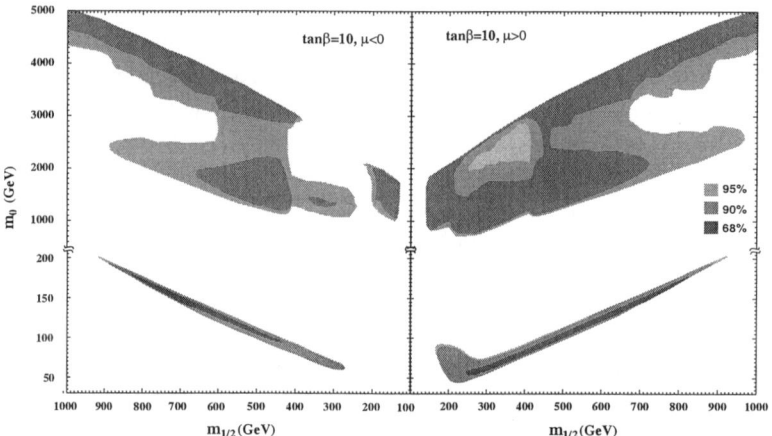

Figure 2. Contours of the likelihood at the 68%, 90% and 95% levels for $\tan\beta = 10$, $A_0 = 0$ and $\mu > 0$ (left panel) or $\mu < 0$ (right panel), calculated using information of m_h, $b \to s\gamma$ and $\Omega_{CDM} h^2$ and the current uncertainties in m_t and m_b.

The bulk region is more apparent in the right panel of Fig. 2 for $\mu > 0$ than it would be if the experimental error in m_t and the theoretical error in m_h were neglected. Fig. 3 complements the previous figures by showing the likelihood functions as they would appear if there were no uncertainty in m_t, keeping the other inputs the same. We see that, in this case, both the coannihilation and focus-point strips rise above the 68% CL.

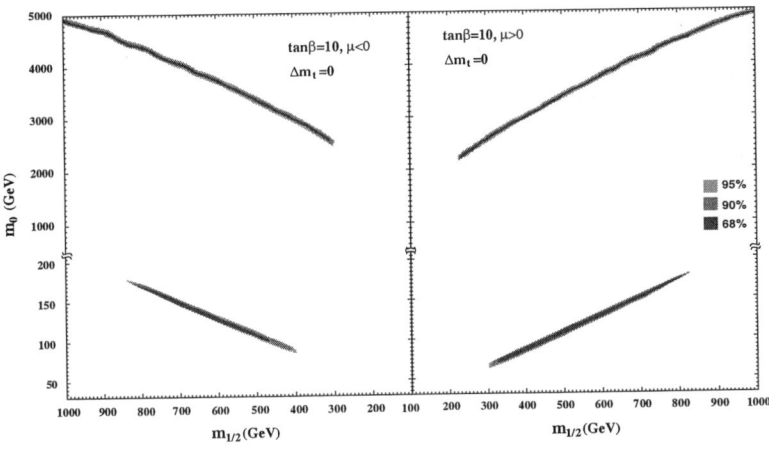

Figure 3. *As in Fig. 2 but assuming zero uncertainty in m_t.*

4. Beyond the CMSSM

The results of the CMSSM described in the previous sections are based heavily on the assumptions of universality of the supersymmetry breaking parameters. One of the simplest generalizations of this model relaxes the assumption of universality of the Higgs soft masses and is known as the NUHM [9] In this case, the input parameters include μ and m_A, in addition to the standard CMSSM inputs. In order to switch μ and m_A from outputs to inputs, the two soft Higgs masses, m_1, m_2 can no longer be set equal to m_0 and instead are calculated from the electroweak symmetry breaking conditions. The NUHM parameter space was recently analyzed [9] and a sample of the results are shown in Fig. 4.

In the left panel of Fig. 4, we see a $m_{1/2}, m_0$ plane with a relative low value of μ. In this case, an allowed region is found when the LSP contains a non-negligible Higgsino component which moderates the relic density independent of m_0. To the right of this region, the relic density is too small. In the right panel, we see an example of the m_A, μ plane. The

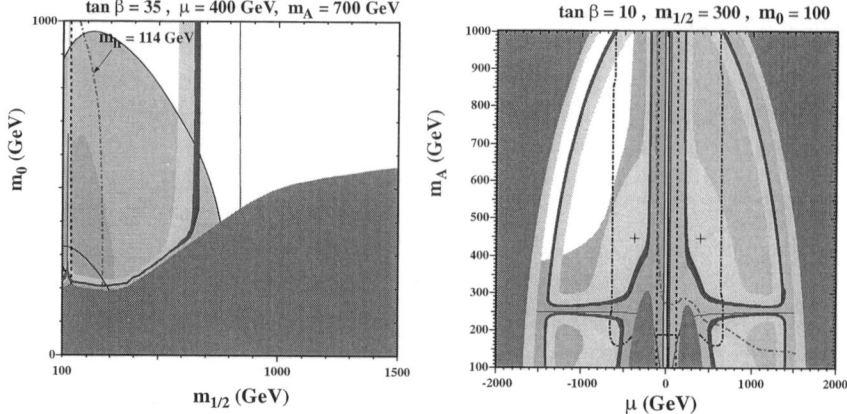

Figure 4. a) The NUHM $(m_{1/2}, m_0)$ plane for $\tan\beta = 35$, (a) $\mu = 400$ GeV and $m_A = 700$ GeV b)the NUHM (μ, m_A) plane for $\tan\beta = 10$, $m_0 = 100$ GeV and $m_{1/2} = 300$ GeV, with $A_0 = 0$. The (red) dot-dashed lines are the contours $m_h = 114$ GeV, and the near-vertical (black) dashed lines are the contours $m_{\chi^\pm} = 103.5$ GeV. The dark (black) dot-dashed lines indicate the GUT stability constraint. Only the areas inside these curves (small μ) are allowed by this constraint. The light (turquoise) shaded areas are the cosmologically preferred regions with $0.1 \leq \Omega_\chi h^2 \leq 0.3$. The darker (blue) portion of this region corresponds to the newer WMAP densities. The dark (brick red) shaded regions is excluded because a charged particle is lighter than the neutralino, and the lighter (yellow) shaded regions is excluded because the LSP is a sneutrino. The medium (green) shaded region is excluded by $b \to s\gamma$. The regions allowed by the $g-2$ constraint are shaded (pink) and bounded by solid black lines. The solid (blue) curves correspond to $m_\chi = m_A/2$.

crosses correspond to CMSSM points. In this single pane, we see examples of acceptable cosmological regions corresponding to the bulk region, co-annihilation region and s-channel annihilation through the Higgs pseudo scalar.

Rather than relax the CMSSM, it is in fact possible to further constrain the model. While the CMSSM models described above are certainly mSUGRA inspired, minimal supergravity models can be argued to be still more predictive. In the simplest version of the theory [10] where supersymmetry is broken in a hidden sector, the universal trilinear soft supersymmetry-breaking terms are $A = (3 - \sqrt{3})m_0$ and bilinear soft supersymmetry-breaking term is $B = (2 - \sqrt{3})m_0$, i.e., a special case of a general relation between B and A, $B_0 = A_0 - m_0$.

Given a relation between B_0 and A_0, we can no longer use the standard CMSSM boundary conditions, in which $m_{1/2}$, m_0, A_0, $\tan\beta$, and $sgn(\mu)$ are input at the GUT scale with μ and B determined by the electroweak symmetry breaking condition. Now, one is forced to input B_0 and instead $\tan\beta$ is calculated from the minimization of the Higgs potential [11]. In Fig. 5, the contours of $\tan\beta$ (solid blue lines) in the $(m_{1/2}, m_0)$ planes for two values of $\hat{A} = A_0/m_0$, $\hat{B} = B_0/m_0 = \hat{A} - 1$ and the sign of μ are displayed [11].

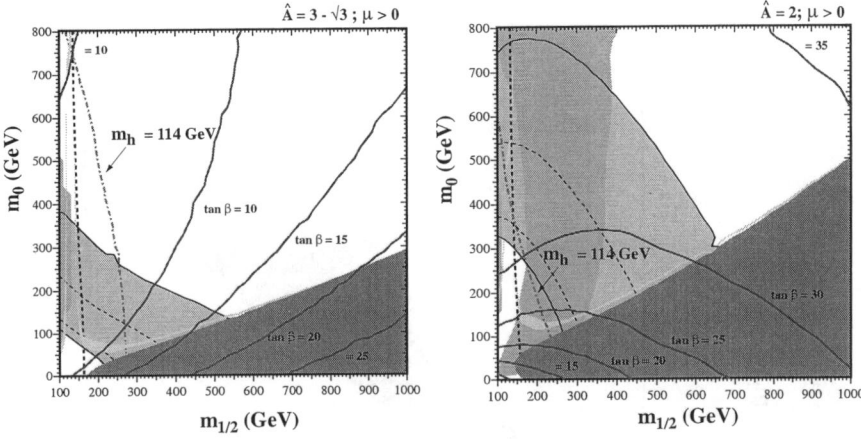

Figure 5. Examples of $(m_{1/2}, m_0)$ planes with contours of $\tan\beta$ superposed, for $\mu > 0$ and (a) the simplest Polonyi model with $\hat{A} = 3 - \sqrt{3}, \hat{B} = \hat{A} - 1$ and (b) $\hat{A} = 2.0, \hat{B} = \hat{A} - 1$. In each panel, we show the regions excluded by the LEP lower limits on MSSM particles, those ruled out by $b \to s\gamma$ decay (medium green shading), and those excluded because the LSP would be charged (dark red shading). The region favoured by the WMAP range has light turquoise shading. The region suggested by $g_\mu - 2$ is medium (pink) shaded.

In panel (a) of Fig. 5, we see that the Higgs constraint combined with the relic density requires $\tan\beta \gtrsim 11$, whilst the relic density also enforces $\tan\beta \lesssim 20$. For a given point in the $m_{1/2} - m_0$ plane, the calculated value of $\tan\beta$ increases as $\hat{A}$ increases. This is seen in panel (b) of Fig. 5, when $\hat{A} = 2.0$, close to its maximal value for $\mu > 0$, the $\tan\beta$ contours turn over towards smaller $m_{1/2}$, and only relatively large values $25 \lesssim \tan\beta \lesssim 35$ are allowed by the $b \to s\gamma$ and $\Omega_{CDM}h^2$ constraints, respectively. For any given value of $\hat{A}$, there is only a relatively narrow range allowed for $\tan\beta$.

5. Detectability

The question of detectability with respect to supersymmetric models is of key importance particularly with the approaching start of the LHC. Clearly the center of mass energy of any future linear collider is paramount towards the supersymmetry discovery potential of the machine. We can emphasize this point in general models by plotting the masses of the two lightest (observable) sparticles in supersymmetric models. For example, in Fig. 6 [12], a scatter plot of the masses of the lightest visible supersymmetric particle (LVSP) and the next-to-lightest visible supersymmetric particle (NLVSP) is shown for the CMSSM.

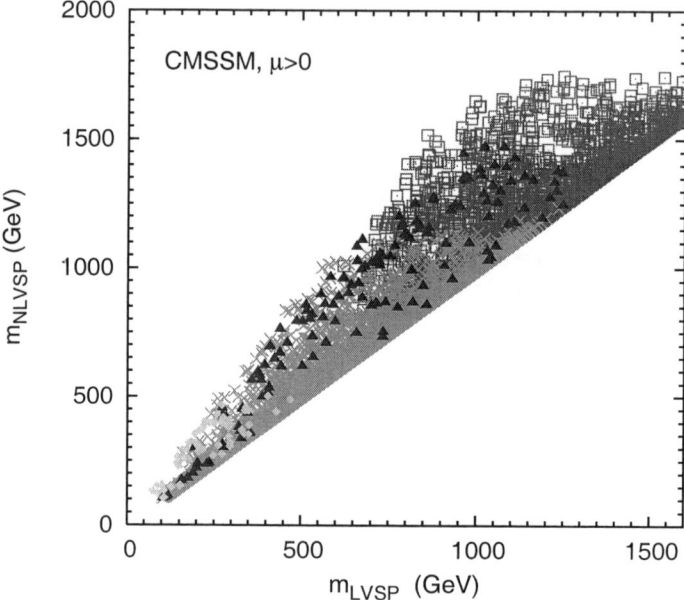

Figure 6. Scatter plots of the masses of the lightest visible supersymmetric particle (LVSP) and the next-to-lightest visible supersymmetric particle (NLVSP) in the CMSSM for $\mu > 0$. The darker (blue) triangles satisfy all the laboratory, astrophysical and cosmological constraints. For comparison, the dark (red) squares and medium-shaded (green) crosses respect the laboratory constraints, but not those imposed by astrophysics and cosmology. In addition, the (green) crosses represent models which are expected to be visible at the LHC. The very light (yellow) points are those for which direct detection of supersymmetric dark matter might be possible.

All points shown in Fig. 6 satisfy the phenomenological constraints discussed above. The dark (red) squares represent those points for which the relic density is outside the WMAP range, and for which all coloured sparticles (squarks and gluinos) are heavier than 2 TeV, and we regard the dark (red) points as unobservable at the LHC. Most of these points have $m_{NLVSP} \gtrsim 1.2$ TeV. Conversely, the medium-shaded (green) crosses represent points where at least one squark or gluino has a mass less than 2 TeV and should be observable at the LHC.

The darker (blue) triangles are those points respecting the cosmological cold dark matter constraint. Comparing with the regions populated by dark (red) squares and medium-shaded (green) crosses, one can see which of these models would be detectable at the LHC, according to the criterion in the previous paragraph. We see immediately that the dark matter constraint restricts the LVSP masses to be less than about 1250 GeV and NLVSP masses to be less than about 1500 GeV. In most cases, the identity of the LVSP is the lighter $\tilde{\tau}$. While pair-production of the LVSP would sometimes require a CM energy of about 2.5 TeV, in some cases there is a lower supersymmetric threshold due to the associated production of the LSP χ with the next lightest neutralino χ_2 [13]. Examining the masses and identities of the sparticle spectrum at these points, we find that $E_{CM} \gtrsim 2.2$ TeV would be sufficient to see at least one sparticle. Similarly, only a LC with $E_{CM} \geq 2.5$ TeV would be 'guaranteed' to see two visible sparticles (in addition to the χ LSP). Points with $m_{LVSP} \gtrsim 700$ GeV are predominantly due to rapid annihilation via direct-channel H, A poles, while points with 200 GeV $\lesssim m_{LVSP} \lesssim 700$ GeV are largely due to χ-slepton coannihilation.

An $E_{CM} = 500$ GeV LC would be able to explore the 'bulk' region at low $(m_{1/2}, m_0)$, which is represented by the small cluster of points around $m_{LVSP} \sim 200$ GeV. A LC with $E_{CM} = 1000$ GeV would be able to reach some way into the coannihilation 'tail', but would not cover all the WMAP-compatible dark (blue) triangles. Indeed, about a third of these points are even beyond the reach of the LHC in this model. Finally, the light (yellow) filled circles are points for which the elastic χ-p scattering cross section is larger than 10^{-8} pb.

Direct detection techniques rely on an ample neutralino-nucleon scattering cross-section. In Fig. 7, we display the allowed ranges of the spin-independent cross sections in the NUHM when we sample randomly $\tan \beta$ as well as the other NUHM parameters [14]. The raggedness of the boundaries of the shaded regions reflects the finite sample size. The dark shaded regions includes all sample points after the constraints discussed above (including the relic density constraint) have been applied. In a random sample, one often hits points which are are perfectly acceptable at low energy scales but when the parameters are run to high energies approaching the GUT scale, one or several of the sparticles mass squared runs negative. This has been referred to as the GUT constraint here. The medium shaded region embodies those points after the GUT constraint has been applied. After incorporating all the cuts, including that motivated by $g_\mu - 2$, we find that the light shaded region where the scalar cross section has the range 10^{-6} pb $\gtrsim \sigma_{SI} \gtrsim 10^{-10}$ pb, with somewhat larger (smaller) values being possible in exceptional cases.

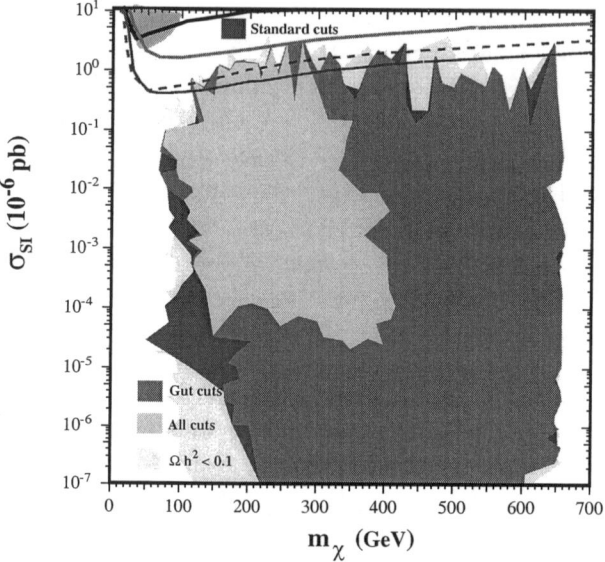

Figure 7. Ranges of the spin-independent cross section in the NUHM. The ranges allowed by the cuts on $\Omega_\chi h^2$, m_h and $b \to s\gamma$ have dark shading, those still allowed by the GUT stability cut have medium shading, and those still allowed after applying all the cuts including $g_\mu - 2$ have light shading. The pale shaded region corresponds to the extra area of points with low relic densities, whose cross sections have been rescaled appropriately. Also shown are the limits from the CDMS[15] and Edelweiss[16] experiments as well as the recent CDMSII result [17] on the neutralino-proton elastic scattering cross section as a function of the neutralino mass. The CDMSII limit is stronger than the Edelweiss limit which is stronger than the previous CDMS limit at higher m_χ. The result reported by DAMA [18] is found in the upper left.

The results from this analysis [14] for the scattering cross section in the NUHM (which by definition includes all CMSSM results) are compared with the previous CDMS [15] and Edelweiss [16] bounds as well as the recent CDMSII results [17] in Fig. 7. While previous experimental sensitivities were not strong enough to probe predictions of the NUHM, the current CDMSII bound has begun to exclude realistic models and it is expected that these bounds improve by a factor of about 20.

References

1. H. Goldberg, *Phys. Rev. Lett.* **50**, 1419 (1983); J. Ellis, J.S. Hagelin, D.V. Nanopoulos, K.A. Olive and M. Srednicki, *Nucl. Phys.* **B238**, 453 (1984).
2. J. L. Feng, S. Su and F. Takayama, arXiv:hep-ph/0404231; arXiv:hep-ph/0404198; J. L. Feng, A. Rajaraman and F. Takayama, Phys. Rev. Lett. **91** (2003) 011302 [arXiv:hep-ph/0302215]; J. R. Ellis, K. A. Olive, Y. Santoso and V. C. Spanos, Phys. Lett. B **588** (2004) 7 [arXiv:hep-ph/0312262]; L. Roszkowski and R. Ruiz de Austri, arXiv:hep-ph/0408227.
3. J. Ellis, T. Falk, and K.A. Olive, *Phys.Lett.* **B444** (1998) 367 [arXiv:hep-ph/9810360]; J. Ellis, T. Falk, K.A. Olive, and M. Srednicki, *Astr. Part. Phys.* **13** (2000) 181 [Erratum-ibid. **15** (2001) 413] [arXiv:hep-ph/9905481].
4. C. L. Bennett *et al.*, Astrophys. J. Suppl. **148** (2003) 1; D. N. Spergel *et al.*, Astrophys. J. Suppl. **148** (2003) 175; H. V. Peiris *et al.*, Astrophys. J. Suppl. **148** (2003) 213.
5. J. R. Ellis, K. A. Olive, Y. Santoso and V. C. Spanos, Phys. Lett. B **565** (2003) 176 [arXiv:hep-ph/0303043].
6. H. Baer and C. Balazs, JCAP **0305** (2003) 006 [arXiv:hep-ph/0303114].
7. A. B. Lahanas and D. V. Nanopoulos, Phys. Lett. B **568** (2003) 55 [arXiv:hep-ph/0303130]; U. Chattopadhyay, A. Corsetti and P. Nath, Phys. Rev. D **68** (2003) 035005 [arXiv:hep-ph/0303201]; C. Munoz, Int. J. Mod. Phys. A **19**, 3093 (2004) [arXiv:hep-ph/0309346] R. Arnowitt, B. Dutta and B. Hu, arXiv:hep-ph/0310103.
8. J. R. Ellis, K. A. Olive, Y. Santoso and V. C. Spanos, Phys. Rev. D **69** (2004) 095004 [arXiv:hep-ph/0310356].
9. J. Ellis, K. Olive and Y. Santoso, *Phys.Lett.* **B539** (2002) 107 [arXiv:hep-ph/0204192].; J. R. Ellis, T. Falk, K. A. Olive and Y. Santoso, Nucl. Phys. B **652** (2003) 259 [arXiv:hep-ph/0210205].
10. J. Polonyi, Budapest preprint KFKI-1977-93 (1977); R. Barbieri, S. Ferrara and C.A. Savoy, Phys. Lett. **119B** (1982) 343.
11. J. R. Ellis, K. A. Olive, Y. Santoso and V. C. Spanos, Phys. Lett. B **573** (2003) 162 [arXiv:hep-ph/0305212]; Phys. Rev. D **70** (2004) 055005 [arXiv:hep-ph/0405110].
12. J. R. Ellis, K. A. Olive, Y. Santoso and V. C. Spanos, arXiv:hep-ph/0408118.
13. A. Djouadi, M. Drees and J. L. Kneur, JHEP **0108** (2001) 055 [arXiv:hep-ph/0107316].
14. J. R. Ellis, A. Ferstl, K. A. Olive and Y. Santoso, Phys. Rev. D **67**, 123502 (2003) [arXiv:hep-ph/0302032].
15. D. Abrams *et al.* [CDMS Collaboration], *Phys.Rev.* **D66** (2002) 122003 [arXiv:astro-ph/0203500].
16. A. Benoit *et al.*, *Phys.Lett.* **B545** (2002) 43 [arXiv:astro-ph/0206271].
17. D. S. Akerib *et al.* [CDMS Collaboration], arXiv:astro-ph/0405033.
18. DAMA Collaboration, R. Bernabei *et al.*, *Phys. Lett.* **B436** (1998) 379.

SEARCHES FOR SUPERSYMMETRY AT HIGH ENERGY COLLIDERS

BEATE HEINEMANN

University of Liverpool, Oliver Lodge Laboratory, Department of Physics,
University of Liverpool, Liverpool L69 7ZE, UK
E-mail: beate@hep.ph.liv.ac.uk

Recent searches for supersymmetric partners of the Standard Model particles at the LEP and Tevatron colliders are presented. In all search analyses the data are found to agree well with the Standard Model background expectation and no evidence for contributions from supersymmetry is found. The data are thus used to constrain the SUSY parameter space. The focus in this report is on analyses taht are sensitive to the lightest supersymmetric particle.

1. Introduction

The Standard Model of particle physics describes all data taken at high energy colliders up to now. However, there are strong theoretical reasons, particularly the hierarchy problem, to believe that new physics should exist at the TeV scale. Also, from cosmology data (particularly the WMAP data) it is known that 95% of the energy and matter in the Universe is not accounted for by SM particles.

There are many theoretical models that attempt to solve these problems by proposing new particles, e.g. Supersymmetry, Extra Dimensions, new gauge groups, compositeness (predicting e.g. leptoquarks, excited fermions) and Technicolor. In this report I will concentrate on Supersymmetry.

Supersymmetry (SUSY) models are among the most promising theories since they address the hierarchy problem and provide a natural dark matter candidate. For each SM particle there is a correspondent super-partner which carries the same quantum numbers apart from the spin: each fermion has a sfermion super-partner of spin 0 and each gauge boson has a gaugino super-partner of spin 1/2. Since no super-partners have yet been observed it is known that SUSY must be a broken symmetry if it exists. The nature of the breaking mechanism largely determines the phenomenology of the models. Furthermore the phenomenology of the models depends strongly on whether R_p is conserved where $R_p = (-1)^{3(B-L)+2S}$ and B is the baryon number, L the lepton number and S the spin. If R_p is conserved the lightest gaugino, the neutralino (χ_1^0) is an excellent candidate for Dark Matter in the Universe.

In this report I will focus on two R_p conserving scenarios since they provide a good candidate for dark matter which differ in the SUSY breaking mechanism: minimal Super-gravity (mSUGRA) and Gauge-Mediated SUSY Breaking (GMSB). Note, that the SUSY models that are used throughout this report should be considered as "benchmark"-models. It is quite likely that none of these *minimal* models is realised exactly in nature but they provide a good guidance for the experimentalists and are useful for comparing the sensitivity of different experiments.

The TeV scale is currently probed by the CDF and DØ experiments at the Tevatron collider at Fermilab. Protons and anti-protons are accelerated to an energy of 980 GeV and brought to collisions at the two experiments at a centre-of-mass energy of $\sqrt{s} \approx 1.96$ TeV. So-called "Run 2" at the Tevatron started in 2001. Until summer 2004 an integrated luminosity of about 700 pb^{-1} has been delivered to the experiments and about 500 pb^{-1} of high quality data are used for physics analyses. In this report most results are based on about 200 pb^{-1} taken up to September 2003.

The LEP electron-positron collider at CERN has stopped operation in 2000. Most data are analysed and published but final combinations from the four experiments (ALEPH, DELPHI, L3 and OPAL) are still ongoing and in many areas of searches for new physics the LEP results provide still the most stringent constraints.

2. mSUGRA Results

In "mSUGRA" models SUSY is broken at the GUT scale (10^{16} GeV). There are five parameters that determine the masses and cross sections for the super-partners: a common scalar mass m_0, a common sfermion mass $m_{1/2}$, the ratio of the vacuum expectation values of the higgs fields $\tan\beta$, the trilinear coupling A_0 and the Higgsino mixing parameter μ. The lightest SUSY particle is the χ_1^0. The characteristic signature of SUSY production is large imbalance in transverse momentum, $\not{E}_T$, due to the outgoing neutralino which is not detected.

2.1. LEP Results

The final LEP results, based on data taken up to the year 2000, on searches for Supersymmetry are nearly all available now. The most severe constraints come from the direct searches for charginos which exclude charginos below 103.5 GeV/c^2 and the search for the Higgs boson. The results are shown in Figure 1 in the plane of the LSP mass versus $\tan\beta$. The limit at low $\tan\beta$ comes mostly from the Higgs search and the limit at high $\tan\beta$ from the chargino searches. Thus, in mSUGRA the mass of the LSP has to exceed 50 GeV [1]. Furthermore, LEP has set stringent direct limits on the masses of sleptons and squarks.

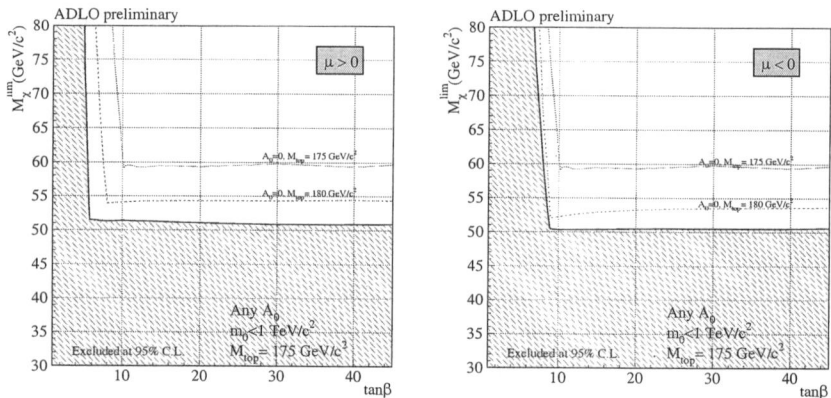

Figure 1. Lower limit on the LSP mass in mSUGRA versus $\tan\beta$ for positive and negative values of μ.

2.2. Trilepton Results

The "trilepton" analysis searches for chargino-neutralino pair production with subsequent cascade decays of the χ_2^0 and χ_1^+ into leptons. The final state then contains three leptons and large $\not{E}_T$ due to outgoing χ_1^0's and neutrinos.

The DØ experiment has analysed four final states: $ee+t$, $\mu\mu+t$, $e\mu+t$ and $\mu^\pm\mu^\pm$ [2]. Here e and μ denote identified electrons and muons, t is a track that is isolated w.r.t. other tracks. The inclusion of a track as 3rd lepton enhances the acceptance since it allows for τ-lepton decays. The $\mu^\pm\mu^\pm$ analysis searches for two muons of the same charge and requires no third lepton. The 1st and 2nd lepton are required to have $p_T > 11$ GeV, the third is required to have $p_T > 5$ GeV. The Standard Model background is further suppressed by requirements on the dilepton invariant mass, the difference in azimuthal angle $\Delta\phi_{ll}$, $\not{E}_T$ and several other variables. Combing all channels 3 events were observed in agreement with the Standard Model expectation of 2.9 events.

The data are thus combined and interpreted as an upper limit on the cross section times branching ratio $\sigma(\chi_2^0 \chi_1^+) \times Br(3$ leptons$)$ shown in Figure 2. The upper limit at 95% C.L. is about 0.4 pb for a parameter choice of $\tan\beta \sim 3$, $A_0 \sim 0$, $\mu > 0$, $M(\chi_1^+) \sim M(\chi_2^0) \sim 2M(\chi_1^0) \sim M(\tilde{l})$. This corresponds to a lower limit on $M(\chi_1^+)$ of 97 GeV which is about 6 GeV lower than the direct model-independent limits obtained at LEP. With only 25% more data the LEP limits will be exceeded in certain regions of the parameter space.

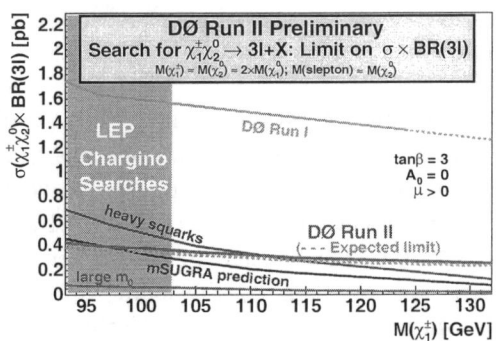

Figure 2. Cross section versus the mass of the lightest chargino χ_1^+. Shown is the experimental upper limit and several theoretical predictions. The line labelled "mSUGRA prediction" shows the theoretical prediction within mSUGRA for $\tan\beta \sim 3$, $A_0 \sim 0$, $\mu > 0$, $M(\chi_1^+) \sim M(\chi_2^0) \sim 2M(\chi_1^0) \sim M(\tilde{l})$. The line labelled "Heavy Squarks" shows the theoretical cross section for a model where the sfermion mass unification is relaxed and the squark masses have very high values. The line labelled "large m_0" is for the case that the sfermion mass scale is larger.

2.3. $B_s \to \mu^+\mu^-$

The SM branching ratio of the B_s meson is only $Br(B_s \to \mu^+\mu^-) = 3.42 \pm 0.54 \times 10^{-9}$ [5]. In SUSY models the branching ratio can be significantly enhanced at high values of $\tan\beta$ [6]. CDF [7] and DØ [8] have searched for this decay in 171 pb^{-1} and 240 pb^{-1} of Run 2 data, respectively. Neither experiment finds any evidence for this decay. CDF observes 1 event compared to 1.1 ± 0.1 expected and DØ observed 4 events compared to 3.7 ± 1.1 expected.

The data are thus used to place an upper limit on $Br(B_s \to \mu^+\mu^-)$. CDF finds $Br(B_s \to \mu^+\mu^-) < 7.5 \times 10^{-7}$ and DØ finds $Br(B_s \to \mu^+\mu^-) < 5.0 \times 10^{-7}$ at 95% C.L.. Note, that this result is complementary to the tri-lepton searches since it constrains the SUSY parameter space at high $\tan\beta$ whilst the tri-lepton search is most sensitive at low $\tan\beta$. It also provides a new stringent constraint for direct Dark Matter detection experiments.

3. GMSB Results

An alternative SUSY scenario with different experimental signatures is the "Gauge Mediated Symmetry Breaking Model" (GMSB). In these models SUSY is broken at a lower scale, ≈ 10 TeV, than in mSUGRA models. The lightest SUSY particle is the gravitino, $\tilde{G}$, with a mass of ≈ 1 keV. In GMSB models the characteristic signature is also $\not{E}_T$ due to the outgoing $\tilde{G}$'s.

If the next-to-lightest SUSY particle is the χ_1^0 it typically decays into a photon and a $\tilde{G}$ resulting in a signature of two photons and large $\not{E}_T$. Both CDF ($L = 202$ pb^{-1}) [3] and DØ ($L = 263$ pb^{-1}) [4] have searched for this signature. CDF observes no event and expects 0.3 ± 0.1 and DØ observes 2 events and expects 3.7 ± 0.6. Thus both experiments the data are consistent with the SM expectation. At high $\not{E}_T$ the main background results from $W\gamma \to e\nu\gamma$ events where the electron is misidentified as a photon due to track misreconstruction or hard Bremsstrahlung.

The data are used to set an upper limit on the cross section. This is then interpreted as a lower limit on the χ_1^0 (χ_1^+) mass of 93 GeV (167 GeV) by CDF and 108 GeV (195 GeV) by DØ at 95% C.L..

4. Conclusions

I have presented a review of the experimental results on searches for supersymmetry at the LEP and Tevatron colliders. At the Tevatron, up to now no significant deviations of the data from the SM have been observed. However, within the next few years the datasets are expected to increase by an order of magnitude leading to a substantial improvement in sensitivity. The LEP experiments still provide the most stringent limits on supersymmetry in many cases, particularly on SUSY, but the new data from the

Tevatron are starting to improve upon those. With the continuously improving performance of the Tevatron and the startup of the LHC many, hopefully exciting, results are expected to appear over the next few years.

Acknowledgments

I would like to thank the local organising committee of IDM 04 for organising the conference and the Particle Physics and Astronomy Research Council and the Royal Society in the UK for their financial support.

References

1. http://lepsusy.web.cern.ch/lepsusy
2. A. Meyer, "SUSY Searches at the Tevatron", Proceedings to 32nd International Conference on High Energy Physics, ICHEP04, Beijing/2004
3. CDF Collaboration, D. Acosta et al., FERMILAB-PUB-04-299-E, submitted to Phys. Rev. D Rapid Communications
4. DØ Collaboration, V. M. Abazov et al., hep-ex/0408146, submitted to Phys. Rev. Lett.
5. A. J. Buras, Phys. Lett. B **566**, 115 (2003).
6. K. S. Babu and C. F. Kolda, Phys. Rev. Lett. **84**, 228 (2000); A. Dedes et al., FERMILAB-PUB-02-129-T; A. Dedes et al., Phys, Rev. Lett. **87**, 251804 (2001); T. Blazek et al., Phys. Lett. B **589**, 39 (2004); R. Demirsek et al. JHEP 0304, 37 (2003); C. Bobeth et al., Phys. Rev. D **66** 074021 (2002); R. Arnowitt et al., Phys. Lett. B **538**, 121 (2002).
7. CDF Collaboration, D. Acosta et al., Phys. Rev. Lett. **93**, 032001 (2004)
8. DØ Collaboration, V. M. Abazov et al., hep-ex/0410039, submitted to Phys. Rev. Lett.

CP PHASES, DARK MATTER AND THE B QUARK MASS

M. E. GÓMEZ*
Dept. de Física Aplicada, U. de Huelva, 21071 Huelva, Spain.
E-mail: mario.gomez@dfa.uhu.es

T. IBRAHIM
Dept. of Physics, Faculty of Science, University of Alexandria,
Alexandria, Egypt†
Dept. of Physics, Northeastern University, Boston, MA 02115-5000, USA.
E-mail: tarek@neu.edu

P. NATH
Dept. of Physics, Northeastern University, Boston, MA 02115-5000, USA.
E-mail: nath@neu.edu

S. SKADHAUGE
Dept. de Fisica Matematica U. de Sao Paulo, 05315-970 Sao Paulo, Brazil.
E-mail: solveig@fma.if.usp.br

When the mSUGRA framework is extended to allow complex soft terms the phases can induce large changes in the SUSY threshold correction to the b quark mass and affect the neutralino relic density predictions of the model as well as the SUSY contribution to the BR($b \to s\gamma$). We present some specific models with large SUSY phases which can accommodate the fermion electric dipole moment constraints and a neutralino relic density within the WMAP bounds. Finally, we discuss the possibility of asymptotic Yukawa unification in these kind of models.

1. Introduction

The recent Wilkinson Microwave Anisotropy Probe (WMAP) data allows a determination of cold dark matter (CDM) to lie in the range[1] $\Omega_{CDM} h^2 = 0.1126^{+0.008}_{-0.009}$. In this analysis we extend the mSUGRA frame-

---

*Speaker.
†Permanent address of T. I.

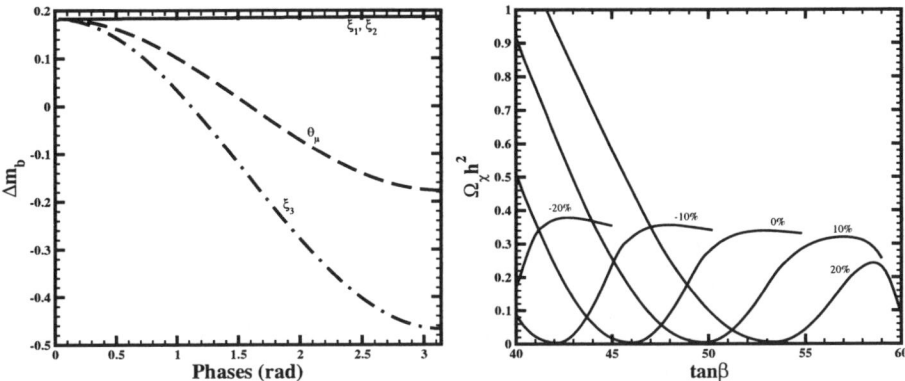

Figure 1. Δm_b as a function of the indicated phases (left) and the variation of the relic density with $\tan\beta$ for several values of Δm_b (right) for the parameters on Eq. (1).

work to include CP phases in the gaugino sector [2], which affects the loop corrections to the b quark mass and also affects the mixing between the neutral Higgs bosons. These corrections then affect relic density computations in important ways. Yukawa unification constraints in mSUGRA are incompatible with WMAP due to the large neutralino mass required to accommodate both the prediction of m_b and the bounds on BR($b \to s\gamma$). The phase dependence of the SUSY contribution to the BR($b \to s\gamma$) requires that one revisit this issue.

2. Phase Generalized mSUGRA and the b Quark Mass

In most of the mSUGRA parameter space [3] the LSP is almost purely a Bino $\tilde{B}$ with a large relic density. However, Ω_χ can lie within the WMAP bounds for certain values of the SUSY parameters such as the coannihilation $\chi - \tilde{\tau}$ region [4], the Hyperbolic Branch/focus-point region (HB/FP)[5] or the regions around the resonances on the Higgs mediated channels[6]. In this section we analyze the impact of enlarging this picture including CP phases using a point in the mSUGRA parameter space which satisfies all the constraints[7], i.e., :

$$\tan\beta = 50, \ m_0 = m_{1/2} = |A_0| = 600\,\text{GeV}. \tag{1}$$

Within mSUGRA there are only two physical phases which can be taken to be the θ_μ, θ_A where θ_μ is the phase of μ and θ_A is the phase of A_0. Naively one expects these phases to be small ($\leq 10^{-2}$) to satisfy the electric dipole moments (EDM) constraints of the electron, of the neutron and of the Hg^{199} atom, i.e.,

$$|d_e| < 4.23 \times 10^{-27}\text{ecm}, \ |d_n| < 6.5 \times 10^{-26}\text{ecm}, \ C_{\text{Hg}} < 3.0 \times 10^{-26}\text{cm}. \tag{2}$$

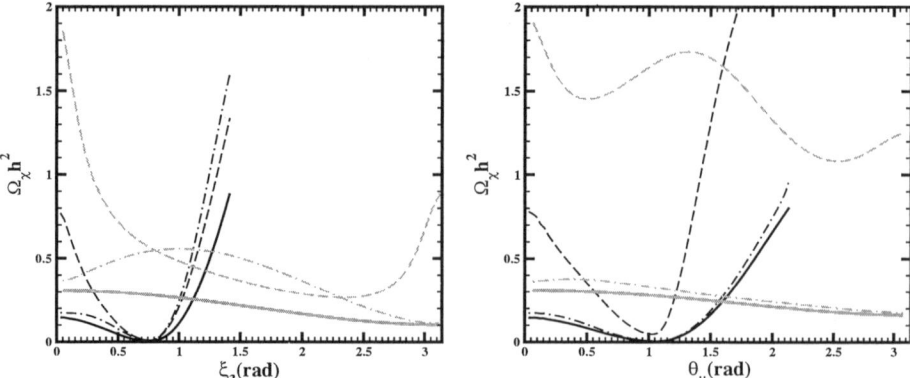

Figure 2. $\Omega_\chi h^2$ as a function of ξ_3 and θ_μ for the parameters on Eq. (1), using the theoretically predicted value of Δm_b (black lines), $\Delta m_b = 0$ (light lines). Solid lines include all contributions, dashed lines (dot-dashed lines) only s-channel H_1 (H_3) mediated annihilation to $b\bar{b}$.

However, large phases can be accommodated in several scenarios such as models with *super heavy sfermions* for the two first generations [8] or models with a non–trivial *soft term* flavor structure[9]. Here, we assume a cancellation mechanism [10] which becomes possible if we assume an extended SUGRA parameter space characterized by the parameters

$$m_0, m_{1/2}, \tan\beta, |A_0|, \theta_\mu, \alpha_A, \xi_1, \xi_2, \xi_3, \quad (3)$$

where, ξ_i is the phase of the gaugino mass M_i. The value of $|\mu|$ is determined by imposing electroweak symmetry breaking.

At the loop level the effective b quark coupling with the Higgs is given by[11]

$$-L_{bbH^0} = (h_b + \delta h_b)\bar{b}_R b_L H_1^0 + \Delta h_b \bar{b}_R b_L H_2^{0*} + H.c. \quad (4)$$

The correction to the b quark mass is then given directly in terms of Δh_b and δh_b so that

$$\Delta m_b = [Re(\frac{\Delta h_b}{h_b})\tan\beta + Re(\frac{\delta h_b}{h_b})]. \quad (5)$$

A full analysis of Δm_b is used[12]. Δm_b depends strongly on ξ_3 and θ_μ and weakly on α_A, ξ_1, ξ_2 as we see from the left panel of Fig. 1. The consequences for $\Omega_\chi h^2$ [a] arising from the changes of Δm_b in this range can be understood from the qualitative analysis on the right panel of Fig. 1, where Δm_b is used as a free parameter.

---

[a]We use *micrOMEGAs* [13] for the computations of $\Omega_\chi h^2$ without phases.

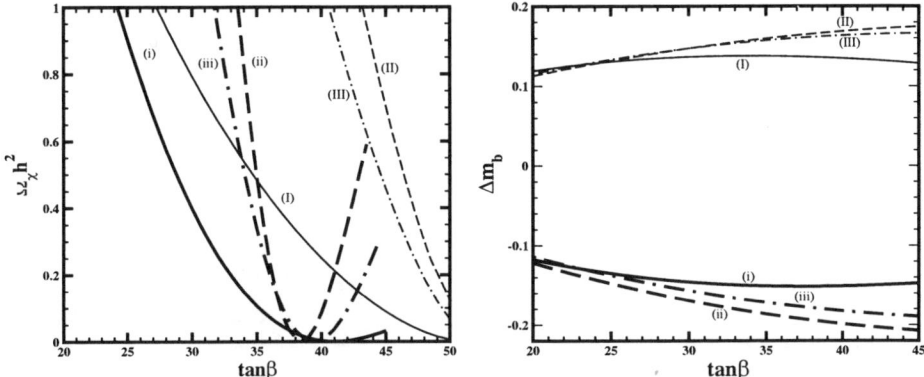

Figure 3. The neutralino relic density as a function of $\tan\beta$ for the three cases (i), (ii), (iii) of the text (left). Lines (I), (II) and (III) correspond to similar set of SUSY parameters for the case of vanishing phases. On the right we present the corresponding values of Δm_b.

CP violating phases induce mixing at one loop of the CP–even, H, h, and CP–odd, A, neutral tree level Higgs bosons: $(H, h, A) \to (H_1, H_2, H_3)$, where H_i (i=1,2,3) are the mass eigen states [b]. Fig. 2 shows the changes of $\Omega_\chi h^2$ with ξ_3 and θ_μ. The light lines are obtained by setting $\Delta m_b = 0$, which implies that the Higgs masses remain almost constant along these lines and hence we see the effects of the CP phases on the vertices without the variation due to Δm_b. Also, $m_{H_1} \sim m_{H_3}$ and $\Gamma_{H_1} \sim \Gamma_{H_3}$, therefore large mixing are possible as we see in the partial contributions of H_1 (dash) and H_3 (dot–dash) mediated s–channels.

In Fig. 3 the neutralino relic density is displayed as a function of $\tan\beta$ for three cases given by: (i) $m_0 = m_{1/2} = |A_0| = 300$ GeV, $\alpha_{A_0} = 1.0$, $\xi_1 = 0.5$, $\xi_2 = 0.66$, $\xi_3 = 0.62$, $\theta_\mu = 2.5$; (ii) $m_0 = m_{1/2} = |A_0| = 555$ GeV, $\alpha_{A_0} = 2.0$, $\xi_1 = 0.6$, $\xi_2 = 0.65$, $\xi_3 = 0.65$, $\theta_\mu = 2.5$; (iii) $m_0 = m_{1/2} = |A_0| = 480$ GeV, $\alpha_{A_0} = 0.8$, $\xi_1 = 0.4$, $\xi_2 = 0.66$, $\xi_3 = 0.63$, $\theta_\mu = 2.5$ where all angles are in units of radians. In all cases the EDM constraints (2) are satisfied for $\tan\beta = 40$ and their values are exhibited in table 1. We also observe that the WMAP bounds are also satisfied in the range of $\tan\beta$ exhibited in Fig.4.

3. CP phases and Yukawa Unification

Fig. 4 summarizes the mSUGRA scenario with Yukawa unification and how the presence of phases can modify this picture. In mSUGRA, the

---

[b] We use *CPsuperH* [14] in our computations.

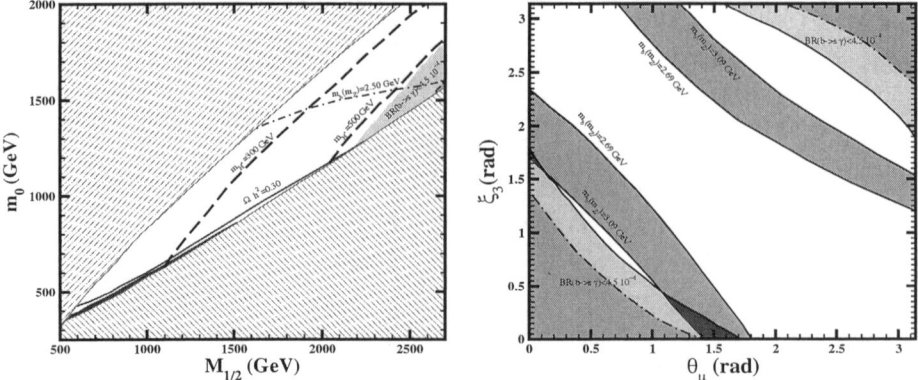

Figure 4. Left, the mSUGRA scenario with Yukawa unification with $\mu < 0$ and $A_0 = -0.5 \cdot M_{1/2}$: on the upper (lower) ruled area $m_A < 120$ GeV (the neutralino is not the LSP), the dark (light) shaded area satisfy WMAP and $(b \to s\gamma)$ constraints. On the right we can see the impact of the phases on the point $m_0 = 820$ GeV, $M_{1/2} = 1400$ GeV and $A_0 = -700$ GeV.

Table 1. The EDMs for $\tan\beta = 40$ for cases of text.

Case	$|d_e|e.cm$	$|d_n|e.cm$	$C_{Hg}cm$
(i)	2.74×10^{-27}	1.79×10^{-26}	8.72×10^{-27}
(ii)	1.29×10^{-27}	1.82×10^{-27}	6.02×10^{-28}
(iii)	9.72×10^{-28}	4.19×10^{-26}	1.41×10^{-27}

prediction of $m_b(m_Z)$ is strongly dependent on its SUSY loop correction which depends on the sign of μ. For $\mu > 0$, $m_b(m_Z)$ becomes too large, while for $\mu < 0$ it is below its lower bound and large values of the SUSY parameters are required to have m_b in the experimental range [15,16]. We observe that the relic abundance prediction satisfies WMAP bounds only when $\chi - \tilde{\tau}$ coannihilations are sizeable and $M_{1/2} < 1500$GeV.

The contribution of the charged Higgs exchange diagram to the $b \to s\gamma$ is large due to small values of m_{H^+}, this contribution adds to the SM prediction while the SUSY contribution takes the opposite sign of μ. To include complex parameters in the SUSY contributions to BR($b \to s\gamma$) we use the computation of [17], which we insert on the code provided by [13]. It was claimed in [18] that the use of $\overline{MS}$ running charm mass instead of the pole mass reduces the NNLO uncertainty in the SM calculation. We present our results using both the pole mass choice, $\frac{m_b}{m_c} = 0.29$, and the $\overline{MS}$ value, 0.22 (dark areas limited by dot-dash lines) [c]. The SM limit becomes 3.33×10^4 in the first case and 3.70×10^4 in the second.

The choice of absolute values for the SUSY parameters on the right panel of Fig. 4 is such that the LSP is almost degenerated with the stau, and light

---

[c]See Ref.[19] for further discussions on this issue.

enough to predict a relic abundance on the WMAP region. Therefore the darker shaded areas on Fig. 4 satisfy all the phenomenological constraints which exclude the Yukawa unification scheme within mSUGRA. Since our result is almost independent of the rest of the phases it is likely to find points such the EDM constraint is also satisfied, this issue is currently under investigation [20].

4. Conclusions

The SUSY threshold correction to m_b, Δm_b, can induce large changes on the two heavier neutral Higgs bosons. For a given m_χ and certain values of Δm_b the neutralino relic density can satisfy the WMAP bounds while a cancellation mechanism among phases can keep the fermion EDM predictions below the current experimental bounds. The presence of phases in the SUSY parameters can reconcile the assumption of Yukawa unification and the observed dark matter density. However, the restriction imposed by the experimental bounds on BR($b \to s\gamma$) remains a severe constraint on the system.

Acknowledgments

MEG acknowledges support from the 'Consejería de Educación de la Junta de Andalucía' and the Spanish DGICYT under contract BFM2003-01266. The research of TI and PN was supported in part by NSF grant PHY-0139967. SS acknowledges support from the European RTN network HPRN-CT-2000-00148.

References

1. C. L. Bennett et al., Astrophys. J. Suppl. **148**, 1 (2003); D. N. Spergel et al., Astrophys. J. Suppl. **148**, 175 (2003).
2. M. E. Gomez, T. Ibrahim, P. Nath and S. Skadhauge, Phys. Rev. D **70**, 035014 (2004); hep-ph/0410007.
3. For a review see, A. H. Chamseddine, R. Arnowitt and P. Nath, Nucl. Phys. Proc. Suppl. **101**, 145 (2001) [arXiv:hep-ph/0102286].
4. J. R. Ellis, T. Falk, K. A. Olive and M. Srednicki, Astropart. Phys. **13**, 181 (2000); M. E. Gomez, G. Lazarides and C. Pallis, Phys. Rev. D **61**, 123512 (2000); Phys. Lett. B **487**, 313 (2000);
5. K.L. Chan, U. Chattopadhyay and P. Nath, Phys. Rev. D **58**, 096004 (1998); J. L. Feng, K. T. Matchev and T. Moroi, Phys. Rev. D **61**, 075005 (2000).
6. A. B. Lahanas, D. V. Nanopoulos and V. C. Spanos, Phys. Rev. D **62**, 023515 (2000); J. R. Ellis, T. Falk, G. Ganis, K. A. Olive and M. Srednicki, Phys. Lett. B **510**, 236 (2001); H. Baer and J. O'Farrill, JCAP **0404**, 005 (2004).
7. D. G. Cerdeño et al, JHEP **0306**, 030 (2003).

8. P. Nath, Phys. Rev. Lett. **66** (1991) 2565. ; Y. Kizukuri and N. Oshimo, Phys. Rev. D **46**,3025(1992).
9. S. Abel, S. Khalil and O. Lebedev, Nucl. Phys. B **606**, 151 (2001); G. C. Branco *et al.*, Nucl. Phys. B **659**, 119 (2003).
10. T. Ibrahim and P. Nath, Phys. Lett. B **418**, 98 (1998) [arXiv:hep-ph/9707409]. ; Phys. Rev. D **57**, 478 (1998) [Erratum-ibid. D **58**, 019901 (1998 ERRAT,D60,079903.1999 ERRAT,D60,119901.1999)] [arXiv:hep-ph/9708456].
11. M. Carena and H. E. Haber, Prog. Part. Nucl. Phys. **50**, 63 (2003)
12. T. Ibrahim and P. Nath, Phys. Rev. D **67**, 095003 (2003)
13. G. Belanger, F. Boudjema, A. Pukhov and A. Semenov, Comput. Phys. Commun. **149**, 103 (2002); for an updated version see hep-ph/0405253.
14. J. S. Lee *et al* Comput. Phys. Commun. **156**, 283 (2004).
15. M. E. Gomez, G. Lazarides and C. Pallis, Nucl. Phys. B **638**, 165 (2002); Phys. Rev. D **67**, 097701 (2003); C. Pallis and M. E. Gomez, hep-ph/0303098.
16. D. Auto, H. Baer, A. Belyaev and T. Krupovnickas, JHEP **0410**, 066 (2004).
17. D. A. Demir and K. A. Olive, Phys. Rev. D **65**, 034007 (2002).
18. P. Gambino and M. Misiak, Nucl. Phys. B **611**, 338 (2001).
19. T. Hurth, E. Lunghi and W. Porod, Eur. Phys. J. C **33**, S382 (2004).
20. M. E. Gomez, T. Ibrahim, P. Nath and S. Skadhauge, work in progress.

NEUTRALINO DARK MATTER IN SUPERGRAVITY THEORIES WITH NON-UNIVERSAL SCALAR AND GAUGINO MASSES

D. G. CERDEÑO*

Institute for Particle Physics Phenomenology (IPPP), Durham DH1 3LE, U.K.

The direct detection of neutralino dark matter in general supergravity theories with non-universal soft scalar and gaugino masses is reviewed. In particular, the neutralino-proton cross section is computed and compared with the sensitivities of dark matter detectors, taking into account the most recent experimental and astrophysical constraints, as well as those coming from charge and colour breaking minima. Gaugino and scalar non-universalities provide a large flexibility in the neutralino sector, and neutralinos close to the present detection limits are possible with a wide range of masses, from over 400 GeV to almost 10 GeV.

1. Introduction

Weakly Interacting Massive Particles (WIMPs) are particularly interesting candidates for the dark matter in the Universe[1], since they can be present in the right amount to explain the observed matter density. Interestingly, supersymmetric extensions of the standard model provide a very natural candidate within this category, the lightest neutralino, $\tilde{\chi}_1^0$.

There are numerous experiments around the world aiming at the direct detection of WIMPs by observing their elastic scattering on target nuclei through nuclear recoils. Due to their increasing sensitivities, it seems very plausible that the dark matter will be found in the near future. Given this situation, and assuming that the dark matter is a neutralino, it is natural to wonder how big the neutralino-proton cross section, $\sigma_{\tilde{\chi}_1^0-p}$, can be, in order to determine the feasibility of its direct detection. In fact, the theoretical predictions for $\sigma_{\tilde{\chi}_1^0-p}$ have been thoroughly analysed during many years[1].

In the usual minimal supergravity scenario, where the soft terms of the minimal supersymmetric standard model are assumed to be universal at the unification scale, $M_{GUT} \approx 2 \times 10^{16}$ GeV, and radiative electroweak

---

*Work supported by the PPARC.

symmetry breaking is imposed, the cross section turns out to be constrained by $\sigma_{\tilde{\chi}_1^0-p} \lesssim 3 \times 10^{-8}$ pb[1]. Clearly, in this case, present experiments are not sufficient and more sensitive detectors producing further data are needed.

The neutralino-proton cross section can be increased in different ways. In particular, it is possible to enhance the scattering channels involving exchange of CP-even neutral Higgses by reducing the Higgs masses (which can be done by increasing $m_{H_u}^2$ and/or decreasing $m_{H_d}^2$), and also by increasing the Higgsino components of the lightest neutralino (through the increase in the value of $m_{H_u}^2$, which entails a reduction in $|\mu|$). The effect of lowering the Higgs masses is typically more important, and it can provide large values for $\sigma_{\tilde{\chi}_1^0-p}$ even in the case of bino-like neutralinos.

The above mentioned effects may be reproduced when the universal structure for the soft terms is relaxed in either the scalar or gaugino masses.

- Non-universal scalars[5,6,7] induce the largest corrections in $\sigma_{\tilde{\chi}_1^0-p}$. The most important effects are due to non-universal Higgs masses, whose values at M_{GUT} can be parameterised as follows:

$$m_{H_d}^2 = m^2(1+\delta_1) , \quad m_{H_u}^2 = m^2(1+\delta_2) , \qquad (1)$$

where m is the common scalar mass parameter.

An increase(decrease) in $m_{H_u}^2$ ($m_{H_d}^2$) at the electroweak scale can be achieved by increasing(decreasing) its value at the GUT scale, i.e., with $\delta_2 > 0 (\delta_1 < 0)$. As above mentioned, these variations in the Higgs mass parameters entail an increase in $\sigma_{\tilde{\chi}_1^0-p}$.

- Non-universal gauginos[6,7,8] can be parameterised as

$$M_1 = M , \quad M_2 = M(1+\delta_2') , \quad M_3 = M(1+\delta_3') . \qquad (2)$$

It is worth noticing that the gluino mass, M_3, appears in the RGEs of squark masses. Thus the contribution of squark masses proportional to the top Yukawa coupling in the RGE of $m_{H_u}^2$ does this less negative if M_3 is small ($\delta_3' < 0$). As discussed above, this enhances the $\sigma_{\tilde{\chi}_1^0-p}$. However, the decrease in M_3 also implies a decrease in the lightest Higgs mass, which therefore constrains this possibility.

2. General case: non-universal scalars and gauginos

Let us now address the general case, with non-universalities in both the scalar and gaugino sectors[9]. We are interested in analysing the conditions under which high values for the cross section are obtained. For this reason, we concentrate on some interesting choices for scalar non-universalities,

$$a)\ \delta_1 = 0,\ \delta_2 = 1;\quad b)\ \delta_1 = -1,\ \delta_2 = 0;\quad c)\ \delta_1 = -1,\ \delta_2 = 1, \qquad (3)$$

and study the effect of adding gaugino non-universalities to these. Non-universalities in both the gauginos and scalars have also been analysed in the context of a SUSY GUT inspired MSSM version[10].

In our computation the most recent experimental and astrophysical constraints are taken into account, as well as those coming from the absence of unbounded from below (UFB) constraints on the Higgs potential. For details on how all these are implemented see reference 9.

Due to the importance of M_3 the possible gaugino non-universalities are grouped in two different cases, depending on whether the ratio M_3/M_1 at the GUT scale decreases or increases with respect to its value in the universal case.

2.1. Decrease in M_3/M_1 ($\delta'_3 < 0$)

In order to satisfy the constraint on the lightest Higgs mass, higher values of M, and therefore of M_1 are necessary. When the neutralino is mostly bino this implies that $m_{\tilde{\chi}^0_1}$ is increased. Thus it is possible to find heavier neutralinos with a relatively high value for their direct detection cross section and a larger Higgsino composition. Another consequence of the decrease in M_3/M_1 is the reduction in the value of the neutralino relic density, $\Omega_{\tilde{\chi}^0_1}$. This may be problematic, since the choices of non-universal scalars (3) already lead to a similar decrease, especially those where the Higgsino components of $\tilde{\chi}^0_1$ increase.

An example with $\delta'_{2,3} = -0.25$ is shown in Fig. 1, where the neutralino-proton cross section is plotted versus the neutralino mass, $m_{\tilde{\chi}^0_1}$, for $\tan\beta = 35$, $A = 0$ and a full scan in m and M for the different choices of non-universal scalar parameters (3). These results are similar to those with only scalar non-universalities[7]. In particular, regions of the parameter space fulfilling all the constraints and with a cross section close to the detection range are attainable for moderate values of $\tan\beta$, entering the DAMA region for $\tan\beta \gtrsim 30$. However, these regions are shifted towards larger M_1 and thus heavier neutralinos are obtained. For instance, points with $m_{\tilde{\chi}^0_1} \gtrsim 400$ GeV and $\sigma_{\tilde{\chi}^0_1-p} \gtrsim 10^{-6}$ pb appear.

Increasing the value of $\tan\beta$ leads to the well known enhancement of $\sigma_{\tilde{\chi}^0_1-p}$. An example with $\tan\beta = 50$ is represented in Fig. 2. In all these cases points with a higher $\sigma_{\tilde{\chi}^0_1-p}$ correspond to those having m_A close to its experimental lower limit, $m_A \gtrsim 100$ GeV.

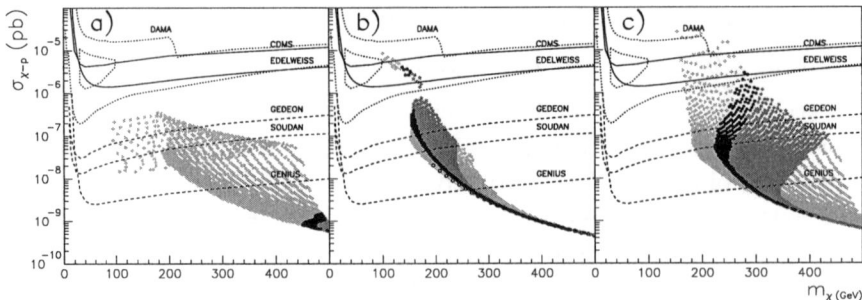

Figure 1. Scatter plot of the scalar neutralino-proton cross section $\sigma_{\tilde{\chi}_1^0-p}$ as a function of the neutralino mass, $m_{\tilde{\chi}_1^0}$, for $\delta'_{2,3} = -0.25$ and the three choices for non-universal scalars (3) in a case with $\tan\beta = 35$ and $A = 0$. The light grey dots correspond to points fulfilling all the experimental constraints. The dark grey dots represent points fulfilling in addition $0.1 \leq \Omega_{\tilde{\chi}_1^0} h^2 \leq 0.3$ and the black ones correspond to those consistent with the WMAP[2] range. Points excluded by UFB constraints are represented with circles. The sensitivities of present and projected experiments[3] are also depicted with solid and dashed lines, respectively. The large (small) area bounded by dotted lines is allowed by the DAMA[4] experiment when astrophysical uncertainties are (are not) considered.

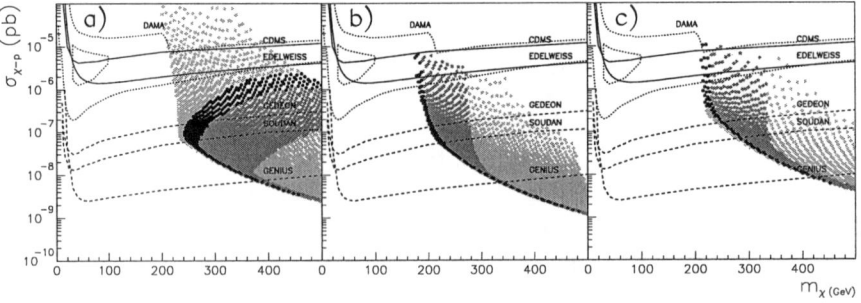

Figure 2. The same as in Fig. 1 but for $\tan\beta = 50$.

2.2. *Increase in M_3/M_1 ($\delta'_3 > 0$)*

In this case, the constraint on the Higgs mass and on $b \to s\gamma$ is satisfied for smaller values of M, and therefore the effective value of M_1 can be smaller than in the universal case. Thus lighter neutralinos appear.

The theoretical predictions for $\sigma_{\tilde{\chi}_1^0-p}$ are represented in Fig. 3 for an example with $\delta'_{2,3} = 1$, $\tan\beta = 50$ and $A = 0$ and the three choices of Higgs non-universalities of (3). As we can see, this choice of gaugino parameters favours the appearance of light neutralinos, which have a large bino component. The predicted cross section is only slightly smaller than in the

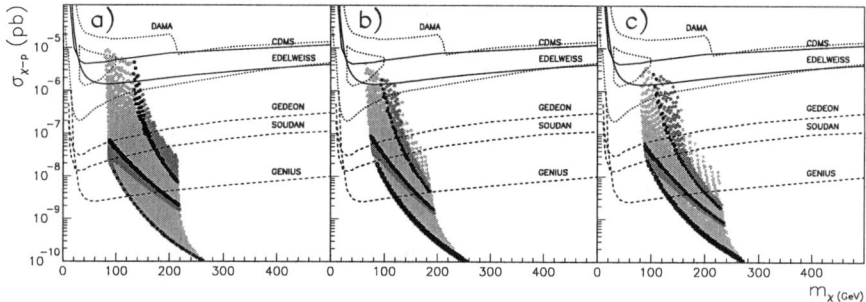

Figure 3. The same as in Fig. 1 but for $\delta'_{2,3} = 1$ and $\tan\beta = 50$

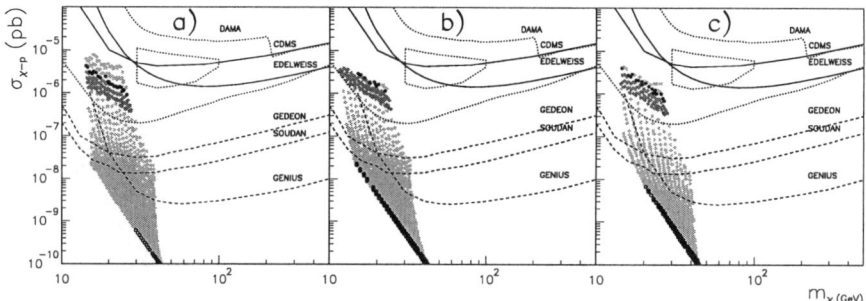

Figure 4. The same as in Fig. 1 but for $\delta'_{2,3} = 10$ and $\tan\beta = 50$.

cases with just non-universal scalars, so these neutralinos can still be close to the sensitivities of dark matter experiments. In particular, points satisfying all the constraints and entering the DAMA region can be obtained for $m_{\tilde{\chi}_1^0} \lesssim 150$ GeV.

The value of M_1 can be further decreased for larger values of δ_3, thus leading to even lighter neutralinos[11,12,13]. Such light $\tilde{\chi}_1^0$ cannot be obtained in SUGRA theories with non-universalities in just the scalar or gaugino sector. One of the requirements for their appearance is to have $M_1 \ll \mu, M_2$ at low energy, almost pure binos. This can be achieved with gaugino non-universalities ($\delta'_{2,3} \gg 1$). However, without a very effective reduction of m_A, the relic density would be too large, inconsistent with observations[13]. Here the presence of non-universal scalars is crucial. Non-universalities as those in the Higgs sector in (3) provide a very effective way of lowering m_A and are thus optimal for this purpose[9].

This is illustrated in Fig. 4 with an example for $\delta'_{2,3} = 10$ with $\tan\beta = 50$ and $A = 0$, where points with $\sigma_{\tilde{\chi}_1^0-p} \gtrsim 3 \times 10^{-6}$ pb are obtained with

$m_{\tilde{\chi}_1^0} \sim 15$ GeV, in agreement with the bound derived in the effMSSM[13].

References

1. For a recent review, see C. Muñoz, *Int. J. Mod. Phys. A* **19** (2004) 3093.
2. WMAP Coll., C. L. Bennett et al., *Astrophys. J. Suppl.* **148** (2003) 1.
3. CDMS Coll., R. Abusaidi et al., *Phys. Rev. Lett.* **84** (2000) 5699; D. Abrams et al., *Phys. Rev. D* **66** (2002) 122003; D. S. Akerib et al., arXiv:hep-ex/0306001; EDELWEISS Coll., A. Benoit *et al.*, *Phys. Lett. B* **513** (2001) 15; *Phys. Lett. B* **545** (2002) 43; A. Morales, arXiv:hep-ex/0111089; GENIUS Coll., H. V. Klapdor-Kleingrothaus et al., arXiv:hep-ph/9910205.
4. DAMA Coll., R. Bernabei et al., *Phys. Lett. B* **480** (2000) 23; *Riv. Nuovo Cim.* **26** (2003) 1. P. Belli, R. Cerulli, N. Fornengo and S. Scopel, *Phys. Rev. D* **66** (2002) 043503.
5. V. Berezinsky, A. Bottino, J. Ellis, N.Fornengo, G. Mignola and S. Scopel, *Astropart. Phys.* **5** (1996) 1; R. Arnowitt and P. Nath, *Phys. Rev.* **D56** (1997) 2820; A. Bottino, F. Donato, N. Fornengo and S. Scopel, *Phys. Rev. D* **59** (1999) 095004; R. Arnowitt and P. Nath, *Phys. Rev. D* **60** (1999) 044002; E. Accomando, R. Arnowitt, B. Dutta and Y. Santoso, *Nucl. Phys. B* **585** (2000) 124; J. Ellis, A. Ferstl and K. A. Olive, *Phys. Rev. D* **63** (2001) 065016; J. Ellis, A. Ferstl, K. A. Olive and Y. Santoso, *Phys. Rev. D* **67** (2003) 123502. R. Arnowitt and B. Dutta, arXiv:hep-ph/0112157. R. Dermisek, S. Raby, L. Roszkowski, R. Ruiz de Austri, *J. High Energy Phys.* **04** (2003) 037; S. Profumo, *Phys. Rev. D* **68** (2003) 015006; *J. High Energy Phys.* **06** (2003) 052; H. Baer, C. Balázs, A. Belyaev and J. O'Farril, *JCAP* **09** (2003) 007.
6. D. G. Cerdeño, S. Khalil and C. Muñoz, Proceedings of CICHEP Conference, Cairo (2001), *Rinton Press* (2001) 214; R. Arnowitt and B. Dutta, arXiv:hep-ph/0210339.
7. D. G. Cerdeño, E. Gabrielli, M. E. Gomez, and C. Muñoz, *J. High Energy Phys.* **06** (2003) 030.
8. A. Corsetti and P. Nath, *Phys. Rev. D* **64** (2001) 125010; V. Bertin, E. Nezri and J. Orloff, *J. High Energy Phys.* **02** (2003) 046; A. Birkedal-Hansen and B. D. Nelson, *Phys. Rev. D* **67** (2003) 095006; A. Birkedal-Hansen, arXiv:hep-ph/0306144; U. Chattopadhyay and D. P. Roy, arXiv:hep-ph/0304108.
9. D. G. Cerdeño and C. Muñoz, *J. High Energy Phys.* **10** (2004) 015.
10. C. Pallis, *Nucl. Phys. B* **678** (2004) 398.
11. M. Drees and X. Tata, *Phys. Rev. D* **43** (1991) 2971; K. Griest and L. Roszkowski, *Phys. Rev. D* **46** (1992) 3309; A. Gabutti, M. Olechowski, S. Cooper, S. Pokorski and L. Stodolsky, *Astropart. Phys.* **6** (1996) 1.
12. D. Hooper and T. Plehn, *Phys. Lett. B* **562** (2003) 18; G. Bélanger, F. Boudjema, A. Pukhov and S. Rosier-Lees, arXiv:hep-ph/0212227; G. Bélanger, F. Boudjema, A. Cottrant, A. Pukhov and S. Rosier-Lees, *J. High Energy Phys.* **03** (2004) 012.
13. A. Bottino, N. Fornengo and S. Scopel, *Phys. Rev. D* **67** (2003) 063519; A. Bottino, F. Donato, N. Fornengo and S. Scopel, *Phys. Rev. D* **69** (2004) 037302; *Phys. Rev. D* **68** (2003) 043506.

RIP
THE MACHO ERA (1974-2004)

N.W. EVANS, V. BELOKUROV

Institute of Astronomy, Madingley Rd
Cambridge CB3 0HA, England
E-mail: nwe@ast.cam.ac.uk, vasily@ast.cam.ac.uk

This article reviews the life and death of a scientific theory

1. The MACHO Ideology

1.1. *The Dawn of Dark Matter*

The hypothesis of dark matter is often ascribed to Fritz Zwicky. Certainly, Zwicky [1] in his book "Morphological Astronomy" noted the discrepancy between masses of clusters inferred from the virial theorem and masses inferred from the visible constituent galaxies. He suggests five possible explanations. The fifth (and most tentative) — after propositions that the clusters may not be in equilibrium or that light may tire on traversal of enormous distances — is: *"Finally, attention must be called to the recent discovery of luminous and of dark intergalactic matter. The existence of this dark matter may seriously affect all previous estimates concerning the distribution of mass in the Universe"*.

The focus of this conference is on the direct and indirect detection of dark matter in the Milky Way galaxy and other nearby galaxies. Even if Zwicky was the first to hypothesise the existence of dark matter in clusters, he did not believe that there was appreciable dark matter in galaxies (in "Morphological Astronomy", he advocated Keplerian fitting to rotation curves to estimate the masses of galaxies). The realisation that galaxies are surrounded by dark matter haloes only came much later. Dark matter on the scales of galaxies became widely accepted after the publication of the rotation curve of the nearby galaxies M31, M81 and M101 by Roberts and collaborators [2]. In an influential paper, Ostriker, Peebles & Yahil [3] brought together a number of lines of evidence to suggest that: *"There are reasons, increasing in number and in quality, to believe that the masses of ordinary galaxies may have been underestimated by a factor of 10 or more... The very large implied mass to light ratios and very great extent of spiral galaxies can perhaps most plausibly be understood as due to a giant halo of faint stars"*

This is the first statement of the MACHO ideology – namely that (some of) the dark matter in galaxy haloes is baryonic and composed of massive objects. The most obvious candidates are faint stars (red dwarfs, white dwarfs, neutron stars), failed stars (brown dwarfs and Jupiters) and massive remnants from an early epoch of Population III stars. The neologism MACHO seems to have been first used in print by Griest [4] as a witty counterpoise to WIMPS (weakly interacting massive particles). MACHO stands for massive compact halo objects.

1.2. The Hey-Day of the MACHO Era (1974-1994)

The Zeitgeist is well documented in the Princeton conference on "Dark Matter in the Universe", which marks the hey-day of the MACHO Era. It was well-known that all the dark matter in galaxies and clusters could conceivably be baryonic without violating constraints from cosmological nucleosynthesis [5]. There even seemed to be arguments in favour of baryonic compact objects as opposed to particle dark matter. For example, Gunn [6] pointed out that; *"There is evidence that the Population II mass function is very steep in the halo and an extension at the low mass-end to quite plausible masses leads to very large mass-to-light ratios... A picture in which the low-mass cut-off progresses smoothly from $0.1\ M_\odot$ to $10^{-3}\ M_\odot$ as one goes from the center of the galaxy outwards makes a qualitatively plausible model... It entails no mystery as to why the amount of dark matter is within an order of magnitude of the visible matter, and makes plausible the fact that rotation curves are flattish from regions where the galaxies are dominated by visible matter out to regions in which they are dominated by dark matter."*

More exuberantly still, Lynden-Bell [7] cited the X-ray data; *"We have rather good evidence that around a number of giant elliptical galaxies, baryonic matter is disappearing from hot, X-ray emitting gas. The place where it disappears is right for the making of dark halos. The rate of its disappearance would build a halo in 10^{10} years. If we want to believe the observations rather than our prejudices, we should take as our best bet that dark halos are baryonic and made from cooling flows... When exotic neutral particles have been found in the laboratory, I shall be happy to postulate them in the* *cosmos, but until then, let us use our observations, not our prejudices."*

But even then, the most important objection to baryonic dark matter as the dominant component of galaxy haloes was clearly understood. It is difficult to understand how such baryonic structures of mass $\sim 10^{12} M_\odot$ could have formed without leaving an imprint in the microwave background [6].

1.3. *The Decline and Fall of the MACHO Era (1994-2004)*

Microlensing as a test for dark, compact objects was suggested very early on (e.g., Zwicky's "Morphological Astronomy" discusses microlensing by neutron stars). But, Paczyński [8] convinced the astronomical community that microlensing could provide a decisive test of the MACHO hypothesis. And so it turned out .... The microlensing experiments led to the decline and fall of the MACHO Era.

Beginning in 1993, large scale monitoring of stars in the Large Magellanic Cloud (LMC) was conducted by two groups (MACHO and EROS) looking for microlensing events. The results of the MACHO experiment are well-known. From 5.7 years of data, Alcock et al. [9] found between 13 to 17 microlensing events and reckoned that the microlensing optical depth (or probability of microlensing) is $\tau \sim 1.2^{+0.4}_{-0.3} \times 10^{-7}$. Interpreted as a dark halo population, the most likely fraction of the dark halo in MACHOs is 20 %, while the most likely mass of the MACHOs is between 0.15 and 0.9 $M_\odot$. After 8 years of monitoring the Magellanic Clouds, the EROS experiment announced 3 microlensing candidates towards the LMC [10]. Although EROS do not report their results in terms of optical depth, they have clearly detected a smaller microlensing signal than MACHO – a discrepancy which could have a number of explanations.

The remainder of this article will argue that Alcock et al. overestimated the microlensing optical depth and that the dark halo has little or no MACHOs.

2. Neural Network Processing

There are two principal difficulties with the microlensing experiments. The first is well-known, the second less so (and thus we concentrate upon it here).

First, just as in direct detection experiments for particle dark matter, there is a background that must be eliminated. In microlensing experiments, stars in the thin disk, thick disk and the LMC all provide possible lenses for microlensing events [11], aside from MACHOs in the dark halo.

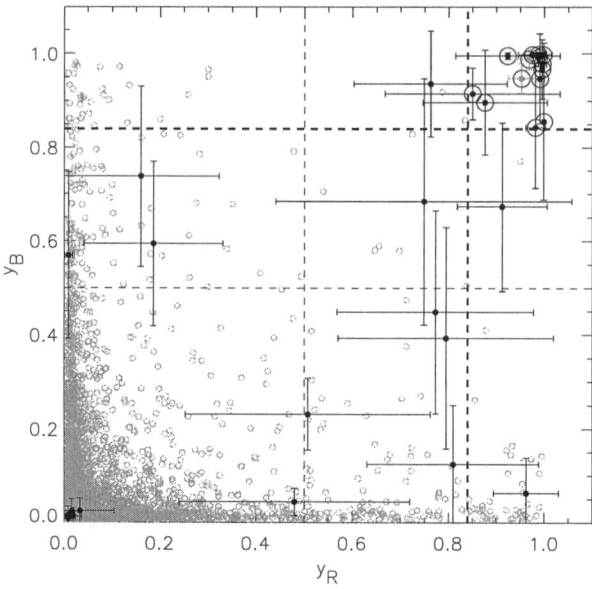

Figure 1. The locations of ≈ 22000 lightcurves as given by the outputs of the neural networks y_R and y_B on processing the red data and the blue data respectively. These include the 29 lightcurves that passed the loose selection of the MACHO collaboration together with ∼ 1000 lightcurves in the vicinity of each candidate. Each point gives the maximum of the moderated output while the error bar gives the network scatter. A large open circles around a point indicates that it lies above the decision boundary ($y_R > 0.87$ and $y_B > 0.87$). Filled black dots represent the 29 lightcurves selected by Alcock et al., while all other lightcurves are represented by open grey dots. [From Belokurov et al. 2004]

The total optical depth due to stellar lensing from known populations [12] is ∼ 0.7×10^{-7}, which is within the 2σ lower bound of Alcock et al.'s claimed detection ($\tau \sim 1.2^{+0.4}_{-0.3} \times 10^{-7}$).

Secondly, the identification of microlensing events (stars that brighten and then fade) takes place against a background of stellar variability that is at least 10^5 times more common. Many varieties of stellar variability are not well-studied or understood. Therefore, the identification of events is much more fraught than usually appreciated. All microlensing groups use a sequence of straight line cuts to identify events (for example, excising chromatic lightcurves or troublesome regions of the colour-magnitude diagram). The decision boundary between microlensing and non-microlensing is therefore polygonal in a multi-dimensional parameter space. Nowadays,

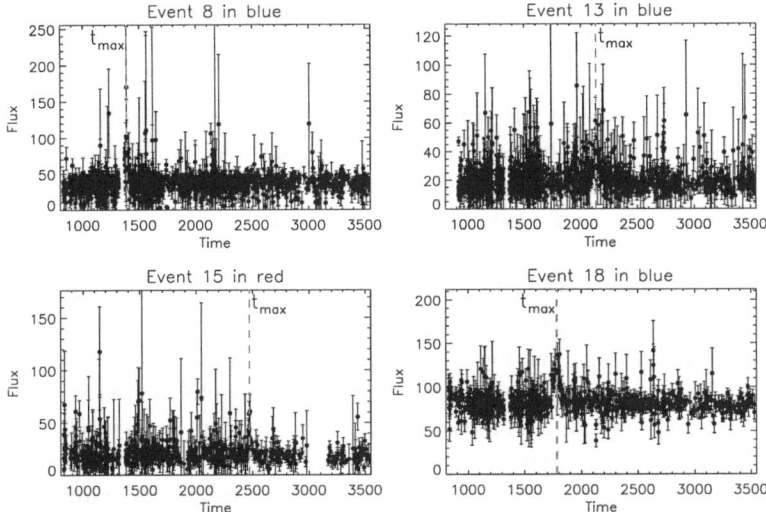

Figure 2. This shows the lightcurves for 4 events which received low probability values y in one or both filters. These are all included in Alcock et al.'s (2000) set of convincing microlensing candidates, but are not confirmed by our neural network analysis. The vertical axis is flux in ADU s^{-1} and the horizontal axis is time in JD-2448000. Vertical lines mark the peak of the event. [From Belokurov et al. 2004]

many high-energy physics experiments prefer to use neural networks for pattern recognition. This is because neural networks permit the construction of complicated decision boundaries.

All this inspired Belokurov, Evans & Le Du [13] to carry out a re-analysis of the MACHO data with neural networks. Microlensing events are characterised by the presence of (i) an excursion from the baseline that is (ii) positive, (iii) symmetric and (iv) single. The event itself is parameterised by (v) a timescale. Motivated by these features, five parameters are extracted from the lightcurves as inputs to the neural networks. Most neural networks require a training set, on which the network is taught to recognise the desired patterns (in this case, microlensing). Here, the training set contains 1500 examples of microlensing and > 2000 examples of other kinds of variability (pre-main sequence stars, Coronae Borealis stars, Miras, Semi-regular variables, Cepheids, Bumpers, Supernovae, novae, eclipsing variables). They are sampled with MACHO sampling and random Gaussian noise is added. All networks are trained using the Netlab package [14]. The output of the network is the posterior probability of microlensing.

Figure 1 shows the locations of $\approx$ 22000 lightcurves. The data for the red and blue passbands are processed separately with neural networks to give outputs y_R and y_B. The decision boundary is shown in the bold broken line – convincing microlensing candidates have $y_{R,B} > 0.84$. This boundary is fixed by insisting that the number of false negatives in the entire MACHO dataset is $\lesssim 1$. This corresponds to a false positive rate of 0.3%. The 29 candidate microlensing lightcurves identified by Alcock et al.[9] are denoted by filled black dots, while all other lightcurves are shown as open grey dots. Twelve of these 29 lightcurves satisfy $y_{R,B} > 0.84$, namely 1a, 1b, 5, 6, 10a, 11, 14, 21-25. There are additionally 2 false positives. Both lie close to the noise/microlensing border in parameter space.

After successfully passing the first tier of neural networks, Belokurov et al.[13] apply a second tier that discriminates against supernovae (SNe) occurring in background galaxies behind the LMC. The colours change dramatically during a supernova explosion as a result of complicated radiation processes inside the ejecta. After a fairly constant pre-maximum epoch with $B - V \approx 0$, a supernova of type Ia typically starts turning red at the time of the maximum light. It reaches $B - V \approx 1$ in about 30 days and then drops back[15]. This can be contrasted with the colour behaviour during gravitational microlensing. Gravity bends light irrespective of its frequency. Therefore, colour does not change during microlensing. However, the achromaticity of the lightcurve only holds good if the source star is resolved and the lens is dark. The presence of other stars within the centroid of light or lensing by a luminous object will result in a colour change during the event. At the baseline, the colour is defined by the combined flux from all the sources. The amplified star will contribute most of the colour around the peak. The colour of a microlensing event can become redder or bluer, depending on the population of the blend, but it usually changes symmetrically about the peak with substantial correlation between passbands[16]. The differing behaviour of colour evolution during SNe and blended microlensing can be quantified as features fed to neural networks, and – as Belokurov et al.[13] show – used to distinguish between the two. This leads to the discarding of a further 3 of the 12 candidates that passed the first tier.

Based on a neural network analysis, Alcock et al.'s sample is seriously contaminated. There are 6 almost certain microlensing events (1, 5, 6, 14, 21 and 25) and two likely ones (9, and 18). Some of the lightcurves rejected by the neural networks, but classified as microlensing by Alcock et al., are shown in Figure 2. The peak of the alleged event is shown as a vertical dashed line. Notice that event 23 – which looks perfect and passes all the

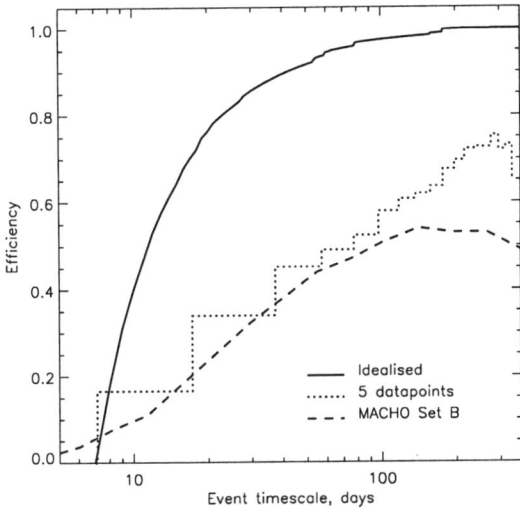

Figure 3. This shows three approximations to the efficiency as a function of Einstein diameter crossing timescale t. The full curve is the idealised efficiency calculated by integration over the distribution of sampling gaps. This is an upper limit. The dashed curve is the published efficiency of Alcock et al. (2000). This is a lower limit. The histogram is a realisation of the true efficiency derived from Monte Carlo simulations of event selection in the training set.

neural networks – has been shown by the EROS collaboration [17] to have a second peak on the lightcurve about 7 years after the first one probed by MACHO and so is an unusual variable star. This is very worrisome for all microlensing experiments.

3. The Optical Depth

The conventional formula for optical depth is

$$\tau = \frac{\pi}{4} \frac{1}{N_\star T} \sum_j \frac{t_j}{\epsilon(t_j)} \qquad (1)$$

where t_j is the Einstein diameter crossing time of the jth event, ϵ is the efficiency, T is the duration of the experiment and $N_\star$ is the number of stars monitored. The summation is taken over the set of microlensing events. There are three major components to the efficiency. The first arises from shortcomings in the cuts used to identify microlensing events. The second arises from blending, which causes both the magnification and the number of stars monitored to be underestimated. The third arises from the temporal sampling, as events are necessarily missed if they fall in a gap in the data-taking. A neural network, properly trained, will all but eliminate any contribution from the first component for the subset of events included in the training set. The second component cancels out to lowest order, as the loss due to the underestimate of the magnification is balanced by the gain due to the fact that an object may contain more than one star [18]. The third component of the efficiency still remains, but fortunately is straightforward to compute.

Figure 3 shows upper and lower bounds to the efficiency as a function of timescale. An upper bound to the efficiency can be found by assuming that events are missed if and only if no data are taken during the bump. We sum the distribution of sampling gaps over the baseline of the experiment and judge an event to be missed if it falls within a gap. The probability of missing an event with timescale t is just

$$P(t) \propto \sum_{t' \geq t} t' n(t') \qquad (2)$$

where $n(t)$ is the number of gaps of duration t. The quantity $1 - P(t)$ is an idealised efficiency which is shown as a full curve in Figure 3. A lower bound to the efficiency is given by the published efficiency results of Alcock et al. [9] for the looser set of candidates. This is because the neural networks necessarily provide a cleaner set of microlensing candidates, uncontaminated by spurious events. A realisation of the actual efficiency is easily found from Monte Carlo simulations of the training set, by finding the fraction of all events that are included (and hence will be inexorably characterised as microlensing by the network). In the simulations, microlensing events are generated with uniform priors. Only those events with five or more datapoints with a signal-to-noise greater than 5 are incorporated into the training set. The efficiency is therefore the ratio of events accepted into the training set to all events. The result is shown as a histogram in Figure 3, and lies between the upper and lower bounds, as expected.

Applying eq. (1) to the set of 9 events found by the neural network, we obtain the following bound on the optical depth to the LMC:

$$3 \times 10^{-8} < \tau < 5 \times 10^{-8}. \qquad (3)$$

Here, the timescales uncorrected for blending given in third column of Table 7 of Alcock et al. [9] are used. This is correct, as the effects of blending cancel out to lowest order.

This is a low value for the optical depth. The optical depths of the thin disk, thick disk and spheroid to be 2.2×10^{-8}, whilst the optical depth of the stellar content of the LMC to be 3.2×10^{-8} on average. In other words, *our total optical depth matches the contribution from the known stellar populations in the outer Galaxy and the LMC. This implies that there is no contribution needed from compact objects in the halo.*

There is supporting evidence for this belief from the exotic events and from the lensing signal towards the Small Magellanic Cloud (SMC). First, the exotic events yield additional information which can break some of the microlensing degeneracies and thus give indirect evidence on the location of the lens. All the exotic lenses belong to known stellar populations in the outer Milky Way or the LMC. Second, the duration of the events towards the SMC is very different from the duration towards the LMC. The EROS collaboration [19] constrain the optical depth towards the SMC to be $< 10^{-7}$ at better than the 90 % confidence level, based on an admittedly small sample. Both these facts militate against the idea that a single population of objects in the Milky Way halo is causing the microlensing events

4. Conclusions

The MACHO Era is over! The dark matter in the halo of the Milky Way is **not** in the form of massive, compact halo objects. The microlensing signal detected by both the MACHO and EROS experiments is entirely consistent with that expected from stellar lenses in the known populations. In particular, the sample of 14 high quality microlensing events in Alcock et al. [9] is contaminated. Realistically, Alcock et al.'s sample has 6 almost certain microlensing events (1, 5, 6, 14, 21 and 25) and two likely ones (9, and 18). This is consistent with expectations from known stellar populations.

Even for the die-hards, the matter will surely soon be settled by the POINT-AGAPE experiment [20]. This is a microlensing experiment towards the nearby Andromeda galaxy (M31), which probes a new line of sight through the Milky Way and M31 dark haloes. It will provide a new estimate of the fraction of the Milky Way and M31 dark haloes that is composed of MACHOs. Two fields north and south of the M31 bulge have been monitored for three years using the Wide Field Camera on the Isaac Newton Telescope. The POINT-AGAPE collaboration have already found a small number of interesting individual microlensing events towards M31 [20], carried out a survey for classical novae [21] and reported the locations, periods and brightness of $\sim$ 35000 variable stars [22]. Very recently, an unrestricted and fully automated search for microlensing events towards M31 has been published [23]. Using a series of seven cuts based on sampling, goodness of fit, consistency, achromaticity, position in the colour-magnitude diagram and signal-to-noise. This leaves just 3 first-level or convincing microlensing candidates and 3 second-level or possible microlensing candidates. The efficiency of this survey is being computed at the moment and will yield an independent estimate of the MACHO fraction.

Die-hards have only a short time to wait.

References

1. F. Zwicky, *Morphological Astronomy*, Springer-Verlag, Berlin (1957)
2. M.S. Roberts, A.H. Rots, *A&A* **26**, 483 (1974); M.S. Roberts, R.N. Whitehurst, *ApJ* **201**, 327 (1975)
3. J.P. Ostriker, P.J.E. Peebles, A. Yahil, *ApJ* **193**, L1 (1974)
4. K. Griest, *ApJ* **366**, 412 (1991)
5. M.J. Rees, In *IAU Symposium 127: Dark Matter in the Universe*, eds. J. Kormendy, G.R. Knapp, Reidel, Dordrecht, p. 396
6. J.E. Gunn, In *IAU Symposium 127: Dark Matter in the Universe*, eds. J. Kormendy, G.R. Knapp, Reidel, Dordrecht, p. 543
7. D. Lynden-Bell, In *IAU Symposium 127: Dark Matter in the Universe*, eds. J. Kormendy, G.R. Knapp, Reidel, Dordrecht, p. 530
8. B. Paczyński, *ApJ* **304**, 1 (1986)
9. C. Alcock et al., *ApJ* **542**, 281 (2000)
10. T. Lasserre et al., *A&A* **355**, L39 (2000)
11. K. Sahu, *Nature* **370**, 275 (1994); N.W. Evans, G. Gyuk, M.S. Turner, J.J. Binney, *ApJ* **501**, L45 (1998); H.S. Zhao, *MNRAS* **294**, 139 (1998); H.S. Zhao, N.W. Evans, *ApJ* **545**, L35 (2000)
12. C. Alcock et al., *ApJ* **479**, 119 (1997)
13. V. Belokurov, N,W. Evans, Y. Le Du, *MNRAS* **352**, 233 (2004)
14. I.T. Nabney, *Netlab*, Springer-Verlag, New York (2002)
15. M.M Phillips, P. Lira, N.B. Suntzef, R.A. Schommer, M. Hamuy, J. Maza, *AJ* **118**, 1766 (1999)
16. R. Di Stefano, A.A. Esin, *ApJ* **448**, L1 (1995)
17. J.F. Glicenstein, Talk at the Hawaiian Gravitational Microlensing Workshop, 2004
18. C. Afonso et al., *A&A* **400**, 951 (2003)
19. C. Afonso et al., *A&A* **404**, 145 (2003)
20. S. Paulin-Henriksson et al., *A&A* **405**, 15 (2003); S. Paulin-Henriksson et al.,*ApJ* **576**, L121 (2002); J. An et al., *ApJ* **601**, 845 (2004)
21. M. Darnley et al., *MNRAS* **353**, 571 (2004)
22. J. An et al., *MNRAS* **351**, 1071 (2004)
23. V. Belokurov et al., MNRAS, in press (astro-ph/0411186)

BARYONIC DARK MATTER — AN OUTSIDER'S VIEW

J. J. QUENBY

Blackett Laboratory, Imperial College, London SW7 2BW, UK

The location and of missing baryonic matter is discussed. Most baryons may lie just outside visible galaxies in the form of a cool or warm gas. Within the Milky Way, the major problem is understanding the dynamics associated with the central bar and non-circular orbits which affect the mesurements in the cusp region. Further out, dense molecular clouds may hold significant baryonic mass.

1. Introduction

The interest from an 'outsider's viewpoint-from one whose primary intererst is in searching for non-baryonic dark matter-is, of course, to eliminate the 'contamination' from known forms of matter in assessing the chance of finding the sypersymmetric particles. In the Milky Way, it is necessary to establishing all known contributions to the gravitational field so that the magnitude of the residual effect of the dark matter halo can be estimated. Further out, a major component of the extragalactic baryonic mass is missing in the sense of not easily being identifyable by photon emission. Here, the doubt is not in the likely presence of a major dark matter contribution to the total Ω value but rather as to where the missing baryons are hiding.

2. Galactic Halo, Baryons Versus Cold Dark Matter

Merrifield (2003) summarises recent information on the local, galactic, matter density. Stellar dynamics, Kuijken and Gilmore (1991), gives $\Sigma \simeq 70 M_\circ pc^{-2}$ for the total mass within 1.1 kpc of the Galactic plane of which there is a visible $\Sigma \simeq 25 M_\circ pc^{-2}$ from stars and $\Sigma \simeq 15 M_\circ pc^{-2}$, from the interstellar medium (Olling and Merrifield,2001). Hence $\Sigma \simeq 30 M_\circ pc^{-2}$, the invisible difference, can be the cold dark matter. For a scale height of the dark matter halo large compared with the distance 1.1 kpc within which this estimate is made, it is found that $\rho_\circ^{DM} \simeq 0.014 M_\circ pc^{-3}$.

If dark and baryonic matter are of similar gravitational importance in the inner galaxy, is there consistency of observed and expected behaviour at a central cusp? A census by Olling and Merrifield (2001) on stars and interstellar material within the inner galaxy concluded they have a combined gravitational attraction equivalent to a spherical distribution of mass

$M_\circ^{baryon} \simeq 5.7 \times 10^{10} M_\circ$. Dynamics, Kerr and Lynden-Bell (1986), gives $M_\circ^{total} \simeq 9.5 \times 10^{10} M_\circ$ within $R_\circ$. Most galaxy formation simulations suggest CDM halos follow a shallow radial power law distribution at small radii going over to a steep distribution far out. Allowing $\rho \sim r^{-\alpha}$ and normalising to $\rho_\circ^{DM}$, Merrifield (2003) finds interior to the Sun, a total CDM mass

$$M_\circ^{DM} = \frac{4\pi}{3-\alpha} \rho_\circ^{DM} R_\circ^3 = \frac{10^{11}}{3-\alpha} M_\circ \quad (1)$$

Subtracting the above $M_\circ^{baryon}$ from $M_\circ^{total}$ yields $\alpha \simeq 0.4$ for the CDM cusp distribution. This low value of α poses an immediate problem for CDM where simulations favour values between 1.0 and 1.5 (Navarro, Frenk and White 1996).

Binney and Evans (2004) consider the minimum baryonic mass in the inner galaxy using microlensing to establish the steller density and take account of the geometry of the galactic bar. Maximising optical depth for minimum number of stars with a high scale height, they plot a steller circular speed rotation curve as observed, a curve generated by gas disk plus stars consistent with microlensing, a cuspy DM halo curve and the combined predicted curve. Any single component will yield a rotation speed $v \sim r^{(1-\alpha/2)}$ and hence the characteristically flat rotation curves, except close the centre, imply $\alpha \sim 2$. The problem with the model with a local CDM $\Sigma \sim 30 M_\circ pc^{-2}$ is that there is so much visible matter contribution at a few kpc that they need $\alpha = 0.3$ for the CDM. Further out, there is room for a substantial dark matter halo. From the baryonic dark matter viewpoint, extra material is required close to the galactic plane within 2 kpc of the centre to maintain the observed rather flat, low α portion of the rotation curve in that region. A rotating bar at the galaxy centre, 3 to 5 kpc long with a speed $\omega = 40 \to 80$km/s/kpc has been established by Debattista et al, (2002). This rotation speed means the bar is travelling close to the speed of the stars at the extremities of it's orbit. If the CDM and baryonic masses are similar in the bar region, some calculations suggest the bar should rapidly slow eg Debattista and Selwood (2004), implying there cannot be a significant CDM cusp but perhaps the CDM material has been flung beyond the solar orbit (Weinberg and Katz 2002, Klypin et al. 2002). Alternative arguments suggest that the bar cannot significantly affect the CDM (Sellwood 2003). Moreover, Merrifield (2003) shows that if the CDM halo follows an ellipsoidal power-law density distribution with a polar to equatorial axis ratio of 0.6, an $\alpha \sim 1.0$ can reproduce the local centripetal acceleration data.

An important contribution to baryonic dark matter away from the galactic centre could be mass 'hidden' in dense cold molecular clouds. Pfeniger et al, (1994) and other suggested cold, self gravitating molecular clouds as a major 'dark matter' component within the galaxy. Ohishi et al (2004), compute with the GEANT 4 code the gamma flux emission from galactic cosmic rays entering dense, uniform clouds of radius $R = 1.5 \times 10^{13}$ cm. A conventional CDM density distribution is adopted for H_2 yielding a cored isothermal sphere

$$\rho = \frac{\sigma^2}{2\pi G(R^2 + z^2 + r_c^2)} \tag{2}$$

with $\sigma = 155$ km s^{-1} and $r_c = 6.2$ kpc. The cosmic ray distribution of Webber et al. (1992) following the cosmic ray source distribution is

$$\frac{J_{cr}(R,z)}{J_{cr}(o)} = (\frac{R}{R_\circ})^{0.6} exp[(R_\circ - R)/L - |z|/h] \tag{3}$$

$R_\circ = 8.54$ kpc, $L = 7$ kpc and $h = 1.5$ kpc, yielding a disk-like character with a central hole. Ohishi et al.(2004) compare EGRET and computed gammma-ray emission at 1 GeV. Towards the centre, $|l| \leq 60°, |b| \leq 10°$, EGRET saw $I = 3 \times 10^{-8} ph, cm^{-2}, s^{-1}, sr^{-1}, MeV^{-1}$. The model above gives a cosmic ray weighted column of $Q = 3.28 \times 10^{-2} g, cm^{-2}$, and an emissivity $E = 1.14 \times 10^{-5} ph, s^{-1}, g^{-1}, MeV^{-1}$ which is appropriate to gas clouds of $\Sigma \leq 1g, cm^2$. For gas cloud column densities $\Sigma \geq 100g, cm^{-2}$ the emissivity $E = 5.5 \times 10^{-6} ph, s^{-1}, g^{-1} MeV^{-1}$ reducing the predicted 1 GeV flux below the observed value. Hence it is possible to 'hide' H_2 from gamma-ray sight. Paolis et al (2000) believe the smooth matching of the nearly flat galaxy rotation curves which they attribute to globular cluster dominated luminous matter close in and a baryonic dark halo further out arises naturally from a single galactic evolutionary model. The Jeans instability, $\lambda = (\frac{\pi kT}{\mu mG\rho})^{0.5}$, allows smaller structures as cooling progresses. UV from AGN formation may halt cooling below ~ 10 kpc for a time, imprinting the $\sim 10^6 M_\circ$ globular mass on the system. Further out, fragmentation continues to 10^{-5} pc scales.

3. Extra-Galactic Hidden Baryonic Matter

Extending the baryonic dark matter hunt beyond the Milky Way, Suto et al (1996) suggested a significant hot gas halo associated with the local group. Sidher, Sumner and Quenby (1998) using ROSAT, limited the electron density from here to 1/10 th the electron density of galactic halo. Their value of electron density, $\leq 4 \times 10^{-4} cm^{-3}$, is consistent with the typical total mass in the cluster gas at the bottom end of the bright X-ray cluster range (ie NGC 4636, data in Reiprich and Bohringer, 2002) The large scale size of the group halo, ~ 350 kpc, compared with the halo scale height of 12 kpc, may therefore allow the intra-group gas to dominate.

To establish the magnitude of the hidden baryonic component, we construct a 'known' mass budget for observed baryons, mainly based on Fukugita (2003), adopting h=0.7, unless explicitly stated as a variable.

For stars in galaxies, the SSDS for 5 colour bands established a global luminosity density $L_z = 3.9 \times 10^8 h L_\circ (Mpc)^{-3}$. A bright galaxy, luminosity function weighted, mass to light ratio $M/L_z \simeq 1.5$ is then found from the analysis of Kauffmann et al. (2003) using a population synthesis model depending on age, metallicity, star formation history and IMF. Finally taking an IMF, flattened at $0.3 M_\circ$, yields $\Omega_{star} = 0.0025 \pm 0.0008$.

HI 21-cm surveys of galaxies give the atomic galactic gas as $\Omega_{HI+HeI} = (6.2 \pm 1.0) \times 10^{-4}$, believed to be a secure value since the column in HI is usually high. The 'conventional' H_2 contribution based on the trace CO temperature integral is $\Omega_{H_2} = (1.6 \pm 0.6) \times 10^{-4}$, neglecting 'hidden' molecular clouds.

There are three independent for determining mass in the hot gas of the largest clusters. First is the determination of dynamical mass from galaxy motions. Typically this involves finding the velocity dispersion σ_8 rms within $8Mpc.h^{-1}$ spheres centred on the cluster. Next, keV X-ray emission assuming hydrostatic equilibrium can define the total mass distribution. More recently, weak gravitational lensing has provided total mass distrbutions. Significant agreement between lensing and X-ray mass distribution for very luminous galaxies has been shown by Allen et al (2004) who worked with Chandra data on RXJ1347.5-1145. The modelling took into account a temperature gradient from 6 to 16 keV and a sub-cluster shock in part of the emission field which accounted for about 5% of the total luminosity. Despite obvious merging which as warned by Quenby et al (1999) could reduce mass estimate due to a non-equilibrium cluster gas, the total estimated mass $M_{200} = 2.29^{+1.74}_{-0.82} \times 10^{15} M_\circ$ is consistent within the error bars of a weak lensing determination. However, an isothermal sphere model proved an unsatisfactory fit. The agreement about total mass seems unaffected by worries about a large 'hidden mass' associated with the mysterious 'soft excess' EUV component, clearly established by Lieu et al.(1996). This excess implies baryonic matter amounts above that deduced from the hot cluster medium. The alternative interpretation that Inverse Compton scatter by cosmic ray electrons in clusters (Sarazin and Lieu, 1998) produces the X-rays runs energy problems, but if proved true, could reduce cluster baryonic mass estimates.

An analysis based on an isothermal temperature model of 106 galaxy cluster observations based on ROSAT and ASCA yielded $\Omega_{cl} = 0.012^{+0.7}_{-0.6}$ for $M > 4.5 \times 10^{13} M_\circ$ within $\rho > 200\rho_{crit}$ (Reiprich and Bohringer, 2002). errors exclude the above caveats. Using the value of the fraction of total mass held in the cluster, X-ray emitting gas, $f_{cl,gas} = 0.14$, Allen et al (2004), $\Omega_{HII,cl} = 0.0017$ for massive clusters.

Arising from Lieu et al (1996) and Quenby et al.(1999), is it worth looking again at the results of Lieu et al. (2000) who found for Abell 2199 a three-phase intercluster medium? They deduced that the inter-cluster gas was a clumped cold, warm and hot medium with the warm medium corresponding to the EUV excess at a temperature $T < 100$ eV as compared with the hot gas $T \sim 4$ keV. A gas mass 43^{+13}_{-29} times that deduced for the hot medium would be required to power the EUV excess if thermal in origin. Is it possible to miss the warm component in the x-ray mass determinations since the hydrostatic equilibrium formula employed shows $M_{tot}(< r) \propto T_{gas}F(r)$? A typical cooling time for a $10^{-3}cm^{-3}$ plasma is 10^9 yr. To supply the required 10^{60} erg energy loss in the soft excess of a large cluster (Lieu 2004), note this author ascibes 10^{64} erg as available in the gravitational energy release of a merger of rich clusters. If a merger rate of $10^{-9}yr^{-1}$ is reasonable for significant sub-clusters, the hypothesis that cluster gas measurements can undershoot the baryonic mass by a factor of a few seems at least energetically plausible.

Although there is some X-ray evidence for warm plasma associated with groups or even perhaps field galaxies. the total contribution to the baryon budget is hard to quantify. Fukugita, Hogan and Peebles (1998) start with a ROSAT survey of groups with $1.2 \times 10^{13}h < M/M_\circ < 8.3 \times 10^{13}$, (Mulchaey et al,1996.) who found $f_{gr,gas} = 0.022h^{-1.5}$, much less than $f_{cl,gas}$ for massive clusters. They then take the mean of three independent estimates of the total gravitational mass. $\Omega_m = 0.12$ comes by extrapolating the Bahcall-Cen (1993), mass function down into the appropriate range. $\Omega_m = 0.14$ arises using a dynamical measure applied outside rich, large systems of galaxies (Bahcal et al 1995) Finally, $\Omega_m = 0.18$ by considering the variation of M/L in different morphological types of galaxies, scaling from great clusters. All figures are quoted at h=0.7. Then using the ROSAT value of $f_{gr,gas}$, the average is $\Omega_{HII,gr} = 0.003h^{-1.5}$

Clearly the nucleosynthesis consensus value of the total baryonic value

$\Omega_b h^2 = 0.021 \to 0.025$ (eg. O'Meara et al., 2001) is the start to establish the missing components. This determination which depends on the simple physical situation setting the deuterium abundance 3 minutes after the big bang is comparerd by Spergel et al, (2003),to WAMP data using the method of Bond and Efstathiou (1984). 372,000 years after the big bang the acoustic peaks seen in the CMB are determined by the adiabatic fluctuation spectrum and the expansion size. Thomson scatter of microwave photons by the electrons at that era in 1st CMB peak which arises from Doppler and gravity fluctuations modifies the oberved peak so that a high z electron density may be found. WAMP gives $\Omega_b h^2 = 0.0224$, in good agreement with nucleosynthesis consensus.

QSO lines by systems at low redshifts near galaxies provides significant evidence for a 'hidden' baryon reservoir. Tripp et al. (2000) report on 5 absorbers showing $OVI\lambda\lambda 1031.92, 1037.62$ doublet resonance line absorption over a z range and also HI Lyman series and other lines in several cases. Typically there are temperatures $T \leq 1.2 \times 10^5 K$ and an occurrence in z of $dN/dz \sim 50$, For solar abundance and from the deduced absorber column densities a limit $\Omega_{b,w} > 0.004 h_{75}^{-1}$ is found.

This contribution to Ω_b maybe a lower temperature variation of the X-ray emitting plasma in groups or the soft excess in clusters or another type of gas existing further out from galaxies than the X-ray/EUV detected gas with a significant total mass.

A useful check on the total of the baryon inventory established within 3 minutes of the big bang but now containing a significant missing component at $z = 0$, is possible at $z \sim 3$ where gas is even more important than stars. Neutral hydrogen at this epoch is mainly in high column density damped Lyman-α absorbers. The number of these absorbers increases with z (Storrie-Lombardi et al, 1996) who give $\Omega_{neutral} = 0.00013 \to 0.0007$. It is believed that this material is continually being converted to stars.

More mass resides in the plasma yielding the Lyman-α forest at $z \sim 3$. To find this plasma mass the detection by trace neutral hydrogen must be modelled based on cloud size, velocity and temperature measurements. Knowledge of the ionising flux is necessary and Zhang et al (1997) work a CDM model with a self-consistent ionising flux. They find $0.01, < \Omega_{HII} < 0.05$. Since this range encompasses the Nucleosynthesis/WAMP consensus, there is no evidence at this level of precision for further missing bayonic mass. The strong hint from from this 'forest' determination is that a copious supply of warm plasma may well be found at $z = 0$, probably on the edge or beyond of galaxies as traced by stars.

4. The Baryonic Mass Budget

Tackling the problem of making the total baryon inventory add up to to expected value, one approach of Fukugita (2003) has been to estimate the total associated with galaxies assuming the WAMP $\Omega_b/\Omega_m = 0.178$ is 'universal'. Then he took the gravity lensing shear value $M/L_r = 170h$ plus L_r for a SDSS sample to give $\Omega_{m,g} = 0.14$ and hence $\Omega_{b,g} = 0.025$ for galaxies in total. There is thus missing mass of the same order as that accounted for which could be in the 'warm' surrounds.

Another approach is to combine all the above estimate at h=0.7, $z \sim 0$ and subtract from the Nucleosynthesis/WAMP value.

$\Omega_b = 0.044$ is the total baryonic mass. This is made up of;

$\Omega_{star} = 0.0025$ due to stars

$\Omega_{HI+HeI+H2} = 0.0008$ due to galaxy atomic and molecular H and to He.

$\Omega_{HII,cl} = 0.0017$ due to hot, $\sim keV$ cluster plasma

$\Omega_{HII,gr} = 0.005$ due to warm, ≥ 100 eV, plasma in groups

$\Omega_{b,w} = 0.004$ due to warm, ≤ 10 eV, plasma near 'field' galaxies

$\Omega_{Known} = 0.014$ is the sum of 'known' baryonic components

$\Omega_{MISSING} = 0.030$ is what's left over.

The greatest contribution to the known mass budget arises from the group and field galaxy $10 \to 100$ keV warm plasmas on the outskirts of the visible objects. Clearly these figures are poorly known and given the apparently adequate supply of plasma at $z = 3$, could account for the majority of the hidden baryonic mass. The speculation that there is a substantial excess EUV component in clusters, although not highly significant in the total budget, nevertheless supports this suggestion.

Cosmological simulations by Cen and Ostriker (1999) and Dave et al. (1999) predict that a substantial fraction of the universe exists in a shock-heated phase, $10^5 \to 10^7$ K, at low z. In further simulations, Ostriker et al (2003) found 20% of mass residing in voids, regions where galaxies do not shine so as to be detectable.

We conclude that improved knowledge of the Milky Way baryonic component depends on better understanding of the Cusp/Bar dynamics near the centre and of the amount of cold, dense molecular clouds far out. Most Baryons may be 'missing', probably in a warm or cool gas just outside galaxies or in clumps which do not shine.

References

S. W. Allen, et al. to appear MNRAS [astro-ph/0111368] 2004
N. A. Bahcall and R. Cen Ap.J. **407** L49 (1993)
J. J. Binney and N. W. Evans to appear, MNRAS (2004) [astro-ph/0108502]
J. R. Bond and G. Efstathiou Ap.J. **285** L45 (1984)
R. Cen and J. R. Ostriker ApJ. **514** p 1 (1999)
R. Dave, L. Hernquist, N. Katz and D. H. Weinberg ApJ. **511** p 521 (1999)
V. P. Debattista and J. A. Sellwood ApJ. **543** p704 (2000)
V. P. Debattista, O. Gerhard and M. N. Sevenster MNRAS **334** p355 (2002)
M. Fukugita [astro-ph/0312517] (2003)
M Fukugita, C. J. Hogan and P. J. E. Peebles ApJ. **503** p 518 (1998)
G. Kauffmann et al. MNRAS **341** p 33 (2003)
F. J.Kerr and D. Lynden-Bell, MNRAS **221** p 1023 (1986)
A. Klypin, H. S. Zhao and R. S. Sommerville ApJ. **573** p 597 (2002)
K. Kuijen and G. Gilmore, Ap. J. **367**, L9 (1991)
R. Lieu et al. ApJ. **458** L5 (1996)
R. Lieu, M. Bonamente and J p. D. Mittaz [astro-ph/0001127] (2000)
R. Lieu, private communication (2004)
M. R. Merrifield, to appear, ASP Conf.Series, [astro-ph/0310497] (2003)
J. S. Mulchaey et al. ApJ. **456** p 80 (1996)
J. F. Navarro, C. S. Frenk and S. D. M.White ApJ. **462** p563 (1996)
M. Ohishi, M. Mori and M. Walker [astro-ph/0403138]
R. P. Ollings and M. R. Merrifield, MNRAS **326**, p164 (2001)
J. M. O'Meara et al. ApJ.**552** p 718 (2001)
J. P. Ostriker, et al. Accepted ApJ. [astro-ph/0305203] (2003)
F. De Paolis,et al. [astro-ph/9906063] (2000)
D. Pfenniger, F. Combes and L. Martinet, A. and A. **285** p79 (1994)
J. J. Quenby, T. J. Sumner, et al 'The Identification Of Dark Matter',(World Scientific) p 137 (1999)
T. H. Reiprich and H. Bohringer ApJ. **567** p716 (2002).
C. L. Sarazin and R. Lieu ApJ. **494** L177 (1998)
J. A. Sellwood ApJ.**587** p 638 (2003)
S. D. Sidher, T. J. Sumner and J. J. Quenby, A. and A. **344** , p333 (1999)
D. N. Spergel et al. ApJ. Suppl. Ser. **148** p 157 (2003)
L. J. Storrie-Lombardi, et al. R. G. MNRAS **283** L79 (1996)
Y. Suto, K. Makishima, Y. Ishisaki et al. ApJ **461** L33 (1996)
T. M. Tripp, B. D. Savage and E. B. Jenkins ApJ.**534** L1 (2000)
M. D. Weinberg and N. Katz ApJ. **580** p627 (2002)
W. R. Webber, Lee, M. and Gupta, M ApJ. **390** p 96 (1992)
Y. Zhang, P. Anninos, M. L. Norman and A. Meilsin ApJ **485** p496 (1997)

EVIDENCE FOR DARK MATTER IN THE FORM OF COMPACT BODIES

M. R. S. HAWKINS

Institute for Astronomy,
University of Edinburgh,
Royal Observatory,
Blackford Hill,
Edinburgh EH9 3HJ,
Scotland, UK
E-mail: mrsh@roe.ac.uk

In this paper we review the evidence for dark matter in the form of compact bodies. This comes largely from the analysis of quasar light curves, where the failure to observe time dilation, the statistical symmetry of the variations, the near achromatic changes in brightness and the cusp like nature of many of the variations all point to a microlensing origin. In addition, where microlensing is observed in gravitationally lensed quasar sytems it is not clear that the variations can be accounted for by the microlensing of stars.

1. Introduction

The possibility that dark matter in the form of compact bodies has been detected from its microlensing effect in quasar light curves (Hawkins 1993,1996) has remained an intriguing possibility. The case has relied on features of quasar light curves which are hard to explain on the basis of intrinsic variation, and the known microlensing effects observed in multiply lensed quasar systems. On the other hand, the case has been weakened by constarints on the halo population of compact bodies derived by the MACHO and OGLE projects.

In the years since 1993, there has been a steady accumulation of evidence favouring the idea that most quasars are being microlensed at a level compatible with the dark matter being entirely composed of compact bodies. The problem of classifying the variations as due to microlensing is made harder by the fact that quasars undoubtedly vary intrinsically at least at some level, and these two modes of variation must be disentangled.

Variations resulting from microlensing have a number of predicable characteristics by which they may be identified. Intrinsic variations on the other hand have until recently been almost completely unconstrained, and thus very hard to pick out from microlensing effects. Over recent years however, there have been a number of attempts to model AGN variations by means of various forms of disc instability. This has at last started to constrain the nature of intrinsic variability in quasars, and thus facilitate the identification of variability due to microlensing.

There is one situation where quasar variation due to microlensing has been established beyond reasonable doubt. This occurs when an intervening galaxy has split the quasar into two or more images. Monitoring of the brightness changes in the system shows evidence for intrinsic variation of the quasar which is seen in all components, with some time lag due to differences in light travel time. In addition, all well observed systems show brightness changes in single images which are not seen in the other components. This is generally accepted as microlensing. There is however a view that these variations can be accounted for by the microlensing effect of stars in the lensing galaxy. There are strong arguments that this cannot be the case, which are beyond the scope of the present paper, but in this case the microlensing can only be due to non-stellar compact bodies in the halo of the lensing galaxy, or a population of compact bodies more generally distributed along the line of sight. Either way, the implication is for dark matter in the form of compact bodies.

Once it can be established that most quasars are being microlensed, the question arises as to the nature of the microlensing bodies. Their mass can be estimated from the timescale of variation, which so far has proved difficult to measure, but the indications are that the light curves imply a mass of around $0.1 M_\odot$. Given that if such bodies make up the missing mass, nucleo-synthesis constraints imply that they must be non-baryonic, the most plausible candidates would appear to be primordial black holes created in the early Universe during the quark/hadron transition.

In the rest of this paper we look at some of the more recent arguments supporting the idea that most quasars are being microlensed.

2. Evidence for microlensing

2.1. Lack of time dilation

Time dilation is a fundamental property of an expanding universe, and is expected to increase rest-frame timescales by a factor of $(1 + z)$ in quasar light curves. As has been mentioned above, timescale is not an easy prop-

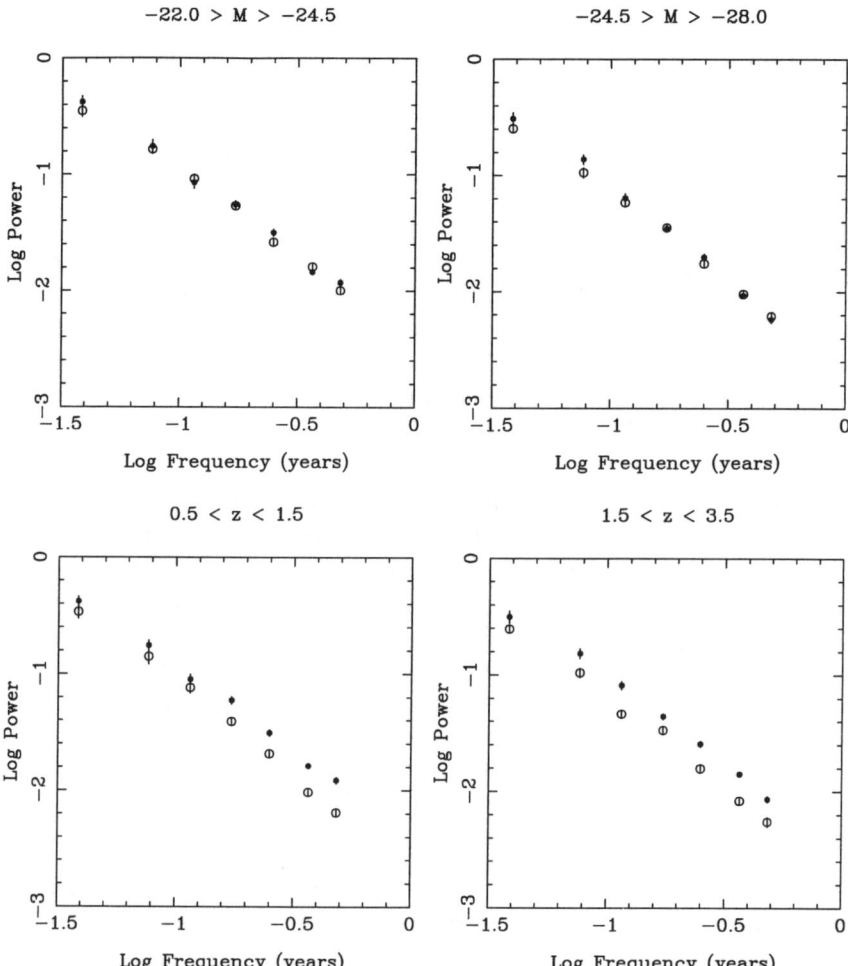

Figure 1. Fourier power spectra for subsamples of quasar light curves in the observer's frame. The top two panels show data for low- and high-luminosity quasars. The filled and open circles are power spectra for low- and high-redshift objects respectively. The bottom two panels show data for low and high-redshift quasars. The filled and open circles are power spectra for low- and high-luminosity objects respectively.

erty to define or measure, but nonetheless one can characterise changes in timescale by movements of the Fourier power spectrum of variations for samples of light curves. This has been done in a recent paper (Hawkins 2001) with the surprising result that time dilation was not observed. The result was based on the observation that in high and low redshift samples of quasars, no movement of the Fourier power spectrum was observed towards longer timescales for higher redshift objects. Fig. 1 shows an updated version of the original figure, where the top two panels show data for high and low luminosity quasars. Within each panel Fourier power spectra for high and low redshift samples of quasars are shown, and in each case they lie exactly coincident with each other. The bottom two panels show a similar comparison for high and low luminosity quasars. In this case it will be seen that in both reshift bins there is more power in the variation of low luminosity quasars.

There appears to be no convincing explanation for this absence of time dilation within the framework of modern cosmology except that the variations are due to microlensing. In this case the variations are mostly caused by lenses at low redshift, and little time dilation effect is to be expected. It is hard to get round this argument as it covers all intrinsic variations regardless of origin, and thus circumvents the difficulty of predicting the nature of variations due to accretion disc instability, starburst or other nuclear activity.

2.2. Colour changes in light curves

It is a well known property of gravitational lensing that it is achromatic in the sense that radiation of all wavelengths from a given source follows the same path. It might therefore be supposed that microlensing of quasars would produce achromatic variation. This is however only the case if the radiation comes from a source small compared with the size of the emitting region. If there is a colour gradient across a large source there will be differences between the blue and red passband light curves. Most astrophysical processes on the other hand produce strong colour changes as they vary, in the sense that they become bluer as they brighten.

With improvements in the quality of light curves from monitoring programmes in more than one passband, it has now become possible to investigate colour changes with a view to distinguishing between different models of variability (Hawkins 2003). Fig. 2 shows some typical quasar light curves in blue and red passbands. It will be seen that although there are no strong

163

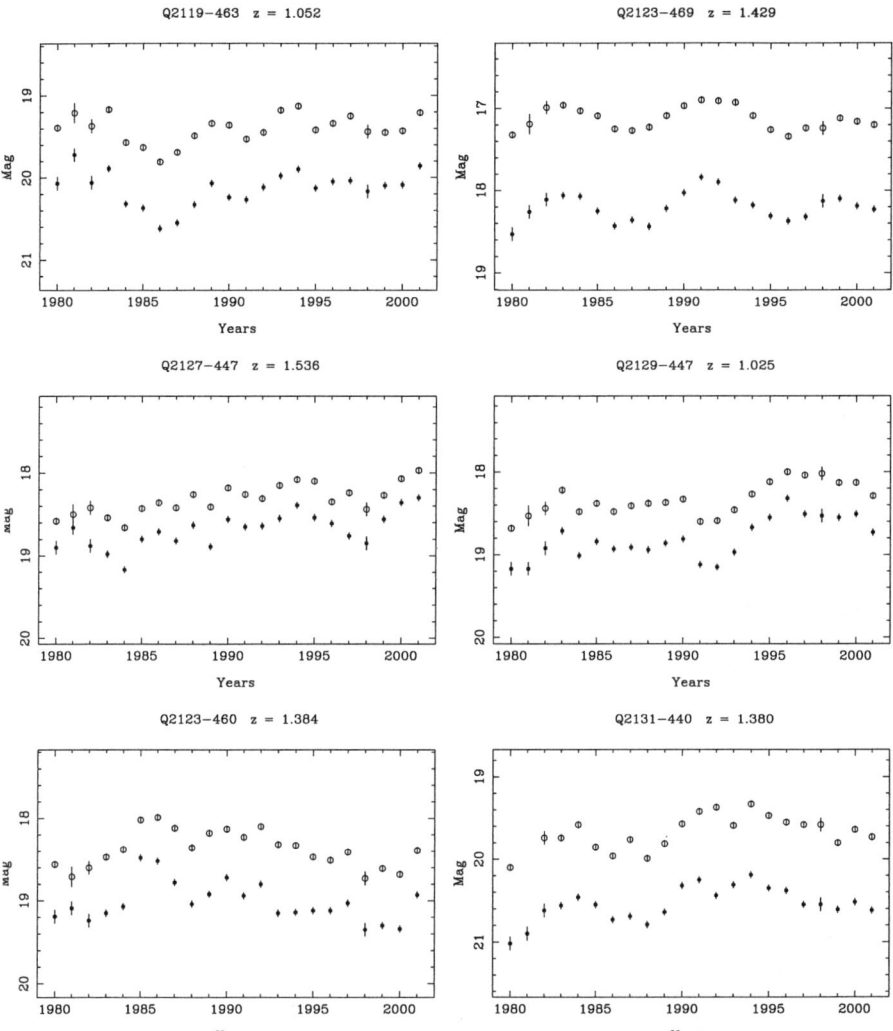

Figure 2. Light curves in B_J (filled circles) and R (open circles) for a typical selection of quasars. The variation is largely achromatic.

colour changes with change in brightness as seen in most intrinsic processes, there is a tendency for sharper features in the blue to be smoothed out in a symmetrical way in the red. This is very characteristic of what would be expected for the microlensing of an accretion disc with a colour gradient, in the sense of a blue nucleus surrounded by a larger redder disc.

2.3. Statistical symmetry of variation

A property of variations caused by microlensing is that the resulting light curves are statistically symmetric, in the sense that it is not possible to tell which way time is running by any statistical analysis of the variations. This is not the case with variations in brightnes caused by most astrophysical processes. For example, accretion disc instabilities tend to show a gradual rise in brightness followed by a sharper fall. Variations attributed to bursts of supernovae would show the opposite effect.

Tests for statistical symmetry on observed light curves imply that there is little deviation from symmetry (Hawkins 1996,2002), and put tight constraints on models of quasar variability, clearly favouring the microlensing interpretation.

2.4. Slope of the structure function

It has recently become possible to compare the observed spectrum of variations for quasars with predictions of theoretical models. This is due to improvements both in the quality of the available data, and to the publication of quantitative predictions from model simulations suitable for comparing with the data (Kawaguchi et al. 1998). Comparisons of the slopes of structure functions for simulated and observed light curves has proved to be a sensitive way of distinguishing models (Hawkins 2002), with microlensing clearly the preferred model.

References

1. M. R. S. Hawkins, *Nature* **366**, 242 (1993).
2. M. R. S. Hawkins, *MNRAS* **278**, 787 (1996).
3. M. R. S. Hawkins, *ApJ* **553**, L97 (2001).
4. M. R. S. Hawkins, *MNRAS* **329**, 76 (2002).
5. M. R. S. Hawkins, *MNRAS* **344**, 492 (2003).
6. T. Kawaguchi, S. Mineshige, M. Umemura and E. L. Turner, *ApJ* **504**, 671 (1998).

IDENTIFICATION OF BLACK-HOLE DARK MATTER

ASANTHA COORAY AND NAOKI SETO

Theoretical Astrophysics
Division of Physics, Mathematics and Astronomy
California Institute of Technology
Pasadena, CA 91125, USA
E-mail: asante@caltech.edu

The high sensitivity of upcoming space-based gravitational wave detectors suggests the possibility that if halo dark matter were composed of primordial black holes (PBHs) with mass between 10^{16} g and 10^{20} g, the gravitational interaction with detector test masses will lead to a detectable pulse-like signal during the fly-by. For an improved version of the Laser Interferometer Space Antenna with a reduced acceleration noise at the low-end of its frequency spectrum, we find an event rate, with signal-to-noise ratios greater than 5, of $\sim$ a few per decade involving black holes of mass $\sim 10^{17}$ g. The detection rate improves significantly for second generation space based interferometers that are currently envisioned, though these events must be distinguished from those involving perturbations due to near-Earth asteroids. While the presence of primordial black holes below a mass of $\sim 10^{16}$ g is now constrained based on the radiation released during their evaporation, the gravitational wave detectors will extend the study of PBHs to a several orders of magnitude higher masses.

1. Introduction

Based on observed rotational velocity measurements of the Milky Way disk, the presence of a substantial dark matter component in the Milky Way halo is now well established [1]. With no unique guidance as to the nature of dark matter, a large number of candidates based on both astronomical arguments, such as baryonic dark matter involving substellar mass remnants [2], and particle physics expectations [3] are now routinely considered to explain the missing mass. An interesting possibility is that the halo dark matter is composed of primordial black holes (PBHs) [4]. While PBHs are expected over a rather wide range of masses [5], the population below $\sim 5 \times 10^{14}$ g is expected to have evaporated by the Hawking radiation over the age of the universe [6]. From the high mass end, constraints come from dynamical arguments such as the potential disruption of galactic clusters and similar bound structures [7] and galactic microlensing, which limits primordial black hole masses to be below roughly few tenths solar masses [8]. In combination, at least over a 10^{16} decade in primordial black hole mass, between evaporating holes with mass below 10^{17} g and ~ 0.1 $M_\odot$, remains yet to be studied.

Though the galactic gamma-ray background and its weak anisotropy suggests an evaporating PBH density of roughly 10^{10} pc^{-3}, PBHs can explain the total halo dark matter if the PBH mass spectrum enhances the abundance above 10^{17} g [9]. These PBHs are not expected to interact or be captured by other massive bodies such as the Sun; they can be considered as another weakly interacting massive particle (WIMPs). They can be detected only through the gravitational interaction. But, as their masses are much larger than standard WIMPs predicted by particle physics, the expected flux of PBHs would be very small and we need detectors with a large effective area for their search. Incidently, these PBHs could in fact be detected directly with planned space-based gravitational wave detectors that have large cross sections to moving massive bodies [10]. These missions include the Laser Interferometer Space Antenna (LISA) [11]. The direct detection simply involves the gravitational interaction between the fly-by PBHs and detector test masses such that with high sensitivity even a small gravitational perturbation can eventually be seen above the detector noise.

2. Calculation

The interaction of a PBH and a gravitational-wave observatory test mass can be considered either as a direct interaction, when the fly-by separation is smaller than the arm length, or as a tidal interaction, in the case where arm length is smaller than the fly-by distance. We first consider the former situation and take the scenario with a PBH mass $M_\bullet$ passing by one of the test masses of the interferometer with a relative velocity V with the closest approach distance R. The time dependent acceleration $a(t)$ of the spacecraft test mass towards the direction of the closest approach has a single pulse-like structure given by

$$a(t) = \frac{GM_\bullet R}{[R^2 + (Vt)^2]^{3/2}}, \qquad (1)$$

where we have set the origin of the time coordinate, $t = 0$, to be at the closest approach.

The Fourier component $a(f)$ of eq. (1) is given in terms of the 1st order modified Bessel function K_1 as

$$a(f) = \int_{-\infty}^{\infty} dt e^{2\pi i f t} a(t) = 2\frac{GM_\bullet}{RV} K_1[2\pi f R/V]. \qquad (2)$$

For the direct detection, the signal-to-noise ratio (SNR) of the pulse $a(t)$ is formally written as $SNR^2 = 2 \int_0^\infty a(f)^2/a_n(f)^2 df$ with the noise spectrum $a_n(f)$ of the detector. In the case of space-based interferometers, the noise has two main contributions involving the proof-mass noise and the optical path noise with the former dominating the low frequency regime. As we will soon discuss, the typical signal $a(f)$ due to the fly-by PBH of mass around $M_\bullet \gtrsim 10^{17}$ g has support in the SNR integral when $f \lesssim 10^{-4}$ Hz. This is in the the low frequency regime of all the proposed future interferometers, including the LISA mission. The proof-mass noise of LISA is estimated to be constant down to $\sim 10^{-4}$ Hz; It might become a factor of two larger at 3×10^{-5} Hz and more at lower frequencies [11].

After some straight forward algebra, we obtain the SNR as

$$SNR^2 = \frac{3\pi}{32} \frac{(GM_\bullet)^2}{VR^3 a^2} \qquad (R < L). \qquad (3)$$

When $R > L$, tidal deformation of the interferometer is a measurable effect, but this involves an additional suppression factor $\sim (L/R)$ in eq. (1). In this limit, we find the SNR to be

$$SNR^2 = \frac{3\pi}{32} \frac{(GM_\bullet L)^2}{VR^5 a^2} \qquad (R > L). \qquad (4)$$

This is the relevant expression for LISA and BBO/DECIGO in the interesting mass range $M_\bullet \gtrsim 10^{17}$g, though, at the low mass end (10^{16} g), the former expression applies for LISA.

Using eq. (4) we can now estimate the maximum length of the closest approach $R_{\max}$ for a given SNR threshold

$$R_{\max} = 5.3 \times 10^{11} \left(\frac{M_\bullet}{10^{17}\text{g}}\right)^{2/5} \left(\frac{V}{350\text{km/s}}\right)^{-1/5}$$
$$\times \left(\frac{SNR}{5}\right)^{-2/5} \left(\frac{a}{a_0}\right)^{-2/5} \left(\frac{L}{L_0}\right)^{2/5} \text{cm} \qquad (5)$$

where $a_0 = 3 \times 10^{-15}\text{m/s}^2\sqrt{\text{Hz}}$ and $L_0 = 5 \times 10^{11}\text{cm}$ are the reference fiducial parameters for LISA in its current design [11]. We have taken the typical velocity dispersion $V = 220\sqrt{5/2} = 350$ km/s of the halo dark matter particles relative to the solar system using the Galactic rotation velocity of 220 km/s and the Galactic radius to the Solar system of $r_g \sim 8$

kpc following Ref. [7]. The gravitational perturbation involves a pulse like signal in the data streams with a characteristic frequency given by

$$V/R_{\max} \sim 6.4 \times 10^{-5} \left(\frac{M_\bullet}{10^{17}\text{g}}\right)^{-2/5} \left(\frac{V}{350\text{km/s}}\right)^{6/5}$$
$$\times \left(\frac{SNR}{5}\right)^{2/5} \left(\frac{a}{a_0}\right)^{2/5} \left(\frac{L}{L_0}\right)^{-2/5} \text{Hz} \quad (6)$$

When we assume that the halo dark matter with density $\rho_{DM} \sim 0.011 M_\odot/\text{pc}^3$ [13] around the Sun is made with PBHs of mass $M_\bullet$ the fly-by event rate, $\dot{\eta}$, above a certain SNR is

$$\dot{\eta} = 0.01 \left(\frac{M_\bullet}{10^{17}\text{g}}\right)^{-1/5} \left(\frac{SNR}{5}\right)^{-4/5} \left(\frac{V}{350\text{km/s}}\right)^{3/5}$$
$$\times \left(\frac{\rho_{DM}}{0.011 M_\odot \text{pc}^{-3}}\right) \left(\frac{a}{a_0}\right)^{-4/5} \left(\frac{L}{L_0}\right)^{4/5} \left(\frac{N}{1}\right) \text{yr}^{-1},$$
(7)

where N represents the effective number of interferometers with sufficient relative distances. While LISA involves a single set of interferometers, some planned missions, e.g. BBO, plans to use multiple interferometer arrays with large separation to improve the localization of binaries sources such that $N \geq 2$. For the reference parameters of LISA, assuming a useful SNR threshold of 5, the detection rate is ~ 0.1 per ten years as shown in eq. (7). The combination (aL^{-1}) that appears in all of the above expressions determine the low frequency sensitivity of the interferometers to the gravitational wave amplitude h. Therefore, once the threshold distances $R_{\max}$ are confirmed to be larger than the arm-length L, we can easily compare the event rate for various interferometers by studying their noise curves at the low frequency regime.

In figure 1, we show the potential detectability of fly-by events involving PBHs with mass $M_\bullet$ using various planned space-based interferometers. Instead of the event rate, our results are presented in terms of the local halo density of PBHs required for a detection, with SNR greater than 5, of one event per decade. For reference, to put these detections in the context of the galactic dark matter density, we also show current observational constraints for the PBH density from the γ-ray background ($M_\bullet \lesssim 10^{17}$ g) [9] and the expected density of dark matter based on a Galactic model ($M_\bullet \gtrsim 10^{17}$g) where the density is now determined to be (0.011 ± 0.005) $M_\odot$ pc^{-3} [13]. Here, we also denote three basic numbers that characterize an interferometer

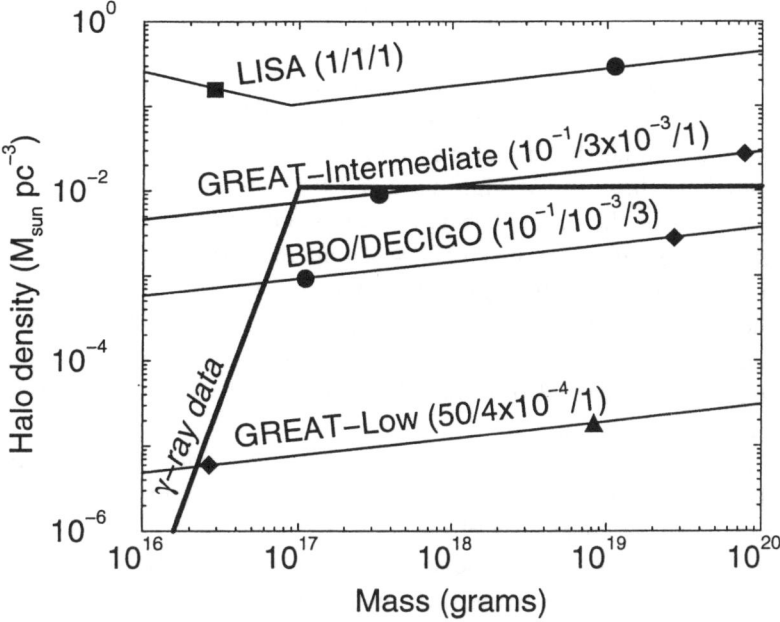

Figure 1. The detection thresholds for PBH fly-by with various upcoming space-based gravitational wave observatories (in solid lines from top to bottom: LISA, GREAT intermediate-frequency-mission, BBO/DECIGO, and GREAT very-low-frequency-mission). The thick solid line shows the expected density of halo dark matter in the form of black holes; at $M_\bullet \simeq 10^{15}$ g, the density is constrained by γ-ray background observations such that the particle density is below $\sim 10^{10}$ pc^{-3}, while no similar constraints exist above $M_\bullet \gtrsim 10^{17}$ g and we take the whole halo to be formed of black holes with the mass given in the horizontal axis (with a density 0.011 $M_\odot$ pc^{-3}). The detection limits assume an event rate of a single detection per decade; if no events are detected, these curves would roughly correspond to the constraint one can put on the black hole contribution to the halo dark matter density. In each of the instruments considered, we label the following three parameters: (arm-length/proof-mass noise/total number of detector arrays) relative to those reference values of LISA (arm length: 5×10^6 km, proof-mass noise: 3×10^{-15} m/s$^2/\sqrt{Hz}$). The symbols on each of the curves represent the average frequency of the gravitational perturbation produced on the interferometer by a black hole of corresponding mass with triangle, diamond, circle and square representing frequencies of $10^{-7}, 10^{-6}, 10^{-5}$ and 10^{-4} Hz, respectively. In the case of LISA, the threshold distance R equals to the arm length L at $M_\bullet = 0.87 \times 10^{17}$ g where the slope of the curve changes.

(arm-length/proof-mass noise/number of detector arrays) normalized by the reference values for LISA. The GREAT-intermediate mission [12] has a marginal possibility for a PBH detection at $M_\bullet \sim 10^{17}$ g while The GREAT-very low frequency mission with parameters $(50/4 \times 10^{-4}/1)$ would detect them even if less than a few percent of the halo dark matter is made with PBHs with mass between 10^{17} g and 10^{20} g.

There is one source of confusion involving the passage of near-Earth asteroids close to detectors (see, section 4.2.5 of [11]). Given the expected flux of minor bodies in the solar system [14], the asteroid perturbations are dominated by those at the high mass end between 10^{13} g to 10^{15} g, with an event rate of the order 0.05 yr^{-1}. Unfortunately, though the relative velocity of asteroid events are small (~ 30 km sec^{-1}) these events could have similar frequencies (as PBHs) due to differences in the maximum distance to which they can be detected. On the positive side, orbital parameters of roughly 10% of such near-Earth asteroids are already known while this fraction is soon expected to grow substantially with dedicated near-Earth asteroid search programs.

While with gravitational wave detectors the presence of PBHs can be established, it is not easy to determine masses of individual events from a single pulse signal that is characterized mainly by two numbers: the amplitude, M/R^3, and the time scale, R/V, made from three variales M, R and V. This is due to the fact that there is an unknown associated with the fly-by distance or the PBH trajectory. If multiple detectors are perturbed, one can establish the trajectory and then used that information to determine mass.

References

1. S. M Faber and J. S. Gallagher, Ann. Rev. Astron. Astrophys. **17**, 135 (1979);
2. V. Trimble, Ann. Rev. Astron. Astrophys. **25**, 425 (1987);
3. G. Jungman, M. Kamionkowski and K. Griest, Phys. Rep. **267**, 195 (1996)
4. S. W. Hawking, Mon. Not. Roy. Astron. Soc. **152**, 75 (1971)
5. B. J. Carr, Astrophys. J. **201**, 1 (1975)
6. S. W. Hawking, Nature **248**, 30 (1974)
7. B. J. Carr & M. Sakellariadou, Astrophys. J. **516**, 195 (1999)
8. C. Alcock et al. Astrophys. J. **550**, 172 (2001); K. Griest, Astrophys. J. **366**, 412 (1991)
9. D. B. Cline, Astrophys. J. Lett. **501**, 1 (1998)
10. N. Seto and A. Cooray, Phys. Rev. D., **70**, 063512 (2004); A. W. Adams and J. S. Bloom, arXiv:astro-ph/0405266.
11. K. V. Danzmann et al., *LISA Pre-Phase A Report*, Second edition, July 1998; http://lisa.jpl.nasa.gov
12. N. J. Cornish, D. N. Spergel and C. L. Bennett, astro-ph/0202001
13. R. P. Olling and M. R. Merrifield, Mon. Not. Roy. Astron. Soc. **326**, 164 (2001).
14. E. M. Shoemaker, Ann. Rev. Earth Planet. Sci. **11**, 461 (1983).

NEWS FROM THE DARK MASS AT THE CENTER OF THE MILKY WAY

A. ECKART, R. SCHÖDEL, C. STRAUBMEIER,
N. MOUAWAD, AND S. PFALZNER

I. Physikalisches Institut
University of Cologne
Zülpicher Str. 77
50937 Cologne, Germany
E-mail: eckart@ph1.uni-koeln.de

The Galactic Center measurements of stellar orbits and strongly variable NIR and X-ray emission from Sagittarius A* at the center of the Milky Way have provided the strongest evidence so far that the dark mass concentrations seen in many galactic nuclei are most likely super massive black holes. As proven by the Keplerian orbits of several of the high velocity stars within the central arcsecond at the position of the compact radio source SgrA* the Galactic Center harbors a $\sim 3.5 \times 10^6 M_\odot$ massive black hole. Simultaneous NIR/X-ray observations of SgrA* in 2003/2004 have revealed first insights into the emission mechanisms of both the powerful near-infrared flares and the 'quiescent' emission from within a few 10 - 100 Schwarzschild radii of the super-massive black hole at the center of the Milky Way. The central source shows synchronous NIR/X-ray flare variations and indications of quasi-periodicity within the NIR flares. In addition the detection of a stellar cusp give evidence for the presence of a spherical potential which is neither Keplerian nor harmonic. In such a potential orbits will precess resulting in rosetta shaped trajectories on the sky and the assumption of non-Keplerian orbits is a more physical approach. It is also the only approach through which cusp mass information can be obtained via stellar dynamics of the cusp members. First results of modeling such a system are now available as well.

1. Introduction

Over the past 10 years, the investigation of the dynamics of stars has provided compelling evidence for the existence of a massive black hole (MBH) at the center of the Milky Way (Eckart & Genzel 1996, Genzel et al. 1997, 2000, Ghez et al. 1998, 2000, 2003a, 2003b, Eckart et al. 2002a, 2002b, Schödel et al. 2002, 2003). The orbit of the high velocity star S2 tells us that a BH mass of $M = 3.6 \times 10^6$ $M_\odot$ is confined within a radius of $R \leq 0.00055$ $pc = 0.655$ ld. However, Sgr A* is remarkably faint in all wavebands other than the radio region. This challenges current theories of matter accretion and radiation surrounding black holes. It is unclear whether the weak emission (10^{-8} of the Eddington rate) is due to a low accretion rate, inefficient angular momentum transport, low radiation efficiency, or a combination of these. Current physical models that explain the emission from SgrA* are based on radiatively inefficient accretion flow models (e.g. RIAFs: Yuan et al. 2004, including ADAFs: Narayan et al. 1995, CDAFs: Ball et al. 2000, ADIOSs: Blandford & Begelman 1999), jet models (Markoff et al. 2001), and Bondi-Hoyle models (Melia & Falke 2001).

The central BH is located in a dense stellar cluster. The Galactic Center (GC) stellar cluster shows some intriguing characteristics: it is extremely dense, with an unusual observed stellar population consisting mainly (80% of all K$\leq$14 stars; Ott et al. 1999) of late-type red giants, many of which are suspected to lie on the asymptotic giant branch (AGB), as well as young massive stars with energetic winds (e.g. Krabbe et al. 1995, Najarro et al. 1997, Horrobin et al. 2004). Spectra of AGB stars show strong 2.3 μm CO bandhead absorption and the massive, hot and windy stars ("He-stars") exhibit He/H emission and strongly interact with the local ISM. The emission line stars appear to dominate the central few arcseconds, where the bright IRS 16 cluster is located. These stars are generally classified as Ofpe/WN9, although some of them might be luminous blue variables (LBV) and a few show characteristics of Wolf-Rayet stars. An additional, less numerous component of the Galactic Center stellar cluster consists of luminous, extended objects with steep, red and featureless (K-band-) spectra and a strong infrared excess. They are likely bow-shock sources with the implication that they are linked to luminous, windy WR or He-stars (Tanner et al. 2002, 2003, Rigaut et al. 2003, Eckart et al. 2004a, Moultaka et al. 2004). There is a small (about 1 arcsend radius) stellar cusp around SgrA* (Genzel et al. 2003). An analysis involving non-Keplerian orbits in a realistic potential (compact BH mass and an extended cusp mass) indicates that the cusp may contain 8000 to not more than a few 10^5 solar masses depending on the mass to light ratio of the dominant stellar constituents (Mouawad et al. 2003, 2004).

2. Variable NIR/X-ray Emission from SgrA*

Additional strong evidence for a massive black hole at the position of Sgr A* came from the observation of interim-quiescent (or IQ) and flare activity from that position both in the X-ray and recently in the near-infrared wavelength domain (Baganoff et al. 2001, 2003, Eckart et al. 2003, 2004b, Porquet et al. 2003, Goldwurm et al. 2003, Genzel et al. 2003, Ghez et al. 2004). Especially the discovery (Genzel et al. 2003b) of powerful NIR flares from SgrA* has now opened the possibility for an improved study of the emission mechanisms. The IR flares occurred at the remarkable rate of 4($\pm$2) times a day, at least twice the rate of X-ray flares detected by Chandra and XMM-Newton between 2000 and 2002.

Eckart et al. (2004b) recently reported on the first successful simultaneous NIR/X-ray campaign using NACO and Chandra as well as quasi-simultaneous mm-data from BIMA. Coincident with the peak of the $\sim 6 \times 10^{33}$ erg/s X-ray flare a fading NIR flare of Sgr A* with >2 times the interim-quiescent flux was detected. The event implies that the NIR/X-ray flare emission was coupled with a time lag not larger than 15 min and probably originated from the same ensemble of electrons. Compared to 8 h before the flare a 10% increased mm-flux density was measured about 8 h after the event.

A remarkable property discovered in two of the brightest K-band flares is a quasi-periodic substructure with a period of 17 minutes (Genzel et al. 2003). If this periodicity is a fundamental property of all flares, it most likely arises from the modulation of gas emission orbiting just outside the event horizon. In that case, the inevitable conclusion is that the Galactic center black hole has at least half of the maximum (Kerr) spin. The X-ray flares have similar durations as the IR-flares and some also do show minute-scale substructure. The reanalysis of the two most powerful ones in the framework of disk modes indicates high spin parameters as well (Aschenbach et al. 2003, 3004). NIR and X-ray time studies will be a powerful tool for exploring the physics and space time structure in the strong gravity regime around a super-massive black hole. The observational data obatined with NACO and Chandra in July 2005 is very much consistent with the previously obtained results.

3. Are there Alternatives to the Black Hole Scenario?

In order to explain the extreme mass concentration at the center of the Milky Way two 'dark particle matter' models have been under discussion as an alternative to a supermassive black hole. These are the so called 'fermion ball' and the 'boson star' scenarios.

The fermion ball as an attempt to explain large compact nuclear masses observed at the centers of galaxies was introduced by (Viollier et al. 1992). In this model the central objects are stabilized by the degeneracy pressure of the corresponding fermion candidates, e.g. neutrinos. The self-gravity of a ball of degenerate fermions can be balanced by the degeneracy pressure of the fermions due to the Pauli principle. The maximum mass of a degenerate fermion ball is given by the Oppenheimer-Volkoff limit M_{OV}. For a given fermion mass m all objects heavier than M_{OV} must be black holes.

For the GC we find from the Oppenheimer-Volkoff limit a maximum mass of a yet unknown, speculative fermion between 351 keV and 417 keV and a minimum fermion mass ranging between 48 keV and 57 keV (depending on the degeneracy g assumed to be g=2 or g=4). The most massive central dark object currently known is located at the center of M87, with a mass of $> 3 \times 10^9$ $M_\odot$. The Oppenheimer-Volkoff limit would allow a maximum neutrino mass of 14 keV in that case. Hence one can exclude the possibility that all compact dark objects at the centers of galaxies can be explained by a neutrino ball model with a single neutrino mass.

The only dark particle matter explanation that cannot be ruled out by the present data is a ball of mini-bosons (Maoz 1998), since such a ball could form a very compact configuration that is difficult to distinguish from a black hole. It would, however, be hard to understand how the bosons managed to cool sufficiently in order to settle down into such a small volume, and did not form a black hole during that process (Maoz 1998). Large boson stars masses require a weak (speculative) repulsive force between bosons (Kaup et al. 1968, Ruffini & Bonazzola 1969, Colpi et al. 1986). For a large range of hypothetical boson masses they can have sizes of only several times their Schwarzschild radii. This makes it difficult to clearly distinguish observationally between boson stars and black holes as candidates for supermassive objects at the nuclei of galaxies (see also Torres et al. 2000, Mielke & Schunck 2000). However, even if a boson star had formed at the center of the Milky Way, it should eventually have collapsed to a black hole through accretion, during its lifetime, of the abundant gas and dust in the central region. As for possibilities of definitely ruling out the boson star scenario, simultaneous multi-wavelength measurements of the emission from Sgr A* (see Eckart et al. 2004b) will allow to constrain the emission mechanism and therefore the compactness of the emitting region around Sgr A* even further. Probably within the next decade it will be possible to image the 'shadow' cast by the putative black hole through deflection of light rays with global radio interferometry at sub-millimeter wavelengths. Such an experiment will involve very long baseline interferometry in the sub-mm regime (Falcke et al. 2000, Melia & Falcke 2001).

Acknowledgments

This work was supported in part by the Deutsche Forschungsgemeinschaft (DFG) via grant SFB 494. We are grateful to all members of the NAOS/CONICA as well as the ESO Paranal teams.

References

Aschenbach, B., Grosso, N., Porquet, D., Predehl, P., 2004 A&A 417, 71
Aschenbach, B., A&A, submitted, astro-ph/0406545
Baganoff, F. K. et al., 2001, Nature, 413, 45
Baganoff, F. K. et al., 2003, ApJ 591, 891
Ball, G. H., Narayan, R. & Quataert, E. 2001, ApJ 552, 221
Blandford, R., & Begelman, M., 1999, MNRAS, 303, L1
Colpi, M., Shapiro, S.L., Wassermann, I., 1986, Phys. Rev. Let. 57, 2485
Eckart, A., & Genzel, R., 1996, Nature 383, 415
Eckart, A., Genzel, R., Ott, T., Schödel, R., 2002a, MNRAS 331, 917
Eckart, A., et al. Proc. Galactic Center Workshop 2002b, Nov. 3-8, 2002, Kailua-Kona, Hawaii, Astron. Nachr., Vol. 324, 557;
Eckart, A., Moultaka, N., Viehmann, T., et al., 2003, Astron. Nachrichten Suppl. 324, 557
Eckart, A., et al. 2004a, ApJ 602, 760
Eckart, A., et al. 2004b, A&A 427, 1
Falcke, H., Melia, F., Agol, E., 2000, ApJ 528, L13
Genzel, R., Eckart, A., Ott, Th., Eisenhauer, F., 1997, MNRAS, 291, 219;
Genzel, R., Pichon, C., Eckart, A., Gerhard, O. E., Ott, T., 2000, MNRAS 317, 348
Genzel, R., Schödel, R., Ott, T., et al., 2003, Nature 425, 934
Ghez, A. M., Klein, B. L., Morris, M., Becklin, E.E., 1998, ApJ 509, 678
Ghez, A.M., Morris, M., Becklin, E. E., Tanner, A., Kremenek, T., 2000, Nature 407, 349
Ghez, A.M., Duchene, G., Matthews, K. et al., 2003a, ApJ 586, L127
Ghez, A.M. et al. 2003b, 2003b, ANS 324, 527
Ghez, A.M. et al. 2004, ApJ 601, L159
Goldwurm, A. et al. 2003, ApJ 584,751
Horrobin et al. 2004, AN 325, 88

Kaup, D.J., 1968, Phys. Rev. 172, 1331
Krabbe et al. 1995, ApJ 447, L95
Maoz, E., 1998, ApJ 447, L91
Markoff, S., Falcke, H., Yuan, F. & Biermann, P.L., 2001, A&A 379, L13;
Melia, F. & Falcke, H. 2001, ARAA 39, 309;
Mielke, E., & Schunck, F., 2000, Nucl. Phys. B, 594, 1985
Mouawad, N., Eckart, A., Pfalzner, S.; Schödel, R., Moultaka, J.; Spurzem, R. 2004 ANS 325, 102
Mouawad, N.; Eckart, A., et al. 2003, ANS 324, 315
Moultaka, J., Eckart, A., Viehmann, T., Mouawad, N., Straubmeier, C., Ott, T., Schödel, R., 2004 A&A 425, 529
Najarro et al. 1997, A&A 325, 700
Narayan, R., Yi, I., & Mahadevan, R., 1995, Nature, 374, 623
Ott et al. 1999, ApJ 523, 248;
Porquet, D., Predehl, P., Aschenbach, B., et al., 2003, A&A 407, L17
Rigaut et al. 2003, ANS 324, 551;
Ruffini, R., &, Bonazzola, S., 1969, Phys. Rev. 187, 1767
Schödel, R., Ott, T., Genzel, R., Eckart, A., Mouawad, N., Alexander, T., 2003, ApJ 596, 1015
Schödel, R., et al., 2002, Nature 419, 694
Tanner et al. 2003, ANS 324, 597
Tanner et al. 2002, ApJ 575, 860
Torres, D.F., Capoziello, S., Lambiase, G., 2000, Phys. Rev. D 62, 104012
Viollier, R.D., Leimbruber, F. R., and Trautmann, D., 1992, Physics Letters B, 297, 132
Yuan, F., Quataert, E. & Narayan, R., 2004, Ap.J. 606, 894;

MICROLENSING EVENTS TOWARDS LMC AND M31

PHILIPPE JETZER AND SEBASTIANO CALCHI NOVATI

Institute of Theoretical Physics University of Zürich, Winterthurerstrasse 190, CH-8057 Zürich, Switzerland
E-mail: jetzer@physik.unizh.ch, novati@physik.unizh.ch

The nature and the location of the lenses discovered in the microlensing surveys done so far towards the LMC remain unclear. Motivated by these questions we computed the optical depth for the different intervening populations and the number of expected events for self-lensing, using a recently drawn coherent picture of the geometrical structure and dynamics of the LMC disk. The most plausible solution is that the events observed so far are due to lenses belonging to different intervening populations: low mass stars in the LMC, in the thick disk, in the spheroid and some true MACHOs in the halo of the Milky Way and the LMC itself. We report also on recent results of microlensing searches in direction of the M31 galaxy, by using the pixel method. The present analysis still does not allow yet to draw sharp conclusions on the MACHO content of the M31 galaxy.

1. Introduction

Since Paczyński's original proposal [1] gravitational microlensing has been proven to be a powerful tool for the detection of the dark matter component in galactic haloes in the form of MACHOs. Searches towards LMC suggest that up to 20% of the halo could be formed by objects of around $M \sim 0.4\,M_\odot$. However, the location and the nature of the microlensing events found so far towards the LMC is still a matter of controversy. The MACHO collaboration found 13 to 17 events in 5.7 years of observations [2], with a mass for the lenses estimated to be in the range $0.15 - 0.9\,M_\odot$ assuming a standard spherical Galactic halo and derived an optical depth of $\tau = 1.2^{+0.4}_{-0.3} \times 10^{-7}$.

The EROS2 collaboration has recently announced the detection of 4 events (one of which being most probably due to a lens located in the disk of our Galaxy) based on 6.7 years of observation but monitoring about three times as much stars as the MACHO collaboration [3]. EROS2 gives a more stringent limit on the MACHO content of the galactic halo of about 5% of the standard halo for masses between 0.001 to 0.1 $M_\odot$. EROS2 made also following up observations of nine events discovered by the MACHO team. One star, MACHO-LMC-23, showed a new variation in Dec 2001, very similar to that observed by the MACHO group in Feb 1995, so that this event should be removed from the candidate list.

The MACHO collaboration monitored primarily the central part of the LMC, whereas the EROS2 experiment covers a larger solid angle but in less crowded fields. The EROS2 microlensing rate should thus be less affected by self-lensing. This might be the reason for the fewer events seen by EROS2.

A detailed analysis [4,5] has shown that probably the observed events are distributed among different galactic components (disk, spheroid, galactic halo, LMC halo and self-lensing). This means that the lenses do not belong all to the same population and their astrophysical features can differ deeply one another.

Some of the events found by the MACHO team are most probably due to self-lensing, such as MACHO-LMC-9 and MACHO-LMC-14. The event LMC-5 is due to a disk lens and indeed the lens has even been observed with the HST. The other stars which have been microlensed were also observed but no lens could be detected, thus implying that the lens cannot be a disk star but has to be either a true halo object or a faint star or brown dwarf in the LMC itself.

Thus up to now the question of the location of the observed MACHO events is unsolved and still subject to discussion. Clearly, with much more events at disposal, as is expected to be the case with the ongoing SUPER-MACHO experiment [6], one might solve this problem by looking for instance at their spatial distribution. To this end a correct knowledge of the structure and dynamics of the luminous part of the LMC is essential, and we take advantage of a new picture of the LMC disk.

2. LMC optical depth

In a series of interesting papers [7,8], a new coherent picture of the geometrical structure and dynamics of LMC has been given. In the following analysis we use this model and adopt the same coordinate system and notations. The center of the disk coincides with the center of the bar and its distance from us is $D_0 = 50.1 \pm 2.5$ kpc. We take a bar mass $M_{\rm bar} = 1/5\, M_{\rm disk}$ with $M_{\rm bar} + M_{\rm disk} = M_{\rm vis} = 2.7 \times 10^9\, M_\odot$.

We consider also the LMC halo contribution to the optical depth. We use two different models to describe the halo profile density: a spherical halo and an ellipsoidal halo. The values of the parameters have been chosen so that the models have roughly the same mass within the same radius. For the spherical model we adopt a classical pseudo-isothermal spherical density profile, with a halo mass within a radius of 8.9 kpc equal to 5.5×10^9 $M_\odot$.

The computation of the optical depth is made following the method outlined in [4]. As expected, for the self-lensing optical depth, i.e. for events where both the sources and the lenses belong to the disk and/or to the bulge of LMC, there is almost no near-far asymmetry and the maximum value of the optical depth, $\tau_{\rm SL,max} \simeq 4.80 \times 10^{-8}$, is reached in the center of LMC. The optical depth then rapidly decreases, when moving, for instance, along a line going through the center and perpendicular to the minor axis of the elliptical disk, that coincides also with the major axis of the bar. In a range of about only 0.80 kpc the optical depth quickly falls to $\tau_{\rm SL} \simeq 2 \times 10^{-8}$, and afterwards it decreases slowly to lower values.

We computed also the optical depth contour maps for lenses belonging to the halo of LMC in the case of spherical model in the hypothesis that all the LMC dark halo consists of compact lenses. The ellipsoidal model leads to similar results [5]. A striking feature of the map is the strong near-far asymmetry, which is not expected, on the contrary, for a self-lensing population of events.

3. LMC self-lensing event rate

We evaluated the microlensing rate in the self-lensing configuration, i.e. lenses and sources both in the disk and/or in the bar of LMC. We have taken into account the transverse motion of the Sun and of the source stars. We assumed that, to an observer comoving with the LMC center, the velocity distribution of the source stars and lenses have a Maxwellian profile, with spherical symmetry.

In the picture of van der Marel et al. [7,8] within a distance of about 3 kpc from the center of LMC, the velocity dispersion (evaluated for carbon stars) along the line of sight can be considered constant, $\sigma_{\rm los} = 20.2 \pm 0.5$ km/s. Summing over all fields we find that the expected total number of self-lensing events is ~ 1.2. Clearly, taking also into account the uncertainties in the parameter used following the van der Marel et al. model for the LMC the actual number could also be somewhat higher but hardly more than our upper limit estimate of about 3-4 events given in [4].

As a further argument, assuming all the 14 MACHO events as self-lensing, we study the scatter plots correlating the self-lensing expected values of some meaningful microlensing variables with the measured Einstein time or with the self-lensing optical depth. In this way we can show that a large subset of events is clearly incompatible with the self-lensing hypothesis.

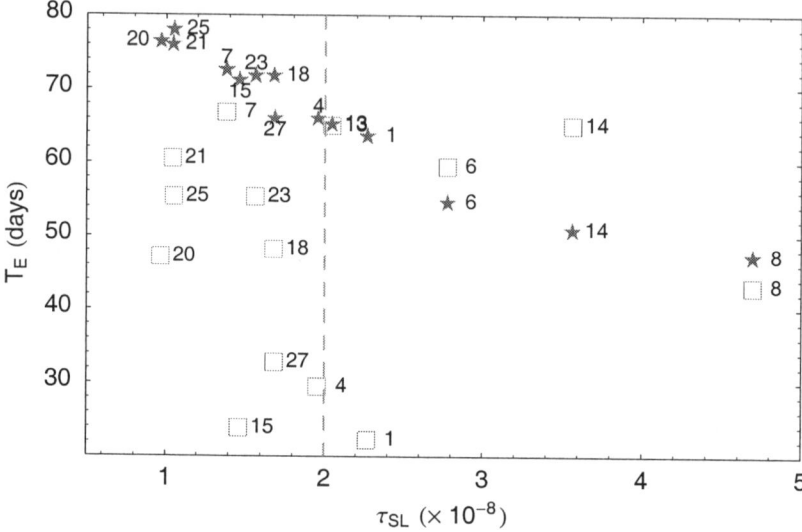

Figure 1. Scatter plot of the observed (empty boxes) values of the Einstein time and of the expected values of the median $T_{E,50\%}$ (filled stars), with respect to the self-lensing optical depth evaluated along the directions of the events.

In Fig. 1 we report on the y–axis the observed values of the Einstein time T_E (empty boxes) as well as the expected values for self-lensing of the median $T_{E,50\%}$ (filled stars) evaluated along the directions of the events. On the x-axis we report the value of the self-lensing optical depth calculated towards the event position; the optical depth is growing going from the outer regions towards the center of LMC. An interesting feature emerging clearly is the decreasing trend of the expected values of the median $T_{E,50\%}$, going from the outside fields with low values of τ_{SL} towards the central fields with higher values of τ_{SL}. The variation of the stellar number density and the flaring of the LMC disk certainly contributes to explain this behaviour.

We now tentatively identify two subsets of events: the nine falling outside the contour line $\tau_{SL} = 2 \times 10^{-8}$ and the five falling inside. In the framework of van der Marel et al. LMC geometry, this contour line includes almost fully the LMC bar, where we expect self-lensing events to be located with higher probability. This plot suggests that the self-lensing events have to be searched among the cluster of events with $\tau_{SL} > 2 \times 10^{-8}$, and at the same time that the cluster of the 9 events including LMC–1 belongs, very probably, to a different population. Moreover, when looking at the spatial distribution of the events one sees a clear near-far asymmetry in the van der Marel et al. geometry; they are concentrated along the extension of the bar and in the south-west side of LMC. Indeed, we have performed a statistical analysis of the spatial distribution of the events, which clearly shows that the observed asymmetry is greater than the one expected on the basis of the observational strategy [5].

4. Pixel lensing towards M31

Searches towards M31 allow to probe a different line of sight in our Galaxy, to globally test M31 halo and, furthermore, the high inclination of the M31 disk is expected to provide a strong signature (spatial distribution) for halo microlensing signals. For extragalactic targets, due to the distance, the sources for microlensing signals are not resolved. This claims for an original technique, the *pixel method*, the detection of flux variations of unresolved sources, the main point being that one follows flux variations of every pixel in the image instead of single stars. Several collaborations have recently presented a handful of microlensing events (SLOTT-AGAPE [9], POINT-AGAPE [10], WeCapp [11] and MEGA [12]).

The SLOTT-AGAPE collaboration has been using data collected on the 1.3m McGraw-Hill Telescope at the MDM observatory, Kitt Peak (USA), and found in a two years campaign three possible candidate events, taking into account also an extension of the light curve using INT data. The POINT-AGAPE collaboration carried out a survey of M31 by using the Wide Field Camera (WFC) on the 2.5 m INT telescope. A first analysis [10] was made with the aim to detect short ($t_{1/2} < 25$ days) and bright variations ($\Delta R < 21$ at maximum amplification), compatible with a Paczyński signal. The first requirement is suggested by the results on the predicted characteristics of microlensing events of a Monte Carlo simulation of the experiment. As an outcome, seven light curves are detected which can be considered viable microlensing events [13].

Once a microlensing event is detected it is important, given the aim to probe the halo content in form of MACHOs, to find out its origin, namely, whether it is due to self-lensing within M31 or to a MACHO. This is not straightforward. The spatial distribution of the events is an important tool.

Among the detected events 4 lie within the bulge region, where the self-lensing contamination is huge with respect to the predicted MACHO signal. Together with the results of a Monte Carlo simulation of the experience an analysis is currently underway with the aim to obtain an upper limit on the halo content in the form of MACHOs [14].

5. Conclusions

We have presented the results of the computation of the optical depth for microlensing towards LMC using a recent picture of the LMC disk. An interesting feature which emerges is a near–far asymmetry of the optical depth for lenses located in the LMC halo, which is not the case for self-lensing. Furthermore, we showed that the timescale distribution of the events and their spatial variation across the LMC disk offers possibilities of identifying the dominant lens population. Through this analysis we have been able to identify a large subset of events that can not be accounted for by self-lensing. As a general outcome of presently available pixel lensing results, we can clearly infer that the detection of microlensing events towards M31 is now established. The open issue to be still explored is the study of the M31 halo fraction in form of MACHOs.

References

1. B. Paczyński, 1986, ApJ **304**, 1
2. C. Alcock, R.A. Allsman, D.R. Alves et al., 2000, ApJ **542**, 281
3. Ph. Jetzer, A. Milsztajn & P. Tisserand, astro-ph/0409496
4. Ph. Jetzer, L. Mancini & G. Scarpetta, 2002, A&A **393**, 129
5. L. Mancini, S. Calchi Novati, Ph. Jetzer & G. Scarpetta, 2004, A&A **427**, 61
6. A. Becker et al., astro-ph/0409167
7. R.P. van der Marel & M.R. Cioni, 2001, AJ **122**, 1807
8. R.P. van der Marel et al., 2002, AJ **124**, 2639
9. S. Calchi Novati et al., 2003, A&A **405**, 851
10. S. Paulin-Henriksson et al., 2003, A&A **405**, 15
11. A. Riffeser et al., 2003, ApJ **599**, L17
12. J. de Jong et al., 2004, A&A **417**, 461
13. S. Paulin-Henriksson & S. Calchi Novati, 2004, to appear in the proceedings of the "Rencontres de Moriond 2004': Exploring the Universe"
14. S. Calchi Novati, S. Paulin-Henriksson et al. (the POINT-AGAPE collaboration) in preparation

SOME CONSEQUENCES OF THE BARYONIC DARK MATTER POPULATION

RUDOLPH E. SCHILD

Center for Astrophysics, 19
60 Garden Street,
Cambridge, MA 02138
Email: rschild@cfa.harvard.edu

Microlensed double-image quasars have sent a consistent message that the baryonic dark matter consists of a dark population of free-roaming planet mass objects. This population has long been predicted to have formed at the time of recombination, 300,000 years after the Big Bang, when the primordial plasma changed to neutral atoms with an attendant large increase in viscosity of the primordial matter. Following a very brief review of the observational basis for this conclusion and some alternative explanations, we review some probable effects of this population. After the particles formed by the usual gravitational condensation - void separation process, they collapsed on a 100 million year Kelvin-Helmholz time scale, and started their inevitable cooling process. Although not yet satisfactorily modeled, this process should have caused significant evaporation of primordial gas and taken them through the condensation and freezing points of hydrogen on their way to the 2.73 K temperature of the present universe. At the 20 K freezing point they should have frozen from the outside in, creating tremendous crushing central pressures that would have easily produced the rocky cores of planets and Kuiper-Oort cloud objects mysteriously over-abundant in the present solar system. The mystery of how did the universe become re-ionized by a Pop III that should have been seen at redshifts 6 to 8, now under scrutiny from direct spectroscopic observation, is cleanly side-stepped. Probably 99% of the baryonic matter in the universe was sequestered away in the dark matter bodies and does not need to be re-ionized for the universe to have its present transparency in the far ultraviolet. And the Dark Energy mystery will evaporate when it is understood how this population reduces the transparency of the universe. It is probably not a coincidence that the "self replenishing dust" model that explains the HST supernova brightness deficits closely matches the known dependence of extinction from $Ly - \alpha$ clouds upon redshift. If these mysterious clouds, that should have diffused away on a short time scale, are reforming from slow evaporation of the planet-mass population, they should produce spherical lenses that refract light out of the supernova images to produce a grey reduction of the transparency of the universe.

1. Introduction and Microlensing Results in Q0957

At the 1996 Sheffield Symposium I reported microlensing results from 15 years of Q0957 monitoring that revealed a persistent, continuous rapid microlensing which indicated that compact planetary mass bodies constitute the baryonic dark matter (Schild, 1996). At the time, the response was, appropriately, "That's nice, but it's just Rudy's data for Rudy's quasar." Today, 8 years on, the rapid microlensing has been confirmed in the 6 additional lens systems having sufficient data for time delay estimation. And new Q0957 observing campaigns have produced evidence for an event with amplitude 1% and time scale of just 12 hours (Colley & Schild 2003). A recent summary of these observational results was presented at the UCLA Dark Matter/Dark Energy 2004 Symposium (Schild, 2004; astro-ph/0406491).

With the observation of rapid microlensing now confirmed, alternative explanations based upon imagined possible quasar structure have fortunately been explored. Thus bright points orbiting in the quasar's accretion disc have been considered (Gould and Escude-Miralde, 1997; Schechter *et al* 2003) and dark clouds swarming around the quasar (Wyithe & Loeb, 2002) have been investigated. These schemes have been frustrated by the extreme physics required, especially for the extremely rapid Colley & Schild (2003) event, and by the implied periodicity. But most important, the simulations do not naturally produce the equal positive and negative events seen in the Schild (1999) wavelet analysis and natural to the planet-mass microlensing explanation (Schild & Vakulik, 2004). These points were more extensively reviewed in Schild (2004).

The above results are based upon reasonably rigorous science, with simulations and confirmed observational results. For the remainder of this contribution I examine more tentative explorations, based largely upon back-of-the-envelope calculations, to examine some immediate applications of the discovery. For example, what would such particles look, feel, smell, and taste like?

2. What do the Particles Look, Feel, Smell, and Taste Like?

Remarkably, these questions were already answered at the 1996 Sheffield symposium by hydrodynamics expert Carl Gibson, who predicted their existence from hydrodynamical analysis of the forces operative in the fluid/gas of the expanding universe at times before, during, and after recombination, 300,000 years after the Big Bang. However this hydrodynamical theory has been largely ignored because of its departures from the prevailing CDM theory, whose failures will need to go to completion before the hydrodynamical theory features receive the scrutiny that they deserve.

In the Gibson (1996, 2000) theory, structure formation limited by viscous-diffusive forces already seeded structures on scales of galaxies, clusters, and superclusters during the plasma epoch, before recombination. After the plasma-gas transition the baryonic gas of the universe fragmented at Jeans mass clumps further seeded with planetary mass "Primordial Fog Particles" (PFP's) limited to sub-stellar mass scales by the viscous and gravitational forces in the fluid. Thus the entire baryonic mass of the universe gravitationally collapsed onto primordial scale-free fluctuations but only the planet mass ones succeeded; they survive today as the baryonic dark matter seen in quasar microlensing. Many or most of these PFP particles would in turn be initially contained in globular cluster scale condensations that are often seen today mysteriously appearing in galaxy interactions.

The PFP's presumably collapsed on the 100 million year Kelvin-Helmholz time scale and then cooled. In their subsequent history they would have swept up dust from supernovae and cool giant stars. In the expanding universe they would presumably have cooled below the hydrogen condensation point, and when passing below the freezing point at 20 K they would have crushed the ices and rocks at their centers to make the solid cores of the planets and the Trans-Neptunian objects seen today.

Gibson (1997) described the PFP's today as "in solid or liquid state, crusted with 14 billion years of accreted dust." It also seems likely that these are the objects at the cores of the Lyman-Alpha clouds seen by the thousand in ultraviolet spectra of quasars.

Today, we do not have an adequate simple model of a planetary mass particle formed at recombination and passively collapsing, reaching its maximum temperature at around 100 million years after recombination, and subsequently cooling. We don't even know if the PFP's would be cool enough at their centers for hydrogen to cool through the hydrogen condensation point at 40 K and the freezing point at 20 K. Because galaxy discs are now measured to have temperatures of 30 K (Bendo *et al*, 2003), there should also have been a point in the asymptotic approach to our 2.73 K background temperature when such Halo PFP objects orbiting galaxy discs went through multiple freezing and melting cycles.

Because the PFP objects are dark, they would be detected by my gravitational microlensing or by chance superpositions against a bright nebular background, as probably seen in the nearest planetary nebula, the Helix (O'Dell & Handron 1996). There they have been interpreted as hydrodynamical or shock instabilities in the expanding shell, even though such interpretations are contradicted by the high masses and densities measured for them (Meaburn et al, 1998). Interpreted as PFP's seen against the bright nebular background, with surface ablation wearing away their outermost layers, their masses are estimated as $10^{-5} M_\odot$ and their diameters are measured to be 10^{16} cm, or 100 times the diameter of Neptune.

These Helix cometary knots are seen in the HST images to occur mostly in clumps, as expected for self-gravitating particles. This would explain why they could not be detected by MACHO microlensing searches, where they would just cause rapid irregular variability rejected by the MACHO search software.

3. The Chemical Enrichment and Re-Ionization of the Universe

Our cosmology today is challenged by the observational result that reionization of the universe occurred at redshifts around 6.5, indicated by the ultraviolet continuum of hi-z quasars (Fan et al, 2004). Thus a vast population of high-redshift galaxies rich in Pop III stars should exist and supply copious $Ly - \alpha$ photons causing the universal ionization. But actual observations of galaxies at the redshift range 6.5 - 7.5 do not show copious $Ly - \alpha$ emission or extreme luminosity; indeed, the searches produced the conclusion that galaxies are x9 less luminous at the highest observed redshifts than at $z = 3$ (Stanway, et al, 2004; Bunker, A. et al, astro-ph/0403223).

We predict that this Pop III will never be found, because only a small fraction of the baryonic matter of the universe needs to be enriched and ionized if 99% of the primordial baryonic matter is sequestered away in our baryonic dark matter particles. A slow PFP evaporation, seen as $Ly - \alpha$ clouds, would slowly enrich the universe with primordial hydrogen/helium to maintain nearly constant elemental abundances. Quasars probably have sufficient ultraviolet radiation to re-ionize the universe (Escude-Miralde, Hashnelt, & Rees, 2000).

4. Relationship to $Ly - \alpha$ Clouds

The primary reason for accepting the concept of Dark Energy is the supernova brightness curves, but in their analysis (Riess et al, 2004) there has been no allowance for the reduction of the transparency of the universe imposed by the detected Baryonic dark matter component. Significant evidence for transmission losses might long have been seen in the quasar number density peak at z = 1.9. Note that the difference between ordinary universe models and the Goobar et al, (2002, Fig 3A) "self replenishing dust" normalized to the supernova flux deficit at z = 0.5, is approximately 1.4 magnitudes for the quasar number density peak at z = 1.9.

We now recognize that the existence of quasar $Ly - \alpha$ forest clouds has evidently long been indicating the source of this transmission loss. The presumption that such clouds must be confined by a hot intercloud medium has been rejected on observational grounds, and the clouds should rapidly dissipate away unless they are being continuously refreshed by the expected evaporation of our PFP objects.

It has long been known (Zuo & Lu, 1993) that the density of $Ly - \alpha$ clouds increases with redshift z as $(1 + z)^{2.8}$. It has also been noticed by Goobar et al, (2002) that an absorption law scaling as $(1 + z)^3$ up to z = 0.5 and constant thereafter, can explain the supernova brightness - redshift relationship of Riess et al, (2004). We are careful not to use the word "grey extinction" because the transmission loss may be dominated by refraction in the spherical lenses, which would contribute to the diffuse background radiation of the universe, a controversial topic.

We are easily left with the following conclusion: before accepting the idea that the Hubble expansion is dominated by a mysterious Dark Energy, it is important to calculate the transmission to distant supernovae and quasars as limited by the baryonic dark matter.

Acknowledgments

I wish to thank Carl Gibson for his extended correspondence related to the expected properties of the microlensing particles.

References

1. G. Bendo, et al, Astron. Journ. **125**, 2361 (2003).
2. W. N. Colley and R. Schild, Astrophys. Journ. **594**, 97 (2003).
3. J. Escude-Miralde, M. Haehnelt, and M. Reese, Astrophys. Journ.. **530**, 1 (2000)
4. X. Fan, et al, Astron. Journ. **128**, 515 (2004).
5. C. H. Gibson, Appl. Mech. Rev. **49**, 299 (1996); astro-ph/9904260.
6. C. H. Gibson, 1997, The Identification of Dark Matter, ed. N. Spooner [London: World Scientific], **114**, (1997); astro-ph/9904283.
7. C. H. Gibson, J. Fluids Eng. **122**, 830 (2000); astro-ph/0003352.

8. A. Goobar, L. Bergstrom, and E. Mortsell, *Astron. & Astrophys.* **384**, 1 (2002).
9. A. Gould, and J. Escude-Miralde, *Astrophys. Journ.* **483**, L13 (1997).
10. J. Meaburn, et al, *Monthly Not. R. A. S.* **294**, 201 (1998).
11. C. R. O'Dell, and K. Handron, *Astron Journ.* **111**, 1630 (1996).
12. A. Riess et al, astro-ph/0402512, 2004.
13. P. Schechter et al, *Astrophys. Journ.* **584**, 657 (2003).
14. R. Schild, *Astrophys. Journ.* **464**, 125 (1996).
15. R. Schild, *Astrophys. Journ.* **514**, 598 (1999).
16. R. Schild, 2004, *New Astronomy Reviews: Proceeding of the Dark Matter/ Dark Energy Symposium [Elsevier]*, in press (2004); astro-ph/0406491.
17. R. Schild and V. Vakulik, *Astron. Journ.* **126**, 689 (2003).
18. E. Stanway, et al, *Astrophys. Journ.* **607**, 704 (2004).
19. S. Wyithe and A. Loeb, *Astrophys. Journ.* **577**, 615 (2002).
20. L. Zuo and L. Lu, *Astrophys. Journ.* **418**, 601 (1993).

DAMA: RESULTS AND PERSPECTIVES

R. BERNABEI, P. BELLI, F. CAPPELLA, F. MONTECCHIA,* F. NOZZOLI
*Dipartimento di Fisica, Università di Roma "Tor Vergata",
INFN, Sezione di Roma II, I-00133, Roma, Italy*

A. INCICCHITTI, D. PROSPERI
*Dipartimento di Fisica, Università di Roma "La Sapienza",
INFN, Sezione di Roma, I-00185, Roma, Italy*

R. CERULLI
INFN - Laboratori Nazionali del Gran Sasso, I-67010 Assergi (Aq), Italy

C. J. DAI, H. H. KUANG, J. M. MA, Z. P. YE[†]
IHEP, Chinese Academy, P.O. Box 918/3, Beijing 100039, China

DAMA is an observatory for rare processes based on the development and use of various kinds of radiopure scintillators; it is operative deep underground at the Gran Sasso National Laboratory of the I.N.F.N.. Several low background set-ups have been realized and many rare processes have been investigated. In particular, the DAMA/NaI set-up ($\simeq$ 100 kg highly radiopure NaI(Tl)) has effectively investigated the model-independent annual modulation signature, obtaining from the data of seven annual cycles (total exposure of 107731 kg × day) a 6.3 σ C.L. model-independent evidence for the presence of a Dark Matter particle component in the galactic halo. Moreover, some of the many possible corollary model-dependent quests for the candidate particle have also been investigated. At present, the second generation DAMA/LIBRA set-up ($\simeq$ 250 kg highly radiopure NaI(Tl)) is in operation deep underground.

1. Introduction

DAMA is an observatory for rare processes based on the development and use of various kinds of radiopure scintillators. Several low background set-ups have been realized; the main ones are: i) DAMA/NaI ($\simeq$ 100 kg of

---

*also: Università "Campus Bio-Medico" di Roma, 00155, Rome, Italy
[†]also: University of Zhao Qing, Guang Dong, China

highly radiopure NaI(Tl)), which took data underground over seven annual cycles and was put out of operation in July 2002 [1,2,3,4,5,6,7,8,9,10,11,12,13]; ii) DAMA/LXe ($\simeq 6.5$ kg liquid Xenon) [14]; iii) DAMA/R&D, which is devoted to tests on prototypes and small scale experiments [15]; iv) the new second generation DAMA/LIBRA set-up ($\simeq 250$ kg highly radiopure NaI(Tl)) in operation since March 2003. Moreover, in the framework of devoted R&D for radiopure detectors and PMTs, sample measurements are regularly carried out by means of the low background DAMA/Ge detector, installed deep underground since $\gtrsim 10$ years and, in some cases, by means of Ispra facilities.

In the following, we will focus our attention only on the DAMA/NaI results on the annual modulation signature. In fact, the model-independent annual modulation signature (originally suggested in [16]) for Dark Matter particles in the galactic halo is very distinctive since it requires the simultaneous satisfaction of all the following requirements: the rate must contain a component modulated according to a cosine function (1) with one year period, T, (2) and a phase, t_0, that peaks around $\simeq 2^{nd}$ June (3); this modulation must only be found in a well-defined low energy range, where WIMP induced recoils can be present (4); it must apply to those events in which just one detector of many actually "fires" (*single-hit* events), since the WIMP multi-scattering probability is negligible (5); the modulation amplitude in the region of maximal sensitivity is expected to be $\lesssim 7\%$ for usually adopted halo distributions (6), but it can be larger in case of some possible scenarios such as e.g. those in refs. [17,18]. To mimic such a signature spurious effects or side reactions should be able both to account for the whole observed modulation amplitude and to contemporaneously satisfy all the requirements; none has been found (see e.g. ref. [2] and references therein).

Main topics on the DAMA/NaI results will be shortly summarized in the following; for a detailed discussion see ref. [2]. For a description of the set-up and its performances see refs. [1,2,10].

2. The model-independent result of DAMA/NaI

A model-independent approach on the data collected by DAMA/NaI over seven annual cycles offers an immediate evidence of the presence of an annual modulation of the measured rate of the *single-hit* events in the lowest energy region. In particular, in Fig. 1 – *left* the time behaviour of the residual rate of the *single-hit* events in the cumulative (2-6) keV energy interval is reported. The data favour the presence of a modulated cosine-like

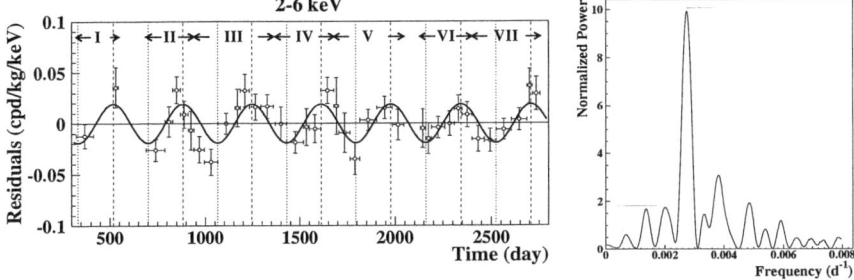

Figure 1. *On the left*: experimental residual rate for *single-hit* events in the cumulative (2–6) keV energy interval as a function of the time over 7 annual cycles (total exposure 107731 kg × day); end of data taking July 2002. The experimental points present the errors as vertical bars and the associated time bin width as horizontal bars. The superimposed curve represents the cosinusoidal function behaviour expected for a WIMP signal with a period equal to 1 year and phase exactly at 2nd June; the modulation amplitude has been obtained by best fit. See ref. [2]. *On the right*: power spectrum of the measured *single-hit* residuals for the cumulative (2–6) keV energy interval calculated including also the treatment of the experimental errors and of the time binning. As it can be seen, the principal mode corresponds to a frequency of $2.737 \cdot 10^{-3}$ d^{-1}, that is to a period of $\simeq 1$ year.

behaviour ($A \cdot \cos\omega(t - t_0)$) at 6.3 σ C.L. and their fit for this cumulative energy interval offers modulation amplitude equal to (0.0200 ± 0.0032) cpd/kg/keV, $t_0 = (140 \pm 22)$ days and $T = \frac{2\pi}{\omega} = (1.00 \pm 0.01)$ year, all parameters kept free in the fit. The period and phase agree with those expected in the case of an effect induced by Dark Matter particles of the galactic halo ($T = 1$ year and t_0 roughly at $\simeq 152.5^{th}$ day of the year). The χ^2 test on the (2–6) keV residual rate disfavours the hypothesis of unmodulated behaviour giving a probability of $7 \cdot 10^{-4}$ ($\chi^2/d.o.f. = 71/37$). The same data have also been investigated by a Fourier analysis as shown in Fig. 1 – *right*, where a clear peak corresponding to a period of $\simeq 1$ year is present. Modulation is not observed above 6 keV [2] [a]. Finally, a suitable statistical analysis has shown that the modulation amplitudes are statistically well distributed in all the crystals, in all the data taking periods and considered energy bins. More arguments can be found in ref.[2]. A careful investigation of all the known possible sources of systematic and side reactions has been regularly carried out and published at time of each data release and quantitative discussions can be found in refs. [2,10]. No

---

[a]We remind that DAMA/NaI took data up to MeV energy region despite the optimization was done for the keV energy range.

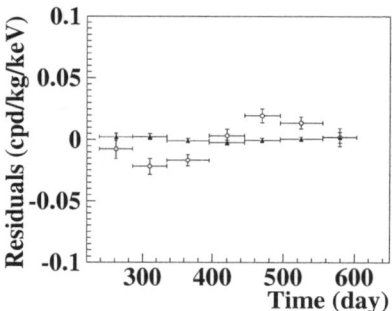

Figure 2. Experimental residual rates over seven annual cycles for *single-hit* events (open circles) – class of events to which WIMP events belong – and over the last two annual cycles for *multiple-hits* events (filled triangles) – class of events to which WIMP events do not belong – in the (2–6) keV cumulative energy interval. They have been obtained by considering for each class of events the data as collected in a single annual cycle and using in both cases the same identical hardware and the same identical software procedures. The initial time is taken on August 7th. See text.

systematic effect or side reaction able to account for the observed modulation amplitude and to satisfy all the requirements of the signature has been found.

As a further relevant investigation, the *multiple-hits* events collected during the DAMA/NaI-6 and 7 running periods (when each detector was equipped with its own Transient Digitizer with a dedicated renewed electronics) have been studied and analysed by using the same identical hardware and the same identical software procedures as for the case of the *single-hit* events (see Fig. 2). The *multiple-hits* events class – on the contrary of the *single-hit* one – does not include events induced by WIMPs since the probability that a WIMP scatters off more than one detector is negligible. The fitted modulation amplitudes are: $A = (0.0195 \pm 0.0031)$ cpd/kg/keV and $A = -(3.9 \pm 7.9) \cdot 10^{-4}$ cpd/kg/keV for *single-hit* and *multiple-hits* residual rates, respectively. Thus, evidence of annual modulation is present in the *single-hit* residuals (events class to which the WIMP-induced recoils belong), while it is absent in the *multiple-hits* residual rate (event class to which only background events belong). Since the same identical hardware and the same identical software procedures have been used to analyse the two classes of events, the obtained result offers an additional strong support for the presence of Dark Matter particles in the galactic halo further excluding any side effect either from hardware or from software procedures or from background.

In conclusion, the presence of an annual modulation in the residual rate of the *single-hit* events in the lowest energy interval (2 – 6) keV, satisfying all the features expected for a Dark Matter particle component in the galactic halo is supported by the data of the seven annual cycles at 6.3 σ C.L.. No systematic effect or side reaction able to account for the observed effect has been found. This is the experimental result of DAMA/NaI; it is model-independent. No other experiment, whose result can be directly compared with this one in a model independent way, is available so far in the field of Dark Matter investigation.

3. Some corollary model-dependent quests for a candidate

On the basis of the obtained 6.3 σ model-independent result, corollary investigations can also be pursued on the nature of the Dark Matter particle candidate. This latter investigation is instead model-dependent and – considering the large uncertainties which exist on the astrophysical, nuclear and particle physics assumptions and on the parameters needed in the calculations – has no general meaning (as it is also the case of exclusion plots and of the WIMP parameters evaluated in indirect detection experiments). Thus, it should be handled in the most general way as we have pointed out with time passing [6,7,8,9,10,11,12,13,2].

Candidates, kinds of WIMP couplings with ordinary matter and implications, cross sections, nuclear form factors, spin factors, scaling laws, halo models, priors, etc. are discussed in ref.[2]. The reader can find in this latter paper and in references therein devoted discussions to correctly understand the results obtained in corollary quests and the real validity of any claimed model-dependent comparison in the field. Here, we just remind that the results briefly summarized here are not exhaustive of the many scenarios possible at present level of knowledge, including those depicted in some more recent works such as e.g. refs. [18,19].

DAMA/NaI is intrinsically sensitive both to low and high WIMP mass having both a light (the ^{23}Na) and a heavy (the ^{127}I) target-nucleus; in previous corollary quests for the candidate, dark matter particle masses above 30 GeV (25 GeV in ref.[6]) have been presented [7,9,11,12,13] for few (of the many possible) model frameworks. However, that bound holds only for neutralino when supersymmetric schemes based on GUT assumptions are adopted to analyse the LEP data [20]. Thus, since other candidates are possible and also other scenarios can be considered for the neutralino itself as recently pointed out[b], the present model-dependent lower bound quoted by LEP for the neutralino in the supersymmetric schemes based on GUT

---

[b]In fact, when the assumption on the gaugino-mass unification at GUT scale is released, neutralino masses down to $\simeq$ 6 GeV are allowed [21,22].

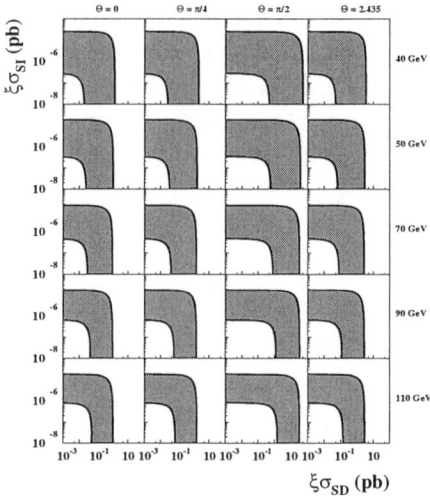

Figure 3. *Case of a WIMP with mixed SI&SD interaction for the model frameworks given in ref.[2]*. Coloured areas: example of slices (of the 4-dimensional allowed volume) in the plane $\xi\sigma_{SI}$ vs $\xi\sigma_{SD}$ for some of the possible m_W and θ values. Inclusion of other existing uncertainties on parameters and models would further extend the regions; for example, the use of more favourable form factors and/or of more favourable spin factors than the ones considered here would move them towards lower cross sections. For details see ref.[2].

assumptions (37 GeV [23]) is simply marked in the following figures. It is worth to note that this model dependent LEP limit – when considered – selects the WIMP-Iodine elastic scatterings as dominant.

For simplicity, here the results of these corollary quests for a candidate particle are presented in terms of allowed regions obtained as superposition of the configurations corresponding to likelihood function values *distant* more than 4σ from the null hypothesis (absence of modulation) in each of the several (but still a limited number) of the possible model frameworks considered in ref.[2]. These allowed regions take into account the time and energy behaviours of the single-hit experimental data and have been obtained by a maximum likelihood procedure (for a formal description see e.g. refs. [6,7,9]) which requires the agreement: i) of the expectations for the modulated part of the signal with the measured modulated behaviour for each detector and for each energy bin; ii) of the expectations for the unmodulated component of the signal with the respect to the measured differential energy distribution and - since ref.[9] - also with the bound on recoils obtained by pulse shape discrimination from the devoted DAMA/NaI-0 data [3]. The latter one acts in the likelihood procedure as an experimental upper bound on the unmodulated component of the signal and – as a matter of fact – as an experimental lower bound on the estimate of the background

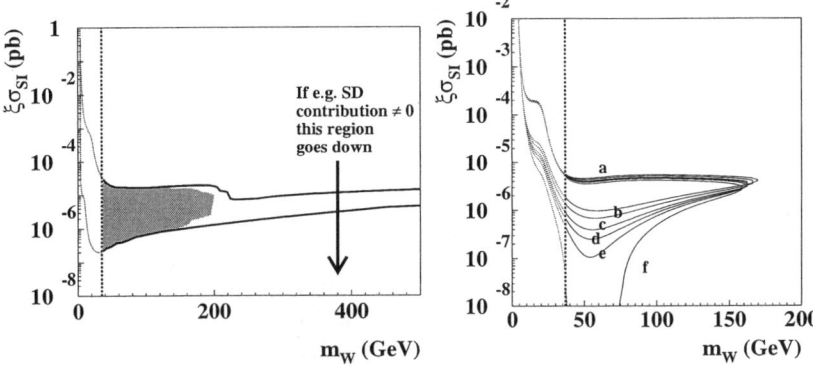

Figure 4. *On the left : Case of a WIMP with dominant SI interaction for the model frameworks given in ref.[2].* Region allowed in the plane $(m_W, \xi\sigma_{SI})$. The vertical dotted line represents a bound in case of a neutralino candidate when supersymmetric schemes based on GUT assumptions are adopted to analyse the LEP data; the low mass region is allowed for neutralino when other schemes are considered (see text) and for every other dark matter particle candidate. While the area at WIMP masses above 200 GeV is allowed only for few configurations, the lower one is allowed by most configurations (the colored region gathers only those above the vertical line). The inclusion of other existing uncertainties on parameters and models would further extend the region; for example, the use of more favourable SI form factor for Iodine alone would move it towards lower cross sections. *On the right: Example of the effect induced by the inclusion of a SD component different from zero on allowed regions given in the plane $\xi\sigma_{SI}$ vs m_W.* In this example the Evans' logarithmic axisymmetric $C2$ halo model with $v_0 = 170$ km/s, ρ_0 equal to the maximum value for this model and a given set of the parameters' values (see ref.[2]) have been considered. The different regions refer to different SD contributions for the particular case of $\theta = 0$: $\sigma_{SD} = 0$ pb (a), 0.02 pb (b), 0.04 pb (c), 0.05 pb (d), 0.06 pb (e), 0.08 pb (f). Analogous situation is found for the other model frameworks. For details see ref.[2].

levels. Thus, the C.L.'s, we quote for the allowed regions, already account for compatibility with the measured differential energy spectrum and with the measured upper bound on recoils. Finally, it is worth to note that the best fit values of cross sections and Dark Matter particle mass span over a large range when varying the considered model framework.

Fig. 3, 4, 6 show some of the obtained allowed regions; details and descriptions of the symbols are given in ref.[2]. Here we only remind that $tg\theta$ is the ratio between the Dark Matter particle-neutron and the Dark Matter particle-proton effective spin-dependent coupling strengths and that θ is defined in the $[0,\pi)$ interval. Obviously, larger sensitivities than those reported in the following figures would be reached when including the effect of other existing uncertainties on the astrophysical, nuclear and particle

Physics assumptions and related parameters; similarly, the set of the best fit values would also be enlarged as well. For details see ref. [2].

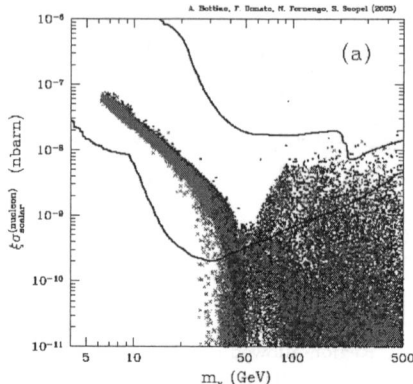

Figure 5. Figure taken from ref.[22]: theoretical expectations of $\xi\sigma_{SI}$ versus m_W in the purely SI coupling for the particular case of a neutralino candidate in MSSM with gaugino mass unification at GUT scale released; the curve is the same as in Fig. 4-*left*.

In Fig. 5 the theoretical expectations in the purely SI coupling for the particular case of a neutralino candidate in MSSM with gaugino mass unification at GUT scale released [22] are shown. The marked curve surrounds the DAMA/NaI purely SI allowed region as in Fig. 4 – *left*.

Specific arguments on some claimed model-dependent comparisons can be found in ref.[2]. They already account, as a matter of fact, also e.g. for the more recent model-dependent CDMS(-II) claim [24] (based on a statistics of 19.4 kg · day, on a data selection and some discrimination procedures), where DAMA/NaI is not correctly and completely quoted and the more recent result of the 7 annual cycles [2] is quoted but not accounted for. In addition, in the single particular scenario considered in [24], uncertainties from the model (from astrophysics, nuclear and particle physics assumptions) as well as some experimental ones (e.g. quenching factor arbitrarily assumed 1, etc.), are not accounted at all. Moreover, the existing interactions and scenarios to which CDMS is largely insensitive – on the contrary of DAMA/NaI – are ignored. Similar arguments also hold for the EDEL-WEISS case, while for Zeplin-I see ref. [2].

On the other hand, some positive hints are present in indirect detection experiments; in fact, an excess of positrons and of gamma's in the space has been reported with the respect to a modellised background; they are not in contraddiction with the DAMA/NaI result. Moreover, recently, it

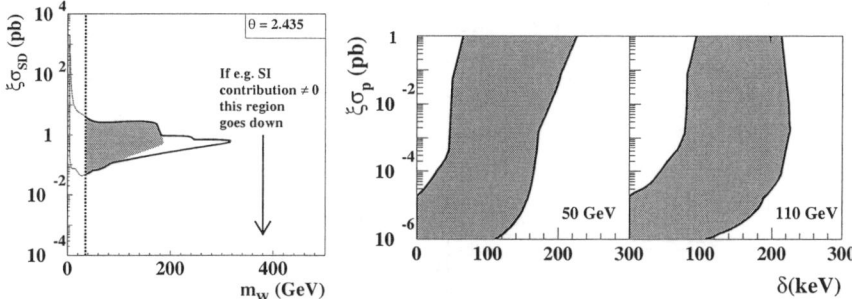

Figure 6. On the left: Case of a WIMP with dominant SD interaction in the model frameworks given in ref.[2]. Example of a slice (of the 3-dimensional allowed volume) in the plane $(m_W, \xi\sigma_{SD})$ at a given θ value (θ is defined in the $[0,\pi)$ range); here $\theta = 2.435$ (Z_0 coupling). For the definition of the vertical line and of the coloured area see the caption of Fig. 4. Inclusion of other existing uncertainties on parameters and models (as discussed in ref.[2]) would further extend the SD allowed regions. For example, the use of more favourable SD form factors and/or more favourable spin factors would move them towards lower cross sections. Values of $\xi\sigma_{SD}$ lower than those corresponding to this allowed region are possible also e.g. in case of an even small SI contribution (see ref.[2]). On the right: Case of a WIMP with preferred inelastic interaction in the model frameworks given in ref.[2]. Examples of slices (coloured areas) of the 3-dimensional allowed volume ($\xi\sigma_p$, δ, m_W) for some m_W values. Inclusion of other existing uncertainties on parameters and models would further extend the regions; for example, the use of more favourable form factors and of different escape velocity would move them towards lower cross sections. For details see ref.[2].

has been suggested [25] that these positive hints and the effect observed by DAMA/NaI can also be described in a scenario with multi-component Dark Matter in the galactic halo, made of a subdominant component of heavy neutrinos of the 4th family and of a sterile dominant component. In particular (see Fig. 7), it has been shown that an heavy neutrino with mass around 50 GeV can account for all the observations, while the inclusion of possible clumpiness of neutrino density as well as new interactions in the heavy neutrino annihilation, etc. can lead to wider mass ranges: from about 46 up to about 75 GeV (see ref. [25] for details).

4. Conclusions and perspectives

DAMA/NaI has been a pioneer experiment investigating as first the WIMP annual modulation signature with suitable sensitivity and control of the running parameters. During seven independent experiments of one year each one, it has pointed out at 6.3 σ C.L. in a model independent way the presence of a modulation satisfying the many peculiarities of an effect induced by Dark Matter particles in the galactic halo; no systematic effect

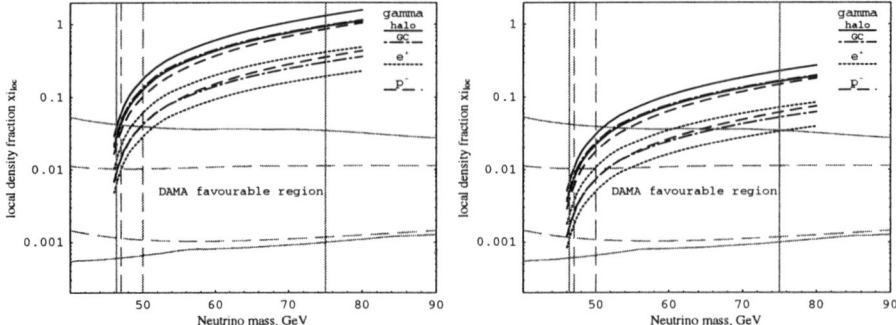

Figure 7. Figure taken from ref.[25]: Case of a subdominant heavy 4th neutrino candidate in the plane local density fraction versus the heavy neutrino mass. The favorable region for this candidate obtained from the DAMA/NaI data (grey dashed line when using the Evan's halo model; solid line when using the other halo models) and the best-fit density parameters deduced from cosmic gamma-radiation (from halo and galactic center), positron and antiproton analysis are shown (left panel). The effect of the inclusion of possible neutrino clumpiness is also reported (right panel). See ref. [25] for details.

or side reaction able to account for the observed effect has been found. As a corollary result, it has also pointed out the complexity of the quest for a candidate particle mainly because of the present poor knowledge on the many astrophysical, nuclear and particle physics aspects. At present after a devoted R&D effort, the second generation DAMA/LIBRA (a $\simeq$250 kg more radiopure NaI(Tl) set-up) has been realised and put in operation since March 2003. It will further investigate with increased sensitivity the model independent result of DAMA/NaI and will improve the corollary quests for the candidate particle. At present, a third generation R&D toward a possible ton NaI(Tl) set-up, we proposed in 1996 [26], is in progress.

References

1. R. Bernabei et al., *Nuovo Cimento* A **112**, 545 (1999).
2. R. Bernabei et al., *La Rivista del Nuovo Cimento* **26** (2003) 1-73 (*astro-ph/0307403*).
3. R. Bernabei et al., *Phys. Lett.* B **389**, 757 (1996).
4. R. Bernabei et al., *Nuovo Cimento* A **112**, 1541 (1999).
5. R. Bernabei et al., *Phys. Lett.* B **408**, 439 (1997); P. Belli et al., *Phys. Lett.* B **460**, 236 (1999); R. Bernabei et al., *Phys. Rev. Lett.* **83**, 4918 (1999); P. Belli et al., *Phys. Rev.* C **60**, 065501 (1999); R. Bernabei et al., *Phys. Lett.* B **515**, 6 (2001); F. Cappella et al., *Eur. Phys. J. direct* C **14**, 1 (2002).
6. R. Bernabei et al., *Phys. Lett.* B **424**, 195 (1998).
7. R. Bernabei et al., *Phys. Lett.* B **450**, 448 (1999).

8. P. Belli et al., *Phys. Rev.* D **61**, 023512 (2000).
9. R. Bernabei et al., *Phys. Lett.* B **480**, 23 (2000).
10. R. Bernabei et al., *Eur. Phys. J.* C **18**, 283 (2000).
11. R. Bernabei et al., *Phys. Lett.* B **509**, 197 (2001).
12. R. Bernabei el al., *Eur. Phys. J.* C **23**, 61 (2002).
13. P. Belli et al., *Phys. Rev.* D **66**, 043503 (2002).
14. R. Bernabei et al., *Nuovo Cimento* A **103**, 767 (1990); *Nuovo Cimento* C **19**, 537 (1996); *Astrop. Phys.* **5**, 217 (1996); *Phys. Lett.* B **387**, 222 (1996) and *Phys. Lett.* B **389**, 783 (1996) (err.); *Phys. Lett.* B **436**, 379 (1998); *Phys. Lett.* B **465**, 315 (1999); *Phys. Lett.* B **493**, 12 (2000); *New Journal of Physics* **2**, 15.1 (2000); *Phys. Rev.* D **61**, 117301 (2000); *Eur. Phys. J. direct* C **11**, 1 (2001); *Nucl. Instrum. Methods* A **482**, 728 (2002); *Phys. Lett.* B **527**, 182 (2002); *Phys. Lett.* B **546**, 23 (2002); in "Beyond the Desert 03", Springer (2004) 541.
15. R. Bernabei et al., *Astrop. Phys.* **7**, 73 (1997); R. Bernabei et al., *Nuovo Cimento* A **110**, 189 (1997); P. Belli et al., *Nucl. Phys.* B **563**, 97 (1999); P. Belli et al., *Astrop. Phys.* **10**, 115 (1999); R. Bernabei et al., *Nucl. Phys.* A **705**, 29 (2002); P. Belli et al., *Nucl. Instrum. Methods* A **498**, 352 (2003); R. Cerulli et al., *Nucl. Instrum. Methods* A **525**, 535 (2004).
16. K.A. Drukier et al., *Phys. Rev.* D **33**, 3495 (1986). K. Freese et al., *Phys. Rev.* D **37**, 3388 (1988).
17. D. Smith and N. Weiner, *Phys. Rev.* D **64**, 043502 (2001).
18. K. Freese et al. *astro-ph/0309279*.
19. G. Prezeau et al., *Phys. Rev. Lett.* **91**, 231301 (2003).
20. D.E. Groom et al., *Eur. Phys. J.* **C15**, 1 (2000).
21. A. Bottino et al., *Phys. Rev.* D **67**, 063519 (2003); A. Bottino et al, hep-ph/0304080; D. Hooper and T. Plehn, MADPH-02-1308, CERN-TH/2002-29, [hep-ph/0212226]; G. Bélanger, F. Boudjema, A. Pukhov and S. Rosier-Lees, hep-ph/0212227.
22. A. Bottino et al., *Phys. Rev.* D **69**, 037302 (2004).
23. K. Hagiwara et al., *Phys. Rev.* D **66**, 010001 (2002).
24. CDMS coll., *astro-ph/0405033*.
25. K. Belotsky et al., *hep-ph/0411093*.
26. R. Bernabei et al.,*Astrop. Phys.* **4**, 45 (1995); R. Bernabei, "Competitiveness of a very low radioactive ton scintillator for particle Dark Matter search", in the volume *The identification of Dark Matter*, World Sc. pub. 574(1997).

CDMS SOUDAN FIRST RESULTS AND STATUS

M. R. DRAGOWSKY

Case Western Reserve University
10900 Euclid Avenue
Cleveland, OH 44106-7079 USA
E-mail: dragowsky@case.edu

The Cryogenic Dark Matter Search (CDMS) Collaboration reports the first results from operations at the Soudan Underground Laboratory. Six cryogenic ZIP detectors were operated to obtain, after analysis cuts, a 19.4 kg-d Ge exposure during the last few months of 2003. A blind analysis yielded no candidate WIMP (Weakly Interacting Massive Particles) interaction events, leading to 90% confidence level upper limits for spin-independent WIMP-nucleon cross section of 4×10^{-43} cm^2 for a 60 GeV/c^2 WIMP assuming a standard halo model. This limit places significant bounds on SUSY parameter space. The CDMS collaboration has a second, more extensive data set from 12 detectors under analysis and 18 additional detectors nearing completion for installation into Soudan during Fall 2004. Running during 2005 is expected to yield $\sim$1000 kg-d Ge exposure before analysis cuts.

1. Context and Strategy

WIMP dark matter is expected to generate nuclear recoils in terrestrial detectors. The energy spectrum would be a falling exponential with mean in the range of 10–50 keV and depending inversely on nuclear mass number. Neutrons having few-MeV kinetic energy will also produce nuclear recoils in the WIMP-search energy range, and therefore constitute a nonrejectable background. Electron recoils arising from radioactive backgrounds are far more common.

The CDMS approach to dark-matter direct detection is to reduce backgrounds through shielding, site location and materials selection and handling. Most significantly, we have developed detectors to achieve event-by-event nuclear- and electron-recoil discrimination in the quest to achieve a background-free experiment.

2. ZIP Detectors

Z-sensitive Ionization and Phonon (ZIP) detectors measure energy from ionizing radiation in two channels: ionization and athermal phonons. The detector is comprised of a single-crystal semiconductor (Ge or Si) as the target mass for WIMP interactions with lithographically patterned sensors on the surface. The substrate is a right circular cylinder with nominal height 1.0 cm and diameter 7.62 cm.

The ionization sensor is patterned on one planar surface and is segmented into inner and outer electrodes to reject particle interactions at the largest radii which suffer from reduced ionization collection.

The crystal surface is covered with four athermal phonons sensors. Particle interactions produce phonons out of thermal equilibrium with the lattice. The majority of energy in phonons will travel diffusively from the interaction site. A small fraction will travel ballistically and these athermal phonons reach the detector surface allowing μs leading-edge timing. Athermal phonons are also generated by electrons and holes drifting in the ionization field and colliding with the lattice, a process known as the Nagonov-Trofimov-Luke (NTL) effect. The ionization sensor is operated at low-voltage bias (3V Ge, 4V Si) so that phonon production by the NTL effect is limited to 50% of the total phonon yield. Low-voltage bias still results in sub-microsecond leading edge timing in ionization and preserves phonon timing information that allows position localization along the longitudinal axis of the crystal.

Event-by-event nuclear- and electron-recoil discrimination is achieved by taking the ratio of the signal in the ionization and phonon sensors. The ratio is termed the ionization yield and normalized to unity for electron-recoil events as determined by gamma-ray calibration. Neutron calibration by means of a ^{252}Cf fission neutron source determines the nuclear-recoil ionization yield response, generally 0.3 but varying slightly with energy, as shown in Fig. 1. Clear separation between the electron- and nuclear-recoil bands is evident in the figure. Statistically, electromagnetic backgrounds are rejected with 99.995% efficiency based on ionization yield and no events in other detectors.

3. Backgrounds & Calibration

Neutrons from cosmic-ray induced muons are a significant problem for direct detection dark matter experiments. Radioactive decay also produces neutrons that can mimic the dark matter signal. Selecting materials for de-

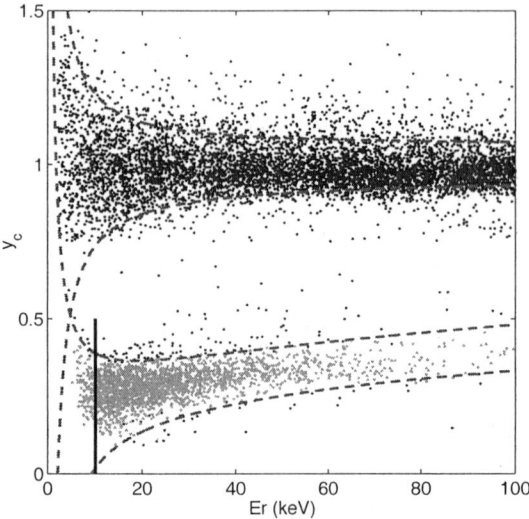

Figure 1. Ionization yield (Y_c) versus recoil energy (E_r) for calibration data with a ^{252}Cf source for Z2, Z3 and Z5 in Tower 1 showing the $\pm 2\sigma$ gamma band and the $\pm 2\sigma$ nuclear-recoil band. The vertical line is the analysis threshold. Events with ionization yield less than 0.75 are only shown if they pass phonon timing cuts to present more clearly the nuclear recoil event distribution.

tectors and the cryogenic enclosure that are radiochemically pure suppresses this rate significantly. Additionally, neutron absorbing material staged in the near-detector volume will reduce this background from environmental neutrons.

The combined strategy to eliminate neutron backgrounds is to operate in a deep site that reduces the primary muon flux and to deploy shielding and a veto system. The deep site for CDMS-II is the Soudan Underground Laboratory located in northern Minnesota in the United States. Suitable infrastructure was established for the experiment over the period 2002-2003. An RF-shielded clean room houses detectors and dilution refrigerator to achieve a radioactively clean and low noise environment.

The shielding material is depicted in Fig. 2. The muon veto is comprised of 40 plastic scintillator panels using 3" photomultiplier tubes for light collection. Internal shielding features 40 cm polyethylene for neutron moderation, 22.5 cm lead for gamma-ray shielding and 10 cm polyethylene immediately surrounding the Cu icebox cans for additional neutron moderation.

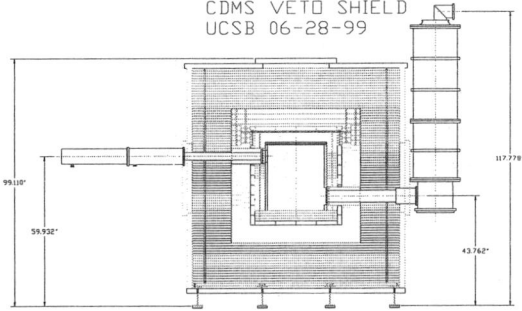

Figure 2. The shielding and veto for the CDMS Soudan apparatus are shown in plan view. The dilution refrigerator is depicted on the right hand side of the diagram. It is coupled to the icebox cans via the cold-stem, which penetrates, in turn, the veto, external polyethylene, Pb (large bricks) and inner polyethylene.

The expected neutron background for this data set has been evaluated for those events coincident with a muon veto signal. Single-scatter and multiple-scatter event rates were obtained and compared to projections obtained through Monte Carlo simulation. These simulations employed detailed geometric modeling of the detectors and shielding material. While these events will be rejected from the WIMP-search data set, they provide a control group to understand our treatment of neutron production and transport. The production is concentrated in the Pb shielding and we expect 1.99 ± 0.44 and 0.89 ± 0.18 neutrons in the WIMP search data set from Ge and Si detectors, respectively.

The veto-coincident data set contained no single- nor multiple-scatter events. This is statistically consistent with the prediction, and does validate the neutron studies to date. Detailed systematic studies must await future data sets with higher exposure.

The relative timing between the leading edge signal from the four phonon quadrants allows for discrimination between low-ionization yield events that are due to β-decay interactions on the detector surface and nuclear recoils occurring in the crystal bulk. These cuts were established on photon calibration events fitting a profile of events with low-yield in two adjacent detectors. The physical model is that an interaction occurs in the surface one detector and causes an electron to be ejected out of the material and into the adjacent detector. Cuts were set to reject almost all such events in the photon calibration data.

These cuts were then applied to the neutron calibration data to establish the nuclear recoil acceptance efficiency. The estimated leakage of surface events into the nuclear recoil band is 0.7±0.3 events in this data set.

Energy calibration is performed *in situ* using ^{133}Ba featuring 356- and 384-keV gamma-rays that penetrate the copper icebox cans. The ionization energy scale in the WIMP-search window below 100 keV is set by this calibration. It is verified by examining a 10.4-keV x-ray resulting from ^{68}Ga and ^{71}Ga beta-decay which is produced in the Ge detectors by cosmogenic activation. The energy resolution is about 5% in the WIMP-search window.

4. First Soudan Analaysis

The results reported in this paper were obtained between 11 October 2003 and 11 January 2004 and were first reported this spring[1]. We obtained 52.6 good live days with the four Ge and two Si detectors of Tower 1 (six close stacked ZIP detectors labeled as Z1(Ge), Z2(Ge), Z3(Ge), Z4(Si), Z5(Ge) and Z6(Si) from top to bottom). Tower 1 in the identical configuration has been operated at both our shallow site Stanford Underground Facility (SUF)[2] and now at our deep underground site (Soudan).

The blind analysis described above yielded no candidate WIMP events. Following the realization that an energy estimation algorithm used with saturated pulses was inadvertently applied to nearly all data the analysis was repeated. The appropriate energy estimator resulted in slightly improved energy resolution and one nuclear recoil candidate event was identified. This finding is consistent with the expected leakage from surface events.

The finding of no candidate WIMP events for this exposure in the blind analysis was converted into a 90% C.L. upper limit on the scalar (spin-independent) coherent WIMP-nucleon cross section making standard assumptions[3]. The minimum cross section is 4×10^{-43} cm^2 associated with WIMP mass 60 GeV/c^2. Limits from the non-blind result were derived using the optimum interval method[4].

5. CDMS II Expectations

The data set reported above was followed by a two-tower data set during the months March–July 2004. Improved noise performance was observed in the detectors of tower 2, which featured 4 Si and 2 Ge detectors. The neutron backgrounds sensitivity will be enhanced by the factor of three increase in Si detector mass in this two-tower run.

The run was ended in July to improve the cryogenic and vacuum systems and to install 3 additional detector towers. These detectors will be commissioned during the last months of 2004 and operated in WIMP-search mode during calendar year 2005. The total mass will be increased to 4.75 kg Ge and 1.1 kg Si. The anticipated data set is expected to provide a factor of 20 improved sensitivity throughout the WIMP mass range 10-1000 GeV/c^2. The expected reach is shown (dotted blue) in Fig. 3 along with the blind (solid blue) and non-blind (dash blue) analysis limits. DAMA[5] (dark green) and Edelweiss[6] (dark red) latest results are shown. Representative supersymmetric models appear as filled regions: Baltz and Gondolo (cyan[7]), Kim et al. (red[8]) and Bottino et al. (yellow[9]).

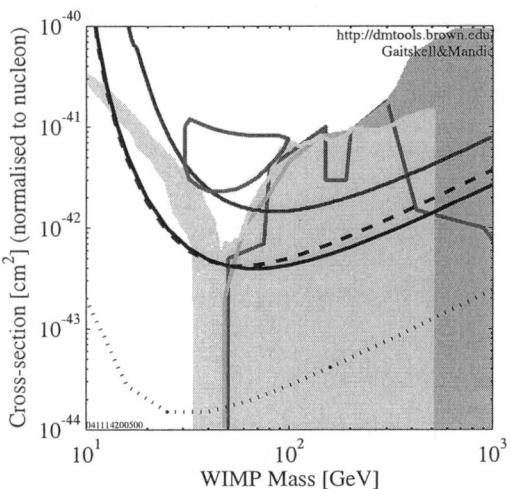

Figure 3. Selected spin-independent WIMP-nucleon scattering results and expectations are shown here. See text for details.

References

1. D. Akerib et al., (CDMS Collab.) Phys. Rev. Lett. **93**, 211301 (2004).
2. D. Akerib et al., (CDMS Collab.) Phys. Rev. D **68**, 082002 (2003).
3. J.D. Lewin and P.F. Smith, Astropart. Phys. **6**, 87 (1996).
4. S. Yellin, Phys. Rev. D **66**, 32005 (2002).
5. R. Bernabei et al., Riv. Nuovo Cim. **26**, 1 (2003).
6. A. Benoit et al., Phys. Lett. **B545**, 43 (2002);
7. E.A. Baltz and P. Gondolo, Phys. Rev. D **67**, 063503 (2003).
8. Y.G. Kim et al., J. High Energy Phys. **0212**, 034 (2002).
9. A. Bottino et al., Phys. Rev. D **69**, 037302 (2004).

FROM EDELWEISS-I TO EDELWEISS-II

V. SANGLARD
FOR THE EDELWEISS COLLABORATION
Institut de Physique Nucléaire de Lyon,
Université Claude Bernard Lyon 1,
43, Bd du 11 Novembre 1918,
69622 Villeurbanne Cedex, FRANCE
E-mail: sanglard@ipnl.in2p3.fr

The EDELWEISS experiment is a Direct Dark Matter Search using 320 g heat-and-ionization Ge cryogenic detectors. The final results obtained by the EDELWEISS-I stage corresponding to a total of 62 kg.day are presented. The status of EDELWEISS-II, involving in a first phase $\sim$ 10 kg of detectors and aiming to gain two orders of magnitude in sensitivity, is also described.

1. Introduction

The EDELWEISS experiment is dedicated to the search for non-baryonic cold dark matter in the form of WIMPs (Weakly Interactive Massive Particles). The direct detection principle consists in the measurement of the energy released by nuclear recoils produced in an ordinary matter target by elastic collisions of WIMPs from the galactic halo.

The EDELWEISS detectors are cryogenic Ge bolometers with simultaneous measurement of phonon and ionization signals. The comparison of the two signals provides an excellent event-by-event discrimination between nuclear recoils (induced by WIMP or neutron scattering) and electronic recoils (induced by β or γ-radioactivity).

The experiment is located in the Modane Underground Laboratory in the tunnel connecting France and Italy under $\sim$1800 m of rock ($\sim$4800 mwe). In the laboratory, the resulting muon flux is 4 $\mu/m^2/d$ and the fast neutron flux has been measured [1] to be $\sim 1.6\times 10^{-6}$ cm^2/s.

During three years, 62 kg.day of data have been accumulated with five 320 g Ge detectors.

2. Experimental set-up

Between 2002 and 2003, three 320 g Ge detectors were operated simultaneously in a dilution cryostat with a regulated temperature of 17 ± 0.01 mK. A passive shielding made of paraffin (30 cm), lead (15cm) and copper (10 cm) surrounded the experiment [2,3].

The detectors [4] are made of a cylindrical Ge crystal with Al electrodes to collect ionization signals and a NTD heat sensor glued onto one electrode to collect the phonon signal. The top electrode is segmented in a central electrode and an annular guard ring and defines a fiducial volume corresponding to 57 ± 2 % of the total volume [5]. On four of the five detectors used in EDELWEISS-I an amorphous layer (either of Ge or Si) was deposited under the electrodes to improve charge collection of near surface events [6].

The resolutions and energy thresholds of the detectors are summarized in Ref. [5]. Detectors with an amorphous layer show a 99.99 % gamma rejection at a recoil energy of 100 keV and a 99.9 % gamma rejection at a recoil energy of 15 keV.

3. Edelweiss-I results

During the EDELWEISS-I stage (2000-2003), four physics runs have been performed with five detectors. In the three first runs, the trigger was the fast ionization signal. For the last run, the trigger was the phonon signal. Thanks to a better resolution and the absence of quenching factor on the phonon signal, the phonon trigger improves the efficiency at low energy for nuclear recoils. In this last configuration, a $\sim$ 100 % efficiency has been reached at 15 keV on the three detectors. With the ionization trigger a $\sim$ 100 % efficiency was reached at 20 keV or 30 keV, depending on the detector. The low-background physics data recorded in the phonon trigger configuration are shown in Fig. 1.

Considering the complete 62 kg.day data set, 60 events compatible with nuclear recoils have been recorded above a recoil energy of 10 keV. The corresponding energy spectrum is shown in Fig. 2, compared with simulations of theoretical spectrum for different WIMP masses, taking into account the recoil energy dependence efficiency[a] of all experimental configurations.

---

[a]The efficiency calculation takes into account thresholds, resolutions, the 90 % efficiency (1.65 σ) for the nuclear recoil band and the 99.9 % rejection (3.29 σ) of the electronic recoils.

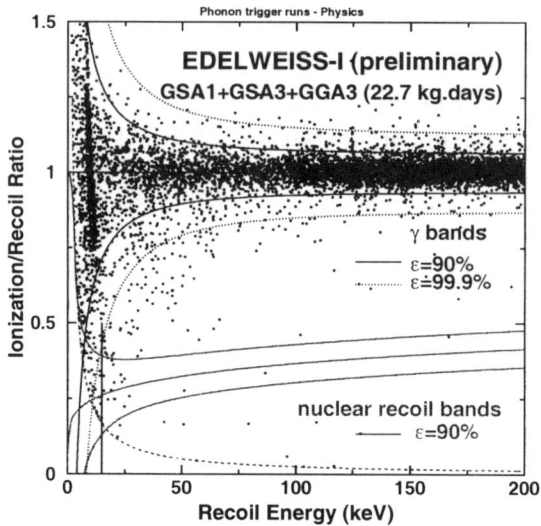

Figure 1. $\frac{E_L}{E_R}$ versus E_R (fiducial volume) for physics runs with the phonon trigger configuration.

The overall shape of the experimental spectrum is incompatible with WIMP masses above 20 GeV/c^2.

Setting our analysis threshold to 20 keV, above which the experimental efficiency is greater than 75 %, 23 events compatible with nuclear recoils have been observed. Considering all these events as possible WIMP interactions and taking into account the efficiency versus recoil energy function of each run[b], a conservative upper limit on the WIMP-nucleon cross-section as a function of the WIMP mass has been derived with the Optimum Interval Method [7]. This method allows to compute an exclusion limit in the presence of an unknown background. Fig. 3 shows the EDELWEISS-I spin independent exclusion limit, assuming a standard spherical and isothermal galactic WIMP halo with a local density of 0.3 GeV/c^2/cm^3, a rms velocity of 270 km/s, an escape velocity of 650 km/s and a relative Earth-halo velocity of 230 km/s. Limits from other running experiments are also shown

---

[b]For example, for the complete data set of Edelweiss-I a 50 % efficiency is reached for a recoil energy of 15 keV.

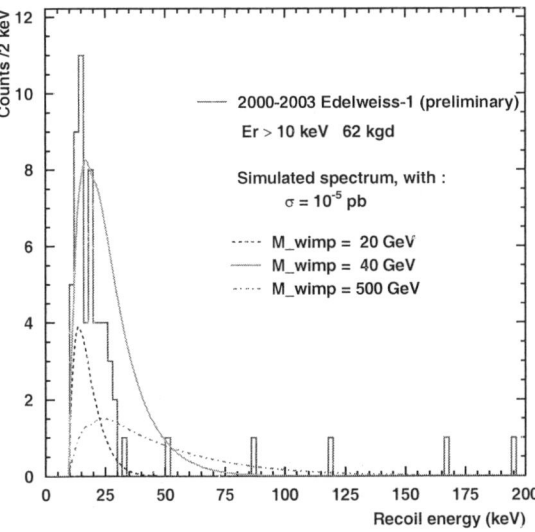

Figure 2. Energy spectrum for the EDELWEISS-I data, for $E_R > 10$ keV, compared to simulated theoretical WIMP spectrum for $M_{WIMP} = 20, 40, 500$ GeV/c^2.

on Fig. 3. With no background subtraction and an extended exposure, the new limit is consistent with the previous published one [8].

Although the EDELWEISS-I limit of Fig. 3 is derived assuming all events as possible WIMP candidates, the experimental data reveal some clues as to the nature of possible backgrounds. For example, a two detector coincidence between nuclear recoils has been recorded. This event is very likely a neutron-neutron coincidence, indicating that a certain fraction of events in Fig. 2 could be due to single hits by neutrons. Miscollected charge events, as indicated by the few events lying between electronic and nuclear recoil bands in Fig. 1, are another possible source of background, because they can simulate nuclear recoils. But with the present statistics, limited largely by the number of detectors (the EDELWEISS-I cryostat could not receive more than 3×320 g detectors), it is not possible to conclude any further.

4. Lessons for EDELWEISS-II

The second phase of the experiment will be EDELWEISS-II, with an expected sensitivity of 0.002 evt/kg/d. Specific improvements are aimed at

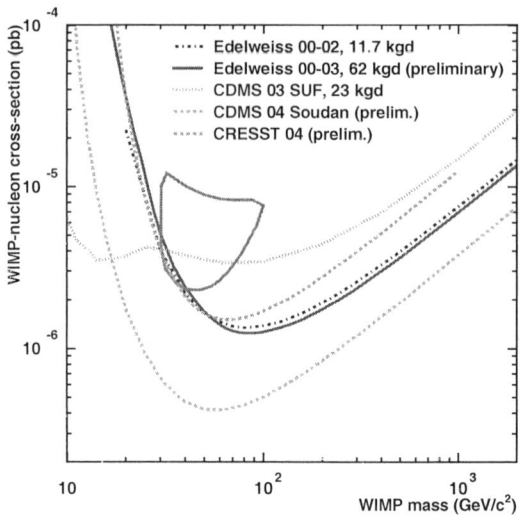

Figure 3. Preliminary spin independent exclusion limit for EDELWEISS-I without background subtraction compared to CDMS limit, when the experiment was operated at Stanford [9] and in the Soudan mine [10], and the CRESST limit [11]. The closed contour is the allowed region at 3σ C.L. from DAMA NaI1-4 annual modulation data [12].

reducing the possible background sources, that may have limited the sensitivity of EDELWEISS-I. In addition, the detector number will be increased up to 28 to achieve a Ge mass of $\sim$ 10 kg in a first stage.

4.1. Neutrons

The low energy neutron background, due to the radioactive surrounding rock, is attenuated by more than three orders of magnitude thanks to a 50 cm polyethylene shielding. In addition, a muon veto [14] surrounding the experiment will tag muons interacting in the lead shielding. The increased number of detectors will improve the possibility of detecting multiple interactions of neutrons.

4.2. Surface events

Surface events, namely interactions near electrodes, show a deficit of the charge collection. One of the R&D goals in EDELWEISS is the event-by-event identification of these miscollected events and their active rejection. A new generation of detectors has been developped with NbSi thin film sensors (instead of the NTD heat sensors for present detectors). They consist in a Ge crystal with two NbSi sensors acting also as electrodes for charge collection. These thin film sensors are sensitive to the athermal component of the phonon signal, acting as near-surface interaction tag [13]. Several tests have been made in the EDELWEISS-I setup with three 200 g Ge detectors showing a reduction by a factor 10 of the surface event rate while retaining a 50 % efficiency. Seven 400 g Ge detectors are being prepared in a first stage for EDELWEISS-II.

Furthermore, improved radiopurity and clean room conditions are expected to reduce the contaminations and the rate of surface electrons.

5. Conclusion

EDELWEISS-I experiment has reached its limit sensitivity near 10^{-6} pb, allowing the exclusion of some optimistic SUSY models. The goal for the future with EDELWEISS-II is to reach more favored models close to 10^{-8} pb. The EDELWEISS-I experiment was stopped in March 2004 to allow the installation of the second phase EDELWEISS-II. The first runs will be performed with 21 × 320 g Ge detectors with NTD heat sensor and 7 × 400 g Ge detectors with NbSi thin film sensor. Data taking in the new setup is scheduled for end-2005.

References

1. G. Chardin and G. Gerbier, in *Proceedings of th 4th International Workshop on Identification of Dark Matter* (IDM2002), eds N.J. Spooner
2. A. Benoit et al., *Phys. Lett. B*, **513**, 15 (2001)
3. Ph. di Stefano et al., *Astropart. Phys.*, **14**, 329 (2001)
   A. de Bellefon et al., *Astropart. Phys.*, **6**, 35 (1996)
4. X.F. Navick et al., *Nucl. Inst. Meth. A*, **444**, 361 (2000)
5. O. Martineau et al., *Nucl. Inst. Meth. A*, **530**, 426 (2004)
6. T. Shutt et al., *Nucl. Inst. Meth. A*, **444**, 340 (2000)
7. S. Yellin, *Phys. Rev. D*, **66**, 032005 (2002)
8. A. Benoit et al., *Phys. Lett. B* **545**, 43 (2002).
9. D.S. Akerib et al., *Phys. Rev. D*, **68**, 082002 (2003)
10. D.S. Akerib et al., astro-ph/0405033, submitted to *Phys. Rev. Lett*
11. G. Angloher et al., astro-ph/0408006, submitted to *Astropart. Phys.*
12. R. Bernabei et al, *Phys. Lett. B*, **480**, 23 (2000)
13. S. Marnieros et al, *Nucl. Inst. Meth. A*, **520**, 185 (2004)
14. L. Chabert, *Ph. D. Thesis (unplubished)*, Université Claude Bernard Lyon 1, France (July 7th, 2004), available on http://tel.ccsd.cnrs.fr

THE CRESST DARK MATTER SEARCH

B. MAJOROVITS, C. COZZINI, S. HENRY, H. KRAUS, V. MIKHAILIK,
A.J.B. TOLHURST, D. WAHL

*Dept. of Physics, University of Oxford, Keble Road, OX1 3RH, England,
email: majorovits@physics.ox.ac.uk*

Y. RAMACHERS

University of Warwick, England

G. ANGLOHER, P. CHRIST, D. HAUFF, J. NINKOVIC, F. PETRICCA,
F. PRÖBST, W. SEIDEL, L. STODOLSKY

Max Planck Institut für Physik, München, Germany

F. V. FEILITZSCH, T. JAGEMANN, W. POTZEL, M. RAZETI, W. RAU,
M. STARK, W. WESTPHAL, H. WULANDARI

Technische Universität München, Germany

J. JOCHUM

Universität Tübingen, Germany

C. BUCCI

Laboratori Nazionali del Gran Sasso, Italy

We present first competitive results on WIMP dark matter using the phonon-light-detection technique. A particularly strong limit for WIMPs with coherent scattering results from selecting a region of the phonon-light plane corresponding to tungsten recoils. The observed count rate in the neutron band is compatible with the rate expected from neutron background. CRESST is presently being upgraded with a 66 channel SQUID readout system, a neutron shield and a muon veto system. This results in a significant improvement in sensitivity.

1. Introduction

Despite persuasive indirect evidence for the existence of dark matter in the universe and in galaxies, the direct detection of dark matter remains one of

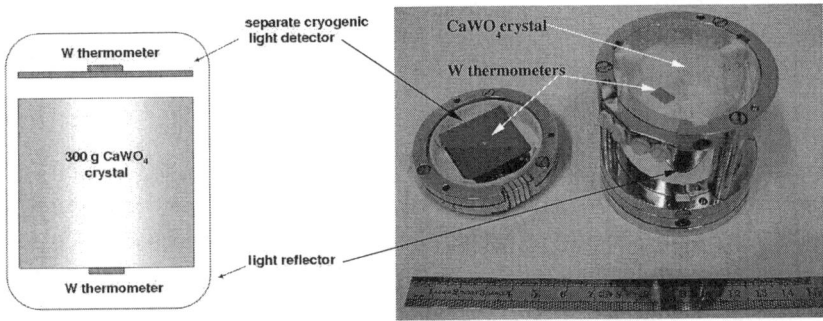

Figure 1. Left: Schematic sketch of the detector for coincident phonon and light measurement. Right: Picture of a detector module, "phonon" (right) and "light channels" (left).

the outstanding experimental challenges of present-day physics and cosmology. The Weakly Interacting Massive Particle (WIMP) is a well motivated candidate for cold dark matter in the form of the lightest supersymmetric particle. It is possible that it can be detected by laboratory experiments, particularly using cryogenic methods, which are well adapted to the small energy deposit expected[1].

2. The CRESST experiment

In the CRESST experiment we attempt to detect WIMP-nucleus scattering using cryogenic methods. Results from the first phase of CRESST using sapphire detectors have been reported previously[2]. For CRESST II[3] we have developed cryogenic detectors based on scintillating $CaWO_4$ crystals. When further equipped with a light detector these provide very efficient discrimination of nuclear recoils from radioactive γ and β backgrounds, down to recoil energies of about 10 keV. The mass of each crystal is about 300 g. Passive background suppression is achieved with a low background installation and the deep underground location at the Gran Sasso laboratory. The overburden of 3500 meter water equivalent reduces the surface muon flux to about $1/(m^2h)$. The detectors themselves are shielded against ambient radioactivity by low background copper and lead. A neutron shield and a muon veto, to be installed for CRESST-II, were not present for the data presented here. A four channel SQUID system allowed the simultaneous operation of only two phonon/light modules.

2.1. Detectors

A single detector module consists of a scintillating $CaWO_4$ crystal, operated as a cryogenic calorimeter (the "phonon channel"), and a nearby but separate cryogenic detector optimized for the detection of scintillation photons (the "light channel"). The phonon channel is designed to measure the energy transfer to a nucleus of the $CaWO_4$ crystal in a WIMP-nucleus elastic scattering. Since a recoil-nucleus differs substantially in the yield of scintillation light from an electron or a γ-quanta of the same energy, an effective background discrimination against γ-particles and electrons is obtained by a simultaneous measurement of the phonon and light signals. The prototype detector modules used here consist of a $300\,g$ cylindrical $CaWO_4$ crystal with $40\,mm$ diameter and height, and an associated cryogenic light detector[4,5]. The light detector is mounted close to a flat surface of the $CaWO_4$ crystal, and both detectors are enclosed by a housing made of a highly reflective polymeric multilayer foil. The arrangement is shown in Fig. 1. The detectors are operated at a temperature of about $10\,mK$ where the tungsten thermometer is in its transition between the superconducting and the normal conducting state, so that a small temperature rise of the thermometer leads to a relatively large rise of its resistance which is measured by means of a two-armed parallel circuit. One branch of the circuit has the superconducting film and the other comprises a reference resistor in series with the input coil of a SQUID, which provides a sensitive measurement of current changes. A rise in the thermometer resistance and so an increase in current through the SQUID coil is then observed as a rise in SQUID output voltage. Incoming pulses are recorded using a 16-bit transient recorder. For a more detailed description of the experimental setup see[6].

2.2. Temperature stability and energy calibration

To monitor the long term stability of the thermometers particle-event like heater test pulses with a range of discrete energies in the energy region of interest were sent every $30\,s$ throughout both dark matter and calibration runs. The response to these proved to be stable within the energy resolution of the detectors. The accuracy of the energy calibration from 10 to 40 keV, as relevant for the WIMP search is in the range of a few percent. This can be inferred from a peak at 47.1 keV which appeared in the energy

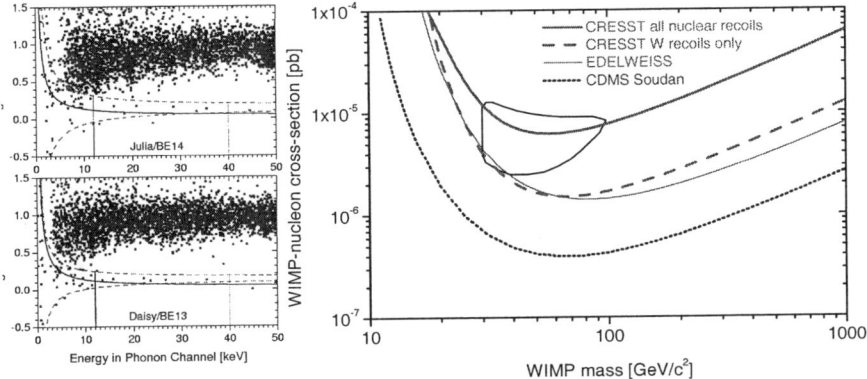

Figure 2. Left: Low energy event distributions in the dark matter run for the two modules. Right: Coherent scattering exclusion limits from the dark matter run. The enclosed region represents the claim of a positive signal by the DAMA collaboration[7].

spectrum of the phonon channel with a rate of (3.2 ± 0.5) $counts/day$. We associate this peak with an external ^{210}Pb contamination resulting in a γ-line at 46.5 keV. The FWHM of the measured peak is 1.0 keV, identical with that for the heater pulses. This good resolution confirms the stability of the response during the dark matter run.

3. Results and discussion

The results shown here were obtained in measurements taken between January 31 and March 23, 2004. The total exposures after all cuts are 9.809 $kg\,d$ and 10.724 $kg\,d$ for the two modules. The low-energy data from the dark matter run is presented in Fig. 2 as a scatter plot. The determination of a nuclear recoil acceptance region in the phonon-light plane is based on a knowledge of the "quenching factor", i.e. the reduction of the light output of a nuclear recoil event relative to an electron/photon event of the same energy. From earlier measurements of recoil neutrons a quenching factor of $Q=7.4$ was determined[4]. From this and the light detector resolution the 90% acceptance band for nuclear recoils is calculated, which is shown in the left panel of Fig. 2 as the area below the upper dashed lines. If we attribute all 16 events from the two detector modules in the acceptance region (corresponding to a count rate of $R = (0.87 \pm 0.22)/(kg\,day)$) to WIMP interactions, we can set a conservative upper limit for the WIMP scattering cross section shown as the full line in Fig. 2. For details on the assumptions made see Ref.[6]. Monte Carlo simulations for our setup without neutron shield yield an estimate for the neutron background of about 0.6 $events/(kg\ day)$ for 12 $keV \leq E_{recoil} \leq 40\ keV$, in reasonable agreement with the observed rate[8].

Kinematic considerations and simulations show that the contribution to the neutron spectrum is dominated by recoil of oxygen nuclei within the $CaWO_4$ crystal[8]. Hence the measured quenching factor will be mostly due to oxygen recoils. However, if one assumes coherent scattering of WIMPs by the nucleus, the interaction cross section will be proportional to A^2. Thus WIMPs are mainly expected to interact with tungsten nuclei. In a seperate measurement the quenching factor of tungsten has been determined to be between $Q_W \geq 40$ [9] for 18 keV and 36 keV tungsten energies, thus WIMP interactions are expected to lie in a seperate band below the one determined from neutron recoils. The discrimination efficiency of the two bands will strongly depend on the energy resolution of the attributed light detector. In Fig. 2 the 90 % acceptance regions for recoil by tungsten are indicated by the area below the solid lines. Since we know that the light detector of one of the two modules did have a slightly deteriorated resolution we discard these data for the extraction of the WIMP coherent cross-section limit. For the better of the two modules there are no recoil events below the full line in the energy range from 12 to 40 keV. Using these data to set a limit we obtain the thick dashed line in the right panel of Fig. 2. As a check we lowered the threshold to 10 keV to include the two events at 10.5 keV and 11.3 keV below the tungsten line and obtained essentially the same curve. At a WIMP mass of 60 GeV/c^2, these tungsten limits, which were obtained without any neutron shielding, are identical to the limits set by EDELWEISS [10]. Very recent results from CDMS at the Soudan Underground Laboratory [11] have improved these limits by a further factor of four.

4. Outlook: Upgrade with neutron shield and a 66-channel SQUID readout system

The limits presented in section 3 were obtained using measurements that were taken without a neutron shield. Thus the sensitivity of these data is limited by the neutron background as well as by the limited exposure. Presently the CRESST setup is being upgraded with a 66-channel SQUID readout system[12]. This will allow the installation of 33 detector modules

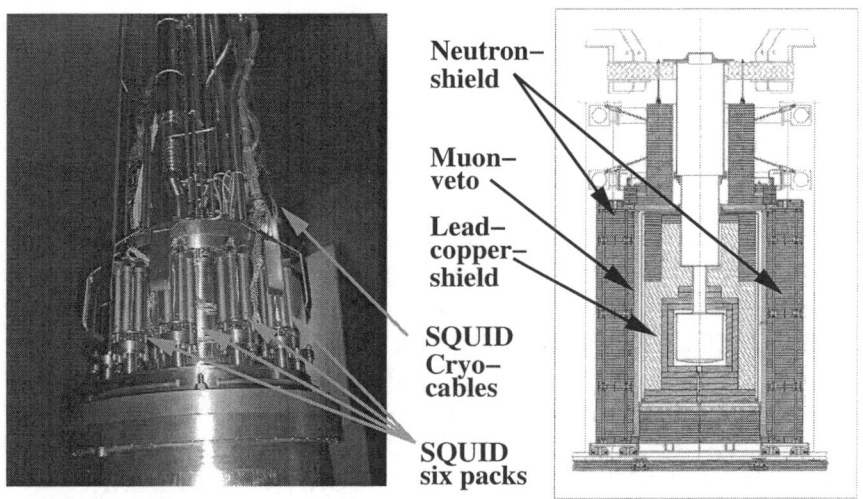

Figure 3. Left: First 24 SQUIDs installed on the cryostat insert at Gran Sasso. Right: schematic view of CRESST cryostat with neutron and muon shield.

providing 10 kg of target material. Together with the neutron shield that has been installed in October 2004 and the muon veto that is presently under construction this shoud improve CRESST's sensitivity to WIMP dark matter interactions by two orders of magnitude. Fig. 3 shows the first 24 SQUIDs that were installed onto the cryostat insert and cooled down to liquid helium temperature at the Gran Sasso underground laboratory. All the tested SQUIDs performed well at liquid helium temperature. The upgrade of the experiment will be completed in early 2005.

References

1. M.W. Goodman and E. Witten, *Phys. Rev.* **D31**, 3059 (1985).
2. G. Angloher et al., *Astropart. Phys.* **18**, 43 (2002).
3. Proposal for a second phase of CRESST, *MPI-PhE/2001-02*.
4. P. Meunier et al. *Appl. Phys. Lett.* **75**, 1335 (1999).
5. G. Angloher et al., *Nucl. Instr. Meth.* **A 520** (2004) 108-111. and F. Petricca, et al., *Nucl. Instr. Meth.* **A 520**, 193 (2004).
6. G. Angloher et al., submitted to *Astropart. Phys.*, **astro-ph/0408006**.
7. R. Bernabei et al., *Phys. Lett.* **B480**, 23 (2000).
8. H. Wulandari et al., in press at *Astropart. Phys.*, **hep-ex/0401032**.
9. J. Ninkovic, PHD thesis, to be published.
10. A. Benoit et. al., Phys. Lett. **B545** (2002) 43.
11. D. S. Akerib et al., submitted to *Phys. Rev. Lett.*, **astro-ph/0405033**.
12. S. Henry et al. *Nucl. Instr. Meth.* **A 520**, 588 (2004).

ZEPLIN I: FIRST LIMITS ON NUCLEAR RECOIL RATE *

UK Dark Matter Collaboration

G. J. ALNER[a], H. M. ARAÚJO[b], G. J. ARNISON [a] J. C. BARTON[c],
A. BEWICK[b], C. BUNGAU[b], B. CAMANZI[a], M. J. CARSON[d],
D. DAVIDGE[b], J. V. DAWSON[b], G. J. DAVIES[b], J. C. DAVIES[d],
T. J. DURKIN[a], T. GAMBLE[d], S. P. HART[a],
R. J. HOLLINGWORTH[d], G. J. HOMER[b], A. S. HOWARD[b],
I. IVANIOUCHENKOV[a,b], W. G. JONES[b], M. K. JOSHI[b],
V. A. KUDRYAVTSEV[d], T. B. LAWSON[d], V. LEBEDENKO[b],
M. J. LEHNER[d], J. D. LEWIN[a], P. K. LIGHTFOOT[d], I. LIUBARSKY[b],
R. LÜSCHER[a,b], J. E. MCMILLAN[d], B. MORGAN[d], A. NICHOLLS[a],
G. NICKLIN[d], S. M. PALING[d], R. M. PREECE[a], J. J. QUENBY[b],
J. W. ROBERTS[a,d], M. ROBINSON[d], N. J. T. SMITH[a], P. F. SMITH[a],
N. J. C. SPOONER[d], T. J. SUMNER[b], D. R. TOVEY[d]

[a] *Particle Physics Department, Rutherford Appleton Laboratory, UK*
[b] *Blackett Laboratory, Imperial College London, UK*
[c] *Department of Physics, Queen Mary, University of London, UK*
[d] *Department of Physics & Astronomy, University of Sheffield, UK*

First preliminary results from the Zeplin I WIMP dark matter detector are presented. They are based on the measurements of scintillation pulse shapes in liquid xenon target with 3.2 kg fiducial mass. The detector was located in the Boulby Underground Laboratory (North Yorkshire, UK) at a depth of 2800 m w. e. The data collected dirung 91.5 days did not reveal the presence of second population of events faster than expected from gamma-induced electron recoils. Based on these results new limits were set on the spin-independent WIMP-nucleon and spin-dependent WIMP-neutron cross-sections.

1. Introduction

It is believed that 20% of the Universe may consist of non-baryonic dark matter. Supersymmetric theories suggest a good candidate – neutralino or

---

*This work is supported by PPARC

Weakly Interacting Massive Particle (WIMP). Liquid and two-phase xenon detectors provide excellent opportunity for direct WIMP searches. The heavy nuclei of xenon have the advantage of having a large spin-independent coupling. The chemical inertness and isotopic composition of liquid xenon provide intrinsically high purity and routes, in principle, to further purification using various techniques. Low-radioactive materials can be used for vessel and internal components to reduce the internal background. Lead and hydrocarbon shielding help to protect the detector against neutrons from rock. Nuclear recoil discrimination in liquid xenon is feasible by measuring both the scintillation light and the ionisation produced during an interaction, either directly or through secondary recombination. Discrimination between nuclear and electron recoils in liquid xenon is expected to be significant based on laboratory tests and theoretical considerations (see, for example, Ref. [1] and reference therein). Any recoil in liquid Xe gives rise to both ionisation and excitation of Xe atoms. The de-excitation result in the emission of 175 nm photons from either singlet (with decay time ~ 3 ns) or triplet (~ 27 ns) states. The ratio single/triplet is several times higher for nuclear recoils compared to electron recoils. In the absence of an electric field, the ions recombine with electrons to produce excited Xe atoms again. The recombination time is smaller for nuclear recoils, than electron recoils.

The UK Dark Matter Collaboration has been running the Dark Matter programme at Boulby Mine (North Yorkshire, UK) at a vertical depth of 2800 m w. e. for more than a decade. One of the key part of the programme in its first stage is the liquid xenon dark matter detector Zeplin I.

2. Zeplin I detector

The ZEPLIN I detector (ZonEd Proportional scintillation in LIquid Noble gases – shown in Fig. 1) consisted of liquid Xe with 3.2 kg fiducial mass (about 5 kg total xenon mass) incased in a copper vessel and viewed by three 8 cm diameter ETL 9265Q PMTs through silica windows. The vessel is lined internally with 5mm thick diffuse PTFE reflector. The detector itself was surrounded from all sides but the top by a 0.93 tonne active scintillator veto, its function being to veto gamma events from the PMTs and other detector components.

PMT signals were digitised with a 1 ns accuracy using a digital oscilloscope driven by a Labview based software at the beginning of the experiment and using an Acqiris CompactPCI based DAQ system later on. The detector was triggered by a 3-fold coincidence of single photoelectron pulses

 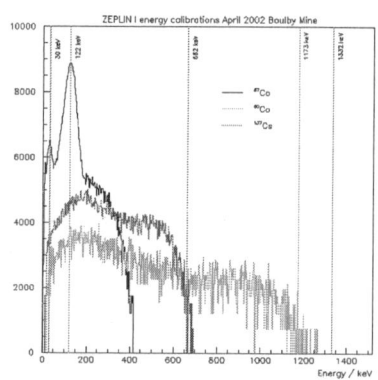

Figure 1. Computer view of ZEPLIN I showing the target viewed by 3 PMTs, encased in an active veto system.

Figure 2. Energy calibration of the Zeplin I detector with various sources.

in each tube. With a light yield of at least 1.5 photoelectrons/keV in the data runs, this gave a 2 keV threshold. A separate digitiser channel was used for the summed signals from the 10 veto PMTs.

3. Calibrations and data analysis

ZEPLIN I had better sensitivity than NaI detectors due to its improved discrimination at low energies. Background discrimination was possible due to the difference in the characteristic time between nuclear and electron recoil pulses. Our standard procedure of data analysis involved calculation of the mean time and amplitude for each scintillation pulse. To eliminate noise pulses and also local PMT background events in the 'turret regions' the following pulse acceptance criteria were imposed: (i) 'trigger cut': requirement of at least one pe in each of the three PMTs; (ii) 'turret cut': rejection of events with more than 67% of photoelectrons in one PMT. (iii) 'noise cut': rejection of events for which, in any single PMT, the mean time of the pulse is less than 10 ns and the number of photoelectrons is more than 2.5 (the latter criteria being set by inspection of a population of fast noise events). These cuts result in some loss of genuine events, and the resulting efficiency factors were obtained as a function of the number of photoelectrons from simulations and data.

Daily energy calibration was performed with a ^{57}Co source automatically placed between target and veto. The 122 keV γs were absorbed within ≈3 mm of their path in the bottom part of the target, making it a calibration point source. A 30 keV K-shell X-ray peak was also observed in the spectrum, its presence was confirmed through the GEANT4 simulation. A full light collection simulation was performed, showing non-uniformity of light collection efficiency. This affected the measured energy of an event and was observed in higher energy gamma calibrations (^{60}Co, ^{137}Cs sources): as different parts of the target were illuminated, the peak position was shifted due to the reduction in light yield. The observations matched well the light collection efficiency simulation. Example spectra from different gamma-sources are shown in Fig. 2.

The mean times of the scintillation pulses followed a gamma density distribution with characteristic times, which were very much different (by about a factor of 2 at low energies) for electron and nuclear recoils. Electron and nuclear recoils thus gave rise to two populations with different characteristic times. Calibration of the detector was done at the surface, prior to moving the detector underground, using neutron source and ambient neutrons, which produce nuclear recoils, and gamma-ray source, which produces electron recoils via Compton scattering. Nuclear recoils were also expected from WIMP-nucleus interactions, whereas electron recoils from gammas constituted the main background. During the neutron source runs, the neutron tagging (using a CsI crystal to detect the simultaneous 4.44 MeV gamma) suppresses the trigger rate by about a factor of 100, making the ambient neutron population negligible. Typical distributions of mean times of scintillation pulses from neutron calibration runs together with fits using the 'gamma' density functions to both gamma and neutron populations are shown in Fig. 3. The additional population resulting from the neutrons can be observed in Fig. 3 but none of the underground runs, at any given energy interval, showed such a population (see Fig. 4).

Minimum χ^2 fits for several energy intervals were obtained for neutron calibration runs assuming two populations of events, thus yielding for each case a best fit mean time ratio $r_{ng} = \tau_{0n}/\tau_{0\gamma}$. The two populations have similar values for the width parameters w, consistent with the expected dependence of w on the number of photoelectrons. The analysis presented here is based on these surface calibrations, owing to technical issues preventing a more extended underground neutron calibration. The best fits to the data showed a decrease of r_{ng} with energy, from ≈ 0.6 at 20-30 keV to 0.42-0.44 in the energy range 3-8 keV relevant to the dark matter limits. However, taking account of the errors in Fig. 3, we used a conservative approach assuming a constant ratio $r_{ng} = 0.5$ below 10 keV.

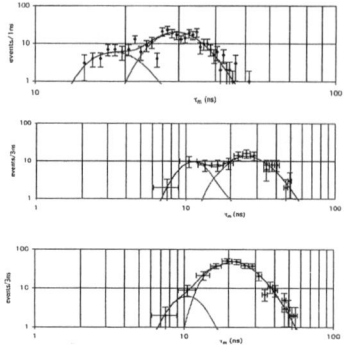

Figure 3. Mean time distributions from neutron calibration runs of the Zeplin I detector: top - neutrons from the source, 20-30 keV, $r_{ng} = 0.64 \pm 0.04$; middle - ambient neutrons, 3-7 keV, $r_{ng} = 0.42 \pm 0.07$; bottom - neutrons from the source, 3-8 keV, $r_{ng} = 0.44 \pm 0.09$.

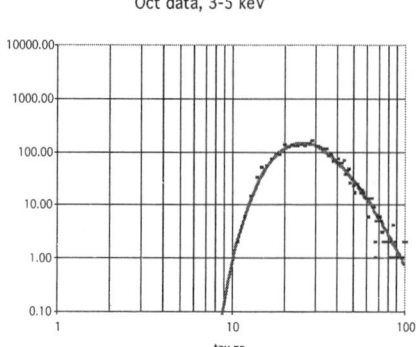

Figure 4. Mean time constant distribution from underground run: measured energy 3-5 keV. Only one population of events is seen unlike runs with neutron source and ambient neutrons.

4. Results

Slopes of the left-hand sides of the mean time distributions from data and calibration (with gamma source) runs were used to demonstrate that the mean nuclear recoil signal was consistent with zero for all runs and all energy bins of interest (within the energy range of 2-10 keV). Our simulations showed that the slopes were very sensitive to the relative intensity of a fast population that could be attributed to nuclear recoils. Comparing the slopes of the distributions from data and gamma calibration (with gamma source) runs we found that they are the same within statistical errors showing no fast population in the data. Based on the absence of nuclear recoils in underground data, the 90% C.L. upper limits on the number of nuclear recoils were extracted by comparing the measured number of events on the left-hand side of the mean time distribution with the expected number for the background events (from the gamma density fit).

The upper limits on nuclear recoil rate for 3 runs and all energy bins, after correction for trigger and cut efficiencies, were converted into the limits on the WIMP-nucleon cross-sections as a function of WIMP mass applying the procedure described in Ref. [2]. A quenching factor of 0.22 was used as a ratio of measured energy to nuclear recoil energy [3]. Energy resolution of $\sigma/E = 1.2/\sqrt{E}$, found to be stable for all runs included in the data analysis, and light collection matrix which described the non-uniformity of the light collection described above, were taken into account. Standard spherical, isothermal halo with Maxwellian velocity distribution of WIMPs with $v_0 = 220$ km/s and WIMP density of 0.3 GeV/cm^3 was assumed. In the spin-dependent case, we adopted a method described in Ref. [4] and use form factors and spin factors calculated in Ref. [5]. The preliminary upper limit on the spin-independent WIMP-nucleon cross-section from 293 kg×days of data is plotted in Fig. 5 in comparison with the new preliminary NAIAD limit and some other world-best limits. Fig. 6 shows the spin-dependent limit on WIMP-neutron cross-section (with form factors for higgsino with Bonne-A potential [5]).

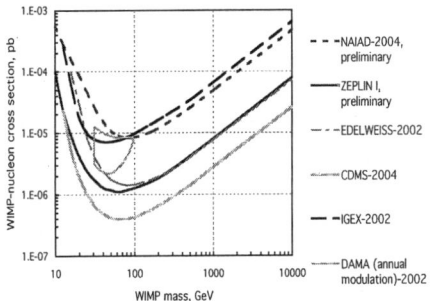
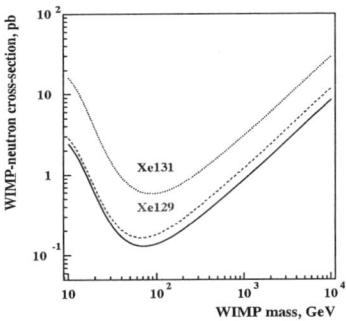

Figure 5. Upper limits (90% C.L.) on spin-independent WIMP-nucleon cross-section from Zeplin I, NAIAD and some other experiments.

Figure 6. Upper limits on spin-dependent WIMP-neutron cross-section from Zeplin I.

5. Conclusions

We presented the first results from WIMP dark matter search with a liquid xenon detector Zeplin I. Analysis of 293 kg×days of data did not reveal fast pulses which could be associated with nuclear recoils expected from WIMP-nucleus interactions. Based on this analysis, preliminary upper limits on spin-independent WIMP-nucleon and spin-dependent spin-neutron cross-sections were reported.

References

1. G. J. Davis et al. *Phys. Lett.* **B320**, 395 (1994).
2. J. D. Lewin and P. F. Smith. *Astroparticle Physics* **6**, 87 (1996).
3. D.Akimov et al. *Physics Letters* **B**, (2002).
4. D. R. Tovey et al. *Physics Letters* **B488**, 17 (2000).
5. M. T. Ressell and D. J. Dean. *Phys. Rev.* **C56**, 535 (1997).

LIMITS ON WIMP CROSS-SECTIONS FROM THE NAIAD EXPERIMENT AT BOULBY *

G. J. ALNER[a], H. M. ARAÚJO[b], G. J. ARNISON[a], J. C. BARTON[c],
A. BEWICK[b], C. BUNGAU[b], B. CAMANZI[a], M. J. CARSON[d],
D. DAVIDGE[b], E. DAW[d], J. V. DAWSON[b], G. J. DAVIES[b],
J. C. DAVIES[d], C. DUFFY[d], T. J. DURKIN[a], T. GAMBLE[d],
S. P. HART[a], R. J. HOLLINGWORTH[d], G. J. HOMER[b],
A. S. HOWARD[b], I. IVANIOUCHENKOV[a,b], W. G. JONES[b],
M. K. JOSHI[b], J. KIRKPATRICK[d], V. A. KUDRYAVTSEV[d],
T. B. LAWSON[d], V. LEBEDENKO[b], M. J. LEHNER[d], J. D. LEWIN[a],
P. K. LIGHTFOOT[d], I. LIUBARSKY[b], R. LÜSCHER[a,b],
J. E. MCMILLAN[d], B. MORGAN[d], A. MURPHY[e], A. NICHOLLS[a],
G. NICKLIN[d], S. M. PALING[d], R. M. PREECE[a], J. J. QUENBY[b],
J. W. ROBERTS[a,d], M. ROBINSON[d], N. J. T. SMITH[a], P. F. SMITH[a],
N. J. C. SPOONER[d], T. J. SUMNER[b], D. R. TOVEY[d], E. TZIAFERI[d]

[a] *Particle Physics Department, Rutherford Appleton Laboratory, UK*
[b] *Blackett Laboratory, Imperial College London, UK*
[c] *Department of Physics, Queen Mary, University of London, UK*
[d] *Department of Physics & Astronomy, University of Sheffield, UK*
[e] *School of Physics, University of Edinburgh, UK*

The NAIAD experiment (NaI Advanced Detector) for WIMP dark matter searches at the Boulby Underground Laboratory (North Yorkshire, UK) ran from 2000 until 2003. We present the results of the analysis of 44.9 kg×years of data collected with 2 encapsulated and 4 unencapsulated NaI(Tl) crystals.

1. Introduction

The UK Dark Matter Collaboration (UKDMC) has been operating various detectors for non-baryonic dark matter searches at the Boulby Underground Laboratory (North Yorkshire, UK) for many years. Limits on the flux of weakly interacting massive particles (WIMPs) were set using the data from the first NaI(Tl) detector [1,2] and later improved with an array of several

---

*This work is supported by PPARC

NaI(Tl) crystals (NAIAD) [3]. NaI has an advantage of having two target nuclei with high and low masses, thus reducing uncertainties related to nuclear physics calculations. The detectors are sensitive to both spin-independent and spin-dependent interactions. The NAIAD experiment is complementary to other dark matter experiments at Boulby, such as ZEPLIN and DRIFT. NaI(Tl) detectors were also used as a diagnostic array to study background and systematic effects for other experiments at Boulby.

NAIAD was constructed in 2000-2001 with 5 unencapsulated crystals to allow better control of the crystal surface, to reduce the background due to surface events and to improve the light collection [3]. In 2002 two more encapsulated crystals were added to the array. Here we present results from the NAIAD experiment based on all data sets collected in 2000-2003.

2. The NAIAD experiment

The NAIAD array was sited in the underground laboratory at Boulby mine at a vertical depth of 1070 metres. It consisted of 7 NaI(Tl) crystals from different manufacturers with a total mass of about 55 kg. Two detectors contained encapsulated crystals, while 5 other crystals are unencapsulated. To avoid their degradation by humidity in the atmosphere, the unencapsulated crystals were sealed in copper boxes filled with dry nitrogen.

One of the crystals (DM70-Saclay) was previously running at Modane and then moved to Boulby in 2001. When encapsulated, it showed the anomalous fast population of events [4,5], which were explained as being due to implanted surface contamination of the crystal by an alpha-emitting isotope from radon decay [5,6]. It was de-encapsulated in 2002, polished and was running as part of the NAIAD array in 2002-2003. No fast population was seen in the crystal after de-encapsulation.

Each crystal was mounted in a 10 mm thick solid PTFE reflector cage and was coupled to 5 inch low background PMTs, ETL type 9390UKB, through 4-5 cm long quartz light guides. Temperature control of the system was achieved through copper coils outside the copper box. Chilled water was constantly pumped through the coils maintaining the temperature of the crystals stable to within a fraction of a degree during a single run. The temperature of the crystal, ambient air, water in the pipes and copper was measured by thermocouples. If, for any reason (for example, chiller failure), the variation of crystal temperature exceeded the predefined limit, the data from these periods were not included in the analysis.

Pulses from individual PMTs were integrated using a buffer circuit and then digitised using an Acqiris CompactPCI based DAQ system. Note that previous results [3] were obtained with a digital oscilloscope and Labview-based DAQ software. The digitised pulse shapes (5 μs total digitisation time, 10 ns digitisation accuracy) were passed to a computer running Linux OS and stored on disk. The gain of the PMTs was set to give about 2.5 mV per photoelectron. Low threshold discriminators were set to about 10 mV threshold, which corresponded to about 1-1.5 keV. A software threshold was set at 4 keV since no pulse shape discrimination was observed below this energy [3].

Copper boxes containing the crystals were installed in lead and copper "castles", to shield the detectors from background due to natural radioactivity in the surrounding rock. Wax and polypropylene neutron shielding (about 10 g/cm^2) around the castles was installed in Spring 2002 and the data collected in 2002-2003 were collected with neutron shielding in place.

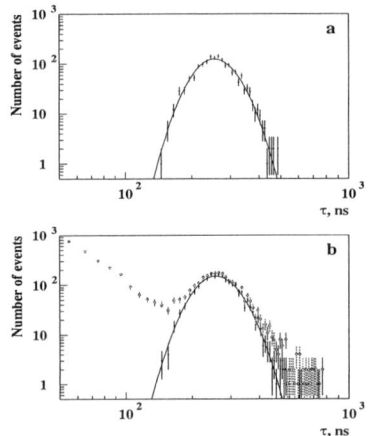
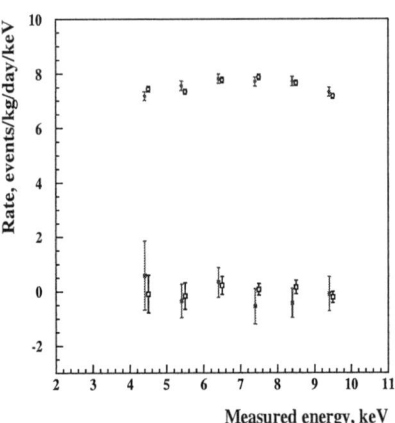

Figure 1. Typical time constant distributions at energies 5-6 keV: a – for gamma calibration run (after cuts), b – for data before and after cuts.

Figure 2. Energy spectra from two runs of one of the crystals (filled and open circles) and the nuclear recoil rate (filled and open squares) with error bars drawn at 90% CL.

3. Analysis procedure

Final analysis was performed on the sum of the pulses from the two PMTs attached to each crystal. The parameters of the pulses from each PMT were used to apply various cuts to remove the noise events from the analysis. Our standard procedure of data analysis involved fitting of a single exponential to each integrated pulse in order to obtain the index of the exponential, τ, the amplitude of the pulse, A, and the start time, t_s. Conversion of the pulse amplitude to the number of photoelectrons and the energy was achieved using pre-determined conversion factors, which come from energy and single photoelectron calibrations. Calibration with various gamma-ray sources provided also the energy resolution of crystals [3].

For each run the "energy – time constant" $(E - \tau)$ distribution was constructed. If all operational settings (including temperature) were the same for several runs, the $(E - \tau)$ distributions for these runs were summed together. To reduce PMT noise and, particularly, events where a spark (flash) in the dynode structure of one PMT was seen by both PMTs, various cuts described in details in Ref. [8] were applied. These cuts are an improvement upon those used in Ref. [3], having been shown to reject spurious events more efficiently while retaining a larger fraction of genuine scintillation pulses [8].

For any small energy bin (1 keV width, for example), the time constant distribution was approximated by a Gaussian in $\ln(\tau)$ (log(Gauss) function) [1,3,7] with three free parameters: mean time constant τ_o, width w and normalisation factor N_o. In experiments where a second population is seen (for example, nuclear recoils from a neutron source or possible WIMP-nucleus interactions), the resulting τ-distribution was fitted with two log(Gauss) functions with the same width w, since the width is determined mainly by the number of collected photoelectrons. The reference τ-distributions were obtained for all energies of interest (2-40 keV) and for all crystals with gamma-ray (^{60}Co) and neutron (^{252}Cf) sources [3,7].

4. Results and discussion

The data from 6 crystals were used to set the limits on WIMP-nucleus cross-section reported here. One crystal was excluded from the data analysis due to its small mass (4 kg) and high background rate.

An energy range from 4 to 10 keV was used in the data analysis. Figure 1 shows typical time constant distributions at 5-6 keV from data (before and after cuts) and a calibration run with gamma-ray source (after cuts). PMT noise events are seen at small values of time constant before cuts but are absent after the cuts were applied. Limits on the nuclear recoil rate at any particular energy were obtained by fitting the measured time constant distribution with two log(Gauss) functions having known parameters: mean time constants and widths, known from gamma and neutron calibrations. Free parameters were the total numbers of electron and nuclear recoils (see Ref. [3] for further details). When both positive and negative values for the numbers of nuclear and electron recoils were allowed, the best fit values for nuclear recoil rates were either positive or negative but normally within

1.5 standard deviations from zero. Total event rates and nuclear recoil rates extracted from the best two log-Gaussian fits to the data are shown in Figure 2 for two runs with crystal DM74. The error bars for nuclear recoil rates correspond to 90% CL. As can be seen, in none of the energy bins does the nuclear recoil rate deviate significantly from zero. Hence no contribution from WIMP-nucleus interactions were observed in these data. This was found to be true for all crystals. Since negative values are non-physical, to set upper limits on nuclear recoils we restricted the rates to non-negative values.

The limits obtained on the nuclear recoil rate for each energy bin and each crystal were converted into limits on the WIMP-nucleon spin-independent and WIMP-proton spin-dependent cross-sections following the procedure described in Ref. [9] and used previously in Ref. [3]. Expected nuclear recoil spectra from WIMP-nucleus interactions were calculated for a standard spherical isothermal halo model. For the spin-dependent case a pure higgsino was assumed. The spin factors and form factors were computed for sodium and iodine nuclei on the basis of nuclear shell model calculations [10] with Bonn A potential. The quenching factors were taken as 0.275 for sodium and 0.086 for iodine recoils [11].

The limits on the cross-section for various energy bins, targets (sodium and iodine) and crystals were combined following the procedure described in Ref. [9] and used previously in Ref. [3], assuming the measurements for different energies and crystals were statistically independent. Figure 3a (3b) shows the NAIAD limits on WIMP-nucleon spin-independent (WIMP-proton spin-dependent) cross-sections as functions of WIMP mass based on the 44.9 kg×years of data. Also shown in Figure 3a is the region of parameter space favoured by the DAMA annual modulation signal [12]. Model-independent limits on spin-dependent WIMP-proton and WIMP-neutron cross-sections, calculated following the procedure described in Ref. [13], are plotted in Figure 4 (see Refs. [14,15] for a compilation of results).

5. Conclusions

Results from the NAIAD experiment for WIMP dark matter search at Boulby mine were presented. Pulse shape analysis was used to discriminate between nuclear recoils, possibly caused by WIMP interactions, and electron recoils due to gamma background. We obtained upper limits on the WIMP-nucleon spin-independent and WIMP-proton spin-dependent cross-sections based on the data accumulated by 6 NaI(Tl) crystals (44.9 kg×years exposure).

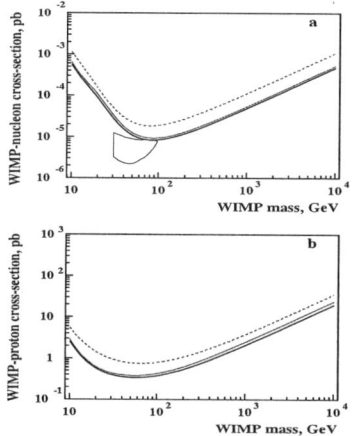
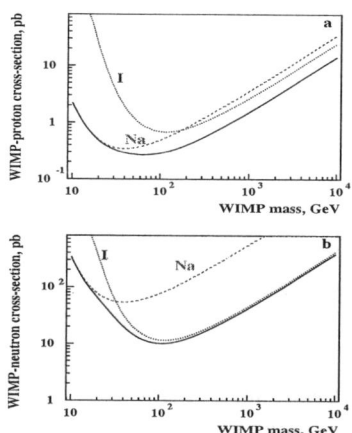

Figure 3. Upper limits (90% CL) on spin-independent WIMP-nucleon (a) and spin-dependent WIMP-proton (b) cross-sections from NAIAD: dashed curve – previously published limit [3], thin solid curve – limit from 2002-2003 data, thick solid curve – combined limit (2000-2003). Closed curve – DAMA allowed region [12].

Figure 4. Model independent upper limits on spin-dependent WIMP-proton (a) and WIMP-neutron (b) cross-sections from NAIAD.

6. Acknowledgements

We would like to thank Prof. G. Gerbier, Drs. J. Mallet, L. Mosca (Saclay), and Dr. C. Tao (CPPM, Marseille) for providing us with one of their crystals for the NAIAD experiment.

References

1. P. F. Smith et al. *Physics Letters B*, **379** (1996) 299.
2. J. Quenby et al. *Astroparticle Physics* **5** (1996) 249.
3. B. Ahmed et al. *Astroparticle Physics*, **19** (2003) 691.
4. G. Gerbier et al. *Astroparticle Physics* **11** (1999) 287.
5. V. A. Kudryavtsev et al. *Astroparticle Physics*, **17** (2002) 401.
6. N. J. T. Smith et al. *Physics Letters B*, **485** (2000) 9.
7. V. A. Kudryavtsev et al. *Physics Letters B*, **452** (1999) 167.
8. M. Robinson et al. *Nucl. Instrum. and Meth. in Phys. Res. A*, in press.
9. J. D. Lewin and P. F. Smith. *Astroparticle Physics* **6** (1996) 87.
10. M. T. Ressell and D. J. Dean. *Phys. Rev. C*, **56** (1997) 535.
11. D. R. Tovey et al. *Physics Letters B*, **433** (1998) 150.
12. R. Bernabei et al. *Physics Letters B*, **480** (2000) 23.
13. D. R. Tovey et al. *Physics Letters B*, **488** (2000) 17.
14. A. Benoit et al. (The EDELWEISS Collaboration), astro-ph/0412061.
15. F. Giuliani. *Phys. Rev. Lett.*, **93** (2004) 161301.

RECENT RESULTS FROM THE SIMPLE DARK MATTER SEARCH

TA GIRARD[1], F. GIULIANI[1], J.I. COLLAR[2], T. MORLAT[1], D. LIMAGNE[3], G. WAYSAND[3,4], M. AUGUSTE[4], D. BOYER[4], A. CAVAILLOU[4], H.S. MILEY[5], M. DA COSTA[6,7], R.C. MARTINS[7], J.G. MARQUES[6], A.R. RAMOS[6,1], A.C. FERNANDES[6], J. PUIBASSET[3], AND C. OLIVEIRA[6]

[1] *Centro de Física Nuclear, Universidade de Lisboa, 1649–003 Lisbon, Portugal*
[2] *Department of Physics, University of Chicago, Chicago IL, 60637 USA*
[3] *Groupe de Physique des Solides (UMR CNRS 75–88), Université Paris 7 & 6, 75251 Paris, France*
[4] *Laboratoire Souterrain à bas Bruit, 84400 Rustrel–Pays d'Apt, France*
[5] *Pacific Northwest National Laboratory, Richland, WA 99352 USA*
[6] *Instituto Tecnológico e Nuclear, Estrada Nacional 10, 2686-953 Sacavém, Portugal*
[7] *Department of Electronics, Instituto Superior Técnico, Av. Rovisco Pais 1, 1049–001 Lisbon Portugal*

SIMPLE is an experimental search for evidence of spin-dependent dark matter, based on superheated droplet detectors using C_2ClF_5. We report preliminary results of a 0.6 kgdy exposure of five one liter devices, each containing $\sim$10 g active mass, in the 1500 mwe LSBB (Rustrel, France). In combination with improvements in detector sensitivity, the results exclude a WIMP–proton interaction above 5 pb at $M_\chi = 50$ GeV/c^2.

1. Introduction

SIMPLE[1,2] is is one of two experiments[3] to search for evidence of spin-dependent dark matter using fluorine–loaded superheated droplet detectors (SDDs). The detector is based on the nucleation of the gas phase by energy deposition in the superheated liquid. The conditions for bubble nucleation imply energy depositions of order $\sim$200 keV/μm, rendering SIMPLE SDDs effectively insensitive to most of the traditional backgrounds which plague the majority of the conventional dark matter search detectors.

In 2000, we reported[1] the first exclusion limits from a prototype SDD with 9.2 g active mass operated for 16 days. Refrigerant-free 'dummy' modules however yielded signals identical to bubble nucleations arising from pressure microleaks in the SDD caps at a rate of 1 event per day[1], which comprised the mahority of the prototype signal.

We here report the preliminary results of a measurement using five detectors of similar construction, with design modifications in the capping of the detectors which result in dummy device rates a factor 10 less than previous, and an overall factor 2 reduction in our previous exclusion limits. Although still limited by statistics, the results clearly demonstrate the experiment competivity with the higher exposure search experiments.

2. Experimental

The detectors, ranging in active freon mass from 9.2–16.6 g with a total of ~60 g, were installed in the GESA area of the LSBB. The SIMPLE SDDs, based on C_2ClF_5 (R-115), are fabricated inhouse with a 1–3% loading, according to previously–described procedures[2].

The set was placed inside a thermally–regulated 700 liter water bath, surrounded by three layers of sound and thermal insulation, resting on a dual vibration absorber. A hydrophone is placed within the detector water bath, and a second acoustic monitor positioned outside the shielding.

A bubble nucleation is accompanied by an acoustic shock wave, which is detected by a piezoelectric microphone embedded in a plastic finger (transducer) immersed in a glycerin layer at the top of the detector. The transducer signal was amplified a factor 10^5; in the case of an event in any of the detectors, the temperature, pressure, and threshold voltage for each device, plus its waveform trace and fast Fourier transform, were recorded in a LabView platform.

The operating pressure (2 atm) and temperature (9 °C) were chosen to reduce the detector sensitivity to background, while preserving WIMP sensitivity. In these conditions, predominant background sources are either neutron or α–particles, or electronics.

Due to their low stopping power, γ–rays below 6 MeV can produce background nucleations only through γ–induced high stopping power electrons (Auger electron cascades following interactions of environmental γ-rays with Cl atoms in the refrigerant[2], visible above 15 °C) or recoil nuclei from γ scattering (kinematically below threshold for $T < 12°C$).

The detector response to α's was described elsewhere[2]. The presence of a small ($\sim 10^{-4}$ pCi/g) ^{228}Th contamination was measured via low–level α spectroscopy, yielding an overall background level of < 0.5 events/kg freon/dy.

The response of smaller SDDs to various neutron fields has been studied extensively and found to match theoretical expectations[4,5]. The SIMPLE detector response to neutrons has been investigated using ^{252}Cf and monochromatic low energy neutron beams generated by filtering the thermal column of the Portuguese Research Reactor[2,6]. In both cases, the expected nucleation rate was calculated as a function of temperature via MCNP4 simulations of the response, following Refs. 5, 7. The ^{252}Cf measurements yielded a detection efficiency of 34% for 2 atm operation, consistent with the filter irradiations[6], and in good agreement with the thermodynamic calculations.

The metastability of a superheated liquid is described by the homogeneous nucleation theory[8], which predicts a spontaneous nucleation rate exponentially decreasing with decreasing temperature. This process should not be significant for rates down to the level of current measurement.

3. Data analysis

As shown in Table 1, the detectors were operated at 9°C for 10 days, 3°C for 15 days. Additional experiments were performed at 14°C in order to assess detector performance. The data record was filtered according to the criteria that (i) one and only one detector had a signal, and (ii) no monitoring detector had a signal.

Table 1. Raw data results, without acoustic detection efficiency or background correction.

run	exposure (dy)	T (°C)	P (bar)	anticoincidence (ev/kgdy)	5.5–6.5 kHz (ev/kgdy)
2704	10	8.9	2.0	197 ± 18	143 ± 16
2805	14	3.2	1.9	64 ± 9	48 ± 8

The fast Fourier transform of the transducer signal comprises a well-defined frequency response, with a primary harmonic at ~6 kHz and a time span of a few milliseconds: only filtered events with a primary harmonic between 5.5–6.5 kHz were accepted.

The 3.2 °C results, not being WIMP–relevant data, were used to estimate a lower limit on the overall background rate. Following the purification studies of Ref. 2, the 9 °C-to-3 °C rate ratio is ~2.5, yielding a difference of 23±26 events/kgdy. At this level of exposure, the signal rate is almost entirely due to contributions from known background origins.

The cosmological parameters and method described in Ref. 9 are used in the calculation of the WIMP elastic scattering rates. A comparison of present with previous results is shown in Fig. 1, indicating the factor 2 improvement over the prototype results. Note that simply assuming *no* WIMPs were detected would yield an order of magnitude better limit on the expected WIMP rate.

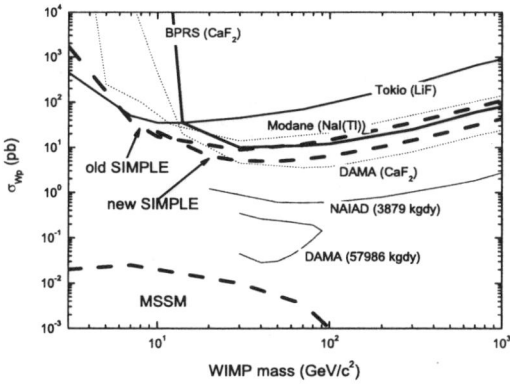

Figure 1. comparison of preliminary present SIMPLE limits with those of Ref. 1; the DAMA/NaI and NAIAD have been updated. The WIMP characteristic velocity is 230 km/s, the earth velocity in May is 257 km/s, and $\rho = 0.3$ GeVc^{-2}cm^{-3}.

Fig. 1 is obtained within a model–dependent formulation. The data were also analyzed using a model–independent formalism[10], in which the odd-group approximation for the WIMP-nucleus cross section σ_A is abandoned and the full spin-dependent interaction with all isoptope nucleons is taken into account: $\sigma_A \propto (a_p <S_p> + a_n <S_n>)^2$, where $a_{p,n}$ ($S_{p,n}$) are the nucleon coupling strengths (spins) and $<S_p>, <S_n> \neq 0$, respectively. Since the phase space is now 3-dimensional, the results can be displayed by projection onto the a_p-a_n plane for each given WIMP mass M_χ, as shown in Fig. 2 at 90% C.L. for $M_\chi = 50$ GeV/c^2 (which is in the DAMA/NaI–preferred range[11]), together with the NAIAD[12], Tokyo/NaF[13], CRESST-I[14] and DAMA/Xe-2[15] experiments. Within this formulation, the SIMPLE results are already seen to eliminate a large part of the parameter space allowed by NAIAD at this mass cut. The CRESST result is assumed to be a result of Al only, since ^{16}O is an even-even, spinless and doubly magic nucleus with no magnetic moment.

We also show the impact of the EDELWEISS[16] and CDMS[17] experiments, customarily considered as spin-INdependent searches, which by themselves are surprisingly even more efficient in reducing the allowed parameter space of the NAIAD–DAMA/Xe-2 intersection.

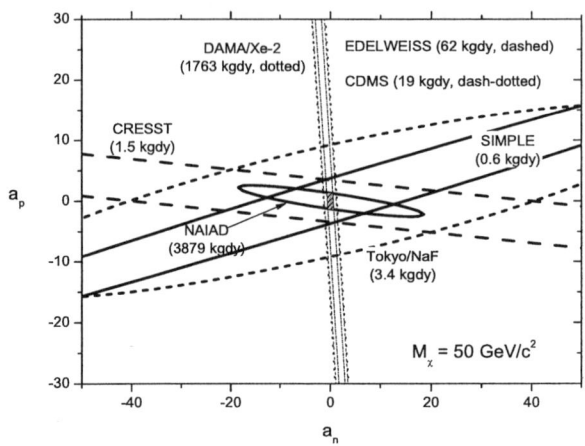

Figure 2. a_p–a_n plots for SIMPLE (0.6 kgdy), NAIAD (3879 kgdy), CRESST-I (1.51 kgdy), Tokyo/NaF (3.38 kgdy) and DAMA/Xe-2 (1763 kgdy) for WIMP mass of 50 GeV/c^2, as indicated. Also shown are the single nuclei spin-INdependent EDELWEISS (62 kgdy) and CDMS (19 kgdy). The unexcluded region of each experiment lies inside its respective contour.

The reason for the large impact of the fluorine–based detectors is (i) the relative sign of $<S_n>$ / $<S_p>$ opposite to I, and (ii) both $<S_n>$ and $<S_p>$ are non–negligible. The near–orthogonality of the fluorine ellipses results from (i); chlorine and sodium, the other nuclei present in the detectors, do not fulfill condition (ii), and are essentially spectators for WIMP–neutron interaction detection. Nevertheless, they make the unexcluded regions of the experiments closed ellipses instead of open conics[10], with the chlorine being the weaker of the two constraint–wise. This weakness is responsible for the high eccentricity of the SIMPLE ellipse, owing to the low spin values of ^{35}Cl and low concentration of ^{37}Cl.

Despite the small active detector mass, the limits reflect the favourable ^{19}F nuclear spins, and the reduced background inherent to a detection method essentially blind to the traditional backgrounds. Furthermore, the temperature–dependent threshold of the detector allows a background estimate at a temperature at which the detector is no longer sensitive to neutralino–induced events. The present SIMPLE limits remain constrained by the large statistical uncertainty associated with the short exposure accumulated so far, as well as continuing "anomalies" in the electronic response, both of which are currently being addressed.

Acknowledgements

This work was supported by grant POCTI/FNU/43683/2002 of the Portuguese Foundation for Science and Technology (FCT), co–financed by FEDER.

References

1. J.I. Collar, et al., Phys. Rev. Lett. **85**, 3083 (2000).
2. J.I. Collar, et al., New Journ. of Phys. **2**, 14.1 (2000).
3. L. A. Hamel et al., Nucl. Instr. Meth. **A388**, 91 (1997).
4. M. Harper and J. Rich, Nucl. Instr. Meth. **A336**, 220 (1993).
5. Y. Ch. Lo and R. Apfel, Phys. Rev. **A38**, 5260 (1988).
6. F.Giuliani, J.I. Collar, D. Limagne et al., Nucl. Instr. Meth. **A526**, 348 (2004).
7. J.I. Collar, Phys. Rev. **D54**, 1247 (1996).
8. F. Seitz, Phys. Fluids **1**, 2 (1958).
9. J.D. Lewin, and P. F. Smith, Astrop. Phys. **6**, 87 (1996); N.J.C. Spooner et. al., Phys. Lett. **B473**, 330 (2000).
10. F. Giuliani, Phys. Rev. Lett. **93**, 161301 (2004); D. R. Tovey et al., Phys. Lett. **B488**, 17 (2000).
11. R. Bernabei et al., Phys. Lett. **B509**, 197 (2001).
12. B. Ahmed et al., Astrop. Phys. **19**, 691 (2003).
13. A. Takeda et al., Phys. Lett. **B572**, 145 (2003).
14. W. Seide et al., Dark Matter in Astro- and Particle Physics, eds. H.V. Klapdor-Kleingrothaus and R.D. Viollier, 517 (Springer-Verlag, Berlin, 2002).
15. R. Bernabei et al., Phys. Lett. **B436**, 379 (1998).
16. A. Benoit et al., Phys. Lett. **B545**, 43 (2002).
17. D.S. Akerib et al., arXiv **Astro-ph/0405033**.

STATUS OF KIMS EXPERIMENT SEARCHING FOR WIMP WITH CSI(Tℓ) CRYSTALS

J. W. KWAK[1],[*] H. BHANG[1], S. Y. KIM[1], H. S. LEE[1], J. LEE[1],
S. S. MYUNG[1], M. J. LEE[1], S. C. KIM[1], S. K. KIM[1], J. I. LEE[2], Y. D. KIM[2],
M. J. HWANG[3], Y. J. KWON[3], I. S. HAHN[4], H. J. KIM[5], J. J. ZHU[6], J. LI[6,7]
(KIMS COLLABORATION)

[1]*DMRC and School of Physics, Seoul National University, Seoul 151-742, Korea*
[2]*Department of Physics, Sejong University, Seoul 143-747, Korea*
[3]*Physics Department, Yonsei University, Seoul 120-749, Korea*
[4]*Department of Science Education, Ewha Woman's University, Seoul 120-750, Korea*
[5]*Physics Department, Kyungpook National University, Daegu 702-701, Korea*
[6]*Department of Applied Physics, Tsinghwa University, Beijing, China*
[7]*Institute of High Energy Physics, Beijing, China*

The Korea Invisible Mass Search(KIMS) collaboration has developed low background CsI(Tℓ) detectors and constructed a new experimental hall in the 700 m deep underground facility in Yangyang, Korea. Successful reduction of the internal background of CsI(Tℓ) crystal has been realized. We constructed a new shield composed of 60 tons of pure materials and the muon detector with full coverage of solid angle. In this letter, the preliminary results of WIMP search using CsI(Tℓ) crystal detector and the construction of new underground laboratory will be presented.

1. KIMS experiment and Experimental hall in Yangyang

According to the recent researches, our universe is composed of 4 % of known matters, 23 % of dark matter and 73 % of dark energy [1]. The rising interest in dark matter is allied to not only the search for the unknown components of universe but also the observation of the new elementary particle. Weakly Interacting Massive Particle(WIMP) is a strong candidate for the dark matter and also corresponds to the neutralino which is the super partner of gauge and Higgs bosons in Supersymmetric models [2].

An underground laboratory, called as "Y2L", has been built in the underground facility of a water pumped storage power plant in Yangyang.

---
[*]e-mail : jwkwak@hep1.snu.ac.kr

Table 1. The list of environment parameters in the Y2L

Item	Values
Depth	Minimum 700 m (length of access tunnel = 2 km)
Temperature	20 ~ 25 °C
Humidity	35 ~ 60 %
Rock Contents	^{238}U less than 0.5 ppm ^{232}Th 5.6 ± 2.6 ppm K$_2$O 4.1%
Muon flux	4.4×10^{-7}/cm^2/s
Neutron flux	8×10^{-7}/cm^2/s
Radon in air	2 ~ 4 pCi/liter

In Table 1, several environmental parameters of the Y2L are listed. In the underground experimental hall, various facilities have been installed to control and monitor the environmental parameters. To keep the uniform temperature and low humidity, Y2L is equipped with an air-conditioning system which controls the temperature and humidity of the detector room and supplies clean air. Also a N$_2$ flowing system has been installed to remove the humidity and Radon in a copper box, in which CsI(Tℓ) crystal detectors are mounted. The temperature and humidity are always monitored using "Environment Monitoring system". From monitoring for a long time, we confirmed that the temperature and humidity inside the copper box are maintained to be stable. The variance of the monitored temperatures of the CsI(Tℓ) crystal is less than 0.2 degree.

The ingredients of the surrounding rocks around the underground facility were inspected to check the isotope amounts which are related to γ and neutron backgrounds. Fortunately, the amount of ^{238}U which is less than the sensitivity of the measurement method is extraordinary small. That results in low Radon contamination in air. To reduce external γ and neutron backgrounds, a 60 ton shielding structure made of pure materials has been constructed. The structure can house up to 250 kg of CsI(Tℓ) crystals. In the outermost part of the shielding structure, there is a 30 cm thick mineral oil layer. The mineral oil layer composed of 5% of liquid scintillator and 95% of mineral oil, works as an active detector for muon detection and a passive shield for neutron reduction. Inside of the mineral oil layer, a 15 cm thick lead and 5 cm thick Polyethelene layers are placed. The innermost part of the shielding structure is a 10 cm thick copper box in which CsI(Tℓ) crystal modules are located. We measured γ rates at both the outside and inside of the shield using a HPGe detector to check the reduction ratio, which is more than 99.99% of external γ's. According to a Monte-Carlo simulation, 99.9% of external neutrons are shielded by the 30 cm thick mineral oil layer.

The expected muon rate in 700 m deep underground is 10^{-5} times of the rate in the ground level. The muon detector, which is composed of 8 modules and 28 PMT channels, covers the full solid angles of the shielding structure. The measured muon flux at Y2L is 4.4×10^{-7} muons/cm^2/s. Using the "Kinematic fitting" technique, we reconstructed the hit positions of muon of which the spacial resolution is about 8 cm. Using the hit information, the polar and azimuthal angles of the muon track were calculated. The distributions of the angles agree well with the mountain shapes at the position of Y2L.

Using the neutron monitoring detector consisting of 1 liter of BC501A liquid scintillator, we measured the fast neutron flux at the inside and outside of the shield. The dominant sources for the fake nuclear recoil event candidates inside the shield are α's from the decay chains of ^{238}U and ^{232}Th existed in the detector itself. The α events can be tagged by the delayed coincidence between $\alpha - \alpha$ or $\beta - \alpha$ decay events. As a result, all neutron candidates were tagged as α's from ^{238}U and ^{232}Th decay chains and the null neutron rate inside the shield was consistent with less than 3 counts/day/liter at 90% CL. The neutron flux outside the shield was measured using the same detector. That makes easy to extract out the real neutron events from the internal α backgrounds by simply subtracting the inside α spectrum normalized by time. The outside neutron flux was measured to be 8×10^{-7} neutrons/cm^2/s. With these measurements, we conclude the effect on the CsI(Tℓ) detector by the background neutron is small enough to be neglected.

For monitoring the Radon amount in air at the experiment hall, we have installed a Radon detector which have 70 liter volume and a 3×3 cm^2 Si-photodiode. To estimate the amount of Radon in the 70 liter of air refreshed every hour, α's from the decays of ^{218}Po and ^{214}Po are tagged by their energies. The Radon amount in air of Y2L is about 2 to 4 pCi/liter.

2. CsI(Tℓ) detector — Calibration and DAQ system

The KIMS collaboration has designed an underground experiment using the CsI(Tℓ) crystals with low intrinsic background for WIMP search. It is well known that CsI(Tℓ) crystal gives light yield comparable to the NaI(Tℓ) crystal, is less hygroscopic and better in Pulse Shape Discrimination(PSD) ability than the NaI(Tℓ) crystal. Since the main background source is intrinsic, the internal background of the CsI(Tℓ) crystal has been extensively studied [4,5,6]. The internal background sources are well understood by the contribution of ^{137}Cs, ^{134}Cs and ^{87}Rb. Through the successful reduction process of the background sources, 6 cpd of the internal background level at 10 keV has been achieved. We have obtained the 237 kg · days data with the CsI(Tℓ) crystal of the 6 cpd internal background level.

To discriminate the nuclear and electron recoil events in the CsI(Tℓ) crystal detector, the reference calibration data have been taken in a calibration facility in Seoul National Univ, where a 300 mCi Be-Am source for the calibration of the nuclear recoil events has been installed. A ^{137}Cs source has been used for the calibration of the electron recoil events. With about 50 % of a branching ratio, a Be and a α from Am sources go to the 4.4 MeV-state of ^{12}C and a neutron. Simultaneously, ^{12}C* deexcites to the ground state of ^{12}C with a 4.4 MeV γ. To tag the neutron events in the CsI(Tℓ) crystal detector, it's required the simultaneous 4.4 MeV γ emitted from ^{12}C* is measured in a liquid scintillator, which covers the Am-Be source fully except of a collimation hole for neutrons. A 3×3×3 cm^3 CsI(Tℓ) detector is located at the end of the collimation hole. In order to identify neutrons scattered from the CsI(Tℓ) crystal, neutron detectors made of BC501A liquid scintillator are located at the various angled positions from the collimation hole. The energy of the neutron incident upon the CsI(Tℓ) crystal can be calculated with the time of flight from the source position to the CsI(Tℓ) detector. ^{137}Cs calibration data were taken both by the small crystal used for neutron calibration and by the full size search crystal installed in Y2L to confirm the mean time characteristics between two crystals is the same. The quenching factor of the CsI(Tℓ) crystal was measured using a tandem accelerator (maximum terminal voltage of 1.7 MV) at the Korea Institute of Geoscience and Mineral resources(KIGAM) [3].

The KIMS DAQ is based on the "ROOT" program in Linux operating system. Various home-made electronics have been developed for KIMS DAQ. Among those, a 500 MHz FADC plays an important role. The 500 MHz FADC has a programable trigger system and a double buffering system to remove DAQ dead time. We take the wave form for the pulse shape within 32 μs of gate window at 500 MHz of sampling rate to separate each single photoelectron cluster of the signal in the CsI(Tℓ) crystal detector. As for hardware trigger condition, we required more than 5 photoelectron clusters within 2 μs which is equivalent to 1 keV energy deposition. That means the energy threshold of a hardware trigger level is about 2 keV.

3. Preliminary results of WIMP analysis and future plan

The KIMS collaboration has collected data with a $8 \times 8 \times 23$ cm^3 CsI(Tℓ) crystal detector with 6 cpd level background since the summer of 2004. Since the discrimination between nuclear and electron recoils is not good enough and the background from PMT is not negligible below 3 keV, the software energy threshold has set to 3 keV. An electron equivalent energy range from 3 to 10 keV is used in this analysis. We used PMT noise events which have been taken without any crystal and ^{137}Cs calibration data to set analysis cut conditions with which PMT noise events are mostly removed. The upper limits of the nuclear recoil events have been obtained for each energy bin by fitting with the reference distributions of "Mean time" for neutrons and Compton γ's from ^{137}Cs decay described in the previous section. The mean time is defined as;

$$<t> = \frac{\sum A_i t_i}{\sum A_i} - t_0, \qquad (1)$$

where A_i is the charge, t_i is the time of the ith cluster, and t_0 is the time of the first cluster.

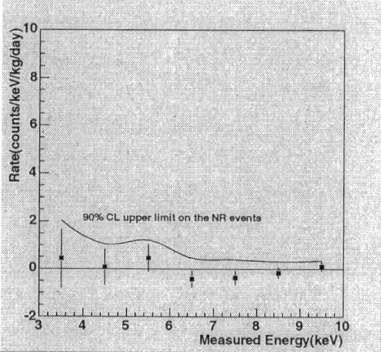

Figure 1. The solid line indicates Upper limits (90 % CL) on the nuclear recoil rate for each energy bin, while the points show the numbers of nuclear recoil events from fitting with the reference distributions.

Fig. 1 shows the 90 % CL upper limits on the nuclear recoil rate and the fitting values with errors for each bin, which are all consistent with zero nuclear recoil event. In order to calculate the limit curve of KIMS experiment, our measured quenching factor [3], WIMP density of 300 MeV/cm^3, average velocity of 230 km/s and escape velocity of 650 km/s were applied.

Fig. 2 shows the preliminary KIMS limits on the WIMP-nucleon cross-section as a function of WIMP mass based on the 237 kg days data, which is comparable to the limits of the other experiments with more than 10

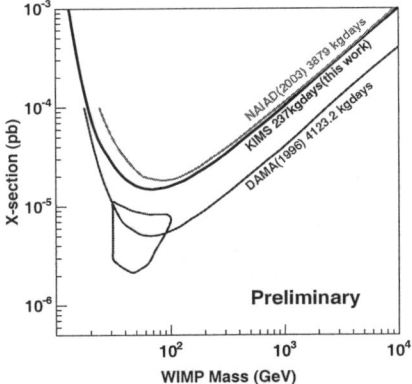

Figure 2. The preliminary limit curve of KIMS experiment using 237 kg · days data of CsI(Tℓ) crystal with the internal background level of 6 cpd.

times of data of NaI(Tℓ) crystals. This is mainly due to better PSD power and lower recoil energy threshold of the CsI(Tℓ) crystal.

The current shielding structure can house 250 kg of CsI(Tl) crystals without any modification. Using 250 kg of CsI(Tℓ) crystals with the internal background level of 1 cpd based on the recent R&D study [6], we can improve the limit on the cross section by two orders of magnitude for one year data taking. That's enough to cover DAMA signal region [7] and the current best limit achieved by CDMS [8].

4. Acknowledgment

This work was supported by Creative Science Research Initiative program of Korea Science and Engineering Foundation. We are very grateful to the Korea Midland Power Co. and their staffs for providing the underground laboratory space at Yangyang.

References

1. C. L. Bennett et al., *ApJS*, **148**, 1 (2003).
2. G. Jungman, M. Kamikowski and K. Griest, *Phys. Rept.* **267**, 195 (1996).
3. H. Park et al., *Nucl.Instrum.Math.* **A491**, 460 (2002).
4. Y. D. Kim et al., *Proceeding paper for IDM2002* (2002).
5. S. K. Kim et al., *Nuclear. Physics.* **B(Proc.Suppl.)**, **124**, 217 (2003).
6. H. S. Lee et al., *Proceeding paper for IDM2004* (2004).
7. R. Bernabei et al., *Phys. Lett.* **B480**, 23 (2000).
8. D. S. Akerib et al., astro-ph/040533(2004).

DRIFT: STATUS AND PROSPECTS

J. C. DAVIES AND N. J. C. SPOONER
ON BEHALF OF THE DRIFT COLLABORATION*
Department of Physics and Astronomy, University of Sheffield, S3 7RH, UK
E-mail: J.C.Davies@sheffield.ac.uk, N.Spooner@sheffield.ac.uk

The DRIFT experiment is the first directionally sensitive dark matter search. The design is based on a gaseous time projection chamber and employs carbon disulphide gas at low pressure. DRIFT-I was installed in the underground laboratory at Boulby mine in 2001 and has acquired more than 1500 hours of data. The second generation detector, DRIFT-II, is undergoing initial testing and will be installed in the new experimental hall at Boulby early in 2005.

1. DRIFT: A directional dark matter detector

The DRIFT experiment design has delivered the world's first directionally sensitive dark matter detector[1,2]. The design concept (Figure 1) employs time projection counters (TPCs) with multi-wire proportional chambers (MWPCs) for readout. The first generations of detector use carbon disulphide (CS_2) gas due to its electronegativity, which allows negative ions to be drifted within the TPCs, rather than electrons, and so avoids large diffusion and greatly improves the spatial resolution.

Most models suggest that the Earth is travelling through a halo of WIMPs in the Milky Way. A dark matter detector should, therefore, observe a directional 'WIMP wind' signal. A time projection chamber is well suited for dark matter detection because it enables directional track information to be recorded, and so can be used to provide the most powerful evidence for the existence of WIMPs in the Galaxy (Figure 2) and to distinguish between postulated halo models[3,4]. The DRIFT design has excellent background rejection potential via range-ionization discrimination[5]. It can probe the WIMP parameter space to sensitivities (defined as a minimum of the sensitivity curve) as shown in Figure 3[4].

---

*UKDMC (University of Sheffield, University of Edinburgh, R.A.L., Imperial College), Occidental College (L.A.), Temple University (Philadelphia), University of New Mexico, University of Boston, University of Thessaloniki (Greece), University of Darmstadt (Germany)

243

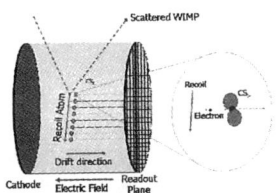

Figure 1. Diagram showing the DRIFT detector concept.

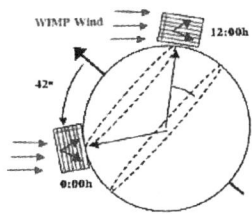

Figure 2. Image of the 'WIMP wind' that may be seen as a directional signal by a detector such as DRIFT.

2. DRIFT-I

DRIFT-I is the first full-scale detector from the DRIFT collaboration[2]. It was installed underground at Boulby mine during the summer of 2001. It consists of a 1 m³ fiducial volume within two back-to-back TPCs, with a shared central cathode and uses two MWPC readouts mounted horizontally. This detector is placed inside a stainless steel vacuum vessel, which is filled with CS_2 gas at low pressure (standard running pressure of 40 torr, giving a target mass of 167 g). The DRIFT-I data acquisition system was designed at SLAC.

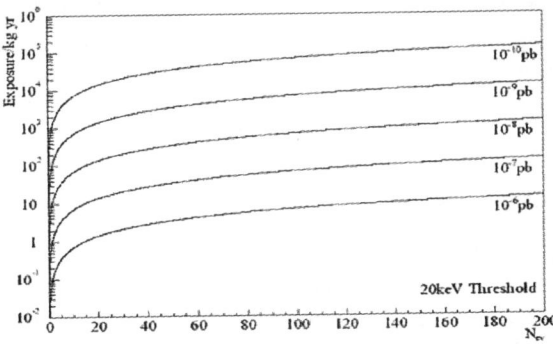

Figure 3. Directional sensitivity that may be achieved by a DRIFT-type detector for a given exposure and number of detected events.

2.1. Simulations

Simulations investigating neutron background sources have been performed for DRIFT-I using both a 'home-grown' Monte Carlo and also the GEANT4 toolkit. The input neutron spectra were obtained from measurements of the uranium and thorium content in the DRIFT cavern and component materials, using a germanium detector in the cavern and mass spectrometry techniques for the other materials. The neutron rate predicted by the simulations for an unshielded detector is 12.6 events / kg / day due to neutrons from rock.

2.2. Operations

During the DRIFT-I (Figure 4) running time, since 2001, there have been a number of operational issues[2], some of which have been found to have solutions that allow the detector to remain at least partially operational and so it has still been possible to take data, although not always with the full detector.

In 2004 new alpha veto hardware was installed, as was $\sim$ 8 tons of CH_2 neutron shielding - 30 g / cm^2 thick on all sides - in the form of polypropylene pellets (see Figure 5).

Figure 4. Photograph of the DRIFT-I inner detector positioned inside the stainless steel vessel with the front panel removed.

Figure 5. Photograph of the partly shielded DRIFT-I vessel in position at Boulby mine with the front section of shielding removed.

2.3. Preliminary results

DRIFT-I has managed to acquire over 1500 hours of data and its response to neutrons, alphas and gammas has been observed.

DRIFT-I has also attained a preliminary background measurement from 37.25 days of livetime of 28 events / kg / day, for an unshielded detector (Figure 6). The high rate, compared to that predicted by computer simulations for neutrons, may be due to neutrons from sources other than the surrounding rock, or alphas that have managed to avoid all the data cuts.

2.4. DRIFT-I summary - Lessons learned

The experience from installing and running DRIFT-I has allowed many invaluable lessons to be learned, both in detector engineering and in performance optimisation. These lessons have been applied to the design of the next generation detector - DRIFT-II.

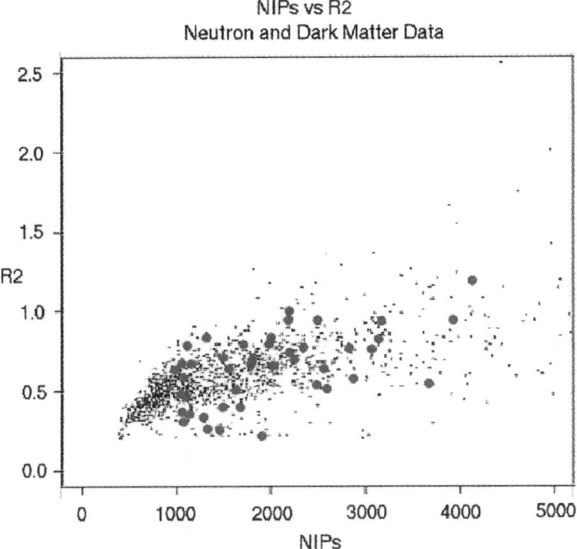

Figure 6. Plot of the Number of Ion Pairs produced in an event against the 2-dimensional range of the event. Data from a neutron source is shown (small dots) along with background data (large dots).

3. DRIFT-II

DRIFT-II is the newest DRIFT detector and will begin taking data underground in 2005. The design is basically an improved and expanded version of DRIFT-I (see Figure 7)[6]. DRIFT-II will be an array of modules, with each module consisting of a stainless steel vacuum vessel of internal dimensions of $1.5 \times 1.5 \times 1.5$ m^3 and with hinged door (see Figure 8). The vessels each contain two back-to-back time projection chambers and two MWPC readouts mounted vertically (see Figure 7), which will reduce the effects of falling debris and allow both MWPCs to operate in identical environments. Initially DRIFT-II will employ CS_2 gas at low pressure again, with a gas system able to maintain various pressures and flow rates. Subsequent gas systems will also be able to maintain different gas mixtures. Care has been taken in choosing the materials used for detector components to ensure they are known to be of at least a minimum purity, such as the low background stainless steel vessel, high purity copper used for the field cages and the radiopure lucite also used inside the vessel. The spatial resolution is improved, as is the 3-dimensional track reconstruction capability, and the background noise level of the data acquisition system is reduced.

Another important consideration that had to be made while designing

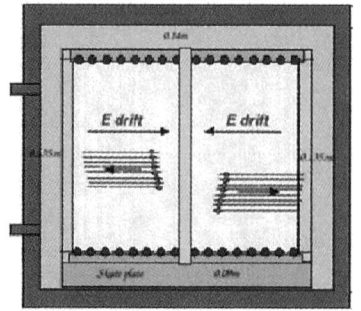

Figure 7. Diagram showing the design of the DRIFT-II detector.

Figure 8. Photograph of the first DRIFT-II vessel at Occidental College, L.A., where commissioning of the first DRIFT-II detector was performed.

the DRIFT-II detector array was the size and layout of the experimental hall at Boulby mine. The limited space available places a constraint on the size of vessel and amount of passive neutron shielding that can be used. By looking at these dimension constraints and taking information from computer simulations, the layout of the DRIFT-II array will be as rows of four modules sharing some shielding, as shown in figure 9.

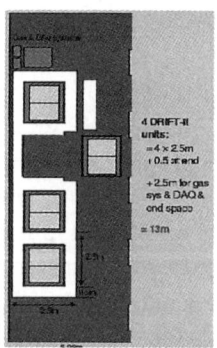
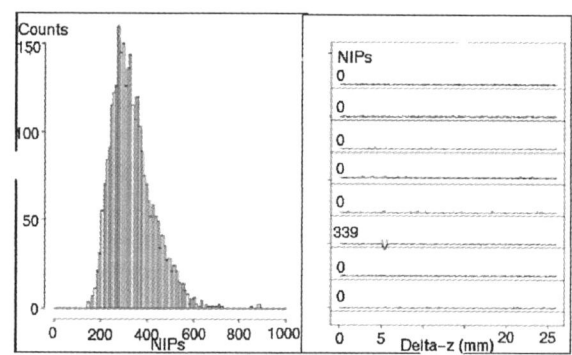

Figure 9. Diagram of the planned layout for a group of four DRIFT-II modules with their shielding.

Figure 10. Initial data plots from the commissioning of the first DRIFT-II detector showing the ^{55}Fe spectrum on the left with an ^{55}Fe dignal shown on the right.

3.1. Data acquisition and commissioning

The data acquisition system for the DRIFT-II detectors has the grid DAq, which uses a grouping technique and so records signals through only eight channels from 512 wires, using Amptek pre-amplifiers. The grid DAq also has wires at the front and back, which are used as alpha vetoes in the DRIFT y-direction. There is also an anode DAq system that simply registers hits and also has alpha vetoes in the DRIFT x-direction.

The initial commissioning of the first DRIFT-II detector module is currently underway at Occidental College, L.A. (Figure 10), where, to date, leak testing has been done and ^{55}Fe calibration data has been taken (Figure 10).

3.2. Simulations

Particle backgrounds are a big issue for dark matter detector. It is important to investigate possible sources of these backgrounds and to use computer simulations to try to understand their effects. The main problem backgrounds for dark matter detectors are sources of neutrons because the signal seen when a neutron interacts inside the detector can be indistinguishable to that of the WIMPs being searched for. Work on simulating the neutron backgrounds for this type of detector has been performed[7] and the results have been used to make decisions on material purity and shielding requirements.

4. Future prospects

The next generation, DRIFT-III, is expected to have a sensitivity down to about 10^{-8} pb, assuming no background and using an array of 100+ units of dimension $(1.5\text{m})^3$. Further expansion up to a 1 ton target would give a potential sensitivity to 10^{-10} pb. This would allow the possibility of distinguishing such objects as the Sagittarius CDM stream[8]. It is also possible for future DRIFT detectors to have the capability of searching for other rare events, such as KK-axions and Universal Extra Dimensions.

While the first DRIFT detectors use CS_2 gas, future detectors may use an alternative gaseous target (single gas or gas mixtures). Research and development work on alternative charge readout devices, focussing on the possibilities of MICROMEGAS, is ongoing, along with halo modeling and computer simulation of the DRIFT detectors.

References

1. C. J. Martoff et al., *Nucl. Instr. & Meth. A* **440** (2000) p355.
2. G. J. Alner et al., *Nucl. Instr. & Meth. A* **535**, (2004), p644.
3. C. J. Copi and L. M. Krauss, *Phys. Rev. D* **63** (2001) 043507.
4. B. Morgan, PhD thesis, University of Sheffield, 2004.
5. D. P. Snowden-Ifft et al., *Phys. Rev. D* **61** (2000) 101301
6. The DRIFT Collaboration, In preparation.
7. M. J. Carson et al. in these proceedings.
8. K. Freese et al., *Phys. Rev. Lett.* **92**, (2004), p. 111301.

XMASS EXPERIMENT

SHIGETAKA MORIYAMA

Kamioka Observatory, Institute for Cosmic Ray Research, University of Tokyo,
Higashi-Mozumi, Kamioka-cho, Hida-city, Gifu-prefecture, Japan, E-mail:
moriyama@icrr.u-tokyo.ac.jp

The XMASS project utilizes ultrapure liquid xenon and aims to detect pp and ^{7}Be solar neutrinos by means of ν + e scattering. It requires low background and a low threshold which will also enable us to search for dark matter in the galactic halo. By using a prototype detector, we have confirmed its feasibility to realize low background and low threshold. We have estimated the sensitivity of an 800 kg liquid xenon detector for a dark matter search experiment based on the experimental results.

1. Introduction

The recent observations by the Super-Kamiokande and SNO experiments have provided evidence that the solar neutrino problem is caused by neutrino oscillations. Furthermore, the results from KamLAND strongly supports this evidence. However, the realtime solar neutrino experiments only observed ^{8}B neutrinos which constitutes a very small fraction (0.17%) of solar neutrinos. Since the rest of the solar neutrinos are low in energy (such as pp and ^{7}Be neutrinos), it is still important to observe this low energy neutrino spectrum.

The XMASS project utilizes liquid xenon as a scintillator and aims to detect pp and ^{7}Be neutrinos. Although there are many advantages for using liquid xenon as a solar neutrino detector, no directional nor coincidence information for solar neutrino signals are available. Therefore it is clear that we must realize an ultralow background in the fiducial volume of the detector. The key idea is to utilize the self-shielding effect of xenon. Just several tens of centimeters of outer layer of liquid xenon can absorb and shield low energy external gamma rays since xenon has a large atomic number. By using a reconstruction method which gives event vertices based on scintillation light patterns, we can extract just the events deep inside the detector. This self-shielding is quite useful for solar neutrino detection as well as dark matter detection. Here we will present experimental results from a prototype detector. They are as expected and are well reproduced by our Monte Carlo (MC) simulations.

Another important issue is radioactive contamination in liquid xenon. ^{85}Kr is a well known radioactive isotope which could not be removed effectively until now. We developed a distillation system for gas xenon to reduce Kr levels and found Kr contamination can be reduced by a factor of 1000 with only one pass.

Based on those encouraging experimental results, we are now designing an 800 kg detector which aims to detect dark matter. Since its sensitivity estimation for a dark matter search is based on a similar MC which was used for the prototype detector, the dark matter detector's feasibility is better demonstrated. The solar neutrino detector will be a next step since it needs a large mass of xenon.

2. Prototype detector

The prototype detector consists of a 31 cm cubic oxygen-free-highpurity-copper (OFHC) chamber, 54 photomultipliers (PMT's), and a heavy shield. The PMT's detect scintillation photons outside the chamber through MgF_2 windows. The heavy shield consists of 15 cm-thick polyethylene, 5 cm-thick boric acid powder, 15 cm-thick lead blocks, 30 μm EVOH sheets, and 5 cm OFHC blocks. The heavy shield reduces external gamma rays and neutrons. The EVOH sheets are important to keep low radon level inside the shield[2].

The detected photons are used to estimate the event vertex and deposited energy. Since the time constant of the scintillation is around 40 ns, event reconstruction cannot be done by timing information. However, a large photon yield enables us to reconstruct events by using the photoelectron patterns of PMT's. Fig. 1 shows a typical hit pattern of a background measurement. The details of the reconstruction methods can also be found in Ref. [2].

3. Demonstration of the self-shielding effect

The key idea to reduce background is to utilize the self-shielding effect of xenon. Since good reconstruction performance is required to realize low background deep inside the detector, we took data to demonstrate its performance. We have three small holes backside of the heavy shield. These holes can be used as collimators of gamma rays from calibration sources. We put ^{137}Cs and ^{60}Co gamma ray sources at the holes and took data. We

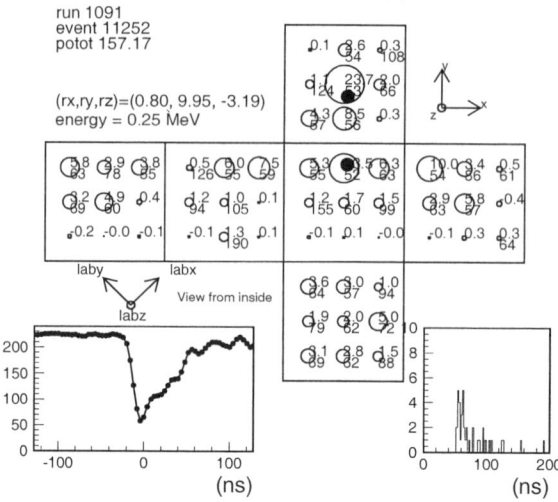

Figure 1. Typical background event with the reconstructed information. The top center figure shows a development of the detector viewed from inside. The area of an open circle is proportional to the detected photo-electrons. By comparing these distributions with expected distributions, we can reconstruct event vertex as well as its deposited energy. The closed circles correspond to the projected position of the reconstructed vertex. The lower left figure shows FADC information of the summed signal of PMT's. The lower right figures shows hit timing distributions of each PMT.

expected exponential dumping of events towards the beam direction. Fig. 2 shows the distributions of the event vertices for ^{137}Cs and ^{60}Co gamma ray sources. The good agreement between data and MC strongly support the validity of our simulations.

4. Background data of the prototype detector

We also took background data without any radioactive sources. The corresponding livetime is 3.9 days. Fig. 3 shows the real data and MC expectations. Since we have the self-shielding effect, we can expect a large reduction of background if we restrict the fiducial volume deep inside the detector. The solid histograms correspond to the most inner, 10 cm cube volume. The background at low energy is order of 10^{-2} keV^{-1} kg^{-1}day^{-1} as expected by MC simulations. This agreement strongly supports the validity of our MC simulations, analysis method, and the background study for the next 800 kg detector.

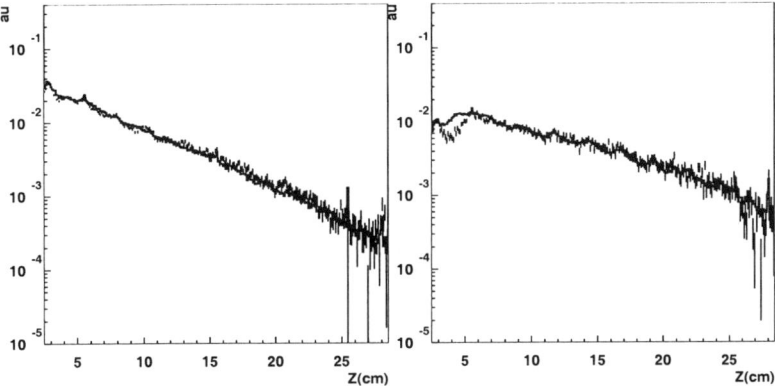

Figure 2. The reconstructed event vertex distributions for ^{137}Cs (left) and ^{60}Co (right) gamma rays. The crosses show real data and solid histograms show Monte Carlo simulations. The horizontal axis corresponds to the depth of the event measured from the injection point. The exponential dumping toward the beam direction can be seen. The discrepancy at 2-5 cm in the right hand figure is due to an incomplete treatment of PMT saturation effect. Good agreement between real data and Monte Carlo simulations demonstrates the self shielding effect.

The MC simulation is based on the measured radioactive contamination of PMT's and other parts which are used in the detector. The discrepancy between data and MC is due to incomplete understanding of photon absorption and scattering in the liquid xenon as well as the reflection coefficient of the inner surface of the detector. We are now tuning those parameters in MC.

The increase of the event rate towards low energy is due to misreconstruction of the events. Since the PMT's are outside the detector, there are some dead angles from the PMT's. This causes the misreconstruction of the events near the wall as if their interaction points are deep inside the detector. The misreconstruction is well reproduced in MC simulations and understood. In the 800 kg detector we will immerse PMT's inside the liquid xenon. Therefore this misreconstruction will be negligible compared with real background.

5. Radioactive contamination and reduction in xenon

Radioactive contamination in xenon is another issue. For U and Th chains, we can select the Bi-Po events in each chain by using FADC information. As a result, we found ^{238}U is $33 \pm 7 \times 10^{-14}$g/g and ^{232}Th is less than 63×10^{-14}g/g. Those values are a factor of 30 larger than the target values

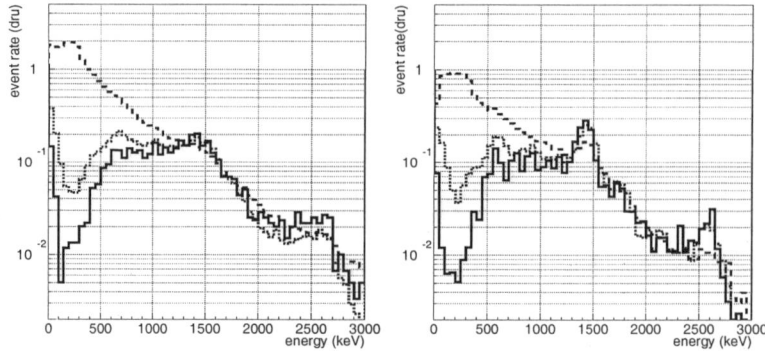

Figure 3. Background spectrum for real data (left) and MC simulations (right). The horizontal axis is the reconstructed energy and the vertical axis is the event rate in unit keV^{-1} kg^{-1}day^{-1} (DRU). The thick dotted line, thin dotted line, and solid line correspond to all volume (31 cm cube), inside 20 cm cube, and the most inside 10 cm cube volume, respectively. One can see a large reduction of background around 100-500 keV region due to the self shielding effect. One can see the reduction factor is as expected by MC simulations. See text for the peak below 100 keV in the 10 cm cube volume data.

for the 800 kg detector. However, since we did not use any filter expect for a simple getter, we have the prospect to reduce those contaminations by using more specialized filters we are now developing.

As for the Kr contamination, we independently developed a distillation system which reduces Kr levels in xenon. The principle is based on the difference of boiling points of xenon and krypton. We built this distillation system, processed our xenon, and confirmed the reduction factor is as expected: It is almost a factor of 1000 since the raw xenon gas has 3 ppb Kr and the processed xenon gas has less than 5 ppt Kr. Though this value is a factor of 5 larger than the target value, we can modify our system to achieve the target value easily by just increasing the height of the distillation tower.

The method to measure 5 ppt Kr in xenon gas is also an interesting technique. We enriched the Kr concentration by a factor of 10 by using a liquid nitrogen trap, and measured it with an API-MS. Further improvement is possible by increasing this enrichment.

6. Expected sensitivity of the 800 kg detector

Based on these encouraging experimental results, we are now designing the 800 kg detector. Fig. 4 shows the schematic view of the detector and its sensitivity for the spin-independent cross section of dark matter.

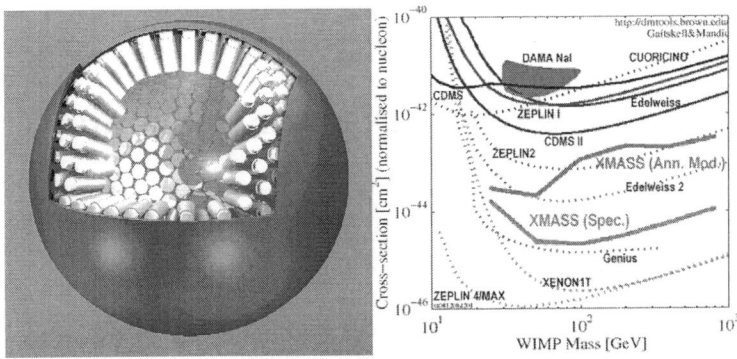

Figure 4. The schematic view of the 800 kg detector and its sensitivity. Lower thick curve corresponds to an analysis to find an increase due to dark matter signal in the raw spectrum. Upper thick curve corresponds to an analysis to look for the annual modulation. Both analysis can give large improvements compared with existing experiments.

7. Summary

The XMASS project aims to detect low energy solar neutrinos, dark matter and neutrinoless double beta decay. A prototype detector provided experimental data which validates our expectations for the future detectors. All experimental results agree with expectations which were studied before the experiment. We are now developing an 800 kg detector to detect dark matter and are going to start the dark matter search in the near future.

References

1. The Super-Kamiokande Collaboration, Phys. Rev. D 69 (2004) 011104; The SNO collaboration, Phys. Rev. Lett. 92 (2004) 181301; KamLAND collaboration, Phys. Rev. Lett. 90 (2003) 021802 and references threrein.
2. S. Moriyama, *In the proceedings of the International Workshop on Technique and Application of Xenon Detectors, Xenon01*, World Scientific, pp 123-135, University of Tokyo, Japan, 3-4 December, 2001; *In the proceedings of the Forth International Workshop on the Identification of Dark Matter, IDM2002*, World Scientific, pp 390-395, York, UK, 2-6 September, 2002; *In the proceedings of The Forth International Workshop on the Neutrino Oscillations and Their Origin, NOON2003*, World Scientific, pp 364-371 Kanazawa, Japan, 10-14 February, 2003; *In the proceedings of BEYOND 2003*, Springer Verlag, pp 385-396, Castle Ringberg, Tegernsee, Germany, 9-14 June, 2003.

STATUS OF ZEPLIN II AND ZEPLIN III

H. WANG[*]

UCLA, Physics and Astronomy Department
475 Portola Plaza
Los Angeles, CA 90095-1547
wangh@physics.ucla.edu

We describe the ZEPLIN II (30-kg) and ZEPLIN III (7-kg) discriminating dark matter detector using two-phase xenon designed for direct detection of cold dark matter in the form of Weakly Interacting Massive Particles. These two detectors are currently being commissioned. Both detector will begin operation in the Boulby Mine, UK in 2005. ZEPLIN II & III are capable of discriminating between nuclear recoils and background events and have a design reach up to two orders of magnitude beyond current limits. These two detectors will also serve as a step in the development program for a next-generation ton-scale detector.

1. Introduction

The ZEPLIN collaboration was formed in the early 90's, after a series of research and development results obtained by ICARUS, to develop large scale liquid xenon detectors for direct WIMP (Weakly Interacting Massive Particles) dark matter search. The key results stimulated the formation of ZEPLIN are: (a) liquid xenon purification to achieve greater than 5-ms free electron life time in liquid xenon [1] and (b) first principle demonstration of gamma and alpha events discrimination [2]. Then xenon as a target for dark matter search were studied extensively [3,4,5,6].

The ZEPLIN collaboration is currently commissioning the ZEPLIN II and ZEPLIN III detector to begin operation in 2005. ZEPLIN II is a large mass (30-kg) two-phase xenon detector while ZEPLIN III is a low mass (7-kg) detector also operates in two-phase. While capable of good physics reach, these two detectors also serve as a step in the development program for a next generation ton-scale detector.

---

[*]Presented by Hanguo Wang for the ZEPLIN Collaboration.

Liquid xenon satisfies the following basic requirements for a dark matter target: (a) It is available in sufficiently large quantities with high purity [1], (b) It scintillates via two mechanisms, and respond differently to nuclear and electron recoil events, (c) It contains both odd and even isotopes, suitable for spin-dependent and scalar interactions, offering the possibility of using enriched odd or even isotopes to identify the type of interaction, (d) Its high atomic number provides a good kinematical match to the theoretically favored particle mass range of 100 − 200 GeV.

Target masses between 100 and 1000 kg may be needed to reach the lowest predicted event rates. The 30 kg detector now under construction as ZEPLIN II represents a major step towards this. The ZEPLIN design principle could subsequently be scaled up to give a total target mass 1000 kg or more. The latter would achieve sensitivity comparable to the lowest predicted neutralino event rates (0.0001 − 0.01/kg/d) and could detect the annual signal modulation that would confirm the Galactic origin of any signal.

With liquid xenon, signal discrimination can be achieved in two basic ways: (1) Scintillation pulse shape - the decay time constant differs by a factor of 2 ∼ 3 for nuclear recoil and background events. (2) Using an electric field to inhibit recombination and measuring (a) the 'primary scintillation' S1 and (b) drifting the ionization into a strong electric field to produce a 'secondary scintillation' signal S2. Method (2) is more powerful, involving a comparison of two distinct signals associated with each individual event.

2. The ZEPLIN II Detector

The key features of the ZEPLIN II detector are:

(a) A shallow cylindrical chamber using 7 PMTs covering the top surface, and with liquid depth (14.1 cm).
(b) Active target mass of 30 kg, to increase the number of events and to provide a realistic module for scale-up to larger masses.
(c) Due to insufficient information about ionization from low energy xenon recoils, the detector was designed using a POLYTETRAFLUOROETHYLENE (PTFE) cone to confine the liquid xenon active volume hence eliminate 'dead' regions, in which a low energy gamma signal could lose its secondary pulse and mimic a nuclear recoil event.
(d) The design of the internal vessel ensures that all of the xenon liquid is active, removing the possibility of a misinterpreted signal from an inactive volume.

(e) PMTs placed inside the detector to maximize the light collection efficiency. Custom made PMTs from Electron Tube Inc. with thin hemispherical quartz profiles and platinum underlay coating ensure high pressure and low temperature operation.

(f) ZEPLIN II is designed capable of operating at high field if the ionization from nuclear recoil is detectable. In that case a full 3-D event reconstruction and perfect background rejection can be achieved (see details in ZEPLIN III design section 3).

Fig. 1 shows some photos of the target assembly and commissioning using Rutherford Appleton Laboratory's dedicated facility.

Figure 1. Internal target installation. Figure shows completed PTFE cone assembly, PMT layout, PMT assembly, field shaping ring assembly with two resistor chains shown, general overview of the ZEPLIN II detector, associated infrastructure and closing up of the target.

3. The ZEPLIN III Detector

ZEPLIN III is designed to get the best possible performance from a two-phase xenon detector of intermediate mass. ZEPLIN III is expected to allow a better ultimate background discrimination than ZEPLIN II by utilizing the ionization yield from xenon nuclear recoil.

Fig. 2 shows the design of ZEPLIN III and the construction in progress. A number of critical features guarantee the competitive edge of the chosen scheme. The placement of 31 PMTs inside the liquid, used to register the scintillation photons from the liquid and subsequent electroluminescence from the gas phase, as well as the thin active liquid xenon volume above the PMTs improves light collection for the primary scintillation, thus improving the energy threshold which allows deeper penetration, by a factor of five, into the cross-section parameter space. The high charge extraction electric field ensures that both primary and secondary signals are present for both nuclear and electron recoils allowing to trigger on a much larger secondary signal and therefore to locate a smaller primary as a precursor to the secondary, which not only makes the signal identification more secure but also again allows the energy threshold to be lowered. In fact these two factors make it possible to achieve a sub-keV energy threshold in detection of the primary recoil. The result is that the separation in secondary signals between electron and nuclear recoils for a given size primary is therefore, maximized. The principle of operation has been demonstrated with a prototype high-field detector, although it should be noted that a definitive nuclear recoil test must await deployment in a sufficiently low background environment due to event confusion at the surface.

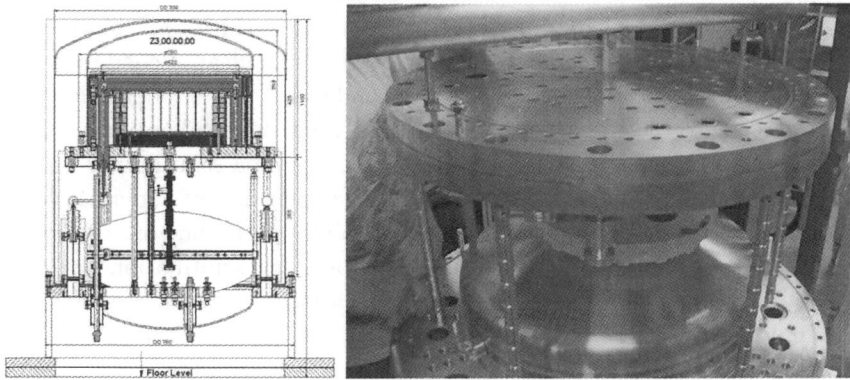

Figure 2. ZEPLIN III side view and trial assembly of the lower base flange, the liquid nitrogen reservoir and the xenon chamber cooling ring

3.1. *Background Estimation and Detector Simulations*

Extensive Simulation studies have been performed for both ZEPLIN II and III detector ANSYS and GEANT4 to assess the target characteristics, including drift field characteristics, primary light yield and light collection uniformity, and the expected neutron background when ZEPLIN II & III is deployed at Boulby. These studies indicate [7,8,9] that the main background are due to U/Th contamination in the PMT array, which can be reduced by conversion of the Compton veto to a neutron veto through Gd loading.

A γ-ray rejection factor of $10^4 - 10^5$ has been demonstrated by simulation for ZEPLIN III. A rejection factor of 10^4 will yield an experimental sensitivity of $\sim 10^{-8}$ pb after a year of running [8]. This is partly enabled by using an all-copper construction as far as possible. However, to reach this sensitivity does require suppression of the neutron background caused by U and Th in the PMTs.

4. ZEPLIN II gas filled tests at RAL

First performance tests on this full scale system, filled with xenon gas, were completed at RAL during the summer of 2004. Performance data obtained from these tests confirms that the ZEPLIN II detector performs as designed. Data were taken with a ^{241}Am gamma source located below the detector. Using the pulse area, timing, and coincidence information, detector performance can be studied carefully.

Fig. 3 shows the histogram of the primary pulse width. The black curve shows all events, the green curve shows events with clean secondaries, and the blue curve shows events without secondaries. Three populations are clearly seen: A, left sharp peak is due to large Cherenkov pulse mixed with gamma events; B, Middle peak is due to gamma scintillation; and C, the peak above 500-ns is due to overlap between primaries and secondaries or miss-triggers on secondaries. The red points at the bottom are events clearly not Cherenkov (97% efficiency on primary scintillation cut) and without secondaries. The energy spectrum of these events are shown in the middle plot where only one event below the interested 10 keV (out of 17,288 total events).

These gas tests confirmed that gamma events produced both primary (scintillation) and secondary (ionization) pulses, as expected from the earlier work with a small test chamber. It was confirmed that some events originating above the top field grid and close to the PMTs, give also a secondary pulse due to strong reverse field. This type of events can be eliminated using the secondary signal. A few rare events very near to the PMT produce very small secondary. These events can be eliminated by pulse arrival time and compare pulse height amount triggered tubes. The gas test run with 17,288 random events show only one event pass all cuts (99.994% rejection above 5 keV). Further reduction of these events are

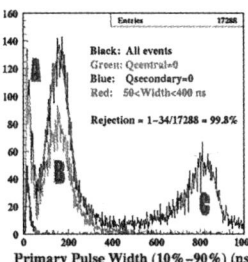 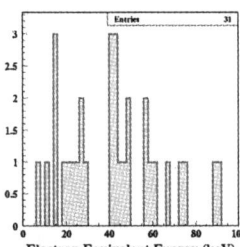 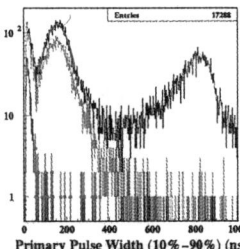

Figure 3. Primary pulse width distribution (left linear scale, right: log scale), middle: spectrum of final single pulse primaries after all cuts. Note that only one event below 10keV.

possible using pulse shape discrimination statistically. Studies will be done when more data are available. Conditions in liquid may vary and test result will soon available when ZEPLIN II operates underground.

5. Conclusion

ZEPLIN II and ZEPLIN III design are based on R&Ds carried out by the ICARUS and the ZEPLIN collaboration during the last decade and both detectors are expected to operate under Boulby mine in 2005 ls -la with expected sensitivity of 10^{-8} Pb.

6. Acknowledgments

The presenter wishes to acknowledge support from the U.S. Department of Energy, grant number DE-FG03-91ER40662, DE-FG03-95ER40917 and the National Science Foundation grant number PHY-0139065.

References

1. P. Benetti et al., Nucl. Instr. and Meth. A **329** 361-364 (1993).
2. P. Benetti et al., Nucl. Instr. and Meth. A **329** 203-206 (1993).
3. F. Arneodo et al., Nucl. Instr. and Meth. A **449** (2000) 147-15 7.
4. D. B. Cline et al., Nuclear Physics B - Proceedings supplements, V**124** (2003) 221-224, .
5. D. Cline et al.: *Astroparticle Physics* 12 (2000)373-377.
6. H. Wang, Physics Reports, Volume 307 Issues 1-4 (1998) 263-267.
7. C Bungau et al. Astroparticle Physics (in Press)
8. T. J. Sumner, 6th Int. Symp. Sources and Detection of Dark Matter/Energy in the Universe (Marina Del Ray, California, feb. 2004).
9. J. V. Dawson, Ph.D Thesis, Imperial College London, 2003

STATUS AND PLANS FOR THE XENON DARK MATTER EXPERIMENT

R. J. GAITSKELL
ON BEHALF OF THE XENON COLLABORATION
Department of Physics,
Brown University,
Providence,
RI 02912, USA
E-mail: gaitskell@brown.edu

The XENON experiment aims at the direct detection of dark matter in the form of WIMPs (Weakly Interacting Massive Particles) via their elastic scattering off Xenon nuclei. With a fiducial mass of 1000 kg of liquid xenon, a sufficiently low threshold of 16 keV recoil energy and an un-rejected background rate of 10 events per year, XENON would be sensitive to a WIMP-nucleon interaction cross section of $\sim 10^{-46} cm^2$, for WIMPs with masses above 50 GeV. A 1 tonne scale experiment (XENON1T) would be realized with an array of ten identical 100 kg detector modules (XENON100). The detectors are time projection chambers operated in dual (liquid/gas) phase, to detect simultaneously the ionization, through secondary scintillation in the gas, and primary scintillation in the liquid produced by low energy recoils. The distinct ratio of primary to secondary scintillation for nuclear recoils from WIMPs (or neutrons), and for electron recoils from background, is key to the event-by-event discrimination capability of XENON. A 3kg dual phase detector with light readout provided by an array of 7 photomultipliers is currently being tested, along with other prototypes dedicated to various measurements relevant to the XENON program. We present some of the results obtained to-date and briefly discuss the next step in the phased approach to the XENON experiment, i.e. the development and underground deployment of a 10 kg detector (XENON10) during 2005.

1. Introduction

The question of the nature of the dark matter in the Universe is being addressed with numerous direct and indirect detection experiments using a variety of methods, detectors and target materials. For a recent review of the field see Gaitskell[1].

The proposed XENON experiment is among the new generation direct searches for dark matter weakly interacting massive particles (WIMPs) with the ambitious goal of a sensitivity reach which is several orders of magnitude better than the lowest exclusion limit currently set by the CDMS experiment[2] To achieve a sensitivity goal of $\sim 10^{-46}$ cm^2 XENON relies on a target mass of 1 tonne of liquid xenon (LXe), with less than about 10 background events per year. Efficient background identification and reduction is based on the distinct ratio of the ionization and scintillation signals produced in LXe by nuclear (from WIMPs and neutrons) and electron (from gamma, beta and alpha backgrounds) recoil events [3]. The main challenge is to accomplish this event-by-event discrimination down to a few tens of keV nuclear recoil energy. Additional techniques used for background suppression are an additional active LXe shield around the sensitive target, passive gamma and neutron shielding, and the detector's 3-D position resolution. The position information is crucial to select single hit events characteristic of a WIMP signal, and to veto multiple hit events associated with neutrons as well as other backgrounds which propagate from the edge of the detector into the fiducial volume.

To test the XENON concept and verify achievable threshold, background rejection power and sensitivity, a detector with a fiducial mass on the order of 10 kg (XENON10), is under development for underground deployment in 2005. The detector exploits several key systems which have been tested and optimized with the 3 kg prototype, but will feature significant improvement in overall performance and sensitivity down to a 16 keV nuclear recoil energy. The experiment will be carried out at the Gran Sasso Underground Laboratory (3500 mwe). The depth and the expected background rejection power will allow us to reach a sensitivity a factor of 20 below the best existing measurements from CDMS II [2], of 2 dark matter events/10 kg/month, without the need of a muon veto for fast neutrons.

Another important goal of the XENON10 phase is to pave the way for the design of a 100 kg scale detector. With 3 months of operation deep underground, at a background level below 1×10^{-5} cts/keVee/kg/day after rejection, XENON100 would provide a sensitivity of $\sim 10^{-45}$cm^2. The full 1 tonne scale experiment (XENON1T) will be realized with ten XENON100 modules.

Fig. 1 shows the sensitivity projected for XENON10 experiment, in comparison to current WIMP searches, which are probing event rates at ~ 0.1 evts/kg/day. The projected performance of XENON100 and XENON1T detectors is also shown. In order to continue progress in dark matter sensitivity, and to obtain further performance data for the XENON Collaboration liquid xenon detectors in a low background environment, it will be important to have the 10 kg scale detector operational and taking science data in early 2006.

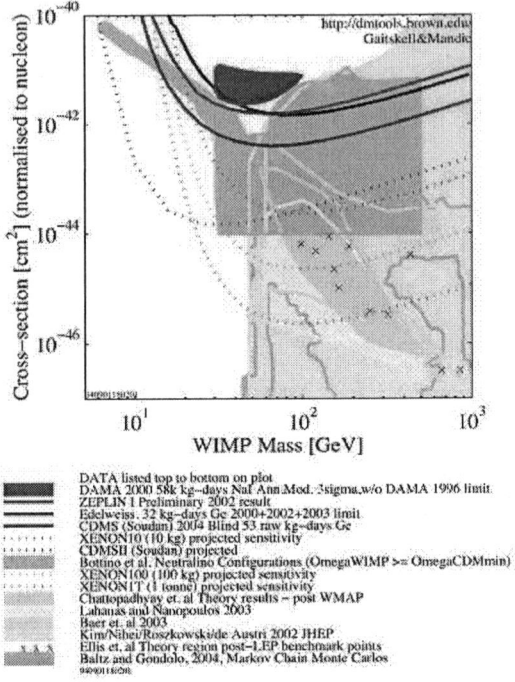

Figure 1. Theoretically predicted regions for SUSY WIMP candidate, along with the best detection dark matter limits from current direct detection experiments. Also shown as four dotted lines (top to bottom at right) are the projected sensitivities for CDMS II at Soudan [2], and for XENON10, XENON100 and XENON1T [11].

2. The XENON Detector Baseline

Figure 2 shows the design of the detector proposed as unit module for the XENON experiment. It is a dual phase TPC, with the active LXe volume defined by a 50 cm diameter CsI photocathode immersed in the liquid, at about 30 cm from the first of three wire grids defining a gas proportional scintillation region. An array of compact, metal channel UV sensitive PMTs developed by Hamamatsu Photonics Co. to work at LXe temperature and recently optimized for low radioactivity content, are used for primary and secondary light detection.

Figure 2. Schematic view of the XENON 100 dual phase detector.

The TPC is enclosed in a leak-tight cylindrical structure made of PTFE and OFHC. The PTFE is used as effective UV light reflector [4] and as electrical insulator. The fraction of direct light heading downward will be efficiently detected with the CsI photocathode [5]. The whole structure is immersed in a bath of LXe, serving as active veto shield against background. The LXe for shielding is readout by PMTs.

An event in the XENON TPC will be characterized by three signals corresponding to detection of direct scintillation light, proportional light from ionization electrons and CsI photoelectrons. Since electron diffusion in LXe is small, the proportional scintillation pulse is produced in a small spot with the same X-Y coordinates as the interaction site, allowing 2D localization with an accuracy of 1 cm. With the more precise Z information from the drift time measurement, the 3D event localization provides an additional background discrimination via fiducial volume cuts. The simulated detection efficiency of the primary scintillation light is about 5 p.e./keV for the XENON100 detector.

3. Results from the 3kg XENON prototype

3.1. *3 kg Prototype Design and Operation*

R&D for the XENON program is being carried out with various prototypes dedicated to test several feasibility aspects of the proposed concept, and to measure the relevant detector characteristics such energy threshold and background discrimination as well as ionization and scintillation efficiency of Xe recoils in LXe as a function of energy and electric field. Here we limit the discussion to the results obtained to-date with a dual phase xenon prototype with $\sim$ 3 kg of active mass. The primary scintillation light (S1) from the liquid, and the secondary scintillation light (S2) from the ionization electrons extracted into the gas phase, are detected by an array of seven PMTs, operating in the cold gas above the liquid.

A drawing of the 3kg XENON prototype is shown in Fig. 3 while the photo of Fig. 4 shows the integrated set-up at the Columbia University Nevis Laboratory. The sensitive volume of the TPC ($7.7 \times 7.7 \times 5.0$ cm^3) is defined by PTFE walls and grids with high optical transmission, made of Be-Cu wires with a pitch of 2 mm and 120μm diameter. Negative HV is applied to the bottom grid, used as cathode. Grids on the top close the charge drift region in the liquid and with appropriate biasing, create the amplification region for gas proportional scintillation. Shaping rings located outside of the PTFE walls and spaced 1.5 cm apart, are used to create a uniform electric field for charge drift.

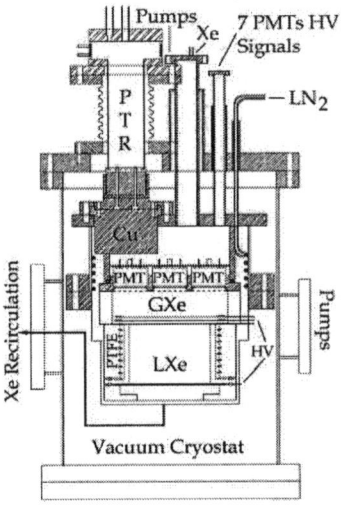

Figure 3. Schematic drawing of the dual phase prototype.

An array of seven, 2 inch PMTs (Hamamatsu R9288), mounted 2.3 cm above the top grid, is used to detect both primary and proportional light.

Figure 4. The detector integrated with the vacuum cryostat, refrigerator, gas/recirculation and DAQ systems.

Figure 5. An array of 7 PMTs on the top of the chamber in the gas phase.

The PMT array mounted on a PTFE frame is shown in Fig. 5. The custom-developed HV divider bases are also clearly visible. We used LEDs to measure the PMTs gain and single photoelectron response.

The LXe detector is insulated by a vacuum cryostat. A Pulse Tube Refrigerator (PTR) optimized for LXe, is used to cool down the detector, liquefy Xe gas and maintain the liquid temperature at the desired value. The typical operating temperature is -100 °C with a stability better than 0.05 °C. At this temperature the Xe vapor pressure is ∼1.8 atm. A reliable and stable cryogenics system is an essential requirement for the XENON experiment since both PMT's gain and the proportional light yield vary with temperature. With a cooling capacity of 100 W at 165 K the same PTR equipment will be used for XENON10 underground.

The XENON experiment requires ultra high purity LXe to enable ionization electrons to drift freely over the 30 cm proposed for the XENON100 unit module. Furthermore, the LXe purity has to be maintained during the long-term detector operation required for statistics and annual modulation analysis. A Xe purification, re-circulation and recovery system was built and operated with the 3 kg prototype [6](see Fig. 6).

For Xe gas purification, a single high temperature SAES getter was used[7]. Electron lifetime longer than 500 μsec is routinely achieved after a few days of continuous purification. A similar gas system with the addition of a Kr removal section will be used for XENON10 underground. We plan to start with commercial Xe gas with a Kr level of roughly 10 ppb, which can be reduced to a level well below 1 ppb by an adsorption-based system currently under construction. Methods for further Kr reduction are being studied within the collaboration using chromatographic separation in charcoal columns[8]. The reliability and efficiency of both the cryogenics and gas purification systems have been tested with repeated experiments lasting for several weeks continuously.

To maximize the information available from time structure and amplitude of the primary (S1), secondary (S2) and possible CsI photoelectron signal (S3), the 7 PMTs are digitized by both fast (1 ns/sample, 8 bit) and slow (100 ns/sample, 12 bit) ADCs. The gain of the fast ADCs is matched to observe signals down to single photoelectrons, whereas the slow ADCs are optimized to observe the longer proportional light signals from S2/S3. The DAQ system has been developed and successfully applied to record S1, S2 and S3 signals from the current prototype (Fig. 7). The coincidence of more than one PMT signals was required to create a trigger.

3.2. Detector Response to Gamma and Alpha Events

The 3kg dual phase detector's operation was tested using low energy gamma-rays from a Co-57 source, alpha particles from Po-210 deposited on the cathode, and neutrons from an AmBe source. Two typical waveforms of the direct light produced in the liquid (S1) and the proportional light produced by electrons extracted in the gas (S2) are shown in Fig.8, for an alpha and gamma event. The S1 signals are prompt while the S2 signals have a width of a few μsec, as expected. The time separation between S1 and S2 is dominated by the drift time in the liquid so that the position of the source along the drift axis is accurately inferred from the known drift velocity at the applied electric field. Since both the Co-57 122 keV gammas and the Po-210 5.43 MeV alphas are very localized in the dense LXe, the

267

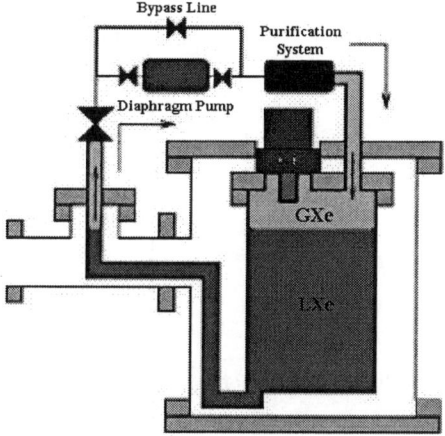

Figure 6. Schematics of the continuous Xe circulation and purification system.

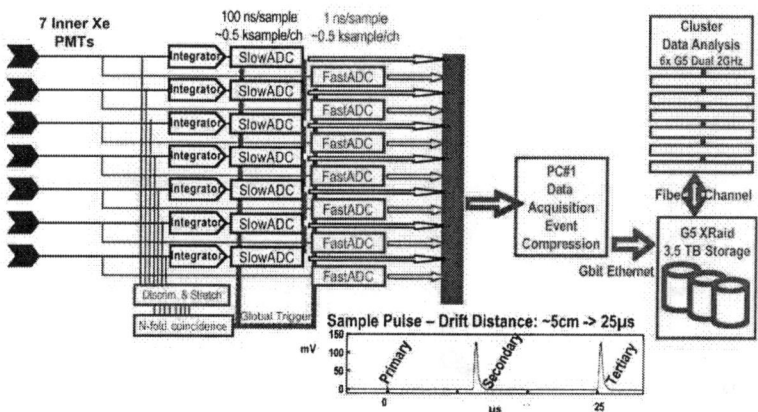

Figure 7. The DAQ system for recording S1, S2 and S3 signals from seven PMTs.

time separation between S1 and S2 is close to the maximum drift time of 25 μsec. The other two coordinates are inferred from the center of gravity of the proportional light emitted near the 7 PMTs. Simulations and a preliminary analysis of the alpha data show that the transverse direction of the source can be localized with an accuracy of 1 cm.

In Fig. 9(left) the S1 and S2 signals, simultaneously recorded with gamma and alpha irradiation, are plotted together. The detector was op-

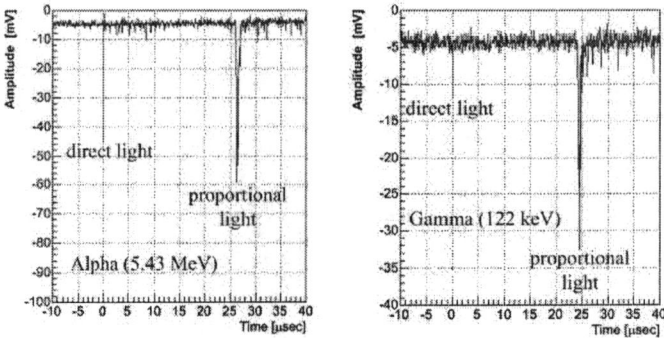

Figure 8. Waveforms of direct and proportional light for an alpha recoil (left) and an electron recoil (right).

erated at 1kV/cm in the drift region and 10 kV/cm in the gas region. The two classes of events are well differentiated. Another visualization of this event separation is shown in Fig. 9 (right) where the distribution of the same events is plotted as a function of the S2/S1 ratio, in logarithmic scale. The peak from gamma rays is normalized to 1.

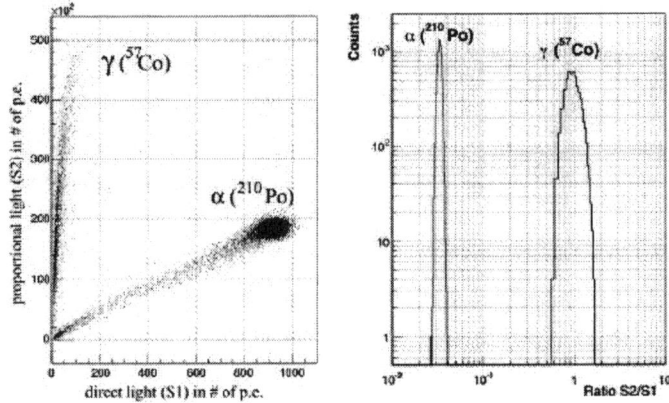

Figure 9. Measured distribution of S2 (proportional light) versus S1 (direct light) for combined alpha and gamma events (left). Distribution of alpha and gamma events plotted as a function of the ratio S2/S1 (right).

The measured ratio $(S2/S1)_\alpha/(S2/S1)_\gamma$ is about 0.03. The separation between alpha recoils and electron recoils is already remarkable, despite the non optimized light collection of the detector at this stage. This ratio is even larger, if we account for the fraction of primary light produced by alpha recoils which is absorbed by the ^{210}Po source disk. Another distinct feature that separates alpha recoils from electron recoils is the dependence of the light on applied electric field. While the primary light from an electron event is strongly quenched by the field because of the reduced recombination rate, this is not the case for a heavily ionizing particle such as an alpha. This means that the primary light is barely affected by the field and the S2/S1 ratio is essentially constant. The dependence of the primary light on the applied field was previously measured by Aprile et al.[9] and has been verified with data from the XENON prototypes.

3.3. Detector Response to Neutron Events

The $S2/S1$ ratio in LXe for nuclear recoil events was established using a 5 Ci ^{241}AmBe source, emitting neutrons in the energy range 0–8 MeV, in conjunction with a BC501A scintillation coincidence counter to detect events scattering from the LXe target. The 1.4 liter BC501A counter was placed at a distance of 50 cm from the center of the LXe chamber, at a scattering angle of 130 deg.

A population of candidate neutron scattering events in the LXe was obtained by identifying events which were (i) tagged as neutron recoils in the BC501A (by pulse shape discrimination) and (ii) also had an implied ToF (time of flight) between the LXe and BC501A in a window of 40–70 ns. The selected events also contained a significant population of accidental coincidences, between gammas scattering in the LXe, and neutrons, emitted separately, which interact in the BC501A in the appropriate time window. Figure 10 shows LXe event data, for the AmBe source, in which there is a population of events with $S2/S1 \sim 1000$, associated with electron recoils, and a second population of events, with $S2/S1 \sim 100$, associated with elastic nuclear recoils. The figure also contains events arising from the inelastic scattering of neutrons from ^{129}Xe (nat. abun. 27%) and ^{131}Xe (nat. abun. 21%) which have excited states of 40 keV, and 80 keV, respectively. A simulation of the predicted event distribution $S2/S1$ vs. $S1$ is shown in Fig. 11 for comparison with the data in Fig. 10.

A histogram comparing the $S2/S1$ distributions for events $S1 < 20$ p.e. for the AmBe source, and separately a ^{137}Cs source, are shown in Fig. 12. The AmBe curve shows the two populations associated with electron and nuclear recoils. The second population is absent in the ^{137}Cs data.

The leakage of electron recoil events in the ^{137}Cs data into the $S2/S1$ region for nuclear recoil events is $< 1\%$. It was established, using separate calibration work, that the the $S1$ signals for electron and nuclear recoils is 0.14 p.e./keV$_{ee}$ and 0.07 p.e./keV$_r$, respectively.

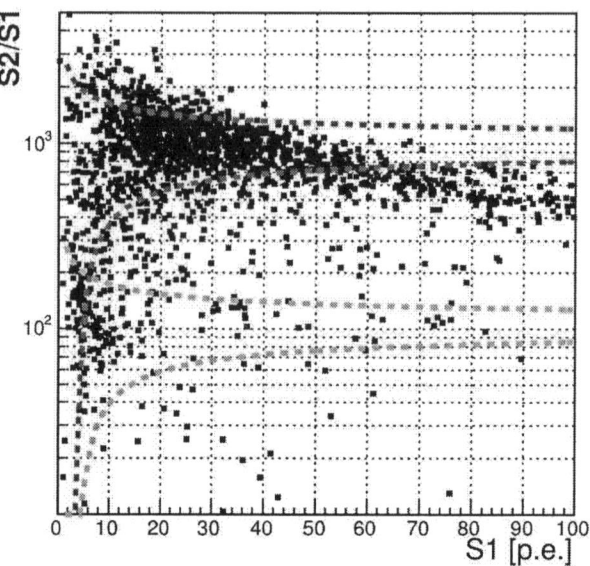

Figure 10. Events in LXe from an AmBe source, tagged using the method described in the text. Events with $S2 > 50$k photoelectrons are rejected since they saturate the acquisition electronics. The region defined by the red dashed lines represents a band with $S2/S1 \sim 1000$ associated with electron recoil events. The region defined by the green dashed lines represents a band with $S2/S1 \sim 100$ associated with elastic nuclear recoil events. Inelastic neutron scattering events are also present. Their predicted event distribution is shown in Fig. 11.

3.4. *Discussion of Primary and Secondary Signal Sizes*

The signal sizes for $S1$ and $S2$ can be estimated from the following expressions.

$$S1 = (\alpha_1 Q_{PMT}) n_p \tag{1}$$

$$S2 = (\alpha_2 Q_{PMT})(\gamma_{gas} \epsilon_{ex} \epsilon_d \epsilon_q) n_e \tag{2}$$

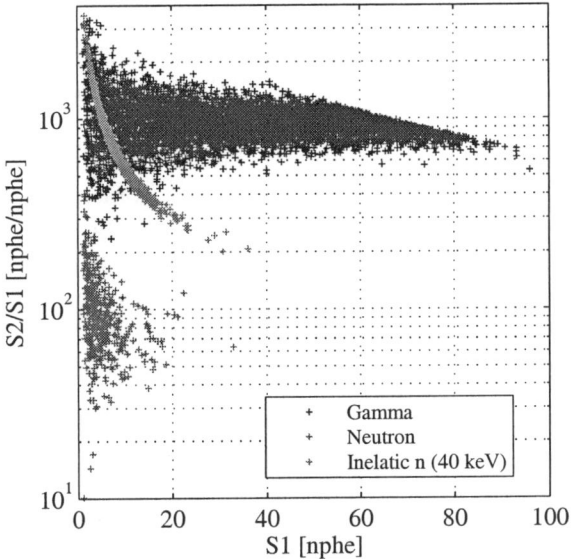

Figure 11. Simulation of detector response, $S2/S1$ vs $S1$, for AmBe neutrons and uniform gamma spectrum. Distributions are shown for electron recoils (blue, $S2/S1 \sim 1000$), neutron elastic recoils (red, $S2/S1 \sim 100$), and neutron inelastic recoils associated with ^{129}Xe 40 keV excited state. The inelastic scattering for the ^{131}Xe 80 keV excited state is not shown. The statistical fluctuations, associated with all steps in the generation of S1 and S2 signals have been considered. The variation in signals with position within the detector is not simulated for this plot.

The ratio of the secondary to primary light is therefore given by

$$S2/S1 = (\alpha_2/\alpha_1)(\gamma_{gas}\epsilon_{ex}\epsilon_d\epsilon_q)(n_p/n_e) \qquad (3)$$

The definitions of these parameters, and some of the values for the operation of the chamber during the AmBe neutron source data taking, are shown in the table overleaf.

Light simulation Monte Carlos[10] were performed to estimate the values of the effective geometric light collection of photons generated in the liquid and the gas, α_1 and α_2 respectively. Previous simulations for other chambers have been found to be in very good agreement with observed collection efficiencies. In the simulations the α values were seen to vary with the position of the light source, however, the mean values are given in table overleaf, and the overall range of collection efficiencies $< \pm 30\%$.

Analysis is currently ongoing to determine the values of the other pa-

Symbol	Value	Comment
α_1	7% (No CsI)	Scintillation light S1 geometrical collection efficiency for all PMTs. This represents the mean prob. that a photon, generated in LXe, strikes the front face of any of the PMTs.
α_2	13% (No CsI)	Scintillation light S2 geometrical collection efficiency for all PMTs. Defined as S1 but for a photon generated in Xe gas above liquid.
Q_{PMT}	12%	The effective efficiency with which a PMT converts a photon arriving at its front face into a photo electron (p.e.). It combines the quantum efficiency of the PMT photocathode ((typ. 20% for $\lambda \sim 175$ nm), with the collection efficiency of electrons from the photocathode inside the PMT (typ. 60%).
γ_{gas}	$\sim$200 ph/elec	Number of photons generated due to electro- luminescence in gas phase per electron extracted from the liquid surface. It is a function of the Xe gas pressure, electron drift length in gas, and the applied field in the gas.
ϵ_{ex}	$\sim 100\%$	Extraction efficiency of electrons from liquid to gas. Close to 100% for extraction field of 9 kV/cm used in this work.
ϵ_d	$\sim 100\%$	Charge drift efficiency. This is a function of the drift distance of the electrons, however, the electron mean free path was measured to be very much greater than the maximum drift length due to the high purity of LXe achieved.
$\epsilon_{q,\gamma}$		Fraction of available charge from the initial
$\epsilon_{q,n}$		interaction site that starts drifting, for electron (γ) or nuclear (n) recoils. It is drift field dependent (4 kV/cm in this work) and parameterizes the recombination of ionization at the primary site.
n_p		Number of scintillation photons created in LXe at the primary interaction.
n_e		Number of ionization electrons created in LXe at the primary interaction.

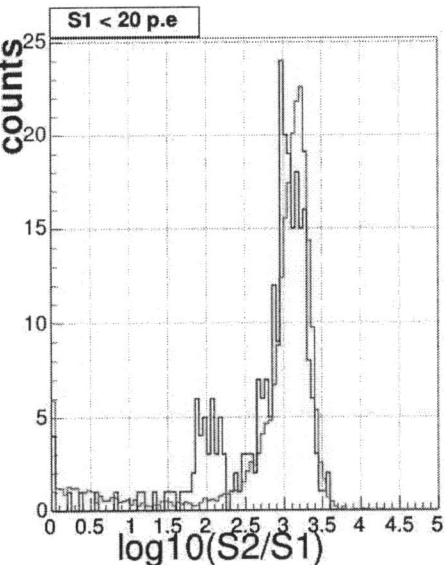

Figure 12. The blue (dark) line shows a histogram of $S2/S1$ for events with $S1 <$ 20 p.e. taken for AmBe source data shown in Fig. 10. For comparison the $S2/S1$ distribution (red (light) line) for Compton electron recoil events, in the same $S1$ range, for a ^{137}Cs source of 662 keV gamma rays is shown. The AmBe curve shows two distinct populations associated with electron and neutron recoils at $S2/S1 \sim 1000$ and $S2/S1 \sim 100$, respectively. The second population is absent in the ^{137}Cs data.

rameters in the Table. It should be noted that the value of the $(S2/S1)$ for elastic neutron scattering events is a factor ~ 3 higher than that seen for α scattering events, although most of this difference can be explained by the suppression of the $S1$ signal for α events since they occur very near the alpha source which has bad light reflection properties.

3.5. *Optimization of Light Detection*

The light collection efficiency of the 3kg prototype, discussed in the text above, is very poor. In particular, the observed $S1$ signal corresponded to ~ 0.35 p.e./keV$_{ee}$ (at zero drift field) for electron recoils. The PMTs were widely spaced, and as a consequence the geometrical coverage in this initial test configuration was low. Also, a large fraction of the UV photons for the $S1$ signal are not detected by the PMTs, which are located in the gas, due to total internal reflection of the photons at the liquid-gas interface. As originally proposed for the XENON baseline detector, a CsI photocathode in place of a common cathode, at the bottom of the liquid, can significantly improve light collection, and lower the minimum energy threshold [11]. Monte Carlo simulations show that the primary light detection efficiency of the 3kg prototype would increase to ~6 p.e./keV$_{ee}$ (at zero drift field), with the cathode grid replaced by a CsI photocathode. Following a similar symbolic convention to that used in Eqns. 1 & 2 the size of $S3$ in the PMTs will be given by

$$S3 = (\alpha_2 Q_{PMT})(\gamma_{gas}\epsilon_{ex}\epsilon'_d)(\alpha_3 Q_{CsI})n_p \qquad (4)$$

where α_3 is the geometric light collection efficiency of the CsI for photons generated in the liquid. Results from recent tests with various photocathodes are very encouraging. We have confirmed the high QE (Q_{CsI}) in LXe (see Fig.13), first measured by the Columbia group more than ten years ago [5]. We have also demonstrated the effective suppression of the photon feedback connected with a CsI in a dual phase detector, using a commercial HV switch unit (PVX-4130 from Directed Energy, Inc). The normal rise and fall time of < 100 nsec was slowed to 1 μsec, and not appreciable noise from the switching was observed on the PMTs in our 3 kg prototype. Proportional scintillation could be stopped as expected by deriving an appropriate gate signal from the light trigger.

From measurements with a CsI photocathode in the 3kg dual phase detector we infer a similar value for QE as that measured with a simple ionization chamber without PMT and switching supply. The QE of the photocathode as a function of field (up to 3 kV/cm) is derived from comparing the measured and theoretical signal size ratio between S4 (Eq. 6) and S2 (Eq. 2). S4 is the proportional light signal from the extracted photoelectrons, which were induced by light from the original S2 signal being collected on the CsI photocathode, such that

$$S4 = (\alpha_2 Q_{PMT})(\gamma_{gas}\epsilon_{ex}\epsilon'_d)(\alpha_4 Q_{CsI})(\gamma_{gas}\epsilon_{ex}\epsilon_d)n_e. \qquad (5)$$

α_4 is the geometric light collection efficiency of the CsI for photons generated in the gas, and is estimated using a Monte Carlo simulation. The photocathodes used in these experiments were deposited at the same time, using the same substrate and thickness of the CsI layer. Combined results are shown in Fig.13. We are finalizing a CsI deposition apparatus at Columbia which will enable us to optimize preparation and test CsI photocathode for maximum performance in LXe.

Interestingly, eqns. 1–5 imply that a combined ratio of the light signal magnitudes

$$\frac{S2S3}{S1S4} = \frac{\alpha_2 \alpha_3}{\alpha_1 \alpha_4} \qquad (6)$$

is dependent only on the α parameters (which are determined only by the geometrical light collection properties of the detector). This relationship was shown to hold experimentally.

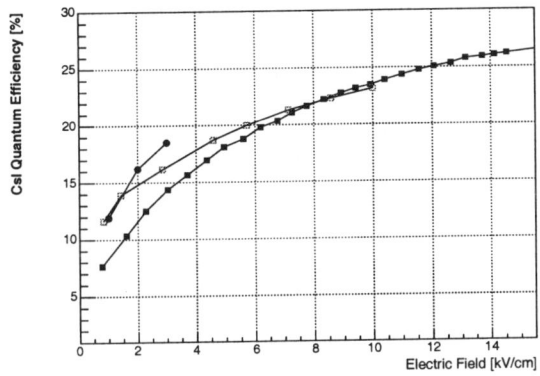

Figure 13. CsI QE as a function of electron extraction field. The solid square data points are from [5]. The open square are from recent measurements with a similar setup as in [5]. The circle data points are inferred from the signals measured with the 3kg dual phase prototype.

4. Conclusion

The experience gained with the 3kg prototype, its performance and results to-date, as well as results obtained with other detectors not presented here[12], are guiding the realization of XENON10. The 10 kg scale detector will use a light readout with a CsI photocathode in the liquid and an array of PMTs in the gas for much improved light detection efficiency and sensitivity to low energy recoils.

This work was supported by grants from the National Science Foundation to the Columbia Astrophysics Laboratory (Grant No. PHY-02-01740) and from the Department of Energy to the Particle Astrophysics Group at Brown University (Grant No. DE-FG-02-91ER40688).

References

1. R.J. Gaitskell: Annu. Rev. Nucl. Part. Sci., 54, (2004) 315-59.
2. CDMS Collaboration: Phys. Rev. Lett. 93, (2004) 211301.
3. M. Yamashita, et al.: Astropart. Phys., 20 (2003) 79-84.
4. M. Yamashita, et al.: Nucl. Inst. and Meth. A, 535 (2004) 692-698.
5. E. Aprile, et al.: Nucl. Inst. Meth. A 338 (1994) 328.
6. S. Mihara, et al.: Cryogenics, 44 (2004) 223-228.
7. SAES Pure GAS, Inc.: http://www.saesgetters.com/.
8. T. Shutt, CWRU, Private Communication.
9. E. Aprile, et al.: IEEE Trans. Nucl. Sci., 37 (1990) 553.
10. K. Ni, Columbia University, (GEANT4) Private Communication.
11. XENON Collaboration: NSF proposal number 0201740, "XENON: A Liquid Xenon Experiment for Dark Matter", proposal submitted to NSF, Particle and Nuclear Astrophysics in Sep. 2001.
12. E. Aprile, et al.:"Scintillation Response of Liquid Xenon to Low Energy Nuclear Recoils", submitted to Phys. Rev. D.

AN ENGINEERING DESIGN STUDY FOR A LARGE-SCALE XENON WIMP DETECTOR [*]

E. J. DAW[†]
*University of Sheffield
Department of Physics and Astronomy
Hicks Building, Hounsfield Road,
Sheffield, South Yorkshire, S3 7RH, United Kingdom
E-mail: e.daw@sheffield.ac.uk*

Some engineering issues relevant to tonne scale liquid xenon based detectors for weakly interacting dark matter particles (WIMPs) are discussed. The vessel design chosen for this study is a pure copper vessel containing 250 kg of liquid and vapour phase xenon at a maximum pressure of 4 atmospheres. Finite element modelling tools are used to calculate the mass and dimensions of the vessel. The physical size of a passive, room temperature gas dump for such a detector is estimated.

1. Introduction

Weakly interacting massive particles of mass in the range $[10, 1000]$ GeV/c^2 could be the cold dark matter thought to form halos around this and other spiral galaxies [1]. Numerous experiments aiming at direct detection and study of WIMPs in our own galactic halo have published results, are currently underway, or are proposed [2,3,4,5,6,7,8,9,10,11,12]. These experiments rely on detection of the energy, typically a few keV, transferred when a WIMP undergoes an elastic collision with a nucleus of the detector material. Since these collisions take place through the exchange of a very massive vector boson, interaction cross sections are very small. Therefore a sensitive detector should be as massive as possible in order to maximize the rate of WIMP interactions. Other critical requirements include deep underground operation to shield against background events from cosmic ray showers, and detector construction from materials of low radioisotope contamination.

---

[*]Work supported in part by the U. K. Particle Physics and Astronomy Research Council.
[†]On behalf of the United Kingdom Dark Matter Collaboration.

2. Why use liquid xenon for large-scale WIMP detectors?

Liquid xenon is a scintillator, emitting 175 nm photons when either nuclei or electrons from xenon atoms are struck by WIMPs or by radioactive background [13]. Liquid xenon detectors therefore consist of large volumes of liquid xenon instrumented with UV-sensitive photon detectors, usually quartz windowed photomultiplier tubes. Liquid xenon has several properties that make it particularly suitable for WIMP detection[14,15].

Firstly, radionuclide and chemical impurity concentrations in liquid xenon can be reduced using techniques unavailable for detectors involving solid crystals such as getters, distillation, or centrifuge. For example, the XMASS collaboration has demonstrated the reduction of ^{85}Kr concentration from 330 parts per billion in xenon gas to 5 parts per trillion in liquid xenon purified by distillation [5].

Secondly, ionization electrons produced in the liquid can be drifted in an electric field through very pure liquid xenon and detected [16]. The size, or simply the presence, of this secondary ionization signal, compared to the size of the primary scintillation signal provides a powerful discriminant between events due to the recoil of a nucleus, which can be due to WIMPs or background neutrons, and events due to the recoil of an electron, which are predominantly due to collisions with gamma rays. In an electron recoil, the struck electron is ionized, and moves off at a high velocity, colliding with and ionizing other electrons in other xenon atoms, each of which also have a relatively high recoil velocity, and hence are unlikely to re-combine with their partner xenon ions. Hence the free electron charge liberated in an electron recoil event is relatively high. In a nuclear recoil, the xenon is again ionized, but the bulk of the recoil energy is transferred to a heavy xenon ion, which has a far smaller recoil velocity than a recoiling electron of the same kinetic energy. A large electric field is necessary to surpress immediate recombination of the electron ion pair, and even when separated, the electron and ion have far smaller velocities than in the case of an electron recoil. Hence the recoil velocities of electrons from subsequent secondary ionization events will also be small, and the chances for recombination of these secondary electron ion pairs, high. Therefore the ionization charge liberated in an nuclear recoil event is significantly smaller than that liberated in an electron recoil event of the same recoil energy.

Finally, the price of liquid xenon is approximately £1k/kg. This compares favourably with other proposed technologies for tonne-scale instruments.

3. Vessel Design for This Study

The strategy for this study is to select constraints that seem reasonable for the design of a real tonne-scale liquid xenon WIMP detector.

- Mass of xenon. Assume one tonne of xenon overall distributed in four 250 kg modules. Simulation studies[17] indicate that such a detector, instrumented with existing technology and readout, could achieve sensitivity to elastic wimp nucleon scattering at the 10^{-10} pb level.
- Vessel material. Assume a copper vessel. Simulation studies[17] indicate that a 500kg copper vessel contributes a background of around 2 events per year.
- Readout. Assume that the readout will detect ionization electrons and scintillation light. Measurements[18] indicate that the level of electronegative impurites such as oxygen in liquid xenon, can be reduced such that the electron mean free path in the xenon significantly exceeds 100 mm. Cleaner xenon may allow greater depths, with the risk of reduced scintillation light collection efficiency.
- Temperature and pressure of xenon. Assume liquid xenon at $-110°C$, contained in a pressure vessel rated for 4 atmospheres, including a safety margin. This configuration has been utilized succesfully on previous experiments[9].

The density of liquid xenon is 3×10^3 kg m^{-3}. A disk of depth 100 mm should be 515 mm in radius to contain 250 kg of xenon. This geometry is illustrated in Figure 1. Above the disk of liquid is a thinner disk of xenon vapour. Electrons drifted from the liquid are detected in this region.

Figure 2 shows the vessel containing the xenon and enough space for readout system. The force on the vessel walls is dominated by the 4 atmosphere pressure requirement. This force is maximal in the regions where the cylindrical walls are joined to the spherical end caps, where the radius of curvature of the walls is minimal. FEMLAB[19], a commercial finite element modelling program, was used to evaluate the von Mises stress [a] in the walls of the copper vessel for a range of wall thicknesses. It was assumed that copper has a density of 8.94×10^3 kg/m^3, a bulk modulus of 0.14×10^{12} N/m^2, a Poisson ratio of 0.32, and a yield stress of 69×10^6 N/m^2.

---

[a] The definition of the von Mises stress σ_v is
$\sigma_v = \sqrt{[(\sigma_1 - \sigma_2)^2 + (\sigma_2 - \sigma_3)^2 + (\sigma_3 - \sigma_1)^2]/2}$, where $\sigma_i | i = 1 \cdots 3$ are the principal components of the stress tensor.

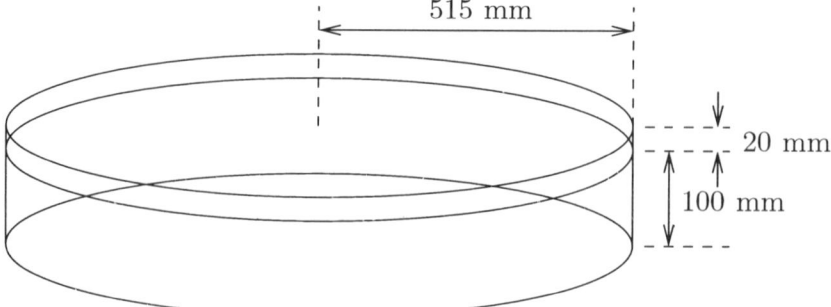

Figure 1. Dimensions of the xenon for this study. The lower, 100 mm thick disk is liquid xenon. The upper 20 mm thick disk is xenon vapour.

4. Results of Finite Element Modelling

Figure 3 is a contour plot of the stress profile in the cross section of the vessel wall indicated by box A in Figure 2, with an end wall thickness of 30mm. The maximum von Mises stress is 2.3×10^7, a factor of 3 less than the yield stress for copper. Based on two ends of this thickness and cylindrical side walls of thickness 27 mm, the total mass of this inner vessel is estimated at 870 kg, comprised of 240kg for the mass of each end and 390 kg for the cylindrical barrel. The high mass of this inner vessel presents an engineering challenge due to the high thermal mass of 870 kg of copper, which will make initial cooldown of the vessel a challenge, and the necessity of mechanically supporting the inner vessel whilst maintaining good thermal insulation of the inner vessel.

5. A Passive Gas Dump System

The safety threat and financial cost in the event of a xenon vent necessitate the construction of a xenon dump system. In the event of a warm up of the inner vessel, the resulting increased gas pressure ruptures a burst disk, releasing the xenon into a volume sufficient to contain the vapourised xenon. One tonne of liquid xenon is 7600 moles, which at 3 atmospheres occupies 57 m^3. A vessel, or assembly of vessels having this overall volume is a challenge in a deep underground laboratory. An alternative is a cryogenic or 'active' dump, perhaps maintained at 77 K with liquid nitrogen. The xenon would freeze to the walls of the dump chamber, thereby requiring a much smaller volume. The challenge with an active dump is to achieve 100% reliability of the dump system and its cryogenics.

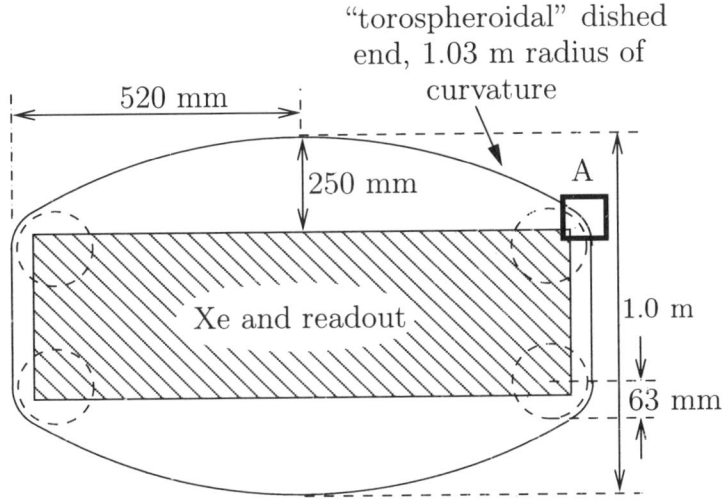

Figure 2. Dimensions and size of the vessel to contain the xenon and readout.

6. Conclusions

The issues highlighted in this article do not preclude the construction of very large scale liquid xenon WIMP detectors. Cooling of tonne-scale masses in vacuum systems has been achieved by many groups, notably those involved in constructing gravitational wave bar detectors. The engineering challenges of experimental physics are a large part of what makes it exciting and rewarding, and we look forward to meeting these and many other challenges on the road to tonne-scale WIMP detectors.

References

1. G. Jungman, M. Kamionkowski, K. Griest, *Phys. Rept.* **267**, 195-273 (1996). Preprint at http://arxiv.org/abs/hep-ph/9506380.
2. Y. Ramachers, *Nucl. Phys. Proc. Suppl.* **118** 341-350 (2003). Preprint at http://arxiv.org/abs/astro-ph/0211500.
3. E. Aprile *et al.* (XENON collaboration), *Proceedings of the Xenon01 International Workshop on Techniques and Applications of Xenon Detectors*, World Scientific (2001). Preprint at http://arxiv.org/abs/astro-ph/0207670.
4. R. Gaitskell (XENON collaboration), *these proceedings (IDM2004)*
5. S. Moriyama (XMASS collaboration), *these proceedings (IDM2004)*
6. D. Akerib *et al.* (CDMS collaboration), *Phys. Rev. Lett.* **93** 211301 (2004). Preprint at http://arxiv.org/abs/astro-ph/0405033.
7. B. Ahmed *et al.* (UKDMC collaboration), *Astropart. Phys* **19** 691-702 (2003).

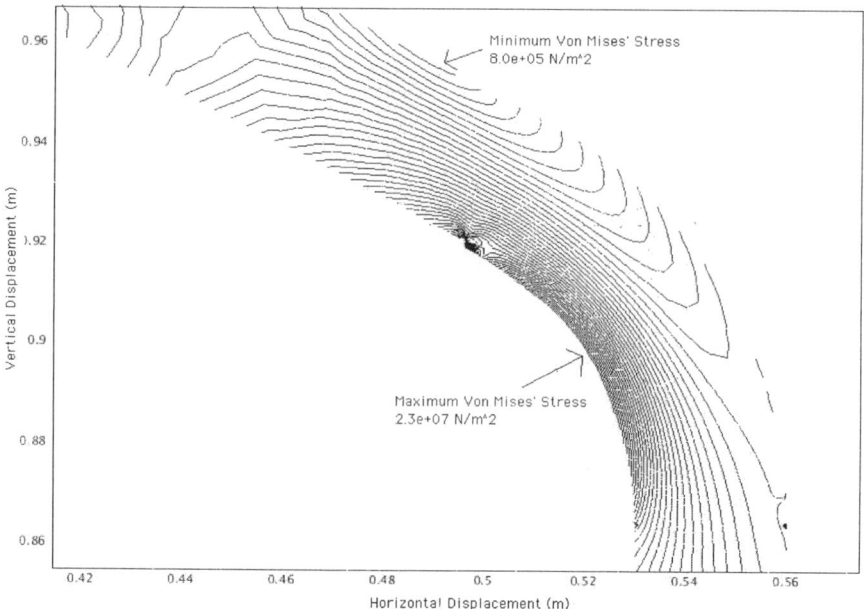

Figure 3. Von Mises' stress in a cross section (Figure 2, box A) through the dished end of a copper vessel having a wall thickness of 30 mm.

8. G. Chardin et al. (EDELWEISS collaboration), Nucl. Instrum. Meth. A **520**, 101 (2004).
9. V. Kudryavtsev (UKDMC collaboration), *these proceedings (IDM2004)*
10. B. Majorovits et al. (CRESST collaboration), *these proceedings (IDM2004)*. Preprint at http://arxiv.org/abs/astro-ph/0411396.
11. P. Belli (DAMA collaboration), *these proceedings (IDM2004)*
12. R. Bernabei et al. (DAMA collaboration), *Riv. Nuovo Cim.* **26** number 1, 1-72 (2003) and references therein.
    Preprint at http://arxiv.org/abs/astro-ph/0307403.
13. A. Hitachi, *Proceedings of the Fourth International Workshop on the Identification of Dark Matter (IDM2002)*, World Scientific, ISBN 981-238-237-2, pages 357-362 (2003)
14. G. Davies et al. *Phys. Lett. B* **320** 395-399 (1994) and references therein.
15. H. Wang, *Phys. Rept.* **307**, 263-267. Preprint at
    http://hepwww.rl.ac.uk/UKDMC/pub/papers/journal/PRep307HW.pdf.
16. T. Takahashi et al. *Phys. Rev. A.* **12** 1771-1775 (1975)
17. M. J. Carson et al., Astropart. Phys. **21**, 667 (2004).
    Preprint at http://arxiv.org/abs/hep-ex/0404042.
18. H. Wang et al., Nucl. Instrum. Meth. A **329** 361-364 (1993)
19. COMSOL, inc, Burlington, MA, USA. http://www.comsol.com

STATUS OF THE ANAIS EXPERIMENT AT CANFRANC

M. MARTÍNEZ, J. AMARÉ, B. BELTRÁN, J.M. CARMONA, S. CEBRIÁN,
E. GARCÍA, I.G. IRASTORZA,* H. GÓMEZ, G. LUZÓN, A. MORALES[†]
J. MORALES, A. ORTIZ DE SOLÓRZANO, C. POBES, J. PUIMEDÓN,
A. RODRÍGUEZ, J. RUZ, M.L. SARSA, L. TORRES,‡ J. A. VILLAR

Laboratory of Nuclear and High Energy Physics,
University of Zaragoza,
50009 Zaragoza, Spain
E-mail: mariam@unizar.es

The status and prospects of the ANAIS experiment (Annual Modulation with NaI's) at the Canfranc Underground Laboratory are presented. After a prototype stage with a single NaI crystal (10.7 kg), which resulted in an average of 1.2 counts/(keV kg day) in the 4-10 keV range after an exposure of 2069.85 kg day, the whole experiment has been designed to accommodate up to 10 crystals (more than 100 kg of NaI). Several R&D works are in progress to reduce both energy threshold and background and are presented here.

1. Introduction

The identification of the non-baryonic dark matter component of the Universe is one of the challenges of the present Particle Physics and Astrophysics, being the WIMPs (Weakly Interacting Massive Particles) one of the favorite candidates. WIMPs are supposed to form a halo with Maxwellian velocity distribution in the most simple models. The variation in relative velocity of the Earth with respect to the halo implies an annual modulation in the interaction rate observed in a detector. This is a small effect ($< 7\%$ of the WIMP expected signal) noticeable only in the low energy region (a few keV). In the past years, a positive result has been reported by the DAMA collaboration, at Gran Sasso National Laboratory [1]: after seven cycles, a modulation has been noticed in the low energy region with a period of one

---

*Present address: CEA, Saclay, France and CERN, Geneva, Switzerland.
†Deceased
‡CEE fellow in the Network of Cryodetectors, under contract HPRN-CT-2002-00322, presently at the Dipartamento di Fisica dell'Università di Milano-Bicocca, Italy.

year and maximum rate around June 2nd.

The ANAIS experiment is a large mass experiment with more than 100 kg of NaI(Tl). Three of the detectors that will comprise the ANAIS set-up were used before in NaI32[2], a pioneer experiment looking for annual modulation with a total mass of 32 kg of NaI. A first prototype of the ANAIS experiment was developed in the past years[3], with a single NaI crystal of 10.7 kg. Working with one photomultiplier and using a pulse shape discrimination technique to reject noise from the PMT, a threshold of 4 keV was achieved. A background of 1.2 counts/(keV kg day) was obtained in the 4-10 keV region. The low energy spectrum before and after noise rejection can be seen in Fig. 1. Figure 2 shows the exclusion plots obtained after an exposure of 2069.85 kg·day for spin independent and spin dependent interacting WIMPs.

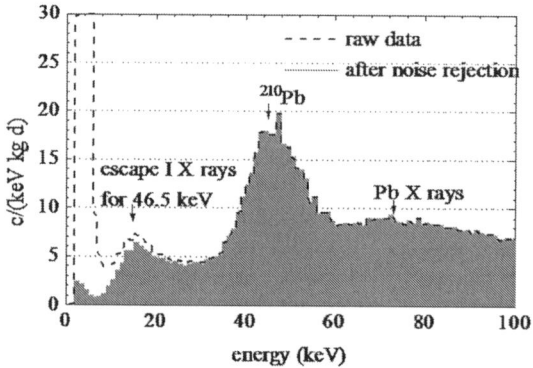

Figure 1. Low energy spectrum obtained with the prototype I after an exposure of 2069.85 kg·day, before and after the noise rejection by PSD.

2. ANAIS experimental set-up

The ANAIS experiment will operate at the new Canfranc Astroparticle Undergrond Laboratory (LSC), presently under construction, that is foreseen to be finished by the summer of 2005[4], as is reported in these Proceedings. The detector will be composed of 10 hexagonal NaI(Tl) crystals of 10.7 kg each. Every crystal will work with two photomultipliers in coincidence. The designed shielding (see Fig. 3) will consist of 10 cm of archaeological lead with an activity less than 9 mBq/kg from ^{210}Pb and an additional shielding of 20 cm of low-activity lead. To avoid radon gas entering the

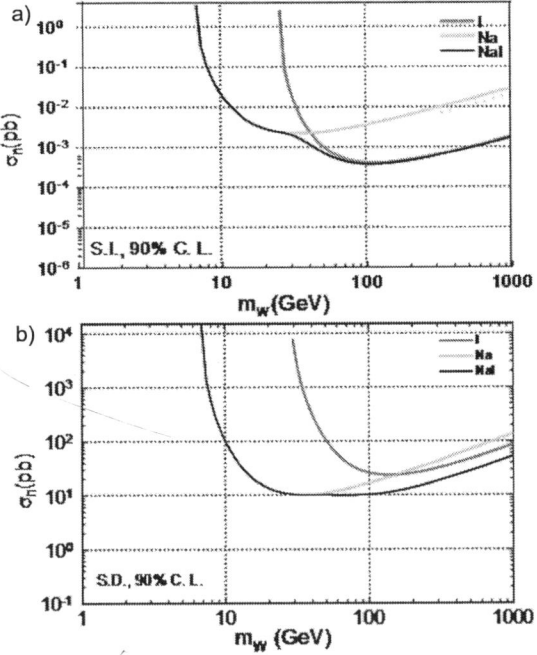

Figure 2. Exclusion plots for a) spin independent and b) spin dependent interacting WIMPs derived with the ANAIS prototype I after an exposure of 2069.85 kg·day, where σ_n represent the WIMP-nucleon cross section and m_W the WIMP mass.

structure a PVC box will keep an overpressure of nitrogen gas coming from evaporation of liquid nitrogen. The neutron shielding will consist of a 2 mm thickness cadmium sheet and 40 cm of a suitable neutron moderator, such as polyethylene or borated water tanks. There will also be plastic scintillators as active muon vetoes.

The data acquisition system, that will combine standard NIM and CAMAC modules controlled by a biprocessor PC with real-time capability under Linux operating system, will record the energy and the digitized pulse shape (sampling rate up to 2 Gs/s, 500 MHz analog bandwidth) for each PMT, as well as the multiplicity of the event. Coincidences between detectors can be removed later by software. A complete slow control will be designed to keep track of effects that could mimic the modulation signal: the temperature inside the structure as well as outside, the high voltage supply to the photomultipliers, the radon air contamination, the flux of the nitrogen injected in the set-up, etc.

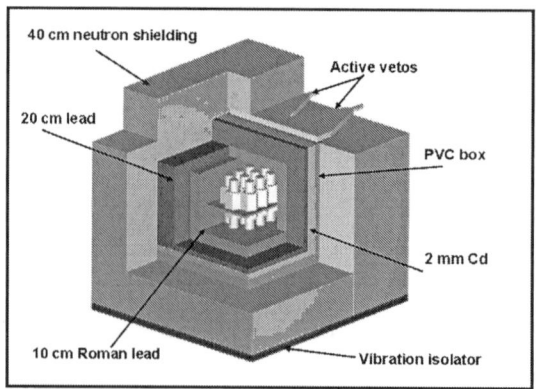

Figure 3. Schematic view of the ANAIS experiment.

3. Threshold and background reduction: Prototype II

As we stated before, the nature of the annual modulation signature makes necessary to reduce the energy threshold and low energy background level as much as possible. Efforts are being made in both directions arriving to a compromise where requirements for the corresponding improvements are in conflict. On this basis, a second prototype of the ANAIS experiment has been constructed (see Fig. 4). It consists of an hexagonal NaI crystal optically coupled by two methacrylate light guides to two ultra-low background PMTs. Two PMTs in coincidence are essential to reject electrical noise and dark pulses. A copper box, sealed to keep inside a dry atmosphere, allows us

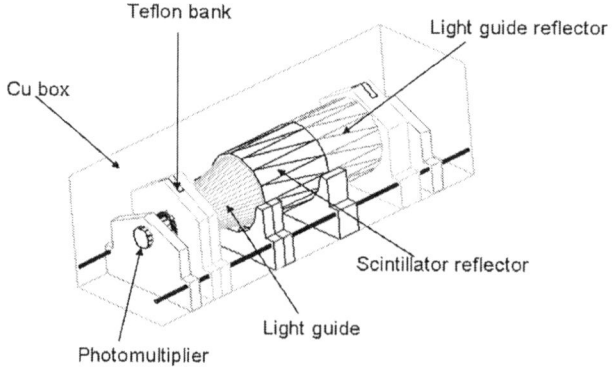

Figure 4. Scheme of the Prototype II of the ANAIS experiment

to operate the NaI crystal without any coating. The box contains a Teflon bank flexible enough to accommodate light guides of different lengths and different thicknesses of specular or diffuse reflectors. The aim is to optimize the experimental set-up in order to improve the light collection. Prototype II is ready for operation.

In parallel, Monte Carlo simulations of the light propagation for different optical configurations have been performed[a]. Some information can be drawn from these simulations: spatial dependence of the light collection, dependence on optical material parameters (reflectivity, attenuation length, ...), dependence on the light guide length (see, for instance, Fig. 5), etc. Further experimental tests of the simulation code are in preparation in order to get a quantitative estimate of the absolute light collection.

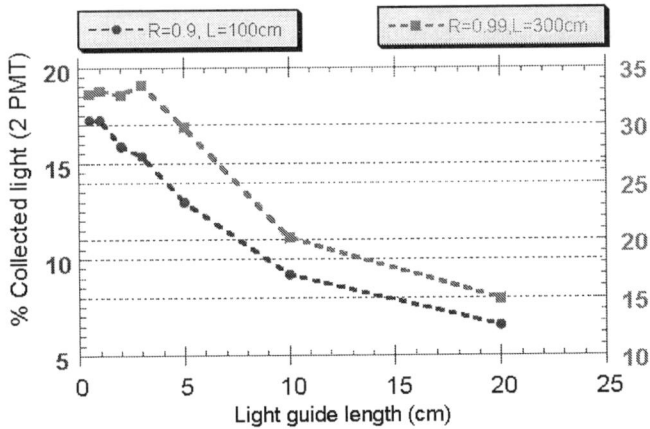

Figure 5. The percentage of collected light at the photomultipliers decreases with the light guide length. The circles and squares correspond to two different choices of the attenuation length in NaI (L) and the reflectivity of the crystal coating (R).

Results from NaI32 and ANAIS prototype I suggest that an important contribution to the low energy background comes from ^{210}Pb (see Fig. 1). In this direction, a pulse shape discrimination of the alpha against beta and gamma events in the high energy region, based on the different time constants of their respective pulses, has been carried out. An analysis is in progress in order to estimate the alpha contaminants in the bulk scintillator

---

[a]With a code adapted from OPTICS[5].

crystal (^{210}Po, for instance). Moreover, in order to improve our knowledge of the background contaminants, underground measurements with the raw crystal inside the copper box are planned.

4. Conclusions and Prospects

A second prototype of the ANAIS experiment (that will look for annual modulation in the signal of galactic WIMPs) is ready for operation, aiming at optimization of the design parameters to reduce low energy background and threshold. The PSD technique has been applied to discriminate alpha particles in the high energy region. An extension of the technique to neutron discrimination at low energies is in progress.

Acknowledgments

The Canfranc Astroparticle Underground Laboratory is operated by the University of Zaragoza. This research was founded by the Spanish Commission of Science and Technology (CICYT) under contract No. FPA2001-2437.

References

1. R. Bernabei et al., *Riv. N. Cim.* **26 n.1**, 1 (2003).
2. M. L. Sarsa et al., *Phys. Rev. D* **56**, 1856 (1997).
3. S. Cebrián et al., *Nucl. Phys. B (Proc. Suppl.)* **114**, 111 (2003).
4. J. Morales et al., Published in this volume.
5. E. Frlêz et al., *Computer Physics Comunications* **134**, 110 (2001).

MODEL INDEPENDENT EXPERIMENTAL LIMITS ON SPIN-DEPENDENT WIMPS

F. GIULIANI

Centro de Física Nuclear,
Universidade de Lisboa,
1649–003 Lisbon, Portugal
E-mail: criodets@cii.fc.ul.pt

Current spin-dependent WIMP searches are analyzed within the model-independent framework of Tovey *et al.*, which itself is extended to the case of positive signal experiments. The results indicate the need of combining experiments with different neutron–to–proton group spin ratios in order to obtain, for a given WIMP mass, reduced allowed areas in the spin-dependent WIMP–proton, WIMP–neutron cross section plane. In the sample survey presented here, and in the case of null signal experiments, these derive from NAIAD and ZEPLIN I, yielding, for a WIMP mass of 50 GeV, $\sigma_n \leq 0.18$, $\sigma_p \leq 0.7$ pb; confirmation of the DAMA/NaI results would imply $\sigma_n \leq 0.17$, $0.01 \leq \sigma_p \leq 0.29$ pb.

1. Introduction

Direct WIMP dark matter searches are generally only able to set an upper limit to the WIMP rate, which is customarily converted to an upper limit to WIMP–nucleon cross sections.

Experiments exploring the spin-dependent channel have to face the problem that, due to spin pairing effects, the total proton (neutron) spin is high only when Z (N) is odd. Since odd–odd nuclei are rare or radioactive, only even–A nuclei are used, dividing the experiments into the two classes of odd–Z and odd–N (nuclei) experiments. No experiment to date combines an odd–Z and an odd–N isotope in the active material. This circumstance is customarily used to simplify the procedures to convert the limits on WIMP–nucleus cross sections to limits on WIMP–nucleon cross sections, by simply stating the experiment is only sensitive to WIMP–proton (WIMP–neutron) interactions.

This odd group approximation has two drawbacks: 1) odd–Z and odd–N experiments do not compare, as they provide limits only on the WIMP–proton or WIMP–neutron cross section respectively. 2) Limits are WIMP-model dependent, since if WIMPs couple mainly with neutrons (protons), odd–Z (odd–N) experiments are not able to constrain theory.

The framework can be "forced" by substituting the assumption of zero even group spin with that of the WIMP couples only with protons (neutrons). This allows to compare odd–Z and odd–N experiments, but extremizes the WIMP–model dependence.

In the model-independent framework both limitations 1) and 2) are removed, and best limits on generic spin-dependent WIMPs are obtained by combining not only an odd–Z and an odd–N experiment, but also an experiment using an isotope, like ^{19}F, whose neutron to proton group spin ratio is between 0.1 and 10.

2. Getting rid of the model dependence

2.1. Basic formulae

As can be shown on general grounds[1], the effective nonrelativistic lagrangian for spin-dependent elastic scattering of a generic spin 1/2 WIMP on nucleons is:

$$L = 4\sqrt{2} G_F \overline{\chi} \vec{\sigma} \chi (a_p \overline{p} \vec{\sigma} p + a_n \overline{n} \vec{\sigma} n) \qquad (1)$$

where p, n and χ are the proton, neutron and WIMP fields respectively, $\vec{\sigma}$ are the Pauli matrices, G_F is the Fermi coupling constant, and $a_{p,n}$ are the effective proton (neutron) coupling strengths (coupling constants in units of $\sqrt{2} G_F$).

At tree level, the resulting (zero momentum transfer) WIMP–nuclide spin-dependent elastic scattering cross section σ_A for a nucleus of mass number A is [1,2,3]:

$$\sigma_A = \frac{32}{\pi} G_F^2 \mu_A^2 (a_p <S_p> + a_n <S_n>)^2 \frac{J+1}{J}. \qquad (2)$$

where $<S_{p,n}>$ are the expectation values of the proton (neutron) group's spin, G_F is the Fermi coupling constant, $a_{p,n}$ are the effective proton (neutron) coupling strengths, μ_A is the WIMP–nuclide reduced mass, and J is the total nuclear spin.

2.2. Traditional odd group approach

The traditional approach to spin-dependent WIMP–nucleon cross section limits consists in observing that experiments usually employ only isotopes with either $<S_n> \approx 0$ or $<S_p> \approx 0$, i.e., satisfying the odd group approximation. This is not a strange choice, since most odd–odd nuclei either are radioactive or have low isotopic abundances. ^{14}N is a notable exception, but practically unique. Since the proton (neutron) cross section is of course $\sigma_p = \frac{32}{\pi} G_F^2 \mu_A^2 \frac{3}{4} a_p^2$ ($\sigma_n = \frac{32}{\pi} G_F^2 \mu_A^2 \frac{3}{4} a_n^2$), it is clear that whenever $a_n <S_n> \approx 0$ ($a_p <S_p> \approx 0$) $\sigma_A \propto \sigma_p$ ($\sigma_A \propto \sigma_n$). Naturally, when this approximation is assumed, an experiment using only odd-Z nuclei cannot constrain σ_n, and reaches conclusions invalid for any candidate such that $a_p \ll a_n$, i.e. some candidates are ignored a priori and without need. In fact, though significantly different, both $<S_n>$ and $<S_p>$ are nonzero.

The same exclusion results of the odd group approximation can also be obtained by setting either a_n or a_p to 0. This way, as clearly shown in Fig. 1, odd-Z experiments can be compared with odd-N, at the price of extremizing the model dependence of the exclusion results. Moreover, in Fig. 1 a) large exposure odd-N experiments like DAMA/Xe-2 give poor results compared to the odd-Z, due to the model restriction $a_n = 0$. On the other hand, if $a_p = 0$ as in Fig. 1 b) odd-N results are clearly the most stringent limits.

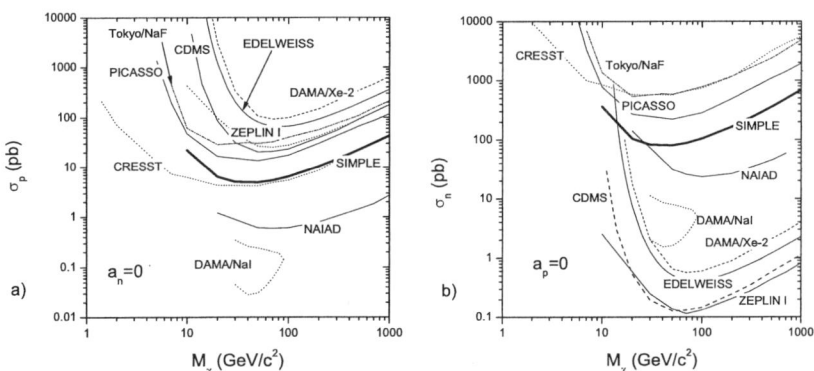

Figure 1. Exclusion plots for $M_\chi = 50$ GeV/c^2 and a) $a_n = 0$ and b) $a_p = 0$. Data are from Ref.s 6–14.

2.3. Model-independent approach

It is possible to display exclusion limits in a model-independent fashion by using the following equations:

$$\sum_A \left(\frac{a_p}{\sqrt{\sigma_p^{lim(A)}}} \pm \frac{a_n}{\sqrt{\sigma_n^{lim(A)}}} \right)^2 \leq \frac{\pi}{24 G_F^2 \mu_p^2} \tag{3}$$

$$\sigma_{p,n}^{lim(A)} = \frac{3}{4} \frac{J}{J+1} \frac{\mu_{p,n}^2}{\mu_A^2} \frac{\sigma_A^{lim}}{<S_{p,n}>^2} \tag{4}$$

where σ_A^{lim} is the upper limit on the cross section of the A-th isotope in the detector, calculated by attributing the whole rate to isotope A. As shown in Fig. 2, Eqn. (3) makes the exclusion plots 3D. Due to the quadratic nature of Eqn. (3), this is constituted by a tube whose axis is the M_χ coordinate

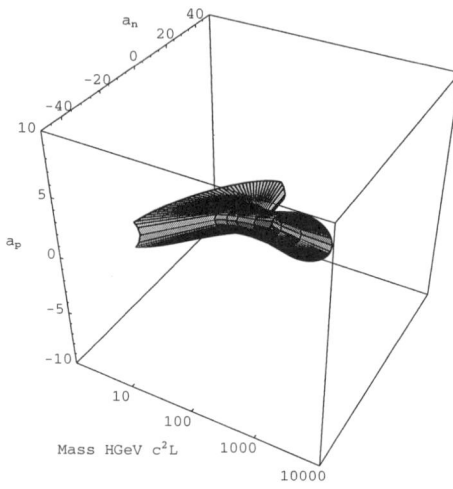

Figure 2. 3D exclusion plot of NAIAD.

axis, while the transversal section is a conic in the a_p–a_n plane[4].

The difficulty of showing in a 2D sheet the many volumes intersecting to the currently allowed volume, the normal choice is[4,5] to display the intersection of selected planes of M_χ = constant with the various experiments' 3D exclusion plots. A typical result is shown in Fig. 3 for $M_\chi = 50$ GeV/c^2.

The overall permitted region is the hatched area, which is the intersection of NAIAD with ZEPLIN I. Note that fluorine–based experiments, like SIMPLE or PICASSO, have the potentiality of further refining this area, due to the orientation of their exclusion plots which is determined by their detector compositions[4].

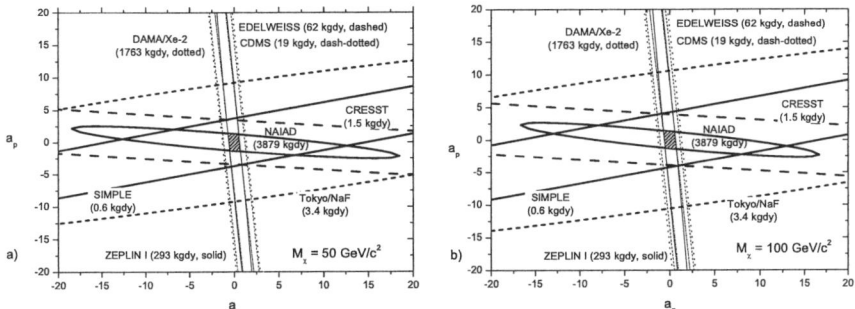

Figure 3. a_p–a_n regions allowed by various experiments for a) $M_\chi = 50$ GeV/c^2 and b) $M_\chi = 100$ GeV/c^2. The ZEPLIN I result is preliminary.

The natural extension of the model-independent framework to the case of a positive signal is based on the simple observation that now there is also a lower bound to the WIMP rate[4]:

$$\sum_A (\frac{a_p}{\sqrt{\sigma_p^{liminf(A)}}} \pm \frac{a_n}{\sqrt{\sigma_n^{liminf(A)}}})^2 \geq \frac{\pi}{24 G_F^2 \mu_p^2} \quad (5)$$

$$\sigma_{p,n}^{liminf(A)} = \frac{3}{4} \frac{J}{J+1} \frac{\mu_{p,n}^2}{\mu_A^2} \frac{\sigma_A^{liminf}}{<S_{p,n}>^2} \quad (6)$$

where σ_A^{liminf} is calculated from the lower bound to the WIMP rate in the same way as σ_A^{lim}.

Figure 4 shows what would happen when various experiments obtain a positive signal: a single experiment, like DAMA/NaI, results in an allowed elliptical shell, still leaving a wide variety of possible combinations of a_p and a_n (in particular, cannot exclude that one of $a_{p,n}$ is zero). Intersecting an odd–Z with an odd–N experiment can greatly improve the determination of the WIMP, because the corresponding allowed regions tend to be perpendicular. But, due to the high elongation of the shells, the intersection is 4 allowed areas, corresponding to two possible values for both $|a_p|$ and $|a_n|$. A third experiment with a significantly different orientation, like an F–based experiment, can establish the value of $|a_p|$ and $|a_n|$ also providing an extra piece of information: the relative sign of a_p and a_n. This is the maximum achievable determination of $a_{p,n}$, and further experiments can only refine error bars, i.e. reduce the two allowed areas in size.

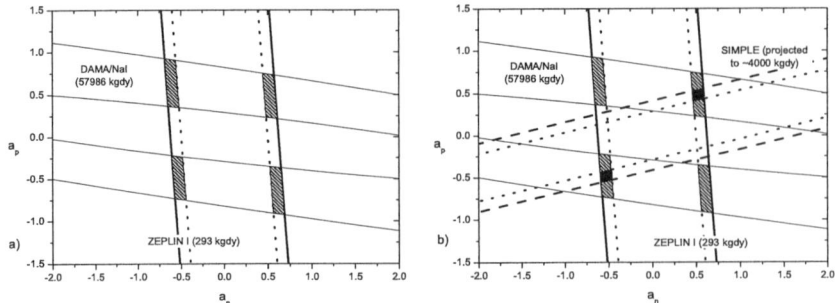

Figure 4. Typical situation (for $M_\chi = 50$ GeV/c^2) expected when various experiments get a positive signal: a) two almost experiments allow four spots, corresponding to two possible cross sections. b) Adding a third experiment with different orientation can eliminate the cross section ambiguity, also providing the relative sign of $a_{p,n}$. For ZEPLIN I the outer limit is the same of Fig. 3, while for SIMPLE it is a projection based on latest results. The inner ellipses of ZEPLIN and SIMPLE are arbitrarily assigned for illustration purposes. The DAMA/NaI region is from 3σ data[7].

Acknowledgements

This work has been supported by grant POCTI/FNU/39067/2002 of the Portuguese Foundation for Science and Technology (FCT), co-financed by FEDER. The author is supported by grant SFRH/BPD/13995/2003 of FCT.

References

1. A. Kurylov and M. Kamionkowski, *Phys. Rev.* **D69**, 063503 (2003).
2. J. D. Lewin and P. F. Smith, *Astrop. Phys.* **6**, 87 (1996).
3. J. Engel et al., *Int. J. Mod. Phys.* **E1**, 1 (1991).
4. F. Giuliani, *Phys. Rev. Lett.* **93**, 161301 (2004).
5. C. Savage et al., astro-ph/0408346.
6. B. Ahmed et al., *Astrop. Phys.* **19**, 691 (2003).
7. R. Bernabei et al., *Phys. Lett.* **B509**, 197 (2001).
8. A. Takeda et al., *Phys. Lett.* **B572**, 145 (2003).
9. W. SeideL et al., *Dark Matter in Astro- and Particle Physics*, eds. H.V. Klapdor-Kleingrothaus and R.D. Viollier, 517 (Springer-Verlag, Berlin, 2002).
10. R. Bernabei et al., *Phys. Lett.* **B436**, 379 (1998).
11. V. A. Kudryavtsev, *arXiv* **astro-ph/0406126**.
12. A. Benoit et al., *Phys. Lett.* **B545**, 43 (2002).
13. D.S. Akerib et al., *arXiv* **Astro-ph/0405033**.
14. N. Boukhira et al., *Astropart. Phys.* **14**, 227 (2000).

WIMP DIRECT DETECTION: HALO MODELLING AND SMALL SCALE STRUCTURE

ANNE M GREEN

Physics and Astronomy, Hicks Building,
University of Sheffield, Sheffield, UK
E-mail: a.m.green@shef.ac.uk

Weakly Interacting Massive Particle (WIMP) direct detection strategies and data analyses are often based on the simplifying assumption of a standard spherical, isotropic halo model, but observations and numerical simulations indicate that galaxy halos are in fact triaxial and anisotropic and contain substructure. The annual modulation of the event rate (due to the motion of the Earth) provides a potential WIMP 'smoking gun', however this signal depends sensitively on the local WIMP velocity distribution. I briefly review observations and numerical simulations of the stucture of dark matter halos and the construction of halo models. I then discuss the effect of two specific models (with parameters chosen to reproduce the properties of observed and simulated halos) on exclusions limits and the annual modulation signal. Exclusion limits undergo a relatively small (halo model and experiment dependent) change of shape. The phase and amplitude of the annual modulation signal can change significantly, however. Finally I discuss the possibility that the dark matter distribution is not completely smooth on the very small (sub-mpc) scales probed by WIMP direct detection.

1. WIMP direct detection signals

WIMP direct detection experiments have just begun to reach the sensitivity required to detect WIMPs via elastic scattering. The event rate depends on the (normalised) WIMP velocity distribution in the detector rest frame of the detector, $f(v)$, and the local WIMP density, ρ_χ as

$$\frac{dR}{dE} \propto \rho_\chi \int_{v_{\min}}^{\infty} \frac{f(v)}{v} \, dv, \qquad (1)$$

where $v_{\min} = [E(m_\chi + m_A)^2/2m_\chi m_A]^{1/2}$ is the minimum WIMP speed that can produce a nuclear recoil of energy E.

The velocity of the detector relative to the Galactic rest frame varies on an annual basis due to the Earth's orbit about the Sun. This provides us with two potential WIMP smoking guns. Firstly the event rate is direction dependent, being greatest in the forward direction [1,2]. Secondly the Earth's velocity, and hence the event rate, varies annually due to the Earth's orbit about the Sun [3]. If the local WIMP velocity distribution is isotropic then the annual modulation is roughly sinusoidal with a maximum in early June (when the Earth's speed with respect to the Galactic rest frame is largest) and amplitude of order a few per-cent.

The DAMA collaboration, using NaI with an exposure of 108 000 kg-days, have detected an annual modulation with the properties described above which they interpret as a WIMP signal [4]. The allowed region of WIMP mass–cross-section parameter space appears at first glance (i.e. assuming standard spin independent interactions and the standard halo model which has a maxwellian velocity distribution) incompatable with the exclusion limits from the CDMS, Edelweiss and Zeplin experiments [5] (see Fig. 1).

2. Halo properties

The standard halo model assumes that the Milky Way halo is spherical and isotropic, with density varying as $1/r^2$ so that the local velocity distribution is given by

$$f(v) = \frac{1}{\sqrt{\pi} v_c} \exp\left[-\left(\frac{v}{v_c}\right)^2\right], \qquad (2)$$

where $v_c \approx 220 \pm 20 \text{kms}^{-1}$ is the asymptotic circular velocity. Probes of the shapes of other galaxies find non-negligible deviations from sphericity [6] while the flattening of the Milky Way halo can be probed using the dynamics of the tidal streams from the Sagitarrius dwarf galaxy (e.g. Ref. [7]). Numerical simulations (see e.g. Refs. [8,9]) produce triaxial and anisotropic halos with the triaxiality and anisotropy varying significantly between halos and as a function of radius.

3. Halo modelling

The steady state phase space distribution function of a collection of collisionless particles is govereneed by the collisionless Boltzmann equation (see e.g. Ref. [10]):

$$\frac{\delta f}{\delta t} + \mathbf{v}.\nabla f - \nabla \Phi . \frac{\delta f}{\delta v} = 0, \qquad (3)$$

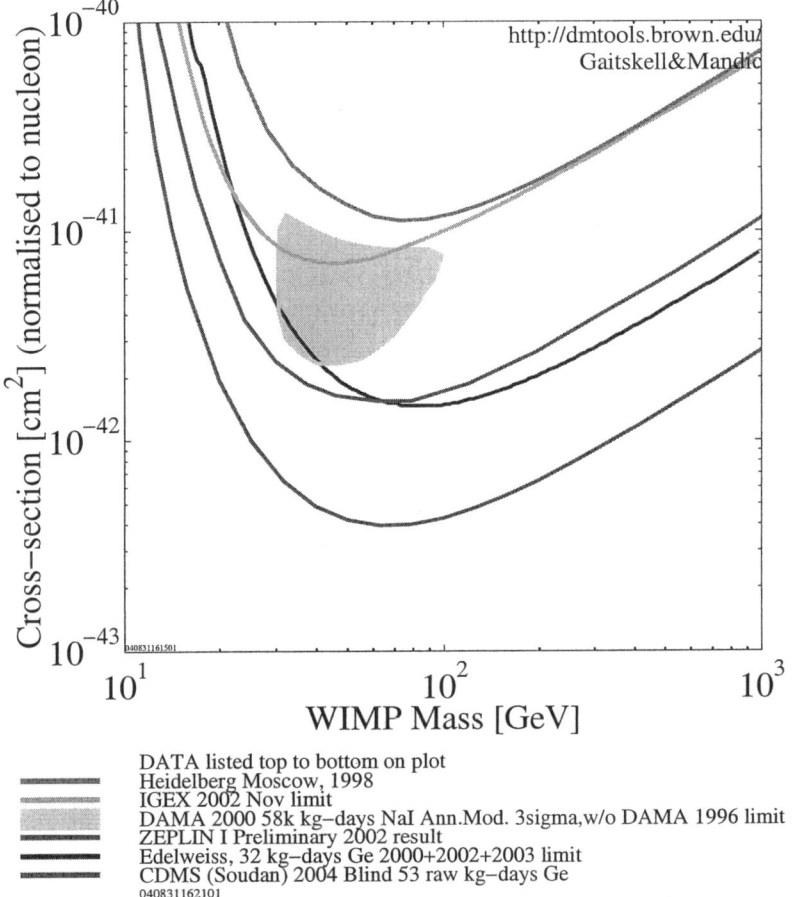

Figure 1. Comparison of the region of WIMP mass cross-section parameter space corresponding to the DAMA annual modulation signal with the exclusion limits from the CDMS, Edelweiss, Zeplin, IGEX and Heidelberg-Moscow experiments (assuming the standard halo model).

and for self-consistent systems (where the potential, Φ, is generated by the density distribution)

$$\nabla^2 \Phi = 4\pi G \rho = 4\pi G \int f(v) \mathrm{d}^3 v \,. \qquad (4)$$

For spherical and isotropic systens for any given $\rho(r)$ there is a unique $f(v)$. Triaxial and anisotropic systems are far more complicated however. There are more degress of freedom than constraints and assumptions have to be made about the form of the velocity distribution and/or the global symmetries of the halo.

We focus on two particular anisotropic and/or triaxial halo models. The Milky Way may well not be well described by either of these models, however they are reasonable 'non-standard' models with which to investigate the uncertainties in the expected WIMP signals due to our ignorance of the detailed structure of the Milky Way and the local phase space distribution function. Ultimately a model independent approach which assumes nothing about the velocity distribution and/or allows to extract information about the local velocity distribution from an observed signal is preferable.

3.1. *Logarithmic ellipsoidal model*

The logarithmic ellipsoidal model [11] is the simplest triaxial generalisation of the standard model. It assumes that the velocity ellipsoid is aligned with conical co-ordinates. In any of the planes of the halo conical co-ordinates coincide with galactic co-ordinates and the local velocity distribution can be approximated by a multi-variate Gaussian. The triaxiality and anisotropy are independent of radius.

3.2. *Osipkov-Merritt*

The Osipkov-Merritt models [12] are spherical and axially symmetric and assume that the distribution function depends on a particular combination of energy and angular momenutum. The resulting velocity anisotropy increases with radius.

4. Effect on signals

Exclusion limits depend on the time averaged speed distribution. We found that the exclusion limits from the IGEX and Heidelberg-Moscow experiments vary by of order tens of per-cent when the models discussed above (with parameters chosen to match the properties of observed and simulated halos) are considered, with the shift being experiment dependent [13]. The annual modulation signal is far more sensitive to the WIMP velocity distribution, and the amplitude and phase can change significantly (see Ref. [14] and references therein for details). It is also important to note that accurate calculation of the shape and phase of the annual modulation signal requires all three components of the Earth's velocity with respect to the Sun, and also the Sun's motion with respect to the Local Standard of Rest, to be taken into account.

Comparing results from different experiments is somewhat non-trivial as the minimum WIMP speed which can cause a recoil of a particular energy depends on the WIMP and target mass, and hence different experiments probe different regions of the WIMP speed distribuion.

5. Small scale structure

The halo models discussed in Sec. 3 assume that the phase space distribution function has reached a steady state. It is not a priori obvious that this assumption is valid however and it is possible that the dark matter distribution could exhibit sub-structure on the sub mpc scales probed by direct detection experiments.

Helmi, White and Springel [15] studied the distribution of 'particles' at the solar radius in a high resolution simulation. They found a smooth distribution, apart from a small number of high velocity particles from a late accreted sub-halo. However the resolution of simulations is of order 100 pc, many orders of magnitude larger than the scales of interest. Moore et al. [9] have argued that the dark matter distribution on small scales depends on the structure (and hence fate) of the first generation of dark matter halos to form, while Stiff and Widrow [16], using an inverse simulation technique, found that the velocity distribution at a single point in a simulated halo consisted of a number of peaks.

Ultimately any calculation of the formation of structure on very small scales requires the Cold Dark Matter power spectrum on small scales as input. This is found by combining the primordial power spectrum and the gravitational growth of perturbations with the microphysics of the CDM particles [17,18].

6. Summary

Direct detection data analyses and experiment comparisons usually assume the standard (spherical, isotropic) halo model, however observations and simulations indicate that the Milky Way halo is unlikely to be perfectly spherical and isotropic. Halo modelling is a non-trivial business; different models with the same macroscopic properties can have very different velocity distributions. For two halo models with parameters chosen to reproduce the range of properties of simulated and observed halos exclusion limits only vary by of order tens of per-cent (with the change in shape being experiment and halo model dependent).The amplitude and phase of the annual modulation signal can however change significantly.

Furthermore the local WIMP distribution is unlikely to be completely smooth. It is very possible that a stream(s) of high velocity particles from a late accreted sub-halo(s) is passing through the solar neighbourhood. It has has been argued that substructure could exist on the sub-mpc scales probed by direct detection experiments. This is currently an open question.

Acknowledgments

AMG was supported by PPARC and the Swedish Research Council and is grateful to Amina Helmi for useful comments and Stefan Hofmann, Ben Morgan and Dominik Schwarz for useful discussions and collaboration.

References

1. D. N. Spergel, *Phys. Rev. D* **37**, 1353 (1988).
2. B. Morgan, A. M. Green and N. J. C. Spooner, astro-ph/0408047; B. Morgan, these proceedings.
3. A. K. Drukier, K. Freese and D. N. Spergel, *Phys. Rev. D* **33** 3495 (1986).
4. R. Bernabei et al. *Riv. N.Cim* **23**, 1 (2003).
5. D. S. Akerib, astro-ph/0405033; A. Benoit et al. *Phys. Lett. B* **545**, 43 (2002); N. J. Smith et al. p302 proceedings of the Fourth International Workshop on the Identification of Dark Matter, ed. N. J. C. Spooner & V. Kudryavtsev (World Scientific), (2003).
6. P. D. Sackett, p393 Galaxy Dynamics, ASP Conference Series 182 eds. D. Merritt, J. A. Sellwood and M. Valluri (1999); M. Merrifield, these proceedings.
7. A. Helmi, these proceedings.
8. B. Moore et al. *Mon. Not. Roy. Astron. Soc.* **310**, 1147 (1999)
9. B. Moore et al. 2001, *Phys. Rev. D* **64**, 063508 (2001).
10. J. Binney and S. Tremaine, *Galactic Dynamics*, Princeton University Press (1987).
11. N. W. Evans, C. M. Carollo and P. T. de Zeeuw, Mon. Not. Roy. Astron. Soc. **318**, 1131, (2000).
12. L. P. Osipkov, Pis'ma Astron, Zh. **55**, 77 (1979); D. Merritt, Astron. J. **90**, 1027 (1985).
13. A. M. Green, *Phys. Rev. D* **66** 083003 (2002).
14. A. M. Green, *Phys. Rev. D* **68**, 023004 (2003).
15. A. Helmi, S. D. M. White and V. Springel *Phys. Rev. D* **66**, 063502 (2002).
16. D. Stiff and L. M. Widrow, *Phys. Rev. Lett* **90**, 211301 (2003).
17. S. Hofmann, D. J. Schwarz and H. Stöcker, *Phys. Rev. D* **64** 083507 (2001); A. M. Green, S. Hofmann and D. Schwarz, *Mon. Not. Roy. Astron. Soc.* **353**, L23 (2004); S. Hofmann these proceedings.
18. V. Berenzinsky, V. Dokuchaev and Y. Eroshenko, *Phys. Rev. D* **68**, 103003 (2003); V. Dokuchaev, these proceedings.

SOME ISSUES RELATED TO THE DIRECT DETECTION OF SUSY DARK MATTER.

J.D. VERGADOS[A,B]

[A] *Theoretical Physics Division, University of Ioannina, Gr 451 10, Ioannina, Greece.*
[B] *T-DO, Theoretical Physics Division, LANL, Los Alamos, N.M. 87545.*

Since the expected rates for neutralino-nucleus scattering are expected to be small, one should exploit all the characteristic signatures of this reaction. Such are: (i) In the standard recoil measurements the modulation of the event rate due to the Earth's motion. (ii) In directional recoil experiments the correlation of the event rate with the sun's motion. One now has both modulation, which is much larger and depends not only on time, but on the direction of observation as well, and a large forward-backward asymmetry. (iii) In non recoil experiments gamma rays following the decay of excited states populated during the Nucleus-LSP collision. Branching ratios of about 6 percent are possible.

1. Introduction

It is now established that dark matter constitutes about 30 % of the energy matter in the universe. The evidence comes from the cosmological observations [1], which when combined lead to:

$$\Omega_b = 0.05, \Omega_{CDM} = 0.30, \Omega_\Lambda = 0.65$$

and the rotational curves [2]. It is only the direct detection of dark matter, which will unravel the nature of the constituents of dark matter. In fact one such experiment, the DAMA, has claimed the observation of such signals, which with better statistics has subsequently been interpreted as modulation signals [3]. These data, however, if they are due to the coherent process, are not consistent with other recent experiments, see e.g. EDELWEISS and CDMS [4].

Supersymmetry naturally provides candidates for the dark matter constituents. In the most favored scenario of supersymmetry the LSP can be simply described as a Majorana fermion (LSP or neutralino), a linear combination of the neutral components of the gauginos and higgsinos [5−8]. We are not going to address issues related to SUSY in this paper, since thy have already been addressed by other contributors to these proceeding. Most models predict nucleon cross sections much smaller the the present experimental limit $\sigma_S \leq 10^{-5} pb$ for the coherent process. As we shall see below the constraint on the spin cross-sections is less stringent.

Since the neutralino is expected to be non relativistic with average kinetic energy $< T > \approx 40 KeV(m_\chi/100 GeV)$, it can be directly detected mainly via the recoiling of a nucleus (A,Z) in elastic scattering. In some rare instances the low lying excited states may also be populated [9]. In this case one may observe the emitted γ rays.

In every case to extract from the data information about SUSY from the relevant nucleon cross section, one must know the relevant nuclear matrix elements [10-11]. The static spin matrix elements used in the present work can be found in the literature [9].

Anyway since the obtained rates are very low, one would like to be able to exploit the modulation of the event rates due to the earth's revolution around the sun [12,13]. In order to accomplish this one adopts a folding procedure, i.e one has to assume some velocity distribution [12,13,14,15] for the LSP. One also would like to exploit other signatures expected to show up in directional experiments [16]. This is possible, since the sun is moving with relatively high velocity with respect to the center of the galaxy.

2. Rates

The differential non directional rate can be written as

$$dR_{undir} = \frac{\rho(0)}{m_\chi} \frac{m}{A m_N} d\sigma(u,v)|v| \qquad (1)$$

where $d\sigma(u,v)$ was given above, $\rho(0) = 0.3 GeV/cm^3$ is the LSP density in our vicinity, m is the detector mass and m_χ is the LSP mass

The directional differential rate, in the direction $\hat{e}$ of the recoiling nucleus, is:

$$dR_{dir} = \frac{\rho(0)}{m_\chi} \frac{m}{A m_N} |v| \hat{v}.\hat{e} \; \Theta(\hat{v}.\hat{e}) \frac{1}{2\pi} \; d\sigma(u, v \; \delta(\frac{\sqrt{u}}{\mu_r v \sqrt{2}} - \hat{v}.\hat{e})$$

where $\Theta(x)$ is the Heaviside function and:

$$d\sigma(u,v) == \frac{du}{2(\mu_r bv)^2}[(\bar{\Sigma}_S F(u)^2 + \bar{\Sigma}_{spin} F_{11}(u)] \qquad (2)$$

where u the energy transfer Q in dimensionless units given by

$$u = \frac{Q}{Q_0} \;,\; Q_0 = [m_p Ab]^{-2} = 40 A^{-4/3} \; MeV \qquad (3)$$

with b is the nuclear (harmonic oscillator) size parameter. $F(u)$ is the nuclear form factor and $F_{11}(u)$ is the spin response function associated with the isovector channel.

The scalar cross section is given by:

$$\bar{\Sigma}_S = (\frac{\mu_r}{\mu_r(p)})^2 \sigma^S_{p,\chi^0} A^2 \left[\frac{1 + \frac{f^1_S}{f^0_S} \frac{2Z-A}{A}}{1 + \frac{f^1_S}{f^0_S}} \right]^2 \approx \sigma^S_{N,\chi^0} (\frac{\mu_r}{\mu_r(p)})^2 A^2 \quad (4)$$

(since the heavy quarks dominate the isovector contribution is negligible). σ^S_{N,χ^0} is the LSP-nucleon scalar cross section. The spin Cross section is given by:

$$\bar{\Sigma}_{spin} = (\frac{\mu_r}{\mu_r(p)})^2 \sigma^{spin}_{p,\chi^0} \zeta_{spin}, \quad \zeta_{spin} = \frac{1}{3(1 + \frac{f^0_A}{f^1_A})^2} S(u) \quad (5)$$

$$S(u) \approx S(0) = [(\frac{f^0_A}{f^1_A} \Omega_0(0))^2 + 2\frac{f^0_A}{f^1_A} \Omega_0(0) \Omega_1(0) + \Omega_1(0))^2] \quad (6)$$

f^0_A, f^1_A are the isoscalar and the isovector axial current couplings at the nucleon level obtained from the corresponding ones given by the SUSY models at the quark level, $f^0_A(q), f^1_A(q)$, via renormalization coefficients g^0_A, g^1_A, i.e. $f^0_A = g^0_A f^0_A(q), f^1_A = g^1_A f^1_A(q)$. These couplings and the associated nuclear matrix elements are normalized so that, for the proton at $u = 0$, yield $\zeta_{spin} = 1$. If the nuclear contribution comes predominantly from protons ($\Omega_1 = \Omega_0 = \Omega_p$), $S(u) \approx \Omega_p^2$ and one can extract from the data the proton cross section. If the nuclear contribution comes predominantly from neutrons (($\Omega_0 = -\Omega_1 = \Omega_n$) one can extract the neutron cross section. In many cases, however, one can have contributions from both protons and neutrons. The situation is then complicated, but it turns out that $g^0_A = 0.1$, $g^1_A = 1.2$ Thus the isoscalar amplitude is suppressed, i.e $S(0) \approx \Omega_1^2$. Then the proton and the neutron spin cross sections are the same.

3. Results

To obtain the total rates one must fold with LSP velocity and integrate the above expressions over the energy transfer from Q_{min} determined by the detector energy cutoff to Q_{max} determined by the maximum LSP velocity (escape velocity, put in by hand in the Maxwellian distribution), i.e. $v_{esc} = 2.84 \ v_0$ with v_0 the velocity of the sun around the center of the galaxy($229 \ Km/s$).

3.1. Non directional rates

Ignoring the motion of the Earth the total non directional rate is given by

$$R = \bar{K}\left[c_{coh}(A,\mu_r(A))\sigma^S_{p,\chi^0} + c_{spin}(A,\mu_r(A))\sigma^{spin}_{p,\chi^0}\,\zeta_{spin}\right] \quad (7)$$

where $\bar{K} = \frac{\rho(0)}{m_{\chi^0}}\frac{m}{m_p}\sqrt{\langle v^2 \rangle}$ and The parameters $c_{coh}(A,\mu_r(A))$, $c_{spin}(A,\mu_r(A))$, which give the relative merit for the coherent and the spin contributions of a nuclear target compared to those of the proton depend on the velocity distribution, the thresh hold energy Q_{min}, the mass of the LSP and A. They have already been tabulated [17]. Using $\Omega_1^2 = 1.22$ and $\Omega_1^2 = 2.8$ for ^{127}I and ^{19}F respectively the extracted nucleon cross sections satisfy:

$$\frac{\sigma^{spin}_{p,\chi^0}}{\sigma^S_{p,\chi^0}} = \left[\frac{c_{coh}(A,\mu_r(A))}{c_{spin}(A,\mu_r(A))}\right]\frac{3}{\Omega_1^2} \Rightarrow \approx \times 10^4\ (A = 127),\ \approx \times 10^2\ (A = 19) \quad (8)$$

It is for this reason that the extracted limit on the spin cross section is less stringent.

If the effects of the motion of the Earth around the sun are included, the total non directional rate is given by

$$R = \bar{K}\left[c_{coh}(A,\mu_r(A))\sigma^S_{p,\chi^0}(1 + h(a,Q_{min})\cos\alpha)\right] \quad (9)$$

and an analogous one for the spin contribution. h is the modulation amplitude (see Fig. 1) and α is the phase of the Earth, which is zero around June 2nd.

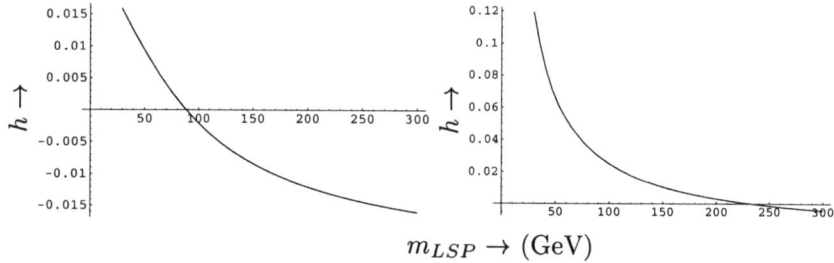

$$m_{LSP} \to (\text{GeV})$$

Figure 1. The modulation amplitude h as a function of the LSP mass in the case of ^{127}I for $Q_{min} = 0$ on the left and $Q_{min} = 10$ keV on the right.

3.2. Directional Rates

Since the sun is moving around the galaxy in a directional experiment, i.e. one in which the direction of the recoiling nucleus is observed, one expects a strong correlation of the event rate with the motion of the sun. In fact the directional rate can be written as:

$$R_{dir} = \frac{\kappa}{2\pi} \bar{K} \left[c_{coh}(A, \mu_r(A)) \sigma^S_{p,\chi^0} (1 + h_m \cos(\alpha - \alpha_m \pi)) \right] \quad (10)$$

and an analogous one for the spin contribution. The modulation now is h_m, with a shift $\alpha_m \pi$ in the phase of the Earth α, depending on the direction of observation. $\kappa/(2\pi)$ is the reduction factor of the unmodulated directional rate relative to the non-directional one. The parameters κ, h_m, α_m strongly depend on the direction of observation. We prefer to use the parameters κ and h_m, since, being ratios, are expected to be less dependent on the parameters of the theory [17].

The asymmetry in the direction of the sun's motion [17] is quite large, ≈ 0.97, while in the perpendicular plane the asymmetry equals the modulation.

For a heavier nucleus the situation is a bit complicated. Now the parameters κ and h_m depend on the LSP mass as well. (see Figs 2 and 3). The asymmetry and the shift in the phase of the Earth are similar to those of the $A = 19$ system.

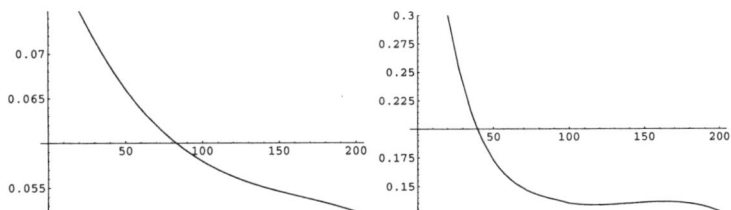

Figure 2. The parameter κ as a function of the LSP mass in the case of the $A = 127$ system, for $Q_{min} = 0$ expected in a plane perpendicular to the sun's velocity on the left and opposite to the sun's velocity on the right.

3.3. Transitions to excited states

Incorporating the relevant kinematics and integrating the differential event rate dR/du from u_{min} to u_{max} we obtain the total rate as follows:

$$R_{exc} = \int_{u_{exc}}^{u_{max}} \frac{dR_{exc}}{du} (1 - \frac{u_{exc}^2}{u^2}) du \ , \ R_{gs} = \int_{u_{min}}^{u_{max}} \frac{dR_{gs}}{du} du \quad (11)$$

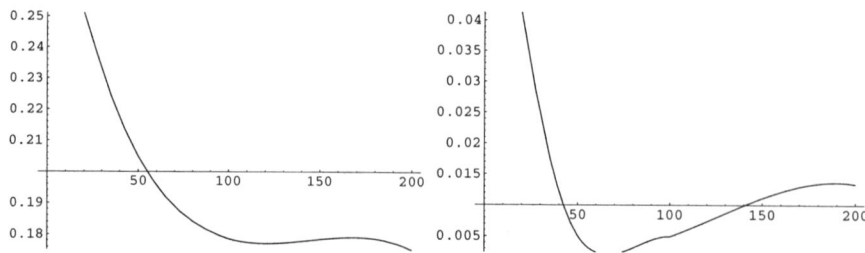

Figure 3. The modulation amplitude h_m in a plane perpendicular to the sun's velocity on the left and opposite to the suns velocity on the right. Otherwise the notation is the same as in Fig 2.

where $u_{exc} = \frac{\mu_r E_x}{A m_N Q_0}$ and E_x is the excitation energy of the final nucleus, $u_{max} = (y/a)^2 - (E_x/Q_0)$, $y = v/\upsilon_0$ and $u_{min} = Q_{min}/Q_0$, Q_{min} (imposed by the detector energy cutoff) and $u_{max} = (y_{esc}/a)^2$ is imposed by the escape velocity ($y_{esc} = 2.84$).

For our purposes it is adequate to estimate the ratio of the rate to the excited state divided by that to the ground state (branching ratio) as a function of the LSP mass. This can be cast in the form:

$$BRR = \frac{S_{exc}(0)}{S_{gs}(0)} \frac{\Psi_{exc}(u_{exc}, u_{umax})[1 + h_{exc}(u_{exc}, u_{max}) \cos \alpha]}{\Psi_{gs}(u_{min})[1 + h(u_{min}) \cos \alpha]} \quad (12)$$

in an obvious notation [17]. $S_{gs}(0)$ and $S_{exc}(0)$ are the static spin matrix elements As we have seen their ratio is essentially independent of supersymmetry, if the isoscalar contribution is neglected. For ^{127}I it was found [9] to be be about 2. The functions Ψ are given as follows :

$$\Psi_{gs}(u_{min}) = \int_{u_{min}}^{(y/a)^2} \frac{S_{gs}(u)}{S_{gs}(0)} F_{11}^{gs}(u) [\psi(a\sqrt{u}) - \psi(y_{esc})] \, du \quad (13)$$

$$\Psi_{exc}(u_{exc}, u_{max}) = \int_{u_{exc}}^{u_{max}} \frac{S_{exc}(u)}{S_{exc}(0)} F_{11}^{exc}(u)(1 - \frac{u_{exc}^2}{u^2})$$
$$[\psi(a\sqrt{u}(1 + u_{exc}/u)) - \psi(y_{esc})] \, du \quad (14)$$

The functions ψ arise from the convolution with LSP velocity distribution. The obtained results, assuming $F_{11}^{exc}(u) \approx F_{11}^{gs}(u)$, are shown in Fig. 4.

4. Conclusions

Since the expected event rates for direct neutralino detection are very low[5,8], in the present work we looked for characteristic experimental signatures for background reduction, such as:

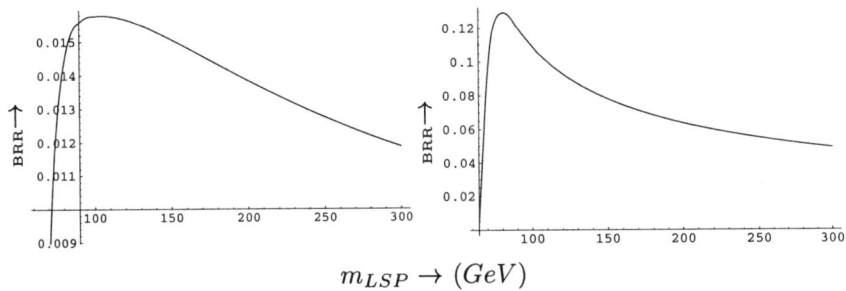

Figure 4. The branching ratio (rate to the excited state divided by that of the ground state) as a function of the LSP mass (in GeV) for ^{127}I. On the left we show the results for $Q_{min} = 0$ and on the right for $Q_{min} = 10$ KeV.

Standard recoil experiments. Here the relevant parameters are c_{coh}, c_{spin} and h. For light targets they are essentially independent of the LSP mass [17], essentially the same for both the coherent and the spin modes. The modulation is small, at most $h \approx 0.2\%$. For heavy targets, unfortunately, even the sign of h is uncertain for $Q_{min} = 0$. The situation improves as Q_{min} increases, but at the expense of the number of counts.

Directional experiments [16]. Here we find a correlation of the rates with the velocity of the sun as well as that of the Earth. One encounters reduction factors $\kappa/2\pi$, which depend on the angle of observation. This factor even in the most favorable case, with the nucleus recoiling opposite to the sun's velocity, is small, $\approx 1/4\pi$. One, however, gets as a bonus a three times larger modulation, $h_m \approx 0.06$. In a plane perpendicular to the sun's direction of motion the reduction factor is close to $1/12\pi$, but now the modulation can be quite high, $h_m \approx 0.3$, and exhibits very interesting time dependent pattern [17] Further interesting features may appear in the case of non standard velocity distributions [15].

Transitions to Excited states. We find that branching ratios for transitions to the first excited state of ^{127}I is relatively high, about 10%. The modulation in this case is much larger $h_{exc} \approx 0.6$. We hope that such a branching ratio and modulation signal will encourage the needed future experiments.

Acknowledgments: This work was supported by the EU contracts RTN No HPRN-CT-2000-00148 and MRTN-CT-2004-503369. Part of this work was performed in LANL. The author is indebted to Dr Dan Strottman for his support and hospitality.

References

1. S. Hanary et al, Astrophys. J. **545**, L5 (2000); J.H.P Wu et al, Phys. Rev. Lett. **87**, 251303 (2001); M.G. Santos et al, ibid **88**, 241302 (2002) P.D. Mauskopf et al, Astrophys. J. **536**, L59 (20002); S. Mosi et al, Prog. Nuc.Part. Phys. **48**, 243 (2002); S.B. Ruhl al, astro-ph/0212229; N.W. Halverson et al, Astrophys. J. **568**, 38 (2002); L.S. Sievers et al,astro-ph/0205287; G.F. Smoot et al, (COBE data), Astrophys. J. **396**, (1992) L1; A.H. Jaffe et al.,Phys. Rev. Lett. **86**, 3475 (2001); D.N. Spergel et al, astro-ph/0302209
2. G. Jungman, M. Kamiokonkowski and K. Griest, Phys. Rep. **267**, 195 (1996).
3. R. Bernabei et al., INFN/AE-98/34, (1998); it Phys. Lett. **B 389**, 757 (1996); Phys. Lett. **B 424**, 195 (1998); **B 450**, 448 (1999).
4. A. Benoit et al, [EDELWEISS collaboration], Phys. Lett. B **545**, 43 (2002); V. Sanglar,[EDELWEISS collaboration]; D.S. Akerib et al,[CDMS Collaboration], Phys. Rev D **68**, 082002 (2003);arXiv:astro-ph/0405033.
5. A. Bottino et al., Phys. Lett B **402**, 113 (1997). R. Arnowitt. and P. Nath, Phys. Rev. Lett. **74**, 4952 (1995); Phys. Rev. D **54**, 2394 (1996); hep-ph/9902237; V.A. Bednyakov, H.V. Klapdor-Kleingrothaus and S.G. Kovalenko, Phys. Lett. B **329**, 5 (1994).
6. M.E.Gómez and J.D. Vergados, Phys. Lett. B **512** , 252 (2001); hep-ph/0012020. M.E. Gómez, G. Lazarides and C. Pallis, Phys. Rev. D **61**, 123512 (2000) and Phys. Lett. B **487**, 313 (2000); M.E. Gómez and J.D. Vergados, hep-ph/0105115.
7. J.D. Vergados, J. of Phys. G **22**, 253 (1996); T.S. Kosmas and J.D. Vergados, Phys. Rev. D **55**, 1752 (1997).
8. A. Arnowitt and B. Dutta, Supersymmetry and Dark Matter, hep-ph/0204187;. E. Accomando, A. Arnowitt and B. Dutta, hep-ph/0211417.
9. J.D. Vergados, P.Quentin and D. Strottman, (to be published)
10. M.T. Ressell et al., Phys. Rev. D **48**, 5519 (1993); M.T. Ressell and D.J. Dean, Phys. Rev. C **56** (1997) 535.
11. P.C. Divari, T.S. Kosmas, J.D. Vergados and L.D. Skouras, Phys. Rev. C **61** (2000), 044612-1.
12. A.K. Drukier, K. Freese and D.N. Spergel, Phys. Rev. D **33**, 3495 (1986); K. Frese, J.A Friedman, and A. Gould, Phys. Rev. D **37**, 3388 (1988).
13. J.D. Vergados, Phys. Rev. D **58**, 103001-1 (1998); Phys. Rev. Lett **83**, 3597 (1999); Phys. Rev. D **62**, 023519 (2000); Phys. Rev. D **63**, 06351 (2001).
14. J.I. Collar et al, Phys. Lett. B **275**, 181 (1992); P. Ullio and M. Kamioknowski, JHEP **0103**, 049 (2001); P. Belli, R. Cerulli, N. Fornego and S. Scopel, Phys. Rev. D **66**, 043503 (2002); A. Green, Phys. Rev. D **66**, 083003 (2002).
15. B. Morgan, A.M. Green and N. Spooner, astro-ph/0408047.
16. D.P. Snowden-Ifft, C.C. Martoff and J.M.Burwell, Phys. Rev. D **61**, 1 (2000); M. Robinson et al, Nucl.Instr. Meth. A **511**, 347 (2003)
17. J.D. Vergados, Phys. Rev. D **67** (2003) 103003; ibid **58** (1998) 10301-1; J.D. Vergados, J. Phys. G: Nucl. Part. Phys. **30**, 1127 (2004);hep-ph/0410378.

CAN WIMP SPIN DEPENDENT COUPLINGS EXPLAIN DAMA?

C. SAVAGE AND K. FREESE

Michigan Center for Theoretical Physics,
Physics Department,
University of Michigan Ann Arbor, MI 48108 USA

P. GONDOLO

Physics Department,
University of Utah,
Salt Lake City, UT 84112 USA

We examine whether the annual modulation found by the DAMA dark matter experiment can be explained by WIMPs with spin-dependent couplings, in light of null results from several other direct and indirect detection experiments. We find that, for WIMP masses above 18 GeV and below 5 GeV, no region of this spin-dependent parameter space is compatible with DAMA and all other experiments. For masses in the range (5-18) GeV, we find acceptable regions of parameter space, including ones in which the WIMP-neutron coupling is comparable to the WIMP-proton coupling.

1. Introduction

Numerous collaborations worldwide have been searching for dark matter in the form of WIMPs. Of these, only DAMA/NaI (hereafter, "DAMA") has found a positive signal, in the form of an annual modulation. It is clear that the spin-independent coupling parameter space that could explain this modulation is severely bounded by the null results of such experiments as Edelweiss and CDMS. In the case of the standard halo model, these results are incompatible.

This work will focus on the alternative of WIMPs coupling to nuclei in a spin-dependent manner. This full analysis is published in Ref. 1 and extends upon work done by Ullio *et al.*[2]; Giuliani applies a similar analysis[3].

2. WIMP Detection

Direct detection experiments are all based upon the same principle– the scattering of a WIMP off a nucleus in a detector will deposit a small amount of energy in that detector as the nucleus recoils. Given a velocity distribution of WIMPs passing by the detector and a WIMP-nuclei cross-section, an energy dependent recoil rate can be can be determined. The differential recoil rate per unit detector mass for a WIMP mass m can be written as:

$$\frac{dR}{dE} = \frac{\rho}{2m\mu^2}\,\sigma(q)\,\eta(E,t) \tag{1}$$

where $\rho = 0.3$ GeV/cm^3 is the standard local halo WIMP density, $q = \sqrt{2ME}$ is the nucleus recoil momentum, $\sigma(q)$ is the WIMP-nucleus cross-section, and the mean inverse speed $\eta(E,t)$ encodes information about the WIMP velocity distribution.

The halo model most frequently assumed for WIMPs in the Milky Way is a simple isothermal sphere ("standard halo model"). In our analysis, we have assumed such a halo, although its validity is by no means guaranteed– halo modelling is actively being studied [4]. The motion of the Earth through this halo throughout the year leads to time variations in the WIMP velocity distribution, resulting in an annual modulation in the count rate of a detector (typically varying by only a couple percent of the average rate). Several experiments, including DAMA, search for WIMPs by exploit this feature.

Alternatively, indirect detection techniques can be used to search for dark matter, of which the most relevant to this paper is WIMP annihilation in the Sun. As WIMPs pass through the Sun and occasionally scatter, some of them lose enough energy to be gravitationally captured. These captured WIMPs eventually fall to the core and annihilate with each other; these annihilations lead to high-energy neutrinos that can be observed by Earth-based detectors such as Super-K and Baksan.

The generic form for spin-dependent WIMP-nucleus cross-sections involves two couplings– the WIMP-proton coupling a_p and the WIMP-neutron coupling a_n,

$$\sigma_{SD}(q) = \frac{32\mu^2 G_F^2}{2J+1}\left[a_p^2 S_{pp}(q) + a_p a_n S_{pn}(q) + a_n^2 S_{nn}(q)\right]. \tag{2}$$

Here the nuclear structure functions $S_{pp}(q)$, $S_{nn}(q)$, and $S_{pn}(q)$ are functions of the exchange momentum q and are specific to each nucleus. Odd-Z nuclei generally have $S_{pp} \gg S_{nn}$, whereas odd neutron nuclei generally have $S_{nn} \gg S_{pp}$. Note that even in an odd-Z, even neutron nuclei, some component of the spin is still effectively carried by the neutrons (and vice versa). The proton has $S_{pp} = \frac{3}{2\pi}$ and $S_{nn} = S_{pn} = 0$, so the WIMP-proton cross-section depends only on a_p; likewise, the WIMP-neutron cross-section depends only on a_n.

3. Experiments

DAMA/NaI, with odd-Z NaI detectors, is the only experiment to claim a positive signal, observing an annual modulation. In our analysis, we have used their observed annual modulation of 0.0200 ± 0.0032 events/kg/day/keVee over electron-equivalent recoil energies of 2-6 keV [5]. DAMA has released results for other recoil bins; our results are generally consistent with these other bins for WIMP masses below 100 GeV.

In addition to DAMA, we have included the null results of several experiments in our analysis. These include: Elegant V (NaI), an annual modulation search similar to DAMA [6]; CRESST I (Al) [7]; CRESST II (Ca,W,O) [8]; DAMA/Xe-2 (Xe) [9]; Edelweiss (Ge) [10]; CDMS I (Si) [11] and CDMS II (Ge) [12], where we combine both data sets to generate a single CDMS limit; and the indirect detection experiments Super-K [13] and Baksan [14]. SIMPLE (F) was not included as its novel detection technique requires a more complex analysis (beyond the scope of this work) [15]. This list is not exhaustive–numerous other experiments provide limits similar to the ones above.

4. Results and Discussion

We will first examine two simple cases– WIMPs couplings to only the proton or only the neutron. For the case of proton only coupling ($a_n = 0$), WIMP-proton cross-section limits are shown in Figure 1. At WIMP masses above 13 GeV, Super-K and Baksan limits rule out cross-sections that would be necessary to generate the annual modulation seen by DAMA. Likewise, at masses below 6 GeV, CRESST I limits rule out DAMA. WIMP masses between 6-13 GeV, however, are consistent with all results in this case.

The case of neutron only coupling ($a_p = 0$) is shown in Figure 2. DAMA/Xe-2 and Edelweiss provide strong limits at masses above 20 GeV and CDMS excludes the DAMA signal at all masses for this case.

We do not *a priori* expect a WIMP to couple solely to the proton or neutron, but instead might have non-zero (and non-equal) couplings to both. Figure 3 shows regions of a_p-a_n allowed by the DAMA modulation signal and by other null results at several different WIMP masses.

The DAMA observed modulation leads to an elliptical ring of finite

312

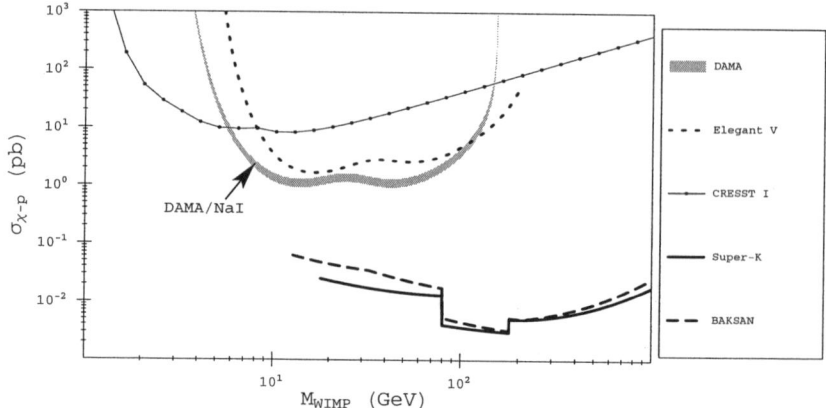

Figure 1. WIMP-proton cross-section limits for the case $a_n = 0$. The DAMA band represents cross-sections necessary to produce their observed signal; remaining lines are exclusion limits. CRESST I rules out the DAMA results below 6 GeV; Super-K rules it out above 18 GeV. WIMPs between 6-18 GeV are consistent with all results in this case.

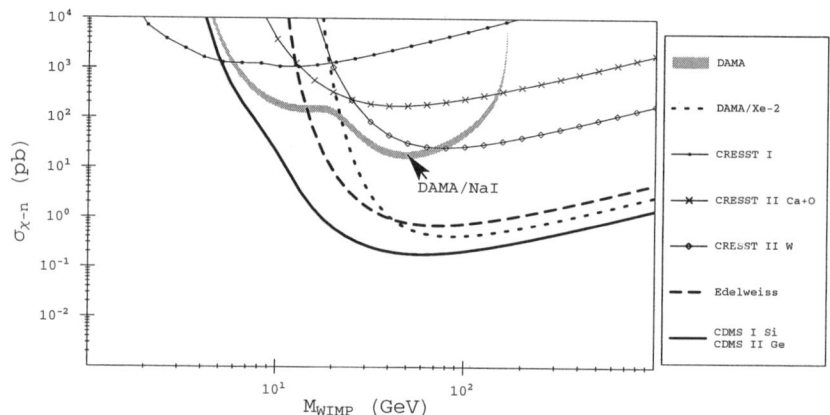

Figure 2. WIMP-neutron cross-section limits for the case $a_p = 0$. CDMS rules out the DAMA result in this case.

thickness (corresponding here to 1σ in their signal), shown in black in the figure. Only couplings *on* this ring would produce the observed modulation– couplings outside this ring would give too large a signal and couplings inside the ring would not give a sufficient modulation amplitude. At some of the masses, the elliptical ring extends beyond the range of the

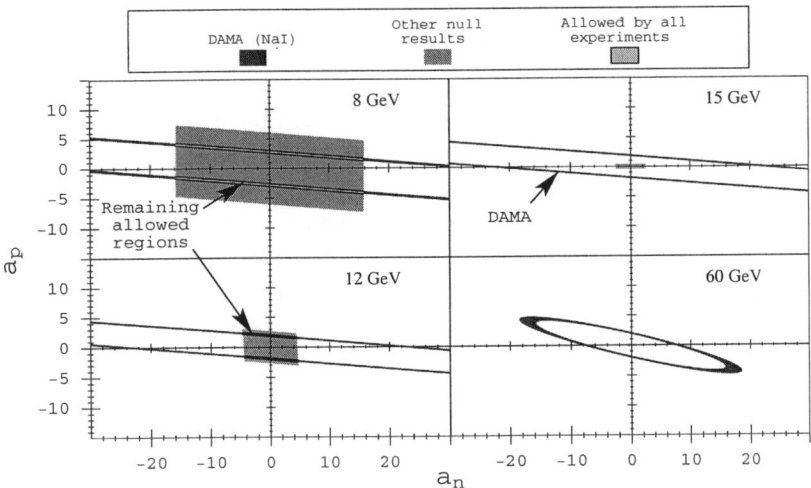

Figure 3. Spin-dependent allowed couplings at several WIMP masses. DAMA allowed couplings are shown in black. The null results of other experiments allow only the couplings shown in medium gray. The overlap of these two regions, where the couplings would be consistent with both DAMA's positive signal and the null results of the other experiments, is shown in light gray – such overlapping occurs only for WIMP masses between 5–13 GeV.

plot and only two bands can be seen.

Null results from Elegant V, CRESST I, CRESST II, DAMA/Xe-2, Edelweiss, CDMS, Super-K, and Baksan yield the allowed coupling region shown in medium gray in Figure 3. Couplings outside this region would have led to an observed signal in at least one of these experiments. At masses above 13 GeV, the allowed region is generated by the combination of limits from Baksan, Super-K, and CDMS. Baksan and Super-K limits depend upon WIMPs interacting with hydrogen in the Sun and thus inherently limit the proton coupling a_p only, whereas the neutron-odd Si-29 and Ge-73 of CDMS give significant limits for a_n. The combination of limits from CDMS and Super-K or Baksan leads to the relatively small allowed parameter regions shown at 15 GeV and 60 GeV. Below the 13 GeV analysis threshold of Baksan, the null result allowed regions are generated from a combination of Elegant V, CRESST I, and CDMS limits.

The region allowed by *all* experiments is that where the DAMA and null result regions overlap, shown in light gray in Figure 3. Only couplings within this region would generate the DAMA signal without generating an observable signal in the other experiments. Such regions can be seen in the 8 GeV and 12 GeV panels of the figure. However, above 13 GeV, with the strong a_p limits from Baksan and Super-K, no such overlap exists– no choice of couplings would be consistent with all the results. At masses below 5 GeV (not shown), CRESST I and CDMS also exclude the DAMA result for any choice of couplings. In the WIMP mass range 5-13 GeV, there still exist regions of parameter space consistent with the modulation seen by DAMA and the null results of others (we remind the reader that these results are based upon a standard halo model).

These compatible coupling regions allow for comparable proton and neutron couplings (e.g. $a_p = a_n = 2.0$ at 10 GeV). All of the allowed regions, however, have $|a_p| > 1$ ($a_n = 0$ is still allowed), which places a lower limit on the WIMP-proton cross-section of $\sigma_{\chi,p} \gtrsim 0.5$ pb for the relevant WIMP masses. Extension of Baksan or Super-K analysis to lower masses could confirm or rule out this remaining spin-dependent parameter space; odd-Z experiments such as CRESST I or SIMPLE could similarly examine this region. Finally, we note that the MSSM does not provide a WIMP candidate for the parameter spaces we have studied here.

Acknowledgments

We acknowledge the support of the DOE and the Michigan Center for Theoretical Physics via the University of Michigan.

References

1. C. Savage, P. Gondolo and K. Freese, *accepted for publication by* Phys. Rev. D, (2004); arXiv:astro-ph/0408346.
2. P. Ullio, M. Kamionkowski and P. Vogel, JHEP **0107**, 044 (2001).
3. F. Giuliani, arXiv:hep-ph/0404010.
4. A. Green et al., *in these proceedings.*
5. R. Bernabei et al. [DAMA Collaboration], Riv. Nuovo Cim. **26N1**, 1 (2003).
6. S. Yoshida et al., SPIN 2000 Symposium, AIP Conf. Proc. **570** 343 (2001).
7. G. Angloher et al. [CRESST Collaboration], Astropart. Phys. **18**, 43 (2002).
8. L. Stodolsky and F. Probst, talk at "The Dark Side of the Universe," Ann Arbor, May 2004.
9. R. Bernabei et al. [DAMA Collaboration], Phys. Lett. B **436**, 379 (1998).
10. V. Sanglard [EDELWEISS Collaboration], arXiv:astro-ph/0306233.
11. D. S. Akerib et al. [CDMS Collaboration], Phys. Rev. D **68**, 082002 (2003).
12. D. S. Akerib et al. [CDMS Collaboration], arXiv:astro-ph/0405033.
13. S. Desai et al. [Super-Kamiokande Collaboration], arXiv:hep-ex/0404025.
14. O. V. Suvorova, arXiv:hep-ph/9911415.
15. J. I. Collar, J. Puibasset, T. A. Girard, D. Limagne, H. S. Miley and G. Waysand, [SIMPLE Collaboration], Phys. Rev. Lett. **85**, 3083 (2000).

DIRECT DETECTION OF NEUTRALINO DARK MATTER IN THE NMSSM

ANA M. TEIXEIRA

Departamento de Física Teórica C-XI and Instituto de Física Teórica C-XVI,
Universidad Autónoma de Madrid, Cantoblanco, E-28049 Madrid, Spain
E-mail: teixeira@delta.ft.uam.es

We address the direct detection of neutralino dark matter in the framework of the Next-to-Minimal Supersymmetric Standard Model. We conduct a detailed analysis of the parameter space, taking into account all the available constraints from LEPII, and compute the neutralino-nucleon cross section. We find that sizable values for the detection cross section, within the reach of dark matter detectors, are attainable in this framework, and are associated with the exchange of very light Higgses, $m_{h_1^0} \lesssim 70$ GeV, the latter exhibiting a significant singlet composition.

1. Introduction

Supersymmetric (SUSY) theories offer some excellent candidates for dark matter. In particular, the lightest neutralino, $\tilde{\chi}_1^0$, is the leading one within the class of Weakly Interacting Massive Particles (WIMPs). WIMPs can be directly detected via elastic scattering on target nuclei and there are currently a large number of experiments devoted to the direct detection of WIMP dark matter[1].

We have studied the theoretical predictions for the direct detection of neutralino dark matter in the framework of the Next-to-Minimal Supersymmetric Standard Model (NMSSM)[2]. Via the introduction of a singlet superfield S, the NMSSM offers an elegant solution to the μ problem of the Minimal Supersymmetric Standard Model (MSSM), while at the same time it also renders the Higgs "little fine tuning problem" of the MSSM less severe. The new fields in the model mix with the corresponding MSSM ones, giving rise to a richer and more complex phenomenology. In particular, a very light neutralino may be present and a very light Higgs boson is not experimentally excluded[3]. The latter aspects, among other features, may modify the results concerning the neutralino-nucleon cross section with respect to those of the MSSM.

2. Neutralino-nucleon cross section in the NMSSM

The NMSSM is defined by the following superpotential

$$W = Y_u H_2 Q u + Y_d H_1 Q d + Y_e H_1 L e - \lambda S H_1 H_2 + \frac{1}{3}\kappa S^3, \quad (1)$$

with S a singlet under the Standard Model (SM) gauge group. After spontaneous electroweak symmetry breaking, the neutral Higgs scalars develop vacuum expectation values, $\langle H_1^0 \rangle = v_1$, $\langle H_2^0 \rangle = v_2$, $\langle S \rangle = s$, and an effective μ term is thus generated, $\mu \equiv \lambda s$.

In the absence of CP violation in the Higgs sector, the CP-even and CP-odd states do not mix. In the NMSSM, we find three scalar and two pseudoscalar Higgs states. Of particular relevance to our analysis is the lightest scalar, h_1^0, which can be written in terms of the original fields as

$$h_1^0 = S_{11} H_1^0 + S_{12} H_2^0 + S_{13} S, \quad (2)$$

where S_{ab} diagonalises the 3×3 scalar Higgs mass matrix[2]. In the neutralino sector, the singlino ($\tilde{S}$) mixes with the Bino, Wino and Higgsinos, and in this model, the lightest neutralino can be expressed as the combination

$$\tilde{\chi}_1^0 = N_{11}\tilde{B}^0 + N_{12}\tilde{W}_3^0 + N_{13}\tilde{H}_1^0 + N_{14}\tilde{H}_2^0 + N_{15}\tilde{S}, \quad (3)$$

with N the matrix that diagonalises the 5×5 neutralino mass matrix[2].

The leading contributions to the neutralino-nucleon cross section ($\sigma_{\tilde{\chi}_1^0-p}$) are associated with the scalar, spin- and velocity-independent term $\alpha_3 \tilde{\chi}_1^0 \tilde{\chi}_1^0 \bar{q} q$ in the effective Lagrangian[1], which receives contributions from squark and Higgs exchange diagrams[2,4]. The term $\alpha_{3i}^{\tilde{q}}$ is formally identical to the MSSM case, differing only in the new neutralino mixings stemming from the presence of a fifth component, and plays a sub-leading role in our analysis. Regarding the Higgs mediated interaction term (α_{3i}^h), the situation is slightly more involved since both vertices and the exchanged Higgs scalar significantly reflect the new features of the NMSSM[2]. It should be emphasised that the exchange of light Higgs scalars in the t-channel might provide a considerable enhancement to the neutralino-nucleon cross section.

3. Results and discussion

In our study[2], we were particularly interested in the various NMSSM scenarios which might potentially lead to values of $\sigma_{\tilde{\chi}_1^0-p}$ in the sensitivity range of detectors which are currently running or in preparation. The analysis of the NMSSM parameter space (minimization of the potential, computation of spectrum and compatibility with LEP experimental constraints) was done using the program NMHDECAY [5].

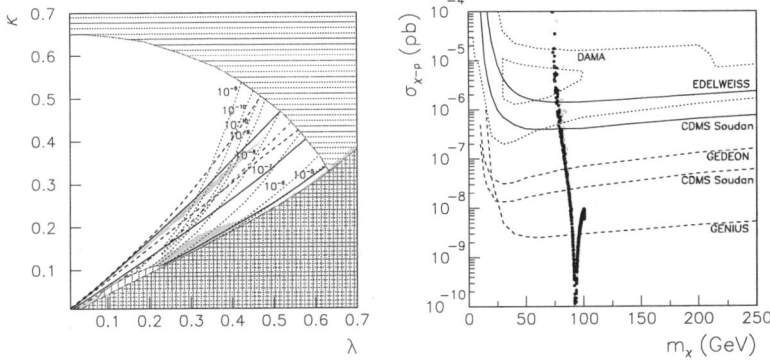

Figure 1. On the left, (λ, κ) parameter space with the corresponding constraints for the case $A_\lambda = 200$ GeV, $A_\kappa = -200$ GeV, $\mu = 110$ GeV and $\tan\beta = 3$. Shaded areas are excluded, while dotted lines in the accepted region represent contours of $\sigma_{\tilde\chi_1^0-p}$. From top to bottom, solid lines indicate different values of lightest scalar Higgs mass, $m_{h_1^0} = 114, 75, 25$ GeV, dashed lines separate the regions where the lightest scalar Higgs has a singlet composition given by $S_{13}^2 = 0.1, 0.9$ and a dot-dashed line reflects the singlino composition of the lightest neutralino (below/above $N_{15}^2 = 0.1$). On the right, and for the same choice of input parameters, scatter plot of $\sigma_{\tilde\chi_1^0-p}$ as a function of the neutralino mass, with the sensitivities of present and projected experiments. Black dots fulfil all constraints, while in grey are those experimentally excluded.

At the electroweak scale, we have the following set of free, independent parameters: λ, κ, $\mu(=\lambda s)$, $\tan\beta$, the soft trilinear terms for the Higgs scalars, A_λ, A_κ, the soft gaugino masses M_i, and a common SUSY scale for the remaining squark masses and trilinear couplings, $M_{\rm SUSY}$.

Not only the cross section itself, but also the allowed regions of the low-energy NMSSM parameter space are strongly sensitive to variations of the input parameters. It proved very illustrative to analyse the relevant features of the model in the plane generated by the Higgs couplings in the superpotential, λ and κ.

As an example, we plot in Fig.1 the (λ, κ) parameter space and the cross section versus the lightest neutralino mass for $\tan\beta = 3$, $A_\lambda = 200$ GeV, $A_\kappa = -200$ GeV and $\mu = 110$ GeV (taking $M_2 = 2M_1 = M_{\rm SUSY} = 1$ TeV). Significant regions of the parameter space are excluded due to theoretical and experimental constraints. The first class comprises the presence of tachyonic CP-even Higgs scalars (gridded area) and the occurrence of false minima and Landau poles (vertically and horizontally ruled regions, respectively). The grey area is associated to points that do not satisfy the LEP constraints. As can be clearly seen from both plots, very large values of $\sigma_{\tilde{\chi}_1^0-p}$ (in fact, even points already excluded by direct searches) can be obtained. From the inspection of the (λ, κ) plane, it is clear that such large values are associated with very light Higgs states (as light as 20 GeV), which are experimentally viable due to their important singlet character ($0.9 \lesssim S_{13}^2 \lesssim 0.95$). In this case, the NMSSM nature is clearly patent in the compositions of the lightest neutralino (a mixed singlino-Higgsino state) and of h_1^0 (light, and mostly singlet-like).

Another example, but for a distinct region in the NMSSM parameter space, is depicted in Fig.2, for $A_\lambda = 300$ GeV, $\mu = 110$ GeV, $A_\kappa = 50$ GeV, and $\tan \beta = 3$. In this case, tachyons arise in both CP-even and CP-odd

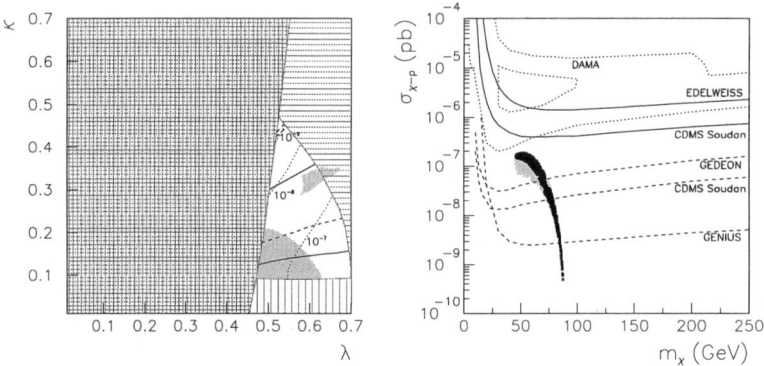

Figure 2. The same as in Fig.1 but for $A_\lambda = 300$ GeV, $\mu = 110$ GeV, $A_\kappa = 50$ GeV, and $\tan \beta = 3$. Only the lines with $m_{h_1^0} = 114, 75$ GeV are represented. Regarding the neutralino composition, only the line with $N_{15}^2 = 0.1$ is shown.

sectors, and close to these areas the experimental constraints from the Higgs sector are more severe. Nevertheless, regions with very light Higgses and $\tilde{\chi}_1^0$ are experimentally viable. In particular, neutralinos with an important singlino composition, $N_{15}^2 \lesssim 0.45$, can be obtained with $m_{\tilde{\chi}_1^0} \gtrsim 45$ GeV. Moreover, the lightest Higgses ($m_{h_1^0} \approx 65 - 90$ GeV) are all singlet-like, and this favours large values of the cross section ($\sigma_{\tilde{\chi}_1^0-p} \lesssim 2 \times 10^{-7}$ pb).

Until here we have only addressed cases where $\tilde{\chi}_1^0$ is essentially a singlino-Higgsino mixture, and this stems from the chosen hierarchy, $\mu < M_1 < M_2$. By relaxing the latter, more general compositions for the neutralino can be found. Let us go back to the example already analysed in Fig.1, but now taking two different values for μ, $\mu = 200, 500$ GeV. Re-

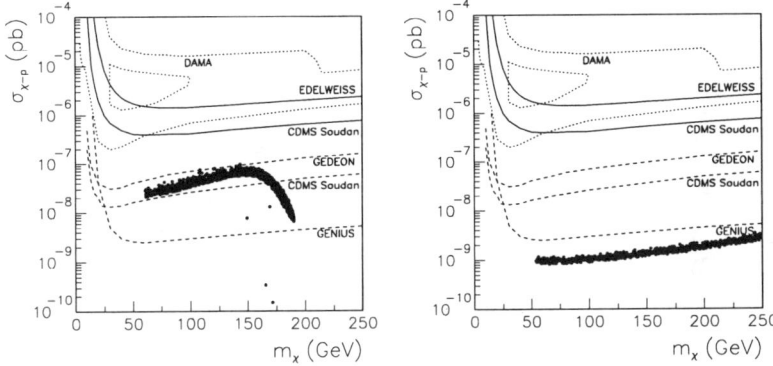

Figure 3. Scatter plot of the scalar neutralino-nucleon cross section as a function of $m_{\tilde{\chi}_1^0}$ for $A_\lambda = 200$ GeV, $A_\kappa = -200$ GeV, $\tan\beta = 3$, $\mu = 200, 500$ GeV. The gaugino masses satisfy the GUT relation, with M_1 in the range 50 GeV $\leq M_1 \leq$ 500 GeV.

garding the gaugino masses, we allow variations in the Bino mass as 50 GeV $\leq M_1 \leq$ 500 GeV, with the GUT relation $M_1 = \frac{1}{2} M_2$. As the value of μ increases, so does the gaugino composition of $\tilde{\chi}_1^0$. Simultaneously, the lightest Higgs becomes heavier and more doublet-like. For $\mu = 500$ GeV, neutralinos lighter than $m_{\tilde{\chi}_1^0} \lesssim 375$ GeV are all Bino-like. The Higgs mediated interaction is now negligible and detection only takes place through the squark mediated interaction. In contrast to the previous examples, Bino-like neutralinos would have $\sigma_{\tilde{\chi}_1^0-p} \lesssim 10^{-9}$ pb, and thus would be beyond the sensitivities of even the largest projected dark matter detectors.

Finally, it is important to conduct a more general survey of the NMSSM parameter space in order to obtain a global view on the theoretical predictions for $\sigma_{\tilde{\chi}_1^0-p}$, and their compatibility with present and projected detectors. We focus on the case $\mu = 110$ GeV, with heavy gaugino masses, $M_1 = \frac{1}{2} M_2 = 500$ GeV, since this choice leads to larger predictions for $\sigma_{\tilde{\chi}_1^0-p}$. The rest of the input parameters are allowed to vary in the ranges -600 GeV $\leq A_\lambda \leq 600$ GeV, -400 GeV $\leq A_\kappa \leq 400$ GeV, and we take $\tan\beta = 2, 3, 4, 5, 10$, with $\lambda, \kappa \in [0.01, 0.8]$. The results of this scan are shown in Fig.4. Points with large predictions for $\sigma_{\tilde{\chi}_1^0-p}$ are found, and correspond to very light singlet-like Higgses, even with $m_{h_1^0} \gtrsim 15$ GeV, which are more easily obtained for low values of $\tan\beta$ ($\tan\beta \lesssim 5$), and typically originate from regions where $\mu A_\lambda > 0$.

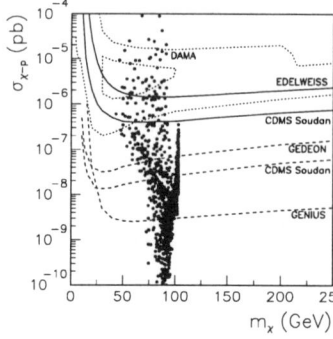
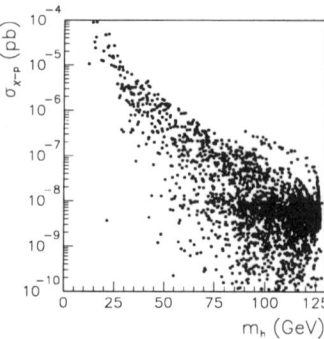

Figure 4. Scalar neutralino-nucleon cross section as a function of the lightest neutralino mass (left) and of the lightest Higgs mass (right). The input parameters are $\mu = 110$ GeV, -600 GeV $\leq A_\lambda \leq 600$ GeV, -400 GeV $\leq A_\kappa \leq 400$ GeV, and $\tan\beta = 2, 3, 4, 5, 10$.

4. Conclusions

We have performed a systematic analysis of the low-energy parameter space of the Next-to-Minimal Supersymmetric Standard Model (NMSSM), studying the implications for the direct detection of neutralino dark matter[2]. In the computation of $\sigma_{\tilde{\chi}_1^0-p}$ we have taken into account the relevant constraints on the parameter space from accelerator data. We have found that large values of $\sigma_{\tilde{\chi}_1^0-p}$, even within the reach of present dark matter detectors (see e.g. Fig.4), can be obtained, and this is essentially due to the exchange of very light Higgses, $m_{h_1^0} \lesssim 70$ GeV. The NMSSM nature is evidenced in this result, since such Higgses have a significant singlet composition, thus escaping detection and being in agreement with accelerator data.

Acknowledgments

A.M. Teixeira acknowledges the support by Fundação para a Ciência e Tecnologia under the grant SFRH/BPD/11509/2002.

References

1. For a recent review see, C. Muñoz, Int. J. Mod. Phys. A **19** (2004) 3093.
2. D. G. Cerdeño, C. Hugonie, D. E. López-Fogliani, C. Muñoz and A. M. Teixeira, arXiv:hep-ph/0408102.
3. U. Ellwanger et al, arXiv:hep-ph/0305109, and references therein.
4. R. Flores, K.A. Olive and D. Thomas, Phys. Lett. **B263** (1991) 425; V.A. Bednyakov and H.V. Klapdor-Kleingrothaus, Phys. Rev. **D59** (1999) 023514.
5. U. Ellwanger, J. F. Gunion and C. Hugonie, arXiv:hep-ph/0406215.

SPHERICAL STATISTICS FOR WIMP DIRECT DETECTION

B. MORGAN* AND A. M. GREEN

*Department of Physics and Astronomy,
University of Sheffield,
Hicks Building,
Hounsfield Road,
Sheffield, S3 7RH, UK*

The direction dependence of the event rate in WIMP direct detection experiments provides a powerful tool for distinguishing WIMP events from potential backgrounds. We consider a variety of of non-parametric statistical tests to examine the number of events required to distinguish a WIMP signal from an isotropic background, taking into account uncertainties in the reconstruction of the nuclear recoil direction and different models for the Milky Way halo.

1. Introduction

Distinguishing a putative WIMP-induced nuclear recoil signal from backgrounds due to, for instance, neutrons from natural radioactivity is a challenge for WIMP direct detection because of the low WIMP-nucleus elastic scattering rate ($\mathcal{O}(10^{-5} - 1)$ events kg^{-1}day^{-1}) expected. The Earth's motion about the Sun provides two potential smoking guns: i) an annual modulation [1] and ii) a strong direction dependence [2] of the event rate.

The direction dependence of the WIMP scattering rate has several advantages over the annual modulation. Firstly, the amplitude of the annual modulation is typically of the order of a few percent[1] while the event rate in the direction of the Earth's galactic motion is roughly an order of magnitude larger than in the antipodal direction.[2] Secondly, it is difficult for the directional signal to be mimicked by backgrounds; in most cases a background which is anisotropic in the laboratory frame will be isotropic in the Galactic rest frame due to the time dependent laboratory to Galactic coordinate transform. Furthermore the mean direction, in the lab, of

---

*Now at the University of Warwick, e-mail:Ben.Morgan@warwick.ac.uk

WIMP induced recoils will vary over a sidereal day due to the rotation of the Earth.[3]

Designing a detector capable of measuring sub-100keV nuclear recoil directions is however a considerable challenge. Low pressure gas time projection chambers (TPCs) such as DRIFT[4] and NEWAGE[5] seem to offer the best prospects. Determining the number of recoils needed to discriminate a WIMP signal from isotropic backgrounds therefore depends on the accuracy with which recoil directions can be reconstructed. Furthermore, the dependence of the directional event rate on the local WIMP velocity distribution[6] means that uncertainties in this distribution must be considered when designing statistical tests for a WIMP signal. In this article we present results on the minimum number of recoil events needed to identify a WIMP signal at high confidence taking into account the resolution of a TPC detector and different models for the local WIMP velocity distribution.

2. Models for the Milky Way Halo

The simplest possible model of the Milky Way (MW) halo is an isotropic sphere with density distribution $\rho \propto r^{-2}$, in which case the velocity distribution at all points within the halo is maxwellian:

$$f_0(\vec{v}) = \frac{1}{(2\pi/3)^{3/2}\sigma_0^3} \exp\left(-\frac{3|\vec{v}|^2}{2\sigma_0^2}\right), \quad (1)$$

where $\sigma_0 = \sqrt{3/2}v_c$ is the velocity dispersion, with v_c the asymptotic circular velocity which we took as 220kms^{-1}.

Observations and simulations indicate that galaxy halos are in fact triaxial and anisotropic. The logarithmic ellipsoidal model[7] (LGE) is the simplest triaxial generalisation of the isothermal sphere, its shape and velocity anisotropy being independent of radius. In any of the planes of the halo, the local WIMP velocity distribution can be approximated by a multi-variate Gaussian:

$$f_0(\vec{v}) = \frac{1}{(2\pi)^{3/2}\sigma_R\sigma_\phi\sigma_z} \exp\left(-\frac{v_R^2}{2\sigma_R^2} - \frac{v_\phi^2}{2\sigma_\phi^2} - \frac{v_z^2}{2\sigma_z^2}\right), \quad (2)$$

where $(\sigma_R, \sigma_\phi, \sigma_z)$ are the radial, azimuthal and polar velocity dispersions respectively. These are related to the halo axis ratios and the radial velocity isotropy $\beta(r)$ through parameters p and q, the parameter set of Ref. 8 being adopted.

Osipkov-Merritt (OM) models[9,10] provide self-consistent radially anisotropic velocity distributions for halos with spherically symmetric density profiles. We assumed that the MW has an NFW profile[11] with scale radius 20kpc, and used the fitting functions in Ref. 12 to give radial velocity anisotropies at the solar radius of $\beta(R_0 = 8.0\text{kpc}) = 0.14$ and 0.31.

We also considered the, somewhat controversial, possibility of a local tidal stream of WIMPs from the Sagittarius dwarf galaxy(Sgr).[13,14] This stream was modelled with a maxwellian velocity distribution with bulk velocity $\vec{v}_{str} = (-65, 135, -249)\text{kms}^{-1}$ and velocity dispersion $\sigma_{str} = 30\text{kms}^{-1}$:

$$f_0(\vec{v}) = \frac{1}{(2\pi/3)^{3/2}\sigma_{str}^3} \exp\left(-\frac{3|\vec{v} - \vec{v}_{str}|^2}{2\sigma_{str}^2}\right), \qquad (3)$$

with the local density of the stream taken to be $0.07\text{GeVcm}^{-3} \approx 0.25\rho_0$.[13]

3. Modelling the Detector Response

TPC based detectors are designed to measure the directions of nuclear recoils by drifting the ionisation produced by recoils in the gas volume to a suitable charge readout plane. To determine the accuracy with which the primary recoil direction can be reconstructed, we considered a TPC filled with 0.05bar CS_2 (to provide low diffusion electronegative drift[4]), a 200μm pitch micropixel readout plane[15] and a 10cm drift length over which a uniform electric field of 1kVcm^{-1} is applied. Primary 5-200keV S recoils were generated using the SRIM2003 package[16], C recoils being ignored due to their $< 1\%$ contribution to the recoil rate.

The drift and diffusion of the ionisation generated by the S recoils to the readout plane was simulated, together with the subsequent avalanche and signal generation on the pixels. Recoil directions were then reconstructed as the principal axis $\pm\vec{r}$ of the resultant, approximately elliptical, pixel hit patterns. Little charge asymmetry was found along the principal axes, indicating that the sense ($+\vec{r}$ or $-\vec{r}$) of the recoil may not be measurable. The RMS deviation between the principal axis and primary direction was $10 - 20°$, decreasing weakly with increasing energy. However, charge diffusion limited accurate axis reconstruction to recoils with primary energies above 20keV due to the diffusion and recoil track length becoming comparable to the pixel pitch. Overall, the primary limitations to accurate direction reconstruction are multiple scattering of the recoil and charge diffusion.

Directional recoil rates above the reconstruction threshold of 20keV were calculated for each halo model via Monte Carlo simulation in order to include this detector angular resolution. The WIMP velocity distributions were transformed to the detector rest frame and then used to generate random incident velocities of 100GeV WIMPs from the WIMP flux. These WIMPs were elastically scattering on S nuclei, with each S recoil weighted by a factor equal to the Helm form factor appropriate for spin-independent scattering. Finally, the primary recoil direction was smeared by the detector angular resolution determined above. Averaging these results over the year gave the overall recoil rate as a function of galactic coordinates.

Table 1. Number of recoil events required to reject isotropy of recoil directions at 95% confidence in 95% of experiments for each test statistic and halo model considered. The statistics are the Rayleigh-Watson($\mathcal{W}^\star$), Beran($\mathcal{A}$), 'F'($\mathcal{F} = \mathcal{A} + \mathcal{G}$), Dipole($\langle \cos\theta \rangle$), Bingham($\mathcal{B}^\star$), Gine($\mathcal{G}$) and modular Dipole($\langle |\cos\theta| \rangle$) tests. Further details on the halo models and statistics can be found in Ref. 17.

Halo Model	Halo Type	N_{iso} for $(R_c, A_c) = (0.95, 0.95)$						
		Vectorial Statistics				Axial Statistics		
		$\mathcal{W}^\star$	$\mathcal{A}$	$\mathcal{F}$	$\langle \cos\theta \rangle$	$\mathcal{B}^\star$	$\mathcal{G}$	$\langle |\cos\theta| \rangle$
1	SHM	18	18	19	11	235	235	131
1 (no)	SHM	13	13	14	9	120	120	65
2	LGE	17	17	18	10	161	163	93
3	LGE	20	20	21	12	225	224	152
4	LGE	18	18	19	11	215	214	120
5	LGE	20	20	21	12	222	22	153
6	LGE	16	16	16	10	96	97	59
7	LGE	19	20	20	12	125	125	93
8	LGE	18	18	19	11	252	252	142
9	LGE	21	21	22	13	378	383	241
10	OM	21	21	21	12	402	405	247
11	OM	17	17	17	10	180	182	102
12	SHM+Sgr	19	19	20	12	316	320	264

4. Statistical Analysis of Recoil Directions

In statistical terms, the question posed by a directional detector is 'How many events are required to reject the null hypothesis of isotropy at a given confidence level?'. Recoil directions constitute vectors, or, if the senses are unknown, axes, and so can be represented as points on the unit sphere. We therefore used standard inference methods for spherically distributed data, as these permit tests of isotropy without any reference to the (uncertain) form of the WIMP velocity distribution.

We chose a set of statistical tests appropriate for vectorial or axial recoil directions, and through Monte Carlo simulation of 10^5 experiments observing our sample halo models determined the number of events, N_{iso}, required to give a 95% confidence rejection of isotropy in 95% of the experiments (hereafter $CL(95,95)$) as described in Ref. 17. These numbers are summarised in Table 1.

It can be seen that if the senses of recoils can be measured, then $N_{iso} \sim 20$ is required to reject isotropy at $CL(95,95)$ over each halo model. However, if the sense cannot be measured, $N_{iso} \sim 100-400$, and there is a significant variation in this number between halo models. This effect is due to the axial statistics combining all the events in a given sample into one hemisphere, reducing the degree of anisotropy in the data. To determine the effect of the recoil reconstruction on these tests, we also simulated a maxwellian halo with a perfect detector (Model 1(no) in Table 1). It can be seen that N_{iso} for the vectorial tests is only reduced by a few events, whereas for the axial tests it is halved.

Translating these numbers to the equivalent exposure required to observe N_{iso} events, we find that isotropy can be rejected for WIMP-nucleon cross sections $\sigma_0 > 3 \times 10^{-9}$pb with 10^5kgday of exposure if the sense of the recoil is known. If recoil sense are not measurable, then isotropy can be rejected for $\sigma_0 > 3 \times 10^{-8}$pb with 10^5kgday.

4.1. Tests for a Triaxial Halo

A generic feature of triaxial halos is a flattening of the recoil distribution towards the galactic plane. A suitable probe for this feature is to test for rotational symmetry about a specified direction, which, for smooth halo models, is the Earth's mean direction of motion through the MW, $\vec{v}_\oplus$.

Using the Kuiper $\mathcal{V}$ statistic[17] we find that the most flattened, and arguably extreme, triaxial model in our sample (Model 7) requires 1710 events to reject rotational symmetry around $\vec{v}_\oplus$ at $CL(95,95)$. Converting this number to an exposure shows that rotational symmetry could be rejected at $CL(95,95)$ for $\sigma_0 > 3 \times 10^{-7}$pb with 10^5kgdays. An advantage of the Kuiper statistic is that it is valid for vectorial or axial recoil directions.

4.2. Tests for the Sagittarius Tidal Stream

The presence of the Sgr tidal stream leads to the peak in the directional recoil rate deviating from $\vec{v}_\oplus$. It is thus possible to test for the presence of this stream in two ways; i) Compatibility of median recoil direction with

$\vec{v}_\oplus$, ii) Rotational symmetry around $\vec{v}_\oplus$.

In the first case, we used the 'X^2' statistic[17] and determined that 294 events are required to reject $\vec{v}_\oplus$ as the median recoil direction at $CL(95,95)$. Thus the Sgr stream could be identified for $\sigma_0 > 6 \times 10^{-8}$pb with 10^5kgdays, but 'X^2' is only valid for vectorial recoil directions.

For rotational symmetry we again used the Kuiper statistic and determined that 574 events are required to reject rotational symmetry around $\vec{v}_\oplus$ at $CL(95,95)$. Thus with 10^5kgday, the Sgr stream could be identified for $\sigma_0 > 10^{-7}$pb, this test being valid for vectorial and axial recoil directions.

5. Summary

We have investigated the effects of finite detector resolution and variations in the MW halo model on the number of events required by a directional detector to discriminate a WIMP signal from isotropic backgrounds. Statistical tests requiring no assumptions about the WIMP velocity distribution were applied, and these indicate that $\sim 20(100-400)$ events are required if the recoil sense is known(unknown). We have also considered tests for identifying the presence of a triaxial halo and the Sgr tidal stream. These tests require $\mathcal{O}(500)$ events, but more powerful tests may be possible.

References

1. A. K. Drukier, K. Freese and D. N. Spergel, *Phys. Rev.* **D33**, 3495 (1986).
2. D. N. Spergel, *Phys. Rev.* **D37**, 1353 (1988).
3. B. Morgan, *Nucl. Inst. Meth.* **A513**, 226 (2003).
4. G. J. Alner et al, *Nucl. Inst. Meth.* **A535**, 644 (2004).
5. T. Tanimori et al, *Phys. Lett.* **B578**, 241 (2004).
6. C. J. Copi and L. M. Krauss, *Phys. Rev.* **D63**, 043507 (2001).
7. N. W. Evans, C. M. Carollo and P. T. de Zeeuw, *Mon. Not. Roy. Astron. Soc.* **318**, 1131 (2000).
8. A. M. Green, *Phys. Rev.* **D68** 023004 (2003).
9. L. P. Osipkov, *Pis'ma Astron. Zh.* **55**, 77 (1979).
10. D. Merritt, *Astron. J.* **90**, 1027 (1985)
11. J. F. Navarro, C. S. Frenk and S. D. M. White, *Astrophys. J.* **462**, 563 (1996).
12. L. M. Widrow, *Astrophys. J. Suppl. S.* **131**, 39 (2000).
13. K. Freese, P. Gondolo, H. J. Newberg and M. Lewis, *Phys. Rev. Lett.* **92**, 111301 (2004).
14. A. Helmi, these proceedings.
15. R. Bellazzini and G. Spandre, *Nucl. Inst. Meth.* **A513**, 231 (2003).
16. J. F. Ziegler, J. P. Biersack and U. Littmark, The stopping and range of ions in solids, Pergamon Press (1985).
17. B. Morgan, A. M. Green and N. J. C. Spooner, *astro-ph/0408047*.

DARK MATTER SIGNAL AND SIGNAL MODULATION FOR A COLD FLOW [*]

F.-S. LING

Service de Physique Théorique,
Université Libre de Bruxelles (U.L.B.),
Brussels, Belgium

We calculate the flow's velocity and density on Earth for a cold flow of dark matter. The main sources of the modulation, including the Sun's and the Earth's gravity, the orbital and rotational motions of the Earth are analyzed. The expected signal modulation in axion and (non-directional) WIMP detectors is derived, and the DAMA result is discussed.

1. Introduction

Cold Dark Matter (CDM) appears as a central pillar in Standard Cosmology[1], and is believed to form haloes around galaxies. However, the shape and the structure of these haloes is still highly controversial[2], which makes any prediction of a CDM potential signature very flimsy.

It has been argued that, for the purpose of galaxy formation, CDM candidates like axions or WIMPs have to be considered as collisionless particles with a negligible (primordial) velocity dispersion[3]. As a result, the dynamical evolution of the halo should be described by the folding of a 3-dimensional sheet in a 6-dimensional phase-space. Interaction of the halo with the baryonic part of the galaxy does not alter this picture[4].

Here, we will therefore assume that CDM flows are cold [a] and collisionless, and examine the consequences of this fact on a CDM signal in direct detection experiments. In particular, we will discuss the relevance of the diurnal and the annual modulation[5].

---

[*]This work is supported by the I.I.S.N. and the Belgian Science Policy (return grant and IAP 5/27).
[a]*i.e.* they have negligible velocity dispersion

2. Solar wakes of cold dark matter flows

The effect of the Sun's gravity on a cold flow which is uniform far upstream with density d_0 and velocity $\vec{v}_0$, is such that there are three flows ($i = 1, 2, 3$) downstream, whereas there is only one flow ($i = 1$) upstream. In addition, there are two caustics associated with the solar gravitational distortion of the flows. One is a line downstream of the Sun, called the 'spike', where the flow is focused by the Sun's gravity. The other one is a conic surface called the 'skirt', with an opening angle close to the maximum scattering angle Θ_{max}, it separates the region upstream with one flow from the region downstream with three flows. In the limit of a point mass Sun, $\Theta_{max} \to \pi$, flows 1 and 2 exist everywhere, while flow 3 disappears (because $d_3 \to 0$ everywhere in that limit).

To describe the flows, it is convenient to adopt cylindrical coordinates (z, ρ, ϕ) such that $\vec{v}_0 = v_0 \hat{z}$, with $0 \leq \phi < \pi$ and $-\infty < \rho < +\infty$ [b]. The flows at a given location (z, ρ, ϕ) have different impact parameters $b_i(z, \rho)$, and the labeling is such that $|b_1| \geq |b_2| \geq |b_3|$. We will also make use of the spherical coordinates (r, θ, ϕ).

The flows velocity formulas follow from energy and angular momentum conservation

$$\vec{v}_{\odot i}(\vec{r}) = \pm \hat{r}\, v_0 \sqrt{1 + \frac{2a}{r} - (\frac{b_i}{r})^2} - \hat{\theta}\, v_0 \frac{b_i}{r} \quad , \tag{1}$$

with $a = GM_\odot/v_0^2$ and the $+(-)$ sign pertains to where the flow is outgoing (incoming), i.e. where $\dot{r} > 0$ ($\dot{r} < 0$). The impact parameters of flows 1 and 2 are given by

$$b_{\frac{1}{2}}(z, \rho) = \frac{\rho}{2}\left(1 \pm \sqrt{1 + Y}\right) \quad , \tag{2}$$

with $Y = 4a(r+z)/\rho^2$ and assuming that the particles in the flows did not go through the Sun at any point in the past, which is satisfied at most locations in the solar neighborhood, but not everywhere. For flows which have previously gone through the Sun, we use the fact that ($i = 2, 3$)

$$|b_i(\theta)| = b_{si}(\Theta = \theta) + O(\frac{R_\odot}{r}) \quad , \tag{3}$$

where $R_\odot$ is the solar radius, and $b_s(\Theta)$ is the impact parameter of the trajectory with scattering angle Θ, which has been derived for a particular model of the mass distribution inside the Sun[6].

---

[b]instead of the usual $0 \leq \phi < 2\pi$ and $0 \leq \rho < +\infty$

The densities of the flows that did not go through the Sun are given by

$$d_{\frac{1}{2}}(\vec{r}) = \frac{d_0}{4}\left(\sqrt{1+Y} + \frac{1}{\sqrt{1+Y}} \pm 2\right) \ . \tag{4}$$

The spike caustic is evident because Y diverges as $8ar/\rho^2$ when $\rho \to 0$. For flows that did go through the Sun, we have $(i = 2, 3)$

$$d_i = \frac{d_0}{r^2|\sin\theta|} \left| \frac{b_{si}(\theta)}{\frac{d\Theta}{db_{si}}(\theta)} \right| (1 + O(\frac{R_\odot}{r})) \ . \tag{5}$$

3. Axion signal modulation

Axion detection is based on the photoconversion of axions in a cavity permeated by a strong magnetic field[7]. The photon energy is given by

$$\hbar\omega = \frac{m_a c^2}{\sqrt{1-(\frac{v}{c})^2}} = m_a c^2 \left(1 + \frac{1}{2}(\frac{v}{c})^2 + \frac{3}{8}(\frac{v}{c})^4 + ...\right) \tag{6}$$

where $m_a \simeq 10^{-5}$ eV is the axion mass and v is the axion speed in the laboratory frame. The conversion is resonantly enhanced if the photon frequency $f = \omega/2\pi$ falls within the cavity bandwidth. Galactic halo axions have velocities of order $10^{-3}c$. Hence the spectrum of microwave photons from $a \to \gamma$ conversion in the cavity detector has width of order $10^{-6}\nu_a$ above $\nu_a \equiv m_a c^2/2\pi\hbar \simeq 2.4$ GHz.

It can be shown that detectors with a high quality factor such as the ADMX detector[7] can achieve a frequency resolution $\Delta f/f$ better than 10^{-11}. With such a resolution, a cold flow of dark matter axions would produce a peak in the frequency spectrum of microwave photons, whose size is proportional to the density of the flow, and whose frequency would modulate as the Earth rotates and orbits around the Sun. In particular, the diurnal and annual modulation can be followed and used to reconstruct the flow direction.

Here, we will only sketch the results for the modulation of the frequency and of the density. The reader is referred to Ref. [5] for detailed calculations. The velocities in the detector rest frame are decomposed as

$$\vec{v}_i = \vec{v}_{\odot i} - \vec{v}_{\rm orb}(t) - \vec{v}_{\rm rot}(t) + \Delta\vec{v}_{i\oplus} \tag{7}$$

where $\vec{v}_{\rm orb}(t)$ is the orbital velocity of the Earth, $\vec{v}_{\rm rot}(t)$ is the velocity component of the laboratory caused by Earth's rotation, and $\Delta\vec{v}_{i\oplus}$ is the effect of the Earth's gravity. We neglect the relativistic effects as well as the gravitational effects due to Jupiter and other planets. The frequency shift $\Delta f_i/\nu_a = v_i^2/2c^2$ will consist of many terms, some of which can be derived from energy conservation. Among these, the diurnal modulation term (typically of order 10^{-9}) is given by

$$v_{rot} \left(\sin \phi_s \; v_{\oplus ix} - \cos \phi_s \; v_{\oplus iy}\right) \frac{1}{c^2} \; , \tag{8}$$

where $v_{rot} = 0.465$ km/s $\cos l$ is the velocity of the laboratory in a reference frame that is co-moving but not co-rotating with the Earth, and depends on the local latitude l. $\vec{v}_{\oplus i} = \vec{v}_{\odot i} - \vec{v}_{\text{orb}}$, and $\phi_s \equiv 2\pi t/t_s$, t is the sidereal time[c] and t_s is the sidereal day. Eq. (8) shows that the measurement of the amplitude and the phase of the daily frequency modulation determines the values of $v_{\oplus ix}$ and $v_{\oplus iy}$ on any day of the year. It can be shown[5] that all three components can be determined by observing the annual modulation[d], which is typically of order 10^{-7} for flow 1 and 10^{-9} for flow 2 and 3.

The density modulation is essentially given by Eqs. (4-5). Away from the spike, $Y \ll 1$, so that $d_1 \simeq d_0$ and flows 2 and 3 have negligible densities $d_2 \simeq d_0 Y^2/16 \sim 10^{-5} d_0$, $d_3 \sim d_2$. Close to the spike ($Y \gg 1$), the densities of flows 1 and 2 are large and nearly equal, producing a density peak,

$$d_1 \simeq d_2 \simeq \frac{d_0}{4} \sqrt{Y} \simeq d_0 \sqrt{\frac{a}{2a_\oplus} \frac{1}{\varepsilon}} \tag{9}$$

where $a_\oplus$ is the semi-major axis of the Earth's orbit and $\varepsilon \ll 1$ is the angular distance to the spike. We can estimate the duration of a density peak in the most favorable case, namely when $\vec{v}_0$ is parallel to the ecliptic plane or $\varepsilon = 0$. A density enhancement by a factor F will last for $\Delta t \simeq (2/\omega F)(\tilde{v}/v_0)$ where $\omega = 2\pi/1$ year and $\tilde{v} \simeq 42$ km/s. For example, with $F = 10$ and $v_0 = 300$ km/s, $\Delta t \simeq 9$ days. The densities of flows 2 and 3 also become large near the skirt. However, the density peak is much thinner. For the same flow velocity, we get $d_2 \simeq d_3 \simeq d_0 1.53\ 10^{-6}/\sqrt{\delta}$ where δ is the angular distance to the skirt.

4. WIMP signal modulation

WIMP (non-directional) direct detection is based on the measure of the recoil energy deposited in an elastic scattering with a nucleus. The differential event rate is usually written as[8]

$$\frac{dR}{dE_r} = \frac{\sigma_0 d}{2m_\chi m_r^2} F^2(|\vec{q}|) \int_{v_{\min}}^{\infty} \frac{f(v)}{v} dv \; , \tag{10}$$

---

[c] i.e. the time elapsed since the Vernal Equinox last crossed the local meridian
[d] i.e. the cross-term $-\vec{v}_{\odot i} \cdot \vec{v}_{\text{orb}}/c^2$

in order to separate the microphysics that enters the standard cross-section σ_0, from the nuclear form factor F, and from the CDM velocity distribution[e] $f(v)$. m_χ is the WIMP mass, m_N is the nucleus mass, and m_r is their reduced mass.

For a cold flow, of density d_F and speed v_F relative to the laboratory, $f(v) = \delta(v - v_F)$, so that the differential rate is a plateau whose height is inversely proportional to v_F, and whose edge is proportional to v_F^2. To be precise, the maximum recoil energy is

$$E_{\max} = \frac{2m_r^2 v_F^2}{m_N} \quad . \tag{11}$$

This particular shape with a sharp edge that occurs for a cold flow makes the annual modulation of the total rate

$$R = \int_{E_1}^{E_2} \frac{dR}{dE_r} \, dE_r \tag{12}$$

very sensitive to the detector thresholds E_1 and E_2. In particular, its sign is not set by the flow direction. For small recoil energies, the event rate modulation follows that of v_F^{-1}, but for higher recoil energies, there could be some events when v_F is highest, but no events at all when v_F is lowest!

4.1. *Predicted WIMP signal for the caustic ring halo model and the DAMA result*

As an illustrative example, we can consider the predicted signal for the so-called caustic ring halo model, and confront it with the WIMP signal modulation claimed by the DAMA collaboration.

The caustic ring halo model assumes a self-similar infall of collisionless CDM flows that are cold. Therefore, at any location, there is a discrete number of flows corresponding to particles that are falling in and out of the galaxy for the nth time[3]. It predicts the appearance of two types of caustics, outer spheres and inner rings with a triangular section. Interestingly, there is some observational evidence for these caustic rings[9], and the proximity of the solar system to one of these makes the local CDM density dominated by one flow, dubbed the 'big flow'. The big flow direction is such that its velocity on Earth becomes maximal in early november[f], which is five months after when the Earth's velocity in the galaxy is maximal.

---

[e]in the laboratory reference frame
[f]we consider the case where the big flow is the flow 5-. See Ref. [5] for details.

On the other hand, the annual modulation seen by the DAMA experiment is such that the rate peaks in early June, May 21 ± 22 days to be precise. The 'electron equivalent' energy range is 2–6 keV, meaning that this is the scintillation energy range, not the recoil energy. Moreover, no significant modulation is observed in the 6–14 keV energy range.

From this it is clear that the caustic ring halo model can be made consistent with DAMA provided that the total rate modulation follows v_F^{-1}, where v_F is the big flow velocity. Inspection of Eq. (12) shows that this condition favors rather massive WIMPs, $m_\chi \geq 100$ GeV. For lighter WIMPs, the annual modulation can have several annual peaks, or is reversed. Finally, the absence of modulation in the 6–14 keV range is troublesome for the caustic ring halo model, as one expects to see the effect of the edge of the plateau in the differential event rate, and therefore a modulation with the opposite sign in general.[g]

5. Conclusions

By considering CDM flows as cold and collisionless as it should be, several important consequences regarding the future detection of these hypothetic particles can be drawn. The signal in a detector will have a characteristic shape, in the form of a narrow peak in the case of axions, and of a plateau for the differential event rate in the case of WIMPs. A precise measurement of this signal will enable to reconstruct the flows direction and obtain some insight about the halo structure. The presence of caustics associated with the Sun's gravitational effect is also crucial. The temporary signal enhancement might be decisive regarding dark matter discovery.

References

1. C.L. Bennett et al., Ap. J. Suppl. 148 (2003) 1.
2. F. Combes, New Astron. Rev. 46 (2002) 755-766.
3. P. Sikivie, I. I. Tkachev, Y. Wang Phys. Rev. D56 (1997) 1863-1878.
4. P. Sikivie and J.R. Ipser, Phys. Lett. B291 (1992) 288.
5. F.-S. Ling, P. Sikivie and S. Wick, astro-ph/0405231.
6. P. Sikivie and S. Wick, Phys. Rev. D66, (2002) 023504.
7. S. J. Astalos et al., Phys. Rev. D69 (2004) 011101.
8. G. Jungman, M. Kamionkowski, K. Griest, Phys. Rept. 267 (1996) 195-373.
9. P. Sikivie, Phys.Lett. B567 (2003) 1-8.

---

[g]This point is also troublesome in the isothermal halo model.

EXPLOITING THE MATERIALS SIGNATURE IN CRYOGENIC WIMP DETECTORS

H. KRAUS, V.B. MIKHAILIK

Department of Physics,
University of Oxford,
Oxford OX1 3RH, UK

D.DAY, K.B.HUTTON AND J.TELFER

Spectra Physics Hilger Crystals,
Margate, Kent CT9 4JL, UK

Y.RAMACHERS

Department of Physics,
University of Warwick,
Coventry CV4 7AL, UK

The mass number dependence of the WIMP–nucleus scattering offers a method for identifying a true WIMP signal over a neutron background. In this paper we present a study on using a combination of $ZnWO_4$ and $CaWO_4$ absorbers to exploit this materials signature for WIMP detection. We show that already modest exposure in the region of 5 kg years should allow the detection of WIMP interaction for cross sections smaller than current experimental sensitivities. The combination of these two tungstates could form the basis of the first multi-target detector capable of WIMP identification through materials signature.

1. Introduction

Most modern experiments aiming to detect WIMP dark matter are employing some form of active discrimination between nuclear recoil (the signal) and electron recoil (backgrounds)[a]. The simultaneous detection of phonons and scintillation with cryogenic detectors operating at temperatures in the milli-kelvin range in dark matter searches[2,3], and in particular the use of calcium tungstate ($CaWO_4$) was pioneered by CRESST. A range of commonly available scintillators were tested for their feasibility as scintillating

---

[a]The following is a shortened version of[1]

cryogenic detectors[2]. The characterization of scintillation materials at very low temperature has not been the mainstream of scintillation studies so far and although many scintillators have been examined at or near room temperature, little information is available at temperatures below that of liquid nitrogen.

A high scintillation yield in the milli-kelvin temperature range is a key criterion in the choice of a scintillating absorber in dark matter searches. The higher the scintillation yield, the lower is the energy threshold for which discrimination between electron and nuclear recoil is possible for a given confidence level. Lower energy threshold translates into less exposure (mass×time) needed to reach certain levels of cross sections for WIMP–nucleon scattering.

$CaWO_4$ is an excellent target material for cryogenic dark matter searches, offering excellent discrimination between nuclear and electron recoil[4]. Nuclear recoils are produced by WIMP scattering, but also by neutron interaction. The various types of events could be distinguished by the dependence of the scintillation yield on the type of recoiling nucleus. It is thus beneficial to have targets with a variety of nuclei, and also argets that are similar with only one nucleus different. This should allow extracting a WIMP signal via the material signature of WIMPs[5]. $ZnWO_4$ is a very interesting material in this regard. It differs from $CaWO_4$ only by having Ca replaced with Zn and the mass number of Zn (A = 65.41) is close to that of germanium (A = 72.64), thereby making cross calibration with detectors based on germanium easier[6,7].

In this paper we present computer simulations on the sensitivity levels for WIMP–nucleon cross sections that could be accessed via a material signature. For a detailed discussion of a comparative characterisation of $CaWO_4$ and $ZnWO_4$ scintillators, see[1].

2. WIMP material signature sensitivity

The operation of two different target materials for particle dark matter detection offers the exciting possibility of exploiting the target material dependence of the scalar WIMP–nucleus scattering cross section as a signature[5]. We carried out Monte Carlo simulations on the material signature for $CaWO_4$ and $ZnWO_4$ detectors operating together in the same experimental setup.

2.1. Background estimate

The key issue in calculating sensitivity levels for discrimination by material composition is knowledge of the spectral shape of possible backgrounds. Discrimination via phonon and scintillation measurement removes backgrounds resulting in electron recoils, which leaves nuclear recoils caused by neutron interaction as the dominant source of background. In the following we assume the background as being caused by neutron elastic scattering in the target material. Monte-Carlo simulations of each resulting recoil spectrum are obtained using the GEANT4 simulation package[8]. These recoil energy spectra are normalised and used as probability density functions in Monte-Carlo simulations of the background energy spectra. For a further discussion we refer the reader to[1].

2.2. Data analysis method for the material signature

The data analysis algorithm for calculating minimum sensitivities for WIMP signature detection is based on a Monte-Carlo technique, for which the assumptions are:

- the exposure (mass×time) for $CaWO_4$ and $ZnWO_4$ detectors is the same,
- both detectors share the same background, operating in the same set-up concurrently,
- signal-to-background ratio is 10 (see comments below on this ratio),
- energy threshold is 10keV,
- the WIMP signal for each detector material is calculated following[9],
- only spin-independent (scalar) interactions of WIMPs are considered,
- WIMP–nucleon scattering scalar cross sections of 10^{-6} pbarn and 10^{-7} pbarn are assumed.

The data analysis algorithm includes of the following steps:

- calculate the WIMP signals in both materials for a range of WIMP masses,
- produce a WIMP spectrum (Poisson-distributed with mean being the expected rate per energy bin) with total count rate s_m for each material M.
- produce Poisson-distributed background spectra containing 10% of the total simulated signals ($b_m = 0.1\ s_m$) and add these to the sig-

nals s_m in order to obtain recoil spectra of signal-plus-background, S_m.
- produce another background spectrum with total count rate B_m equal to S_m.
- fit the expected WIMP signal on each material to the simulated S_m (see description below) and to the simulated background spectra and store the fit results.
- repeat the simulation for 1000 random measurements and analyse the results.

We use a maximum likelihood fit procedure such that the model must be optimised for both materials simultaneously once for the two random signal-plus-background samples and then for the two pure background samples. The model is defined as containing the WIMP signal and background. The WIMP signal contains two fit parameters, WIMP–nucleon scalar cross section and WIMP mass. The background has one free parameter. These three parameters are calculated for each set of S_m and B_m and their distribution is used for the subsequent hypothesis test, which in the end determines the minimum exposure for WIMP detection. For our hypothesis testing, see[1].

In contrast to WIMP signatures based on a stimulus with known time structure, i.e. annual or diurnal modulation, we found that the material signature requires a signal-to-background ratio of at least one for medium and high WIMP masses. Low-mass WIMPs exhibit a spectral shape sufficiently different to background spectral shapes to be still detectable even with a signal to background ratio of one. Nevertheless, for a realistic WIMP search one should aim for an experimental setup which offers a large signal to background ratio for as large a range of WIMP cross sections as possible. Due to the WIMP-mass dependence of this signature, we recommend an 'on/off'-approach to verify a possible detection. A measurement utilising the best possible signal to background ratio should be compared to a previous and/or subsequent measurement with increased background. In our simulations we found that spectra containing neutron background only give rise to a goodness of fit nearly independent of WIMP masses above about 50 GeV. Attempts to fit such a neutron data sample effectively return the start value for the WIMP mass as best fit. If, however, the recoil spectrum contains a contribution from WIMP interaction, the fit procedure behaves in a robust way to variations of the initial parameter values, provided the signal is statistically significant and the signal to background ratio is large enough (we so far checked for a ratio of 10 only). Recoil energy spectra with a lower signal to background ratio should not exhibit a robust solution with regard to WIMP signal identification. In that case the goodness of fit is nearly independent of WIMP mass, indicating neutron background as discussed above.

The minimum exposures for detecting a WIMP signature with the above material combination are shown in Fig. 1. The most remarkable feature should be the absolute scale of exposures. For most WIMP masses, assuming a WIMP–nucleon cross section of 10^{-6} pbarn, exposures of less than 5 kg years are sufficient for detection at 90% C.L. This should be compared with minimum exposure calculations for the time-dependent WIMP modulation signature[10], which are generally higher by at least an order of magnitude. A cross section of 10^{-6} pbarn represents the level of current best sensitivities of WIMP direct detection experiments. Results for the experimentally more challenging cross section of 10^{-7} pbarn can be obtained by scaling the minimum exposures up by a factor of 10. This represents the near or mid-term future sensitivity of experiments. We used separate calculations for cross sections of 10^{-7} pbarn as a numerical cross-check to confirm the expected factor 10 increase in exposure. The steep rise towards higher WIMP masses (greater than about 180 GeV) indicates the limit of applicability of material signature detection for this particular combination of materials. For WIMP masses much greater than that of Zn and Ca the contribution to the recoil energy spectra from scattering by these nuclei tends to be small, making differences in the overall recoil spectra difficult to detect.

3. Conclusion

We have shown that $ZnWO_4$ is a suitable and attractive target material for a cryogenic dark matter search experiment, especially when combined with $CaWO_4$ in order to exploit the materials signature for WIMP–nucleus interaction. Operating both target materials in close proximity, within the same experimental set-up, should allow detection of WIMP–interaction via a material signature already at rather modest exposures in the region of about 5 kg years. This needs to be contrasted with the need for much higher exposures required for detecting an annual modulation signature. The combination of these two tungstates could form the basis for the first multi-target detector capable of WIMP identification through a material signature.

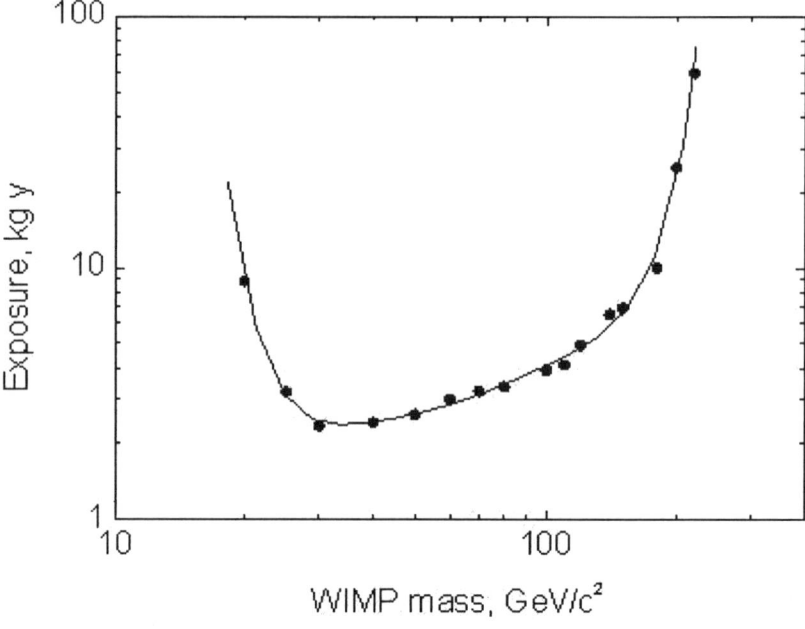

Figure 1. Minimum exposures for WIMP signature detection as a function of WIMP mass. Calculations were carried out for WIMP masses corresponding to the marker symbols; the solid curve is an interpolation shown as guide to the eye. A cross section for scalar WIMP–nucleon elastic scattering of 10^{-6} pbarn was used in the calculations. Exposures below the solid curve are insufficient to detect a WIMP signature. Further details on the calculations resulting in this plot are given in the text.

References

1. H. Kraus, V.B. Mikhailik, Y. Ramachers, D. Day, K.B. Hutton and J. Telfer, *submitted to Phys. Lett. B* (2004).
2. Meunier P. et al. (CRESST collaboration), *Appl. Phys. Lett.*, **75** 1335 (1999)
3. Cebrian S. et al. (ROSEBUD collaboration), *Phys. Lett. B* **563** 48 (2003)
4. Bravin M. et al. (CRESST collaboration), *Nucl. Instr. and Meth. A* **444** 323 (2000)
5. Smith P.F. and Lewin J.D., *Phys. Reports* **187** 203 (1990)
6. Akerib D.S. et al. (CDMS collaboration), *Phys. Rev. D* **68** 082002 (2003)
7. Benoit A. et al. (EDELWEISS collaboration), *Phys. Lett. B* **545** 43 (2002)
8. GEANT4, http://cern.ch/geant4.
9. Angloher G. et al., (CRESST collaboration), *Astropart. Phys.* **18** 43 (2002)
10. Ramachers Y., *Astropart. Phys.* **19** 419 (2003)

THE GENIUS-TEST-FACILITY AND THE HDMS DETECTOR IN GRAN SASSO

H.V. KLAPDOR-KLEINGROTHAUS,[*] I.V. KRIVOSHEINA[†]

Max-Planck-Institut für Kernphysik, P.O. Box 10 39 80
D-69029 Heidelberg, Germany

The first four naked high purity Germanium detectors (10 kg) were installed successfully in liquid nitrogen in the GENIUS-Test-Facility (GENIUS-TF) in the GRAN SASSO Underground Laboratory on May 5, 2003. This is the first time ever that this novel technique aiming at extreme background reduction in search for rare decays is going to be tested underground. First results on the background are presented. The GENIUS-TF experiment, aims to search for the annual modulation of the Dark Matter signal using 40 kg of naked-Ge detectors in liquid nitrogen. It should be able to confirm the DAMA result within two or three years of measuring time.
HDMS (Heidelberg Dark Matter Search) is the only experiment worldwide, operating *an enriched ^{73}Ge detector* and is looking for spin-dependent WIMP-neutron interactions. Results for the measurement Febr. 2001 - July 2003 are presented. They improve the best existing present limits for low WIMP masses.

1. Introduction

The present status of further cold dark matter search, of investigation of neutrinoless double beta decay and of low-energy solar neutrinos all require new techniques of *drastic* reduction of background in the experiments. For this purpose we proposed the GENIUS (GErmanium in liquid NItrogen Underground Setup) project in 1997 [2,3,4,5,6,7]. The idea is to operate 'naked' Ge detectors in liquid nitrogen (as applied **routinely already for more than 20 years** by the CANBERRA Company for technical functions tests [1]), and thus, by removing all materials from the immediate vicinity of the Ge crystals, to reduce the background considerably with respect to conventionally operated detectors. The liquid nitrogen acts both as a cooling medium and as a shield against external radioactivity.

---

[*]Spokesman of Heidelberg-Moscow (and GENIUS-TF and HDMS) Collaborations, E-mail: H.Klapdor@mpi-hd.mpg.de, Home-page: http://www.mpi-hd.mpg.de.non_acc/
[†]on leave of the Radiophysical-Research Institute, Nishnij-Novgorod, Russia, E-mail: Irina.Krivosheina@mpi-hd.mpg.de, http://www.mpi-hd.mpg.de.non_acc/

That the removal of material close to the detectors is the crucial point for improvement of the background, we know from our experience with the HEIDELBERG-MOSCOW double beta decay experiment [11,12,13,14,15], which is the most sensitive double beta experiment for more than 10 years now. Monte Carlo simulations for the GENIUS project, and investigation of the new physics potential of the project have been performed in great detail, and have been published elsewhere [2,3,4,5,6,7,9,10]. We were the first to show (in our HEIDELBERG low-level facility in 1997) that such device can be used for spectroscopy [2,3,4,5,6].

A small scale version of GENIUS, the GENIUS-Test-Facility has been approved by the Gran Sasso Scientific Committee in March 2001. The idea of GENIUS-TF is to prove the feasibility of some key constructional features of GENIUS, such as detector holder systems, achievement of very low thresholds of specially designed Ge detectors, long term stability of the new detector concept, reduction of possible noise from bubbling nitrogen, etc.

Additionally the GENIUS-TF will improve the limits on WIMP-nucleon cross sections with respect to our results with the HEIDELBERG-MOSCOW experiment [23,24] thus allowing for a test of the claimed evidence for WIMP dark matter from the DAMA experiment [25,28]. The relatively large mass of Ge in the full scale GENIUS-TF compared to existing experiments would permit to search directly for a WIMP signature in form of the predicted [27] seasonal modulation of the event rate [20]. A detailed description of the GENIUS-TF project is given in [18,21,22].

In section 2 we will present first spectra measured with GENIUS-TF and in particular discuss the present status of background. In section 3 we discuss our recent results from HDMS experiment operating an enriched ^{73}Ge detector and looking for spin-dependent WIMP-neutron interactions [31,32].

2. Description of the GENIUS-TF Setup and of Present Background

After installation of the GENIUS-TF setup in 2002 between halls A and B in Gran Sasso, opposite to the buildings of the HEIDELBERG-MOSCOW double beta decay experiment and of the DAMA experiment, the first four detectors have been installed in liquid nitrogen on May, 5 2003 and have started operation. This has been reported in [21] and [22].

Fig. 1 shows the contacted crystals after taking them out of the transport dewars, in the holder made from high-purity PA5 (a type a teflon), in which they then are put into liquid nitrogen. Each detector has a weight of 2.5 kg. The depth of the core of the detectors was reduced to guarantee a very low threshold, estimated by ORTEC to be around 0.5-0.7 keV, with only marginal

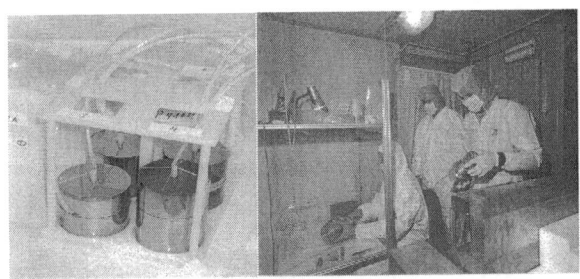

Figure 1. Right: Taking out the crystals from the transport dewars and fixing the electrical contacts in the clean room of the GENIUS-TF building - from left to right: Herbert Strecker, Hans Volker Klapdor-Kleingrothaus. Left: The first four contacted naked Ge detectors before installation into the GENIUS-TF setup.

deterioration of the energy resolution.

The liquid nitrogen (in total $\sim$ 70 l) is kept in a thin-walled (1 mm) box of high-purity electrolytic copper of size 50x50x50 cm^3. Inside this copper box, i.e. also inside the liquid nitrogen, is installed another box with walls of 5 - 10 cm monocristalline Ge bricks ($\sim$300 kg) forming the first highly efficient shield of the Ge detectors.

The copper box is thermally shielded by 20 cm of special low-level styropor, followed by a shield of 10 cm of electrolytic copper (15 tons) and 20 cm of low-level (Boliden) lead (35 tons).

The high-purity liquid nitrogen used, is produced by the BOREXINO nitrogen plant, which has been extended by our group for increase of the production capacity to be able to provide enough nitrogen also for GENIUS-TF (see [22]).

The nitrogen level in the detector chamber is measured by a capacitive sensor.

The data acquisition system we developped recently for GENIUS-TF and GENIUS is decribed in detail in [19]. It uses multichannel digital processing technology with FLASH ADC modules with high sampling rates of 100 MHz and resolution of 13 bits. It allows to capture the detailed shape of the preamplifier signal with high-speed ADC, and then perform digitally all essential data processing functions, including precise energy measurement over a range of 1 keV - 3 MeV, rise time analysis, ballistic deficit correction and pulse shape analysis. Thus we obtain both the energy and the pulse shape information from one detector using one channel of the Flash ADC module.

To allow for regular calibration of the detectors, a source of ^{133}Ba fixed on a wire can be introduced through a teflon tube into the center among the detectors. The source is transported via a magnetic system. Fig. 2 shows a spectrum measured, a few days after installation, with a ^{60}Co source *outside* the setup, and the ^{133}Ba source *inside*.

Also in Fig. 2 are shown two spectra in the range 200-650 keV, measured after the setup has been closed by the final shielding in December 2003. The prominent lines are originating from ^{214}Pb (left three lines) and ^{214}Bi (right line). They are identified as a signature of a ^{222}Rn contamination *diffusing* into the setup (see [17].)

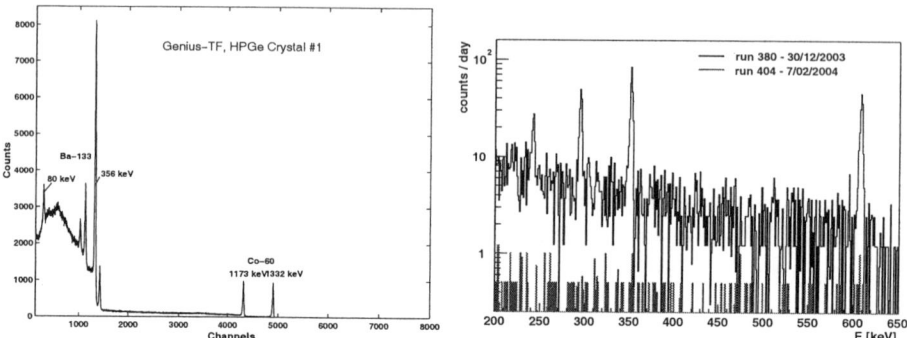

Figure 2. Left: A first spectrum measured with detector 1 with a ^{60}Co source outside, and the ^{133}Ba source inside the setup (see [22]). Right: Comparison between run n.380 (lifetime = 1.807 d), the run showing the highest countrate from ^{222}Rn and run n.404 (lifetime = 1.977 d), one of the latest acquired runs. Both measurements were performed with the GENIUS-TF detector n.2. The comparison clearly shows the reduction in the ^{222}Rn activity (see text and [17]).

We realized that this Rn was introduced largely during the refilling procedure of the liquid nitrogen. With additional measures to avoid this the background reduced to the lower of the spectra shown in Fig. 2 (right side). The remaining count rate still corresponded to a ^{222}Rn activity of ~ 1.2 mBq/m^3. Additional shielding of the whole setup from the external air by a radon-dense foil gave a further reduction by a factor of 10.

The present level of radon activity brings the background index in the energy interval 0-50 keV to a level of 1 count/(kg keV y). This value, calculated from a comparison with the simulation, is by about a factor of 20 higher than foreseen in the GENIUS-TF proposal [18]. This value is *still compatible* (see Table 1 in Ref. [18]) with the goal of the experiment of reaching 2-4 events/kg y keV, namely, **to check the evidence from cold dark matter** found by the DAMA collaboration, by the modulation signal. It would, at the present level, however, **cause very serious problems** for **any full** GENIUS-like experiment.

2.1. GENIUS-TF-II

In October 2004 we have installed a new setup (see Fig 3), containing in addition to the earlier setup (see [22]), a second copper vessel, for further shielding of the Radon (see [16]). First results show a strong reduction of the ^{222}Rn background.

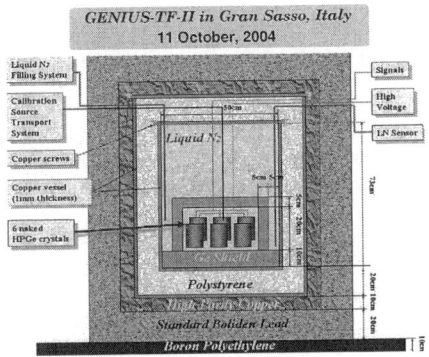

Figure 3. Cross section of the new setup GENIUS-TF-II.

3. The HEIDELBERG Dark Matter Search Experiment (HDMS)

The HDMS project operates two High Purity Germanium detectors in unique configuration at the Gran Sasso National Laboratory. A small p-type enriched ^{73}Ge crystal (enriched to 86%) is surrounded by a well-type natural Ge crystal, both being mounted in the same copper cryostat. This configuration reduces the background by the shielding provided by the outer crystal, and by the anti-coincidence between the two detectors.

The final setup was operated from February 2001 to July 2003 (85.5 kg d). Fig. 4 shows the spectrum collected during this time. The data set is divided into three measuring periods of 30.9, 29.5 and 27.6 kg d, respectively. The latter period is obviously less affected by the background of X-rays from ^{68}Ge.

Fig. 5 (left side), shows the determined limits on WIMP-neutron spin-dependent coupling from the last partial data set, for different values of the spin factors $\langle S_n \rangle$ and $\langle S_p \rangle$. For comparison is shown the current best limit on spin-dependent WIMP-neutron interaction, coming from the odd-neutron nucleus ^{129}Xe (from the DAMA Xenon experiment [26]).

Our result improve these limits in the region of low WIMP masses.

Fig. 5 (right side), shows the sensitivities of HDMS in the framework of

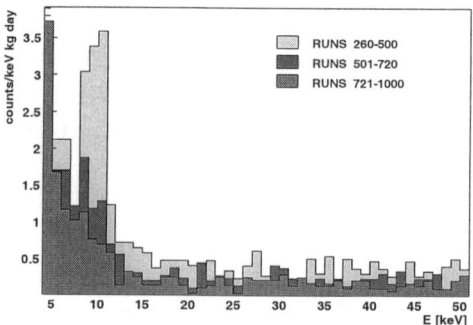

Figure 4. Anti-coincidence spectra from the HDMS experiment (see [29,30]). The result of Runs 721-1400 was used to extract limits on WIMP mass and cross section (see Fig. 5).

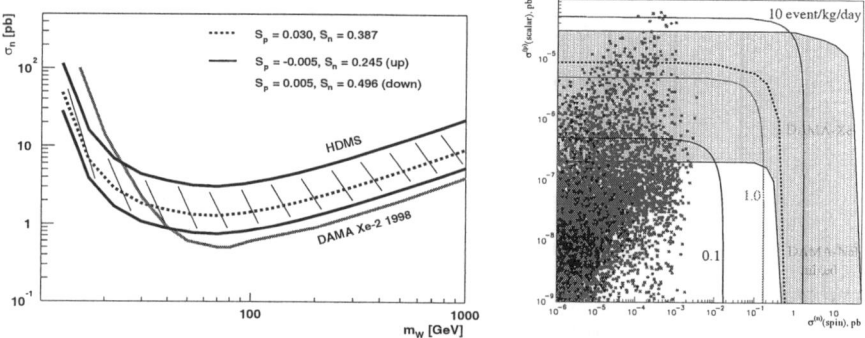

Figure 5. Left: Experimental limits on WIMP-neutron SD coupling from the HDMS experiment (data from runs 721-1000). The result of the DAMA Xenon experiment [26] is shown as comparison (from [29,30]). Right: The solid lines (marked with numbers of $R(15, 50)$ in events/kg/day) show the sensitivities of the HDMS setup with ^{73}Ge in the framework of mixed SD WIMP-neutron and SI WIMP-nucleon couplings together with the DAMA-NaI allowed region for sub-dominant SD WIMP-neutron coupling ($\theta = \pi/2$). The *present HDMS* result **cuts already part of the DAMA allowed range**. The scatter plots give correlations between σ_{SI}^p and σ_{SD}^n in the effMSSM for $m_\chi < 200$ GeV. The squares (red) correspond to sub-dominant relic neutralino contribution $0.002 < \Omega_\chi h_0^2 < 0.1$ and triangles (black) correspond to WMAP relic neutralino density $0.094 < \Omega_\chi h_0^2 < 0.129$. The dashed line from [26] shows the DAMA-LiXe (1998) exclusion curve for $m_{\text{WIMP}} = 50$ GeV (from [32]).

mixed spin-dependent (SD) WIMP-neutron and spin-independent (SI) WIMP-nucleon couplings (see [32]). The present result (sensitivity close to the curve 10 event/kg day - in the energy range 15-50 keV) already cuts part of the DAMA NaI allowed region, which is shown here for subdominant SD WIMP-neutron coupling ($\Theta = \pi/2$).

4. Conclusions

The annual modulation, due to the motion of the Earth with respect to the galactic halo, is the main signature of a possible WIMP signal. A positive indication of this modulation has been found over the past years by the DAMA experiment and it would be of great importance to look for the same effect with another experiment, expecially in the region of the WIMP parameter space indicated by the DAMA results.

The GENIUS-TF experiment [22,18] will be - in addition to DAMA [28] - the *only* experiment which will be able to probe the annual modulation signature in a foreseeable future. The at present much discussed cryo detector experiments, such as CDMS [34], CRESST [35], etc. have at present hardly a chance to look for modulation because the mass used and projected in these experiments is still by far too low (see also [33]).

The first four naked Ge detectors (10 kg) have been installed in liquid nitrogen into the GENIUS-Test-Facility in the GRAN SASSO Underground Laboratory on May 5, 2003. This is the first time that this novel technique is applied under realistic background conditions of an underground laboratory. **With the successful start of GENIUS-TF a historical step has been achieved into a new domain of background reduction in underground physics in the search for rare events.**

We presented a first analysis of the GENIUS-TF background after the complete assembling of the external shielding made of copper and lead. With an appropriate isolation of the setup from the external air and a constant flushing with gaseous nitrogen, the radon contamination was reduced finally to a level of 1 count/(kg keV y) in the low energy region (0-50 keV).

A new GENIUS-TF II setup has been installed in October 2004 with additional shielding against radon and additional two detectors, increasing the total mass to 15 kg.

The HDMS (HEIDELBERG Dark Matter Search) project runs an enriched ^{73}Ge detector in Gran Sasso, looking for *spin-dependent* WIMP-neutron interaction. The measurement over the period 2001-2003 improved the best up to now existing limits, by the ^{129}Xe DAMA measurement, in the range of low WIMP masses. At present efforts are going on to improve the sensitivity of HDMS, to be able to restrict the SUSY prediction region (see Fig. 5).

5. Acknowledgement

The authors would like to thank their collaborators C. Tomei and H. Strecker, and also the technical staff of the Max-Planck Institut für Kernphysik and of the Gran Sasso Underground Laboratory. We acknowledge the invaluable support from BMBF and DFG, and LNGS of this project. We are grateful to the former State Committee of Atomic Energy of the USSR for providing the monocristalline Ge shielding material used in this experiment.

References

1. J. Verplancke, CANBERRA Company, private communication, 5 March 2004.
2. H.V. Klapdor-Kleingrothaus in Proceedings of BEYOND'97, First International Conference on Particle Physics Beyond the Standard Model, Castle Ringberg, Germany, 8-14 June 1997, edited by H.V. Klapdor-Kleingrothaus and H. Päs, *IOP Bristol* (1998) 485 - 531 and in *Int. J. Mod. Phys.* **A 13** (1998) 3953.
3. H.V. Klapdor-Kleingrothaus, J.Hellmig and M.Hirsch, *GENIUS-Proposal*, 20 Nov. 1997.
4. J. Hellmig and H.V. Klapdor-Kleingrothaus, *Z. Phys.* **A 359** (1997) 351 - 359, and nucl-ex/9801004.
5. H.V. Klapdor-Kleingrothaus and M. Hirsch, *Z. Phys.* **A 359** (1997) 361 - 372.
6. H.V. Klapdor-Kleingrothaus, J. Hellmig and M. Hirsch, *J. Phys.* **G 24** (1998) 483 - 516.
7. H.V. Klapdor-Kleingrothaus, CERN Courier, Nov. 1997, 16 - 18.
8. H.V. Klapdor-Kleingrothaus et al. **MPI-Report MPI-H-V26-1999**, and *Preprint: hep-ph/***9910205**, and in Proceedings of the 2nd Int. Conf. on Particle Physics Beyond the Standard Model BEYOND'99, Castle Ringberg, Germany, 6-12 June 1999, edited by H.V. Klapdor-Kleingrothaus and I.V. Krivosheina, *IOP Bristol* (2000) 915 - 1014.
9. H.V. Klapdor-Kleingrothaus, "60 Years of Double Beta Decay - From Nuclear Physics to Beyond the Standard Model", World Scientific, Singapore (2001) 1281 pages.
10. H.V. Klapdor-Kleingrothaus, Springer Tracts in Modern Physics, 163 (2000) 69-104, Springer-Verlag, Heidelberg, Germany (2000).
11. H.V. Klapdor-Kleingrothaus, I.V. Krivosheina, A. Dietz et al., Phys. Lett. B 586 (2004) 198-212.
12. H.V. Klapdor-Kleingrothaus, A. Dietz, I.V. Krivosheina et al., NIM A 522 (2004) 371-406.
13. H.V. Klapdor-Kleingrothaus et al. hep-ph/0201231 and *Mod. Phys. Lett.* **A 16** (2001) 2409-2420.
14. H.V. Klapdor-Kleingrothaus, A. Dietz and I.V. Krivosheina, *Part. and Nucl.* **110** (2002) 57-79.
15. H.V. Klapdor-Kleingrothaus, A. Dietz and I.V. Krivosheina, *Foundations of Physics* **31** (2002) 1181-1223 and Corrigenda, 2003 home-page: http://www.mpi-hd.mpg.de/non_acc/main_results.html.
16. H.V. Klapdor-Kleingrothaus et al., Subm for publ. (2005)
17. H.V. Klapdor-Kleingrothaus, C. Tomei, I. Krivosheina, et al., *Nucl. Instrum. Meth.* **530** (2004) 410-418.
18. H.V. Klapdor-Kleingrothaus et al., *Nucl. Instrum. Meth.* **A 481** (2002) 149-159.
19. T. Kihm, V.F. Bobrakov and H.V. Klapdor-Kleingrothaus, *Nucl. Instrum. Meth.* **A 498** (2003) 334-339 and hep-ph/0302236.

20. C. Tomei, A. Dietz, I. Krivosheina, H.V. Klapdor-Kleingrothaus, *Nucl. Instrum. Meth.* **A 508** (2003) 343-352, and hep-ph/0306257.
21. H.V. Klapdor-Kleingrothaus, CERN Courier, 2003.
22. H.V. Klapdor-Kleingrothaus et al.,*Nuclear Instruments and Methods in Physics Research* **A 511** (2003) 341 - 346 and hep-ph/0309170, and H.V. Klapdor-Kleingrothaus and I.V. Krivosheina, in Proc. of Beyond the Desert 2002, BEYOND02, Oulu, Finland, June 2002, IOP 2003, ed. H.V. Klapdor-Kleingrothaus.
23. HEIDELBERG-MOSCOW collaboration, Phys. Rev. D **59** (1998) 022001.
24. H. V. Klapdor-Kleingrothaus, A. Dietz, G. Heusser, I.V. Krivosheina, D. Mazza, H. Strecker and C. Tomei, Astroparticle Physics, **18**, Issue 5 (2003) 525-530, and hep-ph/0206151.
25. R. Bernabei et al., Phys. Lett. B **424** (1998) 195; Phys. Lett. B **450** (1999) 448; Phys. Lett. B **480** (2000) 23.
26. R. Bernabei *et al. Phys. Lett.* **B436** (1998) 379.
27. K. Freese, J. Frieman and A. Gould, Phys. Rev. D **37** (1988) 3388.
28. R. Bernabei et al., Riv. Nuovo Cim. 26 (2003) 1-73.
29. H.V. Klapdor-Kleingrothaus et al., in Gran Sasso Annual Rep. 2003, INFN (2004).
30. C. Tomei, Diss. L'Aquila University, 2004
31. H.V. Klapdor-Kleingrothaus et al., in Press (2004).
32. V. Bednyakov, H.V. Klapdor-Kleingrothaus, hep-ph/0404102 and in Press (2004) in Phys. Rev. D.
33. H.V. Klapdor-Kleingrothaus, *Int. J. Mod. Phys.* **A17**, 3421-3431 (2002), in Proc. of *Internat. Conf. LP01*, Rome, Italy, July 2001.
34. T. Saab et al., Nucl. Phys. Proc. Suppl. 110 (2002) 100-102.
35. F. Probst et al., Nucl. Phys. Proc. Suppl. 110 (2002) 67-69.

WARP: A WIMP DOUBLE PHASE ARGON DETECTOR

R. BRUNETTI, E. CALLIGARICH, M. CAMBIAGHI, C. DE VECCHI,
R. DOLFINI, L. GRANDI, A. MENEGOLLI, C. MONTANARI, M. PRATA,
A. RAPPOLDI, G.L. RASELLI, M. RONCADELLI,
M. ROSSELLA, C. RUBBIA*AND C. VIGNOLI

*INFN and Department of Physics at University of Pavia,
Via Bassi 6, I-27100 Pavia (PV), Italy*

F. CARBONARA, A. COCCO, A. EREDITATO, G. FIORILLO,
G. MANGANO AND R. SANTORELLI

*INFN and Department of Physics at University of Napoli,
Via Cintia, I-80126 Napoli (NA), Italy*

F. CAVANNA, N. FERRARI AND O. PALAMARA

*INFN Gran Sasso Laboratory,
INFN, Laboratori Nazionali del Gran Sasso (LNGS),
S.S. 17/bis Km 18+910, I-67010 L'Aquila - Italy*

PRESENTED BY L. GRANDI
E-mail: luca.grandi@pv.infn.it

The *WARP* programme for dark matter search with a double phase argon detector is presented. In such a detector both excitation and ionization produced by an impinging particle are evaluated by the contemporary measurement of primary scintillation and secondary (proportional) light signal, this latter being produced by extracting and accelerating ionization electrons in the gas phase. The proposed technique, verified on a 2.3 *liters* prototype, could be used to efficiently discriminate nuclear recoils, induced by *WIMP's* interactions, and measure their energy spectrum. An overview of the 2.3 *liters* results and of the proposed 100 *liters* detector is shown.

---

*also at ENEA, Presidenza, Lungotevere Thaon Di Revel 76, I00196 Roma, Italy

1. Introduction

The very small interaction cross section makes $WIMP$-nucleus scattering a very rare event. Such a rate is not easily predicted, since it depends on many variables which are poorly defined. In practice, uncertainties may encompass many orders of magnitude, although the minimal $SUSY$ leaves open the optimistic possibility of very significant rates [1]. Any new experiment must therefore reach an ultimate sensitivity which is several orders of magnitudes higher than the one of the presently ongoing searches [2]. To achieve such a goal, both sensitive mass and background discrimination should be as large as possible. From this point of view the cryogenic noble liquids technology (Ar and Xe) is one of the most promising techniques since it provides both a highly efficient discrimination and the potentiality to be extended to multi-ton sensitive volumes. Already in 1993 [3] the $ICARUS$ collaboration pointed out, for the first time, that the simultaneous measurement of scintillation and ionization produced in noble liquids by an impinging particle permits to efficiently discriminate the nature of the particle itself. In particular this makes it possible to discriminate a nuclear recoil, eventually induced by a $WIMP$ in the typical energy range $\approx 10 - 100 \ keV$, from the dominant gammas and electrons background induced by materials radioactivity.

The $WARP$ (Wimp ARgon Programme) collaboration has focused on argon since its technology is already fully operational, well supported at industrial level and it has low cost, providing at the same time a sensitivity similar to xenon for a reasonable energy threshold ($E_{rec} > 30keV$): the less effective coherent effect is compensated by the less steeper nuclear form factor [4].

To verify the proposed technique a 2.3 $liters$ double phase argon detector has been realized and an intensive set of measurements and calibrations with various radioactive sources has been conducted (see Sec. 2). At the end of this feasibility study the argon chamber, opportunely refurbished, has been installed at the underground Gran Sasso National laboratory (LNGS) to study the background (see Sec. 3). On the basis of the obtained results a 100 liters sensitive volume argon detector, to be installed at LNGS, has been proposed [4]. The construction of the 100 $liters$ detector has been approved and funded by INFN starting from 2004 (see Sec. 4).

2. 2.3 liters chamber: feasibility study

The 2.3 *liters* drift chamber consists of a lower liquid volume and an upper region with Argon in the gaseous phase, kept in thermal equilibrium by immersing the hole chamber in a liquid argon bath. LAr electro-negative impurity concentration is kept at a level ≤ 0.1 *ppb* (O_2 equiv.) with the use of standard $Hydrosorb^{TM}$ and $Oxisorb^{TM}$ filters. The prototype is equipped with a single 8" cryogenic photomultiplier coated with TPB to wave-shift VUV scintillation photons [4] and installed in the gaseous phase. In such an experimental setup the ionization electrons, produced in the liquid by an interacting particle, are drifted towards the liquid-gas boundary, extracted with the help of an electric field and accelerated in a high electric field region to produce proportional scintillation light. Extraction and multiplication fields are generated through the use of grids. A high diffusive reflector layer surrounds the inner volume to increase light collection efficiency. The typical light signal generated by an interaction in the liquid volume and recorded by the PMT is then constituted by a prompt primary peak (primary signal $S1$), produced by de-excitations and recombination processes, followed after a drift time (depending on actual location of the interaction) by a secondary peak (secondary signal $S2$), associated to the ionization electrons drifted in the liquid and accelerated in the gas phase.

As a first test the chamber has been used as a standard scintillation counter with extraction and multiplication fields turned off. In this way only primary signal survives since the eventually drifted electrons reaching the liquid-gas boundary do not produce proportional light. To evaluate the collected light yield and hence the detection efficiency, the chamber has been exposed to a variety of radioactive sources: in particular a ^{109}Cd source, providing several x-rays peaks in the region $20 - 25$ keV, has been placed inside the sensitive volume. The measured photo-electron yield, deduced from several acquired scintillation spectra, is $\Gamma(0) = 2.9$ phe/keV at zero drift field and $\Gamma(1000) = 2.35$ phe/keV at 1 kV/cm drift field: this decrease is due to the fact that the applied drift field avoids ionization electrons-ions recombination.

With a similar setup but the fields on, the electrons extraction process has been studied as function of the extraction field (maintaining a fixed multiplication field). The extraction probability is obviously function of the electric field applied at the liquid-gas interface through the use of two grids, one placed in liquid ($g1$) and the other in gas ($g2$). The measured extraction efficiency is in excellent agreement with experimental results found in literature[5]. At fields of the order of 2.5 kV/cm almost the totality of the electrons reaching the interface is extracted in less than 0.1 μs differently from what occurs in xenon for which higher fields are required.

Intensive tests on multiplication process has been conducted too. Providing a fixed extraction field, the amplitude of the secondary signal as function of the applied field has been investigated. Proportional light production starts at fields of the order of 1 kV/cm (significantly lower than the one needed for xenon). At 5 kV/cm the gas gain is about 32 $photons/electron/cm$ for a saturated gas pressure of 1 bar at 87 K).

After having verified the possibility of extracting the ionization electrons and producing proportional light emission with a double phase argon detector, the response of the chamber to different kind of particle has been investigated. Since different particles produce excitation and ionization in a different amount, the ratio $S2/S1$ should be function of the nature of the interacting particle. In particular, we have explored the signatures of an hypothetical *WIMP*, exposing the chamber to a $D-T$ 14 MeV neutron generator and to an Am-Be neutron source: fast neutrons scattering elastically on nuclei behave like "strong interacting" *WIMP's*, producing nuclear recoils in the energy range of interest. At the working electric fields a strong correlation between primary and secondary signals is evident: differently from gamma-like signals (electrons mediated) for which $S2/S1 \gg 1$, the observed nuclear recoil events are characterized by $S2/S1 \ll 1$ populating a completly different region of the scatter plot $S2/S1$ versus $S1$. The obtained results are in good agreement with the predictions of the so-called Box Model of Thomas and Imel[7]: due to local density ionization a smaller amount of electrons, if compared to minimum ionizing particles, is able, under the effect of drift field, to escape from recombination (for details see *WARP* proposal[4]). The measured photo-electron yield for nuclear recoils, deduced from their primary scintillation spectrum, is 0.66 phe/keV (at 1 kV/cm) to be compared to 2.9 phe/keV for gammas at zero drift field.

3. Run in underground laboratory

The 2.3 *liters* chamber, opportunely refurbished, has been installed at the Gran Sasso National Laboratory to work in a low background environment. It has been equipped with seven 2 *inch* cryogenic photo-multipliers and several technical improvements (for a better field uniformity) have been applied.

As a first step the primary scintillation spectrum has been measured

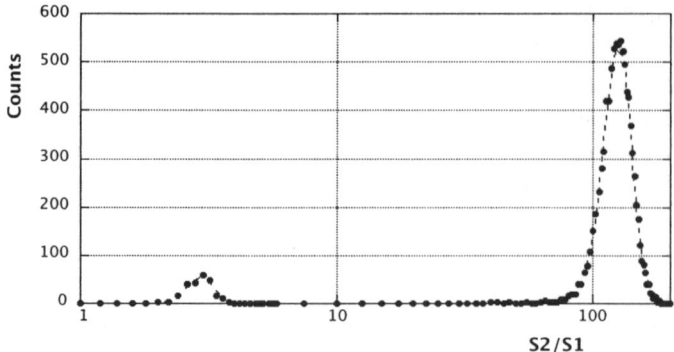

Figure 1. $S2/S1$ distribution. Two well separated peaks are evident. One due to minimum ionizing particles and the other to $^{222}Rn \rightarrow ^{218}Po + \alpha + 5.489\ MeV$ decays (half-life of 3.825 $days$).

and compared to the one acquired at surface level. As expected the gamma spectrum at LNGS ends at 3 MeV. The dominant contributions comes from ^{232}Th and ^{238}U chains and ^{40}K.

With this new internal setup and different fields value a new measurement of the discriminating power has been conduced. Figure 1 shows the $S2/S1$ distribution. Two well separated families are visible: one, associated to minimum ionizing particles (gammas), centered around 125.7 and another, due to ^{222}Rn α-decay contained in the argon, around $S2/S1 = 2.9$. The gamma-α suppression factor is about 44/1. As expected alpha particles, due to their strong recombination, behaves similarly to nuclear recoils providing a depleted ionization signal. Obviously the value of $S2/S1$ is function of the applied electric fields and of the argon gas pressure (equal to the atmospheric pressure and different according to the measurement site).

At the moment the chamber has been surrounded by a 10 cm thick lead shield and it is running in a low background condition to study the effects of shield and internal contamination.

4. The proposed 100 liters detector

On the basis of the comfortable results obtained with the small prototype, a new argon based detector has been proposed [4]. Its basic scheme, shown in Figure 2, foresees a fiducial volume of LAr (about 100 $liters$), tracing the layout of the 2.3 $liters$ chamber, with a uniform electric field drifting

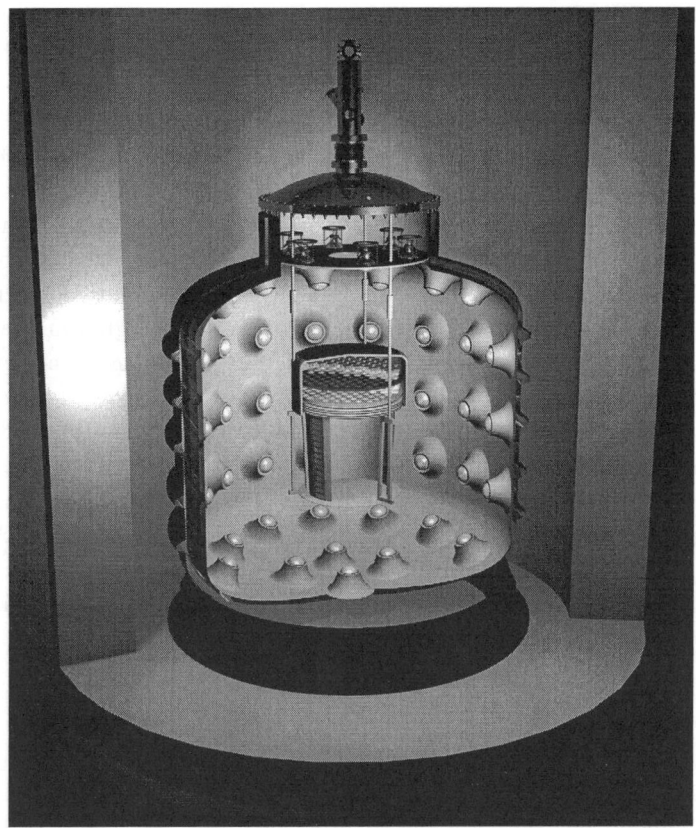

Figure 2. 3D view of the 100 *liters* prototype. The core of the detector is represented by the double-phase 50 *cm* drift sensitive volume, completely immersed in a 60 *cm* thick VETO region readout by 400 3 *inch* photo-multipliers. Through the use of a stainless steel cap and some heaters a gas pocket is maintained in the central volume equipped with 61 2 *inch* PMTs. The external passive shield is also shown.

ionization electrons towards a liquid to gas interface. A set of grids with an appropriate voltage arrangement provides then the extraction of ionization electrons from the liquid phase and their acceleration in the gas phase for the production of the secondary light pulse. A set of photomultipliers placed above the grids sense both the primary scintillation signal in the liquid argon and the delayed secondary pulse in the gas phase. *PMTs* granularity allows reconstruction of event position in the horizontal plane with about 1 *cm* resolution. Position along the drift coordinate is given by the drift time (position reconstructed in 3D). The whole detector has been designed trying to minimize the weight and therefore the amount of materials (and radioactive contamination) to be placed around the inner active volume.

The detector is completely submersed in a LAr volume that works as an anti-coincidence (*Active VETO*), which is also readout by a set of phototubes, the two volumes are optically separated. The *VETO* region is used to reject the events due to neutrons or other particles penetrating from outside or travelling out from the central part. Dimensions of the outer LAr volume are chosen in such a way that the probability for a neutron to interact in the inner detector without producing a signal in the *VETO* system is negligible. Only events with no signals in the *VETO* are potential *WIMP's* candidates.

An external shield is added to adequately reduce environmental neutron and gamma background. With 100 *liters* sensitive volume, which corresponds to an active mass of about 150 kg, thanks to the rejection/identification power of the two-phase technique, we could reach a sensitivity about two orders of magnitude better than the present experimental limit from *CDMS* (or to the presently indicated hint from the *DAMA* experiment).

References

1. Y.G. Kim, T. Nihei, L. Roszkowski and R. Ruiz de Austri, J. HEP 0212:034 (2002).
2. The liquid Xenon set-up of the DAMA experiment, Nucl. Instr. and Meth. **A482** (2002), 728; CDMS Coll., Phys. Rev. **D66**, 122003 (2002); A. Benoit et al., Phys. Lett. **B545** (2002) 43.
3. P. Benetti et al., Nucl. Instr. and Meth. **A327** (1993) 203.
4. *WARP: Wimp ARgon Programme - Experiment Proposal.* See also webpage *http://warp.pv.infn.it.*
5. E.M. Gushchin et al., Sov. Phys. JETP **55** (5) (1982).
6. J. Lindhard et al., Mat. Fys. Medd. Dan. Vid. Selsk. 33 n. 10-14 (1963).
7. J. Thomas and D.A. Imel, Phys. Rev. **A38** (1988) 5793.

COUPP, A HEAVY-LIQUID BUBBLE CHAMBER FOR WIMP DETECTION

J. BOLTE§, J.I. COLLAR§, M. CRISLER¶, D. HOLMGREN¶,
D. NAKAZAWA§, B. ODOM§, K. O'SULLIVAN§, R. PLUNKETT¶,
E. RAMBERG¶, A. RASKIN§, A. SONNENSCHEIN§, J. VIEIRA§ *

¶ *Fermi National Accelerator Laboratory, Batavia, IL, USA*
§ *KICP and Enrico Fermi Institute, University of Chicago, IL, USA*

The capabilities and reach of the first phase of COUPP (the Chicago Observatory for Underground Particle Physics) are described. During this first phase of the experiment a 2 kg CF_3I bubble chamber sensitive to WIMPs will be operated at the ~300 m.w.e. of the Minos-near gallery at FNAL. Prospects for larger devices are briefly discussed.

Our group is presently investigating the application of bulk superheated liquids to Weakly Interacting Massive Particle (WIMP) detection. It has been possible to demonstrate [1] that relatively large volumes of heavy refrigerants can be kept in a radiation-sensitive metastable state for long enough to perform rare-event searches. For certain choices of operating pressure and temperature, the vaporization of the liquid (Fig. 1) can be produced exclusively by particles having a high stopping power (e.g., nuclear recoils like those expected from WIMPs or neutrons), making the detector insensitive to minimum ionizing backgrounds. The devices are operated at near room temperature and the industrial refrigerants used are inexpensive, non-flammable and non-toxic, with a chemical composition that maximizes sensitivity to neutralino interactions through both the spin-dependent and independent channels [2]. For these reasons, the technique seems to be ideally fitted for the goal of building ton or even multi-ton WIMP detectors, devices able to probe most of the supersymmetric phase space where the supersymmetric dark matter may abide.

---

*Work supported by the Kavli Institute for Cosmological Physics via NSF grant PHYS-0114422 and CAREER award 0239812. E-mail: collar@uchicago.edu

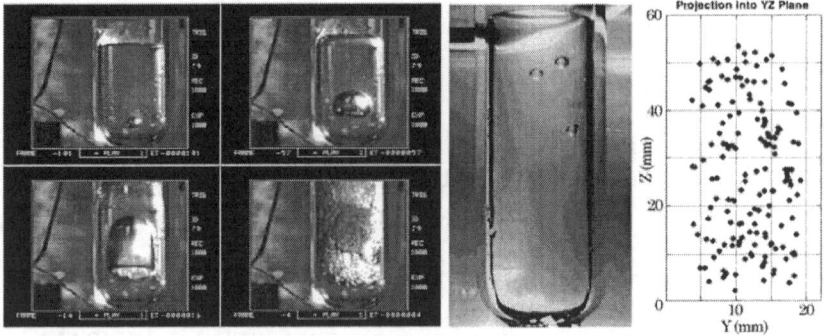

Figure 1. Left: High-speed footage of boiling induced by a neutron source in superheated CF_3Br. The bubble diameter expansion rate is ~1mm/ms in normal operating conditions (the depicted sequence spans 0.1 seconds). This footage can be viewed at http://cfcp.uchicago.edu/~collar/bubble.mov. Center: Multiple-bubble event produced by a neutron triple-scatter. Right: Automatic image analysis using LabVIEW's Vision Development Module allows precise bubble positioning and counting in 3-D (two perpendicular cameras are used).

Several techniques have been identified and exploited to maximize the stability of small bubble chamber prototypes containing CF_3Br and CF_3I. Namely, avoidance of contact with rough metallic surfaces, use of an immiscible liquid lid above the active volume, outgassing of surface imperfections in the presence of a buffer liquid, surface cleaning techniques and wetting improvement via vapor deposition [3]. Small prototypes (~30 g) remain superheated for periods of 15 minutes on the average at the shallow 6 m.w.e. depth of the EFI underground laboratory, a nucleation rate compatible with the measured neutron flux and energy spectrum in the site [4]. The insensitivity (rejection factor) to minimum ionizing particles (MIPs) in operating conditions at which the liquids are fully responsive to low energy nuclear recoils has been measured to be $> 10^9$ using strong gamma sources [4]. This guarantees the ability to build much larger prototypes in the ton or multi-ton regime, essentially without any concern for MIPs. Calibrations using neutron sources with a well-defined endpoint energy (11.1 MeV for Am/Be and 152 KeV for ^{88}Y/Be) allowed testing the response of the liquids to nuclear recoils down to 4 keV in the case of CF_3I and the establishment of agreement with theoretical models of this response [4].

The construction of a 1-liter (2 kg) active volume bubble chamber is well advanced (Fig. 2): at the time of this writing, it has been operated at 6 m.w.e. for several weeks, with a nucleation rate (~1/min) again compatible with the modelled contribution from the measured environmental fast neutron flux. The purpose of this prototype and experiment (the Chicago Observatory for Underground Particle Physics, COUPP) is to study the ultimate limits to the stability of the superheated liquid in a deeper location, with reduced neutron backgrounds. However a device of this mass can already be an extremely competitive WIMP detector, given the optimal choice of target nuclei and insensitivity to most backgrounds (Fig. 3).

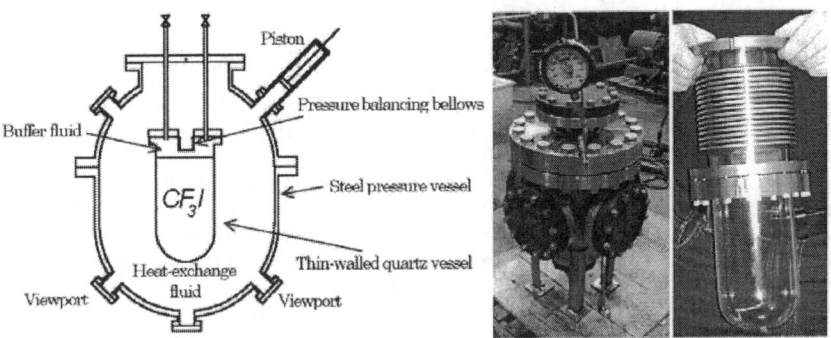

Figure 2. Left: Conceptual design of the 2 kg CF_3I chamber to be used in the preliminary phase of COUPP. Center: the recompression vessel in this prototype. Right: inner quartz vial and pressure-compensation bellows for the same. Footage of the response of this chamber to neutrons can be seen at http://cfcp.uchicago.edu/~collar/triple_bubble(0.4s).mov

While the ultimate goal is to deploy a large bubble chamber-based dark matter search in the Soudan Mine, there are considerable logistical benefits to having the initial commissioning and testing of this prototype device take place in the MINOS-near detector gallery at FNAL. Siting the initial commissioning work at Fermilab allows ready access to the detector, a critical need during the first few months of operation, until the behavior of the detector at this depth is completely understood and operation is fully automated. In this FNAL site, the experiment will profit from ~300 ft of rock overburden. A preliminary estimate of the cosmic-ray associated backgrounds at this depth reveals that the nucleation rate assumption used to generate the projections in Fig. 3 could in principle be met there. In the presence of 30 cm of polyethylene shielding, muon-induced energetic neutrons producing a few nucleations per kg/day are expected to be dominant, with only a small additional component from beam-related backgrounds. After a few months of preliminary tests at FNAL, the chamber will be

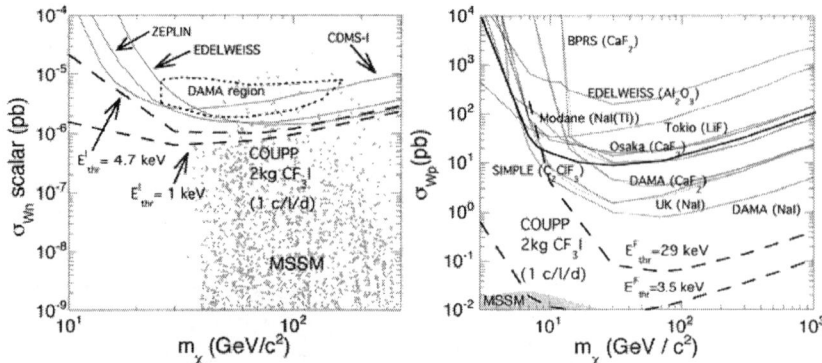

Figure 3. Sensitivity limits in the spin-independent (left) and spin-dependent (right) neutralino parameter space achievable during the first phase of COUPP (dashed lines), compared with other experiments. The neutron background rate used for these estimates is representative of what can be expected in the Minos near detector gallery and, according to Monte Carlo calculations, *a factor of ~100 too conservative* for the Soudan depth. The limits are plotted for two different energy thresholds, one already demonstrated with the ^{88}Y/Be neutron source calibrations and a second one (the best that can be expected before gamma background rejection is lost) soon to be tested with a ^{124}Sb/Be neutron source. NOTE: *the recently released CDMS-II limits surpass those from CDMS-I in the figure by a factor ~4.*

transported to a final emplacement in the Soudan mine (MN).

MCNP-Polimi [5] Monte Carlo simulations of the response to a typical underground neutron flux indicate that large enough bubble chambers (few hundred liters) would have ideal features as WIMP detectors. For instance, a sizeable inner fiducial volume would be shielded against events produced by "punch-through" neutrons able to penetrate any reasonable thickness of neutron moderator. These represent the ultimate challenge for next-generation WIMP detectors. Their interactions would nevertheless be revealed in these chambers by multi-bubble events which WIMPs cannot produce (Figs. 1 and 4). To this unique feature, one can add the ability to easily exchange liquids from those containing fluorine as the heaviest atom (e.g. C_3F_8) to those containing also iodine or bromine (CF_3Br, CF_3I). For targets like these, the expected WIMP and neutron induced bubble-nucleation rates can be radically different, a signature that could be exploited for WIMP discovery. The 2 kg prototype will also serve the purpose of studying the feasibility of building much larger chambers. New challenges will certainly arise during its operation, besides those already envisioned (for instance, Radon emanation from metallic parts in the in-

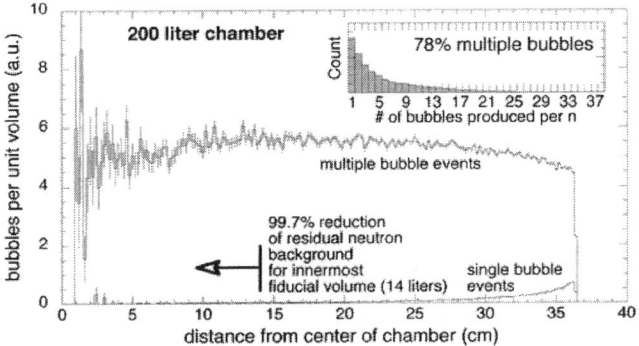

Figure 4. Simulated distribution of bubble sites and their multiplicity, expected from a typical underground neutron spectrum (LNGS). The probability of having a fast neutron reach the center of a large volume of target liquid, produce a single bubble and exit without further interaction is seen to be extremely small. This self-shielding effect provides an additional background rejection mechanism for the innermost fiducial mass, already considerable (∼30 kg) in a 200 liter chamber (∼4 kg for 50 l). WIMPs should produce a uniform distribution of singles: irreducible alpha backgrounds can do the same, but with a markedly different rate as a function of temperature [2,6].

ner vessel can give rise to a recoiling-daughter background [4], but prospects based on BOREXINO measurements of cleaned steel parts are reassuring at $< 5\mu Bq/m^2$) [7]. However, rapid upscaling to the ton or even multi-ton target mass regime seems feasible, in view of the simplicity of the method, optimal choice of target nuclei and excellent *intrinsic* background rejection ability.

References

1. L. Bond *et al.*, *Bubble Chambers with Sensitivity to WIMPs*, Proc. *Topics in Astroparticle and Underground Physics* (**TAUP 03**) University of Washington, Seattle, Washington, 2003. *Nucl. Phys. B (Proc. Suppl.)* in press.
2. J.I. Collar *et al.*, Phys. Rev. Lett. **85**, 3083 (2000) (astro-ph/0001511); J.I. Collar *et al.*, New J. Phys. **2**, 14.1 (2000) (http://www.njp.org).
3. M.A. Grolmes and H.K. Fauske, Proc. 5th Intl. Heat Transfer Conf., Tokyo 1974; M.G. Buivid and M.V. Sussman, Nature 275 (1978) 203; P. Reinke, Exp. Heat Transfer 10 (1997) 133; P. Reinke, *Surface Boiling of Superheated Liquid*, Ph.D. Thesis no. 11598, Swiss Federal Institute of Technology, Zurich, Switzerland 1996.
4. J.I. Collar *et al.*, in preparation, to be submitted to Phys. Rev. Lett.
5. S.A. Pozzi *et al.*, Nucl. Instr. Meth. **A513** (2003) 550.
6. M. Barnabe-Heider *et al.*, (hep-ex/0311034).
7. Borexino collaboration, Astrop. Phys. **18**, 1 (2002)

MIMAC-HE3: A NEW DETECTOR FOR NON-BARYONIC DARK MATTER SEARCH

DANIEL SANTOS

Laboratoire de Physique Subatomique et de Cosmologie (CNRS-IN2P3/Université Joseph Fourier) 53, Av. des Martyrs
38026 Grenoble, France

EMMANUEL MOULIN

Laboratoire de Physique Subatomique et de Cosmologie (CNRS-IN2P3/Université Joseph Fourier) 53, Av. des Martyrs
38026 Grenoble, France

The project of a micro-TPC matrix of cells of 3He for direct detection of non-baryonic dark matter is presented. The privileged properties of 3He for this detection are highlighted. The double detection: ionization – projection of tracks is explained and its rejection evaluated. The specific capabilities of this project with respect to other experiments are mentioned and its complementarities concerning the supersymmetric phenomenology explicitly showed.

1. Why ^{3}He?

In the last years, our work on ^{3}He as a target for detecting WIMPs allowed us to confirm its privileged properties for direct detection [1].

These properties can be enumerated as follow: i) its fermionic character opens the axial interaction with fermionic WIMPs, as the neutralinos, ii) the extremely low Compton cross section reduces in several order of magnitude the natural radioactive background with respect to other targets, iii) the high neutron capture cross section gives a clear signature for neutron rejection, iv) its light mass allows a higher sensitivity to light WIMP masses than other targets, v) the elastic energy transfer is bounded to a very narrow range of energy (a few keV) allowing to have all the interesting events concentrated in a narrow energy range offering a high signal to noise ratio.

The extremely low Compton cross section and the possibility to detect events in the keV range (< 5.6 keV) have been demonstrated by the electron conversion 57Co detection recently reported [2]. The fact that electrons of 7 keV could be detected in the MACHe3 [3] prototype with the source emitting the 122 keV

gamma rays embedded in the ^{3}He is a clear demonstration of the virtual transparency of this medium to the electromagnetic radiation.
The work developed on the MACHe3 prototype concerning the simulations of the interaction of the electromagnetic radiation and cosmic particles and the rejection estimation based on the properties above mentioned [3,4] are applied to a new TPC detector offering the electron – recoil discrimination.

2. Micro-TPC

The micro temporal projection chambers with an avalanche amplification using a pixelized anode [5] present the required features to discriminate electron – recoil events by the double detection of the ionization energy and the track projection onto the anode. A schematic structure of the chamber is shown on fig1. In order to get the electron-recoil discrimination, the pressure of the TPC should be such that the tracks of the electrons having energies less than 6 keV could be well resolved with respect to those corresponding to the recoils of the same energy convoluted by the quenching factor.

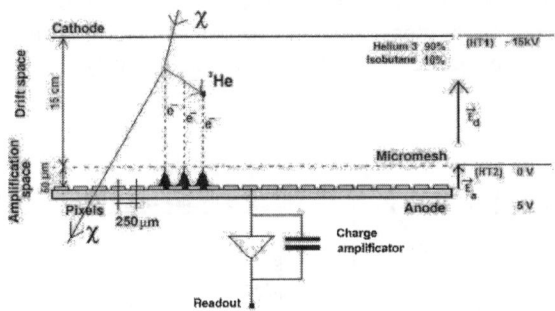

Figure 1. A typical cell chamber of MIMAC-He3 filled with gaseous ^{3}He. The pixellised anode allows a 2D projection of the recoil track.

Simulations have been done, as a function of the pressure, for electrons using Geant 4 [6] and for recoils using SRIM [7]. The results are shown on fig 2. The electrons produced by the primary interactions will drift to the grid in a diffusion process following the well known distribution characterized by a radius of $D \sim 200 \mu m \sqrt{L[cm]}$ where L is the total drift in the chamber up to the grid. This process has been simulated with Garfield [8] and the drift velocities estimated as a function of the pressure and the electric field. A typical value of $26\mu m/ns$ is obtained for 1kV/cm at a pressure of 1 bar.

To prevent confusion between the electron tracks projection and the recoil ones the total drift length should be limited to L~15 cm. This fact defines the elementary cell of the detector matrix and the simulations performed on the ranges of electrons and recoils suggest that with an anode of 250 μm the electron-recoil discrimination required can be obtained.

As mentioned above, the quenching factor is an important point that should be addressed to quantify the amount of the total recoil energy recovered in the ionization channel. No measurements of the quenching factor (QF) in ^{3}He have been reported. However, an estimation can be obtained applying the Lindhard calculations [9]. We have plotted on fig. 3 the estimated quenching factors for different pairs of nuclei (^{132}Xe, ^{74}Ge, and ^{3}He) of the same atomic and mass numbers as a function of the recoil energy. The Ge curve has been validated by a neutron induced measurement more than ten years ago [10]. The ^{3}He curve shows that we can expect up to 60% of the recoil energy in the range of interest (<6 keV) going through ionization.

In order to measure the QF in ^{3}He at such low energies, we have designed, at the LPSC laboratory, an ion source to accelerate the helium ions to be coupled to the micro-TPC chamber. The ions accelerated by the source will pass through a thin foil of polypropylene that will neutralize them before enter to the chamber. The measurement of the energy of the atoms of ^{3}He entering to the chamber will be made by a time of flight measurement. The determination of the QF is one important step of the project and it will be performed shortly.

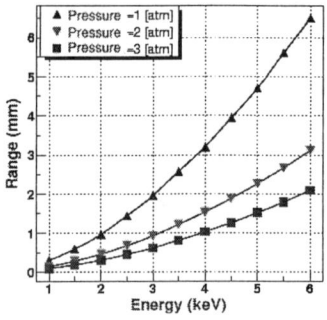 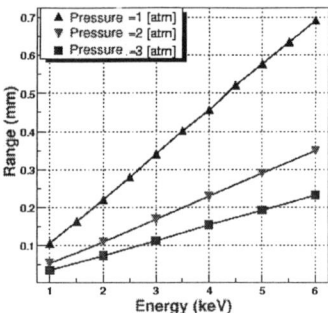

Figure 2. The left (resp. right) side plot shows the electron (resp. ^{3}He) simulated range versus the kinetic energy for 1, 2 and 3 atm pressure.

3. Double Detection: Ionization-Track Projection

In order to characterize the distribution of pixels on the anode produced by the different kind of trajectories we define the ratio between perpendicular symmetry axis of the pixel distribution (a/b) where a is the larger axis of the distribution.
We plot on fig. 4 the histograms of the ratio a/b distribution for the simulations performed at different energies of electrons and recoils at different pressures. An isotropic spherical emission of electrons and recoils at L~10 cm from the grid has been injected as the input of the simulation. For the recoils a very concentrated distribution around 1 is expected, and for the electrons a very wide one.
The rejection of events using the a/b ratio is a strong function of the energy and the pressure of the chamber, but even at 1 keV and 3 bar only a small number of the total events can be confused.

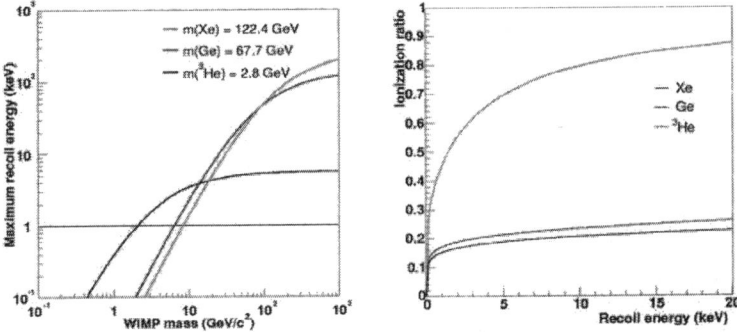

Figure 3. The plot on the left shows the maximum recoil energy of the target nucleus versus the incident WIMP mass for different target nuclei. In the case of ^{3}He, the energy range where all the searched events have to fall in is from the energy threshold up to 5.6 keV. For heavier nuclei, the energy range is much wider. The plot on the right side shows the ionization ratio predicted by Lindhard [9] for different sensitive media. For ^{3}He, up to 60% of the recoil energy is expected to be released in the ionization channel.

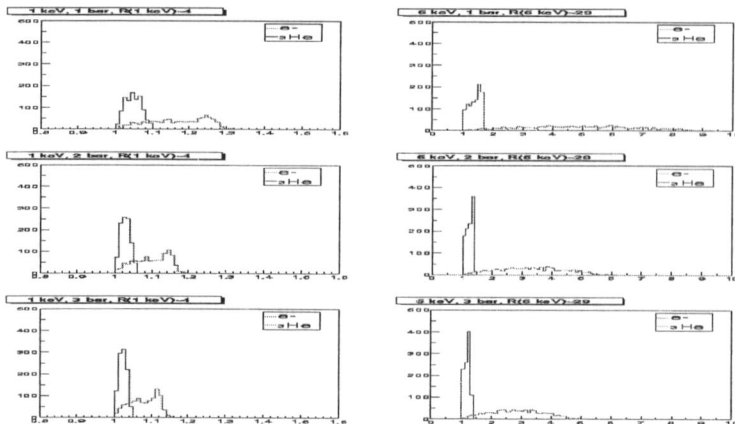

Figure 4. The histograms of the a/b distribution, for a recoil energy of 1 and 6 keV, and a pressure of 1, 2 and 3 bar.

4. Supersymmetry Models with MIMAC-He3

We have shown, in the last years, the complementarities of the axial interaction searches with those using mainly the scalar interaction [11].
These complementarities arise from the fact that the relative position of the different supersymmetric models with respect to the exclusion curves are very different when we plot the axial or scalar cross-section as a function of the neutralinos mass. On fig.5 (left) we show the models generated by the DarkSUSY code [14] accessible to MIMAC-He3 and on fig.5 (right) the same models in the scalar cross-section plot. The models are distributed over a wider range of cross-section, and some of them present an extremely small scalar cross-section. If the neutralino nature is described by these models it will be very difficult to detect them by scalar interaction.
A particular interesting point recently addressed in the frame of non universal gaugino masses [12,13] is that neutralino lighter masses than the LEPII limit could be allowed. In such scenarios neutralinos masses of ~10 GeV are proposed requiring an experimental challenge concerning the direct detection threshold. The fact that ^{3}He has a light mass allows us to expect to have substantial signal over the threshold of 500 eV.

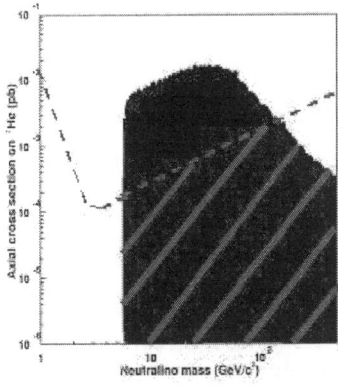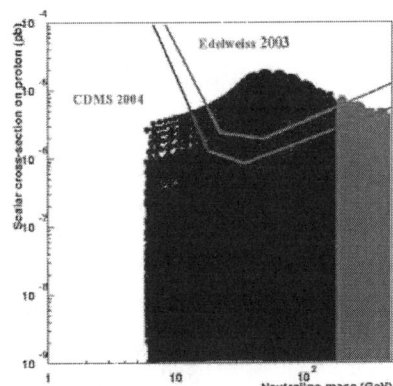

Figure 5 (left): Axial cross-section on ^{3}He versus the neutralino mass. The black points correspond to SUSY models allowed by collider as well as cosmological constraints. The projected exclusion curve is displayed for a 10^{-2} day^{-1} background level in a 10 kg ^{3}He based detector. A lot of models are above the exclusion curve. (right): Scalar cross-section on proton versus the neutralino mass. The black points are the models giving a rate in MIMAC-He3 higher than the 10^{-2} day^{-1} background level. The red points are the models giving a rate lower than 10^{-2} day^{-1}. The exclusion curves from Edelweiss(2003) and CDMS(2004) are shown as a reference. A lot of models accessible to MIMAC-He3, are far below the scalar detection prospects.

References

1. D. Santos, F. Mayet, E. Moulin et al. Proc. of the 4th International Workshop on the Identification of Dark Matter (IDM2002), York, Sept.2002.
2. E. Moulin et al, to be published in Phys. Rev. D.
3. F. Mayet et al., *NIM*. A455, 554 (2000).
4. E. Moulin et al, astro-ph/0309325.
5. I. Giomataris et al, NIM A376,29 (1996)
6. S. Agostinelli et al, NIM A506,250 (2003)
7. J.F. Ziegler, www.srim.org.
8. R. Veenhof, Garfield (1999), http://consult.cern.ch/writeup/garfield.
9. J. Lindhard, et al , *Mat. Fys. Medd. K. Dan. Vidensk. Selsk.*. 33, 10 (1963).
10. T. Shutt et al, *Phys. Rev. Lett.* 69, 3425 (1992).
11. F. Mayet et al., *Phys. Lett*. B538, 257 (2002).
12. G. Bélanger et al, hep-ph/021227.
13. A. Bottino et al, hep-ph/0212379.
14. P. Gondolo et al, Dark SUSY, astro-ph/0406204.

THE ULTIMA PROJECT: ULTRA-LOW TEMPERATURE INSTRUMENTATION FOR MEASUREMENTS IN ASTROPHYSICS

YU. M. BUNKOV, C. B. WINKELMANN AND H. GODFRIN

Centre de Recherches sur les Très Basses Températures, CNRS,
38042, Grenoble cedex 9, France

We describe our project of applying superfluid ^{3}He as an extremely sensitive detector for the search of Dark Matter. Different prototypes of this detector have been developed and used in our Grenoble facility, demonstrating a real potential of ^{3}He as a sensitive medium with unique properties for direct detection of non-baryonic dark matter. The new project will associate the efforts of the ultra-low temperature laboratories of Grenoble, Helsinki and Kyoto, with theoretical support from groups in Moscow and Grenoble.

1. Introduction

A substantial body of astrophysical evidence supports the existence of non-baryonic Dark Matter (nBDM) in the halo of our galaxy, in particular in the form of new, yet undiscovered, weakly interacting massive particles. One of the leading candidates is the neutralino predicted by supersymmetric extensions of the Standard Model of particle physics. Direct observation of nBDM is one of the main challenges of modern astrophysics.

The most performing detectors for direct detection are currently based on Ge and Si crystals[1,2] and explore the spin-independent interaction channel with the neutralino. While these experiments are now achieving sufficient sensitivity for probing the first supersymmetric models, a substantial part of the supersymmetric parameter space remains out of reach even of the most optimistic projections. Direct detection experiments are mainly limited by contamination due to the neutron background and to the target material's intrinsic radioactivity.

We propose a direct detection method based on the bolometric detection of nBDM using superfluid ^{3}He at ultra-low temperatures ($\sim$100 μK) as the target material. The diffusion of a neutralino on a ^{3}He nucleus is expected to release a maximum energy of about 5.6 keV, rather independently on the neutralino mass. In ^{3}He, the background contamination due to neutrons is strongly suppressed owing to the high cross-section of the nuclear capture reaction $n + {}^3\text{He} \rightarrow {}^3\text{H} + p + 764$ keV. An incident neutron will therefore produce a large energy release, well above the expected energy range for the neutralino scattering. In addition, the absence of free electrons makes ^{3}He rather insensitive to γ-ray contamination.

Furthermore, the fermionic nature of the ^{3}He nucleus allows the spin-dependent interaction channel with the neutralino. As discussed by Mayet et al.[3,4], a large bolometric detector based on ^{3}He could be sensitive to the neutralino within a wide class of supersymmetric models which are beyond the reach of current detectors.

2. Superfluid ^{3}He at Ultra Low Temperatures

^{3}He, together with its isotope ^{4}He, is the only material to remain liquid down to the lowest temperatures, under moderate pressures. It therefore presents the considerable advantage over crystal detectors not to display any coherent recoil of the material after a particle impact. Furthermore, the purity of ^{3}He at ultra-low temperatures is virtually absolute, since no other material can dissolve in it at these temperatures, not even ^{4}He.

Superfluid ^{3}He at ultra-low temperatures is essentially a renormalized quantum vacuum containing a dilute gas of quasiparticles with energy of about $\Delta = 10^{-7}$ eV each. Owing to the isotropic gap of ^{3}He in it's superfluid B phase, the heat capacity decreases exponentially with T^{-1}. At 130 μK, the thermal enthalpy of a detector cell containing 0.1 cm^3 of ^{3}He is therefore as low as a few hundred keV.

A direct method of thermometry of the superfluid is achieved by measuring the damping of a Vibrating Wire Resonator (VWR) immersed in the superfluid. A VWR is a fine superconducting wire bent into semi-circular shape and oscillating perpendicularly to it's plane (Fig. 1). The damping of the mechanical resonator is dominated by the collisions with the surrounding quasiparticles of the superfluid[5]. The damping factor of the VWR provides therefore a very sensitive and direct probe of the superfluid's temperature since no thermal surface impedance is involved.

3. Principle of detection

A schematic view of a superfluid ^{3}He bolometer is shown in Fig. 1. The target ^{3}He is confined into a copper cell of a fraction of a cm^3 which is in thermal contact with the surrounding ^{3}He bath through a small orifice in one of the cell walls[6]. Energy deposited to the superfluid inside the cell from whatever source (for instance by an incident particle) generates an excess of quasiparticles. The increase of the quasiparticle density (determined by the volume of the detector) increases the damping of the Vibrating Wire (which is monitored continuously by measuring the amplitude of the wire motion at the VWR's mechanical resonance frequency). The VWR response to temperature variations is relatively fast, and the signal rises within a second. This sharp rise is followed by a slow recovery as the excess quasiparticles 'leak out' through the detector orifice. The time constant for the recovery is determined simply by the size of the detector orifice. The detector is therefore particularly adapted for the detection of rare events of small energy.

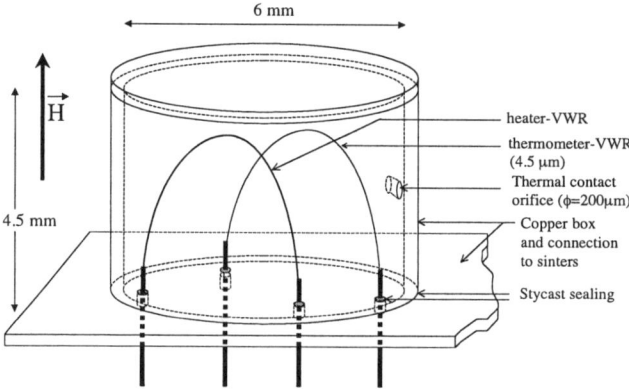

Figure 1. Bolometric cell for particle detection using superfluid ^{3}He as the target material. The fine VWRs are sensitive thermometers in the submilliKelvin range. The superfluid inside the cell is in weak thermal contact with the outer ^{3}He heat bath via the small orifice. A second VWR in the cell can be used for calibrating the bolometric cell (see text).

The amplitude of the signal is proportional to the energy deposited within the box. In order to establish an accurate relation between the increase of the VWR damping and the deposited energy, it is essential to calibrate the detector. Owing to the large thermal boundary resistance between a solid surface and the superfluid ^{3}He, it is difficult to transfer heat to the superfluid (i.e. to create quasiparticles) by conventional Joule heating. However, it is possible to use a second VWR present for heating by driving it with a large excitation current at its mechanical resonance and heating thus the liquid via the momentum transfer from the heater VWR to the quasiparticles. Measuring the voltage component in phase with the current yields directly the dissipation, thus providing an accurate way to apply known amounts of energy into the superfluid. We have used continuous and pulsed heating techniques in order to achieve a calibration of the detector[7]. These measurements, which also constitute the basis for the development of a sub-milliKelvin temperature scale, provide us with an accurate energy calibration over the range 10 keV to 1 MeV.

4. Previous results

In Grenoble, we have designed and tested three generations of superfluid ^{3}He detector prototypes. In a first experience, the energy release by the nuclear neutron capture reaction was observed[8]. In the following years, the detection sensitivity could be lowered by an order of magnitude, which allowed to identify the peak due to cosmic muons crossing the cell, at about 60 keV, in agreement with numerical estimations[9].

In recent measurements on a detector prototype of three adjacent bolometric cells, the detection threshold was lowered to the keV level[10]. In the same experiment, the coincident detection of muons in two or three cells simultaneously allowed to illustrate the possibility of efficient muon background rejection by a large matrix of bolometric cells. Furthermore, in one of the three detector cells, we embedded a small amount of radioactive ^{57}Co. The decay of ^{57}Co produces γ-rays of 122 and 136 keV as well well as electron emission lines, mainly at 7 and 14 keV. These low energy events are clearly detected, and are well above the threshold (Figs. 2 and 3). The latest detector prototype therefore already displays the required sensitivity for observing bolometrically the neutralino signal.

5. ULTIMA outlook

The potential of a superfluid ^{3}He detector using a mass of 10 kg target material has been analyzed by Mayet et al.[4]. We are currently constructing a nuclear demagnetization refrigerator of a new generation, in order to cool down to 100 μK a target mass of a few hundred grams. This new detector will have to be tested in an underground laboratory, since at the earth surface the detector is largely saturated by muons.

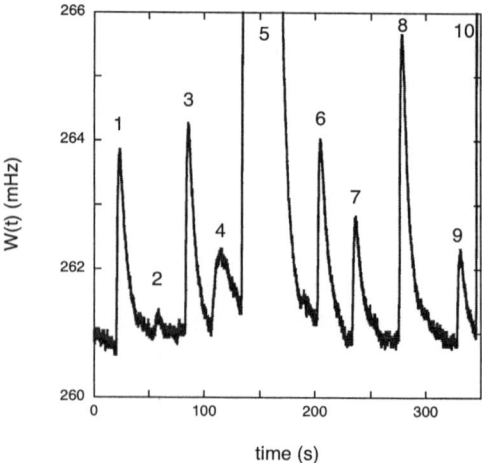

Figure 2. VWR damping as a function of time. Low energy electrons (events marked 1, 3, 6, 7 and 9) of 7 and 14 keV observed in the cell containing the ^{57}Co source. The high energy events (5 and 10) are most likely due to cosmic muons. Energy depositions below 1 keV can be distinguished (2).

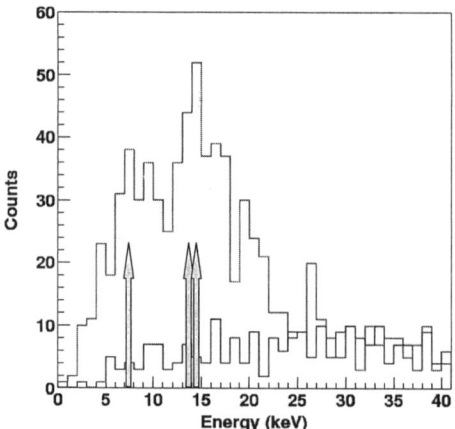

Figure 3. Detection spectrum of the main low energy electron emission lines (arrows) in the cell containing the ^{57}Co source (upper line) and in a neighbouring cell without source (lower line).

A final version of a detector looking for nBDM will have to integrate a method of parallel discrimination of the γ-ray background. This issue is still under study, but the irradiation dependent scintillation rate of helium could provide an interesting method for fiurther discrimination of incident γ-rays and muons from nuclear recoils.

An alternative method of thermometry, based on Nuclear Magnetic Resonance (NMR) measurements of the superfluid ^{3}He, is also under study. Superfluid ^{3}He-B responds to a magnetic field gradient parallel to the main applied magnetic field by separating into two domains. In one of these, the equilibrium magnetization of the superfluid is deflected by an angle $\theta_L = 104$ degrees, producing thus a Homogeneous Precession Domain (HPD) of the magnetization[11]. The dissipative part of the large NMR signal associated to a HPD is a sensitive probe of the quasiparticle density. Such a method of thermometry would allow to decrease the response times of the detector to about 1 μs.

References

1. A. Benoît et al., *Phys. Lett. B* **545**, 43 (2002).
2. D. Abrams et al., *Phys. Rev. D* **66**, 122003 (2002)
3. F. Mayet et al., *Nucl. Instr. and Meth. A* **455**, 554 (2000).
4. F. Mayet, D. Santos, Yu. M. Bunkov, E. Collin, H. Godfrin, *Phys. Lett. B* **538** 257 (2002).
5. S. N. Fisher, A. M. Guénault, C. J. Kennedy, and G. R. Pickett, *Phys. Rev. Lett.* **63** (1989) 2566.
6. D. I. Bradley, Yu. M. Bunkov, D. J. Cousins, M. P. Enrico, S. N. Fisher, M. R. Follows, A. M. Guénault, W. M. Hayes, G. R. Pickett, and T. Sloan, *Phys. Rev. Lett.* **75**, 1887 (1995).
7. C. Bäuerle, Yu. M. Bunkov, S. N. Fisher, H. Godfrin, *Phys. Rev. B* **57**, 14381 (1998).
8. C. Bäuerle, Yu. M. Bunkov, S. N. Fisher, H. Godfrin, G. R. Pickett, *Nature* **382**, 332 (1996).
9. E. Collin, E. Moulin, C. B. Winkelmann, Yu. M. Bunkov, H. Godfrin, M. Krusius, and D. Santos, *submitted to Astroparticle Physics*.
10. C. B. Winkelmann, E. Moulin, Yu. M. Bunkov, J. Genevey, H. Godfrin, J. Macias-Pérez, J. A. Pinston, and D. Santos, "MACHE3, a prototype for non-baryonic matter search: keV event detection and multicell correlation", Proceedings of the XXXIXth. Rencontres de Moriond Exploring the Universe, La Thuile/Italy April 2004, to appear.
11. A. S. Borovik-Romanov, Yu. M. Bunkov, V. V. Dmitriev, and Yu. M. Mukharskii, *JETP Lett.* **40**, 1033 (1984).

R & D STATUS OF THE NEWAGE EXPERIMENT *

KENTARO MIUCHI, KAORI HATTORI, HIDETOSHI KUBO,
TSUTOMU NAGAYOSHI, HIRONOBU NISHIMURA, YOKO OKADA,
REIKO ORITO, HIROYUKI SEKIYA, ATSUSHI TAKADA,
ATSUSHI TAKEDA, TORU TANIMORI,

*Cosmic-Ray Group, Department of Physics, Graduate School of Science,
Kyoto University, Kitashirakawa-oiwakecho, Sakyo-ku, Kyoto, 606-8502, Japan
E-mail: miuchi@cr.scphys.kyoto-u.ac.jp*

NEWAGE (NEw generation WIMP search with an Advanced Gaseous tracking dEvice) is an experiment seeking for the WIMPs with a gaseous micro time projection chamber(μ-TPC). We have developed a prototype μ-TPC with an original two-dimensional imaging detector, or the μ-PIC and investigated its performance as a WIMP detector.

1. Introduction

Direction-sensitive detection of the WIMPs (Weakly Interacting Massive Particles) is said to be a very convincing signature of the halo dark matter. There has been an intense effort for the directional detection with gaseous detector (DRIFT project[1]). We have developed an advanced gaseous tracker, or the μ-TPC (micro time projection chamber) based on our original two-dimensional imaging detector, or the μ-PIC. The outstanding properties of our detector, which are 1) the μ-PIC is a fine pitch (400 μm) device and 2) the μ-TPC can detect real three-dimensional tracks, would provide advantages over the DRIFT projects. We, therefore, started the studies on its application for the WIMP detection[2].

---

*This work is supported by a grant-in-aid for the 21st century coe "center for diversity and universality in physics"; a grant-in-aid in scientific research of the japan ministry of education, culture, science, sports, technology.

2. μ-PIC and μ-TPC

We developed several μ-PICs of a practical size $(10 \times 10$ cm$^2)$[3,4]. The schematic structure of the μ-PIC is shown in the left panel of Figure 1. Orthogonal anode and cathode strips are formed with a pitch of 400 μm on the rear and front side of a thick(100μm) polyimide substrate, respectively. Because μ-PIC is made by the print circuit board (PCB) technology, large area detectors ($> 30 \times 30$ cm^2) are in principle made at a relatively low cost, which is an important feature for a WIMP detector. Another outstanding feature of the μ-PIC is that all structures for the multiplication and the readout are on one board and would provide a very stable and robust long term operation without the deformation. The gas multiplication occurs near an anode pillar and same amount of charges are read from corresponding anode and cathode strips. This is also important for a direction sensitive WIMP detector, because the three-dimensional recoil tracks would provide much more reliable and concrete results than the projected tracks do. Two-dimensional images as shown in the right panel of Figure 1 are detected by the μ-PIC. Two dimensional position resolution was measured to be 120 μm from the edge image.

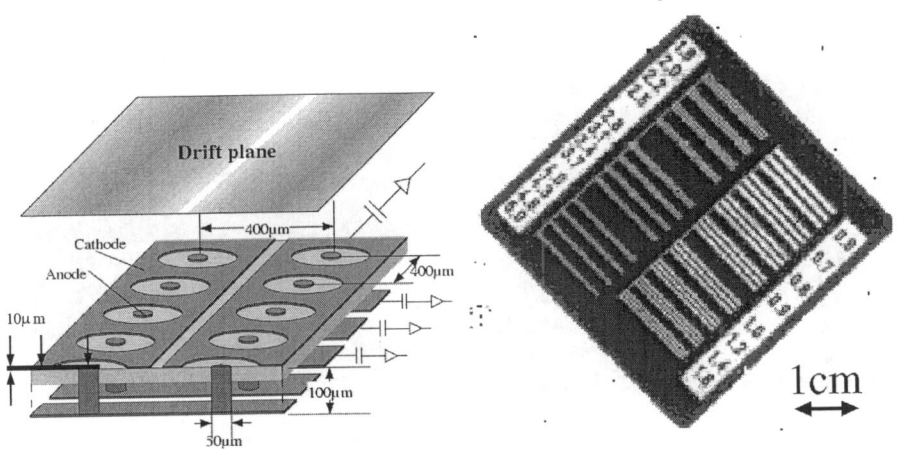

Figure 1. Schematic structure of the μ-PIC (left) and the a two-dimensional X-ray image detected by the μ-PIC (right).

One of the important, but often avoided, task we need to work on with any micro-structure detectors is the data acquisition of the many channels (more than 500 channels a for 10×10 cm^2 detector). We developed a

FPGA-based fast data acquisition system. Anode and cathode signals are discriminated into the LVDS-level digital signals in the preamplifier chip[5]. Then the anode-cathode coincidence within an internal clock(50 MHz) of a FPGA are taken. The anode and cathode positions (X and Y) and the timing (Z) are recoded for each coincidence. A track of a charged particle is consequently recorded as successive points. The data size of each point is 32 bit. We also record the waveform of the summed analog signals with a 8 bit 100 MHz flash ADC (FADC). This summed waveform is used to measured the deposited energy and also to know the track sense because these summed signals basically carry the Bragg curve shapes.

We developed a prototype μ-TPC (micro time projection chamber) with a detection volume of $10 \times 10 \times 10$ cm^3. A picture of the μ-PIC(left) on the mother board and a picture of the prototype μ-TPC is shown in the left and right panel of the Figure 2, respectively. A field cage of 10 cm drift length which consists of a drift electrode, nine copper wires of 0.2mm diameter, and glass fiber reinforced plastics(GFRP) rods was set on the μ-PIC. The wires are set around the field area of 15×15 cm^2 with 1cm pitch, which forms a uniform electric field in the detection volume. We set the μ-TPC in a 6mm-thick aluminum vessel of 20cm(height) $\times$ 60cm(diameter) and measured its performance.

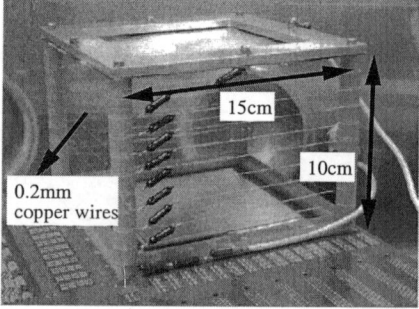

Figure 2. A picture of the μ-PIC(left) on the mother board(left) and a picture of the prototype μ-TPC(right)

3. Performance

Performance of the μ-TPC had been measured with a normal pressure gas flow in the most of the former development phase. Because the required gas pressure for the WIMP search is 0.05 - 0.2 atm, we operated the μ-TPC with low-pressure chamber gas. As this is a first measurement with a low pressure gas, we put the whole mother board which has resistances and decoupling capacitances without any out-gas suppression. As expected, the out-gas made the μ-TPC unstable below 0.1 atm. We, therefore, performed the measurement with a gas mixture of Ar-C_2H_6(10%) at 0.2 atm. We are going to work on the out-gas suppression in the near future.

We operated the μ-PIC(TOSHIBA S/N 040223-2) with a gas gain of about 3000 and irradiated the μ-TPC with neutrons from ^{252}Cf source. Typical recoil proton tracks with energies between 100 keV and 300 keV are shown in left panel of Figure 3, while one of the carbon track candidates is shown in the right panel of Figure 3. Track length known from the digital track data and the energy measured from the FADC waveform reconcile with SRIM2003 [7] calculations. We are going to analyze these data statistically in order to measure the energy resolution and tracking spatial resolution.

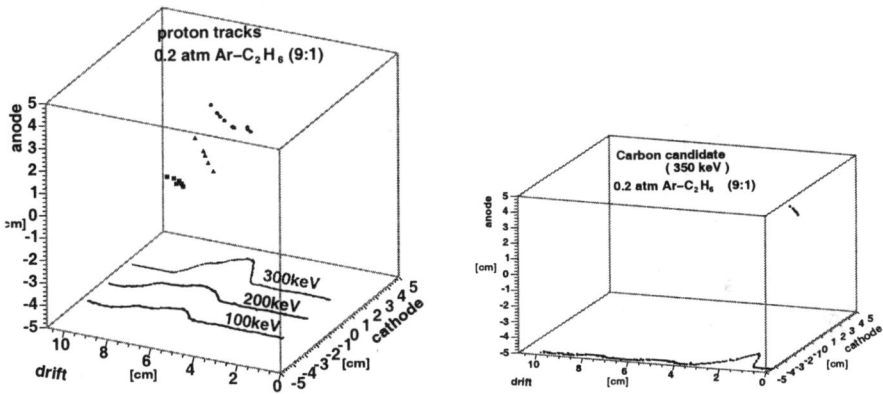

Figure 3. Detected proton(left) and carbon tracks. Proton tracks of 100 keV(square), 200 keV(triangle) and 300 keV(circle) are shown.

4. Neutron background to the gaseous detectors

Neutron background is known to be a serious problem in the next generation WIMP search experiment. In fact, the comparison of the simulation codes for the neutron background calculation was intensely discussed in this conference[8]. Gaseous detectors have an advantage for the neutron background measurement and monitoring because the neutron flux inside of the shield is easily monitored by a same type detector (μ-TPC) with an addition of neutron targets to a WIMP target gas: *i.e.* hydrocarbon for the fast neutron and ^{3}He for the thermal neutron. We estimated the fast neutron background at Kamioka Observatory with the WIMP and the neutron target gas (CF_4 and C_2H_6, respectively). As we need to demonstrate the performance of this new device before developing a large volume ($\sim$ 1m^3) WIMP detector, we are going to measure the neutron background at Kamioka Observatory with a small-size (30×30×30cm^3) prototype. We, therefore, simulated the response of a μ-TPC with a detection volume of 30×30×30cm^3 filled with 0.3 atm gas without any shield at Kamioka Observatory. We assumed an exponential neutron spectrum above 0.5 eV scaled to the measured flux of 1.15 ×10^{-5}nn · cm^{-2} · s^{-1} and used the neutron scattering cross section data of the ENDF/B-VI[9]. We found that the neutron background can be measured with a short running time($\sim$ 100 days) with a prototype detector.

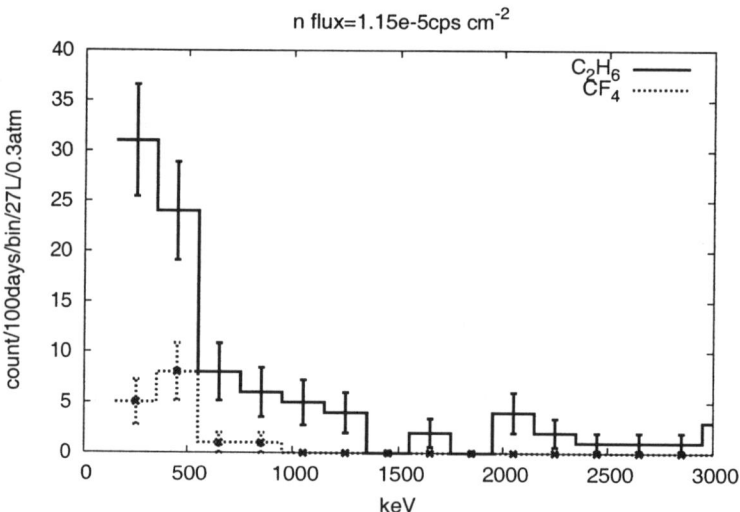

Figure 4. Expected neutron background without any shield at the Kamioka Observatory. The solid and the dotted lines represent the spectra with C_2H_6 and CF_4 target gas, respectively.

5. Prospects

A larger area (30×30 cm^2) μ-PIC is now being manufactured and is expected to be ready at the end of 2004. We are going to develop a prototype μ-TPC with a volume of 30×30 ×30cm^3 and measure its performance. We are going to measure the gamma-ray rejection power, energy and tracking resolution, and absolute efficiency by the end of 2005 and we hope to start the underground run with a neutron background measurement in 2006.

6. Summary

We started investigating the perfromance of the μ-TPC for the direction sensitive WIMP search. We operated the μ-TPC with low pressure (0.2 atm) chamber gas and detected low energy proton and carbon recoil tracks. We hope to start the underground run in 2006 after a development and performance study of a 30 cm size μ-TPC.

References

1. N. Spooner et al., these proceedings and the references therein.
2. T. Tanimori, et al., Phys. Lett. B 578 (2004) 241
3. T. Nagayoshi, et al., Nucl. Instr. Meth. A 513(2003)277,
4. T. Nagayoshi, et al., Nucl. Instr. Meth. A 525(2004)20, A. Takeda, et al., IEEE Trans. Nucl. Sci, in press.
5. R. Orito, et al. IEEE Trans. Nucl. Sci., 51 (2004) 1337
6. K. Miuchi, et al., IEEE Trans. Nucl. Sci., 50 (2003) 825, K. Miuchi, et al., Nucl. Instr. Meth. A in press.
7. J. F. Ziegler, J. P. Biersack and U. Littmark, SRIM - The Stopping and Range of Ions in Matter, Pergamon Press, New York, 1985.
8. V. Kudryavtsev, et al.
9. ENDF/B-VI the Evaluated Nuclear Data File/Version B, the U.S. national nuclear data file.
10. K. Miuchi, et al. Nucl. Instr. Meth. A 517(2004)219, T. Tanimori, et al. Nucl. Instr. Meth.

DARK MATTER SEARCH WITH DIRECTION SENSITIVE SCINTILLATORS

H. SEKIYA

Department of Physics, School of Science, Kyoto University
Kitashirakawa, Sakyo, Kyoto 606-8502, Japan
E-mail: sekiya@cr.scphys.kyoto-u.ac.jp

M. MINOWA, Y. SHIMIZU, W. SUGANUMA

Department of Physics, School of Science, University of Tokyo
7-3-1 Hongo Bunkyo-ku, Tokyo, 113-0033

Y. INOUE

International Center for Elementary Particle Physics, University of Tokyo
7-3-1 Hongo Bunkyo-ku, Tokyo, 113-0033

We have carried out the dark matter search with a 116g direction-sensitive stilbene crystal in Kamioka Observatory. With the crystal fixed to the earth, we searched the modulation of the light output. No modulation signal was found due to the small size of the detector crystal and the higher background rate yet to be eliminated. However, it demonstrated the effectiveness of the method of direction sensitive search for the dark matter with an implementation of the anisotropic organic scintillation crystal.

1. Introduction

The most convincing signature of the WIMPs appears in the direction of nuclear recoils induced by WIMPs[1]. Although studies on detecting the signature by measuring the recoil directions have been carried out ever since it was indicated to be a reliable method, no dark matter search had been conducted with direction sensitive detector because of its difficulties. Recoil energy is only a few tens of keV and the track length of recoil nucleus should be short. Consequently, low pressure TPC have been studied as the directional WIMP detector principally, such as DRIFT and NEWAGE[2]. However, realistic experiments with gaseous detectors are very challenging because they are required the large fiducial volume and the great stability

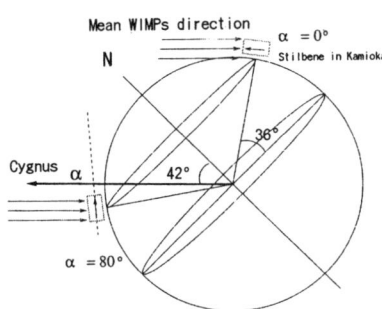 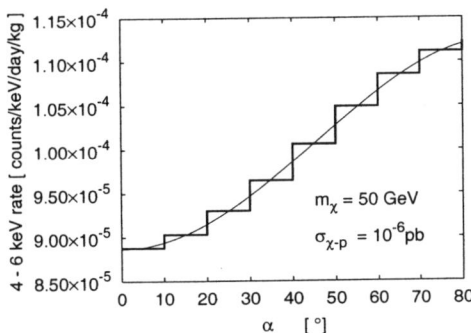

Figure 1. Left: Schematic drawing of the experimental approach mentioned in the text. Right: The expected event rate for 4-6 keV as a function of α. The thin line shows $A - S\cos 2\alpha$, where $S = 1.20 \times 10^{-5}$ counts/keV/day/kg and $A = 1.01 \times 10^{-4}$ counts/keV/day/kg in this case. The parameters that we used in the calculation are $\rho_0 = 0.3$ GeV/cm^3, $v_0 = 220$ km/sec, WIMP-proton spin independent cross section $\sigma_{\chi-p} = 10^{-6}$ pb, and $m_\chi = 50$ GeV.

with very fine resolution. Therefore, it is significant to explore alternative experimental approach.

It is known that scintillation efficiency of organic crystals to heavy particles depends on the direction of the particles with respect to the crystallographic axes. This property makes it possible to propose a WIMP detector sensitive to the recoil direction of the nucleus[3].

We measured the carbon recoils in a stilbene crystal for recoil energies of 30 keV to 1 MeV and shown that the scintillation efficiency does vary by 7% depending on the direction of the recoil carbon with respect to c' axis[4].

Then, we estimated the response to WIMPs when stilbene crystals installed in Kamioka[5]. As illustrated in Fig. 1, a suitable arrangement for the stilbene crystal is to fix the detector with the c' axis in parallel to the horizontal plane and towards the North assuming the WIMP halo is an isothermal sphere. In that case, the mean incident angle of the WIMP with respect to c' axis, α, varies about 80° within a sidereal daily period.

The expected event rate for 4-6 keV region as a function of α calculated by Monte Carlo method is shown in Fig. 1. As indicated, the variation can be well fitted by a function $A - S\cos 2\alpha$.

As the next step, we have performed a pilot experiment at Kamioka to prove the feasibility of this method. In this paper, we report on the measurement and its results.

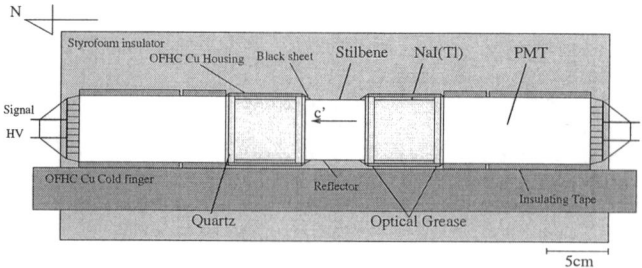

Figure 2. Schematic view of the detector setup.

2. The Experimental Setup

The schematic view of the detector assembly is shown in Fig. 2. The $\phi 50$ mm $\times$ 50 mm (116 g) cylindrical stilbene crystal is viewed by two Hamamatsu R8778MOD low background PMTs through two Horiba low background NaI(Tl) active shields. Self coincidence of two PMTs are required and both PMTs are cooled down at about $-7°C$ to reduce dark current further.

The detector assembly is shielded with 10cm OFHC copper, 15 cm Lead, and 20cm polyethylene. The EVOH sheets are formed into air tight bags filled with nitrogen gas for purging the radon gas.

The whole setup is laid with the c' axis of the stilbene crystal parallel to the north-south direction.

3. Measurement Results

With the detector system, we started the measurement in October 25, 2003 and it was halted in December 11, 2003[5]. The obtained energy spectrum with the stilbene is shown in Fig. 3.

The background event rate is as high as 2000 counts/keV/day/kg for 4-6 keV region, however, the event rate of WIMPs should change in a cycle of one sidereal day —i.e. 23.934 hours— independent of halo models. Therefore, in order to search the modulation signal in frequency domain, we derived the power spectrum of the time data of the event rate for 4-6 keV region during the measurements[5] using Lomb's periodogram method[6]. Fig. 4 shows the power spectrum for the frequency interval 0 - 0.3 hour^{-1}.

The signal with the frequency of 1/23.934 hours cannot be discerned from Fig. 4. That means the isotropic "white noise" events dominates the high rates of background events which should be eliminated.

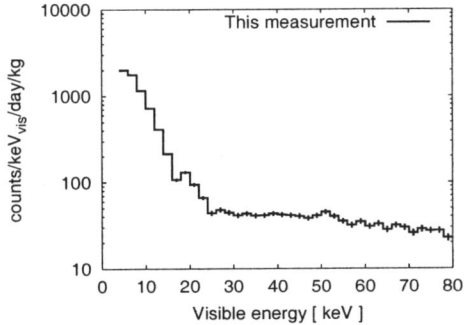

Figure 3. Low energy spectrum obtained with the stilbene crystal.

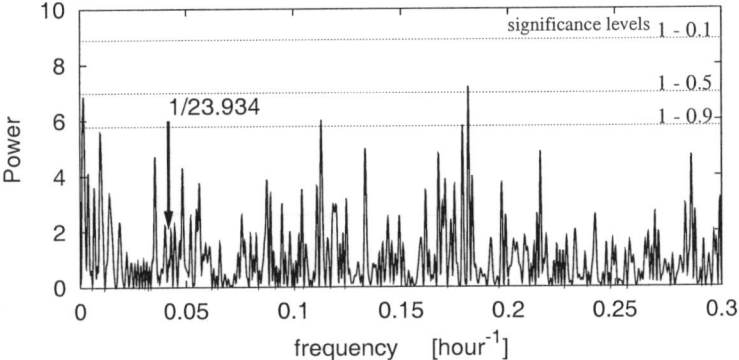

Figure 4. Power spectrum of the time data of event rate for 4-6 keV in the stilbene.

For all that, we derived limits on cross sections from this isotropic results. Fig. 5 shows the measured event rate for 4-6 keV as a function of α. As indicated in Fig. 1, the event rate should vary as $A - S\cos 2\alpha$, and both A and S are in proportion to $\sigma_{\chi-p}$. Accordingly, from the measured unmodulated part, A, conventional limits can be derived, and from the measured modulation amplitude, S, limits of direction sensitivity can be derived. The obtained limits on $\sigma_{\chi-p}$ are shown in Fig. 6. The limit is far looser than the contemporary non-directional limits, however, it is the limit from the directional signature of WIMPs. In addition, as the directional limit is derived from the "signal of WIMPs", the better limits will be obtained with the higher statistics.

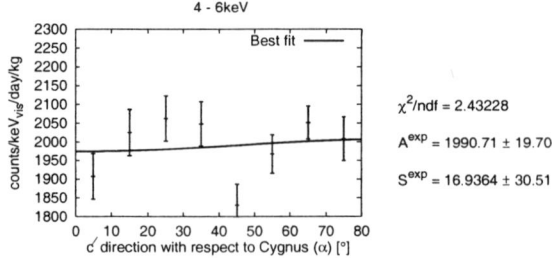

Figure 5. The measured event rate for 4-6 keV as a function of α. Expected $A - S \cos 2\alpha$ is fitted.

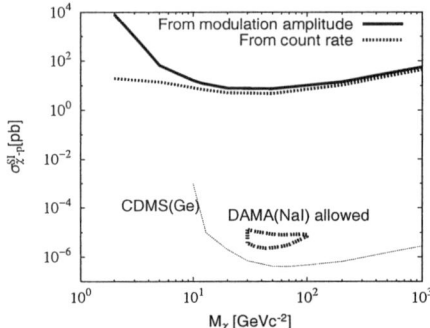

Figure 6. The obtained $\sigma_{\chi-p}$ limits as a function of WIMP mass M_χ.

Fig. 7 indicates the background and exposure dependence of the achievable limits on $\sigma_{\chi-p}$. We see from Fig. 7 that the directional limit will be comparable to contemporary non-directional limit if we achieved the background level as low as 10^{-3} counts/keV/day/kg. Stilbene crystals could potentially detect the robuster WIMP signal arising from the earth's rotation in our galaxy as compared with the annual modulation signal.

4. Discussions and Prospects

It is obvious that rather high background rate due to the radioactivity in PMTs limits the sensitivity. The small light yield of the stilbene is another essential problem. Therefore, in order to overcome the difficulties, highly radio-pure, high quantum efficiency, high gain photon detector is indispensable. In that respect, we focused on Avalanche photodiodes(APD).

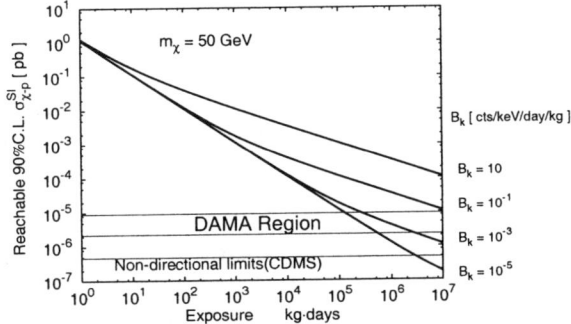

Figure 7. Background rate dependence of the expected sensitivity.

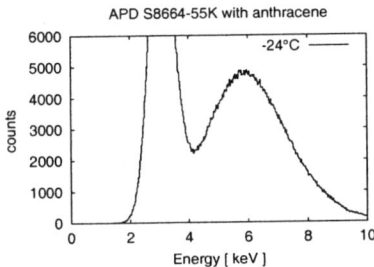

Figure 8. An example of the performance of the HAMAMATSU S8664-55K. 5.9 keV X-ray spectrum measured at $-24°C$ with anthracene crystal($15 \times 15 \times 15$ mm^3).

Fig. 8 shows an example of the performance of an APD (HAMAMATSU S8664-55K) with anthracene crystal. 5.9keV X-rays from ^{55}Fe are clearly resolved. This result suggests that APDs will be promising devices for organic scintillators and will pull up the potential of the direction sensitivity.

References

1. D.N. Spergel, *Phys. Rev.* **D 37**, 1353 (1988).
2. N.J.C. Spooner – these proceedings; K. Miuchi – these proceedings.
3. Y. Shimizu et al., *Nucl. Instr. and Meth.* **A 469**, 347 (2003); R. Bernabei et al., *Eur. Phys. J.* **C 28**, 203 (2003); N.J.C. Spooner et al., *International Workshop on Identification of Dark Matter*, (World Scientific 1997), p481.
4. H. Sekiya et al., *Phys. Lett* **B 571**, 132 (2003).
5. H. Sekiya et al., astro-ph/0405598, To appear in the proceedings of the 5th Workshop on "Neutrino Oscillations and their Origin" (NOON2004)
6. Y. Ramachers, *Astropart. Phys* **19**, 419 (2003).

PERFORMANCE OF A SCINTILLATING SAPPHIRE BOLOMETER FOR THE ROSEBUD EXPERIMENT

J. AMARÉ, B. BELTRÁN, J.M. CARMONA, S. CEBRIÁN, E. GARCÍA,
I.G. IRASTORZA,* H. GÓMEZ, G. LUZÓN, M. MARTÍNEZ, A. MORALES[†]
J. MORALES, A. ORTIZ DE SOLÓRZANO, C. POBES, J. PUIMEDÓN,
A. RODRÍGUEZ, J. RUZ, M.L. SARSA, L. TORRES[‡] J.A. VILLAR

Laboratorio de Física Nuclear y Altas Energías,
Facultad de Ciencias, Universidad de Zaragoza,
50009 Zaragoza, Spain
E-mail: Lidia.Torres@unizar.es

N. CORON, G. DAMBIER, J. LEBLANC, P. DE MARCILLAC

Institut d'Astrophysique Spatiale (IAS)
Bâtiment 121
91405 Orsay Cedex, France

The performance of a scintillating sapphire bolometer has been estimated and its suitability for dark matter detection has been studied. Characterization of the detectors to be used in the ROSEBUD (Rare Objects SEarch with Bolometers UndergrounD) experiment has shown high discrimination power (electron versus nuclear recoils) down to about 10-15 keV. The light-heat anticorrelation observed in the γ events is used to improve significantly the energy resolution and the particle discrimination threshold. Prospects for future runs at the Canfranc Underground Laboratory, planned for the coming months, are also presented.

1. Introduction

The identification of the nature of the events in bolometric experiments by means of the simultaneous measurement of two different magnitudes related to the deposited energy has been successfully used in the rejection of background, for example in the study of very rare decays of nuclei [1,2,3]

---

*Present address: DAPNIA, CEA, Saclay, France and CERN, Geneva, Switzerland.
†Deceased.
‡CEE fellow in the Network on Cryodetectors, under contract HPRN-CT-2002-00322, presently at the Dipartimento di Fisica dell'Università di Milano-Bicocca, Italy.

and in dark matter experiments, both for charge-heat [4,5] and light-heat measurement [6,7]. In particular, the sensitivity of bolometric experiments to WIMPs (Weakly Interacting Massive Particles) has been enhanced by this feature allowing them to give some of the best results in these searches [4,5,7].

The ROSEBUD Collaboration is presently working on the development of light-heat discrimination techniques. Adequate scintillation yield at low temperature has been observed in several materials at IAS [8], which provides a high flexibility in the choice of absorbers. Scintillation produced by an event is particle dependent : γ and β particles give higher scintillation yield than nuclear recoils, induced by neutrons or WIMPs and it is possible to disentangle γ/β events from nuclear recoils by measuring simultaneously the energy converted into heat and light.

A scintillating $CaWO_4$ bolometer prototype was operated at the Canfranc Underground Laboratory in a previous phase of the experiment [6]. In this paper we will report on a preliminary analysis of the recent progress achieved with a scintillating sapphire bolometer and its potential for a future WIMPs search underground.

2. Experimental set-up

The detector module consists of a double bolometer arrangement: a 50 g massive sapphire bolometer and an optical bolometer (a 0.25 g germanium disc that collects the light produced in the sapphire; see reference [9] for details), placed inside a reflecting silver-coated cavity, facing at each other. Both bolometers use NTD germanium thermistors as temperature sensors and are operated at a base temperature of 20 mK attained with a small dilution refrigerator. Details of double bolometer arrangement and dilution refrigerator have been published elsewhere [1,6,10].

When an interaction occurs in the scintillating crystal, the heat signal is drawn from the temperature increase of the sapphire absorber and the light signal from the heat produced by the scintillation photons absorbed in the optical bolometer.

Calibrations of both, germanium bolometer as light detector and sapphire bolometer, are performed with external gamma (^{57}Co, ^{60}Co, ^{137}Cs) and neutron (^{252}Cf) sources; also an internal ^{241}Am source is used as a low energy gamma source (^{237}Np recoils and alphas were blocked by a 50 microns thick copper foil). An internal ^{55}Fe X ray source placed at the back of the optical bolometer allows absolute calibration of energy collected in the form of light. Stability is monitored by using short square infrared pulses sent periodically to the bolometers.

3. Performance of the scintillating sapphire bolometer

The sapphire bolometer (labelled B213) had been previously used in the ROSEBUD dark matter experiment (phase I and II [10]) at Canfranc Underground Laboratory with single collection of heat. In these new runs heat and scintillation light have been simultaneously measured at IAS.

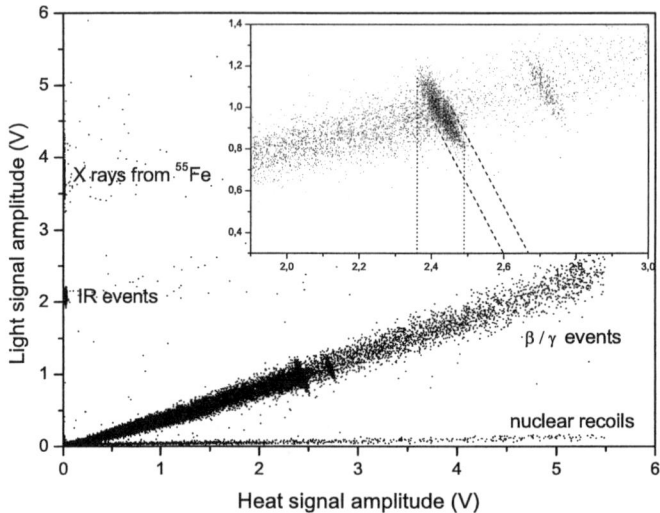

Figure 1. Light vs. heat discrimination plot for a γ calibration with ^{57}Co external source. The 122.1 keV and 136.5 keV γ lines from ^{57}Co are clearly seen at values around 2.4 and 2.7 V for heat signal amplitudes and the 59.5 keV from the internal ^{241}Am source is hardly distinguished from the underlying background at 1.2 V. The insert shows in detail the light-heat anticorrelation observed in the 122.1 keV and 136.5 keV lines. The different widths of the 122.1 keV line in the heat channel and in the new variable obtained through a rotation of light-heat axis are illustrated by the dotted and dashed lines, respectively.

In Fig. 1 we plot light vs. heat signal amplitudes obtained in a calibration with the external ^{57}Co source. Two bands of different slope are clearly distinguished, the γ/β events band, with greater light response, and the band of nuclear recoil events induced by environmental neutrons, with lower light output. A light yield of 12.7 keV/MeV for γ/β events and a ratio $(light/heat)_\gamma : (light/heat)_{rec}$ of about 16 at 122.1 keV have been estimated. The latter shows the excellent discrimination capability of sapphire (typical values for other scintillators are 10 and 12-15 for $CaWO_4$ and BGO, respectively).

The lines at 59.5, 122.1 and 136.5 keV, corresponding to γ events show anticorrelation between light and heat signals (see Fig. 1). This effect suggests the convenience of measuring the energy of an event through a new variable chosen in order to minimize the energy resolution. Since the anticorrelation slope is not exactly the same for the 59.5 keV line and the 122.1 and 136.5 keV lines, a compromise has been done to determine the rotation angle. When projecting on this new energy axis, an improvement of a factor ~ 2 is obtained in energy resolution (see Table 1).

Table 1. Energy resolution obtained before and after taking into account the light-heat anticorrelation.

Energy (keV)	FWHM before (keV)	FWHM after (keV)	Factor gained
59.5	1.6	1.1	1.5
122.1	3.7	1.6	2.3
136.5	3.3	1.6	2.1

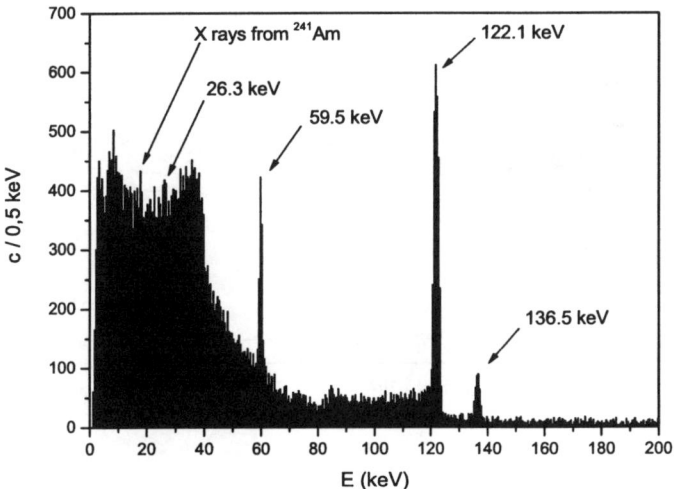

Figure 2. Energy spectrum obtained after considering the light-heat anticorrelation.

The appropriateness of this procedure is confirmed by the appearance of some more peaks at lower energies: the 26.3 keV γ peak and X-rays at ~18 keV from ^{241}Am (see Fig. 2). This has allowed us to improve the low energy calibration adding a fourth point (26.3 keV), and to assess the linearity of the detector within this energy range. The particle discrimination energy threshold has also improved after the anticorrelation analysis (see Table 2 and Fig. 3).

Table 2. Particle discrimination energy threshold at different C.L. before and after considering the anticorrelation effect.

C.L.	E_{th} before (keV)	E_{th} after (keV)
1σ	10.4	8.8
2σ	14.0	12.5
3σ	19.0	17.5

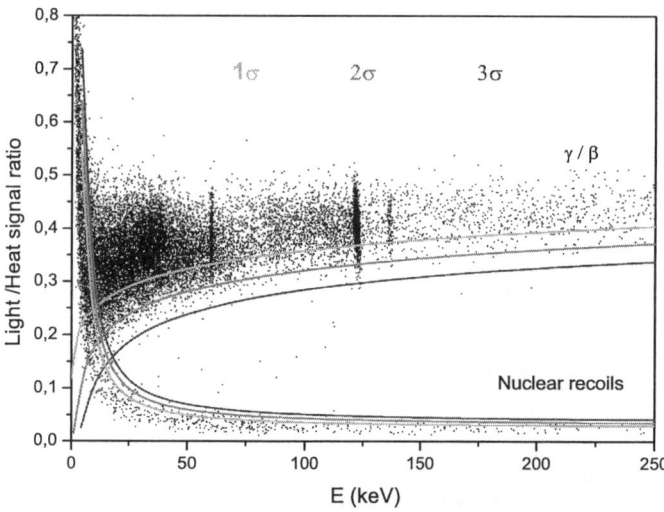

Figure 3. Discrimination plot obtained in a γ calibration with ^{57}Co. The lines delimit the γ/β and nuclear recoils zones at 1σ, 2σ and 3σ determined after separate calibrations with ^{57}Co and ^{252}Cf.

4. Conclusions and prospects for ROSEBUD

The consideration of the light-heat anticorrelation in the off-line analysis of data has led us to improve energy resolution, calibration at low energy and particle discrimination threshold. A full understanding of this effect can give us valuable information about the physics underlaying the interaction. For this reason, further research is in progress. The light-heat anticorrelation could be partly responsible for the poor resolution obtained with this bolometer in previous runs with collection of heat alone, where we had a FWHM of 3.2 keV at 122.1 keV.

The low particle discrimination energy threshold achieved with sapphire proves its suitability for dark matter experiments. We have estimated from background spectra obtained in previous runs at Canfranc (assuming γ/β background) the reduction that would be obtained with simultaneous collection of heat and light. For a 90% C.L. nuclear recoils acceptance band, a background rejection of 95% and 99.9% at 10 and 15 keV, respectively, is attained, reaching background levels of about 5 and 0.1 c/(keV kg day). Future runs underground will consist of simultaneous light and heat measurements on sapphire and BGO.

Acknowledgments

This work has been supported by the Spanish Commission for Science and Technology (CICYT, grant FPA2001-2437), by the French CNRS/INSU (MANOLIA and BOLERO projects) and the EU Network Contract HPRN-CT-2002-00322. The Canfranc Underground Laboratory is operated by the University of Zaragoza.

References

1. P. de Marcillac et al., Nature **422**, 876 (2003).
2. S. Cebrián et al., Phys. Lett. **B556**, 14 (2003).
3. C. Cozzini et al., CRESST Collaboration, published in this volume.
4. M.R. Dragowski et al., CDMS Collaboration, published in this volume.
5. V. Sanglard et al., EDELWEISS Collaboration, published in this volume.
6. S. Cebrián et al., Phys. Lett. **B563**, 48 (2003).
7. B. Majorovits et al., CRESST Collaboration, published in this volume.
8. N. Coron et al., Nucl. Inst. Meth. **A520**, 159 (2003).
9. N. Coron et al., Opt. Eng. **43**, 1568 (2004).
10. S. Cebrián et al., Astrop. Phys. **21**, 23 (2004).

DEVELOPMENT OF LOW BACKGROUND CsI(Tℓ) CRYSTALS

H.S.LEE[1],* H.BHANG[1], S.Y.KIM[1], J.LEE[1], J.W.KWAK[1], S.S.MYUNG[1],
M.J.LEE[1], S.C.KIM[1], S.K.KIM[1], J.I.LEE[2], Y.D.KIM[2], M.J.HWANG[3],
Y.J.KWON[3], I.S.HAHN[4], H.J.KIM[5], J.J.ZHU[6], J.LI[6,7]

(KIMS COLLABORATION)

[1] *DMRC and school of Physics, Seoul National University, Seoul 151-742, Korea*
[2] *Department of Physics, Sejong University, Seoul 143-747, Korea*
[3] *Physics Department, Yonsei University, Seoul 120-749, Korea*
[4] *Department of Science Education, Ewha Womans University, Seoul 120-750, Korea*
[5] *Physics Department, Kyungpook National University, Daegu 702-701, Korea*
[6] *Department of Applied Physics, Tsinghwa University, Beijing, China*
[7] *Institute of High Energy Physics, Beijing, China*

Searches for weakly interacting massive particle(WIMP) is being carried out at the underground laboratory, Yangyang, Korea. Characteristics and internal background of CsI(Tℓ) crystal have been investigated. In our extensive R&D, we developed a technique to reduce internal background in the CsI(Tℓ) crystal. With the latest CsI(Tℓ) crystal, we have achieved 6 counts/keV/kg/day level of background. Further reduction of internal background is foreseen with the CsI powder lately produced.

1. Introduction

Scintillation crystals are frequently used for WIMP search. NaI(Tℓ) crystals have been adopted by several experiments [1,2]. Because of its high hygroscopicity, one needs encapsulation of the crystal and it could be introduce unwanted surface background contamination. The CsI(Tℓ) crystal is a good candidate for WIMP search because its hygroscopicity is much lower than NaI(Tℓ) and it has a good pulse shape discrimination(PSD) power with a high light yield[3,4,5,6]. But, internal background from radioisotopes of cesium (^{137}Cs, ^{134}Cs) and rubidium (^{87}Rb) can cause high rate of

---

*e-mail : hslee@hep1.snu.ac.kr

gamma background[7]. Therefore reduction of internal radioisotopes is a critical issue for low background WIMP search experiments. We already identified major source of internal radioisotopes[7,8]. We have studied the internal radioisotopes in CsI(Tℓ) crystals by measurement inside a heavy shield at the underground laboratory and by calculations with GEANT4 Monte Carlo(MC) simulation.

2. Internal Backgrounds

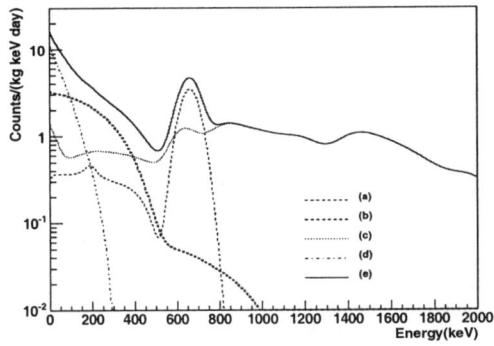

Figure 1. Background spectra obtained using GEANT4 simulation for the 8x8x30 cm^3 CsI(Tℓ) crystal with 10mBq/kg ^{137}Cs contamination, 30mBq/kg ^{134}Cs contamination, and 10ppb ^{87}Rb contamination (a) spectrum of $^{137}Ba^*$ (b) beta-ray spectrum of ^{137}Cs (c) ^{134}Cs spectrum (d) ^{87}Rb spectrum (e) total summed spectrum

^{137}Cs ($t_{1/2} = 30.07$ y, Q=1175.6 keV) is a beta-emitter decaying with a 95% branching ratio to the metastable state($t_{1/2} = 2.55$ minutes) of ^{137}Ba at 661.7 keV. Therefore, the beta electron and the subsequent gamma ray are not correlated in time, and the background at 10keV region is not negligible. ^{134}Cs ($t_{1/2} = 2.065$ y, Q = 2058.7 keV) is also a beta-emitter decaying to ^{134}Ba, but the subsequent gamma ray is correlated with the beta electron. Therefore the background level from ^{134}Cs decays are not significant at low energies. ^{87}Rb ($t_{1/2} = 4.75 \times 10^{10}$y Q = 282 keV) is also beta-emitter decaying to the ground state of ^{87}Sr without gamma-ray emission. Since the Q value is small and it has very long half-life, the background level from ^{87}Rb decay is relatively high at low energies. Figure 1 shows background spectrum of each radioisotope which is obtained by GEANT4 simulation. This spectrum is well matched with previous

CsI(Tℓ) crystal measurement[7]. As a result, we can quantify contribution of each isotope to the background level at 10keV. 1 mBq/kg of ^{137}Cs and ^{134}Cs contribute 0.3 counts/keV/kg/day(cpd) and 0.07 cpd respectively and 1ppb of ^{87}Rb does 1.07 cpd to the background rate.

We already reported contamination level of ^{137}Cs, ^{134}Cs, and ^{87}Rb in CsI(Tℓ) crystals commercially available from several companies[7]. The contaminations in CsI(Tℓ) crystal are 18~210 mBq/kg for ^{137}Cs, 38~586 mBq/kg for ^{134}Cs, and 3.2~816 ppb for ^{87}Rb, respectively. Background level of commercially available crystals at 10keV are higher than 65 cpd. Also, we had grown a 20 cpd level crystal at 10keV with careful powder selection, which contains 15.5 ±2.6 mBq/kg of ^{137}Cs, 27.4 ±4.6 mBq/kg of ^{134}Cs and 20 ± 1 ppb of ^{87}Rb [9].

3. Reduction of Internal Background

3.1. ^{137}Cs

Table 1. Activity of ^{137}Cs isotopes (unit in mBq/kg)(error is statistical only).

	Water	CsI powder	CsI(Tℓ) Crystal
'Regular' Water	0.35~1.16	24.5~81.2	
'Pure' water	0.052±0.01	5.2±1.0	7.83±1.70
'Ultra pure' water	<0.01	2.2±0.3	

There was an indication that the water for Cs extraction might be a main source for ^{137}Cs contamination in CsI powder[8]. We have measured the ^{137}Cs contamination of three different kinds of water samples; one 'Regular' water routinely used by the company for the processing of Cs extraction, one 'Pure' water, and one 'Ultra pure' water. Figure 2 shows the spectra of the three samples measured by HPGe detector in the underground laboratory. As one sees in this Figure and Table 1, it is clear that the purification of water reduces ^{137}Cs amounts. It has been tried to produce CsI powder using only one kind of water at a time in the powder company. We measured samples from the products and the results are shown in Table 1. Crystal grown with CsI powder only using the 'Pure' water have 14 cpd background level at 10keV region as shown in Figure 4 (c) and Table 2. The ^{137}Cs amounts in this crystal is 7.83 ± 1.70 mBq/kg which can contribute 2.74 ± 0.60 cpd at 10keV. The crystal with powder produced by the 'Ultra pure' water is on going. We can estimate background level of this crystal using the result of powder measurement. As

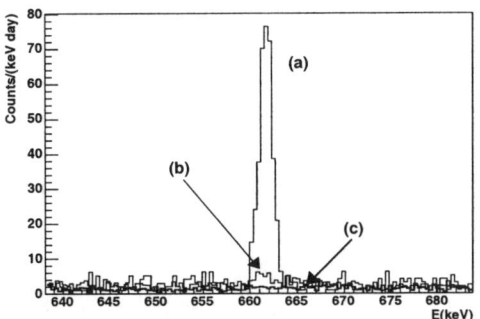

Figure 2. HPGe spectra for Cs-137. (a) 'Regular' water from company (b) 'Pure' water from company (c) 'Ultra pure' water

shown in Table 1, the ^{137}Cs amounts in the powder using the 'Ultra pure' water is 2.2 ± 0.3 mBq/kg, which can contribute about 0.8 cpd at 10keV region.

3.2. ^{87}Rb

Figure 4 (c) and Table 2 show that the ^{87}Rb contamination of 14cpd level crystal is about 10ppb, which can contribute 10.7 cpd at 10keV. It is known that ^{87}Rb can be reduced by the recrystallization method. We have achieved less than 1 ppb contamination of the ^{87}Rb in CsI powder using the method. Figure 4 (d) and Table 2 show background level of the crystal made of this powder. By using the 'Pure' water and the recrystallization method, we achieved 6 cpd level background at 10keV. The ^{87}Rb amounts in this crystal is about 0.9 ppb, which can contribute 0.96 cpd at 10keV region.

3.3. ^{134}Cs

In the 6cpd crystal, the ^{134}Cs contamination is 25.9 ± 2.06 mBq/kg, which contributes 1.81 ± 0.144 cpd at 10keV. But suppression of ^{134}Cs contamination in CsI powder is very difficult because it was mainly made by neutron capture process of ^{133}Cs, which is stable isotope. But gammas from ^{134}Cs correlated with the beta electron, as described in section 2, we can suppress low energy beta with gamma tagging. Figure 3 shows the simulated background spectra from ^{134}Cs for the center crystal in 3x3 array of full size crystals with and without gamma tagging by the surrounding crystals. We can reduce background from ^{134}Cs about factor 20 if we use

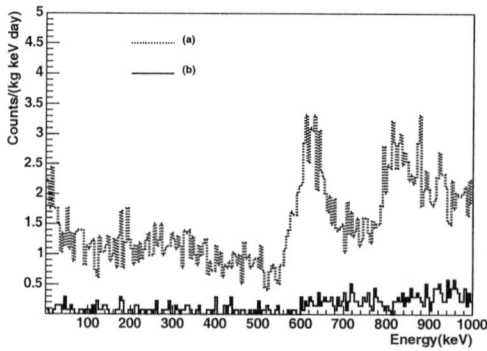

Figure 3. Background spectrum of ^{134}Cs obtained using Geant4 simulation at center of 3×3 array in the case of 30mBq ^{134}Cs amounts and 8cm × 8cm × 30cm size (a) without compton veto (b) with rejection of gamma tagged event in surrounding crystals

surrounding crystals as veto detectors for gamma tagging.

Table 2. Activity of Cs and Rb isotopes of various crystals which is correspond with Figure 4(unit in mBq/kg)(error is statistical only).

radioisotope	Crystal a	Crystal b	Crystal c	Crystal d
^{87}Rb (ppb)	3.2	10.4	10	0.9
^{137}Cs (mBq/kg)	193.1±4.1	23.3±3.3	7.83±1.70	7.96±1.53
^{134}Cs (mBq/kg)	38.0 ±3.2	34.4±2.6	13.7±2.09	25.9±2.06
total at 10keV (cpd)	65	20	14	6

4. Conclusion

We have studied internal background and its reduction of CsI(Tℓ) crystals. As a result of our extensive R&D, we have grown the purified crystal which is about 6cpd in 10keV region with low ^{137}Cs contamination by using the 'Pure' water and with low ^{87}Rb contamination by recrystallization method. The ^{137}Cs, ^{134}Cs, and ^{87}Rb contamination levels of this crystal are 7.96±1.53 mBq/kg, 25.9±2.06 mBq/kg, and about 0.9ppb, respectively. We also have developed further purified CsI powder using the 'Ultra pure' water and crystal growing is on process. The expected ^{137}Cs contamination level at CsI(Tℓ) crystal is less than 3 mBq/kg, which can contribute 1cpd at 10keV region. The ^{87}Rb contamination level may be less than 1ppb with further recrystallization because we had further repeat of recrystallization, contribution of ^{87}Rb at 10keV will be less than 1cpd. The ^{134}Cs contamination level will be similar with crystal (d) of Table 2. Using the neigboring

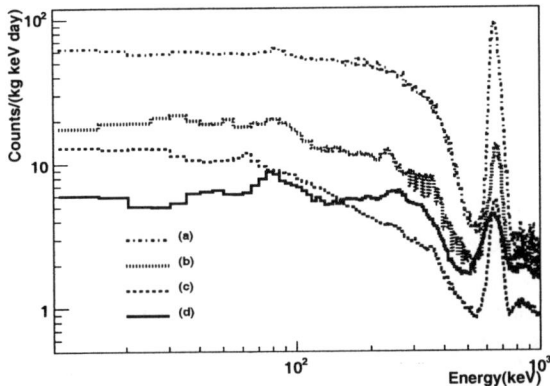

Figure 4. Background spectra of various crystals with different reduction status at R&D: (a) Best crystal in the market (b) Crystal was made by careful powder selection (c) Crystal was made by ^{137}Cs reduced powder with 'Pure' water (d) Crystal was made by ^{87}Rb and ^{137}Cs reduced powder

crystals as a compton suppressor, contribution from ^{134}Cs can be less than 0.1 cpd at 10keV as discussed in section 3.3. In conclusion, we can have low background CsI(Tℓ) crystal which is less than 2cpd level at 10keV.

Acknowledgments

This work was supported by Creative Science Research Initiative program of Korea Science and Engineering Foundation. We are very grateful to the Korea Midland Power Co. and their staffs for providing the underground laboratory space at Yangyang.

References

1. R. Bernabei et al, Physics Letters B **480**, 23 (2000).
2. N. J. C. Spooner et al, Physics Letters B **473**, 330 (2000).
3. S. Pecourt et al, Astroparticle physics **11**, 457 (1999).
4. V. A. Kudryavtsev et al, Nucl. Instru. Method. **A456**, 272 (2001).
5. H. J. Kim et al, Nucl. Instru. Method. **A457**, 471 (2001).
6. H. Park et al, Nucl. Instru. Method. **A491**, 460 (2002).
7. T. Y. Kim et al, Nucl. Instru. Method. **A500**, 337 (2003).
8. Y. D. Kim et al, Journal of Korean Physical Society **40**, 520 (2002).
9. Y. D. Kim et al, Proceedings of IDM 2002.

PROPERTIES OF LIQUID RARE GAS SCINTILLATION FOR WIMP SEARCHES

AKIRA HITACHI

Kochi Medical School, Nankoku, Kochi 783-8505, Japan

Scintillation in liquid rare gases, Ar and Xe is discussed in relation with WIMP detectors. Although the model is approximate, it shows good agreement with neutron scattering experiments. The RC/γ ratio in liquid Ne is also estimated. A simple way to estimate the RC/γ ratios in organic and inorganic scintillators is proposed and demonstrated for CsI(Tl) crystal.

1. Introduction

The search for WIMP (weakly interacting massive particle), candidate for galactic dark matter, is the most exciting issue in cosmology and particle physics. Theories indicate that WIMPs have masses of $\sim 1\text{-}10^3$ GeV with a mean velocity of ~270 km/s. The signals which WIMPs may generate will be very weak and rare. Elastic scatterings of WIMPs by the nuclei of scintillators would produce nuclear recoils, with energy up to a few tens of keV, of $\sim 10^{-4}$ to 1 event/kg/day. The neutron scattering experiments can produce recoil ions of such energies in the detector, however these are not easy to perform. Information on the scintillation yield is crucial for the selection of scintillators and design and development of WIMP detectors. We will discuss simple methods to estimate scintillation efficiencies for recoil ions.

Among rare gases, liquid Xe has been considered for rare event detectors mainly because of high Z and A [1]. Recently, other rare gases are also came into consideration and are being investigated [2,3]. Condensed rare gases are divided into two groups. Scintillation and quenching mechanisms in He and Ne are different from those in Ar, Kr and Xe [4]. The quenching factors and the recoil-ion to gamma ratios, RC/γ, in liquid rare gases and CsI(Tl) crystal will be discussed.

2. Nuclear and electronic quenching: RC/γ ratio

Since the recoil ions are slow, considerable amounts of their energy E are expended to atomic motion [5] and do not produce scintillation or ionization signals. We define the "nuclear" quenching factor $q_{nc} = T/E$ where T is the energy given to electronic motion. The RC/γ ratio in Si show good agreement with q_{nc}.

Semiconductors have good linearity from electrons to fission fragments except for small defects. All T can be used. However, this is not the case for liquid and solid scintillators. Those scintillators show "electronic" quenching (q_{el}) for high LET (Linear Energy Transfer) particles. The total quenching factor q_{TTL} is given by $q_{nc} \times q_{el}$ ($q=1$ for no quenching). For fast ions such as αs and γs, $q_{TTL} = q_{el}$.

q_{TTL} is not always equal to the RC/γ ratio. βs and γs show no quenching but have escaping electrons in liquid rare gases [6]. These show reduced efficiency for electron–ion recombination also in inorganic crystals [7], producing less photons which means equivalent to $q_{el} <1$. Furthermore, the scintillation efficiency for γs has a considerable energy dependence and even has some structures [7,8]. These facts should be considered.

3. High excitation density quenching

3.1. Liquid Ar and Xe

The quenching mechanism in LAr and LXe has been discussed before [9]; therefore it is only briefly discussed here. The electronic energy T delivered to the liquid is divided in the core (T_c) and in the penumbra (T_p) of the ion track. Quenching takes place only in the high-ionization-density core. Then the energy T_s available for scintillation is, $T_s=q_{el}T=q_cT_c+T_p$. Here q_{el} and q_c are the overall electronic quenching factor and that in the core, respectively.

We use a diffusion-reaction model of free exciton with a specific reaction rate [9]. We take a cylindrical track core and a Gaussian initial radial distribution of excitons and electron-ion pairs. The proposed quenching mechanism is biexcitonic collisions, $R^*+R^* \rightarrow R+R^++e^-$(K.E.), where, R^* is the "free" exciton. The ejected electron e^- carries away the excess energy close to an excitation energy and loses it before recombination. The electron-ion recombination on heavy ion tracks is very quick. Thus, the recombination component also undergoes this quenching processes.

The range R of recoil ions was obtained using the stopping power found in HMI tables [10]. The nuclear processes follow the usual procedure of screened Rutherford scattering. The electronic process is based on a Thomas-Fermi treatment. Then the electronic LET for recoil ions (LET$_{el}$ =dT/dR) is calculated. When a slow ion enters a medium, heavy charge-transfer processes; capture and loss, take place and the charge state is soon determined by the velocity and Z. The stopping power calculations take this into account. For 40 keV recoil Ar-ions in LAr, we have a q_{nc}=0.37 [5], i.e., more than 60% of the energy is spent in atomic motion. q_{nc} are listed in Table 1 for typical maximum energy of ions recoiled by WIMP collisions. The WIMP mass was assumed to be much heavier than those of atoms.

The energy partition T_c/T is given by the equipartition principle of stopping power contribution between glancing and head-on collisions, and by the energy left by the δ-rays in the core. The value of T_c/T in LAr was estimated to be 0.51 for relativistic ions and 0.72 for the αs [9].

We assume $T_c/T = 1$ for recoil ions since most δ-rays produced by recoil ions do not have sufficient energy to effectively penetrate the core, forming a fairly undifferentiated core [9]. So obtained radial distribution of excited species for recoil-ion track is similar to that for α track core in condensed rare gases [11]. The proposed method for estimation of the RC/γ ratio exploits this similarity and assumes that q_c for the αs is the same as q_{el} for the recoil-ions. Using $q_{el}=0.71$ measured for αs, we have $T_s/T=0.71=0.72q_c+0.28$ then $q_c(\alpha)=0.6$, thus obtain $q_{el}=0.6$ for recoil-Ar ions. The total quenching factor is $q_{TTL} = q_{nc} \times q_{el} \approx 0.37 \times 0.6 = 0.22$. The q_{nc} and q_{TTL} are shown in Fig. 1 as functions of recoil energy. These values are relative to the relativistic heavy ions (RHI), which show basically no quenching in LAr and LXe.

To obtain more practical quantity, RC/γ ratio, one needs γ/RHI ratio. No direct measurements have been reported for γ/RHI. We use $q_{el}=0.71$ for α and γ/α~ 1.1 and obtain RC/γ= 0.28 for 40 keV recoil-Ar ions.

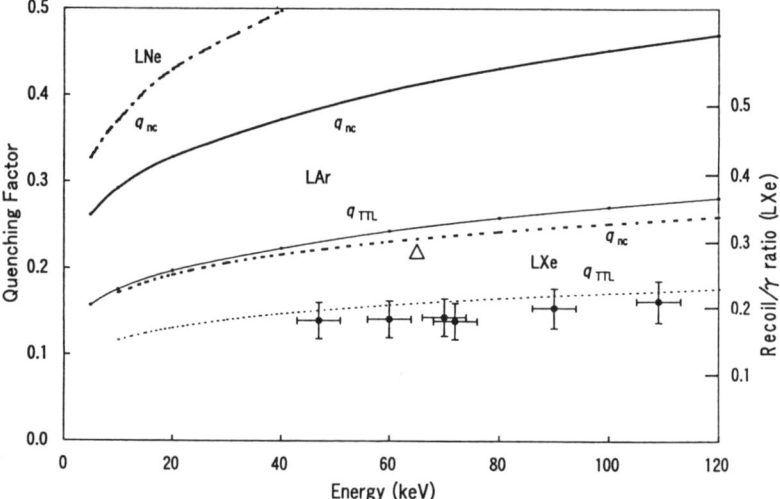

Figure 1. The nuclear q_{nc} [5] and the total $q_{TTL} = q_{nc} \times q_{el}$ quenching factors as a function of recoil ion energy in LNe (dot-dashed line), LAr (solid lines) and LXe (broken lines). •; neutron-recoil measurements in LXe [12], Δ; in LAr [3]. NB: For RC/γ ratio, the values should be divided by factors, 0.8 and 0.77 for LAr and LXe, respectively (shown in the right hand axis for LXe).

Table 1. Estimated scintillation for recoil ions in liquid rare gases.

liquid		LNe	LAr	LXe
Energy	keV	20	40	60
Range	μm	0.1	0.13	0.07
q_{nc}	= T/E	0.43	0.37	0.23
q_{TTL}	= $q_{nc} \times q_{el}$	(0.3)	0.22	0.16
RC/γ		(0.3)	0.28	0.21
N_{ph}	photon/keV	(2.3)	11	12

For LXe, we assume T_c/T to be the same as in LAr. Using q_{el} =0.77 for the αs in LXe [6], we have 0.77=0.72q_c +0.28. We obtain q_{el} = 0.68 and q_{TTL} = $q_{nc} \times q_{el}$ ≈ 0.23 × 0.68 = 0.16 for 60 keV recoil-Xe ions. With q_{el} (α) = 0.77 and α/γ ~ 1, we have RC/γ=0.21. The results in LAr and LXe [11] agree well with the nuclear recoil experiments [3,12] as shown in Fig. 1.

3.2. Liquid Ne

The scintillation mechanism in LNe is different from that in LAr and LXe. No satisfactory quenching theory exists. Farther more, experimental investigations are scare [2]. The electronic energy T is about 43% of recoil ion energy for 20 keV recoil-Ne ions in LNe [5].

We assume q_{el} for recoil Ne is also the same as q_c in the α-track and the energy partition for αs in LNe is the same as that in LAr, i.e., q_{el}(RC) = q_c (α) and T_c/T =0.72. No information is available for q_{el} for αs in LNe. We assume no quenching for γs and take the α/γ ratio = 5.9/7.4=0.8 [2] for $q_{el}(α)$ in LNe. Then we have, T_s/T=0.8=0.72q_c+0.28 for αs in LNe and we calculate q_c = 0.72 for α track core. Taking T_c/T =1 for recoil–Ne ions gives q_{el}(RC)=0.72. Then, q_{TTL} = $q_{nc} \times q_{el}$ ≈ 0.43 × 0.72 = 0.3. Also, we obtain RC/γ = (RC/α)·(α/γ) = q_{nc} × q_{el}(RC)/q_{el}(α)·(α/γ) ≈ 0.43× 0.72/0.8 × 0.8 = 0.3. For the number of photons produced, we have N_{ph}(RC) = N_{ph}(γ) ×RC/γ ≈ 7.4 × 0.3 = 2.3 photons/keV.

3.3 Other Scintillators

The scintillation efficiency or q_{el} is more or less a function of LET for complex compounds particularly for slow ions in which contributions by energetic δ–rays can be ignored. One uses the Bragg rule for stopping power calculation. The main difference between organic scintillators and inorganic crystals such as NaI(Tl) is that the former do not show the reduced efficiency of electron-ion recombination at low LET while the latter does. One can use scintillation data as a function of LET and/or Birk's formula [7].

Z and A for Cs and I atoms are very close and similar to Xe, These are treated similarly in the stopping calculations and values for LXe can be used for CsI(Tl) crystals. The range of recoil ions were obtained from stopping powers for Xe ions in HMI tables. The energy given to electronic excitation was obtained from Lindhard. Then LET_{el} was calculated for recoil ions. The LET dependence of scintillation efficiency in CsI(Tl) crystals is given in the literature [7,13]. The LET_{el} of recoil ions with energies of concern here are in the same range in LET as the αs. We use the scintillation efficiency for the αs at the same LET for that for the recoil ions. The RC/γ ratio obtained in this way is shown in Fig. 2 together with values for neutron recoil measurements [14]. No parameter fittings has been made. The present estimation gives good agreement with the experimental values for $\gtrsim 40$ keV, but gives smaller values in the low energy. However, neither theories nor measurements at the low energy range is considered very good.

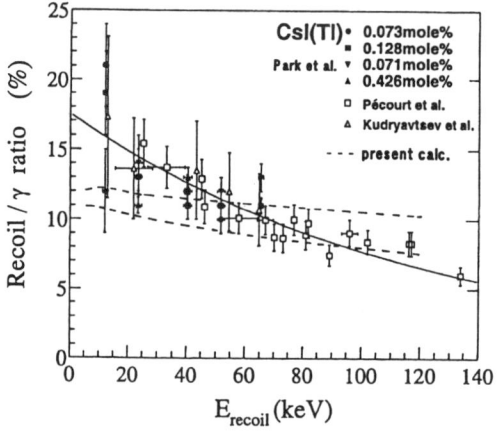

Figure 2. The recoil-ion to γ ratio reported in CsI(Tl) [14]. The broken lines are present estimates for Tl=0.31% (upper) and 0.046% (lower). The solid line is fitting to an exponential fn. by Park [14].

4. Discussion

Table 1 shows the quenching factor, RC/γ ratio and N_{ph} calculated for liquid rare gases using α/γ ratios and W-values in the literature [6]. q_{nc} for LAr is much larger than that in LXe; however, N_{ph} are expected almost the same. This is because of relatively heavy electronic quenching in LAr and the difference in W_{ph} values (19.5 eV and 14 eV, respectively, for LAr and LXe).

The value of N_{ph} for LNe is much smaller than thate for LAr and LXe. Some of the excited states responsible for scintillation in LNe have metastable character; therefore non-radiative processes may exist. The addition of Ar or Xe may be useful, apart from the wavelength shift and lengthening of the scattering length. The mixing may transfer the energy to the dopants before the system undergoes non-radiative processes. Here, the concentration of the dopants should be large so that energy transfer is collisional in order to maintain desirable characteristics for recoil-γ discrimination [15].

The RC/γ ratios for LXe and CsI(Tl) are shown in Fig. 1 and Fig. 2, respectively, as a function of recoil-ion energy. The values show the opposite trends in each other. The RC/γ ratios for liquid rare gases decrease slowly as the energy of the recoil ion decreases whereas that for CsI increases. This is because that LAr and LXe are the most strong among solid and liquid scintillators against high LET quenching. On the other hand, quenching in inorganic crystals is rather large. As the energy of recoil ions increases the LET_{el} increases as a result q_{TTL} and RC/γ ratios for CsI fall off rapidly. Pécourt [8] fitted their experimental values by the Birks-Lindhard model with single free parameter. The fitting shows rapid increase in the q_{TTL} at low energy; however, the increase seems to be too steep since q_{nc} eventually goes down.

Acnowledgements

The author would like to thank Dr. A. Mozumder for reading of the manuscript.

References

1. T. J. Sumner, in *The Identification of Dark Matter* IV, 396 (2002).
2. R. A. Michniak et al. Nucl. Instr. Meth. **A 482**, 387 (2002).
3. WARP Proposal (2004), http://warp.pv.infn.it/index.php
4. *Rare Gas Solids*, ed. by K.L.Klein and J.A. Venables (Academic, N.Y. 1976) vol. 1.
5. J. Lindhard et al. Mat. Fys. Medd. Dan. Vid. Selsk. **33**, no.10 (1963).
6. T. Tanaka et al., Nucl. Instr. Meth. **A 457**, 454 (2001) and refs. therein.
7. J. B. Birks, *The theory and practice of scintillation counting*, (Pergamon, Oxford, 1964).
8. S. Pécour et al. Astroparticle Phys. **11**, 457 (1999).
9. A. Hitachi, T. Doke and A. Mozumder, Phys. Rev. **B 46**, 11463 (1992).
10. J. P. Biersack et al. Hahn-Meitner Inst. Pub. No. HMI-B 175 (1975).
11. A. Hitachi, in *The Identification of Dark Matter* IV, 357 (2002); Proc. 46th Jpn. Soc. Radiation Chemistry, 148 (2003).
12. F. Arneodo et al. Nucl. Instr. Meth. **A 449**, 147 (2000).
13. R. Gwin, Oak Ridge National Lab. Rpt. ORNL 3354 (1962).
14. H. Park et al. Nucl. Instr. Meth. **A 491**, 460 (2002) and refs. therein.
15. A. Hitachi, Nucl. Instr. Meth. **A 327**, 11 (1993).

LOW TEMPERATURE TESTS OF PHOTOMULTIPLIERS FOR USE IN LIQUID XENON EXPERIMENTS

R.J. HOLLINGWORTH,[*] J.E. MCMILLAN

*Department of Physics and Astronomy, University of Sheffield,
Hicks Building, Hounsfield Road, Sheffield,
S3 7RH, UK*

Eleven developmental 128mm hemispherical quartz photomultipliers were cooled to liquid xenon temperatures, to determine their response (combination of photo-cathode quantum efficiency and dynode gain) as a function of temperature. The responses were measured at two wavelengths of light; 460 and 370nm. The tests were performed using light sources at varying distance from the photo-cathode, to investigate the effect of resistivity changes on tube response. Single electron noise was also studied as a function of temperature. The equipment used is described in detail and results are presented.

1. Introduction

ZEPLIN II[1] is a Dark Matter detector currently being constructed for operation at the Boulby mine in the UK. It uses liquid xenon as the target medium. Interactions in the xenon cause recoil events, generating scintillation light. This light is detected by a set of seven photomultipliers. The tubes need to detect the 175nm scintillation light, and to operate at liquid xenon temperatures, $\approx -110°C$. Previous to these tests, the tubes selected had only been operationally tested at room temperature, and it was necessary to qualify their behaviour at low temperatures before they could be installed. Some of the difficulties that could have been encountered included, gross failure of the photomultiplier, changes in dynode gain, and changes in photo-cathode resistance causing electrostatic field distortions and consequent loss of collection efficiency.

---

[*]corresponding author: r.j.hollingworth@sheffield.ac.uk

2. Tube Description

The photomultipliers that were tested were developmental devices from Electron Tubes, model no. D742QKFLB, with the following properties:

- 128mm (5") in diameter, giving large detection area.
- Hemispherical bulb.
- UV transmissible quartz window.
- Bi-alkali photo-cathode.
- High ^{40}K content, due to the presence of a graded glass seal.

3. Experimental Set-up

The set-up used was a modification of that used by McMillan and Reid[2]. Similar tests have been performed by Araujo et al[3].

The photomultiplier under test was installed inside an improvised cryostat, based a large wooden box. The joints of the box were sealed with rubber sealant, and the entire box coated on the inside and out with black paint to completely isolate it from the light. The box was covered on the outside with approximately 100mm of thermal insulation foam. The photomultiplier was installed vertically, through a hole in a shelf inside the box, with the photo-cathode facing downward. The photomultiplier was held in place by two rubber O-rings that were fastened around its base. The photomultiplier was connected to an external power supply. All tests were performed at a potential difference of 1,200V. All previously tested photomultipliers were kept in the box, either on the bottom or in specially installed baskets around the sides. This was done so that they would experience many temperature cycles, and any cyclic failure could be identified. The box was fitted with a plastic right-angled tube to allow the electrical cables into the box without letting any light in. The box was cooled via an external liquid nitrogen dewar that supplied a constant flow of cold nitrogen gas. The gas entered the box via a metal pipe located at the top of the box, then into the inner volume of the box through two tubes located at the front right and rear left of the volume. The gas exited via the right-angled tube and the non-gas-tight joint between the door and the main box. The photomultipliers were tested using light from an external LED pulser array. The light was carried into the box via plastic optical fibres. The fibres entered the box through the same plastic right-angled tube as the electrical cables.

The configuration of the box changed significantly over the duration of the experiment. Initially the tubes were only tested using one wavelength of light, 460nm, and the temperature of the tube was measured using a platinum resistance thermometer placed in the box on a shelf above the photomultiplier under test. During the process of experimentation, light from a 370nm LED and a second 460nm LED were introduced into the box. The 370nm LED was introduced to provide a light source that was closer in wavelength to that to be detected in the ZEPLIN II experiment. The second 460nm LED was introduced so that tests could be performed to detect any changes in the photomultiplier photo-cathode resistivity. No tests were performed using 175nm light. This is because the light used was pulsed from LEDs (see below), and there are no currently available LEDs that produce light with that wavelengths that short. A Xe emission lamp has been recently acquired, and plans have been made for retesting of the tubes. A second platinum resistance thermometer was also introduced to the box. This thermometer was glued to a thin sheet of copper, and then strapped to the side of the photomultiplier under test. This was done to improve the accuracy of the temperature measurements of the photomultipliers. All tubes initially tested with light from just the one LED, were retested once the modifications had been made. Figure 1 shows the final configuration of the cold-box.

The optical fibres that carried light from the first 460nm and the 370nm were terminated by plastic cylindrical prisms that were cut at a 45° angle. These prisms projected a cone of light up onto the photomultiplier. The prisms were located at the bottom of the box, approximately six inches below the photomultiplier. The second 460nm optical fibre was placed so that it terminated $\sim$ 1mm away from the centre of the photomultiplier bulb.

4. Experimental Procedure

After a tube had been installed in the cold-box, it was left unpowered for a period of approximately twenty four hours. This was done so that any photoluminescence in the glass, caused by its exposure to daylight, would have decayed away and would not contribute to the output signals.

As described previously, the tubes were cooled by cold nitrogen gas from an external liquid nitrogen dewar. The dewar was equipped with a heater, so that the rate of nitrogen flow could be increased as the temperature of the box decreased. This was done so that the necessary increase in the cooling rate with lower temperatures could be achieved. The tubes were cooled at a rate of between 20 and 30 Khr^{-1}, so that thermal equilibrium between

Figure 1. Photograph of the final configuration of the cold-box.

the outside of the photomultipliers, the thermometer position, and inside, where the internal components under test were, could be maintained.

The response of the photomultiplier was firstly taken at room temperature, and then at decreasing intervals of 10K, until liquid xenon temperatures were reached. The response of the photomultiplier was measured by pulsing light from each LED at a frequency of 10kHz and measuring the response signal. The duration of each of the LED pulses was $\sim$ 4ns. The LEDs were pulsed using a circuit based on that of Kapustinsky et al[4]. The response of the photomultiplier was defined as the amount of charge produced in the photomultiplier from a single input light pulse; this quantity is a combination of the quantum efficiency of the photo-cathode and the dynode gain. The anode output signal from the photomultiplier was fed into a LeCroy WaveRunner LT342 oscilloscope. The signal was then integrated to give the charge per pulse and this value was histogrammed to give a measurement of the photomultiplier response. The value of the response was taken to be the value of the median of the histogram distribution, approximately the peak of the distribution. 5,000 signal pulses were recorded for each data histogram.

The light from the first 460nm LED and the 370nm LED illuminated the whole of the photomultiplier photo-cathode. Light from the second 460nm LED only illuminated the very central region of the photo-cathode. If there were any changes to the efficiency of the photo-cathode due to changes in its resistivity, then the ratio of the responses from the near and far 460nm

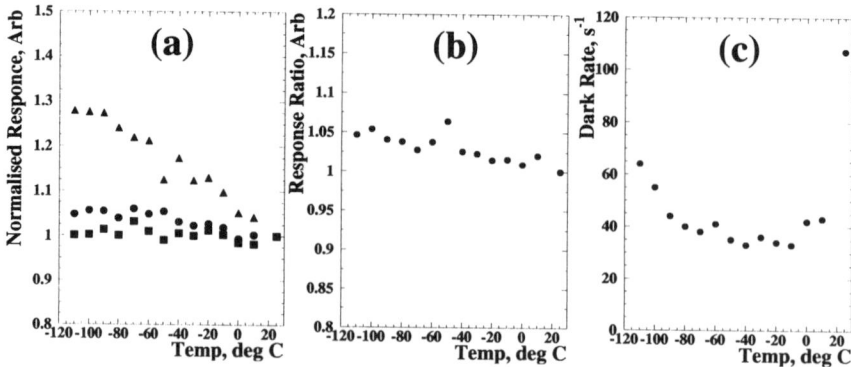

Figure 2. Typical results for, (a) photomultiplier response (● - far 460nm, ■ - near 460nm, ▲ - 370nm), (b) photomultiplier response ratio, and (c) photomultiplier dark rate.

light pulses would also change (see below).

Before the response measurements were taken for each temperature value, the dark rate of the tube was first measured.

5. Results and Analysis

Figure 2(a) shows the variation of the response to the three LED light pulses with temperature. The results are normalised, since it was the change in the response that was of interest, and not the absolute magnitude of the signal. Since each photomultiplier had a different operating voltage the size of the signal from each, when run at a uniform voltage, was different. These results show that the response did not diminish as the temperature was decreased. For the 460nm light pulses, the response either remained approximately constant or increased slightly. For the 370nm light pulses, the response increased significantly between that at room and liquid xenon temperatures.

Figure 2(b) shows the ratio of the responses to light pulses from the two 460nm LEDs, R_{Far}/R_{Near}, with temperature. If the resistivity of the photo-cathode did increase, then it would mean that the photo-electrons, emitted due to incident photons, would be harder to replenish. This would lead to a deficiency of available photo-electrons, and a local positive charge. Both of these effects would cause a decrease in the quantum efficiency of the photo-cathode at that site. Another result could be a loss in collection efficiency of the first dynode, due to the non-uniform potential of the photo-cathode. The photomultipliers are designed assuming that the photo-cathode is all at one potential. Any changes in the potential of could cause the photo-electrons to miss the first dynode or even return to the photo-cathode. Since light from the near 460nm LED only illuminated the central region of the photo-cathode, the effect would be most pronounced for photons incident there. The effect would be evident as an increase in the ratio of the response to the light from the far LED compared with the near. These results show that the ratio of the photomultiplier responses did not increase significantly as the temperature was decreased, thus there could not have been a significant change in the resistivity of the photo-cathode.

Figure 2(c) shows the variation of the photomultiplier dark rate with temperature. Since single photo-electron signals coming from the dark current would be a background to low energy recoil events, it was necessary to also measure the dark rate. As expected, the dark rate initially fell quickly with temperature. A plateau region was reached, but the dark rate started to increase again before liquid xenon temperatures were reached.

6. Conclusions

Low temperature tests were performed on eleven 128mm hemispherical photomultipliers. No results were found that would cause any of the photomultipliers to be rejected for use in the ZEPLIN II experiment. The response of the photomultipliers as temperature decreased increased slightly for 460nm light, and significantly for 370nm light. Analysis of this difference is difficult, and it is intended to perform additional measurements at a range of wavelengths.

Acknowledgements

RJH acknowledges the Particle Physics & Astronomy Research Council (PPARC) for funding.

References

1. Lüscher R et al. Nucl. Phys. B-Proc. Sup. **95**: 233-236 (2001).
2. McMillan JE, Reid RJO J. Phys. E-Sci. Inst. **22(6)**: 377-382 (1989).
3. Araujo HM et al. IEEE Trans. Nucl. Sci. **45(3)**: 542-549 (1998)
4. Kapustinsky JS et al. Nucl. Inst. Meth. A **241**: 612-613 (1985)

DETECTION OF DARK ELECTRIC MATTER OBJECTS FALLING OUT FROM EARTH-CROSSING ORBITS

E. M. DROBYSHEVSKI

Ioffe Physico-Technical Institute, Russian Academy of Sciences
St.Petersburg, 194021 Russia
e-mail: emdrob@mail.ioffe.ru

If the DM consists of elementary Planckian black holes, their number (and flux) should be fairly low. If, however, they carry an electric charge corresponding to their mass (up to $Ze \approx 10e$), such DArk Electric Matter Objects, daemons, should interact strongly with matter. They should be slowed down somewhat when crossing celestial bodies, build up in them, and in multiple systems, in *close lying orbits* too (e.g., in Earth-crossing orbits). Capture, say, of a Fe nucleus by a *negative* daemon releases >100 MeV of energy, i.e., cause ejection of ~10 nucleons. The detector consisting of two ZnS(Ag) scintillation screens stacked one upon the other (four modules 0.25 m² each) detects at CL > 99% events with a time shift corresponding to velocities $V \sim 30\text{-}5$ km/s (in both down- and upward crossings). Such velocities are typical for objects trapped into helio- and geocentric orbits (with the latter crossing the Earth's surface to become finally confined to its interior). Of particular significance ($> 3\sigma$) is a group with $V \approx 10\text{-}15$ km/s, which is characteristic of objects falling from near-Earth almost circular heliocentric orbits. Their flux is $>10^{-9}$ cm^{-2}s^{-1} and varies with $P = 0.5$ year.

1. Our Ideology

Our search for DM particles was based on a hypothesis that as long as the Universe started from Planckian scales, the most of its mass should also have remained in Planckian objects. Their simplest version is elementary black holes, whose gravitational radius is one fourth of the Compton wavelength. We assumed also that such DArk Electric Matter Objects, daemons, carry an electric charge, whose repulsive action is counterbalanced by gravitation. Then $m_\delta \approx 3 \times 10^{-5}$ g, $r_g \approx 2 \times 10^{-33}$ cm, and $Ze \approx 10e$. Such objects can be stable. They were discussed by many authors (e.g., [1-7]), but no one apparently attempted their search for a number of reasons.

First, the daemon having a giant mass, their flux from the Galactic halo onto the Earth can barely reach $\sim 10^{-12}$ cm^{-2}s^{-1}.

However, if we assume that the Galactic disk has a DM population comparable to the stellar density [8], and that the velocity dispersion of DM objects ~4-20 km/s [8], then the Sun should exert a strong focusing action on the daemons when moving through their background. When traversing the Sun, some of the disk daemons are decelerated to the extent where they can be captured into elongated orbits with perihelia lying inside the Sun. In subsequent Sun crossings by the daemons these orbits contract and lose orientation (remaining however in planes containing the Solar apex direction), with the daemons descending into the Sun [9]. If, however, in moving along such an orbit a daemon traverses the Earth's sphere of action, its orbit changes such that its perihelion will leave with a high probability the Sun's volume, to make this strongly elongated Earth-crossing heliocentric orbit (SEECHO) stable until the next time it enters the Earth's sphere of action. Such events transfer the objects into near-Earth almost-circular heliocentric orbits (NEACHOs) crossing the Earth's orbit mainly in close to equinoxes regions. Estimates reveal that the daemon flux from similar orbits onto the Earth may be as high as ~10^{-7} cm^{-2}s^{-1} [10].

Second, a daemon moving with a velocity ≤ 100 km/s cannot produce scintillations, that aroused skepticism concerning their detection [1].

However, in their passage through matter negative daemons should capture nuclei with an energy release $W \approx 1.8ZZ_nA^{-1/3}$ MeV, which for Fe is > 100 MeV. A nucleus is excited and ejects ~10 protons, neutrons, and their clusters (the first to be emitted are the atomic electrons). These particles are capable of producing a strong scintillation. For $Z_n \geq 2$, the ground state orbit of the daemon lies within the nucleus (a similar situation occurs in muon physics), and for $Z_n \geq 24/Z$, even inside a proton in the nucleus. It appears plausible that, on entering a proton, the daemon will cause its decay [11] (recall the monopole catalysis of proton decay [12]). So, the charge of the nuclear remainder dragged along by the daemon should decrease continuously until $Z_n < Z$. Then the daemon will be capable of capturing another nucleus on getting rid of the very neutron-rich remainder. Thus, daemon's traversal of condensed matter with $Z_n > 10$ is accompanied by fast events of successive capture and evaporative de-excitation of a nucleus (at $V \approx 10$ km/s, the capture of a nucleus occurs on a path ~1 µm [13]) and slower processes of digestion of the nuclear remainder.

2. Detection System

Thus, we invoke as a working hypothesis the existence of daemons (in a more general sense, of particles with $m_\delta \sim M_{Planck}$ carrying a large negative charge $Z \approx 10$; these may be also monopoles, dyons etc.). This assumption implies inevitably (i) their electromagnetic interaction with matter, including capture of nuclei with their strong excitation, (ii) proton decay catalysis, as well as (iii) advantages of looking for objects of a slow concentrated population captured in the Solar system from the Galactic disk.

Based on this assumption, we made several approaches to the problem [14] which culminated in development of a simple detector consisting of four identical modules [13,15–16]. Each module represents a cubic case 51 cm on a side made of tinned sheet iron. Its top side is a sheet of black paper. At the case's center, two transparent polystyrene screens 4 mm thick are arranged horizontally at a distance of 7 cm from one another, which are coated on the underside by a ZnS(Ag) layer ~3.5 mg/cm^2 thick (we deliberately made our detector asymmetric to stress the contrast between the signatures of up- and downward moving objects). The screens are separated by black paper. Each screen is viewed by a PM tube. The tube signals are fed to a digital double-trace oscilloscope. The oscilloscope is triggered by signals from the top PM tube. If the second trace also carries a signal, such a paired event is sent to computer memory. Note the advantage of a thin (~10 μm) ZnS(Ag) layer. While relativistic particles like cosmic-ray muons leave an energy of <10 keV in crossing it, heavy charged particles ejected from excited nuclei release in it ~1 MeV. Therefore preliminary experiments can be performed at ground level with a not very sensitive equipment.

We are interested in events with signals whose origins are displaced with respect to one another by Δt. The most frequent paired events are not displaced in time ($\Delta t < 0.5$ μs). They are caused by cosmic rays. Such signals feature a short leading edge (~1 μs) and do not differ in shape from intrinsic PM tube noise. Sometimes, signals with a longer leading edge and a fairly flat, 2.5-3 μs long, maximum are observed. Such signals are characteristic of scintillations produced by α-particles. We called them Heavy Particle Scintillations (HPSs) [13,16].

3. Main Results

Figure 1a displays the signal distribution $N(\Delta t)$ (with the events with $\Delta t \leq$ 0.5 μs excluded) collected during March 2000 [15]. A peak in the region Δt = 30 μs is clearly seen with a sgnificance of 2.6σ. If we accept only events with HPSs in the top screen, the number of events drops nearly to one half (from 413 to 212), but the significance of the peak rises to 2.85σ (CL = 99.5%). This suggests that the peak is caused by heavy particles rather

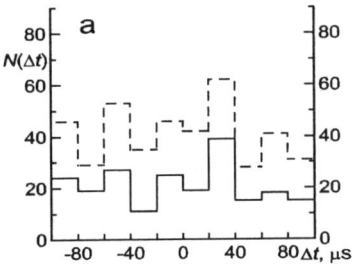 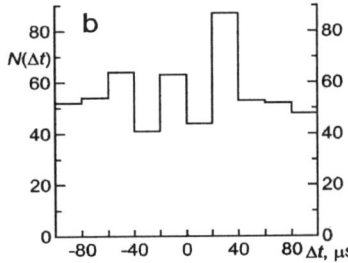

Figure 1. Distributions $N(\Delta t)$ of pair events on their time shift relative to the upper screen scintillations. (a) March, 2000; (b) Sum of Marches of 2000, 2003, and 2004. (- - -) all type events in the upper screen; (——) HPS events only in the upper screen.

than to instrumentation malfunctions or some noise. The sum distribution $N(\Delta t)$ of March events of three years is presented graphically in Fig.1b. The significance of the 30-μs peak increased to 3.33σ (CL = 99.9%).

Knowing the detector dimensions, one can determine the velocity of the particle from the time it needs to cross it. Our detector was designed assuming the lifetime of the daemon-containing proton to be $\Delta \tau_{ex} \sim 10^{-7}$ s [11]. This is what accounts for our choosing 7 cm for the separation between the two scintillator screens. The time $\Delta t = 30$ μs yields, however,

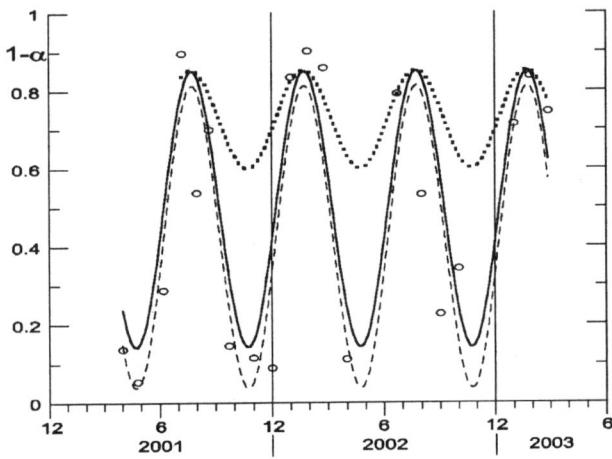

Figure 2. Seasonal variation of 1 - α (see text). (- - - -) Weights of all the points are equal, the correlation coefficient of the sine curve ($P = 0.5$ yr) with the points is $r = 0.86$, its CL > 99.9%. (———) Weights of the points are proportional to 1 - α; $r = 0.75$, its CL = 98.7%. (..........) Points with 1 - α > 0.5 are considered only; their weights are equal to 1 - 2α; $r = 0.36$, its CL $\approx$ 50%.

$V = 2.5 - 3$ km/s only. It appears more reasonable to choose for the base length the 29-cm distance between the top screen and the bottom tinned-iron case lid. Then the transit velocity increases to $V \approx 10\text{-}15$ km/s, a physically sound figure corresponding to the fall of objects from NEACHOs.

We may conceive now of the following sequence of events [13,15-16]. A downward moving daemon captures a S or Zn nucleus in the top scintillator, where an HPS is created. While digesting the remainder of the nucleus, it crosses without interaction the bottom scintillator. Only when the system approaches the bottom lid of the case, the charge of the nuclear remainder drops to $Z_n < Z$, after which the daemon is again capable of capturing a new Sn or Fe nucleus in the lid. The numerous electrons (and, possibly, some protons) ejected in the process can traverse the 22-cm distance to the bottom scintillator and excite in it a not very strong scintillation in $\Delta t \approx 30$ μs. This yields also the estimate $\Delta \tau_{ex} \sim 10^{-6}$ s. It is obvious that an upward moving daemon with $V \approx 10\text{-}15$ km/s is in an unfavorable position; indeed, it can excite the bottom scintillator even twice, but the probability for it to initiate the top scintillator located not far from the bottom one is slim. This is why in place of a maximum at $\Delta t \approx -30$ μs one observes rather a minimum, which may be employed to refine the value of $\Delta \tau_{ex}$ (for more details, see [13]). The number of events in the 30-μs peak corresponds to a flux of $>10^{-9}$ cm^{-2}s^{-1}.

Interestingly, the April $N(\Delta t)$ distributions reveal maxima at $\Delta t = \pm(40\text{-}80)$ μs. They correspond to a $V < 5$ km/s, i.e., to objects captured into geocentric Earth-surface-crossing orbits and moving both down- and upward [13].

We see the daemon flux onto the Earth changes continuously its parameters. Fig.2 shows a seasonal variation of $1 - \alpha$, the confidence level that the $N(\Delta t)$ cannot be represented by a constant, which was derived by the χ^2 criterion [16]. The totality of the data collected may be fitted by a sine curve with $P = 0.5$ yr and maxima falling about on the January-March and July-September periods. This is consistent with the above ideas on accumulation of daemons in NEACHOs crossing the Earth orbit in regions where velocities of the SEECHO objects are in the greatest compliance with the Earth's orbital velocity, i.e. (accidentally) in the near-equinox regions.

4. Conclusion

The above results characterize apparently the first attempt at a direct detection of super-massive (Planckian?) DM particles. Because of their giant mass, their flux is not large, which is a serious impediment to accumulation of good statistics. We made use of the fairly obvious phenomenon of their buildup in Earth-crossing orbits from the slow population of the Galactic disk due to a combined action of the Earth and the Sun. Nevertheless, accumulation of a satisfactory monthly statistics is a hard task, even in these conditions. As follows from our experiments, the smallest possible detector area is ~ 1 m^2, and seasonal variations permit one to collect reasonable data in a time of one year. Add here the difficulties involved in adjustment of a detector based on basically novel principles (for instance, the use of daemon-stimulated proton decay, a concept postulated by us), where some of the parameters governing its operation are poorly controllable or not known at the present time altogether (indeed, sometimes it is unclear to which side should the "tuning knob" be turned, but in either case one will have to wait a year to see whether the decision was right!). Nevertheless, the results thus far obtained are encouraging and have a CL > 99.9%.

I am greatly indebted to M.E.Drobyshevski for calculating many-loop daemon trajectories passing through the Sun, to N.N.Nikonov for processing March, 2003 and 2004, data, and especially to Dr. N.G.Bourova for supporting my attendance at the IDM2004.

References

1. M.A.Markov, *ZhETF*, **51**, 878-890 (1966).
2. K.P.Stanyukovich, *Doklady Acad. Scis. USSR*, **168**, 781-784 (1966).
3. M.Turner, in *"Dark Matter in the Universe"* (IAU Symp. No117), J.Kormendy & G.R.Knapp (eds.), Dordrect: Reidel, pp.445-488 (1987).
4. J.H.MacGibbon, *Nature*, **329**, 308-309 (1987).
5. J.D.Barrow, E.J.Copeland, A.R.Liddle, *Phys. Rev.*, **D46**, 645-657 (1992).
6. A.D.Dolgov, P.D.Naselsky, I.D.Novikov, *astro-ph*/0009407 (2000).
7. S.Alexeyev et al., *Class. Quantum Grav.*, **19**, 4431-4443 (2002).
8. J.H.Bahcall, C.Flynn, A.Gould, *Astrophys. J.*, **389**, 234-250 (1992).
9. E.M.Drobyshevski, *A&A. Trans.*, **23**, 173-183 (2004).
10. E.M.Drobyshevski, in: *Dark Matter in Astro- and Particle Physics*, H.V. Klapdor-Kleingrothaus & Y.Ramachers (eds.), World Sci., pp.417-424 (1997).
11. E.M.Drobyshevski, *Mon. Not. Royal Astron. Soc.*, **282**, 211-217 (1996).
12. V.A.Rubakov, *Pis'ma v ZhETF*, **33**, 658-660 (1981).
13. E.M.Drobyshevski et al., *A&A Trans.*, **22**, 19-32 (2003).
14. E.M.Drobyshevski, in: *Proc. of the 2nd Intnl. Workshop on Identification of Dark Matter*, N.J.C.Spooner & V.Kudryavtsev (eds.), World Sci., pp.643-648 (1999).
15. E.M.Drobyshevski, *A&A Trans.*, **21**, 65-73 (2002).
16. E.M.Drobyshevski, *astro-ph*/0402367 (2004).

FIRST RESULTS FROM THE CERN AXION SOLAR TELESCOPE (CAST)

S. ANDRIAMONJE[b], V. ARSOV[m], S. AUNE[b], D. AUTIERO[a],
F. AVIGNONE[c], K. BARTH[a], A. BELOV[k], B. BELTRÁN[f],
H. BRÄUNINGER[e], J. M. CARMONA[f], S. CEBRIÁN[f], E. CHESI[a],
J. I. COLLAR[g], R. CRESWICK[c], T. DAFNI[d], M. DAVENPORT[a],
L. DI LELLA[a], C. ELEFTHERIADIS[h], J. ENGLHAUSER[e],
G. FANOURAKIS[i], H. FARACH[c], E. FERRER[b], H. FISCHER[j], J. FRANZ[j],
P. FRIEDRICH [e], T. GERALIS[i], I. GIOMATARIS[b], S. GNINENKO[k],
N. GOLOUBEV[k], M. D. HASINOFF[l], F. H. HEINSIUS[j],
D.H.H. HOFFMANN[d], I. G. IRASTORZA[b,*] J. JACOBY[m], D. KANG[j],
K. KÖNIGSMANN[j], R. KOTTHAUS[n], M. KRČMAR[o], K. KOUSOURIS[i],
M. KUSTER[e], B. LAKIĆ[o], C. LASSEUR[a], A. LIOLIOS[h], A. LJUBIČIĆ[o],
G. LUTZ[n], G. LUZÓN[f], D. W. MILLER[g], A. MORALES[f], J. MORALES[f],
M. MUTTERER[d], A. NIKOLAIDIS[h], A. ORTIZ[f], T. PAPAEVANGELOU[a],
A. PLACCI[a], G. RAFFELT[n], J. RUZ[f], H. RIEGE[d], M. L. SARSA[f],
I. SAVVIDIS[h], W. SERBER[n], P. SERPICO[n], Y. SEMERTZIDIS[d],
L. STEWART[a], J. D. VIEIRA[g], J. VILLAR[f], L. WALCKIERS[a],
K. ZACHARIADOU[i], K. ZIOUTAS[h]
(CAST COLLABORATION)

[a] *European Organization for Nuclear Research (CERN), Genève, Switzerland*
[b] *DAPNIA, Centre d'Études Nucléaires de Saclay (CEA-Saclay), Gif-sur-Yvette, France*
[c] *Department of Physics and Astronomy, U. of South Carolina, Columbia, SC, USA*
[d] *Institut für Kernphysik, Technische Universität Darmstadt, Darmstadt, Germany*
[e] *Max-Planck-Institut für Extraterrestrische Physik, Garching, Germany*
[f] *Instituto de Física Nuclear y Altas Energías, Universidad de Zaragoza, Zaragoza, Spain*
[g] *Enrico Fermi Institute and KICP, University of Chicago, Chicago, IL, USA*
[h] *Aristotle University of Thessaloniki, Thessaloniki, Greece*
[i] *National Center for Scientific Research "Demokritos", Athens, Greece*
[j] *Albert-Ludwigs-Universität Freiburg, Freiburg, Germany*
[k] *Institute for Nuclear Research (INR), Russian Academy of Sciences, Moscow, Russia*
[l] *Department of Physics and Astronomy, U. of British Columbia, Vancouver, Canada*
[m] *Johann Wolfgang Goethe-Universität, Institut für Angewandte Physik, Frankfurt am Main, Germany*
[n] *Max-Planck-Institut für Physik (Werner-Heisenberg-Institut), Munich, Germany*
[o] *Rudjer Bošković Institute, Zagreb, Croatia*

---

*attending speaker, e-mail: igor.irastorza@cern.ch

Hypothetical axion-like particles with a two-photon interaction would be produced in the Sun by the Primakoff process. In a laboratory magnetic field ("axion helioscope") they would be transformed into X-rays with energies of a few keV. Using a decommissioned LHC test magnet, CAST has been running for about 6 months during 2003. The first results from the analysis of these data are presented here. No signal above background was observed, implying an upper limit to the axion-photon coupling $g_{a\gamma} < 1.16 \times 10^{-10}$ GeV^{-1} at 95% CL for $m_a \lesssim 0.02$ eV. This limit is comparable to the limit from stellar energy-loss arguments and considerably more restrictive than any previous experiment in this axion mass range.

1. Introduction

Axions and other hypothetical axion-like particles with a two-photon interaction have been invoked in a number of well-motivated scenarios. In particular, they may provide a solution for the strong CP problem and are viable dark matter candidates[1,2]. They can transform into photons in external electric or magnetic fields[3]; an effect that may lead to measurable consequences in laboratory or astrophysical observations[1,4,5,6,7,8,9,10]. For example, axions would contribute to the magnetically induced vacuum birefringence, interfering with the corresponding QED effect[11,5]. The PVLAS experiment[12] apparently observes such an effect far in excess of the QED expectation, although an interpretation in terms of axion-like particles requires a coupling strength far larger than existing limits.

Stars could produce these particles by transforming thermal photons in the fluctuating electromagnetic fields of the stellar plasma[13,14]. Anomalous stellar energy loss by axion emission is constrained by the observed properties of globular cluster stars, implying[14] $g_{a\gamma} \lesssim 10^{-10}$ GeV^{-1} for the axion-photon coupling, where the axion-photon interaction is written in the usual form $\mathcal{L}_{a\gamma} = -\frac{1}{4} g_{a\gamma} F_{\mu\nu} \tilde{F}^{\mu\nu} a = g_{a\gamma}\, \mathbf{E} \cdot \mathbf{B}\, a$. Therefore, the Sun would be a strong axion source and thus offers a unique opportunity to actually detect these particles by taking advantage of their back-conversion into X-rays in laboratory magnetic fields[4]. The expected solar axion flux at the Earth due to the Primakoff process[a] is $\Phi_a = g_{10}^2\, 3.67 \times 10^{11}$ cm^{-2} s^{-1} (where $g_{10} \equiv g_{a\gamma} 10^{10}$ GeV) with an approximate spectrum $d\Phi_a/dE_a = g_{10}^2\, 3.821 \times 10^{10}$ cm^{-2} s^{-1} keV^{-1} $(E_a/\text{keV})^3/(e^{E_a/1.103 \text{ keV}} - 1)$ and an average energy of 4.2 keV [15].

---

[a]Axion interactions other than the two-photon vertex would provide for additional production channels, but in the most interesting scenarios these channels are severely constrained, leaving the Primakoff effect as the dominant one[14]. In any case, it is conservative to use the Primakoff effect alone when deriving limits on $g_{a\gamma}$.

The conversion probability in a B-field in vacuum is[4] $P_{a\to\gamma} = (g_{a\gamma}B/q)^2\sin^2(qL/2)$, where L is the path length and $q = m_a^2/2E_a$ is the axion-photon momentum difference. For $qL \lesssim 1$ where the axion-photon oscillation length far exceeds L we have $P_{a\to\gamma} = (g_{a\gamma}BL/2)^2$, implying an X-ray flux of

$$\Phi_\gamma = 0.51 \text{ cm}^{-2} \text{ d}^{-1} g_{10}^4 \left(\frac{L}{9.26 \text{ m}}\right)^2 \left(\frac{B}{9.0 \text{ T}}\right)^2. \quad (1)$$

For $qL \gtrsim 1$ this rate is reduced due to the axion-photon momentum mismatch. The presence of a gas would provide a refractive photon mass m_γ so that $q = |m_\gamma^2 - m_a^2|/2E_a$. For $m_a \approx m_\gamma$ the maximum rate can thus be restored[18].

The Tokyo axion helioscope[19] of $L = 2.3$ m and $B = 3.9$ T has provided the limit $g_{10} < 6.0$ at 95% CL for $m_a \lesssim 0.03$ eV (vacuum) and $g_{10} < 6.8$–10.9 for $m_a \lesssim 0.3$ eV (using a variable-pressure buffer gas)[20]. Limits from crystal detectors[21,22,23] are much less restrictive.

2. CAST experiment

In order to detect solar axions or to improve the existing limits on $g_{a\gamma}$ an axion helioscope has been built at CERN by refurbishing a de-commissioned LHC test magnet[24] which produces a magnetic field of $B = 9.0$ T in the interior of two parallel pipes of length $L = 9.26$ m and a cross-sectional area $A = 2 \times 14.5$ cm^2. The magnet is mounted on a platform with $\pm 8°$ vertical movement, allowing for observation of the Sun for 1.5 h at both sunrise and sunset. The horizontal range of $\pm 40°$ encompasses nearly the full azimuthal movement of the Sun throughout the year. The time the Sun is not reachable is devoted to background measurements. A full cryogenic station is used to cool the superconducting magnet down to 1.8 K[25]. The hardware and software of the tracking system have been precisely calibrated, by means of geometric survey measurements, in order to orient the magnet to any given celestial coordinates. The overall CAST pointing precision[26] is better than 0.01°. At both ends of the magnet, three different detectors have searched for excess X-rays from axion conversion in the magnet when it was pointing to the Sun. Covering both bores of one of the magnet's ends, a conventional Time Projection Chamber (TPC) is looking for X-rays from "sunset" axions. At the other end, facing "sunrise" axions, a second smaller gaseous chamber with novel MICROMEGAS (micromesh gaseous structure – MM)[27] readout is placed behind one of the magnet bores, while in the other one a focusing X-ray mirror telescope is

Table 1. Data sets included in our result.

Data set	Tracking exposure (h)	Background exposure (h)	$g_{a\gamma}$(95%) $(10^{-10}$ GeV$^{-1})$
TPC	62.7	719.9	1.55
MM set A	43.8	431.4	1.67
MM set B	11.5	121.0	2.09
MM set C	21.8	251.0	1.67
CCD	121.3	1233.5	1.23

working with a Charge Coupled Device (CCD) as the focal plane detector. Both the CCD and the X-ray telescope are prototypes developed for X-ray astronomy[28]. The X-ray mirror telescope can produce an "axion image" of the Sun by focusing the photons from axion conversion to a ~ 6 mm^2 spot on the CCD. The enhanced signal-to-background ratio substantially improves the sensitivity of the experiment. A detailed account of the technical aspects of the experiment will be given elsewhere.

3. Data Analysis and First Results

CAST has been in operation for about 6 months from May to November in 2003, during which time most detectors were taking data. The results presented here were obtained after the analysis of the data sets listed in Table 1. An independent analysis was performed for each data set. Finally, the results from all data sets are combined.

For a fixed m_a, the theoretically expected spectrum of axion-induced photons has been calculated and multiplied by the detector efficiency curves. These spectra, which are proportional to $g_{a\gamma}^4$, are directly used as fitting functions to the experimental subtracted spectra (tracking minus background) for the TPC and MM. For these data, the fitting is performed by standard χ^2 minimization. Regarding the CCD data, the analysis is restricted to the small area on the CCD where the axion signal is expected after the focusing of the X-ray telescope. The resulting low counting statistics in the CCD required the use of a likelihood function in the minimization procedure, rather than a χ^2-analysis. For more details on the analysis we refer to [29].

Each of the data sets is individually compatible with the absence of any signal. The 95% CL limits on $g_{a\gamma}$ for each of the data sets are shown in the last column of Table 1. They can be statistically combined by multiplying the Bayesian probability functions to find the global result for the 2003 CAST data:

$$g_{a\gamma} < 1.16 \times 10^{-10} \text{GeV}^{-1} (95\% \text{CL}).$$

Thus far our analysis was limited to the mass range $m_a \lesssim 0.02$ eV where the expected signal is mass-independent because the axion-photon oscillation length far exceeds the length of the magnet. For higher m_a the overall signal strength diminishes rapidly and the spectral shape differs. Our procedure was repeated for different values of m_a to obtain the entire 95% CL exclusion line shown in Fig. 1.

4. Summary

Our limit improves the best previous laboratory constraints [19] on $g_{a\gamma}$ by a factor 5 in our coherence region $m_a \lesssim 0.02$ eV. This result excludes an important part of the parameter space not excluded by solar age considerations [30] and is comparable, in this range of masses, to the limit derived from stellar energy-loss arguments. A higher sensitivity is expected from the 2004 data with improved conditions in all detectors, which should allow us to surpass the astrophysical limit. In addition, starting in 2005, CAST plans to take data with a varying-pressure buffer gas in the magnet pipes, in order to restore coherence for axion masses above 0.02 eV. The extended sensitivity to higher axion masses will allow us to enter into the region shown in Fig. 1 which is especially motivated by axion models [31].

References

1. S. Eidelman et al., Phys. Lett. B **592**, 1 (2004).
2. R. Bradley et al., Rev. Mod. Phys. **75**, 777 (2003).
3. H. Primakoff, Phys. Rev. **81**, 899 (1951).
4. P. Sikivie, Phys. Rev. Lett. **51**, 1415 (1983) [Erratum-ibid. **52**, 695 (1984)].
5. G. Raffelt and L. Stodolsky, Phys. Rev. D **37**, 1237 (1988).
6. D. S. Gorbunov, G. G. Raffelt and D. V. Semikoz, Phys. Rev. D **64**, 096005 (2001).
7. C. Csaki et al., JCAP **0305**, 005 (2003).
8. C. Csaki, N. Kaloper and J. Terning, Phys. Rev. Lett. **88**, 161302 (2002).
9. E. Mörtsell and A. Goobar, JCAP **0304**, 003 (2003).
10. B. A. Bassett, Astrophys. J. **607**, 661 (2004).
11. L. Maiani, R. Petronzio and E. Zavattini, Phys. Lett. B **175**, 359 (1986).
12. G. Cantatore et al., these proceedings.
13. D. A. Dicus et al., Phys. Rev. D **18**, 1829 (1978).
14. G. G. Raffelt, Ann. Rev. Nucl. Part. Sci. **49**, 163 (1999).
15. The spectrum in [18] has been changed to that proposed in [16], however with a modified normalization constant to match the total axion flux used here, which is predicted by a more recent solar model [17].
16. R. J. Creswick et al., Phys. Lett. B **427**, 235 (1998).

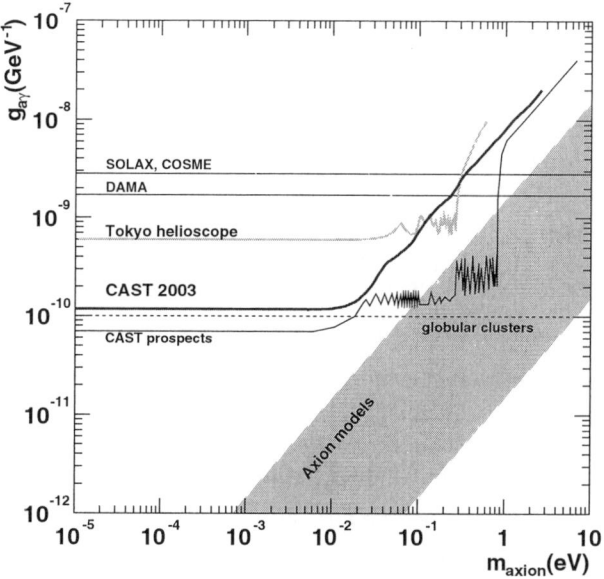

Figure 1. Exclusion limit (95% CL) from the CAST 2003 data compared with other constraints discussed in the introduction. The shaded band represents typical theoretical models. Also shown is the future CAST sensitivity as foreseen in the experiment proposal.

17. J. N. Bahcall, M H. Pinsonneault and S. Basu, Astrophys. J. **555**, 990 (2001).
18. K. van Bibber et al., Phys. Rev. D **39**, 2089 (1989).
19. S. Moriyama et al., Phys. Lett. B **434**, 147 (1998).
20. Y. Inoue et al., Phys. Lett. B **536**, 18 (2002).
21. F. T. Avignone et al., Phys. Rev. Lett. **81**, 5068 (1998).
22. A. Morales et al., Astropart. Phys. **16**, 325 (2002).
23. R. Bernabei et al., Phys. Lett. B **515** 6 (2001).
24. K. Zioutas et al., Nucl. Instrum. Meth. A **425** 480 (1999).
25. K. Barth et al., Proc. 2003 Cryogenic Engineering Conference (CEC) and Cryogenic Materials Conference (ICMC).
26. http://cast.web.cern.ch/CAST/edited_ tracking.mov
27. Y. Giomataris et al., Nucl. Instrum. Meth. A **376** 29 (1996).
28. J. Altmann et al., in *Proceedings of SPIE: X-Ray Optics, Instruments, and Mission*, 1998, edited by Richard B. Hoover and Arthur B. Walker, p. 350; J. W. Egle et al., *ibid.*, p. 359; P. Friedrich et al., *ibid.*, p. 369.
29. S. Andriamonje et al. [CAST collaboration], arXiv:hep-ex/0411033.
30. H. Schlattl, A. Weiss and G. Raffelt, Astropart. Phys. **10** 353 (1999).
31. D. B. Kaplan, Nucl. Phys. B **260**, 215 (1985).

PVLAS RESULTS ON LASER PRODUCTION OF AXION-LIKE DARK MATTER CANDIDATE PARTICLES

E. ZAVATTINI[1], G. ZAVATTINI[4], G. RUOSO[2], E. POLACCO[3], E. MILOTTI[1],
M. KARUZA[1], U. GASTALDI[2], G. DI DOMENICO[4], F. DELLA VALLE[1],
R. CIMINO[5], S. CARUSOTTO[3], G. CANTATORE[1,*], M. BREGANT[1]

1 - Università e I.N.F.N. Trieste, Via Valerio 2, 34127 Trieste, Italy

2 - Lab. Naz.di Legnaro dell'I.N.F.N., Viale dell'Università 2, 35020 Legnaro, Italy

3 - Università e I.N.F.N. Pisa, Via F. Buonarroti 2, 56100 Pisa, Italy

4 - Università e I.N.F.N. Ferrara, Via del Paradiso 12, 44100 Ferrara, Italy

5 - Lab. Naz.di Frascati dell'I.N.F.N., Via E. Fermi 40, 00044 Frascati, Italy

Microscopic processes in the quantum vacuum, such as photon-photon scattering and the Primakoff effect, where light, neutral, scalar/pseudoscalar particles are produced from a two-photon vertex, could give rise, in the presence of an external magnetic field, to macroscopically observable optical phenomena, such as vacuum birefringence and dichroism. Such particles are candidate constituents of the Cold Dark Matter. The PVLAS collaboration is operating a high-sensitivity ($\sim 10^{-7}$ 1/$\sqrt{Hz}$) optical ellipsometer capable of detecting very small changes in the light polarisation state induced by a strong transverse magnetic field on a linearly polarised laser beam. This ellipsometer is capable of measuring vacuum birefringences and dichroisms, in an independent way, down to levels below 10^{-8} rad, for about one hour of data taking time. Preliminary results, involving the observation of candidate signals in vacuum, will be discussed.

1. Introduction

The existence of zero point fluctuations in the energy of quantum vacuum opens the possibility of considering vacuum itself as material medium with dielectric properties. In the presence of an external magnetic field, quantum vacuum can behave as an optically anisotropic medium and induce both an ellipticity and a dichroism on a linearly polarised light beam propagating transversely with respect to the field [1,2,3]. A determination of the total photon-photon scattering cross section at low energies, and of mass and coupling constant of hitherto undetected spin-zero particles, produced via the Primakoff effect, can be extracted from a positive observation of magnetically induced vacuum

---

*Corresponding author

birefringence and dichroism. Such particles are considered good candidate constituents of the Cold Dark Matter [4,5].

The PVLAS [6,7,8,9] collaboration is operating, at the INFN Legnaro National Laboratory, Legnaro, Padova, Italy, a high-sensitivity ($\sim 10^{-7}$ $1/\sqrt{Hz}$) optical ellipsometer capable of detecting, using the heterodyne technique, both ellipticities and dichroisms in an independent way, down to levels below 10^{-8} rad, for about one hour of data taking time.

2. PVLAS apparatus and experimental technique

Figure 1 shows a horizontal schematic layout of the PVLAS ellipsometer: the actual set-up extends vertically. The interaction region, where laser photons interact with the external magnetic field, is contained within a high-finesse Fabry-Perot optical resonator consisting of a pair of dielectric, multilayer, high-reflectivity mirrors (M_1 and M_2) placed 6.4 m apart. Vacuum ($P < 10^{-7}$ mbar) is maintained within a 4.6 m long, 25 mm diameter quartz tube (the tube and the additional vacuum chambers containing the optical elements are omitted in Figure 1 for clarity) traversing the room temperature bore of a 1.1 m long superconducting dipole magnet. A turntable, actuated by a hydraulic motor, can rotate the liquid–He cryostat housing the magnet [10], around a vertical axis, to provide a time-varying signal for heterodyne detection. During data taking, the turntable rotates at frequencies $f_M = 0.3$–0.5 Hz. The heterodyne ellipsometer proper consists of a pair of crossed polarising prisms P and A, together with an ellipticity modulator (Stress Optic Modulator, or SOM, see [11]).

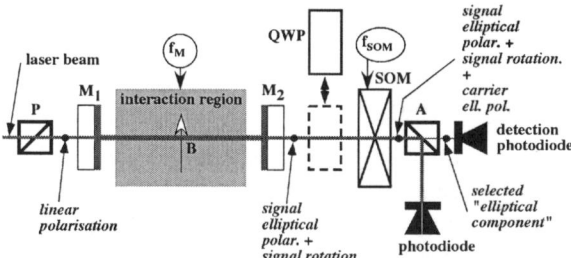

Figure 1 - Schematic layout of the PVLAS apparatus. The figure shows only the elements relevant to the discussion of the heterodyne ellipsometer (see text) and omits details on the frequency-locking feedback loop necessary to keep the laser and the FP cavity at resonance, the vacuum chambers, the cryogenic system, the magnet rotation assembly and the data acquisition and digitisation system. The quarter–wave plate (QWP), which transforms rotation into ellipticity (and ellipticity into rotation), is inserted when performing rotation measurements.

The ellipticity modulator provides an ellipticity carrier signal for the purposes of heterodyne detection and is driven at a frequency f_{SOM} = 506 Hz. A quarter–wave plate QWP can be inserted between FP cavity and SOM: when properly aligned it transforms apparent rotations (dichroisms) into ellipticities [12], which can then beat with the SOM carrier ellipticity signal and be detected. The laser beam, coming from a 1064 nm, 100 mW CW output power Nd:YAG laser, is kept at resonance with the FP cavity by means of an electro-optical feedback loop. This has the effect of amplifying the optical path within the interaction region by a typical factor of N = 63700. Details on the PVLAS cavity and on the frequency locking system can be found in [13]. Light transmitted trough the analyser is detected by a photodiode and the resulting voltage signal is both directly fed to a computer for digitisation at an 8.2 kHz sampling rate ("fast" acquisition) and to a lock-in amplifier referenced at the 506 Hz SOM carrier frequency: the demodulated signal is frequency analysed on-line by a spectrum analyser, and digitised for off-line analysis ("slow" acquisition). The triggering and gate signals for the slow acquisition are obtained from a series of 32 marks placed around the circumference of the turntable: in this way, for every acquired data point the direction of the rotating magnetic field with respect to the fixed initial polarisation direction is known, and absolute phases can be determined for all acquired signals.

The detection photodiode signal is analysed off–line giving amplitude and phase of quantities of interest. In particular, ellipticity and dichroism signals generated by interactions with the magnetic field should appear, in a Fourier spectrum, at twice the magnet rotation frequency (since data acquisition is synchronous with magnet rotation, this is the convenient unit for measuring frequencies).

3. Measurements and results

Recall that, with the PVLAS set-up, both ellipticities and dichroisms induced in the laser beam polarisation state can be detected in an independent way by inserting, in the beam path, a quarter–wave plate (QWP) before the ellipticity modulator. To determine the behaviour of vacuum as a material medium, several measurement runs, each lasting about 10^3 s of data taking, were conducted in the following conditions:
- magnet at 5.5 T and rotating at a frequency around 0.3 Hz
- FP cavity at resonance with $F \sim 10^5$
- QWP inserted (dichroism) or QWP removed (ellipticity)

As a separate check, test gases such as Neon and Nitrogen, were preliminarily inserted in the vacuum chamber in order to measure amplitude and phase of their magnetic birefringence (Cotton-Mouton effect [14]). Since the phase of "true" physical signals is determined by the choice of the direction of the initial linear polarisation of the laser beam, when the measured phases of gas signals reproduce this phase the apparatus is functioning properly. In our case, signals corresponding to positive birefringences (Neon) have a phase around 195°, while negative birefringences (Nitrogen) appear around 15° indicating proper operation of the ellipsometer.

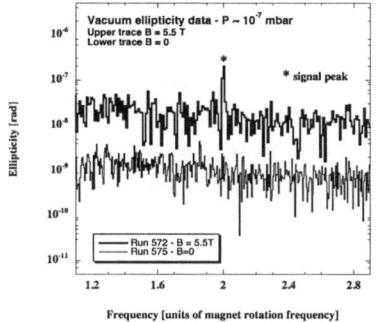

Figure 2 - Sample Fourier spectra of the detection photodiode signal around twice the magnet rotation frequency (see text). Spectra correspond to data taken with vacuum (residual pressure P < 10-7 mbar) in the interaction region and cavity present. The upper and lower traces refer to field intensities of 5.5 T and 0 T, respectively. Notice the peak appearing exactly at twice the magnet rotation frequency in the upper trace.

Figure 2 shows sample Fourier spectra of the current generated by the detection photodiode when B = 0 (lower trace) and B = 5.5 T (upper trace), and the interaction region is in vacuum. Notice the signal peak in the upper trace appearing at the frequency 2 in units of magnet rotation frequency. The peak is absent in the lower trace, where the field intensity is zero. These spectra where obtained with the QWP removed, and represent therefore ellipticity data. Dichroism signal peaks, observed when the QWP is inserted, show a similar appearance.

An extensive series of tests has been conducted on the observed ellipticity and dichroism signals in order to determine their nature and origin. We can summarise the preliminary results of these tests as follows.
- For both QWP inserted and removed:

- signal peaks appear at the expected frequency when B ≠ 0 and the FP cavity is present
  - spurious effects such as electrical pick-ups and diffusion from magnetised surfaces have been excluded
  - signals are generated within the Fabry-Perot cavity
- When the QWP is removed (ellipticity measurement):
  - magnetic birefringence due to residual gas excluded by mass spectrometer measurements
  - magnetic birefringence due to stray magnetic fields induced by reflection on mirror surfaces excluded by direct measurement on test apparatus [15]
- When the QWP is inserted (dichroism measurement):
  - signal peak phase changes by 180° when QWP is rotated by 90°

The observed signals are therefore compatible with a "true" ellipticity and a "true" dichroism generated within the FP cavity.

There are, however, issues which we have not been yet able to clarify. First of all, when measuring ellipticity peaks we find that signal phases do not group well around the direction of the "physical axis" as should be expected from a signal having a clear physical origin (as opposed to an instrumental one). Secondly, statistics on dichroism measurements should be increased. If we set aside, for the moment, the discrepancies on the phase measurements we can take an average of the ellipticity peak amplitude across the entire available data set giving, for B = 5.5 T and F = 97000, a value of $\Psi \sim 6 \cdot 10^{-7}$ rad. Similarly, we can give an average value for the dichroism peak amplitude of $\varepsilon \sim 3 \cdot 10^{-7}$ rad.

4. Discussion and conclusions

Measurement runs conducted in vacuum up to now show the presence of signal peaks, well above background, at the expected frequency, both when measuring ellipticities and when measuring dichroisms. Diagnostic tests show that these signals, which appear only when the magnetic field intensity B ≠ 0 and the FP cavity is present, are optical effects generated inside the FP cavity itself. The distribution of ellipticity signal phases, however, when compared to the expected phase distribution of a physical signal, does not exclude the presence of an unknown, albeit subtle, instrumental effect mimicking a true anisotropy of vacuum. The PVLAS apparatus can be still exploited to its full potential in upcoming measurement runs by increasing statistics on dichroism measurements and by conducting studies on the effects of low–pressure pure gases in the interaction region. Furthermore, medium term improvements of the apparatus as a whole are already planned, including a change in the probe laser wavelength from 1064 nm to 532 nm, and the installation of a new amagnetic access structure to the apparatus.

Finally, as an exercise, one might assume that the observed PVLAS signals correspond to true physical effects and try to interpret the observed signal amplitudes. Regarding the ellipticity, it is clear that such a signal could not possibly come from QED processes alone. In this case, where an ellipticity $\Psi \sim 6 \cdot 10^{-7}$ rad would correspond to a measured index of refraction difference of $\Delta n = 3.4 \cdot 10^{-18}$, the expected number from QED alone would be $\Delta n_{QED} = 1.2 \cdot 10^{-22}$ (for B = 5.5 T, λ = 1064 nm, F = 97000) and such processes could not possibly account for the observed value. It is always possible to use Ψ to give a limiting value for the total photon-photon scattering cross section of $3 \cdot 10^{-21}$ pb, to be compared to the previously available experimental value of $1.5 \cdot 10^{-12}$ pb [16]. Carrying the exercise further, the ellipticity peak amplitude could be interpreted as due to virtual production of neutral, light, pseudoscalar particles coupled to two photons [4,5]. Following this line of reasoning, the dichroism signal could then be ascribed to real production, and the two amplitudes together, $\Psi \sim 6 \cdot 10^{-7}$ and $\varepsilon \sim 3 \cdot 10^{-7}$ rad, used to extract mass m_p and inverse coupling M to two photons of the produced particles giving $m_p \sim 10^{-3}$ eV and $M \sim 4 \cdot 10^{-5}$ GeV.

References

1. W. Heisenberg, and H. Euler, *Z. Phys.*, **38**, 714 (1936)
2. J. Schwinger, *Phys. Rev.* **82**, 664 (1951)
3. E. Iacopini, and E. Zavattini, *Phys. Lett.* **B85**, 151 (1979)
4. L. Maiani, E. Petronzio, E. Zavattini, *Phys. Lett.* **B175**, 359 (1986)
5. E. Massó and R, Toldrà, *Phys. Rev.* **D52**, 1755 (1995)
6. D. Bakalov et al., *Quantum and Semicl. Optics*, **10**, 239 (1998).
7. E. Zavattini et al., AIP Conf. Proc., **564**, 77 (2001)
8. F. Brandi et al., *Nuclear Instr. and Meth.,* **A461**, 329 (2001),
9. M. Bregant et al., in *IDM 2002 – Proc. of the Fourth Int. Work. on the Identification of Dark Matter*, N.J.C. Spooner and V. Kudryavtsev eds., World Scientific, Singapore (2003)
10. R. Pengo et al., in *"Frontier Tests of QED and Physics of the Vacuum"*, E. Zavattini, D. Bakalov and C. Rizzo eds., 59, Heron Press, Sofia (1998)
11. F. Brandi et al., *Meas. Sci. Technol.*, **12**, 1503 (2001)
12. M. Born and E. Wolf, *Principles of Optics*, Pergamon, New York (1980).
13. G. Cantatore et al., *The Rev. of Scient. Instr.*, **66**, no.4, 2785 (1995)
14. C. Rizzo et al., *Intern. Rev. in Phys. Chem.* **16**, no.1, 81 (1997)
15. G. Bialolenker et al., *Appl. Phys.* **B68**, 703 (1999).
16. D. Bernard et al., *The European Physical Journal D*, **10**, 141 (1999)

STATUS OF THE AXION DARK-MATTER EXPERIMENT (ADMX)

D. KINION

Lawrence Livermore National Laboratory,
7000 East Ave.,
Livermore, CA 94550, USA
E-mail: kinion1@llnl.gov

The Axion Dark Matter Experiment (ADMX) at Lawrence Livermore National Laboratory has been searching for dark-matter axions using the Sikivie microwave cavity technique. Axions are light pseudoscalar particles which arise from the Peccei-Quinn solution to the Strong-CP problem. The ADMX collaboration includes LLNL, the University of Florida, the National Radio Astronomy Observatory, and the University of California at Berkeley.

1. Introduction

Axions, a promising cold dark matter candidate, arise from a minimal extension of the Standard Model to enforce Strong-CP conservation. The Peccei-Quinn solution to the Strong-CP problem in QCD[1] involves an approximate $U_{PQ}(1)$ global symmetry which is spontaneously broken at some unknown symmetry-breaking scale f_a, and the axion is the associated pseudo-Goldstone Boson.[2]

The properties of the axion depend mainly on the symmetry breaking scale f_a, for instance its mass is inversely proportional to f_a and given by

$$m_a[eV] \approx 0.6 \; eV \frac{10^7 \; GeV}{f_a \; [GeV]}. \qquad (1)$$

All of the axion couplings are proportional to m_a including the relevant two-photon coupling described by

$$L_{a\gamma\gamma} = g_\gamma \frac{\alpha \phi_a}{4\pi f_a} F_{\mu\nu} \tilde{F}^{\mu\nu} = -g_{a\gamma\gamma} \phi_a \mathbf{E} \cdot \mathbf{B} \qquad (2)$$

where α is the fine structure constant, ϕ_a is the axion field, g_γ is a model-dependent constant of order unity, and $g_{a\gamma\gamma} = (\alpha g_\gamma/\pi f_a)$. For the two most important axion models, KSVZ[3] and DFSZ,[4] $g_\gamma \sim 0.97$, and $g_\gamma \sim -0.36$ respectively.

Since f_a is unknown and arbitrary, m_a could have any value. Astrophysical and cosmological considerations help constrain m_a resulting in the allowed mass range, or axion window:[5] $10^{-6} < m_a < 10^{-2}$ eV.

2. The Sikivie Microwave-Cavity Technique

To date, the most efficient method of searching for axions is the microwave cavity technique originally proposed by Sikivie.[6] In a static background magnetic field, axions will decay into single photons via the Primakoff effect. The energy of the photons is equal to the rest mass of the axion with a small contribution from its kinetic energy, hence their frequency is given by $hf = m_a c^2 (1+O(10^{-6}))$. At the lower end of the axion window, the frequency of the photons lies in the microwave regime. A high-Q resonant cavity, tuned to the axion mass serves as the detector for the converted photons. The expected signal power varies with the experimental parameters as[6,7]

$$P_{a \to \gamma} \propto B^2 V C Q f \rho_a \qquad (3)$$

where B is the background magnetic field, V is the cavity volume, C is a mode dependent form factor, Q is the loaded quality factor, f is the resonant frequency, and ρ_a is the local halo axion density. Axions couple most strongly to the TM_{010} cavity mode ($C \sim 0.5$), so it is the only mode used in most searches. For the parameters of the ADMX experiment, the power from KSVZ axions is typically 5×10^{-22} W.

Since the axion mass is unknown, the frequency of the cavity must be tunable. For a given signal-to-noise ratio, the scan rate scales as:

$$\frac{df}{dt} \propto \frac{f_o^2 Q_u C^2 B^4 V^2}{T_s^2}. \qquad (4)$$

3. Current ADMX Hardware and Limits

Figure 1 is a schematic of the ADMX detector. The magnet employed in this search is an 8 T superconducting NbTi solenoid.

The microwave cavity is a right-circular cylinder 50 cm in diameter and 1 m long constructed from stainless steel and plated with ultra-high purity, oxygen-free copper. The resonant frequency of the empty single-cavity is 460 MHz and the unloaded Q is approximately 200000. Moving a combination of metal and dielectric rods, running the full length of the cavity, changes the resonant frequency.

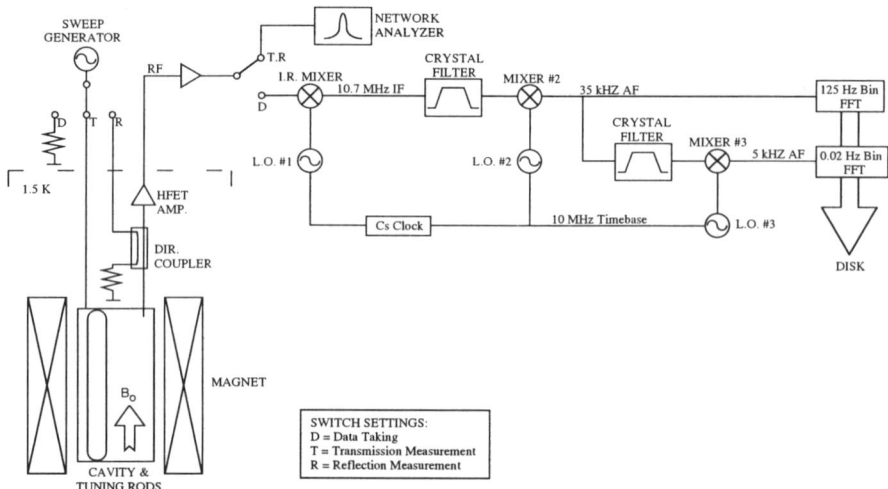

Figure 1. Schematic of the ADMX experiment.

Superfluid ^{4}He maintains the physical temperature of the cavity near 1.5 K. The cryogenic amplifiers currently used are double-balanced GaAs HFET amplifiers supplied by NRAO.[8] The *in situ* measured noise temperatures range from 1.7 - 4.5 K. Cascading two of these amplifiers achieves sufficient gain (35 dB) to render downstream noise contributions negligible.

Before data is taken at a given frequency, a transmission measurement is made to determine the resonant frequency and Q. Next, the double-heterodyne receiver shown in Figure 1 mixes a small bandwidth centered on the cavity frequency down to 35 kHz. This audio frequency signal is then sent to medium and high-resolution spectrum analyzers.

The medium-resolution search channel consists of a commercial FFT spectrum analyzer with a frequency resolution of 125 Hz. These data are coadded and the result searched for Maxwellian peaks a few bins wide (about 700 Hz) characteristic of thermalized axions in the halo.[9]

An independent, high-resolution search channel operates in parallel to explore the possibility of fine-structure in the axion signal [10,11]. The 35 kHz signal passes through a third mixing stage to shift the center frequency to 5 kHz. A PC based DSP takes a single 50 second spectrum and performs an FFT with 20 mHz frequency resolution, about the limit imposed by the Doppler shift due to the earth's rotation.

Positive fluctuations in the power spectrum are identified as candidate

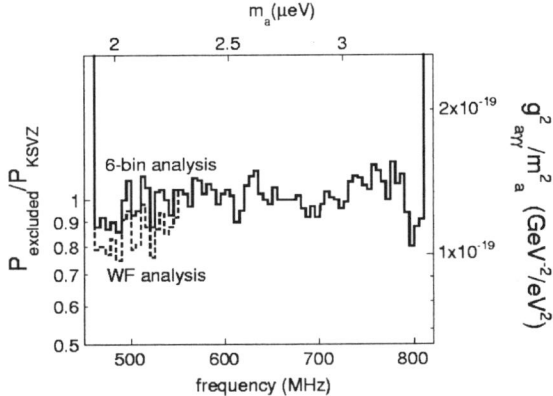

Figure 2. Axion couplings and masses excluded at the 90% confidence level by ADMX normalized to the KSVZ model predictions. The dotted line is the exclusion from a Wiener filter analysis.[12]

peaks and rescanned. Peaks which are statistical in nature will not reappear and can be eliminated as axion signals. Candidates which survive the rescan are considered persistent, and must be checked in other ways. Those few that remained have all been linked to external sources by using an antenna in the room. If a peak were to survive all of these checks, the definitive test would be to see if it appears only when the magnetic field is on.

So far, no axion signal has been detected. Based on these results, we exclude at 90% confidence a KSVZ axion of mass between 1.9 and 3.3 μeV, assuming that thermalized axions comprise a major fraction of our galactic halo ($\rho_a = 450$ MeV/cm^3). This exclusion region is shown in Figure 2. For more details see Ref. 13.

4. Phase I Upgrade

A major upgrade to ADMX began in the summer of 2004 with the goal of increasing the scan rate of the existing experiment. As seen in Equation 4, the scan rate depends strongly on the parameters B, V, Q and T_S. B and V are both constrained by the current hardware, and the cavity Q is already very close to the theoretical limit for copper in the anomalous skin-depth regime. There is, however, room for significant improvement in the system noise temperature T_S.

The NRAO HFET amplifiers are state-of-the-art, however their noise temperatures are still more than an order of magnitude higher than the

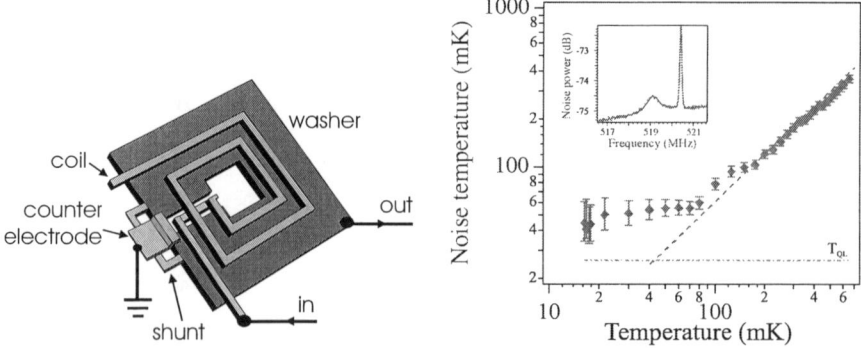

Figure 3. (Left) Schematic of a microstrip SQUID amplifier. (Right) Measured noise temperature versus physical temperature for a microstrip SQUID amplifier. The inset shows the Nyquist noise peak as well as a calibration signal.

Standard Quantum Limit ($T_N = h\nu/k_B$).[5] In the past several years a group at Berkeley led by John Clarke has developed dc SQUID amplifiers in the 100 - 3000 MHz range specifically for the ADMX experiment.

As seen on the left of Figure 3, a dc SQUID consists of two Josephson junctions connected in parallel on a superconducting loop. The SQUID produces an output voltage in response to a small input flux, and is a very sensitive flux-to-voltage transducer. Flux is coupled into the SQUID through a microstrip input coil. Near the fundamental frequency of the stripline, the gain of the amplifier is strongly enhanced. The highest frequency amplifier built so has a resonant frequency greater than 8 GHz. More details can be found in Ref. 5.

The dominant source of noise in these devices is the Johnson noise from the resistive shunts across the Josephson junctions. This noise scales linearly with temperature, so the noise temperature of microstrip amplifiers is expected to be proportional to the physical temperature until either the Quantum Limit is reached or hot electron effects in the shunts become dominant. This behavior is seen in Figure 3 which shows the result of a noise temperature measurement at 520 MHz.[14] Work is continuing to demonstrate quantum-limited noise performance.

The second, almost equal, contribution to the current value of T_S comes from the 1.5 K physical temperature. This can be reduced below 100 mK by cooling the entire experiment with a dilution refrigerator.

These two pursuits of lowering the system noise temperature have been divided into a two-phase upgrade. The goal of the Phase I upgrade which is now underway is getting a SQUID amplifier working in the current experiment at 1.5 K. This will result in a modest improvement in noise temperature, but will demonstrate the feasibility of using microstrip SQUID amplifiers. The real improvement comes in Phase II which will involve cooling the entire system to 100 mK by installing a dilution refrigerator. The ultimate result should be an order-of-magnitude reduction of the system noise temperature.

5. Conclusion

After twenty five years, the axion remains the most elegant solution to the Strong-CP problem in QCD. The ADMX experiment has been operating since February 1996 and has excluded KSVZ axions in the mass range 1.9 to 3.5 μeV with greater than 90% confidence. An upgrade is underway to incorporate SQUID amplifiers which will eventually allow us to reach sensitivity to DFSZ axions.

Acknowledgments

This work was performed under the auspices of the U.S. Department of Energy by University of California, Lawrence Livermore National Laboratory under Contract W-7405-Eng-48.

References

1. R. Peccei and H. Quinn, *Phys. Rev. Lett.* **38**, 1440 (1977).
2. S. Weinberg, *Phys. Rev. Lett.* **40**, 223 (1978); F. Wilczek, *ibid.* 279 (1978).
3. J.E. Kim, *Phys. Rev. Lett.* **43**, 103 (1979); M.A. Shifman, A.I. Vainshtein, and V.I. Zakharov, *Nucl. Phys.* **166**, 493 (1980).
4. M. Dine, W. Fischler, and M. Srednicki, *Phys. Lett.* **104**, 199 (1981); A.R. Zhitnitsky, *Sov. J. Nucl. Phys.* **31**, 260 (1980).
5. R. Bradley *et al.*, *Rev. Mod. Phys.* **75**, 777 (2003).
6. P. Sikivie, *Phys. Rev. Lett.* **51**, 1415 (1983).
7. L. Krauss *et al.*, *Phys. Rev. Lett.* **55**, 1797 (1985).
8. E. Daw and R.F. Bradley, *J. Appl. Phys.* **82**, 1925 (1997).
9. M.S. Turner, *Phys. Rev.* D **42**, 3572 (1990).
10. P. Sikivie and J. Ipser, *Phys. Lett.* B **291**, 288 (1992).
11. P. Sikivie *et al.*, *Phys. Rev. Lett.* **75**, 2911 (1995).
12. D. B. Yu *et al.*, *Phys. Rev.* D **69**, 011101 (2004).
13. S. Asztalos *et al.*, *Phys. Rev.* D **64**, 092003 (2001).
14. Mück, M., J. B. Kycia, and J. Clarke, *Appl. Phys. Lett.* **78**, 967 (2001).

LABORATORI NAZIONALI DEL GRAN SASSO AND THE ILIAS[†] INITIATIVE

N. FERRARI

Laboratori Nazionali del Gran Sasso
S.S. 17bis km 18+910 – 67010 Assergi, Italy

Laboratori Nazionali del Gran Sasso (LNGS) is actually the largest deep underground laboratory in the world completely devoted to fundamental science. In this talk I give a review of the main characteristics of the infrascructure and of the past and ongoing scientific activities. LNGS is taking part in the EU initiative ILIAS (Integrated Large Infrastructures for Astroparticle Physics). Within ILIAS LNGS is collaborating with the other three european deep underground sites: Laboratoire Soterrain de Modane (France), Laboratorio Subterraneo de Canfranc (Spain), and Boulby Mine Underground Laboratory (UK).

1. The Laboratori Nazionali del Gran Sasso (LNGS)

1.1. *The infrastructure*

Laboratori Nazionali del Gran Sasso are located in the Gran Sasso highway tunnel connecting L'Aquila to Teramo, in central Italy, at about 120 km from Roma. The laboratory is operated by Istituto Nazionale di Fisica Nucleare (INFN).

The underground area includes 3 main halls (called A, B, and C) with dimensions of about 100 m x 20m x 15 m, and a number of service tunnels, for a total volume of about 180000 m^3 and a total surface exceeding 6000 m^2. A sketch of the underground area is shown in Figure 1. The excavation of the underground lab started in 1982 and was completed in 1987.

The lab is located on the vertical of Monte L'Aquila (2600 meters high), and has an average rock coverage of 1400 meters. The muon flux is reduced by a factor 10^6 with respect to the surface; the neutron flux is also reduced by a factor

---

[†] The ILIAS initiative is supported by European Union under the contract RII3-CT-2004-506222

1000, thanks to the low content of U and Th in the dolomite rocks of the mountain.

Outside facilities are located near the village of Assergi (on the L'Aquila side of the tunnel) at a height of about 1000 m. They include offices for the lab staff and for host researchers and technicians, mechanical workshop, chemical laboratory, electronic workshop, computing center, library, canteen, conference rooms, large assembly rooms, and the administration department. A view of the external facilities is shown in Figure 2.

The permanent staff of the laboratory is composed by 60 people (physicists, technicians, engineers, and administration staff). The scientists involved in LNGS experiments includes more than 700 researchers from 24 countries.

The laboratory structure is organized into Divisions and Services including research division, technical and general services division, directorate, administration, public affairs, prevention and protection service. Two separate bodies, the Laboratory Council and the Scientific Committee, assist the director in the guidance of the laboratory.

The geographical location (inside the Gran Sasso-Monti della Laga National Park) and the special operating conditions (near the highway tunnel and in proximity to water basins) requires special attention to safety and environmental aspects.

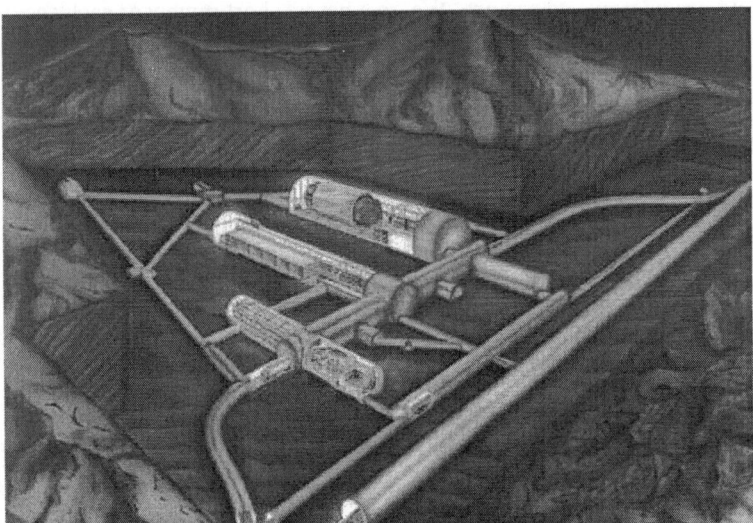

Figure 1. Sketch of the LNGS underground area. The three Halls (A, B and C) are linked by several interconnecting tunnels and directly accessible from the highway connecting Teramo to L'Aquila.

Figure 2. The LNGS outside facilities.

1.2. *The scientific programme*

The experimental activities ongoing at LNGS include all major research topics in the field of underground science. We give here a short review of each sector. For a complete review and bibliography of the experiments described in this section see [1] [2] and the references quoted therein.

1. *Neutrino astrophysics*: thanks to the large areas available underground, LNGS is an ideal infrastructure for large experiments designed for the detection of astrophysical neutrinos. Many experiments have been and are being carried out for the study of neutrinos from the sun, from supernovae, and from the atmosphere.

   The Gallex/GNO experiment has been measuring low energy solar neutrinos with a radiochemical technique using a 30 t gallium target. The experiment was successfully taking data between 1991 and 2003, and detected for the first time the low energy "pp neutrinos"; moreover it gave evidence at the beginning of 90s for neutrino oscillations, and was monitoring the low energy solar neutrino flux for a complete solar cycle.

   After the success of GNO, the solar neutrino observations at LNGS are expected to continue in the next future with the Borexino detector, made of 300 tons of ultrapure liquid scintillator (+ 1000 tons of buffer). The scintillator is contained in a stainless steel sphere surrounded by water; the detection of solar neutrino interactions via elestic scattering off electrons, requires ultra-low level radiupurity in all the components of the apparatus. The aim of Borexino is to study in real-time the ^{7}Be component of the solar neutrino flux. Borexino is now ready for filling after a partial stop of the activities due to an accident occurred in august 2002. Besides solar neutrinos, Borexino will also be able to detect supernova ν, geophysics ν, and to test ν magnetic moment.

The MACRO experiment was taking data on atmospheric neutrinos between 1991 and 2001 with a massive detector made of streamer tubes and liquid scintillator modules. The results after 10 years of successful data taking supported a strong evidence for neutrino flavour oscillations, in agreement with the japanese experiment of Superkamiokande. Other results by MARCRO are a complete and precise characterization of the muon energy spectrum and angular distribution underground, and the best upper limit in the world on magnetic monopole parameters.

The LVD detector, made of 1000 tons of liquid scintillator in 840 counters is looking for neutrinos and antineutrinos from a galactic supernova. The detector is continuously taking data since 1992 with a very high duty cycle waiting for the next galactic supernova to explode.

2. *Long baseline neutrino detection*: the CNGS project has the aim to study the neutrino oscillation parameters with a neutrino beam produced at CERN and shot to LNGS. Two experiments will detect at LNGS neutrinos produced at CERN after travelling a 720 km distance.

OPERA is a 1.8 kton detector made of Pb sheets and nuclear emulsions in the form of 230000 emulsion cloud chambers, and two big magnetic spectrometers (RPC and scintillating fibers). The main goal of OPERA is to detect for the first time in the world the appearance of tau neutrinos from a muon neutrino beam. The emulsion chamber technique will allow identification of the tau emitted by ν interactions with an almost zero background. The experiment is under construction: the magnetic spectrometers are expected to be completed in 2005 and data taking should start in 2007.

ICARUS is a 3 kton detector based on the use of liquid argon as a large time projection chamber. The first 600 ton module of ICARUS was built and tested above ground, and is going to be transported to LNGS before the end of 2004. Installation of the complete 3 ktons requires major works in the underground infrastructure, so it is still not clear if the complete detector will be ready for the neutrino beam commissioning in 2007. In any case ICARUS is a general-purpose innovative detector with a broad programme not limited to the CNGS project.

3. *ββ-decay search*. At LNGS a lot of efforts are ongoing on this issue, crucial for the determination of the absolute neutrino mass. Different and complementary techniques are being employed.

   The ββ Heidelberg-Moscow experiment operated 11 kg of enriched ^{76}Ge crystals in the form of HP-Ge detectors at liquid nitrogen temperature. Data taking was going on regularly in the period 1993-2003; this experiment is presently the most sensitive in the world in the ββ decay sector. Evidence for a possible ββ decay signal is claimed, corresponding to a neutrino mass in the range 0.1-0.9 eV. This evidence calls for furher confirmation possibly using different isotopes.

   Cuoricino (upgrade of the Mibeta experiment) has recently started to operate 40.7 kg of TeO_2 crystals as thermal detectors at the temperature of a few millikelvin. Cuoricino is expected to reach a sensitivity of the order of 0.3 eV on the neutrino mass after 3 years of data taking. In a few years Cuoricino will be upgraded to Cuore: the TeO_2 mass will be increased to 750 kg and the expected sensitivity on the neutrino mass will go down to about 30 meV.

   The aim of the recently approved GERDA experiment is to build a setup of HP-Ge detectors enriched in ^{76}Ge with a total mass of about 20 kg and improved background reduction.

4. *Dark matter search*; due to the extreme importance of this subject for cosmology and particle physics, many experiments are ongoing at LNGS looking for WIMPs dark matter candidates. Detailed reports from all the experiments have been presented at this conference.

   Dama/NaI was operating a 100 kg detector of ulta pure NaI crystals with the aim to detect the scintillation light produced by elastic scattering of WIMPs. The experiment was taking data between 1995 and 2002 with increasing sensitivity. Data from 7 annual cycles show a modulation compatible with WIMPs interactions. The DAMA/NaI setup was recently upgraded to 250 kg of sensitive mass, and the new detector (LIBRA) started data taking in 2003.

   CRESST is operating a thermal detector made of $CaWO_2$ crystals at low temperature. The readout of both the thermal and scintillation signals produced by particle interactions in the crystals allows a powerful discrimination of WIMPs signals against background.

   WARP is an argon double phase (liquid+gas) detector planned for installation at LNGS in the next years. Particles interacting in the liquid Ar phase give a double signal, the first from the primary scintillation light, and the second from scintillation in the gas originated by multiplication of ionization electrons drifted and extracted into the gas phase by an electric field. A 2.3 liters prototype is being successfully operated; the installation of the 100 liters detector will start in 2005.

The Cuore and GERDA experiments, mainly designed for ββ decay search, will also be potentially sensitive to dark matter. Special investigations have been carried out to study the sensitivity of Ge detectors to dark matter search by the heidelberg-Moscow collaboration with the HDMS and GENIUS-TF experiments.

5. *Nuclear astrophysics.* LNGS hosts one of the best facilities in the world for the study of nuclear reactions relevant for astrophysics. The facility consists of two electrostatic accelerators (50 kV and 400 kV) operated by the LUNA collaboraton. In almost 10 years of measurements, LUNA obtained very important results from precise measurements of the cross sections of the reactions $^{3}He(^{3}He,2p)^{4}He$ (relevant for the pp chain inside the stars), $d(p,\gamma)^{3}He$ (relevant for the pp chain and reaction rates in proto-stars), $^{14}N(p,\gamma)^{15}O$ (the slowest reaction of the CNO cycle in the stars). The location of the accelerator and detectors underground in absence of backgrounds from cosmic rays makes possible to measure the extremely low cross sections at stellar energies.

6. *Geophysics, biology and environmental sciences.* The low background environment inside the Gran Sasso laboratory and its location on a particularly active seismic area, is ideal for a number of interesting research projects in the fields of geophysics and environmental sciences.

Operating in the field of geophysics, GIGS is a laser interferometer for geophysical purposes operating inside the LNGS area since 1994 and monitoring the microseismic movements of the Gran Sasso fault. The TELLUS project is designed to carry out a continuous tilt monitoring to detect aseismic creep strain episodes associated with earthquakes preparation. UNDERSEIS is an underground seismic array aimed to monitor seismic radiation with vey high sensitivity by short period seismometers.

In the field of environmental sciences several activities are ongoing at the LNGS low background facilities. For example ERMES is a project for the monitoring of radioisotopes in the seabed and seawater: extremely low levels of radioactivity in selected samples can be measured in the LNGS facilities by HP-Ge and liquid scintillation detectors.

In the field of biology PULEX is an ongoing experiment whose aim is to investigate the effects of background radiation on the methabolism of cells.

2. ILIAS (Integrated Large Infrastructures for Astroparticle Physics)

2.1. *What is ILIAS*

ILIAS [3] is an Integrated Infrastructure Initiative funded by European Union within the 6th framework Programme. It is based on the cooperation of many EU institutions operating in the sector of Astroparticle Physics. ILIAS was proposed under the coordination and review of APPEC (AstroParticle Physics European Coordination).

The ILIAS project is based on three groups of activities: Networking, Transnational Access, and Joint research activities.

- Networking activities have the objective to favour the contacts and collaborations among researchers working on the same fields in different EU countries. Five different networks are active in ILIAS: deep underground science laboratories, direct dark matter detection, search on double beta decay, gravitational wave research, and theoretical astroparticle physics.
- Transnational access activities have the aim to support the access of research teams to the major EU research infrastructures in the field of astroparticle physics to carry out research activities. The support includes travels and subsistence, and technical support on site. Within ILIAS a coordinated transnational access to the four EU deep underground sites is active.
- Joint research activities (JRA): the objective of these activities is to carry out joint R&D projects among different EU instutution working on common subjects in astroparticle physics and underground sciences. Three JRA are active within ILIAS: Low background techniques for deep undergroun science, double beta decay european observatory, and study of thermal noise reduction in gravitational wave detectors.

2.2. *ILIAS and the EU Deep Underground Laboratories*

Four deep Underground Laboratories in Europe are hosting important experiments on underground sciences: Laboratori Nazionali del Gran Sasso (LNGS, Italy), Laboratoire Souterrain de Modane (LSM, France), Laboratorio Subterraneo de Canfranc (LSC, Spain), and Boulby Mine Underground Lab (UK). The LNGS infrastructure and scientific programme was discussed in the previous section; a similar discussion for LSM, LSC, and Boulby is given with specific talks at this conference.

The four labs are major infrastructures where important experiments on fundamental rare-event and astroparticle physics are underway. ILIAS is offering for the first time the opportunity to start an effective collaboration among the labs.

The underground labs are deeply involved in three activities of ILIAS:

The network of the Deep Underground Laboratories is aimed to support the management of common issues relevant in the operation of the underground sites, such as communication and coordination, service and facilities improvement, extension of underground sites, safety problems and accident prevention, communication and outreach, scientific coordination.

Transnational Access to the underground laboratories gives the opportunity to EU researchers to access the underground sites outside their home country in order to carry out experiments, with priority given to new users and less favoured countries.

The Joint Research Activity on low background techniques gives the opportunity to strengthen and enlarge the low background facilities in the underground labs with a common R&D, and to share the know-how.

References

1. "LNGS Annual report 2003", LNGS Internal report LNGS-EXP/01-04 available online at http://www.lngs.infn.it.
2. A. Bettini et al., ""The Gran Sasso Laboratory 1979-1999" edited by R. Antolini, ISBN 88-86409-20-6.
3. For a complete description of the ILIAS initiative see the web page http://ilias.in2p3.fr

LABORATOIRE SOUTERRAIN DE MODANE : STATUS AND PROJECTS AT FRÉJUS SITE

GILLES GERBIER

Laboratoire Souterrain de Modane
CEA/DSM and CNRS/IN2P3
90, rue Polset,
73500 Modane, France
E-mail: gilles.gerbier@lsm.in2p3.fr

The "Laboratoire Souterrain de modane", with a rock coverage of 4800 mwe, is one of the deepest underground laboratory in the world. It is currently sheltering experiments dedicated to fundamental research in physics and low radioactivity measurements applied to many fields. In this paper, the main characteristics, activities and performances of the lab are given together with a brief history. Short, medium and long term projects are sketched. More informations can be found on dedicated web sites.

1. Introduction

The "Laboratoire Souterrain de Modane"[1] has been running for more than 20 years. This underground laboratory is located in France in Savoie, almost exactly in the middle of the 12.6 km road tunnel between France (Modane) and Italy (Bardoneccia), at an altitude of 1263 m under the Fréjus peak culminating at 3000 m.

Excavated in 1981, it has been opened on december 1982 to host a single experiment, TAUP, a 900 T detector dedicated to the search of the decay of the proton. The size of the cavern was adjusted to the size of the planned experiment, namely a hall of about 30 m long by 10 m wide by 10 m high, plus additional smaller rooms. This German French experiment has been running for 5 years without detecting evidence for proton decay, but setting constraints on proton decay models.

Then the experiment was dismounted in 1988 and progressively, the lab hosted more and more activities requiring protection against cosmic rays and/or dedicated low radioactive environmental background.

The lab became a facility for many experimental programs. Since the beginning, it has been funded equally by IN2P3 (Institut National de Physique

Nucléaire et de Physique des Particules) of CNRS (Centre National de la Recherche Scientifique) and DSM (Direction des Sciences de la Matière) of CEA (Commissariat à l'Energie Atomique). Today, the lab is one of the 20 laboratories of IN2P3 in France. Three structures, the Steering Committee, the Scientific Council, the User group Committee constitute the bodies allowing proper running of the LSM.

The laboratory is currently hosting experiments in particle and astroparticle physics -dark matter and double beta decay searches, super heavy elements search- and low level gamma spectroscopy activities performed with 13 very low activity Germanium detectors.

As explained in the N. Ferrari talk in same proceedings, four european underground labs, Gran Sasso (Italy), Canfranc (Spain), Boulby (UK) and LSM laboratories are involved in a European Community program, ILIAS[4], dedicated to the coordinated development of underground infrastructures, low level measurements and background reduction techniques. This program has three main activities : networking, research and developmment and exchange of research teams. The LSM runs a dedicated program[5], centered on development of gamma ray Germanium detectors. This is in view of preparing next generations of dark matter and double beta decay experiments.

2. Main features and equipments of infrastructure

Main technical features are given in table 1.

Table 1. Main features of lab.

Depth	4800 mwe
Rock	Schistes lustrés
U content	about 1 ppm
Density	2.7 g/cm^3
Temperature of rock	32 deg
Total volume of lab	3300 m^3
Power consumption	60 kW
Air renewal rate	4800 m^3/h
Muon flux	4/m^2/day
Neutron flux	1.6 10^{-6} /cm^2/s (E 1MeV)
Radon concentration	5 to 15 Bq/m^3

All the surfaces of the lab were covered with a low U/Th content concrete. This, together with the high renewal rate of the air allowed to obtain the very low rate of radon inside the lab. The air used to ventilate the lab comes from one of the 4 big road tunnel ventilation units drawing air through a 700 m high chimney from 2000 m altitude mountain.

A special radon reduction device has been installed recently in the laboratory. Though the radon content is quite low, it is still too high for the NEMO3 experiment, who is searching for double beta decay events. A tent has been built around the detector and de-radonised air is sent within the tent to reduce the migration of radon through the joints inside the detector volume. The 150 m^3/h flushed through NEMO3 are produced by compressing, drying, cooling and pushing air in two 500 kg active carbon towers. This system is similar to the one used by SuperKamiokande to produce the air above the water inside the 50kT tank.

Thanks to the optmised cooling unit, which temperature drives the performance of the radon trapping, radon concentration was brought from 5-10 Bq/m^3 down to about 10 mBq/m^3.

All parameters of the lab -temperature, pressure, input air flux, O_2 N_2 CO NO concentrations, radon level and various controls- are now continuously monitored and available on a dedicated web page.

The lab is equipped with an early fire detection system, allowing a prompt detection and triggering the intervention of the firemen within 6 mins.

Access to the lab is done with dedicated vehicules equipped with safety lights and external paintings. According to safety rules defined by the company running the tunnel, the SFTRF, only two cars can park in front of the lab and 17 people can be simultaneously present in the lab at any moment.

Outside facilities are situated in the town of Modane : offices, computer network, storage and mechanical workshop, liquid nitrogen tank, small housing unit run by the lab.

The operational staff is composed of a team of about 10 people, 1 director and deputy, 1 post doc, 1 technical and administrative manager, 1 administrative assistant and 4 technicians.

3. Scientific activities

The figure 1 shows the implantation of the experiments in the main hall and in the dedicated rooms for the Germanium detectors.

Current status of running experimental programs is described in the

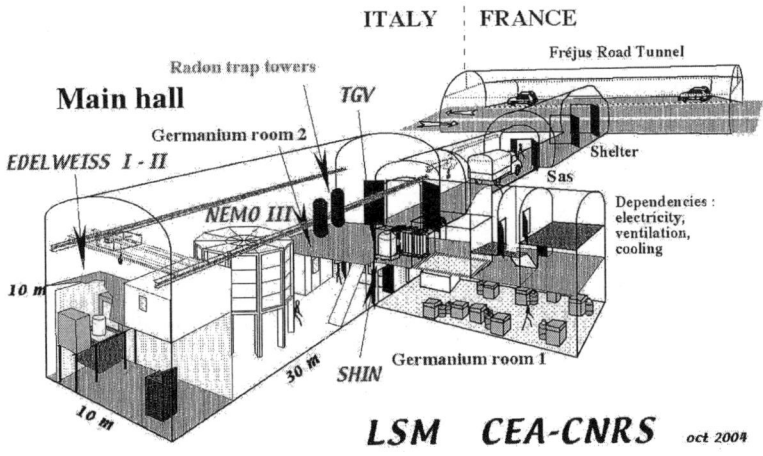

Figure 1. Implantation of experiments in laboratory.

following.

Dark Matter search under the form of WIMP particles is performed by the EDELWEISS[2] collaboration, gathering laboratories from France, Germany and Russia. The first phase of this experiment, EDELWEISS 1, used three Germanium heat-ionisation bolometers of masss 320 g each operated in a dilution refrigerator at 20 mK, made with low radioactivity materials inside shields of lead copper and parrafin. With such twin measurements, most of the residual electromagnetic background can be differentiated and rejected. This allowed to obtain the best world wide sensitivity in 2002 to spin independant coupling WIMP's. The set up has been dismounted these last months to allow for the installation of the second version, EDELWEISS II, which will involve in a first step, 28 detectors in a much larger volume cryostat. End of installation and start of tests are expected to take place in summer 2005. Ultimateley, this refrigeretor will be able to accomodate up to 30 kg of detectors, the largest capacity among the competitors in the 4 to 5 years to come.

NEMO3[3] is the third and largest avatar of a series of "traco-calo" detectors for the double beta decay search. The NEMO collaboration involves laboratories from France, Russia, Czek republik, UK, Finland, USA, Japan. The tracking detector, composed of drift wire chambers operated in Geiger mode allows to identify electron tracks while plastic scintillators measure their total energy. With about 7 kg of Molybden 100 and 0.9 kg of Selenium 82, and other isotope small masses samples, the detector is triggered by "allowed" 2 neutrino double beta decay events at a rate of 0.7 evt/min !

Sensitivity to double beta decays without neutrino was limited until recently radon induced background, from the air surrounding the detector. Now equipped with an air tight "tent" flushed with radon free air -see section 2-, NEMO3 is in the best position to take data and hope to set an upper limit of .2-.35 eV to the effective mass of the neutrino, in a five year run.

TGV (Telescope Germanium Vertical) is a detector designed and built by a collaboration involving Russian, French, Czech, Slovakia laboratories. Combining advantages of excellent energy resolution of semiconductors and identification power of tracking devices, this "telecope", composed of a stack of 32 Germanium detectors of 60 mm diameter can only accomodate 10 to 20 grams of materials in between the detectors. Very rare isotopes are thus the target of such a detector. After setting a limit on Calcium 48 with the TGV1 set up with a 1 g sample, the detector has been upgraded for low energy event detection with the goal of observing for the first time double electron capture process in Cadmium 106, with a 10g sample. The data taking should start in the coming months.

SHIN experiment is a new experiment recently installed in the lab to search for the presence of super heavy elements in nature. Designed and built by a Russian team from Dubna, this detector will search for multi neutron production from fission of superheavy element 108 in a 500 g sample of Osmium (chemically close to the searched element. The detector is composed of 60 ^{3}He counters in a 50 cm diameter, 1 m long cylinder filled with paraffin. With this set-up, a sensitivity to the presence of such element of 10^{-22} g/g in earth material can be achieved in one year, assuming that its half life is 10^9 years.

In addition to these particle physics experiments, a 15 year long tradition in low level gamma spectroscopy has led the lab to shelter 13 ultra low level gamma ray measurement Germanium detectors. These detectors, belonging to 5 differents users, are used for the material selection for physics experiments (NEMO and EDELWEISS), for the environmental control (2 national institutions), and for various scientific and dating applications (Oceanography, climatology, geology, Bordeaux wine dating ...). Among them are 400 CC detectors with background of order of 1 count/day at most natural isotopes lines and two 1000 CC well type detectors.

4. Projects

The next version of the Edelweiss set-up will require a much larger electric consumption, 80 kW instead of 15 kW. This is essentially due to the fact that this innovative cryogenic device will generate its own cold by pulse tubes and ^{4}He liquefier, without, in principle requiring feeding with cryogenic liquids. The operation of the radon trapping device induces also a large increase in the power consumption of the lab. This has led us to redefine the actual cooling system, unable to handle such a dissipation of heat. The new cooling system will use as cold source the air of the tunnel and will have a 300 kW cooling power. This is going to be installed in 2005.

Another project for next years is the building of a new single infrastructure in Modane, which will shelter the offices, the storage hall, the mechanics and chemistry workshops, accomodations for a few guests and a sizeable hall for outreach and mediation, part of a local scientific tourist network.

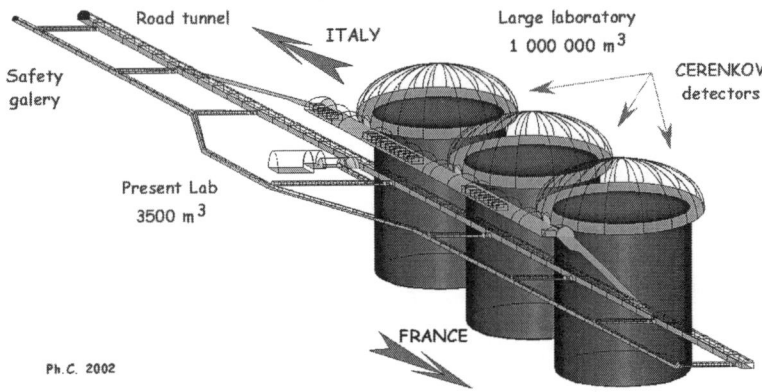

Figure 2. An artist view of what could a Megaton detector along the road tunnel.

A longer term project of a water megaton size detector installed under the Frejus mountain has received special attention these last years. Such a multipurpose detector for rare event search -proton decay, supernova watch- would have also the virtue of contributing to a very sensitive measurement of the mixing angle theta 13, by detecting neutrinos from with a few hundred MeV neutrino beam sent from CERN. Beside the scientific interest and ideal distance from CERN, the Frejus location offers specific advantages : known and good quality of rock, horizontal access by car through possibly shared security tunnel, central location in Europe, fast access by TGV train, highway, closeby airports, nice surroundings for snowboarding in winter and climbing in summer.

This very short introduction does not reflect the amount of work already done by the working groups[6] at CERN and in Europe. The coming NNN05 workshop[7], held in Aussois, 5 km away from Modane, on april 5th to 9th 2005 will be an occasion to gather the world wide community interested in nucleon decay and neutrino studies with large detectors.

References

1. http://www-lsm.in2p3.fr/
2. http://edelweiss.in2p3.fr/
3. http://nemo.web.lal.in2p3.fr/
4. http://ilias.in2p3.fr/
5. http://www-lsm.in2p3.fr/ilias/ilias.html
6. http://nuspp.in2p3.fr/
7. http://nnn05.in2p3.fr/

THE CANFRANC UNDERGROUND LABORATORY. PRESENT AND FUTURE

J. MORALES, B. BELTRÁN, J.M. CARMONA, S. CEBRIÁN, E. GARCÍA, I.G. IRASTORZA, H. GÓMEZ, G. LUZÓN, M. MARTÍNEZ, A. MORALES[*], A. ORTÍZ DE SOLÓRZANO, C. POBES, J. PUIMEDÓN, A. RODRÍGUEZ, J. RUZ, M.L. SARSA, L. TORRES, J.A. VILLAR.

Laboratory of Nuclear and High Energy Physics
University of Zaragoza, 50009 Zaragoza, Spain
E-mail: jmorales@unizar.es

A brief history of the Canfranc Underground Laboratory is presented together with its current status. The description of a new, enlarged underground facility with a main experimental hall of $40 \times 15 \times 11$ m^3 and a total surface of about $1,500$ m^2 at a depth of 2,450 m.w.e. is outlined as well.

1. Introduction

Underground science includes a wide range of exciting disciplines in both fundamental and practical subjects in fields as diverse as Elementary Particle Physics, Nuclear Physics, Astrophysics, Cosmology, Geology, Biology, Material Science, etc.

By using underground laboratories one reaches the low radioactive background environment needed to look for rare event physics which allows us to study, amongst many other amazing research fields, the fundamental laws of physics such as those governing the stability of the proton or the particle-antiparticle properties of the neutrino; we can probe the interior of the Sun or determine the nature of the dark matter.

Figure 1 shows an incomplete list of the underground laboratories around the world, the most outstanding ones in Europe being the L*aboratori Nazionali del Gran Sasso (LNGS)* in Italy, the *Laboratoire Souterrain de Modane (Frejus) (LSM)* in France and *The Boulby Mine Underground Physics Facility* in UK.

---

[*]Deceased

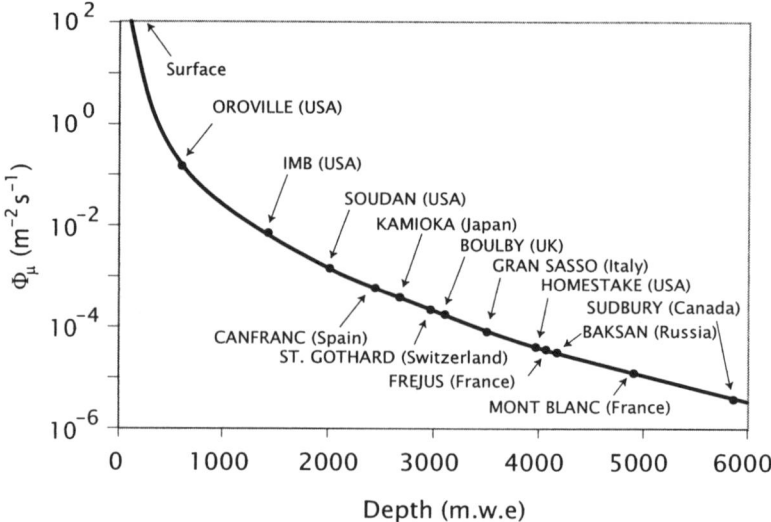

Figure 1. Underground laboratories around the world

In the following, we will briefly draft the history of the Canfranc Underground Laboratory and summarize its experimental program. Finally, the new underground facility which is being constructed at the same location with a total surface of nearly 1,500 m^2 is shortly described.

2. The Canfranc Underground Laboratory (LSC)

In the mid eighties, there was no infrastructure in the Canfranc railway tunnel[a] except for two small halls (about 12 m^2 each plus two 35 m^2, 1 m high galleries) located 780 m from the Spanish entrance below a rock overburden of 675 meters of water equivalent (m.w.e.). In 1986, preliminary measurements of radioactive contamination and cosmic radiation were carried out with a mobile platform along the railway; conditioning works started shortly after (cleaning, electrical power installation, telephone, humidity and temperature control, and ventilation) and the two halls became what we now call Lab-1.

By 1988, the LSC consisted of the two small halls plus a prefabricated cabin, especially reinforced, of about 15 m^2 and installed over the railway

---

[a]LSC is located in a railway tunnel of $\sim$6 km long (not in use) crossing the Pyrenees (entrance at $\sim$1080 m above sea level close to the Canfranc Station, in the Spanish side, 175 km from Zaragoza).

Table 1. Experimental parameters and infrastructure of the site

Depth (max.)	2,450 meters of water equivalent (m.w.e.)
Composition of the rock	limestone, mainly calcium carbonate
Average density	2.7 g·cm^{-3}
Muon flux	2×10^{-7} cm^{-2}·s^{-1}
Radon concentration	20 to 70 Bq·m^{-3}
Neutron flux	a few$\times10^{-6}$ cm^{-2}·s^{-1} (depending on energy)
Ambient photon flux	2×10^{-2} cm^{-2}·s^{-1}
Infrastructure	30 kW independent electric power supply
	Air conditioning and thermalization
	Air extraction and forced ventilation in/from outside
	Low temperature facility (12-20 mK)
	Antivibrational cabin and Faraday cages
	Floor reinforced for supporting heavy shielding
	Tons of archaeological lead
	Ultra-low bkg HpGe detectors for radiopurity measurements

next to the two halls. In August 1988, as a result of a collaboration with the University of South Carolina (USC) and the Pacific Northwest National Laboratory (PNNL), the first ultralow background detector came to Canfranc: a germanium hyperpure detector. With this detector preliminary measurements of materials (lead, copper, polyethylene, etc.) were carried out and fifteen months later (November 1989) the first experiment started in the Canfranc Underground Laboratory, a multidetector system with 14 NaI scintillators and the germanium detector looking for the double beta decay of ^{76}Ge. Since then, many other experiments have been operating in the Canfranc Underground Laboratory, which has undergone important modifications and improvements.

In 1991 a new prefabricated cabin (about 27 m^2) was added to the one already in operation and they were both moved to a new location (1,200 m from the Spanish entrance below an overburden of 1,380 m.w.e.). In this location and with the convenient conditioning of both cabins (electrical power installation, telephone, ventilation), they became Lab-2. In 1994, taking advantage of the excavation works for the new Somport road tunnel, a new experimental hall, 118 m^2, at 2,520 m from the Spanish entrance and below an overburden of 2,450 m.w.e. was excavated. It was named Lab-3 (or main Lab), and has been in operation since the beginning of 1995. This new hall, much larger and deeper underground, has provided for a quantitative and qualitative progress in the research activities of the group; it allowed to start new experiments and to reach a clear improvement in the radioactive backgrounds. With the setting-up of this new hall, Lab-2 was dismounted. The small cabin was installed inside Lab-3 and the large one outside the tunnel for remote control of the experiments and communications. Lab-1 is used now only to store detectors and other materials (which are so kept shielded from the cosmic radiation). Table 1 shows the most relevant experimental parameters of the laboratory as well as the current infrastructure of the site.

The scientific program of the Laboratory includes:

- Neutrino Physics, looking for the double beta decay of ^{76}Ge by using both natural and enriched germanium detectors,
- Dark matter searches, looking for the direct detection of galactic WIMPs by comparing the expected signal rate with the recorded background (using scintillation, ionization and cryogenic detectors) and by searching distinctive signatures as the annual modulation of the WIMP signal, and
- Solar Axion searches through Bragg-scattering in the detector.

A list of experiments already performed or being currently in operation at LSC is presented in Table 2.

Table 2. Experiments already performed or in operation at LSC

Name	Description	Ref.
$2\beta/\gamma$	Decay of ^{76}Ge to excited states of ^{76}Se (coincidence exp.)	[1]
^{78}Kr exp.	Double positron decay of ^{78}Kr	[2]
COSME 1	Looking for WIMPs of low mass	[3]
NaI 32	Search for annual modulation of WIMPs signals with scintillators	[4]
COSME 2	Detection of solar axions through Bragg-scattering	[5]
COSME 2	Looking for WIMPs with a small natural Ge detector	[5]
IGEX	Double Beta Decay of ^{76}Ge	[6]
IGEX-DM	Direct search for WIMPs with an enriched Ge detector	[7,8]
ROSEBUD-I	Direct search for WIMPs with thermal detectors	[9]
ROSEBUD-II	Search for WIMPs with scintillating bolometers	[10 - 12]
ANAIS	Search for annual modulation of WIMP signals with large masses of NaI	[13,14]
GEDEON	Set of GErmanium DEtectors in ONe cryostat (in project)	[15]

In the last fifteen years more than fifty scientists from twelve institutions from eight countries (Argentina, Armenia, France, Italy, Portugal, Russia, Spain and USA) have participated in the LSC Scientific Program. The research infrastructures built in the tunnel, the investments on experimental equipment, the running costs and other scientific activities of the LSC are funded by the Spanish National Programs of High Energy Physics, and of Particle Physics and Accelerators of the Ministry of Science and Technology.

Other funding contributions come from the University of Zaragoza, the TMR Program of the European Union and the Regional Government of Aragon. Occasional contributions are those of the DOE (USA) and NSF (USA), the INFN (Italy) and CNRS (France) and INR and ITEP (Russia). Since 2004, the European Community is contributing to the development of the Laboratory through the ILIAS project within the 6th framework programme.

3. The Future: The new Canfranc Underground Laboratory

Civil works for the construction of the new Canfranc Underground Laboratory are underway. Two experimental halls are being excavated 50 m from the old facility: the main one ($40 \times 15 \times 11$ m^3) oriented towards CERN to allow the possibility to use neutrino beams coming from there and an ultra low background facility ($15 \times 10 \times 8$ m^3). An access corridor housing offices, clear room and workshops for a total of about 1,000 m^2 completes the facility which should be finished next summer 2005. The first call for proposals will be announced soon.

Fig. 2 shows an drawing of the new facility and Table 3 summarizes its characteristics.

Table 3. Characteristics of the new LSC.

Depth (max.)	2,450 meters of water equivalent (m.w.e.)
Main experimental hall	600 m^2 (40x15), 11 m high (oriented to CERN)
Low-background lab	150 m^2 (15x10), 8 m high
Clean room	45 m^2 (100/1000 type)
General services	135 m^2
Offices	80 m^2
Ventilation	11,000 m^3/h (air from outside of the tunnel)
	25,000 m^3/h (air conditioning and filtered)
Electric power	500 kW + 50 kW generator + 15 kW UPS
Infrastructure	Low Temperature facility
	Tons of archaeological lead
	Sensors for NO, NO$_2$ and CO concentration
	Automatic control of temperature and humidity
	Active control of Rn
	Management of dangerous chemical products

There is no room here to describe with a minimum detail the safety regulations established for the new Laboratory. A complete protocol for access, stay in the lab, fire control and evacuation procedures has been included in the safety regulations of the road tunnel.

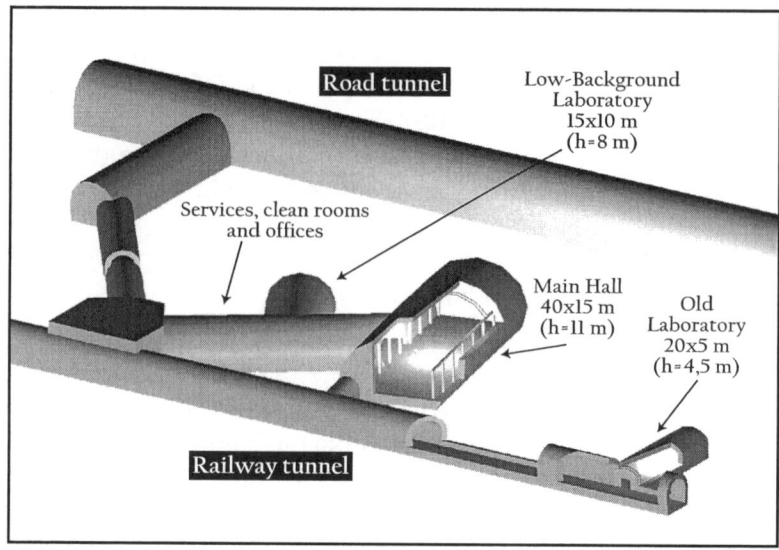

Figure 2. Artistic view of the new facility

Acknowledgments

The Canfranc Astroparticle Underground Laboratory is operated by the University of Zaragoza under contract No. FPA 2001-2437 funded by the Spanish Commission for Science and Technology (CICyT).

References

1. A. Morales et al., Il Nuovo Cimento **104A**, 1581 (1991).
2. C. Sáenz et al., Phys. Rev. **D50**, 1170 (1994).
3. E. García et al., Phys. Rev. **D51**, 1458 (1995).
4. M.L. Sarsa et al., Phys. Rev. **D56**, 1856 (1997).
5. A. Morales et al., Astrop. Phys. **16**, 325 (2002) and references therein.
6. C. E. Aalseth et al., Phys. Rev. **D65**, 092007 (2002).
7. A. Morales et al., Phys. Lett. **B532**, 8 (2002).
8. A. Morales et al., Phys. Lett. **B489**, 268 (2000).
9. S. Cebrián et al., Astrop. Phys. **15**, 79 (2001).
10. S. Cebrián et al., Astrop. Phys. **21**, 23 (2004).
11. S. Cebrián et al., Phys. Lett. **B563**, 48 (2003).
12. S. Cebrián et al., Phys.Lett. **B556**, 14 (2003).
13. S. Cebrián et al., Nucl. Phys. (Proc. Suppl.) **B114**, 111 (2003).
14. M. Martínez et al., this volume.
15. I. G. Irastorza et al., Proceedigs of the 4th International Workshop on the Identification of Dark Matter, York, (World Scientific), p. 308.

THE BOULBY UNDERGROUND LABORATORY

A.S. MURPHY, FOR THE UK DARK MATTER COLLABORATION
*School of Physics,
University of Edinburgh,
Mayfield Road,
Edinburgh EH8 7TU
E-mail: a.s.murphy@ed.ac.uk*

The Boulby Mine is situated near Whitby in the north-east of England. Since 1990 it has been the site of experimental underground science studies, and is currently the home of the UK Dark Matter Collaboration. The physical properties of the mine relevant to low background measurements such as infrastructure and background radiation levels will be discussed.

1. Introduction

Deep underground laboratories provide an essential environment for several types of research in the physical sciences. In 1990, a commercial rock salt and potash producing mine in North Yorkshire, England, made resources available for small scale particle astrophysics studies. Since then, the science activities at the Boulby mine have enlarged, most recently with the installation of major surface and subterranean laboratories to support the UK Dark Matter Collaboration funded through PPARC and the Joint Infrastructure Fund. Although a preliminary report on these developments was presented at the 2002 IDM conference, the formal opening occurred only in April 2003 [1].

2. Facilities

A collection of images depicting major aspects of the surface and underground facilities available are shown in Figure 1. Surface facilities are housed in the purpose built John Barton Building, and include computing, shower, changing, workshop and meeting room facilities as well as storage areas. In addition, Cleveland Potash Limited provide valuable support for health and safety issues, chemical handling, transport and power supply.

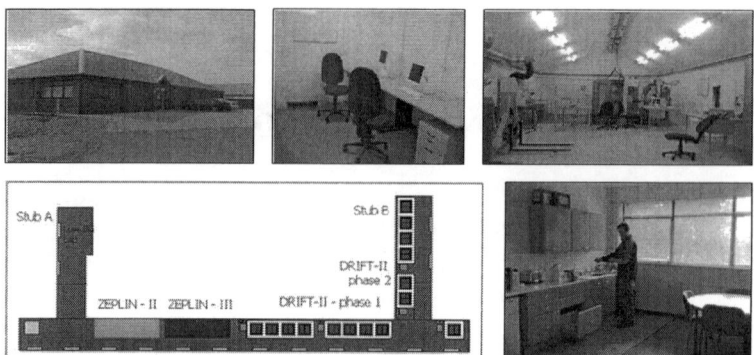

Figure 1. A montage of photographs of the current facilities at the Boulby mine, together with the planned floor space allocations for each of the projects.

The underground areas, at a depth of between 850 m and 1.3 km, are accessed via lift shafts each equiped with a 7000 HP winch capable of lifting ∼22 Tonnes. Shafts have a diameter of 5.5 m, with transportation between surface and underground usually being made within a roughly cubical cage of sidelength 2 m.

Underground, there exist now over 1000 km of tunnels, extending at a rate of about 40 km per year. Areas dedicated to scientific research are maintained within ≈1 km of the main shaft, at a physical depth of ≈1100 m, and are housed within caverns of ∼5 m height and 8 m width, although in principle the local rock would be able to support larger caverns. Ambient rock temperatures are typically 28°C, although within the main scientific laboratories the air conditioning, required for dust control, reduces this to a more comfortable level. Infrastructure underground includes purpose built wooden frame laboratories with thermal insulation and radon proof membranes. Care is taken to allow maintenance of at least a class 10000 clean-room quality environment for operational detectors, and class 5000 clean-room for final assembly of components. Rooms are amply equiped with electrical power, cranes running the full length of the rooms, high speed (100 Mbit/s) fibre-optic links, and telephones.

3. Radiological background

The characteristic of a deep underground laboratory that makes it attractive to certain scientific efforts is that it will have a much reduced radiological background. Such radiation arises from several sources: penetrating high energy muons produced in cosmic ray showers, secondary radiations produced when those high energy muons interact nearby, and radiations emanating from naturally occurring long lived isotopes in the local environment. Recently, extensive efforts have been made at Boulby to provide accurate measurement of all forms of background radiation. These studies suggest that the background levels can be sufficiently attenuated for a tonne-scale dark matter detector to achieve a sensitivity approaching 10^{-10} pb [2]. The following sections provide details of these efforts.

3.1. *Muons*

The muon flux reaching the underground laboratory at Boulby has been directly measured [3], using the ZEPLIN-I liquid scintillator veto illustrated in Figure 2. A total of 106 days-worth of data were collected. The event rate of muons depositing greater than 30 MeV was found to be 53.5 events per day. A typical energy spectrum of these events is shown in Figure 3. Detailed Monte Carlo simulations were then performed that included the geometry of the mine and the detector, and the properties of the rock, liquid scintillator and hardware and software trigger requirements. The resulting simulated energy spectrum is also shown in Figure 3, scaled vertically to provide a best fit to the data. The scaling required corresponds to a muon flux within the mine of $(4.08\pm0.08(\text{stat})\pm0.13(\text{syst}))\times10^{-8}cm^{-2}s^{-1}$. In turn, this corresponds to the scientific laboratory being at vertical depth of 2805±45 m w.e..

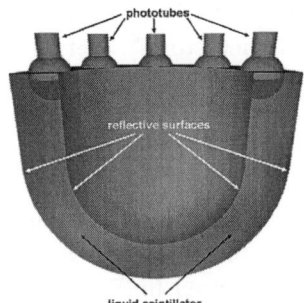

Figure 2. Schematic drawing of the ZEPLIN-I veto detector. Only 5 of the 10 photomultiplier tubes are shown. The height of the detector is 0.65 m, and the hemispherical lower part has inner and outer radii of 0.35 and 0.65 m respectively.

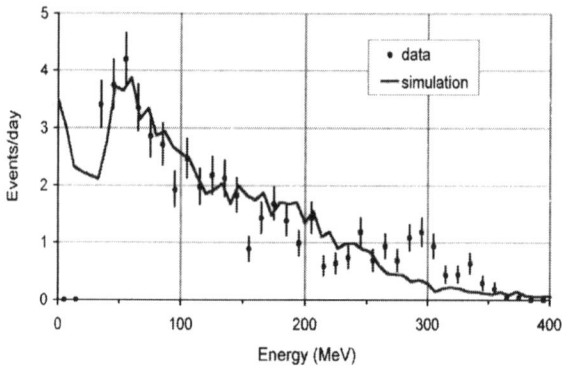

Figure 3. Muon data collected in the ZEPLIN-I veto device.

3.2. Ambient gamma background

Together with the muon flux, it is the long lived isotopes of K, U and Th that are of primary concern in the Boulby laboratory, since it is these that lead to almost all of the locally produced gamma and neutron backgrounds. The mine rock is predominantly composed of halite, known to be relatively low in U and Th, but with significant concentration of K. Also found in the scientific environment in significant quantity are wood and polypropylene, used in construction of the laboratory. We have performed the following measurements to determine the impact these have on the capability to perform low counting rate scientific endeavours.

Direct measurement of the ambient gamma-ray background in the mine has been made using HPGe detectors [4]. Detectors placed centrally within the essentially empty experimental area were allowed to accrue data over a period of 188 hours. Resulting spectra are shown in Figure 4, and peaks corresponding to the uranium, thorium and potassium are clearly visible. The concentrations of U, Th and K that must be present in the rock to generate the obserevd gamma ray fluxes were then estimated with a FAUST Monte Carlo simulation of gamma-ray emission from the rock into the laboratory [5]. Results are summarised in Table 1. Overall, the total flux of gamma-rays at the rock–cavern boundary is approximately 0.09 $cm^{-2}s^{-1}$. In addition, we have commissioned radiological assays of the wood and polypropylene, the results again being shown in Table 1. Analyses were performed by cascadescientific [6]. Similarly, radon levels within the experimental areas have been measured as below 5 Bq/m^3.

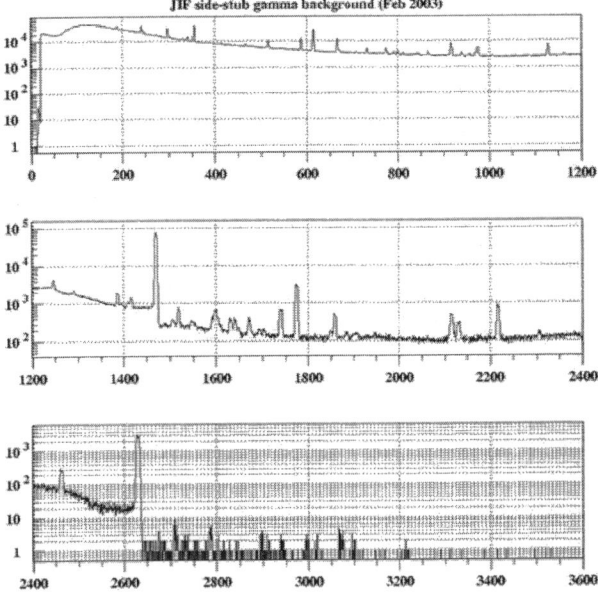

Figure 4. HPGe detector energy spectrum of gamma-rays within the Boulby mine.

Table 1. Abundances of U, Th and K in the Boulby mine environment.

Material	K	U	Th
Polypropylene solid sheet	0.21 mg/kg	0.00012 mg/kg	<0.00005 mg/kg
Wood (sample 1)	29 mg/kg	42 mg/kg	38 mg/kg
Wood (sample 2)	73 mg/kg	4 mg/kg	8 mg/kg
Rock	1130ppm	70ppb	125ppb

3.3. *Neutron backgrounds*

A direct measurement of the neutron flux within the underground laboratory has yet to be made[a]. However, using the known U, Th and K concentrations an updated SOURCES code was used to determine the neutron production rate within the cavern rock [2,7]. The GEANT 4 toolkit was then used to propagate the neutrons and events crossing the cavern–rock interface recorded to estimate the ambient neutron flux. The predicted energy spectrum of neutrons so produced is shown as the solid line in Figure 5,

---

[a]At the time of writing such a measurement is being undertaken

and corresponds to a flux of 2.45×10^{-6} $cm^{-2}s^{-1}$ neutrons above 10 keV, 1.9×10^{-6} $cm^{-2}s^{-1}$ above 100 keV, and 0.9×10^{-6} $cm^{-2}s^{-1}$ neutrons above 1 MeV. Also shown in Figure 5 are the effects on the ambient neutron flux of a variety of thicknesses of both hydrocarbon and lead passive shielding. The uppermost curve, dashed, shows the neutron flux visible at the surface of any such shielding. It is higher than the neutron flux being emitted from the rock wall due to multiple scattering within the cavern. The remaining curves show the neutron flux observed within various shielding structures, indicating reductions of a factor of 10^6 are achievable. The FAUST code has also been employed to estimate the ambient neutron flux [8], giving thoroughly consistent results.

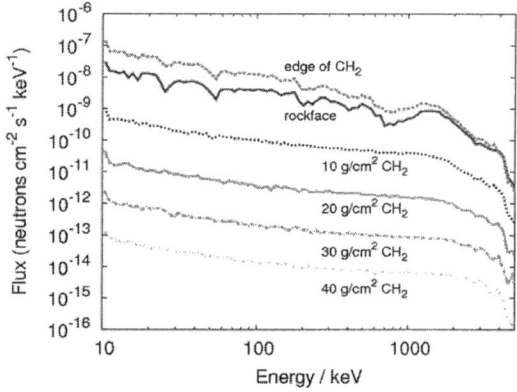

Figure 5. Monte Carlo simulated energy spectra of neutrons emanating from the rock surrounding the Boulby mine experimental laboratory, and for the flux expected within alternative shielding schemes.

4. Management

With the growth of the experimental programme at Boulby, especially the interest being shown from groups outwith the UK Dark Matter Collaboration, it has become necessary to implement a more coherent management plan, ensuring the best use of the facilities available. A robust framework has been developed for new proposals to be reviewed and the implications for the laboratory as a whole to be assessed. Statements of Interest may be submitted at any time, and if these receive support from the Boulby Science committee, a well defined feasibility study will be initiated.

5. Current Programme

The current research programme is dominated by the projects of the UKDMC. The sodium iodide based NAIAD project has been completed and the liquid xenon ZEPLIN-I and negative-ion gas DRIFT-I TPC detectors are now being replaced with second generation detectors. The liquid xenon programme features the development of two-phase readout with ZEPLIN-II and ZEPLIN-III addressing various aspects for optimising scintillation light collection and photoluminescence charge collection. The DRIFT-II programme builds on the its predecessor, now allowing 3-D readout of tracks and more robust construction techniques, while at the same time delivering a substantial cost reduction per modules. In addition, a proposal for a dedicated low background experimental facility utilising a suite of HPGe detectors is proceeding, as are additional measurements of the radiological background through the ILIAS funded by through the FP6 mechanism. Preliminary studies relating to neutrinoless double beta decay measurements have also begun. The expected expected arrangement of the new projects within the main scientific laboratory of the Boulby Mine is shown in Figure 1.

6. Summary and Future Outlook

A comprehensive assessment of the radiological backgrounds present in the Boulby Mine is nearing completion. This suggests that it will be suitable for both the current generation of dark matter detectors, and the next generation of tonne-scale devices. In addition, recent investments in infrastructure provide a secure, well resourced and comfortable environment. The UKDMC is imminently to install and commission several new dark matter detectors, and projects of wider interest are under development.

References

1. See http://hepwww.rl.ac.uk/ukdmc/databases/pr/JIF_opening_030428.html
2. M.J. Carson et al., *Astroparticle Phyiscs,*, **21** (2004) 667
3. M. Robinson et al., *Nucl. Inst. and Meth. in Phys. Res. A*, **511** (2003) 347
4. V. Kudryavtsev (2004). See
   http://hepwww.rl.ac.uk/ukdmc/Radioactivity/Index.html
5. P.F. Smith et al., *Astroparticle Phyiscs,*, in press.
6. Private communication, Cascade Scientific, Brunel Science park, Uxbridge, UB8 3PH.
7. M.J. Carson et al., *Nucl. Inst. and Meth. in Phys. Res. A,*, submitted.
8. C. Bungau et al., *Astroparticle Phyiscs,*, in press.

SNOLAB

DAVID SINCLAIR
Carleton University
Ottawa, Canada

Abstract: An International facility for Underground Science is under construction at Sudbury in Canada. The facility will expand the existing SNO experiment to allow several concurrent experiments to be carried out. All of the space will be established as a connected clean room at a depth of 2000 m (6000 MWE). Proposals for projects that would exploit this low background environment are welcome at any time.

1. Introduction

A new international facility for underground science, SNOLAB, is being established at the Creighton Mine near Sudbury in Canada. It will provide new experimental space adjacent to the existing SNO experiment for both large and small experiments and it will provide a surface building to support the underground experiments.

2. Scientific Opportunities

The scientific case made for the development of SNOLAB centered on the need for deep facilities for the study of low energy solar neutrinos, the search for dark matter, and the search for neutrino-less double beta decay. In order to understand the infrastructure needs for these projects a series of workshops were held. At the most recent meeting, groups were invited to submit Letters of Interest detailing the needs of the experiments. Sixteen letters were received. One project required a space beyond the capacity of SNOLAB to provide but all of the remaining letters described exciting projects that could be accommodated in modest sized caverns and which required the deep location and clean conditions planned for SNOLAB. On the basis of these workshops and letters, a design for the surface and underground facilities has emerged.

The next phase of the scientific definition process will be the determination of the initial round of experiments. Because projects are at various stages of development, the next request will be for expected schedules for the preparation of firm proposals.

The experiment selection is being made with the advice of an international panel of experts who form our Experiments Advisory Committee. Further letters of Interest can be submitted to the committee at any time.

3. The Underground Facility

The layout of the new underground facility is shown in Figure 1. There are two major new areas being excavated. These are a large rectangular experimental hall and an area for smaller experiments and support facilities known as the Ladder area. A second phase of the project will add a 'cryopit' area specifically designed to deal with the hazards associated with large underground cryogenic experiments. The underlying philosophy for the design has been to allocate a dedicated room to each experiment. This approach has been taken rather than the successful model of for example the Gran Sasso laboratory where several experiments may share a single room, in order to make it easier to control the risks associated with experiments that may involve large volumes of hazardous materials such as suffocant gases, flammable or poisonous materials. It also reduces the risk of interferences between experiments. The cost is a reduction in flexibility.

The rectangular hall will be 15m wide, 18 m long and 15m high plus a domed space at the top. The ladder area has a back drift 6m wide and 4m high with a wider section which is 8m wide and 5m high. The total length is 60m. Smaller spaces are provided in the rungs of the ladder. The cryopit would be 15 m diameter and 15m high plus a domed space. Both of the large spaces can be isolated from the rest of the laboratory with high pressure bulkhead doors and relieved to a high capacity return air raise to deal with noxious gas release.

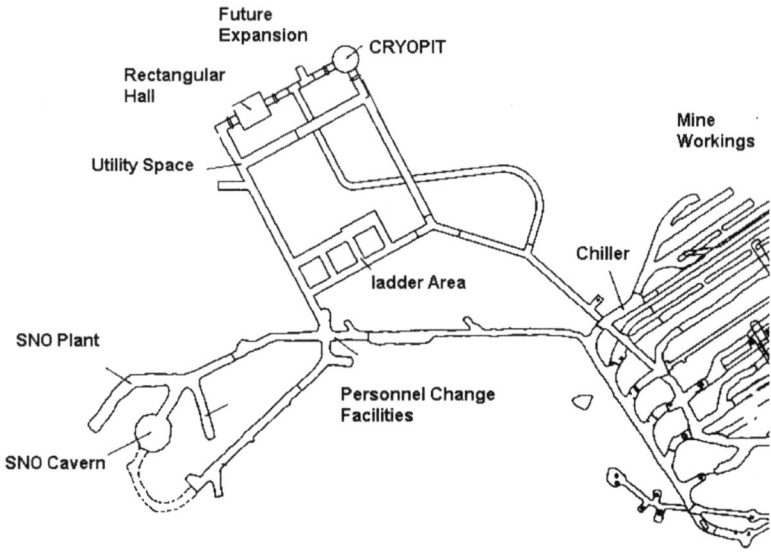

Figure 1. Outline of the SNOLAB underground facilities being developed at a depth of 6800 feet in the Creighton Mine near Sudbury, Canada. The new space will be connected to the existing SNO experiment clean environment.

A major part of the infrastructure is associated with establishing the environment. The system will build on the experience gained with the construction and operation of SNO. The space will be contiguous with the present SNO detector facility. The present personnel change area and materials cleaning facilities will be moved to a common entry area. All of the space will be maintained as a common clean environment. There will be an extensive system of air handlers which will bring air into the laboratory through a series of activated charcoal and fine/HEPA filters. One of the design criteria for this system was to establish a pressure gradient in the laboratory such that the cleanest areas (the experimental halls) have the highest pressure so that all air leakage is away from these sensitive areas. The pressure in the laboratory is maintained positive with respect to the local mine environment again to stop the ingress of mine dust which contains traces of uranium and thorium. A recirculating flow of about 5 air changes per hour is maintained throughout the facility.

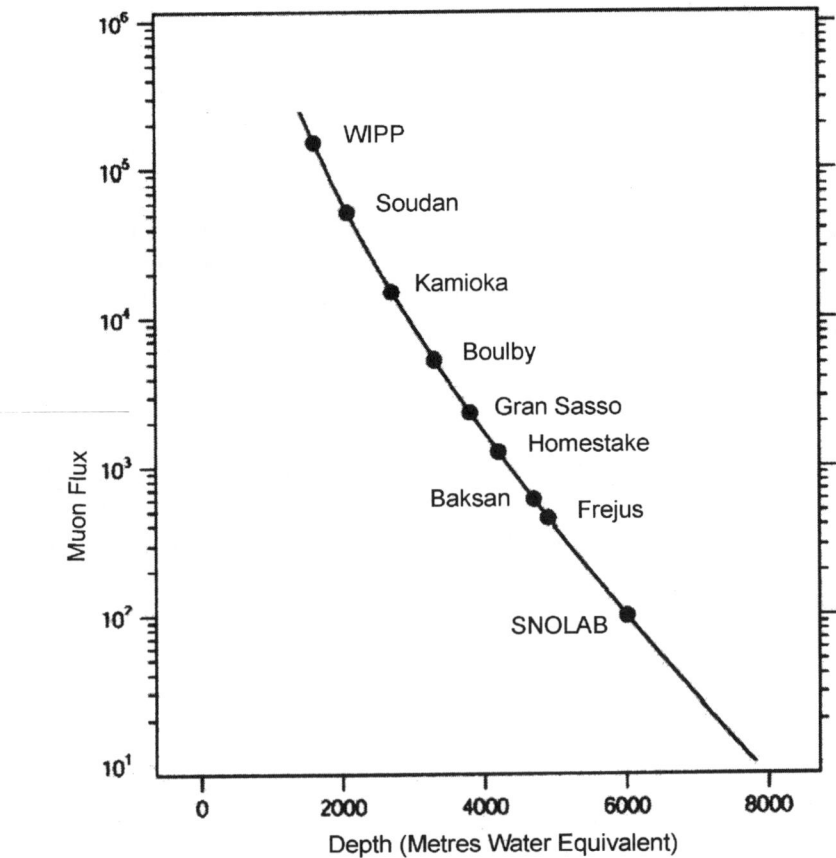

Figure 2. This shows the depth of major existing underground laboratories and the corresponding muon flux. SNO's site at Sudbury is the deepest of these.

Cooling of the laboratory is achieved by operating a chiller which can discharge its waste heat into the return air. The SNOLAB facility will have 300 T of chilling capacity requiring an air flow of 100,000 cfm. The other main infrastructure provided with the laboratory include ultrapure water (about 10 T per hour), power (up to ~1 MW) and high speed networking (high speed fiber within the lab and 1Gb/s off site)

In addition to the newly excavated space, the cavern in which the SNO experiment is located will become available after completion of that experiment in 2007. This cavern is the largest, deep cavern in the world designed for human occupancy with a span of 22 m and a height of 35 m.

4. Surface Facilities

To support the underground experiments, a building is being constructed on surface which will provide 33,000 square feet of clean laboratory space, change rooms, meeting rooms, office and computer facilities. One of the unusual features of the design is the need to allow detector components to be assembled in a clean room on surface and then be transported cleanly through the mine to the laboratory without contamination by the very dusty environment. Four clean rooms will be provided. An auditorium for presentations to up to 100 people is included as well as a number of smaller meeting rooms.

5. Schedule

The completion of the surface building is scheduled for June of this year. The underground excavation has started with the re-conditioning of some of the old mine workings and development of a new access drift. The first new underground space should be available in 2006 with the full facility completed by 2007.

NEUTRON STUDIES AT THE CANFRANC UNDERGROUND LABORATORY

G. LUZÓN, J. AMARÉ, B. BELTRÁN, J.M. CARMONA, S. CEBRIÁN,
E. GARCÍA, I.G. IRASTORZA,* H. GÓMEZ, M. MARTÍNEZ, A. MORALES†
J. MORALES, A. ORTIZ DE SOLÓRZANO, C. POBES, J. PUIMEDÓN,
A. RODRÍGUEZ, J. RUZ, M.L. SARSA, L. TORRES‡ J.A. VILLAR

*Laboratory of Nuclear and High Energy Physics,
University of Zaragoza, E-50009 Zaragoza, SPAIN
E-mail: luzon@unizar.es*

Four sets of the IGEX-DM low energy data, obtained with different neutron-shielding conditions, have been compared to simulations to quantify the neutron populations from different sources: the flux of neutrons coming from the radioactivity of the surrounding rock, $(3.82 \pm 0.44) \times 10^{-6}$ cm^{-2}s^{-1}, the flux of muon-induced neutrons in the rock, $(1.73 \pm 0.22(\text{stat}) \pm 0.69(\text{syst})) \times 10^{-9}$ cm^{-2}s^{-1} and the rate of neutron production by muons in the lead shielding, $(4.8 \pm 0.6(\text{stat}) \pm 1.9(\text{syst})) \times 10^{-9}$ cm^{-3}s^{-1}. It can be concluded that a suitable neutron shielding practically eliminates the main contribution (rock radioactivity neutrons) to the IGEX background at the present level of sensitivity, while the remaining neutron populations (muon-induced neutrons in the rock or in the shielding) are below the present background level thanks to the veto system. These neutron studies are extremely useful to understand the effect of neutrons in other current and future experiments at the Canfranc Underground Laboratory.

1. Introduction

The mountain over the Canfranc Underground Laboratory (LSC) (2450 m.w.e.) serves as a natural shield against cosmic rays and only very energetic muons are able to traverse the rock. Along their path they produce energetic neutrons in the rock and also in the lead shielding of the experiment. However, the main neutron contribution to the background of

---

*Present address: CERN, Geneva, Switzerland and DAPNIA, CEA-Saclay, Gif-sur-Yvette, France.
†Deceased.
‡Present address: Dipartimento di Fisica dell'Università di Milano-Bicocca, Milano, Italy.

our experiments, though much less energetic, comes from the radioactivity of the rock. For deep underground sites, such as the LSC, the neutron component of the background used to be below the typical gamma levels, but the high radiopurity of materials and the expertise achieved in low background techniques have reduced the raw background to levels where neutrons can be of crucial importance. Neutrons have become the real worrisome background for WIMPs since they can produce nuclear recoils in the detector target nuclei which would mimic WIMP interactions. Consequently, a program to understand the neutron background of the LSC has been undertaken [1].

The IGEX-DM [2] experiment was intended for the detection of WIMPs through its coherent scattering off germanium nuclei producing a nuclear recoil. The 2 kg enriched Ge detector is fitted into a precision-machined lead chamber where a flux of nitrogen gas (140 l/hour) creates an overpressure to minimize radon intrusion. The shielding from inside to outside can be described as follows. About 2.5 tons of 2000-year-old archaeological lead form a cubic block of 60 cm side around the central chamber. The archaeological lead block is surrounded by 20 cm of lead bricks made from 70-year-old low-activity lead (~ 10 tons). Two layers of plastic seal this central assembly against radon intrusion and a 2-mm-thick cadmium sheet surrounds the ensemble. Covering the top and three sides of the shield are the cosmic muon vetoes (BC408 plastic scintillators). Finally, an external neutron moderator (made of polyethylene bricks and borated water tanks) surrounds the whole set-up. Its thickness has been changed in several occasions. In such a way, independent sets of data (A, B, C and D) have been obtained (in the case of the set B, two of the shielding sides were left without moderator). These sets of data have been compared with simulations (see Table 1) where the GEANT[3] package has been used to estimate the contributions to the IGEX background levels of a known flux of neutrons and the FLUKA[4] code to study the neutron production by muons.

Table 1. Main features of the data sets A, B, C and D: thickness of the neutron moderator, statistics time and measured average background level (in anticoincidence with the veto system) from 4 to 10 keV.

	A	B	C	D
thickness of moderator (cm)	0	20	40	80
statistics time (days)	17	118	97	41
background [counts/(kg keV day)]	0.74(6)	0.39(2)	0.22(1)	0.24(2)

2. Muon-induced neutrons in the shielding and in the rock

The spectrum of neutrons induced by muons in lead has been obtained using FLUKA, in quite satisfactory agreement with the analytical formulae used[5] for evaporation and direct emission neutrons (Medium Energy spectrum $E < 20$ MeV, ME) and for neutrons produced by high energy hadronic processes (High Energy spectrum $E > 20$ MeV, HE). The obtained neutron production rate, 1.7×10^{-3} (g/cm^2)$^{-1}$ neutrons per muon, is pretty close to the value estimated in [6,7] for the total neutron yield. These neutron spectra have served as input for a GEANT4 simulation whose comparison to experimental muon-vetoed events (see Fig. 1) results in a neutron production by muons in lead of $(4.8 \pm 0.6(stat) \pm 1.9(syst)) \times 10^{-9}$ cm^{-3} s^{-1}. The systematic error comes from the estimated rates of veto events not corresponding to neutrons, and the unknown fraction of HE and ME neutrons in the complete spectrum. Taken into account also the previous neutron yield, the muon flux crossing the shielding can be deduced to be $(2.47 \pm 0.31(stat) \pm 0.98(syst)) \times 10^{-7}$ cm^{-2} s^{-1}.

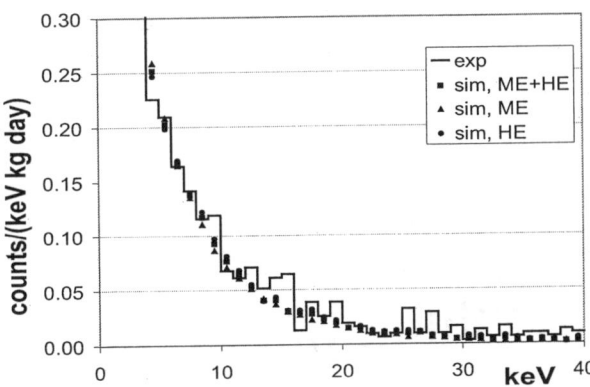

Figure 1. Measured spectrum of vetoed events (labelled *exp*) fitted to simulated spectra, assuming the HE spectrum, the ME spectrum and a weighted combination (50% ME, 50%HE).

The obtained muon flux is of the expected order of magnitude at 2450 m.w.e., and in fact it is in perfect agreement with previous scintillator measurements at the LSC site, which gave a flux[2] of 2×10^{-7} cm^{-2} s^{-1}, therefore showing the reliability of our numerical simulations.

Note that thanks to the veto system this non-negligible contribution is significantly reduced. For the IGEX-DM veto counting rate, 0.16 counts/(keV kg day) in the low energy region (4-10 keV), an veto efficiency of around 92%, reduces this contribution to 1.3×10^{-2} counts/(keV kg day), a value more than one order of magnitude lower than the present background level. However, it might become an important contribution in future more sensitive experiments where a more efficient veto system should be designed.

Muons also induce neutrons in the surrounding rock. Most of the muon-induced neutrons in the rock are not detected by the veto system and become as important as neutrons induced in the shielding itself, even though the latter are produced much closer to the detector. A FLUKA simulation allowed us to obtain the spectra and total yield for neutrons produced in the rock, 4.6×10^{-4} $(g/cm^2)^{-1}$ per muon, compatible with values shown in other references [7], and also the neutrons per muon exiting the walls, 0.01 (ME) and 0.007 (HE). Known the muon flux in Canfranc, the flux of muon induced neutrons in the rock can be deduced to be $(1.73 \pm 0.22(stat) \pm 0.69(syst)) \times 10^{-9}$ $cm^{-2} s^{-1}$.

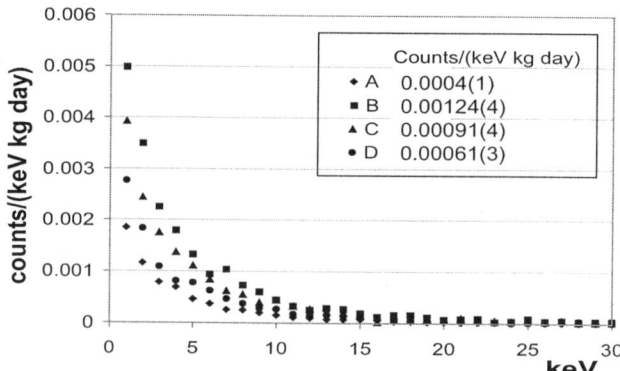

Figure 2. Spectra of the deposited energy in the detector by muon-induced neutrons in the rock: simulation for the different set-ups and background rates (4-10 keV).

A further GEANT4 simulation serves to estimate the contribution of neutrons induced by muons in the rock to the IGEX low energy background in the four different shielding conditions (see Fig. 2). The contribution is maximum for the intermediate value of 20 cm of neutron moderator, decreasing for thicker walls or in the case of total absence of polyethylene: very energetic neutrons (with some hundreds of MeV), which produce nuclear recoils corresponding mostly to energies higher than our range of interest in the absence of moderator walls, can be slowed down in the presence of the polyethylene shielding, inducing in this way nuclear recoils in the low-energy region. In any case, the contribution of these neutrons to the IGEX-DM background is much lower than the present background level.

3. Neutrons from the radioactivity in the rock

Most of the neutrons in the LSC comes from the radioactivity of the rock due to spontaneous fission of uranium and (α, n) reactions. Given their much lower energy compared to muon induced neutrons it can be assumed that 40 cm of polyethylene are enough to moderate down this population. Additional simulations show a reduction of 99.2% for fission neutrons (and 98.3% for (α,n) neutrons) using 20 cm and 99.99% for both populations using 40 cm of moderator.

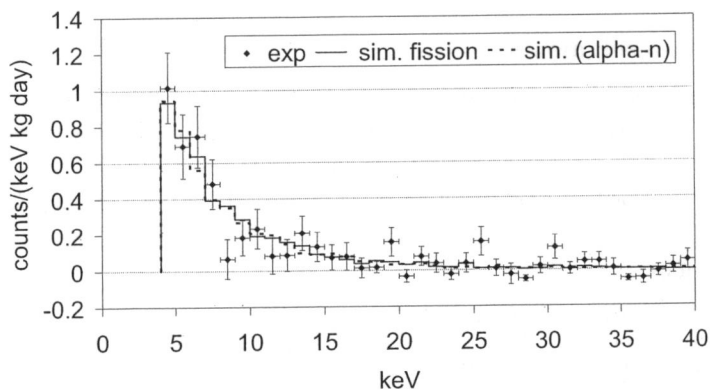

Figure 3. Energy deposited in the detector due radioactive processes —no neutron moderator— compared to the difference between sets A and C (labelled exp).

Simulations of the propagation of these neutrons through the shielding of the set A (no moderator) compared to the difference between measured spectra for sets A and C (see Fig. 3) allow us to deduce the radioactive neutron flux, $(3.82 \pm 0.44) \times 10^{-6}$ cm^{-2} s^{-1}, which is within the expected range. It is remarkable the similar spectral shape of the deposited energy for simulated neutrons coming from fission or from (α, n) reactions.

4. Conclusions

A complete quantitative study of the neutron environment in the LSC has been performed. A set of simulations have been compared with several sets of the IGEX detector data in different conditions of neutron shielding. In decreasing order of importance, the neutron populations studied, whose estimated fluxes are summarized in Table 2, are: neutrons coming from radioactivity in the laboratory rock, neutrons induced by muons in the lead shielding, and neutrons induced by muons in the laboratory rock. Neutrons produced by radioactivity in the lead shielding have been excluded from the analysis since their contribution is negligible at the present level of background of IGEX-DM.

Table 2. Results for the estimates of the fluxes of different neutron populations reaching the IGEX-DM experimental setup in the LSC.

from radioactivity of the rock	$(3.82 \pm 0.44) \times 10^{-6}\,\text{cm}^{-2}\,\text{s}^{-1}$
μ-induced in the rock	$(1.73 \pm 0.22(stat) \pm 0.69(syst)) \times 10^{-9}\,\text{cm}^{-2}\,\text{s}^{-1}$
μ-induced in the shielding lead	$(4.8 \pm 0.6(stat) \pm 1.9(syst)) \times 10^{-9}\,\text{cm}^{-3}\,\text{s}^{-1}$

5. Acknowledgments

The Canfranc Underground Laboratory is operated by the University of Zaragoza. This research was partially funded by the Spanish Commission of Sciences and Technology (CICYT) under contract No. FPA2001-2437. We also acknowledge the funding from the EU FP6 project ILIAS.

References

1. J.M. Carmona *et al.*, Astr. Phys. 21 (2004) 523.
2. A. Morales *et al.* [IGEX Collaboration], Phys. Lett. B 532 (2002) 8.
3. S. Agostinelli *et al.* [GEANT4 Collaboration], NIM A 506 (2003) 250.
4. A. Fassò, A. Ferrari, P.R. Sala, "Electron-photon transport in FLUKA: status", Proceedings of the MonteCarlo 2000 Conference, Lisbon, October 23–26 2000, A. Kling, F. Barao, M. Nakagawa, L. Tavora, P. Vaz eds., Springer-Verlag Berlin, p. 159-164 (2001); A. Fassò, A. Ferrari, J. Ranft, P.R. Sala, "FLUKA: Status and Prospective for Hadronic Applications", ibid, p. 955-960 (2001).
5. T.A. Perera, PhD thesis.
   http://cosmology.berkeley.edu/preprints/cdms/Dissertations/tap_thesis.pdf
6. G. Chardin, IDM2002 Proceedings, York (England), World Scientific, p. 470.
7. V.A. Kudryavtsev, N.J.C. Spooner, and J.E. McMillan, NIM A 505 (2003) 683; V. A. Kudryavtsev, IDM2002 Proceedings, York (England), World Scientific, p. 477.

STUDY OF THE MUON-INDUCED NEUTRON BACKGROUND WITH THE LVD DETECTOR

H. MENGHETTI*

Bologna University and INFN
Via Irnerio 46
Bologna 40126 Italy
E-mail: menghetti@bo.infn.it

High energy neutrons, generated as a product of cosmic muon interaction in the rock or in the detector passive material, represent the most dangerous background for a large list of topics like reactor neutrino studies, the search for SN relic neutrinos, solar antineutrinos, etc. [1] Up to now there are few measurements of the muon-produced neutron flux at large depth underground. Moreover it is difficult to reproduce the measured data with Montecarlo simulation because of the large uncertainties in the neutron production and propagation models. We present here the results of such a measurement with the LVD detector, that is well suited for the detection of neutrons produced by cosmic-ray muons, reporting the neutron flux at various distances from the muon track, for different neutron energies ($E > 20\ MeV$) and as a function of the muon track length in scintillator.

1. Detector description

LVD [2] is a neutrino telescope located in the Hall A of the Gran Sasso INFN laboratory, in data acquisition since 1992, whose main goal is the observation of neutrinos from a Supernova core collapse. It is composed of three independent towers, each of them made of 280 scintillator counters, 1.5 m^3 each, for a total scintillator (C_nH_{2n}, $<n> = 9.6$) mass of about 1000 tons. The detector is also equipped with a tracking system made of limited streamer tubes, interleaved to the scintillator counters, which allows to reconstruct muons; the depth of the LVD site corresponds to a mean muon energy of about 270 GeV.
The observation of neutrinos is made mainly trough the inverse beta decay

---

*On behalf of the LVD collaboration.

reaction of electron anti neutrinos on protons:

$$\overline{\nu_e} + p \to n + e^+ \quad (1)$$

$$n + p \to d + \gamma \quad (2)$$

The prompt signal from the positron and the delayed signal from the 2.2 MeV gamma from the neutron capture (mean lifetime $\tau \sim 185$ μs) constitute the double signature for this reaction. The positron signal is detected with a high energy threshold level (HET), set at 7 MeV for the external counters and 4 MeV for the "core" counters, while the 2.2 MeV gamma is detected with a 1 MeV low energy threshold level (LET), activated by the high energy signal for a time duration of about 1 ms.

2. Neutron signature

The LVD apparatus can detect neutrons with the same signature of the inverse beta decay reaction. High energy neutrons could cause a liquid scintillator proton to recoil (prompt HET signal), and are then thermalized and finally captured by the liquid scintillator protons with the emission of the 2.2 MeV gamma (delayed LET signal). Taking into account the energy transfer in the interaction between neutron and proton, the proton quenching and the value of the high energy threshold in the core of the detector, we estimate that the neutrons detected in this way have energies greater than about 20 MeV.

The background to the neutron detection is due to the accidental coincidences between the high energy signals and the low energy ones. This background, however, has a flat distribution of the delay between the two signals and can be estimated by fitting the time delay distribution. An example of this distribution is shown in figure 1; we can fit the data with the curve

$$\frac{dN}{dt} = P1 \cdot e^{\frac{-t}{\tau}} + P2$$

where $\tau = 185$ μs. From the first parameter we obtain the number of neutron interactions, while the second takes into account the number of accidental coincidences.

3. Analysis and results

We have analyzed the neutron production in association to single muon events, that is events with only one reconstructed track, from 1994 to 2002,

for a total sample of more than 7 millions of single muons events.
First we have evaluated the production of neutrons per counter per event at various distances from the muon track; the distance is defined as the distance between the reconstructed muon track and the center of the counter where the neutron is detected. Notice that in the counters traversed by the muon track we require, in addition to the high energy signal associated to the muon a second one associated to the recoiling proton. Neutron candidates are selected with the procedure described in the previous section; at each distance the background contribution has been evaluated by fitting the time delay distribution between the HET signal and the LET ones. The result obtained is shown in figure 2; we were able to evaluate the neutron

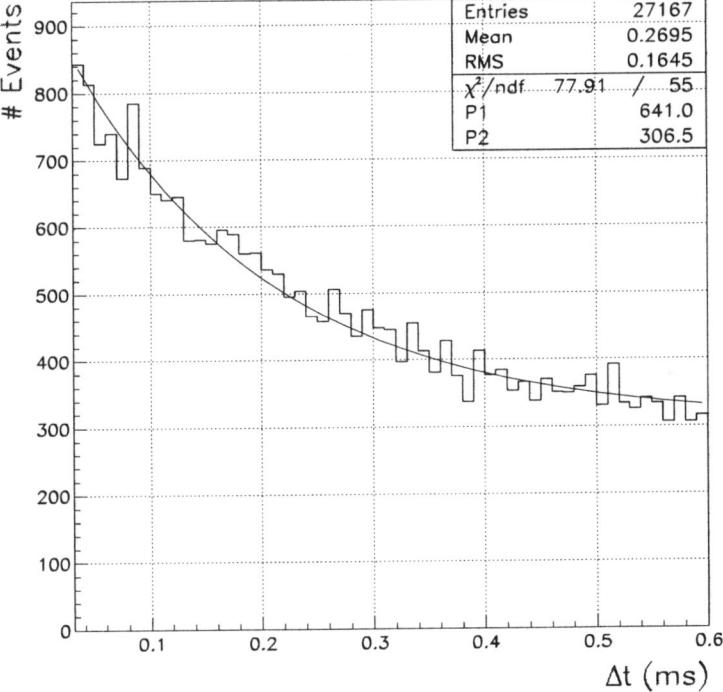

Figure 1. Time delay distribution between the high energy signals and the following low energy one detected in the same counter of the detector. From the distribution we can separate the neutron interaction from the accidental coincidences (see text).

production up to 22 m from the muon track. Due to the non homogeneous distribution of the scintillator in the LVD detector the behavior observed in figure 2 has to be studied with a detailed Montecarlo simulation which is under development.

To estimate the neutron energy spectrum, we studied the number of neutrons detected as a function of the energy released in the scintillator from the recoiling proton. The result is shown in figure 3; the data are well fitted by the power law spectrum:

$$\frac{dN}{dE} = A \times E^{-\alpha}$$

where $A = (1.6 \pm 0.1) \cdot 10^{-5}$ and $\alpha = (1.18 \pm 0.02)$; the errors are statistical only.

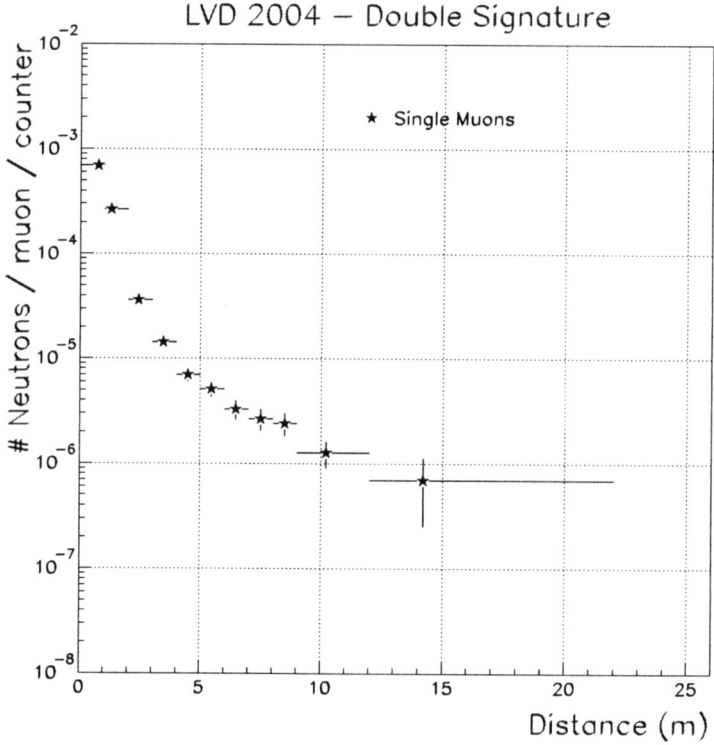

Figure 2. Number of neutrons detected per muon per counters as a function of the distance from the muon track.

Finally we evaluate the neutron production as a function of the muon track length in scintillator. The preliminary result is shown in figure 4. The data are well fitted by:

$$y = 0.13 \cdot 10^{-3} + 0.13 \cdot 10^{-2} \cdot L$$

where the first parameter takes into account the neutron production in the rock, as it is independent from the muon track length inside the liquid scintillator, while the second parameter takes into account the increase in the neutron production with the muon track length in scintillator. Comparing the two values we can conclude that the neutron production in the core of the experiment is mostly due to the interaction of muons with the detector nuclei (Fe,C).

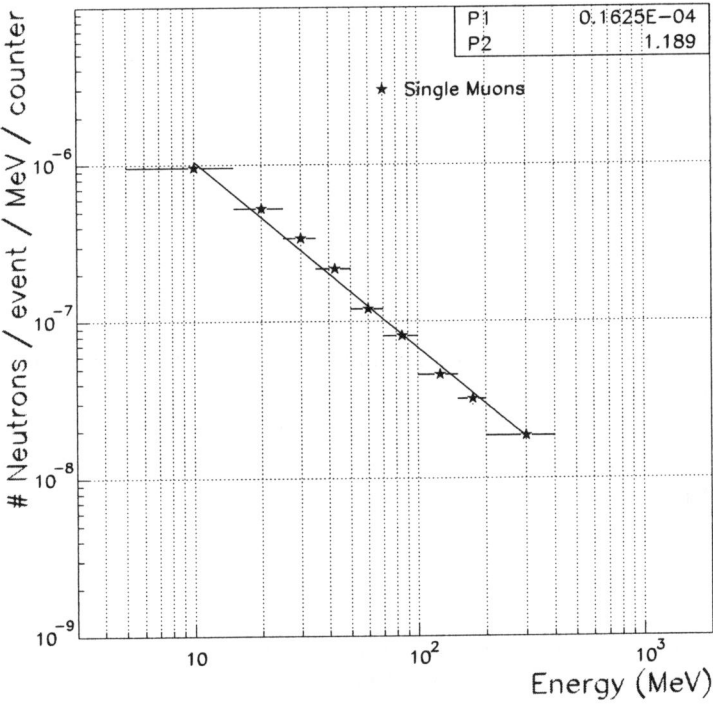

Figure 3. Number of neutrons detected per muon per counters as a function of the proton recoil energy.

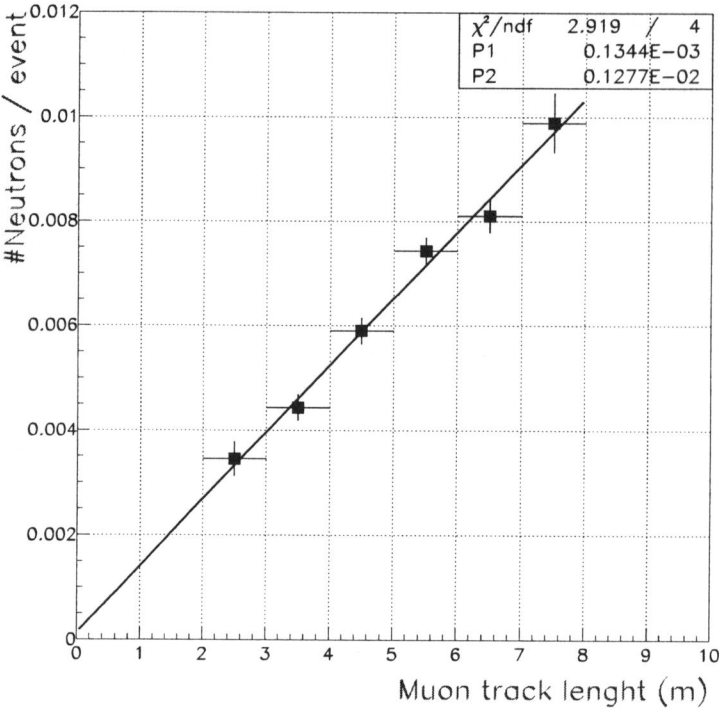

Figure 4. Number of neutrons per events detected as a function of the muon track length in scintillator; the main production of neutron is due to the muon interaction with the detector nuclei.

4. Conclusion

We report the neutron production from muon interaction in the LVD detector for single muon events and selecting neutrons with a double signature, that is neutrons of energies larger than 20 MeV. For this sample of events we were able to measure the neutron flux up to 22 meter from the muon track and the neutron energy spectrum. We also find out that the neutrons selected are produced in the muon interaction with the detector material.

References

1. Khalchukov et al, *Il Nuovo Cimento C* **6**, 320 (1983).
2. Aglietta et al., *Il Nuovo Cimento A* **105**, 1793 (1992).

STATUS OF NEUTRON BACKGROUND STUDY IN CRESST

H. WULANDARI, W. RAU AND F. VON FEILITZSCH

Physikdept. E-15, Technische Universität München,
James-Franck-Str., D-85748 Garching, Germany
E-mail: Hesti.Wulandari@ph.tum.de

J. JOCHUM

Physikalisches Institut I, Eberhard Karls Universität Tübingen,
Auf der Morgenstelle 14, D-72076 Tübingen, Germany
E-mail: josef.jochum@uni-tuebingen.de

We present results of neutron background studies based on Monte Carlo simulations for the direct dark matter search experiment CRESST II. Simulations suggest that the events measured by CRESST II without any neutron moderator are due to the neutrons induced by the local radioactivity of the rock/concrete surrounding the experimental setup. The contributions of neutrons from other origins and how they would affect CRESST are also discussed.

1. Introduction

CRESST (Cryogenic Rare Event Search using Superconducting Thermometers) employs cryogenic detectors for the direct detection of WIMPs (Weakly Interacting Massive Particles)[1]. The CRESST setup is located at hall A of the Gran Sasso underground laboratory (3600 mwe) in Italy.

The sensitivity of the direct search experiments using cryogenic detectors has improved rapidly in the last few years, thanks to the applications of event-by-event discrimination techniques which enable an efficient rejection of the dominant electron recoils induced by radioactive gamma-ray background from nuclear recoils expected from WIMP interactions with matter. In the case of CRESST II this rejection technique is performed by measuring scintillation light and a phonon signal simultaneously. However, neutrons are hardly distinguishable from WIMPs, because they interact with matter in a similar way. Therefore, suppression of neutron background is mandatory for experiments aiming for high sensitivity. Simulations of neutron background are needed for the prediction of sensitivity an experiment can achieve and which measures have to be taken to reach the projected sensitivity of the experiment.

In the following we present the results of our Monte Carlo simulations of neutrons from local radioactivity (fission and (α,n) reactions) as well as muon-induced neutrons in the rock/concrete and shielding materials, and their contributions to the expected background rate in CRESST II.

1.1. *Neutrons from local radioactivity*

Neutron flux at large depth underground is dominated by neutrons from the local radioactivity. We have simulated the flux of neutrons induced by radioactivity in the rock and concrete surrounding the Gran Sasso laboratory using MCNP4B[2]. Details of our simulations are discussed elsewhere[3]. We found from our study that the contribution of (α,n) reactions makes the spectra in the laboratory halls different from the spectrum expected for neutrons produced by fission reactions only, especially at high energies. The flux is dominated by neutrons produced in the concrete layer and therefore does not vary much from hall to hall. It can be expected that as well for other underground laboratories the neutron flux originates mainly from the concrete and not from the rock material. A more detailed spectrum compared to that from measurements has been obtained. We also have shown the dependence of the neutron flux on the humidity of the concrete. Our results for the case of hall A with dry concrete (8% water content) are in good agreement with the experimental data from[4]. This simulated neutron spectrum in hall A was then used for further simulations to get the neutron-induced recoil spectrum in a $CaWO_4$ target crystal. A simplified CRESST setup, which consist of Pb/Cu shields, was used in the simulations. The outer dimension of the simulated lead shield is $(130 \times 130 \times 136)\,cm^3$ and the thickness is 20 cm. Inside the lead is a 15 cm thick copper layer housing the experimental cavity, in which a single cube detector crystal of $(4 \times 4 \times 4)\,cm^3$ is placed. The simulation yields a count rate of (0.56 ± 0.11) cts/kg/day in the detector in the energy range of 12-40 keV. A 50 cm polyethylene shield will reduce the background count rate by about four orders of magnitude.

CRESST II data from early 2004 (without neutron moderator) with two 300 g prototype detector modules and a net exposure of 20.5 kg days give a rate for nuclear recoils between 12 and 40 keV of (0.87 ± 0.22) cts/kg/day, which is compatible with the rate expected from our simulation for the neutron background. The WIMP sensitivity appears to be limited by the neutron background if neutron-induced events and WIMP-induced events are indistinguishable. However, neutron induced signals are mainly due to recoils of the light elements of the target. WIMPs on the other hand scatter predominantly on the tungsten. Measurements indicate a higher quenching factor for tungsten, which would allow a partial discrimination of the neutron background and such lead to a better WIMP sensitivity[5].

We also performed simulations to study the effect caused by neutrons from the radio impurity in the lead shield of CRESST. Only neutrons induced by spontaneous fission of ^{238}U were considered in our simulations, because the contribution of (α,n) reactions in lead is not significant. We found that neutrons from this origin would give 1.5×10^{-2} cts/kg/day in the energy range of 12-40 keV for a ^{238}U concentration of one ppb. Hence, for the CRESST setup without a neutron shield a contamination of some ppb ^{238}U is still acceptable. However, when a 50 cm polyethylene shield is put in place, reducing the contribution of low energy neutrons from the rock significantly, only a few ppt ^{238}U are already a limiting neutron source.

The typical amount of radio impurity in lead commonly used in rare event search experiments is reported by some groups. Allesandrello et al.[6] measured a ^{238}U contamination of $<$ 2 ppb in roman lead and $<$ 12 ppb in low activity lead. Assuming an equilibrium of ^{238}U with its daughter products, the EDELWEISS collaboration found an upper contamination limit of 0.7 ppb ^{238}U in the most recent measurements of its lead. Previous measurements of a different lead sample gave 0.1 ppb[7]. To know the real contribution of neutrons from this origin for CRESST, the contamination in the shielding materials used in CRESST needs to be measured.

1.2. Muon-induced neutrons

The muon flux at the depth of the Gran Sasso laboratory is a factor of $\sim 10^6$ smaller compared to the surface flux. However, neutrons produced by muons in the rock and in the shielding materials can be important for experiments aiming for high sensitivity like WIMP searches.

To calculate differential and integral muon intensities at the depth of the Gran Sasso laboratory we used a special code called SIAM[8]. In this code the differential muon intensity underground was determined using the following equation:

$$I_\mu(E_\mu, X, \cos\theta) = \int_0^\infty P(E_\mu, X, E_{\mu0}) \frac{dI_{\mu0}(E_\mu, \cos\theta^*)}{dE_{\mu0}} dE_{\mu0} \qquad (1)$$

where $\frac{dI_{\mu0}(E_\mu, \cos\theta^*)}{dE_{\mu0}}$ is the muon intensity at the sea level and zenith angle

θ^*:

$$\frac{dI_{\mu0}(E_\mu, \cos\theta^*)}{dE_{\mu0}} = A\frac{0.14 E_\mu^{-\gamma}}{\text{cm}^2\text{s sr GeV}} \times \left\{ \frac{1}{1+\frac{1.1 E_{\mu0}\cos\theta^*}{115\,\text{GeV}}} + \frac{0.054}{1+\frac{1.1 E_{\mu0}\cos\theta^*}{850\,\text{GeV}}} + R_c \right\} \quad (2)$$

The relation between the zenith angle at the Earth's surface, θ^*, and the zenith angle underground, θ, was determined taking into account the curvature of the Earth. R_c denotes the ratio of prompt muons to pions. The parameters in Eq. (2) were taken either according to Gaisser's parameterization[9] ($A=1, \gamma=2.70$) which is modified for large zenith angles and prompt muon flux[10], or following the best fit to the depth-vertical μ intensity relation measured by the LVD experiment[10]. LVD reported the normalization constant $A = 1.84 \pm 0.31$, $\gamma = 2.77 \pm 0.02$ and the upper limit $R_c \leqslant 2 \times 10^{-3}$ (95% C.L.)[11]. In this work we have chosen 10^{-4} for the ratio of prompt muons to pions. $P(E_\mu, X, E_{\mu0})$ is the probability for a muon with energy $E_{\mu0}$ at the surface to have the energy E_μ at depth X[8] and was obtained by propagating muons with various energies at the Earth's surface using MUSIC (Muon Simulation Code)[12].

The absolute muon intensity underground depends in fact on the surface relief, which is very complex for Gran Sasso. We assumed a flat surface as an approximation and plan to take into account the detailed mountain profile of Gran Sasso in the further study. We used the muon spectrum with parameters following the LVD best fit and generated muons at the surfaces of a cube of rock with a size of $(20 \times 20 \times 20)\,\text{m}^3$. Inside the rock cube, the experimental hall was taken to be of a size of $(6 \times 6 \times 5)\,\text{m}^3$. The top of the hall was placed 10 m below the top of the rock cube. The size of the experimental hall used in these simulations was chosen smaller than the real hall at the Gran Sasso laboratory to save computing time. But some test simulations have been done to ensure that the results do not change significantly if a larger size is used. The simplified geometry of the CRESST setup with 50 cm polyethylene is placed inside the hall. We used MUSUN (Muon Simulation Underground)[8] to sample muon energy and angular distribution outside the rock cube and FLUKA[13] to simulate neutron production by muons.

We found that neutrons produced by muons in the shield contribute $\sim 95\%$ to the neutron flux at the detector level. The rest is the contribution of muon-induced neutrons in the rock. Among the different shielding materials, lead is the most effective target of muons to produce neutrons. The expected count rates in the detector are $(1.39 \pm 0.13) \times 10^{-1}$ cts/kg/day and $(1.09 \pm 0.10) \times 10^{-2}$ cts/kg/day in the energy range of 12-40 keV for muon-induced neutrons from the shield and from the rock respectively. If an internal polyethylene shield of 10 cm were inserted in the current experimental cavity (after the copper shield), a reduction of the count rates of muon-induced neutrons from the shields and from the rock by a factor of 30 and 60 respectively could be achieved. In Figure 1 the recoils spectra of muons-induced neutrons are shown together with those of neutrons from the radioactivity of the rock/concrete. It is seen that the count rate of muon-induced neutrons in the shields is not negligible compared to the contribution of neutrons from the radioactivity of the rock/concrete even without neutron moderator.

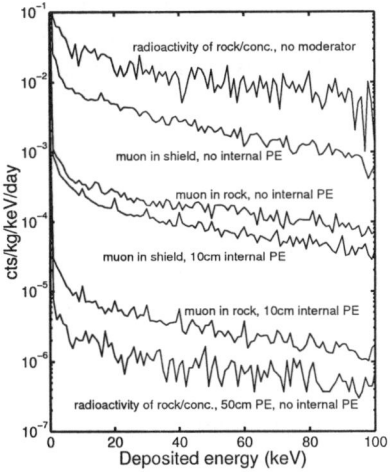
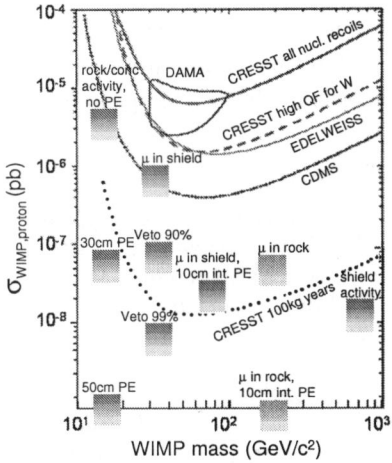

Figure 1. Recoil spectra of neutrons induced by muons and neutrons from the radioactivity of the surrounding rock/concrete.

Figure 2. Limit of WIMP sensitivity due to background neutrons from different origins.

The current WIMP-nucleon cross section exclusion limit (90% CL)[5] and the projected sensitivity of CRESST are shown in Figure 2 together with the present EDELWEISS[14] and CDMS[15] limits and the region of DAMA positive signal[16]. We also show in the figure how neutrons from different origins would limit the sensitivity of CRESST. To reach the projected sensitivity of some 10^{-8}pb a neutron moderator of 30-50 cm polyethylene and a muon veto with an efficiency of more than 90% will be installed for CRESST

II. A 10 cm internal polyethylene could also help, but unfortunately there is not enough place for it in the current setup. The contribution of fission-induced neutrons in the lead shield shown here is assuming a contamination of 0.1 ppb ^{238}U.

1.3. Conclusions

Without a neutron moderator the number of events measured by CRESST II is in agreement with the expected rate of background neutrons from the activity of the rock/concrete. A 50 cm polyethylene shield will reduce the background count rate in the $CaWO_4$ detector by by about four orders of magnitude. Then the background will be dominated by neutrons from other origins. To reach the projected sensitivity, a muon veto will be installed for CRESST. Multiple scattering should be studied and the radio impurity of the shielding materials need to be measured to determine a more realistic contributions of muon-induced neutrons in the rock and fission-induced neutrons in the shielding materials.

References

1. G. Angloher et al., *Astroparticle Physics* **14**, 43 (2002).
2. J.F. Briesmeister, Ed., "MCNP-A General Monte Carlo N-Particle Code, Version 4B", LA-12625-M, Los Alamos National Laboratory (March 1997).
3. H. Wulandari et al., *hep-ex/0312050*, accepted for publication in *Astroparticle Physics*.
4. P. Belli et al., *Il Nuovo Cim.* **101A**, 959 (1989).
5. G. Angloher et al., astro-ph/0408006, submitted to *Astroparticle Physics*.
6. A. Allesandrello et al., *Nucl. Instr. Meth.* **B61**, 106 (1991).
7. G. Gerbier, private communication.
8. V.A. Kudryavtsev, private communication.
9. T.K. Gaisser, "Cosmic Rays and Particle Physics", Cambridge University Press (1990)
10. M. Aglietta et al., *Phys. Rev.* **D58**, 092005 (1998).
11. M. Aglieta et al. (The LVD-Collaboration), *Phys. Rev.* **D60**, 112001 (1999).
12. P. Antonioli et al., *Phys. Lett.* **B471**, 251 (1999).
13. A. Fasso, A. Ferrari and P.R. Sala, in Proceedings of the Monte Carlo 2000 Conference, (Lisbon, October 23-26, 2000), Eds. A. King, F. Barao, M. Nakagawa, L. Tavora, P. Vaz, Springer-Verlag, Berlin (2001) 159; A. Fasso, A. Ferrari, J. Ranft and P.R. Sala, ibid. 995.
14. A. Benoit et al., *Phys. Lett.* **B545**, 43 (2002).
15. D.S. Akerib et al., *astro-ph/0405033*, accepted for publication in Phys. Rev. Lett.
16. R. Bernabei et al., *Phys. Lett.* **B480**, 23 (2000).

NEUTRON BACKGROUND IN A TIME PROJECTION CHAMBER FOR WIMP SEARCHES

M. J. CARSON, J. C. DAVIES*, E. DAW, R. J. HOLLINGWORTH,
J. KIRKPATRICK, V. A. KUDRYAVTSEV, T. B. LAWSON,
P. K. LIGHTFOOT, J. E. MCMILLAN, B. MORGAN, S. M. PALING,
M. ROBINSON, N. J. C. SPOONER, D. R. TOVEY, E. TZIAFERI

Department of Physics and Astronomy
University of Sheffield
Hicks Building
Hounsfield Road
Sheffield, S3 7RH
UK

Neutron background is significant in limiting the sensitivity of dark matter detectors. Presented here are results of simulations performed for a time projection chamber acting as a particle dark matter detector in an underground laboratory. The investigated background includes neutrons from rock and detector components, generated via spontaneous fission and (α,n) reactions, as well as those due to cosmic-ray muons. Also examined are methods of neutron background supression.

1. Introduction

Neutron background is an important issue for future underground dark matter detectors, since neutrons can produce nuclear recoils in the same way as WIMPs. Other backgrounds that have the capability to generate false signals include sources of alphas and gammas. Simulations of alphas and gammas, however, have been neglected in this study as they can be removed by appropriate active vetoes and energy threshold cuts[1,2].

The simulations use either the GEANT4 toolkit[3] or FLUKA[4] (versions FLUKA-2002 and FLUKA-2003) to simulate geometry, particle production, tracking and detection, with the neutron production energy spectra from spontaneous fission and (α, n) reactions being calculated using SOURCES[5]. The background investigated is that of neutrons produced via spontaneous

---

*corresponding author, email: j.c.davies@sheffield.ac.uk

fission and (α, n) reactions in and around the detector, along with muon-induced neutrons from deeply penetrating cosmic rays. These simulations attempt to determine the rate of nuclear recoils that will be observed by the detector due to the neutron background, to estimate the amount of passive neutron shielding required to keep this rate sufficiently suppressed and to formulate requirements for the purity of materials used in the detector construction in order to gain a high sensitivity to WIMPs.

2. Detector design

For the GEANT4[3] simulations a long laboratory area, similar in size to the Boulby Underground Laboratory (North Yorkshire, UK; Boulby mine is run by Cleveland Potash Ltd.), filled with air approximated at 1 atmosphere of an 80 : 20 nitrogen to oxygen mixture, was used. Surrounding the laboratory is rock salt (pure NaCl in these simulations) 3 m thick.

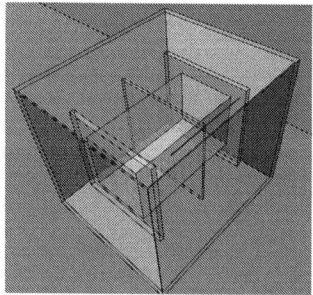

Figure 1. GEANT4 image, using VRML, of the detector geometry used for the simulations. The steel vessel containing the perspex frames and resistor volumes can be seen.

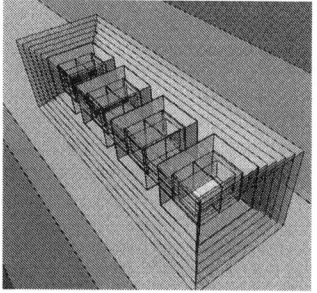

Figure 2. GEANT4 image, using VRML, showing the geometry for simulating 4 detectors with CH_2 neutron shielding both in between surrounding the vessels.

The detector itself consists of a 2 cm thick stainless steel vessel with internal dimensions of $1.2 \times 1.2 \times 1.4$ m^3 and a mass of 1590 kg (Figure 1). Within the vessel the detector includes a cathode frame with dimensions $0.8 \times 0.8 \times 0.02$ m^3 made of perspex and mounted vertically in the centre of the detector volume. To either side of this frame are another two frames, which are used to support the two multi-wire proportional chambers (MWPCs). The dimensions of these frames are $0.8 \times 0.8 \times 0.04$ m^3. The remaining volume within the vessel is filled with CS_2 gas at a pressure of 160 torr, with a density of 0.668 kg m^{-3}. The fiducial volume within the

MWPC frames is $0.5 \times 0.5 \times 1$ m^3, which gives a target mass of 0.167 kg. This is a possible design for the second generation DRIFT-type detectors[6].

An approximation of the resistor chain in the field cage of the detector is also included. The dimensions of the resistors were approximated to a cuboid of ceramic material positioned alongside the fiducial volume. The volume of ceramics was $3.28 \times 3.28 \times 700$ mm^3.

Varying amounts of passive neutron shielding (CH$_2$) were added to the simulation geometry in order to evaluate the quantity required to reduce the rate of nuclear recoils due to the rock neutron background to less than one per year for a target mass of 3.33 kg (20 of the modules described above).

Multiple detectors were also simulated to assess the effects of other neutron background sources and shielding between detectors. To investigate this, four detectors were grouped and positioned in a row with varying amounts of CH$_2$ neutron shielding between the vessels (see Figure 2 for illustration).

3. Neutron simulation

The sources of neutron background investigated in this work are due to uranium and thorium contamination in rock, stainless steel, ceramics and hydrocarbon shielding, along with those due to cosmic ray muons.

Simulations of neutrons being produced in the rock surrounding the detector laboratory were performed chiefly in order to establish the amount of hydrocarbon shielding required to reduce the recoil rate observed in the detector to well below one per year. The neutron energy spectrum assigned to neutrons being generated isotropically within a 2.5 m thickness of rock at the edge of the laboratory volume is shown in Figure 3. The U and Th content presented in Table 1[7,8] gives a total neutron production rate of 6.32×10^{-8} neutrons cm^{-3} s^{-1} with a mean energy of 1.73 MeV.

It is likely that the main limitation on the sensitivity of a dark matter detector (after shielding from rock neutrons) is the rate of the neutron background originating in the detector itself. The component materials expected to produce the largest neutron backgrounds are the stainless steel vessel, the ceramic material of the resistor chain and the hydrocarbon shielding. The neutron production spectra generated by SOURCES for all three components are shown in Figure 4, assuming the U and Th content given in Table 1[7,8].

In all runs simulating neutrons from rock and detector components the number of events processed was large enough to ensure that statistical uncertainties were ≪ 1%.

Table 1. Values for the uranium and thorium content in materials used for these simulations taken from measurements.

Material	U Contamination (ppb)	Th Contamination (ppb)
Rock	60	130
Stainless Steel	0.5	0.5
Ceramics	500	2000
CH_2	0.27	0.05

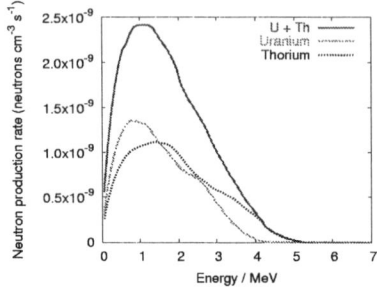

Figure 3. Energy spectrum of neutrons produced in the rock salt surrounding the detector. Spectra due to 60 ppb uranium contamination, 130 ppb thorium contamination and U and Th combined are shown.

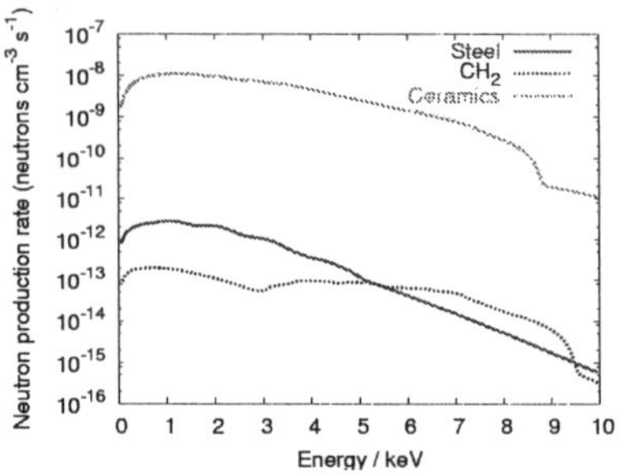

Figure 4. Energy spectra of neutrons produced in the detector components. Spectra are shown for the ceramics of the resistor chain, the stainless steel vessel and the CH_2 shielding surrounding the vessel.

For the case of muon-induced neutrons the muon spectrum and angular distribution was simulated using the MUSUN Monte Carlo code[9]. Normalisation of the muon (and neutron) spectrum was achieved using the measured value for the muon flux at the Boulby Underground Laboratory: $(4.09 \pm 0.15) \times 10^{-8}$ cm^{-2} s^{-1}, which corresponds to a rock overburden at vertical of 2805 ± 45 m w. e.[10].

Simulations of muon propagation and interactions, development of muon-induced cascades, neutron production, propagation and detection were performed with FLUKA[4]. Figure 5 shows the neutron spectrum entering the TPC volume (vessel-gas boundary).

4. Preliminary results and conclusions

The results of the simulations described here are summarised in Table 2 and the nuclear recoil spectra for muon-induced neutrons, and neutrons

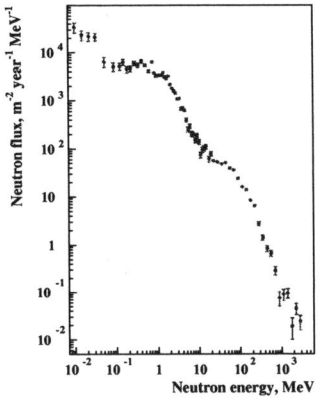

Figure 5. Energy spectrum of muon-induced neutrons entering the TPC (vessel/gas boundary).

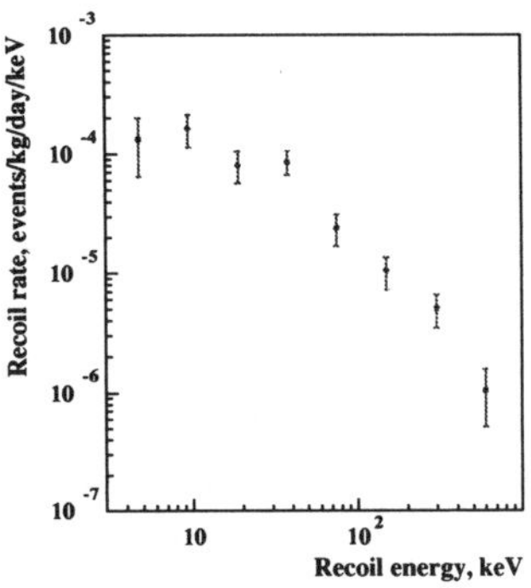

Figure 6. Nuclear recoil energy spectrum in a TPC filled with 167 g of CS_2 from muon-induced neutrons. TPC is shielded with 40 g/cm^2 of CH_2.

originating in rock and detector components are plotted in Figures 6, 7 and 8 respectively.

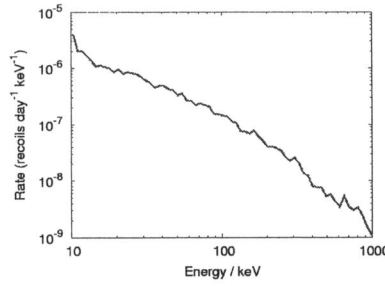
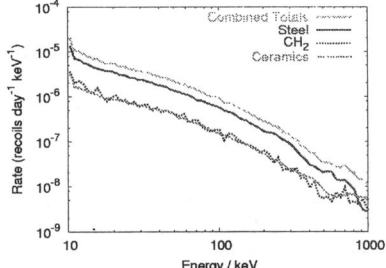

Figure 7. Energy spectrum of nuclear recoils in the 167 g of CS_2 behind 40 g/cm^2 of CH_2 shielding, produced by neutrons originating in the surrounding rock.

Figure 8. Energy spectra of nuclear recoils in the 167 g of CS_2 produced by neutrons originating in detector components.

Table 2. Neutron background rates per year at 10-50 keV recoil energies from different sources in a 167 g, 3.33 kg and 10 kg TPC (40 g/cm^2 of CH_2 shielding against rock neutrons and 50% efficiency to detect electromagnetic component of the muon-induced cascades were assumed).

Detector mass	Nuclear recoil rates per year at 10-50 keV			
kg	Rock	Muons	Detector	Total
0.167	0.01	0.12	0.06	0.19
3.33	0.2	2.4	1.3	3.9
10.0	0.6	7.2	3.8	11.6

We conclude that to suppress the nuclear recoil rate due to neutrons produced in the rock to a reasonable level in a low pressure CS_2 TPC, 30–40 g cm^{-2} of hydrocarbon shielding is required around the modules. It is also advisable, due to the rate of neutrons produced in the steel vacuum vessels, to place at least 5 g cm^{-2} of hydrocarbon shielding between modules that are positioned side by side. Adding low-background hydrocarbon shielding around the sensitive volume of the detector to protect it from neutrons from detector components, and installing an active veto around the detector against muons and their secondaries, would also reduce the background rate by an order of magnitude.

Acknowledgments

This work was performed in affiliation with the UK Dark Matter and DRIFT Collaborations. This work is funded by PPARC. We also acknowledge the funding from EU FP6 programme – ILIAS.

References

1. G. J. Alner et al. *Nuclear Instrum. & Meth. in Phys. Res. A*, in press.
2. D. P. Snowden-Ifft et al. *Nucl. Instrum. & Meth. in Phys. Res. A* **498** (2003) 155;
3. GEANT4 Collaboration. *Nucl. Instrum. & Meth. in Phys. Res. A*, **506** (2003) 250. http://geant4.web.cern.ch/geant4/
4. A. Fassò, A. Ferrari and P. R. Sala. *Proceedings of the MonteCarlo 2000 Conference* (Lisbon, October 23-26, 2000), Ed. A.Kling, F.Barao, M.Nakagawa, L.Tavora, P.Vaz (Springer-Verlag, Berlin, 2001), p. 159; A. Fassò, A. Ferrari, J. Ranft and P. R. Sala, *ibid.* p. 995.
5. W. B. Wilson et al. SOURCES-4A, Technical Report LA-13639-MS, Los Alamos (1999).
6. S. M. Paling (for the Boulby Dark Matter Collaboration). Talk given at the 6th UCLA Symposium on Sources and Detection of Dark Matter and Dark Energy in the Universe (Marina del Rey, CA, USA, 18-20 February, 2004), http://www.physics.ucla.edu/hep/dm04/talks/paling.pdf.
7. P. F. Smith, D. Snowden-Ifft, N. J. T. Smith, R. Luscher and J. D. Lewin. To be published in *Astroparticle Physics*.
8. http://hepwww.rl.ac.uk/ukdmc/Radioactivity/Index.html
9. V. A. Kudryavtsev, N. J. C. Spooner and J. E. McMillan. *Nuclear Instrum. & Meth. in Phys. Res. A*, **505** (2003) 688.
10. M. Robinson et al. *Nuclear Instrum. & Meth. in Phys. Res. A*, **511** (2003) 347.

SIMULATION OF LOW ENERGY NEUTRON RECOILS WITH GEANT4

S. SCHOLL, M. BAUER AND J. JOCHUM

Physikalisches Institut I, Eberhard-Karls Universität Tübingen
Auf der Morgenstelle 14,
72076 Tübingen, Germany
E-mail: scholl@pit.physik.uni-tuebingen.de

For the development and the analysis of high sensitivity dark matter detection experiments, it is important to understand and reduce the background of neutron induced nuclear recoils. For this task, monte–carlo simulations written in GEANT4 are employed to understand the signal due to the neutron background.

1. Motivation

Today, a central problem of modern cosmology is the determination of the value and the composition of the energy density of the universe. From a variety of observations, i.e. the CMB anisotropy [1], the high–z supernova search [2] and the analysis of galaxy redshifts [3], it is obvious that the bulk of the matter in the universe is dark and non–baryonic. To determine the nature of the dark matter, direct detection would be most desireable to determine its nature. Since a hypothetical dark matter particle must be neutral, neutrons constitute a background for these experiments. For the understanding of the neutron background at energies of a few keV, simulations of the interactions of neutrons in the detector are nescessary.

2. Direct Detection of Dark Matter

The CRESST[4], EDELWEISS[5] and CDMS[6] experiments combine phonon readout with scintillation or ionisation readout to discriminate nuclear recoils from electron recoils, respectively. Thus for these experiments the remaining principal background is given by neutrons of radioactive or cosmogenic origin. The neutrons themselves cannot be eliminated by active background rejection techniques and thus provide a limit to the sensitivity of the experiments.

The recoil energy of a target nucleus elastically hit by a neutron is given by the following equation:

$$E_{Recoil}(neutron) = \frac{2E_{neutron}M_{nucleus}M_{neutron}}{(M_{nucleus}+M_{neutron})^2}(1-cos\theta) \quad (1)$$

For a dark matter particle, the elastic recoil energy of the hit nucleus is then given by:

$$E_{Recoil}(DMParticle) = \beta^2 \frac{M_{nucleus}M_{DMParticle}^2}{(M_{nucleus}+M_{DMParticle})^2}(1-cos\theta) \quad (2)$$

In the formulae above E and M are the kinetic energy and the mass incident particle, θ is the scattering angle, β is the velocity of the dark matter particle in units of c and $M_{nucleus}$ is the mass of the recoiling nucleus.

For head on collisions with a typical kinetic energy of the neutron of $1 MeV$, and a dark matter particle with a mass $60 GeV/c^2$ and a typical velocity of $250 km/s$ we get maximum recoil energies as shown in Table 1. For the heavier nuclei, the recoils induced by neutrons and dark matter particles are more difficult to distinguish. Since the spin–independent cross–section between dark matter particles and nuclei raises with A^2, the importance of the neutron background becomes obvious.

Table 1. The maximum recoil energy of a given nucleus for a head on incident neutron ($m_{neutron} \approx 1 MeV$) and a incident dark matter particle of 60 GeV/c^2 mass and an assumed velocity $\approx 10^{-3}c$.

Nucleus	Mass(GeV)	Maximum recoil(keV) Neutron	Dark Matter Particle
O	16	221	20
Si	28	133	26
Ca	40	95	29
Ge	73	53	30
Xe	131	30	26
W	184	22	22

3. Neutron Scattering in GEANT4

For our simulations we use GEANT 4 which is an object–oriented toolkit for simulating the passage of particles through matter. For our purposes, the object–oriented design allows an easy exchange of a given physical model by another while the rest of the simulation stays unaffected. In addition, we can also build our own physics lists, employing several different models in a single simulation. Since the recoil energy is in the order of tens of keV, the accuracy of the simulation must be able to cope with such small energies in the presence of multi MeV neutrons. For a suitable simulation of low energy neutrons in the detector, we must get our data on a single event basis. Getting this event–by–event basis data in MCNP[7] is quite difficult. The most complete simulation for hadronic and electromagnetic interactions, FLUKA, does not treat nuclear recoils individually and low–energy neutron interactions ($< 20 MeV$) are dealt with in an average way[8]. Because of these reasons, we chose GEANT4 as tool for our simulations. The interactions of low energy neutrons are split into four different parts which treated as different models in GEANT4:

Elastic Scattering

In the elastic scattering (n,n) the neutron scatters of the nucleus without exciting the nucleus. In GEANT4, the final state is given by sampling the differential neutron cross–sections of the nuclei in question. The data is given in two representations, a tabulation of the differential cross–section as function of scattering angle and kinetic energy of the incoming neutron. This process makes up the bulk of interactions in the detectors we are interested in.

Radiative Capture

In the radiative capture process (n,$x\gamma$) the incident neutron is captured and the daughter nucleus deexcited through the emission of one or more photons. In GEANT4, the resulting final state is described by either photon multiplicities, photo production cross–sections and the continuous contributions to the photon energy spectra, along with the angular distribution of the emitted photons. The data is given either by the full transition probability array if known, or as function of the kinetic energy of the neutron for each discrete photon plus eventual continuum contributions. For the materials we are interested in, this process is rather unimportant.

Fission

In the fission process, the neutron is captured by the nucleus and the resulting daughter nucleus splits up in two or more fragments. In GEANT4 the first to fourth most common fission channels are considered. The neutron yield is given as function of the incoming and the outgoing neutron energy, the angular distribution is given as expansion in legendre polynomials or as a tabulation. For the background in the detectors this process is unimportant.

Inelastic Scattering

In the inelastic scattering process (n,n'γ) the neutron excites the recoiling nucleus and the final state is given the neutron, the recoiling nucleus plus its daughter particles like γ,n or ^{4}He. In GEANT4 several final states are included. For photons, the energy and the angular distribution is described as in the case of radiative capture. For the final state particles, the energy and angular distribution can be decribed as in fission process, but normally only the tabulation of secondary energies is applicable.

4. Simulation in GEANT4

We found that the elastic scattering (n,n) in treated correctly in GEANT4, for mono-energetic neutrons we found that the elastic scattering yields the correct recoil energy spectra and angular distribution. Radiative capture nor fission do not really contribute to the neutron background, so we did not investigate this type of processes. Additionally, for the detector materials we used, fission is of no interest for the neutron background. However, there are some problems with the way GEANT4 deals with inelastic scattering. In the inelastic process in GEANT4, the neutron and the recoiling nucleus form a compound particle, which decays isotropically in the nucleus and a neutron again. In our testing of the inelastic scattering processes (n,n'γ) and (n,xn) we have observed the fact that GEANT4 fails to simulate these nuclear recoils correctly on the single event basis. On the statistical level, they are treated correctly, but on a single event basis, momentum and energy conservation is violated. In reality, the recoil energy depends on the scattering angle, whereas our simulations have shown no angular dependence.

5. Summary and Outlook

Apart from the inelastic scattering, GEANT4 seems suitable for the calculation of neutron–induced nuclear recoils. However, in the different physics lists the inelastic scattering may be implemented in different ways. This feature has to be understood and corrected if we want to understand the neutron interactions in the detector properly. As a next step, we are going to try to fix the inelastic scattering.

Acknowledgements

We also acknowledge the funding from the Network on Direct Dark Matter Detection N3 of the EU FP6 project ILIAS.

References

1. E. L. Wright, *astro-ph/0306132* (2003).
2. S. Perlmutter et al.,*Astrophys. J.* **517**, 517 (1999).
3. W. J. Percival et al.,*MNRAS* **327**, 129 (2001).
4. G. Angloher et al.,*astro-ph/0408006* (2004).
5. O. Martineau et al.,*NIM A* **530**, 426 (2004).
6. CDMS Collaboration *astro-ph/0405033* (2004).
7. J. F. Briesmeister, Ed. *MCNP–A General Monte Carlo N–Particle Code, Version 4B*,LA–12625–M, Los Alamos National Laboratory (1997).
8. A. Fasso et al. *Proceedings of the Monte Carlo 2000 Conference*, Springer–Verlag, Berlin 159 (2001).

SIMULATIONS OF MUON-INDUCED NEUTRON BACKGROUND WITH GEANT4

M. BAUER, J. JOCHUM, S. SCHOLL

Physikalisches Institut I, Universität Tübingen,
Auf der Morgenstelle 14,
72076 Tübingen, Germany
E-mail: bauer@pit.physik.uni-tuebingen.de

The production of high energy neutrons by muons in rock and in shielding materials typically used in direct dark matter search experiments has been simulated using the GEANT4 Monte Carlo toolkit. The results obtained agree within 25 % for different selections of physics models and within a factor of 2 with FLUKA simulations. Therefore we will use it for our simulations to check the relevance of muon-induced backgrounds.

1. Motivation

CRESST (Cryogenic Rare Event Search with Superconducting Thermometers) is a dark matter search experiment aiming at the direct detection of WIMPs (Weakly Interacting Massive Particles).[1] They are identified by their nuclear recoils, which can be discriminated from electron recoils with an efficiency of at least 99.7 % for energies above 15 keV by the use of cryogenic $CaWO_4$ detector modules which allow the simultaneous measurement of phonons and scintillation light.[2] Thereby most of the background is eliminated. However, unless segmented detectors are used, WIMPs cannot be distinguished from neutrons since they also produce nuclear recoils. Neutrons are therefore the main background to be considered in the experiment.

2. Sources of background

CRESST is carried out in the Gran Sasso underground laboratory at a depth of about 3600 m w.e. At this depth, neutrons originate mainly from the following sources:[3]

(1) Low energy neutrons (< 10 MeV) from fission and (α, n) reactions in the surrounding rock and concrete of the laboratory

(2) Low energy neutrons from fission in the shielding material and experimental setup
(3) High energy neutrons induced by muons in the rock
(4) High energy neutrons induced by muons in the shielding material

At this depth, the flux of high energy neutrons is about 3 orders of magnitude lower than that of low energy neutrons. Nevertheless, it is extremely important and may be the limiting factor for highly sensitive experiments due to the following reasons:[4]

(1) These neutrons have a hard spectrum and can therefore reach a detector far away from the muon track.
(2) Because of their high energies the energy transferred to a nucleus in a reaction can be large, so the nuclear recoils are usually above detector threshold, while the recoils induced by the low energy neutrons often go undetected.
(3) The usual hydrogen-rich shielding material used against low energy neutrons doesn't protect from high energy neutrons, but is itself a target for high energy muons in which additional neutrons can be produced.

Because of the importance of the high energy neutrons I will concentrate on them in the following.

3. Existing simulations

Simulations of the neutron flux induced by high energy muons in the rock and concrete of the laboratory and in the shielding materials of the detector as well as of the nuclear recoil in the detector have already been done.[3,4,5] In these simulations FLUKA[6] was used to simulate neutron production and transport, while the nuclear recoils were simulated using MCNP.[7] The use of several different simulation codes was necessary since no single code was able to do the whole task: FLUKA doesn't treat individual nuclear recoils but averages over them in a statistical way, while MCNP is unable to describe muon-nuclear interactions and is therefore unsuitable for the simulation of neutron production. However, splitting the simulation into different parts imposes some difficulties:

(1) There could be correlations between the production of neutrons and of charged particles which are lost when only pre-calculated neutron spectra are fed into the recoil simulation. However, neutrons which

are accompanied by charged particles do not contribute to the neutron background of the experiment since the charged particles are seen in the muon veto, therefore it is important to know these correlations.
(2) Unlike WIMPs, neutrons can be multiply scattered in the detector, therefore a distinction between WIMPs and neutrons is possible when using a segmented detector.
(3) Neutrons can also be produced in the experimental setup, and these neutrons are not taken into account in this splitted simulation approach.

Therefore it would be preferable to use one single simulation code for the whole task.

4. GEANT4 Simulations

GEANT4 is an object-oriented C++ toolkit for the simulation of the passage of particles through matter.[8] It seems to be the most promising choice for creating a unified muon background simulation. However because of its high flexibility it is extremely important to use it correctly since results depend crucially on the selected configuration. Also the reliability of the relevant hadronic physics still has to be verified and some technical difficulties have to be solved before it can be considered reliable.

As a check of GEANT4, a simulation of muon-induced neutron production has been created as close as possible to an existing FLUKA simulation.[4] Monoenergetic 270 GeV muons (which is the mean muon energy at the Gran Sasso laboratory) were shot onto a wide slab of material of thickness ≈ 5000 g/cm^2, which was chosen such that showers are well developed but the muon doesn't lose too much energy in the material. The number of neutrons produced was counted. Since all physics processes and the models describing them have to be selected by the user of GEANT4,[9] simulations have been carried out using different selections of models for comparison:

(1) Predefined "educated guess" physics list QGSP_GN[10] as well as muon deep inelastic scattering for muons above 1 GeV.
(2) The same as (1), and additionally data-driven high precision models for interactions of neutrons below 20 MeV (NHP).
(3) The same as (2), and additionally the GEANT4 binary cascade model (BIC), which is an improved model for the interaction of protons and neutrons below 10 GeV with nuclei.

(4) The same as (3), but with the Bertini cascade model (BERT) instead of the binary cascade model.

Results are presented in tables 1 and 2. When determining neutron production rates one has to make sure that neutrons are counted only once. Double counting could happen if scattered neutrons, e.g. in (n,n') reactions, are not really scattered, but killed and produced again by the simulation code. In FLUKA, this is handled via the concept of "stars". In GEANT4 the behaviour is unspecified in principle;[11] however, considering the results of our tests, double counting is assumed to happen in all models relevant for this work. Therefore the results given in table 2 (with double counting taken into account) should be considered as the physically meaningful ones. Results obtained with different physics models agree within about 25 % where the models using the binary or the Bertini cascade code are expected to be best.[11] The values are in agreement with FLUKA results[12] within a factor of 2, with a much higher degree of agreement for most materials simulated in this work.

Table 1. Simulated neutron production rates by 270 GeV muons in units of 10^{-5} n/μ/(g/cm^2) for different selections of physics models ((1) QGSP_GN, (2) QGSP_GN + NHP, (3) QGSP_GN + NHP + BIC, (4) QGSP_GN + NHP + BERT, see text) without double counting reduction. For physically meaningful results see table 2.

Material	G4 (1)	G4 (2)	G4 (3)	G4 (4)
Copper	114.1	178.2	183.7	208.3
Lead	251.5	517.3	669.5	620.3
LNGS concrete	44.6	47.3	56.2	58.7
LNGS rock	59.8	52.6	65.8	53.4
PE	26.0	29.1	30.3	29.8

5. Summary and Outlook

Neutron production rates by high energy muons have been simulated with GEANT4 using different selections of physics models. Results are in agreement with each other and with FLUKA results within a factor of 2 (at most), indicating that GEANT4 actually can be used to estimate background levels.

Table 2. Simulated neutron production rates by 270 GeV muons in units of 10^{-5} n/μ/(g/cm^2) for different selections of physics models ((1) QGSP_GN, (2) QGSP_GN + NHP, (3) QGSP_GN + NHP + BIC, (4) QGSP_GN + NHP + BERT, see text) with double counting reduction (see text) compared to FLUKA "precision" results[12]

Material	G4 (1)	G4 (2)	G4 (3)	G4 (4)	FLUKA
Copper	77.1	82.6	76.8	90.3	135.6
Lead	212.2	224.5	281.6	268.9	421.0
LNGS concrete	31.1	34.3	33.4	36.0	35.1
LNGS rock	36.0	36.2	36.9	31.4	40.4
PE	21.2	23.9	21.6	21.1	26.1

Acknowledgements

We acknowledge the funding from the Network on Direct Dark Matter Detection N3 of the EU FP6 project ILIAS.

References

1. G. Angloher et al., Limits on WIMP dark matter using scintillating CaWO$_4$ cryogenic detectors with active background suppression, astro-ph/0408006
2. B. Majorovits for the CRESST collaboration, talk given at IDM2004 conference, Edinburgh, Scotland, http://www.shef.ac.uk/physics/idm2004.html (see also these proceedings)
3. H. Wulandari et al., Neutron Background Studies for the CRESST Dark Matter Experiment, hep-ex/0401032
4. H. Wulandari, Study On Neutron-Induced Background in the Dark Matter Experiment CRESST, PhD thesis, TU Munich 2003
5. H. Wulandari et al., Neutron Flux Underground Revisited, hep-ex/0312050
6. A. Fasso, A. Ferrari and P. R. Sala, in Proceedings of the Monte Carlo 2000 conference (Lisbon, October 23-26, 2000), Eds. A. King, F. Barao, M. Nakagawa, L. Tavora, P. Vaz, Springer-Verlag Berlin (2001) 159; A. Fasso, A. Ferrari, J. Ranft and P. R. Sala, ibid., 995
7. J. F. Briesmeister, Ed., "MCNP – A General Monte Carlo N-Particle Code, Version 4B", LA-12625-M, Los Alamos National Laboratory (March 1997)
8. S. Agostinelli et al., GEANT4 — a simulation toolkit, NIM A 506 (2003) 250
9. GEANT4 Physics Reference Manual: http://wwwasd.web.cern.ch/wwwasd/geant4/G4UsersDocuments/UsersGuides/PhysicsReferenceManual/html/PhysicsReferenceManual.html
10. GEANT4 Hadronic Physics Working Group Homepage: http://www.geant4.com/hadronics/GHAD/HomePage/
11. H.-P. Wellisch, private communication
12. H. Wulandari, private communication

MUON-INDUCED NEUTRON PRODUCTION AND DETECTION WITH GEANT4 AND FLUKA

H. M. ARAÚJO
Blackett Laboratory, Imperial College London, SW7 2BW, UK

V. A. KUDRYAVTSEV
Department of Physics & Astronomy, University of Sheffield, S3 7RH, UK

A comparison study of the Monte Carlo codes GEANT4 and FLUKA for simulating neutron production by muons penetrating deep underground has been carried out. GEANT4 was found to generate fewer neutrons at muon energies above ∼100 GeV, by at most a factor of 2 in some materials, which we attribute mainly to lower neutron production in hadronic cascades. The muon-induced neutron background expected in a 250 kg liquid xenon WIMP dark matter detector was calculated with the two codes and good agreement was found for the recoil event rates.

1. Introduction

The knowledge of muon-induced neutron fluxes in underground laboratories is crucial in experiments searching for and detecting rare events associated with neutrino interactions, double-beta decay and dark matter WIMPs. This is achieved through measurements and simulations using available Monte Carlo codes. FLUKA[1] is very well established for neutron simulations. However, it does not treat the elastic scattering of neutrons at low energies with sufficient detail for the purpose of dark matter experiments. So-called 'Kerma' factors are used to generate energy deposition from nuclear recoils (other than protons), which are equivalent to the average recoil energy for a certain neutron energy. The GEANT4[2] toolkit can potentially be used for end-to-end simulations of experiments, from background calculations down to detailed detector characterisation. Its object-oriented design and open-source nature make it rather flexible. Critically, it generates and tracks in a realistic way the recoiling nuclei from neutron elastic scattering. In this paper we compare GEANT4 (release 6.2) and FLUKA (FLUKA-2003 and previous releases) simulations of neutron production by muons and recoil rates in a conceptual xenon dark matter detector.

2. Muon-induced neutron production

To study neutron production by muons we considered a μ^- beam incident on a block of material with length 3200 g/cm^2 and comparable transverse dimensions. We calculated the muon energy spectra at the end of the block from GEANT4, FLUKA and muon propagation code MUSIC[3] and found them to be in good agreement. Calculated neutron yields were corrected for the muon energy loss, double-counting of inelastically scattered neutrons and edge effects due to the progressive build-up of muon-induced cascades.

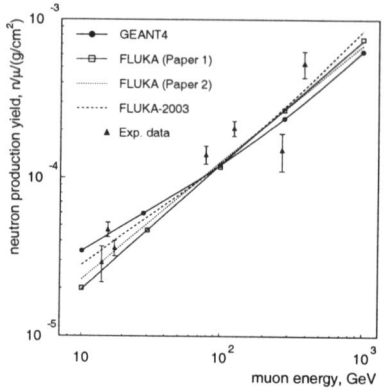 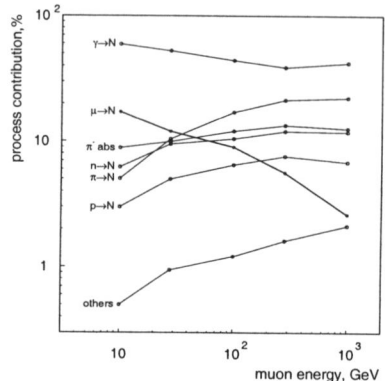

Figure 1. Dependence of the neutron yield per unit muon track length on muon energy for $C_{10}H_{20}$ scintillator (see Paper 1 and Paper 2 for references to the original experiments).

Figure 2. Relative contribution of various processes to the neutron yield in C_nH_{2n} from GEANT4: γ-N interaction ($\gamma \to N$), muon, proton and pion spallations ($\mu \to N$, $p \to N$, $\pi \to N$), π^- absorption (π^- abs) and neutron inelastic scattering ($n \to N$).

The total neutron yields in a generic hydrocarbon with composition C_nH_{2n} (density ρ=0.8 g/cm^3) are shown in Fig. 1 as a function of muon energy. Although GEANT4 results agree with the power law $E^{0.74-0.79}$ predicted by FLUKA-1999 at high energies[4,5] (hereafter Papers 1 and 2, respectively), there is an enhancement with decreasing energy relative to those FLUKA simulations. Present results with FLUKA-2003 show a similar enhancement of neutron production at low energies over the simple power law. Both codes are consistent with the experimental data shown, which have been measured at various depths around the world, keeping in mind the spread in experimental results.

The relative contribution of the most important production mechanisms to the total neutron rate in C_nH_{2n} is shown in Fig. 2 for GEANT4. Both GEANT4 and FLUKA (Paper 2) predict neutron production in electromagnetic (EM) cascades (real photonuclear interaction) to dominate at lower energies and to decrease in importance with increasing muon energy. Both results confirm that most neutrons are not produced in direct muon-induced spallation, but rather in the cascades muons initiate, and more so at higher energies. However, the GEANT4 results reveal a greater dominance of EM cascades at low energies, and this scenario is not significantly reverted at high energies, where FLUKA predicts production in hadronic cascades to take over. In this material, GEANT4 appears to overproduce neutrons in EM cascades and underproduce in hadronic cascades compared to FLUKA.

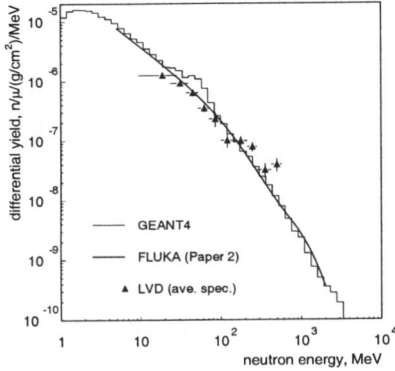
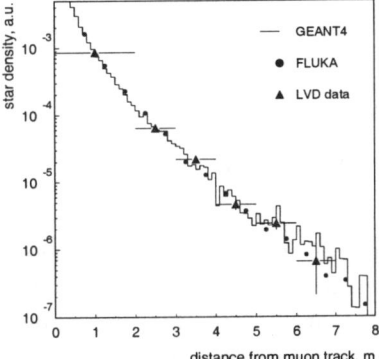

Figure 3. Energy spectrum of muon-induced neutrons in C_nH_{2n} normalised to the GEANT4 spectrum for visual agreement.

Figure 4. Lateral distribution of neutron inelastic interactions from the muon track at about 2.8 km w. e., normalised to the GEANT4 distribution for visual agreement.

The neutron production spectrum in C_nH_{2n} is shown in Fig. 3 for 280 GeV muons, a value close to the mean energy at several underground labs. The GEANT4 result is compared with a parameterised FLUKA spectrum from Paper 2 and measurements with the LVD detector.[6] The lateral distribution of neutron inelastic interactions for the muon spectrum at a depth of 2.8 km w.e. (280 GeV mean energy) is plotted in Fig. 4. Both simulations agree well with the LVD data.[6]

The variation of the total neutron yield with (average) atomic weight of the material is shown in Fig. 5, along with FLUKA results reported in Paper 1. The codes differ at most by a factor of 2, with FLUKA predicting consistently higher yields. The rates can be fitted by a power-law of the atomic weight A: $R = bA^\beta$. We obtain $b=(3.0\pm0.4)\times10^{-5}$ and $\beta=0.82\pm0.03$ for the GEANT4 simulation, while Paper 1 reports $b=(5.33\pm0.17)\times10^{-5}$ and $\beta=0.76\pm0.01$ for FLUKA. Real photonuclear interactions in GEANT4

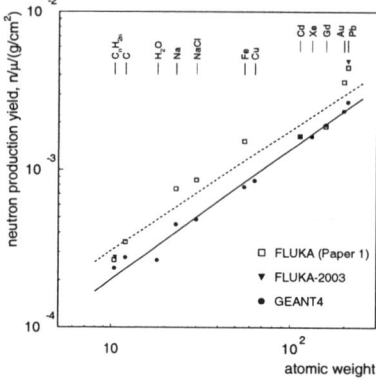
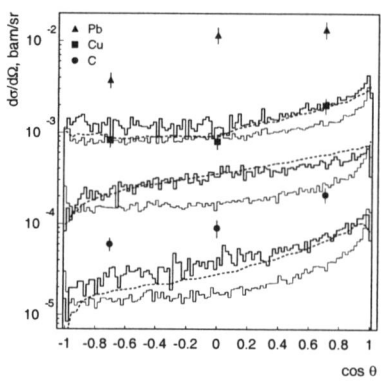

Figure 5. Dependence of the neutron yield on the average atomic weight of the material in FLUKA and GEANT4 for 280 GeV muons.

Figure 6. Differential cross-section of neutron production by 190 GeV muons: data – NA55; dashed curve – FLUKA; thick histogram – GEANT4; thin histogram – GEANT4 (muon-nucleus interaction only).

dominate for all materials, increasing linearly with A. Photoproduction in FLUKA gives less neutrons in hydrocarbon and more neutrons in lead. Secondary neutron production in GEANT4 (from neutron inelastic scattering), increases slightly slower ($A^{0.85}$) than photoproduction. Yields from pion, proton and muon spallations vary approximately as $A^{0.5}$.

Differential cross-sections for fast-neutron production by 190 GeV muons are shown in Fig. 6 for the set-up of the CERN NA55 experiment,[7] which measured mainly direct production in μ–N interactions above a 10 MeV threshold in neutron energy. Both packages agree quite reasonably, but the experimental points lie far above the Monte Carlo data. The reasons behind this discrepancy are still being investigated, but other experimental data do not show such a large difference with simulations, although accurate comparison requires detailed modelling of the set-ups.

3. Neutron background rates in xenon detectors

The calculation of the muon-induced background in a 250 kg liquid xenon target is the case study for a further comparison of the Monte Carlos. The muon flux experimentally measured at the Boulby Mine, and used to normalise the simulations, is $(4.09 \pm 0.15) \times 10^{-8}$ muons/cm^2/s.[8] Muons were sampled according to the energy spectrum and angular distribution at 2800 m w.e. using the MUSUN code,[4] on 5 sides of a 20×20×20 m^3 cube of pure NaCl (ρ=2.2 g/cm^3). A cavern with dimensions 6× 6×5 m^3

is located in the salt cube at a depth of 10 m. A 250 kg liquid xenon target, contained in a copper vessel 2 cm thick, is surrounded by 50 cm of $C_{10}H_{20}$ scintillator to form an active veto followed by 30 cm of lead shielding. The neutron spectra across various boundaries in this set-up are shown in Fig. 7 for GEANT4 and FLUKA simulations[9] (hereafter Paper 3). The figure confirms that most neutrons are produced in the lead shielding. This lead/hydrocarbon combination provides effective shielding only for neutrons below 10 MeV. The two spectra for neutrons emerging across the veto are in good agreement, especially for energies below a few tens of MeV.

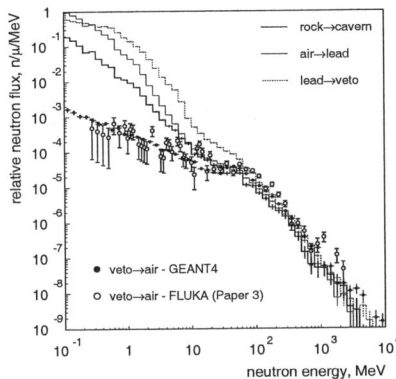
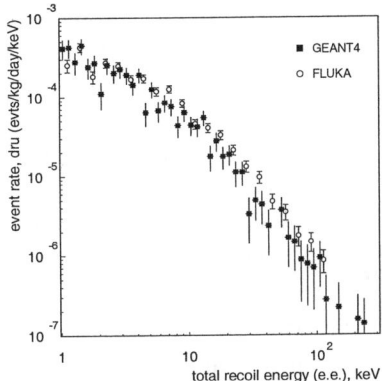

Figure 7. Energy spectra of muon-induced neutron fluxes across several boundaries.

Figure 8. Nuclear recoil rate in the liquid xenon detector as a function of the visible energy from all nuclear recoils in an event.

Table 1. Muon-induced neutron background (events/year) in 250 kg xenon target for several detection thresholds.

Event type	E_{ee}, keV	FLUKA	GEANT4
All NR events	>0	234±8	217±9
Pure NR events	>0	13.6±1.9	10.3±1.9
	>2	7.3±1.4	4.0±1.2
	2–10	3.9±1.0	3.3±1.1
Pure single NR events	>0	6.6±1.3	6.6±1.6
	2–10	2.1±0.7	2.6±1.0
...In anti-coincidence	>0	0	0

The nuclear recoil spectra in the xenon target are plotted in Fig. 8 and compared in Table 1. They include both 'pure' and 'mixed' recoil events, the latter term representing events involving both electromagnetic energy and nuclear recoils. An exposure time of nearly 3 years was simulated with GEANT4 ($\simeq 20 \times 10^6$ muons) and about 4 years with FLUKA-2003. A quenching factor of 0.2 is applied to the energy deposited in neutron elastic scattering events. A threshold of 100 keV e.e. is assumed for detection in an organic scintillator. FLUKA and GEANT4 arrive at a similar total number of neutron events, as well as similar rates in the range 2–10 keV e.e. Note that the numbers presented in Table 1 for FLUKA agree with the previous simulations (Paper 3) for the total recoil rate, but are a factor of 2 smaller for 'pure' nuclear recoil events. This is due to the lower energy cuts for gammas and electrons used in the present simulations.

4. Conclusions

The total muon-induced neutron yields, neutron energy spectrum and lateral distribution from the muon track in hydrocarbon material obtained with GEANT4 and FLUKA are consistent with the available measurements, with GEANT4 predicting fewer neutrons than FLUKA above $\sim$100 GeV muon energy by a small factor (<30%). In most other materials studied, GEANT4 also generates lower neutron fluxes for muon energies in the hundreds of GeV. The simulated total fluxes of fast neutrons (>1 MeV) entering the underground cavern and emerging after 30 cm of lead and 50 cm of hydrocarbon shielding were found to agree within 20%. The nuclear recoil spectrum observed in the xenon target is slightly harder in the FLUKA simulation, which also predicts a small excess of 'pure' recoil events. Both codes predict a similar number of recoils with visible energy of 2–10 keV.

5. Acknowledgements

This work has been undertaken within the framework of the UKDMC and supported by PPARC. The authors wish to thanks H. P. Wellisch (CERN) for his invaluable assistance. We also acknowledge the funding from the EU FP6 project ILIAS.

References

1. A.Fassò, A.Ferrari and P.R.Sala, in: *Proceedings of the MonteCarlo 2000* (Lisbon, 2000), p. 159; A.Fassò, A.Ferrari, J.Ranft and P.R.Sala, *ibid.* p. 995.
2. Geant4 Collaboration, *Nucl. Instrum. Meth. in Phys. Res.* **A506**, 250 (2003).
3. P.Antonioli et al. *Astroparticle Phys.* **7**, 357 (1997).
4. V.A.Kudryavtsev et al., *Nucl. Instrum. Meth. in Phys. Res.* **A505**, 688 (2003).
5. Y.-F.Wang et al., *Phys. Rev.* **D64**, 013012 (2001).
6. M.Aglietta et al., in: *Proc. 26th Int. Conf. Cosmic Rays* (Salt Lake City, 1999), v. 2, p. 44 (1999); hep-ex/9905047; see also H.Menghetti, these Proceedings.
7. V.Chazal et al., *Nucl. Instrum. Meth. in Phys. Res.* **A490**, 334 (2002).
8. M.Robinson et al., *Nucl. Instrum. Meth. in Phys. Res.* **A511**, 347 (2003).
9. M.J.Carson et al., *Astroparticle Phys.* **21**, 667 (2004).

VETO PERFORMANCE FOR LARGE-SCALE XENON DARK MATTER DETECTORS

M. J. CARSON,* J. C. DAVIES, V. A. KUDRYAVTSEV,
M. ROBINSON, N. J. C. SPOONER.

Department of Physics and Astronomy, University of Sheffield, UK

H. M. ARAÚJO, T. J. SUMNER.

Blackett Laboratory, Imperial College London, UK

Monte Carlo simulations of the veto performance for large-scale liquid xenon dark matter detectors are presented. A number of possible veto configurations are considered with the aim of maximizing the neutron and gamma rejection efficiency. Also discussed are the shielding requirements for reducing the external gamma background.

1. Introduction

Future dark matter detectors aim to achieve sensitivities to the WIMP-nucleon cross-section down to 10^{-10} pb, sufficient to probe the region of interest for neutralinos (a favoured dark matter candidate) in SUSY models [1,2]. A number of groups are now planning tonne-scale detectors which, it is hoped, will be capable of reaching this sensitivity [3,4].

WIMPs are expected to interact with ordinary matter primarily by single elastic scattering events producing nuclear recoils which can then be detected by the ionisation, scintillation and/or phonons produced in a suitable target. However, identical signals can be produced by the elastic scattering of neutrons in the target. Further, α, β, and γ radiation can cause signals which, at keV energies, can be misidentified as nuclear recoil events. Therefore, sophisticated techniques capable of distinguishing between WIMP-induced events and those due to background radiations are needed. Here we report on detailed Monte Carlo simulations of veto performance for a model large-scale liquid xenon dark matter detector.

---

*corresponding author, email: m.j.carson@sheffield.ac.uk

Table 1. Radioactive contamination levels in the materials used in the simulations. Note: PMT Potassium contamination level is for ^{40}K.

Material	U (ppb)	Th (ppb)	K (ppb)	^{60}Co (ppb)
PMT	4	4	3.1×10^{-1}	1.9×10^{-9}
Copper vessel	0.02	0.02	1	-
Boulby rock (NaCl)	60	300	1.3×10^{6}	-

Most simulations were carried out with Geant4 version 5.1-p01 [5]. Input neutron spectra were calculated with the Sources 4A [6] code. Gamma spectra for the U and Th chains were generated from reference spectra [7,8].

The detector is modeled as a cylindrical liquid xenon target 10 cm in height and 50 cm in diameter with a mass of approximately 250 kg. The target is viewed by an hexagonal array of 169 Hamamatsu R8778 photomultipliers (PMTs). The target and PMTs are contained in a 1 cm thick copper vessel which is in turn surrounded by a CH_2 veto (contained in a 0.5 cm thick stainless steel vessel) and 10 cm of lead shielding. For the purpose of shielding simulations the detector is housed in a laboratory $30\times6.5\times4.5$ m^3 with walls made of NaCl with an assumed thickness of 1 m. The detector itself is placed in the centre of one face of the laboratory. In all simulations the statistical errors are about 1%.

2. Veto performance for neutrons

Internal contamination is probably the dominant source of background in dark matter experiments if sufficient shielding is installed and the depth of the laboratory is large enough. All materials and components used in any detector are contaminated with one or more of U, Th, K and Co which are the primary sources of background neutrons and gammas. Radioactive Rn is ubiquitous in air and can be a concern if present inside the shielding in significant quantities. Further, xenon itself is contaminated with ^{85}Kr, the β-decay of which produces a gamma-like background of low energy events in the target.

Veto efficiencies were calculated for neutrons generated in the copper vessel with contamination levels given in Table 1. Neutrons were then propagated throughout the detector, veto and shielding and tracked until absorbed either within the detector or surrounding veto. Any neutron depositing energy in the target volume in coincidence with a signal in the veto can in principle be tagged as a neutron event. Efficiency is defined as the fraction of nuclear recoil events in the target with energies between 10-50 keV (2-10 keV electron equivalent energy, quenching factor of 0.2 [9,10])

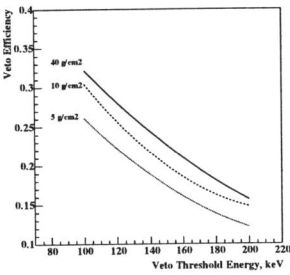

Figure 1. Efficiency of neutron rejection as a function of veto threshold energy for increasing veto thicknesses, assuming the detection of proton recoils only (without gammas from neutron capture).

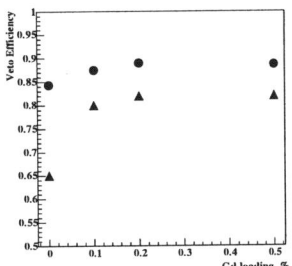

Figure 2. Efficiency as a function of Gd-loading for veto thickness 40 g cm^{-2}. Circles show efficiency of neutron rejection due to the detection of either gammas or proton recoils or both. Triangles show efficiencies for gammas from neutron capture only.

which is detected in the veto. For a veto thickness of 5 g/cm^2 (density of CH$_2$ is assumed to be 1 g/cm^3) and threshold energy of 100 keV the efficiency of neutron rejection is 26% if only proton recoils are detected (see Figure 1). The quenching factor for protons in the veto is $0.2 \times E^{1.53}$ with E in MeV [11]. Greater efficiency can be achieved by increasing the veto thickness but even at 40 g/cm^2 it does not exceed 33%.

Greater neutron rejection efficiency can be achieved by the addition of Gd to the veto scintillator. With the assumption that we only detect the gamma signal from thermal neutron capture the efficiency reaches a maximum of $\sim$ 82%. If we detect the signal from either proton recoil or neutron capture or both, the efficiency for neutron rejection increases to 89% (see Figure 2). This is the maximum efficiency which can be achieved with this detector configuration.

These efficiencies are calculated under the assumption of an infinite time window in which to tag neutron capture gammas. The mean capture time of thermal neutrons on protons is about 200 μs. Increasing the Gd-loading in the veto allows a shorter time window to be used. If the time window is shortened to 100 μs the efficiency drops from 89% to 82% for 0.2% of Gd-loading.

3. Internal and External Gammas

In addition to the U and Th chains, we also have to consider two additional sources of gamma-rays: ^{40}K, which emits a 1.46 MeV photon following electron capture and ^{60}Co which emits two photons (1.17 MeV and 1.33 MeV). Radioactivity levels were estimated from activity measurements presented in [12] and [13]. The β-decay of ^{85}Kr in liquid xenon can also be a significant contribution to the background (depending on the contamination level). However, these β's will only deposit energy in the target and not in the veto. Figure 3 shows the energy deposition from ^{85}Kr β-decay in the liquid xenon target at the level of 5 ppb contamination.

Gammas originate in the vessel and PMTs, propagate outwards isotropically and are tracked until absorbed. ^{60}Co contributes about 50% of the total rate of gammas from the PMTs (which is about 7×10^5 gammas/day). Copper typically has low levels of radio-impurity, less than 0.02 ppb for U and Th. Radioactivity in the copper vessel gives a total rate of about 2×10^4 gammas/day. Figure 3 shows the energy deposition spectrum in liquid xenon of gammas from PMTs and copper vessel. In the 2-10 keV energy range the event rate from the PMTs is about 0.9 events/kg/day, more than 30 times greater than the contribution from copper.

^{222}Rn arises in bulk materials from the decay chain of primordial ^{238}U and emanates slowly from surfaces. A potential background comes from the radon decay in the air surrounding the detector, producing alpha, beta and gamma radiation. The detector vessels will shield against the alpha and beta particles produced outside, but high-energy gamma rays can scatter in the xenon. Another concern is the deposition and nuclear-recoil implantation of daughter radioisotopes in detector elements prior to assembly. This problem can be avoided by careful surface treatment of detector parts after manufacture, followed by storage in a radon-free environment before assembly.

The main gamma production in the ^{222}Rn chain is due to the beta decays of ^{214}Pb and ^{214}Bi. Due to their metallic, non-gaseous nature, it is likely that a large fraction of these will deposit inside the detector shielding (e.g. on the outer copper vessel) or in the hydrocarbon shielding (veto). To assess if this source of background can compromise a sensitivity of 10^{-10} pb a simple simulation was set up for a 250 kg liquid xenon module surrounded by a 1 cm thick copper vessel. An activity of 10 Bq/m^3 in ^{214}Pb and ^{214}Bi was uniformly generated in 10 m^3 of air inside a cubic lead castle. The result, shown in Figure 3, points to rates of a fraction of 1 dru at

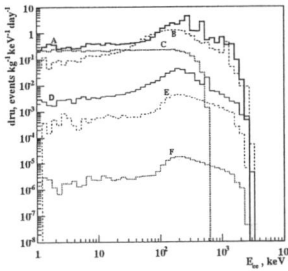
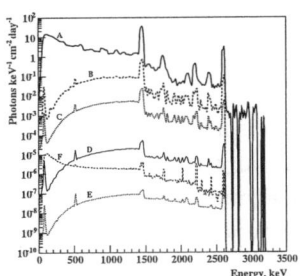

Figure 3. Energy deposition spectra in liquid xenon due to gammas from ^{222}Rn (A), PMTs (B), ^{85}Kr (Coulomb corrected) (C) and copper vessel (D). Also shown is the energy deposition due to gammas from rock activity after 10 cm Pb + 40 g cm^{-2} CH$_2$ (E) and 20 cm Pb + 40 g cm^{-2} CH$_2$ (F).

Figure 4. Gamma spectrum from Boulby rock/cavern interface (A) and after 5 cm (B), 10 cm (C), 20 cm (D) and 30 cm (E) of Pb shielding. Line (F) is the spectrum after 20 cm Pb + 40 g cm^{-2} of CH$_2$.

low energy. This approximate calculation suggests that the radon activity inside the shielding enclosure should not be overlooked.

External shielding will be required to reduce the rate from the cavern rock to below the lowest internal background level. The rock is modeled as pure NaCl with contamination levels shown in Table 1. The total flux of gammas from the rock/cavern boundary is about 0.09/cm^2/s and the spectrum is shown in Figure 4. Varying thicknesses of lead shielding were then placed around the target and veto and the resulting rock spectrum after shielding was calculated. Rock spectra after 10 cm and 20 cm of lead shielding were then propagated through 40 g/cm^2 of CH$_2$ into the detector volume producing the energy deposition spectra shown in Figure 3. It can be seen that only 10 cm of lead is required to reduce the external background to a level below that of the internal background from PMTs. Previous studies [14] have shown that 35-40 g/cm^2 of hydrocarbon and 30 cm of lead shielding are sufficient to suppress the neutron flux from rock activity to less than 1 event per year. This is more than enough to shield against external gamma activity.

4. Veto performance for gammas

Veto efficiencies for gammas from both PMTs and copper vessel were calculated. Again, efficiency is defined as the fraction of events in the target (between 2-10 keV) detected in the veto. For gammas from PMTs or copper vessel the veto efficiency is about 35-40% (with 100 keV veto energy threshold). This relatively low value is attributed to the fact that gammas can be absorbed within the detector (copper vessel, veto container, PMTs) without any energy deposition in the veto. If we consider the veto efficiency against all signals associated with electron recoils in the target then the efficiency will be further reduced if Compton energy deposition does not occur. For example, β's from ^{85}Kr-decay will only deposit energy in the target. Since no subsequent energy deposition occurs in the veto the overall veto efficiency for gammas will be reduced if there is significant ^{85}Kr contamination in the xenon.

5. Acknowledgements

This work was undertaken within the framework of the UK Dark Matter Collaboration and is funded by PPARC. The authors are grateful to Prof. P.F. Smith and Dr. C. Bungau for valuable discussions and to Dr. J.D. Lewin for compiling the radioactive contamination database. We acknowledge financial support from Eu FP6 project ILIAS.

References

1. G. Jungman, M. Kamionkowski and K. Griest, *Phys. Rep.* **267** (1996) 195.
2. L. Roszkowski, *Nuclear Physics B (Proc. Suppl.)* **124** (2003) 30.
3. H. Wang, 6th *UCLA Symposium on Sources and Detection of Dark Matter and Dark Energy in the Universe*, Marina del Rey 2004, http://www.physics.ucla.edu/hep/dm04/dm04.htm
4. E. Aprile, 6th *UCLA Symposium on Sources and Detection of Dark Matter and Dark Energy in the Universe*, Marina del Rey 2004, http://www.physics.ucla.edu/hep/dm04/dm04.htm
5. Geant 4 Collaboration, *Nucl. Instrum. Meth. A* **506** (2003) 250.
6. W. B. Wilson et al., SOURCES-4A, Technical Report LA-13639-MS, Los Alamos (1999).
7. P. F. Smith and J. D. Lewin, *Phys. Reports* **187** No.5 (1990) 203.
8. UKDM website, http://hepwww.rl.ac.uk/ukdmc/Radioactivity/index.html
9. D. Akimov et al., *Phys. Lett. B* **524** (2002) 245.
10. F. Arnedo et al., *Nucl. Instrum. Meth. A* **449** (2000) 147.
11. M. Anghinolfi et al., *Nucl. Instrum. Meth. A* **165** (1979) 217-224.
12. Hamamatsu Photonics KK Ltd., private communication.
13. M. Nakahata for XMASS Collaboration, talk given at LowNu2003 Workshop, http://cdinfo.in2p3.fr/LowNu2003/
14. M. J. Carson et al., *Astropart. Phys.* **21** (2004) 667.

MEASUREMENT OF LOW LEVEL NEUTRON FLUXES: STATUS AND PROSPECTS

J. E. MCMILLAN*

Department of Physics and Astronomy, University of Sheffield
Sheffield S3 7RH, Great Britain
E-mail: j.e.mcmillan@sheffield.ac.uk

In underground laboratories, the identification of WIMPs is limited by the ambient neutron background flux. The measurement of this flux is essential for the thorough understanding of detector performance. This paper reviews previous low-background thermal neutron detector systems and proposes a new detector based on lithium salicylate

1. Introduction

One of the important parameters characterizing low-background underground laboratories is the ambient neutron flux. This flux forms a major background in searches for dark matter WIMPs and ultimately limits the attainable sensitivity of such searches. It is also of importance in the application of neutral current detectors for neutrino oscillation studies and in planned supernova observatories.

Considerable effort has been put into monte-carlo simulations of low-level underground neutron fluxes, and these now require confirmation by experiment. Measurements of low-level neutron fluxes are notoriously difficult, requiring detectors whose energy responses and efficiencies are well understood. Wulandari et al. [12] review five independent measurements performed at Gran Sasso which differ from each other by orders of magnitude. The recent measurements of Borio di Tigliole et al. [3], which indicate that the underground neutron flux can be significantly modulated by annual variations in surface rainfall, further complicate the issue.

In WIMP searches, neutrons with energies greater than 10keV are the main cause for concern. Consequently many of the measurements have con-

---

*Work supported by PPARC and EU FP6 programme ILIAS.

centrated on high-energy neutrons, however, if thermal neutron detectors are used, detection efficiencies can be higher since the interaction cross-sections are several orders of magnitude greater. The detectors can be physically smaller and counting times reduced such that statistically significant results can be obtained in reasonable times. Efficiency calibration of thermal neutron detectors is also simpler. The disadvantage is that no information about the original energy spectrum can be obtained.

Thermal neutron detectors must use one of the following capture reactions:

$$^3\text{He} + \text{n} \rightarrow {}^3\text{H} + {}^1\text{H} + 0.764\text{MeV}$$
$$^6\text{Li} + \text{n} \rightarrow {}^4\text{He} + {}^3\text{H} + 4.8\text{MeV}$$
$$^{10}\text{B} + \text{n} \rightarrow {}^7\text{Li} + {}^4\text{He} + 2.3\text{MeV} + 0.48\text{MeV}(\gamma)$$

While the lithium and boron reactions have been used in a large number of detector designs using scintillators, no solid or liquid compounds or forms of helium are easily available, so this nuclide is only usable in gaseous detectors. The reactions liberate energetic protons, α-particles and tritons (^{3}H ions). The energies specified, unless otherwise marked, appear as the kinetic energy of the reaction products. At these energies, the reaction products have ranges of several microns (μm) in solids and this can be used to effect when high immunity to background γ-rays is required.

Thermal neutron detectors based on capture on gadolinium have also been used. The reactions liberate conversion electrons and γ-rays which are normally detected by scintillation. Gadolinium based detectors consequently have poorer neutron-gamma discrimination than those based on helium, lithium or boron. Where unambiguous registration of neutron flux is required, gadolinium detectors must be operated in time-coincidence with high-energy proton recoil detection.

2. ^{3}He Proportional Counters

Proportional counters filled with ^{3}He have provided high efficiency thermal neutron detection with relatively low background on many experiments. Typical tubes, such as those manufactured by Reuters-Stokes, can be made 630mm long and 25mm diameter with a stainless steel casing. They have a filling of 0.6MPa ^{3}He and Kr(11%) CO_2(1%) are added to increase the stopping power of the gas. The maximum efficiency is restricted by the pressure achievable, while the background is limited by radioactive contamination of the tube walls.

2.1. Neutral Current Detectors at SNO

^{3}He proportional counters were chosen as the Neutral Current Detectors at the Sudbury Neutrino Observatory [4,5,6,11]. For this application, an extremely stringent radiopurity criterion had to be met, namely that the counters should contain <12 parts per 10^{12} thorium by weight.

The counters were constructed with the casing formed from 50mm diameter chemical-vapour-deposited nickel tube; a technique which yields excellent radiopurity. The nickel tube was then electropolished, acid etched to remove radon progeny and laser welded. The filling gas was ^{3}He with CH_4(15%) at 0.25MPa pressure. Figure 1 shows a longitudinal cross-section of one of the counters. The alpha activity of the first detectors was 48 counts/m^2/day [6]

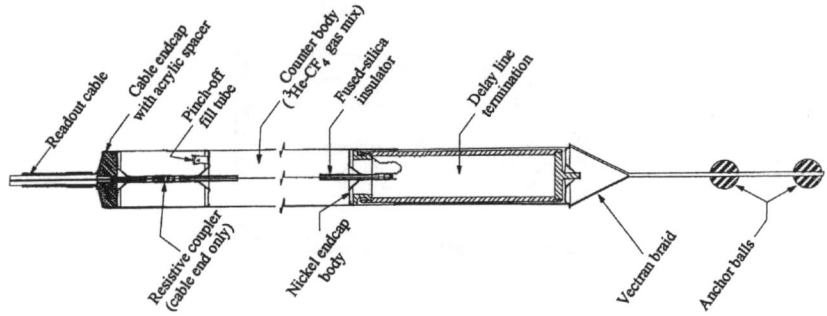

Figure 1. SNO Neutral current detector

The SNO Neutral Current Detectors are currently the world leaders in low-background thermal neutron detection. Their only disadvantage is that of cost, which must be extremely high.

3. ^{6}LiF-ZnS(Ag) Detectors

Low-cost low-background thermal neutron detectors were developed by Barton et al. [2] as part of the instrumentation for a search for rare fission events associated with the decay of superheavy elements. The detectors used ^{6}Li-fluoride powder mixed with ZnS(Ag) scintillator, together with a transparent silicone binder. This mixture was spread in thin (100μm) layers on aluminized mylar sheets which were then overcoated with clear silicone to prevent contamination by radon progeny. These layers were arranged in a stack, interleaved with polypropylene moderator and wavelength shifting lightguides which conveyed the scintillation light to photomultipliers at the ends of the detectors. This is shown in figures 2 and 3. The detectors were constructed largely from plastic materials which had low background contamination. The resulting detectors had high efficiency; eight detectors in cylindrical geometry gave 37% for ^{252}Cf fission neutrons. They were completely insensitive to gammas with energies of a few MeV or less.

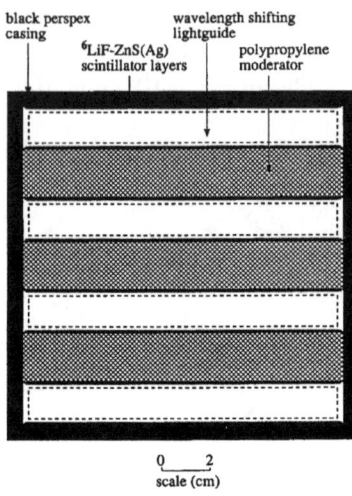

Figure 2. ^{6}LiF-ZnS(Ag) detector cross section.

The main disappointment in these detectors was that the background count was not as low as expected. This did not affect the neutron multiplicity experiment, but when one was subsequently operated at Boulby Mine, a background rate of ~ 0.01Hz was encountered, irrespective of whether the detector was in an open cavern, surrounded by cadmium sheet or immersed to a depth of one metre in a water tank. The conclusion was that there was an internal source of alpha emitters, presumably in the scintillating layers. Further investigation[1] of the time structure of the output pulses revealed that the most probable cause was uranium contamination of the zinc sulphide. This may have been in the zinc, the sulphur or in the silver and nickel used as activators. The removal of the contamination would require working with the producers of the ZnS(Ag) scintillator over an extended period.

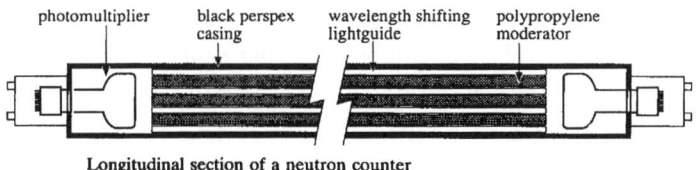

Longitudinal section of a neutron counter

Figure 3. ^{6}LiF-ZnS(Ag) detector longitudinal section.

4. ^{6}Li-Salicylate; a rediscovered neutron scintillator

In order to avoid the background problems associated with ZnS(Ag), a search was made for lithium compounds that were intrinsic scintillators. The most promising of these is lithium salicylate, which when prepared with ^{6}Li provides an efficient thermal neutron scintillator [7,10]. The material has an efficiency 10-15% of anthracene for alpha particles and a peak emission wavelength of 421nm. Mandzhukov et al. [8,9] found that it could be used in polycrystalline layers with optimal thickness at 250μm; thicker layers were found to absorb their own light. Such layers (95% ^{6}Li) had 5% total efficiency for detecting thermal neutrons while being totally insensitive to gammas.

4.1. Synthesis of ^{6}Li-Salicylate

^{6}Li-salicylate is not currently available as a commercial product. The basic recipe[7] for its production is to react isotopically enriched lithium metal with anhydrous methanol, then to neutralize the solution with salicylic acid. The resulting solution can then be dehydrated and crystallized.

Enriched lithium metal is currently easily and relatively cheaply available; Aldrich offer lithium metal chunks (95% ^{6}Li) at 10g for €300. Since the chemistry of lithium is markedly different from uranium or thorium and since the material has been isotopically separated, it is presumably free from radioactive contamination.

Pure salicylic acid is widely available, being a base material for pharmaceuticals, particularly aspirin. Commercially it is prepared using phenol, derived from petroleum, and the reagents NaOH, CO_2 and H_2SO_4. If pharmaceutical grade salicylic acid proves to be radioactively contaminated, it can be further purified by sublimation at 76°C.

Polycrystalline layers of ^{6}Li-salicylate can be deposited from solution in methanol by evaporation[8,9]. An alternative approach would be to incorporate finely ground ^{6}Li-salicylate into a transparent binder and spread the mixture at a controlled thickness. In either case, the layers would need overcoating with a transparent epoxy or silicone to prevent contamination by radon progeny.

5. Conclusion

The use of ^{6}Li-salicylate in polycrystalline layers combined with wavelength shifted readout offers the prospect of constructing large-area, high-efficiency, low-background thermal neutron detectors at low cost. Being entirely solid, they also offer excellent long-term stability. The proposed detectors will give unambiguous measurements of thermal neutron fluxes in underground experiments.

References

1. J.C Barton. Background in ZnS-LiF neutron counters. Internal Report, Birkbeck College, University of London, 1999.
2. J.C. Barton et al. A novel neutron multiplicity detector using lithium fluoride and zinc sulphide scintillator. *J Phys G*, 17:1885–1899, 1991.
3. A. Borio di Tigliole et al. Variability of fast neutron yield in underground environment. *Europhys Lett*, 67(6):1045–1049, 2004.
4. M.C. Browne. *Preparation for Deployment of the Neutral Current Detectors (NCDs) for the Sudbury Neutrino Observatory*. PhD thesis, North Carolina State University, 1999.
5. M.C. Browne et al. Low-background ^{3}He proportional counters for use in the Sudbury Neutrino Observatory. *IEEE Trans Nucl Sci*, NS-46(4):873–876, 1999.
6. S.R. Elliot et al. ^{3}He neutral current detectors at SNO. Technical Report DOE/ER/41020-42, Univ Washington, Seattle, 1998.
7. L.R. Greenwood et al. ^{6}Li-salicylate neutron detectors with pulse shape discrimination. *NIM*, 165:129–131, 1979.
8. I.G. Mandzhukov et al. Possibilities to use polycrystalline lithium salicylate as a thermal neutron scintillator. *Bulg J Phys*, 8(4):349–354, 1981.
9. I.G. Mandzhukov et al. Properties of a new class of organic scintillators: derivatives of salicylic acid. *Instrum Exp Tech*, 24(3):605–611, 1981.
10. H.H. Ross and R.E. Yerick. A new liquid scintillator for thermal neutron detection. *Nucl Sci Eng*, 20:23–27, 1964.
11. P. Thornewell. *Neutral Current Detectors for the Sudbury Neutrino Observatory*. PhD thesis, Oxford University, 1999.
12. H. Wulandari et al. Neutron flux at the Gran Sasso underground laboratory revisited. *Astropart Phys*, 22:313–322, 2004.

CRESST II BACKGROUND DISCRIMINATION: DETECTION OF ^{180}W NATURAL DECAY IN A PURE α-SPECTRUM

C. COZZINI,[*] S. HENRY, H. KRAUS, B. MAJOROVITS, V. MIKHAILIK,
Y. RAMACHERS,[†] A. J. B. TOHLRUST

*Department of Physics, University of Oxford,
Oxford OX1 3RH, U.K.*

G. ANGLOHER, D. HAUFF, J. NINKOVIC, F. PETRICCA, F. PRÖBST,
W. SEIDEL, L. STODOLSKY

*MPI für Physik, Föhringer Ring 6,
80805 Munich, Germany*

C. BUCCI

*Laboratori Nazionali del Gran sasso,
67010 Assergi, Italy*

F. VON FEILITSCH, TH. JAGEMANN, W. POTZEL, W. RAU, M. RAZETI,
M. STARK, W. WESTPHAL, H. WULANDARI

*Physikdepartment E-15, TU München,
85748 Garching, Germany*

J. JOCHUM

*Eberhard-Karls-Universität Tübingen,
D-72076 Tübingen, Germany*

For the first time the natural alpha decay of ^{180}W has been unambiguously detected in a (γ, β and neutron)-free background spectrum. This has been obtained by simultaneously measuring phonon and light signals with CRESST II cryogenic detectors. Results on the radio purity of the detectors and on the measured half-life of ^{180}W are presented.

---

[*]Corresponding author; e-mail: c.cozzini@physics.ox.ac.uk
[†]Present Address:University of Warwick, Coventry CV4 7AL, U.K.

1. Introduction

The high sensitivity and excellent energy resolution of low temperature detectors (see for example Ref. 1, 2) and the great efforts made to suppress and understand the radioactive background in rare event searches make such apparatus sensitive to rare nuclear decays. CRESST (Cryogenic Rare Event Search with Superconducting Thermometers) is an ultra-low background facility operating at the Gran Sasso underground laboratories. Detailed descriptions can be found in Ref. 1, 3.

The CRESST experiment is primarily devoted to the direct detection of WIMP dark matter particles via their scattering by nuclei. Such nuclear recoils resulting from WIMP interaction can be discriminated from electron background (caused by photons or electrons) by measuring phonons and scintillation light simultaneously (see Fig. 1). The energy detected via phonons in a cryogenic detector is, to first order, independent of the nature of the particle while the scintillation yield varies significantly. WIMP and neutron induced nuclear recoils give considerably less scintillation light than electrons of the same energy while α particles can be clearly discriminated since they interact with both nuclei and electrons. This results in a pure α spectrum, i.e. one without contributions from β, γ and neutron events as in the middle band of Fig. 2. We remark that the energy measured following an internal α-decay in a cryogenic detector corresponds to the sum of the energies of the α particle and of the recoiling nucleus, i.e. the total decay energy Q.

Here we present clear evidence for the α-decay of ^{180}W from early runs

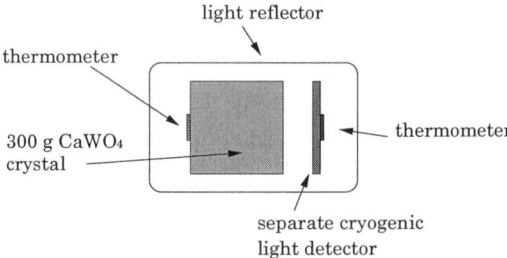

Figure 1. Schematic of a detector module. It consists of a scintillating 300 g CaWO$_4$ crystal (phonon channel) and a Si wafer (light channel), both read out by a SPT. The set-up is surrounded by a reflective foil

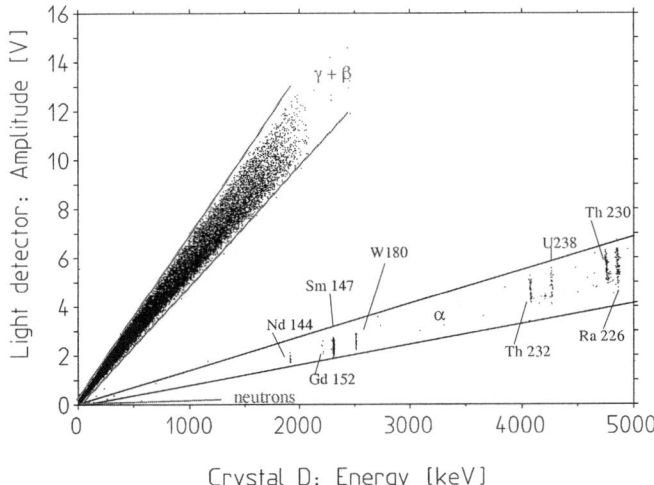

Figure 2. Pulse amplitude in the light channel (scintillation) versus energy in the CaWO$_4$ crystal D (phonons) from run 28. Three clearly separated event populations appear. The solid lines border the three different bands: the upper band is due to γ's and β's, the lowest one to neutrons and the middle one is due to α's. Each group of events in the α band has been identified (see text) and is here labelled accordingly.

of the CRESST II dark matter detectors, originating from the tungsten in the CaWO$_4$ crystals used as the dark matter target[4].

1.1. Experimental set-up

The set-up for the simultaneous detection of scintillation light and phonons is described in Ref. 3 and is shown schematically in Fig. 1 for one absorber module. It consists of two independent detectors, each one with its own thermometer. The main detector consists of a $\sim$ 300g cylindrical CaWO$_4$ crystal ($\varnothing$ = 40 mm, h = 40 mm). The scintillation light produced in each target crystal is detected via an associated calorimeter consisting of a silicon wafer of $(30 \times 30 \times 0.45)$mm^3 volume with a 20 nm thick SiO$_2$ layer on both surfaces. In CaWO$_4$ only a few percent of the absorbed energy is transformed into light. To minimise light losses the whole module is enclosed in a highly reflective foil.

1.2. Background

Due to the powerful discrimination technique of CRESST II detector modules, contaminations from natural decay chains and other α-unstable isotopes have been identified by their α-decays with a sensitivity of ∼ 1μBq/kg. The results of this analysis have shown very good agreement with the results obtained using different techniques commonly employed to determine the presence of impurities in crystals. These include ICPMS[a] (Inductively Coupled Plasma Mass Spectrometry), HPGe[b] (High Purity Germanium γ-spectroscopy) and X-ray luminescence[c] techniques.

Evidence for contaminations due to α-unstable rare earth elements has been found in all the measured detectors. In particular an α-peak at 2.31 MeV has been detected and identified as the α-decay of ^{147}Sm. This isotope is naturally occurring with 15% isotopic abundance and is known to be a pure α-unstable radionuclide with $T_{1/2} = 1.06 \times 10^{11}$ years. The intensity of this peak varies for the different detectors, implying different levels of contamination in each crystal.

ICPMS analyses were performed in the Gran Sasso facility on crystals B and E[4] giving results in agreement with our analysis of the α-peak intensities. Evidence of ^{144}Nd (Q = 1905.1 keV) and ^{152}Gd (Q = 2205 keV)[5] α-decays have also been observed and the results are consistent with the ICPMS analysis. Furthermore, X-ray luminescence analysis[6] on crystal B showed the presence of the characteristic emission peak of Gd and Er, the two rare earth contaminants most abundant in our crystals. Finally, alternative candidates such as ^{174}Hf (Q =2496 keV) and ^{186}Os (Q = 2822)[5] were considered to cause the 2.31 MeV peak. These hypotheses were discarded, however, due to the unreasonably high atomic concentration needed to produce the detected rate.

Once unambiguously identified, ^{147}Sm was used to calibrate the full α spectrum. This allowed a precise identification of the U-Th peaks, which were then used to determine the accuracy of the calibration (see Ref. 4).

2. Results

In the data presented here, three different $CaWO_4$ crystals (named B, D and E) were used in the Gran Sasso set-up for a total of four different runs (22, 23, 27, 28). Crystal D showed the best performance as a detector and its spectrum is shown in Fig. 3. The position in energy for the ^{180}W peak is (2516.5 ± 1.4) keV. A half-life $T_{1/2} = (1.7 \pm 0.2) \times 10^{18}$ y is obtained. This is consistent with the value from run 27 on the same crystal and on crystals

---

[a]Gran Sasso Laboratory and Durham University
[b]Modane Underground Laboratory and Gran Sasso Underground Laboratory
[c]Durham University

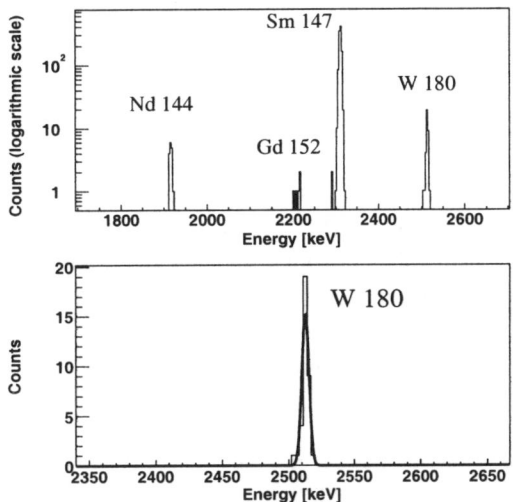

Figure 3. Alpha events measured in run 28. In the top figure the energy spectrum between 1.7 and 2.7 MeV is shown with 3 keV bins. On the vertical axis the number of counts are displayed on a logarithmic scale. In the bottom figure the fit to the ^{180}W α-decay is shown. No background counts are present in a wide energy interval around the peak.

B and E. The consistency of the results from different crystals tends to confirm that the signal originates from the ^{180}W decay and not from unknown impurities which may be expected to vary for each individual crystal. Furthermore ICPMS measurements performed on these two crystals exclude alternative explanations for the tungsten peak, such as ^{174}Hf.

Finally, the four measurements were added, resulting in a spectrum for

Table 1. Half-life for the α-decay of ^{180}W obtained from the present work. The results are given independently for the four different runs and for the total exposure. For the runs 27, 28, and the total exposure, the number of counts is determined by a Gaussian fit to the peak. Errors are given at 1σ. For run 22 and 23, due to the small number of counts, the total number of events contained within $\pm 3\sigma$ of the expected energy position is given.

Crystal	Run number	Exposure [kg days]	Number of counts	Half-life [y]
D	28	12.268	35± 4.3	$T_{1/2} = (1.7 \pm 0.2) \times 10^{18}$
D	27	11.745	28.5 ± 5.6	$T_{1/2} = (1.9 \pm 0.4) \times 10^{18}$
E	23	3.467	9	$T_{1/2} \geq 1.1 \times 10^{18}$ (90% C.L.)
B	22	1.14	4	$T_{1/2} \geq 6.8 \times 10^{17}$ (90% C.L.)
	All Crystals	28.62	75.5±8.7	$T_{1/2} = (1.8 \pm 0.2) \times 10^{18}$

a total exposure of 28.62 kg days. The data from each run have been calibrated individually and then summed. This results in a half-life of $T_{1/2} = (1.8 \pm 0.2) \times 10^{18}$y, which is consistent with previously published limits[7,8] and the indication reported in Ref. 7. Including the uncertainty on the position of the ^{147}Sm line, a peak of energy (2516.4 ± 1.1 (stat.) ± 1.2 (sys.)) keV is found with a resolution $\Delta E = (18 \pm 2)$ keV (FWHM).

Limits on the half-lives for the non-observed α-decays of the other four naturally occurring tungsten isotopes were also calculated. Due to a surface contamination on the Ag reflective foil, the data from run 27 could not be used (see Ref. 4). The limits obtained from the sum of runs 22, 23 and 28 (in which a polymeric foil was used) are reported in Table 2.

Table 2. Lower limits (90% C.L.) on the half-lives of the non-observed α-decays of the other four naturally occurring tungsten isotopes, obtained with a total exposure of 16.875 kg days. For comparison the best previously published limits are also shown.

Isotope	Half-life [y]		
	This work	Previous [7]	Previous [8]
^{182}W	$T_{1/2} \geq 7.7 \times 10^{21}$	$\geq 1.7 \times 10^{20}$	$\geq 2.5 \times 10^{19}$
^{183}W	$T_{1/2} \geq 4.1 \times 10^{21}$	$\geq 0.8 \times 10^{20}$	$\geq 1.3 \times 10^{19}$
^{184}W	$T_{1/2} \geq 8.9 \times 10^{21}$	$\geq 1.8 \times 10^{20}$	$\geq 2.9 \times 10^{19}$
^{186}W	$T_{1/2} \geq 8.2 \times 10^{21}$	$\geq 1.7 \times 10^{20}$	$\geq 2.7 \times 10^{19}$

Acknowledgments

This work was supported by PPARC, BMBF, the EU Network HPRN-CT-2002-00322 on Applied Cryodetectors, the EU Network on Cryogenic Detectors (contract ERBFMRXCT980167), the DFG SFB 375 on Particle Astrophysics and two EU Marie Curie Fellowships.

References

1. G. Angloher et al., *Astroparticle Physics* **18**, 43-55 (2002).
2. C. Arnaboldi et al., *Astroparticle Physics* **20**, 91-110 (2003).
3. G. Angloher et al., submitted to Astroparticle Physics, astro-ph/0408006.
4. C. Cozzini et al., *Physical Review* **C**, accepted for publication, nucl-ex/0408006
5. *Table of Isotopes*, edited by R. B. Firestone et al. (1996) Wiley, New York.
6. J. Ninkovic et al., *Nucl. Instr. and Methods* **A**, in press.
7. F. A. Danevich et al., *Physical Review* **C67**, 014310 (2003).
8. S. Cebrian et al., *Physics Letters* **B556**, 14-20 (2003).

INDIRECT DETECTION OF NEUTRALINOS

J. EDSJÖ*

Department of Physics, Stockholm University,
AlbaNova University Center,
SE-106 91 Stockholm, Sweden
E-mail: edsjo@physto.se

There is compelling evidence for the existence of dark matter in the Universe. One of the favourite candidates is a Weakly Interacting Massive Particle (WIMP). We will here focus on indirect ways to search for WIMPs and compare their advantages and disadvantages. As a concrete WIMP example, we will focus on the neutralino that arises in supersymmetric extensions of the standard model.

1. Introduction

Weakly Interacting Massive Particles (WIMPs) are very good dark matter candidates as they naturally give a relic density of the right order of magnitude. There are many different WIMP candidates (see e.g. [1]), and as a concrete example, we will here use the neutralino, that naturally appears as the lightest supersymmetric particle in many supersymmetric extensions of the standard model (see e.g. [2]).

We will work mostly in the Minimal Supersymmetric Standard Model (MSSM), with the usual low-energy parameters: μ, M_2, $\tan\beta$, m_A, m_0, A_b and A_t. In one example, we will also use the mSUGRA model. See [3,4] for more details.

We will use DarkSUSY [5] to calculate the various rates in indirect and direct searches. The relic density of neutralinos is calculated including coannihilations [4,6,7]. We will here only include cosmologically interesting models, where the neutralinos can make up a major part of the dark matter in the Universe without overclosing it.

---

*Supported by the Swedish Research Council.

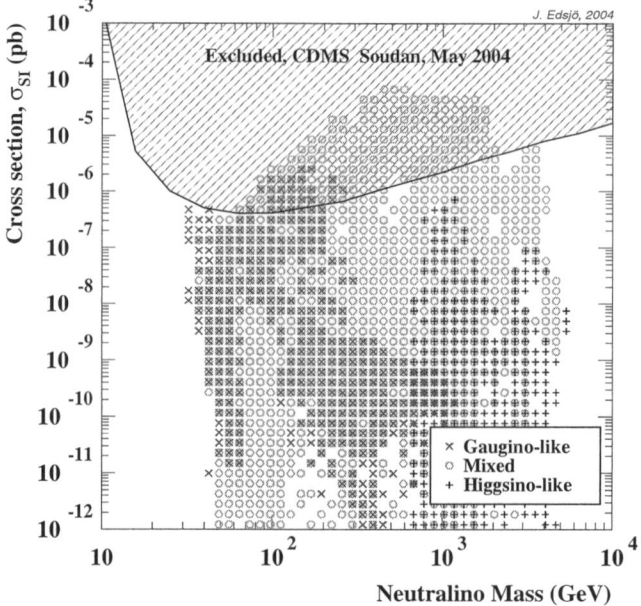

Figure 1. The spin-independent scattering cross section versus the neutralino mass and the current limits from CDMS at Soudan.

2. Indirect searches and comparison with direct searches

There are many different ways to search for neutralino dark matter (for a review see e.g. [1,2]). We will here focus on a few of the more common/promising ways to search for neutralinos.

2.1. Direct searches

The most stringent direct detection limits to date comes from the CDMS experiment at Soudan [8], that has already started to exclude some of our MSSM models. In Fig. 1 we show our set of supersymmetric models and the exclusion limits from CDMS (assuming standard halo parameters). In some of the coming plots, the models excluded by CDMS will be indicated by green filled circles.

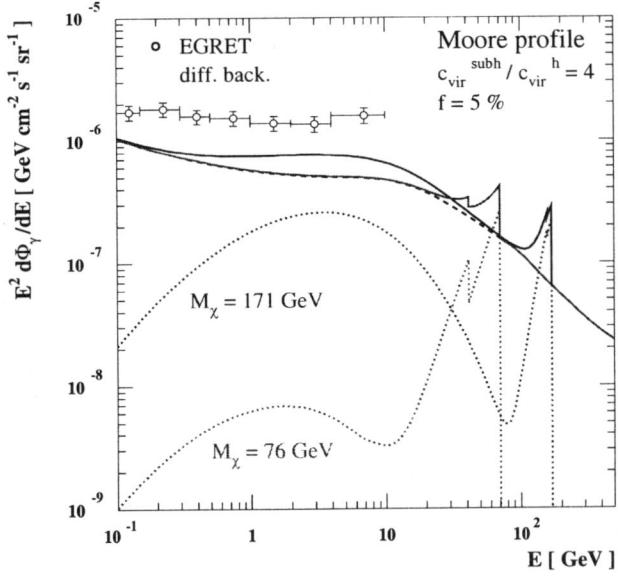

Figure 2. Two example MSSM models of the diffuse gamma ray flux from annihilation in external halos. The monochromatic gamma lines are clearly seen (together with the redshift of the lines from far away halos).

2.2. *Gamma rays*

Neutralinos can annihilate to gamma rays that can be searched for in the cosmic rays. Annihilation can occur either to monochromatic gamma rays at loop level [9] or lower energy continuous gamma rays from pion decays. The monochromatic gamma ray lines are a striking signature that can be probed with future detectors like GLAST [10] or Air Cherenkov Telescopes looking e.g. towards the galactic center of at the diffuse extragalactic flux from external halos. In Fig. 2, an example of such a signature is shown [11]. The figure shows the diffuse gamma ray flux from neutralino annihilation in external halos. The monochromatic gamma lines are clearly seen (together with the redshift effect from far away halos). Even if the MSSM model chosen is an optimistic one, the signature is very clear and should be easily observable as long as the flux is high enough.

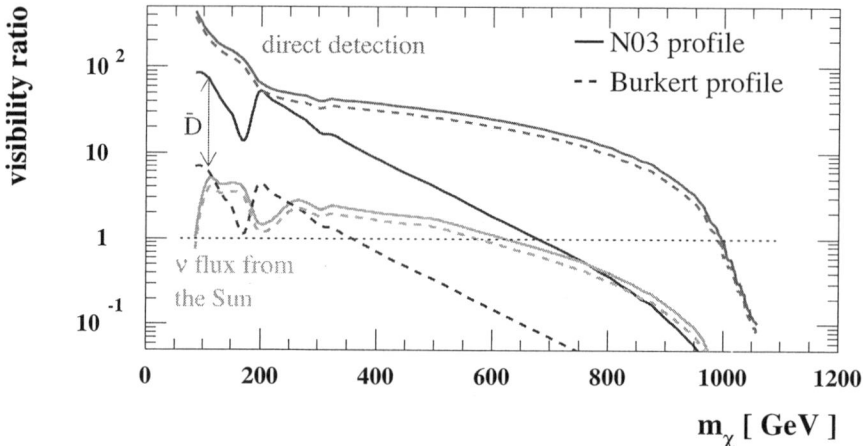

Figure 3. The visibility ratio for future cosmic ray searches for a neutralino signature. The figure shows curves for $\Omega_\chi h^2 = 0.103$ in the focus point region in mSUGRA (see text for more details).

2.3. Antiprotons

Neutralinos can also produce antiprotons from quark jets. The fluxes can be rather high, but unfortunately, the shape of the spectra does not differ significantly from that expected from the background. Hence, discriminating the background from the signal is hard. However, if the predicted signal is significantly higher than the measured flux, the antiproton fluxes can be used to constrain supersymmetric models (or rather supersymmetric models in conjunction with astrophysical models).

2.4. Antideuterons

Antideuterons can also be produced in neutralino annihilations [12]. Compared to the antiprotons, the background of antideuterons is expected to be very low at low energies. Hence, searching for antideuterons at low energies is a nearly background free way to search for a neutralino signature. In Fig. 3, the visibility ratio (roughly the sensitivity to observe a signal), is shown for the focus point region in mSUGRA [13]. A ratio larger than 1 is observable in future searches. The results are shown for two extreme halo models: a cuspy N03 profile and a cored Burkert profile. As shown, the antideuteron fluxes can be a good way to search for these models. These results are for a proposed antimatter probe, GAPS [14].

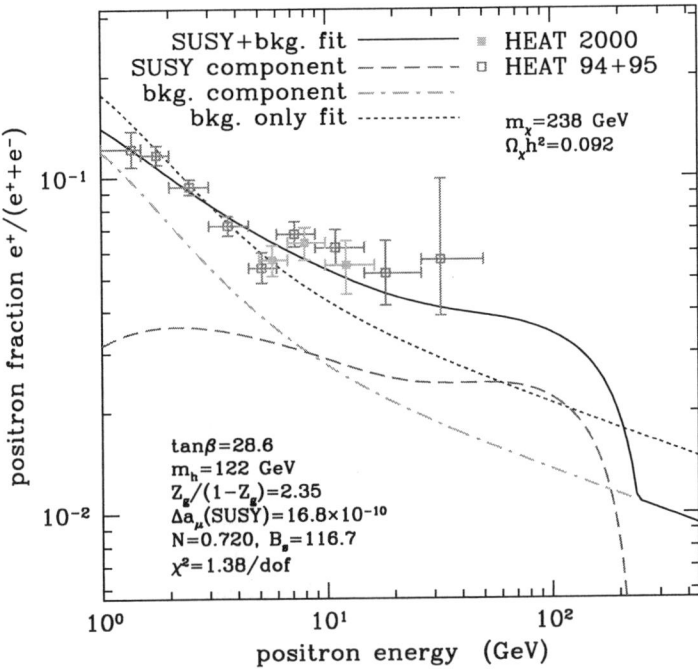

Figure 4. The measured positron fractions by the HEAT experiment. Compared to the estimated background, an MSSM example model is shown to produce a better, but not very good, fit to the data.

2.5. *Positrons*

The HEAT experiment [15] has shown an indication of a positron excess around 8 GeV. In Fig. 4 the HEAT data is shown together with an example MSSM model. Compared to the background estimate, the MSSM model gives a better, but not very good, fit to the data. The model shown has strong annihilation to W^+W^-, whose decay gives the sharp edge at higher energies. Such a feature in the spectrum can be searched for with future detectors, like AMS [16], Pamela [17] or Calet [18].

2.6. *Neutrino-induced muons in neutrino telescopes*

WIMPs can be gravitationally trapped by e.g. the Sun and the Earth, where they can annihilate and produce e.g. muon neutrinos. These can produce muons that neutrino telescopes like Super-Kamiokande, Macro, Baksan, Amanda, Antares or IceCube could search for. The Sun captures WIMPs directly from the halo, while this is not very efficient for the Earth, being

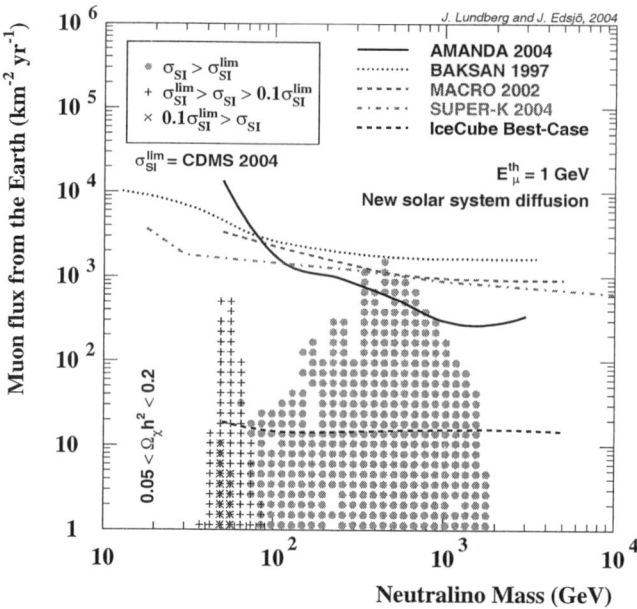

Figure 5. Expected neutrino-induced muon fluxes from neutralino annihilation in the Earth. The current experimental limits are also shown together with the expected future limits from Antares and IceCube. Models that are excluded by the current CDMS limit are shown with green filled circles (see text for details).

deep inside the potential well of the Sun. Instead the Earth captures from a population of WIMPs that have been gravitationally scattered from the halo and diffuse around in the solar system [19]. In [19], it was shown that the velocity distribution at the Earth effectively is the same as if the Earth was in free space due to these diffusion processes. However, in recent years [20], a concern about the effects of solar capture has been raised. In [21], the effects of solar capture were studied in detail via extensive numerical simulations. The main result of these simulations is that if it were not for solar capture, the results of [19] would hold, i.e. the velocity distribution at the Earth would be as if the Earth was in free space. However, solar capture reduces the density of low-velocity WIMPs by almost an order of magnitude. Since these low-velocity WIMPs are the ones that could be captured by the Earth, especially for heavier WIMPs, the corresponding capture rates will be reduced by up to an order of magnitude. As capture and annihilation

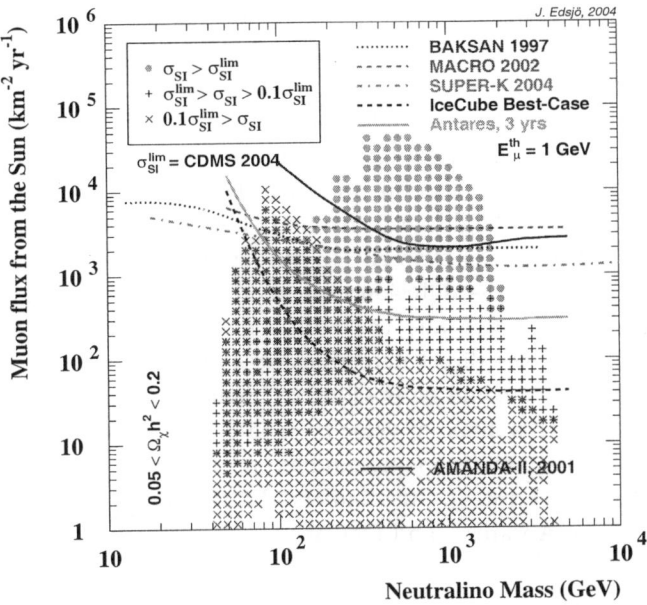

Figure 6. Expected neutrino-induced muon fluxes from neutralino annihilation in the Sun. The current experimental limits are also shown together with the expected future limits from Antares and IceCube. Models that are excluded by the current CDMS limit are shown with green filled circles (see text for details).

is typically not in equilibrium in the Earth, the annihilation rates will be suppressed further (by up to two orders of magnitude). In the following we will use this new estimate of the capture and annihilation rates in the Earth.

In Figs. 5–6 we show the expected fluxes from neutralino annihilation in the Earth and the Sun [21]. We also show the current limits from Baksan [22], Macro [23], Super-Kamiokande [24] and Amanda [25]. All limits have been converted to limits on the muon flux above 1 GeV (without an angular cut-off). We see that these neutrino telescopes have already started to explore the MSSM parameter space. We also indicate which models are excluded by the direct search by CDMS [8]: green filled circles are excluded by CDMS, blue crosses would be excluded with a factor of ten increased sensitivity and red crosses would require an even higher sensitivity. We clearly see that for the Earth, CDMS has already excluded those models that would produce the highest fluxes in neutrino telescopes, whereas for the Sun, the correlation is not as high. The reason for the strong correlation for the Earth is that both the signal in CDMS and the signal from the Earth depends strongly on the spin-independent scattering cross section. For the Sun, on the other hand, the signal depends strongly also on the spin-dependent scattering cross section, on which the limits are not as good from direct searches. One should keep in mind, however, that the correlation seen in Figs. 5-6 depends strongly on the assumed halo velocity profile (and for the Earth on the diffusion of neutralinos in the solar system [19]). The capture by the Earth and the Sun is most efficient for low-velocity neutralinos whereas the direct detection rates are higher for high-velocity neutralinos. Hence, it could be possible to break this strong correlation in signal strengths with a different velocity profile. Figs. 5-6 is produced with the assumption of a standard gaussian halo profile (and the effects of solar capture on the Earth rates). Also indicated in the figure are the expected future limits from Antares[26] and IceCube.

3. Future searches

If we look 5–10 years into the future we can compare the expected sensitivities of future searches by indicating which parts of the MSSM parameter space they will probe. We will here use the gaugino fraction of the neutralino, Z_g, and project the MSSM parameter space onto the $Z_g/(1 - Z_g)$ versus neutralino mass plane. In Fig. 7 we show the regions of the parameter space where future direct detection experiments, neutrino telescopes and gamma ray searches are sensitive. For the direct detection sensitivity, a generic experiment like e.g. GENIUS sensitive down to $\sigma_p = 10^{-9}$ pb is assumed. For neutrino telescopes, the expected (optimistic) sensitivity of IceCube is used. Finally, for the gamma rays, we show the sensitivity of upcoming Air Cherenkow Telescopes (ACTs) and GLAST looking towards the galactic center. That signal depends strongly on the halo profile, and in this figure an NFW profile is assumed. As can be seen, these different searches complement each other fairly well, except for the signal from the Earth, that at least for standard assumptions of the halo profile, will be better probed by direct detection experiments.

4. Conclusions

WIMPs are natural dark matter candidates and a concrete example of a WIMP would be the neutralino, which in many cases is the lightest super-

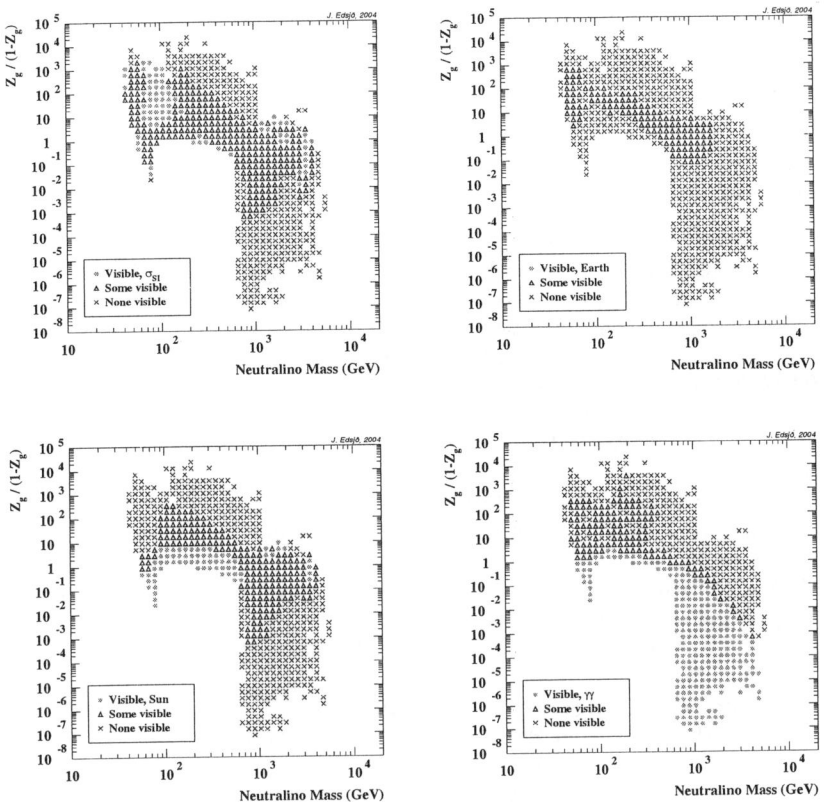

Figure 7. The $Z_g/(1-Z_g)$ versus neutralino mass plane and the expected sensitivity of future experiments. In the top-left panel, the sensitivity of future direct detection experiments is shown. In the top-right, the sensitivity of future neutrino telescopes looking towards the Earth is shown. In the bottom-left, the same, but for the Sun, is shown. In the bottom-right panel, the sensitivity of future gamma ray experiments is shown. Regions where all models can be probed are shown with green dots, regions where some models can be probed are shown with blue triangles, and, finally, regions where no models can be probed, are shown with red crosses. See text for more details.

symmetric particle in supersymmetric extensions of the standard model. The current direct detection experiments have already started to explore the MSSM parameter space. The same is also true for neutrino telescopes searching for neutrinos from the Earth and the Sun, even if the expected rates from the Earth have decreased due to a new calculation of the effects of solar capture on WIMPs diffusing in the solar system. Future searches are going to probe substantial parts of the MSSM parameter space, with a good complementarity between different searches. Clear signatures can, in optimistic scenarios, be expected in e.g. the positron fluxes, the gamma ray fluxes and the antideuteron fluxes.

References

1. L. Bergström, Rept. Prog. Phys. **63** (2000) 793.
2. G. Jungman, M. Kamionkowski and K. Griest, Phys. Rep. **267** (1996) 195.
3. L. Bergström and P. Gondolo, Astrop. Phys. 5(1996)263.
4. J. Edsjö and P. Gondolo, Phys. Rev. **D56** (1997) 1879.
5. P. Gondolo, J. Edsjö, P. Ullio, L. Bergström, M. Schelke and E. Baltz, JCAP **0407**(2004) 008. [astro-ph/0406204]
6. P. Gondolo and G. Gelmini, Nucl. Phys. **B360** (1991) 145.
7. J. Edsjö, M. Schelke, P. Ullio and P. Gondolo, JCAP **0304** (2003) 001. [hep-ph/0301106]
8. D.S. Akerib et al, astro-ph/0405033.
9. L. Bergström and P. Ullio, Nucl. Phys. **B504**199727; P. Ullio and L. Bergström, Phys. Rev. **D57** (1998) 1962.
10. GLAST Collaboration, A. Moiseev et al., astro-ph/9912139; homepage http://www-glast.stanford.edu.
11. P. Ullio, L. Bergström, J. Edsjö and C.G. Lacey, Phys. Rev. **D66** (2002) 123502.
12. F. Donato, N. Fornengo and P. Salati, Phys. Rev. **D62** (2000) 043003.
13. J. Edsjö, M. Schelke and P. Ullio, JCAP **09** (2004) 004.
14. K. Mori, C. J. Hailey, E. A. Baltz, W. W. Craig, M. Kamionkowski, W. T. Serber and P. Ullio, Astrophys. J. **566** (2002) 604.
15. S. W. Barwick et al. (HEAT Collaboration), Astrophys. J. **482** (1997) L191; S. Coutu et al. (HEAT-pbar Collaboration), in Proceedings of 27th ICRC (2001).
16. S. Ahlen et al. (AMS Collaboration), Nucl. Instr. Meth. **A350** (1994) 351.
17. O. Adriani et al. (PAMELA Collaboration), Proc. of the 26th ICRC, Salt Lake City, 1999, OG.4.2.04.
18. K. Yoshida et al., these proceedings.
19. A. Gould, Astrop. J. **368** (1991) 610.
20. A. Gould and S.M.K Alam, Astroph. J. **549** (2001) 72. [astro-ph/9911288]
21. J. Lundberg, J. Edsjö, Phys. Rev. **D69** (2004) 123505. [astro-ph/0401113]
22. M. Boliev et al, proceedings of *Dark Matter in Astro and Particle Physics*, 1997; O. Suvorova, hep-ph/9911415.
23. I. de Mitri (Macro Collaboration), proceedings of idm2002, York, England, September, 2002.
24. S. Desai et al., Phys. Rev. **D70** (2004) 083523. [hep-ex/0404025]
25. K. Woshnagg, these proceedings.
26. S. Cartwright (Antares Collaboration), proceedings of idm2002, York, England, September, 2002.

EGRET EXCESS OF DIFFUSE GALACTIC GAMMA RAYS INTERPRETED AS A SIGNAL OF DARK MATTER ANNIHILATION

W. DE BOER

Institut für Experimentelle Kernphysik
University of Karlsruhe
Postfach 6980
76128 Karlsruhe, Germany
E-mail: wim.de.boer@cern.ch

The EGRET excess in the diffuse galactic gamma ray data above 1 GeV shows all the features expected from Dark Matter WIMP Annihilation: a)it is present and has the same spectrum in all sky directions, not just in the galactic plane. b) The intensity of the excess shows the $1/r^2$ profile expected for a flat rotation curve outside the galactic disc with additionally an interesting substructure in the disc in the form of a doughnut shaped ring at 14 kpc from the centre of the galaxy. At this radius a ring of stars indicates the probable infall of a dwarf galaxy, which can explain the increase in DM density. From the spectral shape of the excess the WIMP mass is estimated to be between 50 and 100 GeV, while from the intensity the halo profile is reconstructed. Given the mass and intensity of the WIMPs the mass of the ring can be calculated, which is shown to explain the peculiar change of slope in the rotation curve at about 11 kpc. These signals of Dark Matter Annihilation are compatible with Supersymmetry and have a statistical significance of more than 10σ in comparison with a fit of the conventional galactic model to the EGRET data. The statistical significance combined with all features mentioned above provide an intriguing hint that the EGRET excess is indeed a signal from Dark Matter Annihilation.

1. Introduction

Cold Dark Matter (CDM) makes up 23% of the energy of the universe, as deduced from the WMAP measurements of the temperature anisotropies in the Cosmic Microwave Background, in combination with data on the Hubble expansion and the density fluctuations in the universe [1]. The Dark Matter has to be much more widely distributed than the visible matter, since the rotation speeds do not fall off like $1/\sqrt{r}$, as expected from the visible matter in the centre, but stay more or less constant as function of distance. For a "flat" rotation curve the DM has to fall off slowly like $1/r^2$

instead of the exponential drop-off for the visible matter. The fact that the DM is distributed over large distances implies that its properties must be quite different from the visible matter, since the latter clumps in the centre owing to its rapid loss of kinetic energy by the electromagnetic and strong interactions after infall into the centre. Since the DM apparently undergoes little energy loss, it can have at most weak interactions. In addition its mass is probably large, since it cannot be produced with present accelerators. Therefore it is generically called a WIMP, a Weakly Interacting Massive Particle.

According to the rules of particle physics the weakly interacting particles can annihilate, yielding predominantly quark-antiquark pairs in the final state, which hadronize into mesons and baryons. The stable decay and fragmentation products are neutrinos, photons, protons, antiprotons, electrons and positrons. From these, the protons and electrons disappear in the sea of many matter particles in the universe, but the photons and antimatter particles may be detectable above the background, generated by particle interactions. Such searches for indirect Dark Matter detection have been actively pursued, see e.g the review by Bergström[2] or more recently by Bertone, Hooper and Silk [3].

The present analysis on diffuse galactic gamma rays differs from previous ones by considering simultaneously the complete sky map *and* the energy spectrum, which allows us to constrain both the halo distribution *and* the WIMP mass. The constraint on the WIMP annihilation cross section from WMAP is discussed in Section 2, while the constraints on the mass and the DM halo profile from the EGRET excess are discussed in Sections 3. The summary is given in Section 4.

2. Annihilation Cross section Constraints from WMAP

In the early universe all particles were produced abundantly and were in thermal equilibrium through annihilation and production processes. At temperatures below the mass of the WIMPS the number density drops exponentially. The annihilation rate $\Gamma = <\sigma v> n_\chi$ drops exponentially as well, and if it drops below the expansion rate, the WIMP's cease to annihilate. They fall out of equilibrium (freeze-out) at a temperature of about $m_\chi/22$ [4] and a relic cosmic abundance remains.

For the case that $<\sigma v>$ is energy independent, which is a good approximation in case there is no coannihilation, the present mass density in units of the critical density is given by [5]:

$$\Omega_\chi h^2 = \frac{m_\chi n_\chi}{\rho_c} \approx (\frac{2 \cdot 10^{-27} cm^3 s^{-1}}{<\sigma v>}). \quad (1)$$

One observes that the present relic density is inversely proportional to the annihilation cross section at the time of freeze out, a result independent of the WIMP mass (except for logarithmic corrections). For the present value of $\Omega_\chi h^2 = 0.113 \pm 0.009$ the thermally averaged total cross section at the freeze-out temperature of $m_\chi/22$ must have been around $2 \cdot 10^{-26} cm^3 s^{-1}$. The observed annihilation rate will be compared with this generic cross section, which basically only depends on the value of the Hubble constant in absence of resonances and if coannihilation with other particles can be neglected.

3. Indirect Dark Matter Detection

The neutral particles play a very special role for indirect DM searches, since they point back to the source. The charged particles change their direction by the interstellar magnetic fields, energy losses and scattering. Therefore the gamma rays provide a perfect means to reconstruct the intensity (halo) profile of the DM by observing the intensity of the gamma ray emissions in the various sky directions. Of course, this assumes that one can distinguish the gamma rays from DM annihilation from the background, mainly from proton-proton interactions. Both for DMA and pp collisions the gamma rays originate mainly from the decay of neutral pions, a light particle produced abundantly in the hadronization process of quarks into hadrons. However, the protons in the galaxies and consequently the quarks inside the protons have a steeply falling energy spectrum ($N \propto E^{-2.7}$). In contrast, the quarks from DM annihilation are monoenergetic, since the WIMPS annihilate almost at rest, so their mass is converted completely into kinetic energy of the much lighter quarks. Each quark thus obtains an energy corresponding to the mass of the WIMP. The gamma rays from such monoenergetic quarks will not be monoenergetic, but they will be smeared towards lower energies as a result of the fragmentation and subsequent decays. However, they still have a considerably harder energy spectrum than the gamma rays from nuclear interactions. This allows on a statistical basis to determine the contribution from DM annihilation in each sky direction. In addition, the monoenergetic quarks from WIMPs lead to a sharp decrease in the energy of the gamma rays towards the WIMP mass, which allows to constrain the WIMP mass. The spectral shape of the gamma rays from either the backgrounds or the monoenergetic quarks are well known from accelerator experiments and can be obtained from the well-known PYTHIA code for quark fragmentation[6].

A very detailed gamma ray distribution over the whole sky was obtained by the Energetic Gamma Ray Emission Telescope EGRET, one of the four instruments on the Compton Gamma Ray Observatory CGRO, which collected data during nine years, from 1991 to 2000. The EGRET telescope was carefully calibrated in the energy range of 0.1 to 30 GeV, but using Monte Carlo simulations the energy range was recently extended up to 120 GeV[7] with a correspondingly larger uncertainty, mainly from the self-vetoing of the detector by the back-scattering from the electromagnetic calorimeter into the veto counters for high energetic showers.

It was already noticed in 1997 that the EGRET data showed an excess of gamma ray fluxes for energies above 1 GeV if compared with conventional galactic models[8]. This analysis was repeated recently using a different analysis technique on the publicly available EGRET data, namely by comparing the data not with the absolute fluxes from galactic models, but only with the shape of the gamma energy spectra from the galactic background, which is much better known[9,10]. Simultaneously to this galactic background the shapes of Dark Matter Annihilation and the extragalactic background are fitted for many different sky directions.

Fitting these three contributions of galactic background, extragalactic background and DMA to the energy spectra of 360 independent sky directions yielded astonishingly good fits with the free normalization of the background agreeing reasonably well with the absolute predictions of the galactic models[11,12] for the energies between 0.1 and 0.5 GeV. Above these energies a clear contribution from Dark Matter annihilation is needed, but the excess in different sky directions can be explained by a single WIMP mass, as shown in Fig. 1 for 3 different sky directions. Alternative explanations for the excess have been plentiful. Among them: locally soft electron and proton spectra, implying that in other regions of the galaxy the spectra are harder, thus producing harder photon spectra, local distortions of the spectra by Supernovae explosions, spatial variation of the diffusion parameters etc. A summary of these discussions have been given by Strong et al.[7], who find that hard proton spectra are incompatible with the antiproton yield and hard electron spectra are incompatible with the EGRET data up to 120 GeV, which they analyzed. However, they find that by modifying the electron and proton injection spectra simultaneously, they can improve the description of the data. The problem is of course, if the local spec-

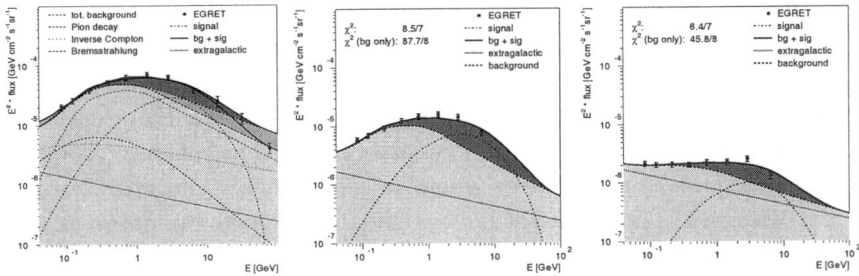

Figure 1. The diffuse gamma-ray energy spectrum of 3 angular regions: from left to right: towards the galactic centre (latitudes $0° < |b| < 5°$; longitudes $0° < |l| < 30°$), the galactic anticentre ($0° < |b| < 10°$; $90° < |l| < 270°$) and the pole regions ($60° < |b| < 90°$; $0° < |l| < 360°$), as measured by the EGRET space telescope. In the two panels on the right the solid straight line represents the fitted contribution from the extragalactic background, while the dotted line indicates the contribution from the annihilation from 65 GeV WIMPs. The total background (DMA) is indicated by the light (yellow) (dark (red)) shaded area, respectively. In the panel on the left the various contributions to the background are indicated as well, while the uncertainties from the background are indicated by the medium shaded (blue) area.

tra of protons and electrons cannot be used to fit the galactic parameters, why should one trust e.g. the B/C ratio, which is used to determine the diffusion coefficient. So the predictive power of models assuming the local neighborhood is not representative for the galaxy is rather limited. In addition, it is difficult to describe the excess in all sky directions simultaneously, since the matter densities are highest in the disc, while the DM has a much broader distribution. So the fit of the combination of background with DM always yields a much better fit than a modified background alone, if all sky directions are considered and if the correlations between the errors are taken into account, which implies a much better knowledge of the shape because of the small statistical and point-to-point errors.

As an example of alternative explanations the analysis of Kamae et al.[13] is considered. They use a harder proton spectrum than locally observed (a power law with index 2.5 instead of 2.7 observed locally) and an updated pp cross section including diffractive scattering and scaling violation. They claim this can describe the EGRET data towards the galactic centre. However, there is a clear overshoot at low energies. Fitting only their shape to the EGRET data, i.e. with a free normalization, still leaves a significant excess, as shown in the left panel of Fig. 1. Here the upper edge of the medium shaded (blue) area corresponds the hardest possible spectrum from Kamae et al.[13] with the power index of 2.5, while the lower edge corresponds to the conventional GALPROP model[7]. Note that the hard spectrum overshoot the highest EGRET point, which was not yet available during the analysis by Kamae et al. In summary, also for the "conventional" explanations[7,?] the fit to *all* sky directions can be much improved, if DM is added, since then both the low and high energy range can be perfectly described. Thus different backgrounds just change the normalization of the DM contribution.

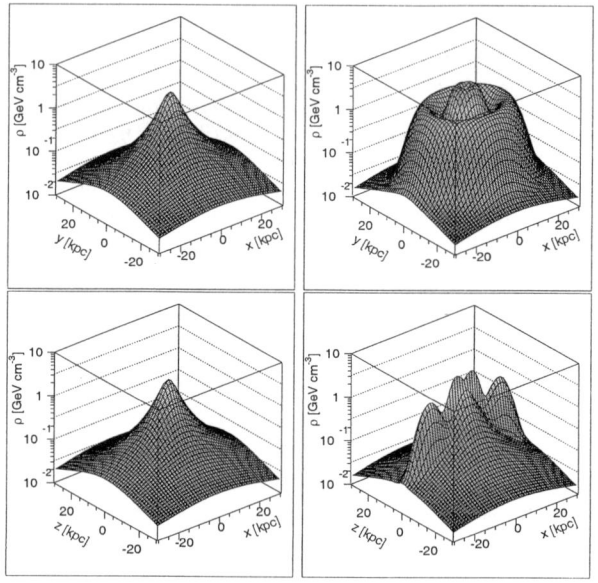

Figure 2. 3D-distributions of the $1/r^2$ haloprofile in the galactic xy-plane (top row) and xz-plane (bottom row) without (left) and with (right) rings.

From the excess in the various sky directions one can obtain the halo profile under the assumption that the clustering of the DM is similar in all sky directions. The result is surprising: in addition to the $1/r^2$ profile expected for a flat rotation curve the EGRET excess show a substructure in the form of toroidal rings at 4 and 14 kpc, as shown in Fig. 2: on the left hand side the contribution from the $1/r^2$ profile is shown, while for the right hand side the ring structure is added. The need for these additional rings is most easily seen by comparing the longitudinal profiles in the galactic plane and towards the galactic poles. As shown in Fig. 3 the pole regions are described reasonably well without rings, but for the galactic plane the $1/r^2$ profile only describes the data towards the centre. For the larger latitudes one needs the rings, as indicated by the right top panel. Note that for each bin only the flux integrated for data above 0.5 GeV has been plotted. The normalization of the background has been obtained from a fit to the flux integrated between 0.1 and 0.5 GeV.

The position and shape of the inner ring coincides with the ring of molecular hydrogen. Molecules form from atomic hydrogen in the presence of dust or heavy nuclei. So a ring of neutral hydrogen suggests an attractive gravitational potential. The position and shape of the outer ring coincides with the ring of stars, discovered in 2003 by two independent groups[14,15]. This ring is thought to originate from the infall of a dwarf galaxy, so additional DMA is expected there. To prove that the enhanced gamma ray density is indeed connected to non-baryonic mass the rotation curve was reconstructed from the excess of the diffuse gamma rays in the following

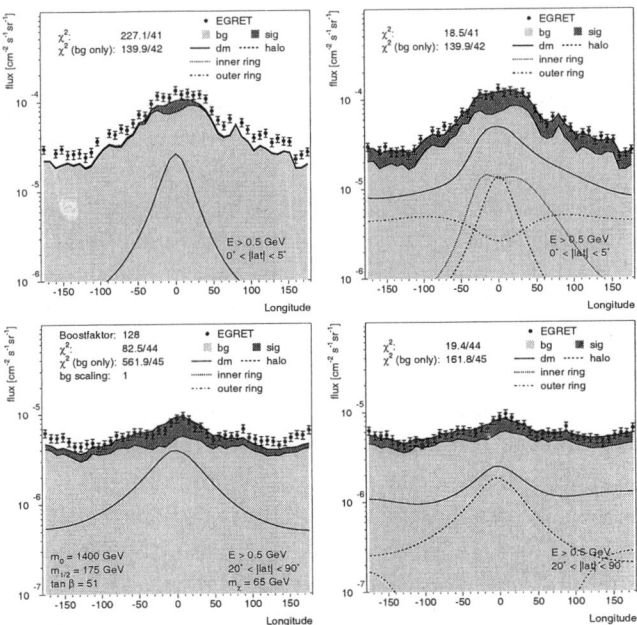

Figure 3. Top row: the longitude distribution of diffuse gamma-rays in the disc of the galaxy (latitudes $0° < |b| < 5°$) for the $1/r^2$ profile without (left) and with rings (right). The points represent the EGRET data. Bottow row: as above for the polar regions of our galaxy (latitudes $20° < |b| < 90°$).

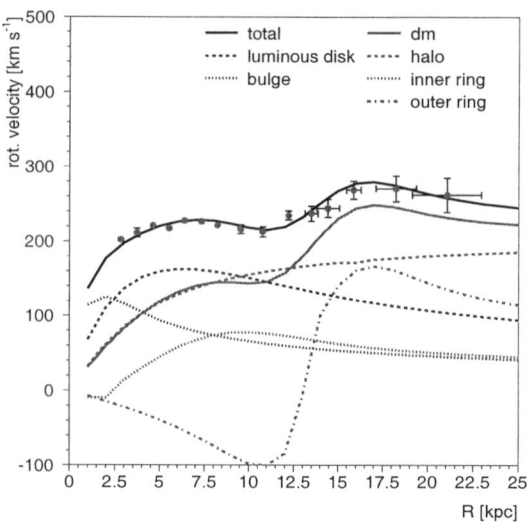

Figure 4. The rotation curve from our galaxy with the DM contribution determined from the EGRET excess of diffuse gamma rays. The data are averaged from Ref. [9].

way: since the flux determines the number density of DM for a given boost factor and since the mass of each WIMP is between 50 and 100 GeV, one can determine the mass in the ring and consequently predict the rotation curve[a]. The two ring model describes the peculiar change of slope at 11 kpc well, as shown in Fig. 4. The contributions from each of the mass terms have been shown separately. The basic explanation for the negative contribution from the outer ring is that a tracer star at the inside of the ring at 14 kpc feels an outward force from the ring, thus a negative contribution to the rotation velocity. It has often been argued that the outer rotation curve cannot be taken seriously, because the errors are large due to the fact that the absolute values of the rotation velocities strongly depend on the value of R_0, the distance between the solar system and the galactic centre. This is true, as shown by Honma and Sofue[16], but they show that the *change in slope* at about $1.1R_0$ is independent of R_0. In addition, it has been argued that the inner and outer rotation curve are difficult to compare, since the methods are completely different. The methods are indeed different, but the first 3 data points from the outer rotation curve (between 8 and 11

---

[a]For the outer ring a total DM mass of a few times 10^{10} solar masses was found in comparison with about 10^9 solar masses in the form of stars.

kpc) show the same slope as the ones from the inner rotation curve, so there seems to be no systematic effect related to the different methods.

4. Summary and Outlook

In summary, the EGRET data shows an intriguing hint of DM annihilation, since it explains many unrelated facts simultaneously:

a) An excess of diffuse galactic gamma rays which shows a *spectrum* consistent with the expectation from WIMP annihilation into quarks.

b) The excess is present in *all* sky directions with the same spectrum, thus excluding that it originates from anomalous contributions in the centre of the galaxy.

c) The excess shows an strongly increased intensity at positions where extra DM is expected, namely at two doughnut shaped structures at radii of 14 and 4 kpc from the centre of the galaxy. At 14 kpc one has observed a ring of stars thought to originate from the infall of a dwarf galaxy, while at 4 kpc one finds an enhanced concentration of molecular hydrogen thought to form from atomic hydrogen in the presence of dust or heavy nuclei, which can be collected in the gravitational potential of a ring of DM.

d) The enhanced excess of gamma rays cannot be due to additional gas in these rings as proven by the rotation curve calculated from the gamma ray excess: the mass in the rings perfectly describe the hitherto unexplained change of slope in the rotation curve at a distance of about 11 kpc. The amount of visible matter is far too low to have such an impact on the rotation curve.

In our analysis we only fit the known spectral shapes of the various processes with arbitrary normalizations, so the analysis becomes largely model independent. Interestingly, the normalization factors come out to be in excellent agreement with expectations, both for the WIMP signal and the background.

Alternative models trying to explain the EGRET excess have to assume that the locally measured fluxes of protons and electrons are not representative for our galaxy, in which case these spectra outside our local bubble can be tuned to obtain the more energetic gamma rays needed for the EGRET excess, although these models provides significantly worse fits to the data, if one takes the strong correlations in the errors between the different energy bins into account. Of course such models cannot explain simultaneously the stability of the ring of stars at 14 kpc and the change of slope in the rotation curve at $r = 1.1 R_0$.

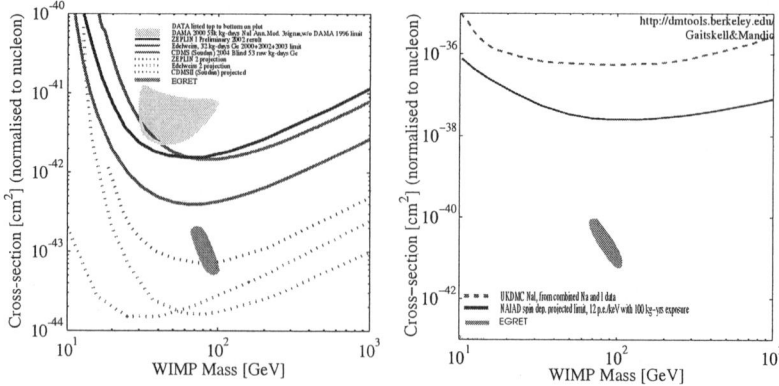

Figure 5. The spin-independent (left) and spin-dependent (right) neutralino-nucleon cross section as function of the neutralino mass for the SUSY parameters from this analysis[9] (oval shaded (brown) area in comparison with results from present and future direct DM detection experiments.

The results mentioned above make no assumption on the nature of the Dark Matter, except that its annihilation produces hard gamma rays consistent with the fragmentation of monoenergetic quarks between 50 and 100 GeV. WIMPs produce such monoenergetic quarks with energies equal to the WIMP mass. WIMP masses in this range and the observed WIMP self annihilation cross section are consistent with the Lightest Supersymmetric Particle predicted in the Minimal Supersymmetric Model with supergravity inspired symmetry breaking, called the mSUGRA model, if one assumes the enhancement of the annihilation by the clustering of DM to be of the order of 50, which is the right order of magnitude[17].

Within this supersymmetric model one finds a spin-independent cross section for elastic scattering of a WIMP on a proton of about 10^{-43} cm^2, which is within reach[19] of future experiments as shown in Fig. 5. This elastic scattering cross section was calculated with Darksusy[18].

Direct and indirect detection experiments do not prove the supersymmetric nature of the WIMPs. If the WIMPs are indeed the lightest supersymmetric particle, then this will become clear at the future LHC collider under construction at CERN in Geneva, where supersymmetric particles of the mass range deduced from the EGRET data[9] should be observable from 2008 onwards, if they exist.

The statistical significance of the EGRET excess of at least 10 σ, if fitted to the shape of the diffuse gamma ray background only, combined with all features mentioned above provides an intriguing hint that this excess is indeed indirect evidence for Dark Matter Annihilation.

I thank my close collaborators A. Gladyshev, D. Kazakov, C. Sander and V. Zhukov for their contributions to this interesting project. Furthermore I thank V. Moskalenko, A. Strong and O. Reimer for numerous discussions on galactic gamma rays.

This work was supported by the DLR (Deutsches Zentrum für Luft- und Raumfahrt) and a grant from the DFG (Deutsche Forschungsgemeinschaft, Grant 436 RUS 113/626/0-1).

References

1. D.N. Spergel et al, 2003, ApJS, 148, 175;
   C.L. Bennett et al., 2003, ApJS, 148, 1; See also: http://map.gsfc.nasa.gov/m_mm/pub_papers/firstyear.html
2. L. Bergström, Rept. Prog. Phys. **63** (2000) 793 [arXiv:hep-ph/0002126].
3. G. Bertone, D. Hooper and J. Silk, arXiv:hep-ph/0404175.
4. E. Kolb, M.S. Turner, The Early Universe, Frontiers in Physics, Addison Wesley, 1990.
5. G. Jungman, M. Kamionkowski and K. Griest, Phys. Rep. **267** (1996) 195.
6. T. Sjöstrand, P. Eden, C. Friberg, L. Lönnblad, G. Miu, S. Mrenna and E. Norrbin, Computer Phys. Commun. **135** (2001) 238.
7. A. W. Strong, I. V. Moskalenko and O. Reimer, Astrophys. J. **613**, 962 (2004); [arXiv:astro-ph/0406254].
8. Hunter, S. D. et al., Astrophysical Journal **481**, 205 (1997)
9. W. de Boer, M. Herold, C. Sander, V. Zhukov, A. V. Gladyshev and D. I. Kazakov, arXiv:astro-ph/0408272.
10. W. de Boer, arXiv:hep-ph/0408166.
11. A. W. Strong and I. V. Moskalenko, Astrophys. J. **509**, 212 (1998); [arXiv:astro-ph/9807150].
12. I. V. Moskalenko and A. W. Strong, Astrophys. Space Sci. **272** (2000) 247; [arXiv:astro-ph/9908032].
13. T. Kamae, T. Abe and T. Koi, arXiv:astro-ph/0410617.
14. B. Yanny et al., Astrophys. J. **588** (2003) 824 [Erratum-ibid. **605** (2004) 575]; [arXiv:astro-ph/0301029].
15. R. A. Ibata, M. J. Irwin, G. F. Lewis, A. M. N. Ferguson and N. Tanvir, Mon. Not. Roy. Astron. Soc. **340** (2003) L21; [arXiv:astro-ph/0301067].
16. M. Honma and Y. Sofue, Publ. of the Astronomical Society of Japan, v.48, p.L103-L106; arXiv:astro-ph/9611156.
17. V. Dokuchaev, these proceedings. V. Berezinsky, V. Dokuchaev and Y. Eroshenko, Phys. Rev. D **68**, 103003 (2003); [arXiv:astro-ph/0301551].
18. P. Gondolo, J. Edsjo, P. Ullio, L. Bergstrom, M. Schelke and E. A. Baltz, JCAP **0407** (2004) 008 [arXiv:astro-ph/0406204] and http://www.physto.se/~edsjo/darksusy/.
19. The curves were calculated with the interactive web-based program on http://dmtools.berkely.edu.

INDIRECT DARK MATTER SEARCHES WITH THE AMS-02 DETECTOR IN SPACE

W. DE BOER
FOR THE AMS COLLABORATION

Institut für Experimentelle Kernphysik
University of Karlsruhe, Postfach 6980
76128 Karlsruhe, Germany
E-mail: wim.de.boer@cern.ch

The Alpha Magnetic Spectrometer (AMS-02) is a large acceptance ($\approx 0.8 m^2 sr$) particle physics detector, planned to perform accurate, long duration measurements of charged cosmic rays (up to the TeV energy range) and gamma rays in space. The detector is approved by NASA to be installed on the International Space Station and is expected to take data for at least 3 years. AMS-02 is designed for superb particle identification, so that it will provide precise spectra for particles and antiparticles and all nuclear isotopes up to Fe. This will allow to constrain the propagation models of our galaxy with unprecedented precision and start to investigate new phenomena, usually not included in such models, like the presence of antimatter in our universe or the production of antimatter and gamma rays by Dark Matter Annihilation. This paper concentrates on the latter topic.

1. Introduction

The Alpha Magnetic Spectrometer AMS-02 detector is a large acceptance detector under construction to study cosmic rays and gamma rays outside the atmosphere up to energies in the TeV range. It was approved by the NASA to be installed by means of a space shuttle flight on the International Space Station (ISS) and take data there for at least 3 years. The Ready-for-Flight date for the final detector AMS-02 is set for September 2007. It will be the major scientific instrument on the ISS and is designed to operate without human interference for several years in space. To ensure that the concept works and e.g. the delicate silicon tracker survives the shuttle start (acceleration 3g) and landing (deceleration 10g), an engineering unit AMS-01 was flown with the shuttle Discovery in February 1998 for 160 orbits. Even this short flight delivered already data with much higher statistics than most of the data from balloon flights[1]. Of course, installing a complex high-tech particle spectrometer in space is not an easy task. Among the requirements to be fulfilled:

a) *Temperature variations* : The ISS will circle the earth in 90 minutes, thus being often exposed to the sun followed by a short "night". Therefore the electronics has to withstand temperatures from -20 °C up to +50°C.

b) *Vacuum* : In the vacuum of space the heat transport by diffusion is negligible. The heat of the electronics can only be removed by an elaborate system of radiators.

c) *Cosmic radiation* : The high flux of cosmic rays outside the atmosphere requires radiation-hard electronic components. In addition protection against micrometeorites is required.

d) *Vibration* : During start and landing vibrations and accelerations of several g will occur, requiring a high mechanical stability. All components have to be tested on vibration tables to ensure that the eigenfrequencies will be high enough in order to avoid resonances.

e) *Long term conditions* : All components have to work reliably without external maintenance. These requires space qualified components where possible. In addition, all but the front-end electronics will be twofold redundant with a fourfold redundance for the central computers.

f) *Powersupply* : AMS is allowed to take only 2 kW power from the solar panels of the ISS as a DC voltage varying between 109-126 V with a nominal value of 120 V. The small available power requires special low power electronics (no standard bus systems for the electronics!) and highly efficient DC/DC converters for transformation to lower voltages for the electronics.

g) *Data communication* : A low 20 kbit/s data link, as widely used in aircraft, is used for commanding and telemetry, while the AMS data can be partially transferred via a high rate data link (43 Mbit/s). The latter has at best a duty cycle of 70%, so online data storage is necessary. With the expected trigger rate of about 1 kHz all data will be stored locally and taken to earth by the astronauts, while part of the data will be transferred online for monitoring and first analysis.

A large fraction of the testing and development work on the detector hardware aims at qualifying it for all the above conditions, like thermal vacuum tests, vibration tests and EMI (Electromagnetic Interference) tests.

The complete detector with its many subdetectors is shown in Fig. 1 with its position on the ISS in Fig. 2. The basic idea is to have a high

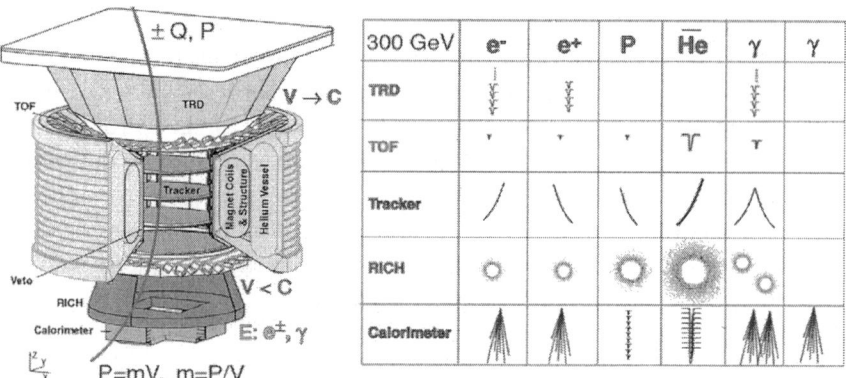

Figure 1. A sketch of the AMS-02 detector (left) and its particle identification capabilities.

energy spectrometer for charged particles - in this case a high precision silicon tracker[2] in a 0.86 T magnetic field provided by a superconducting magnet[3] - complemented by an electromagnetic calorimeter (ECAL)[4], a time-of-flight system (TOF)[5], a ring image Cherenkov counter (RICH)[6] and a transition radiation detector (TRD)[7]. These additional detectors allow perfect particle identification, as summarized on the right hand side of Fig. 1. The momentum and sign of the charge Q is measured by the tracker, while the absolute value of the charge is measured independently by the tracker, RICH and TOF. The velocity of the particle is measured by the TOF, TRD and RICH, thus allowing for redundant particle identification, once the charge and momentum are known. Gamma rays are detected either by the calorimeter or by the pair conversion outside the tracker.

The large acceptance of the detector ($\approx 0.8 \text{m}^2\text{sr}$) combined with rigidity measurements of charged particles up to the multi-TeV range and excellent particle identification will allow for a much better understanding of the cosmic rays in our galaxy than obtained hitherto by means of balloon and lower energy satellite experiments. Such an understanding is of primary importance for studying Dark Matter Annihilation (DMA), which is expected to be a strong source of energetic gamma rays, energetic positrons and low energy antiprotons. This paper will concentrate on this so called indirect detection of DM with the AMS-02 detector, i.e. searching for the stable decay products of DMA in our galaxy.

Figure 2. A computer simulation of the AMS-02 detector on the International Space Station ISS.

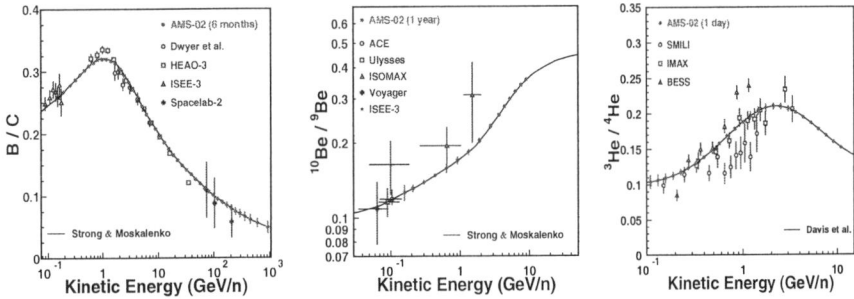

Figure 3. The present data on nucleon ratios in comparison with expectations from AMS-02.

2. Expectation of AMS-02 for Indirect DM Detection

As shown in a separate contribution to this conference[8], diffuse gamma ray data are an interesting tool to study DMA. In addition, DMA is expected to be a strong source of antiparticles, like positrons and antiprotons. Future improvements expected from AMS-02 to this subject include:

- A gamma ray spectrum up to much higher energy (TeV range) with high angular resolution.

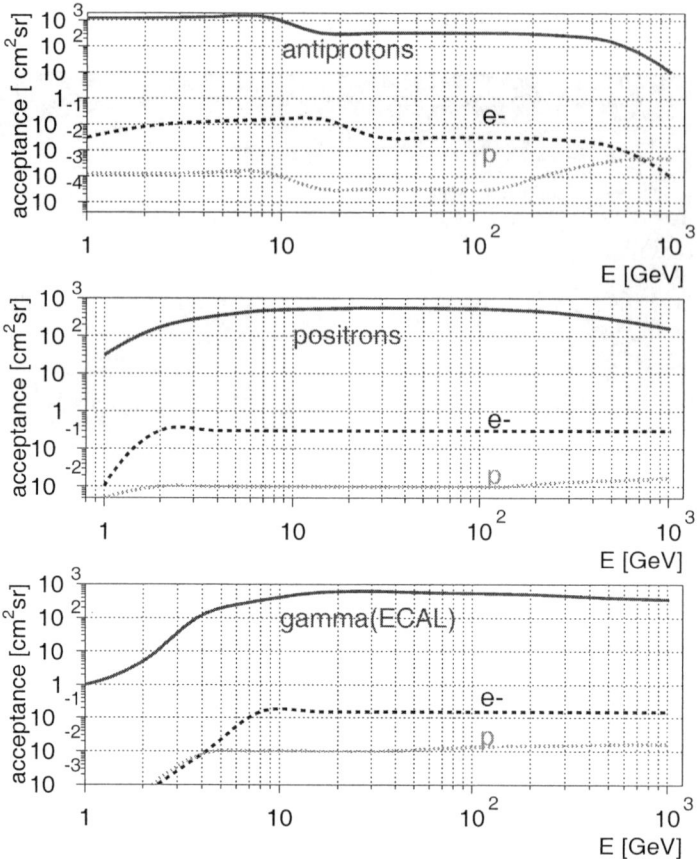

Figure 4. The acceptances of particles of interest for indirect Dark Matter detection in comparison with the small acceptances of the backgrounds.

- Precise spectra for the antimatter particles expected from DMA, i.e. the positron fraction and the antiproton spectrum.
- Precise spectra for all heavier nuclei, which provide strong constraints on the background from nuclear interactions to any DMA signal.

AMS-02 will provide precise spectra of *all* charged particles and gamma rays. A global fit to this information allows one to build a model of our galaxy. Analytical and semi-analytical models often fail when compared with all data. Therefore advanced models incorporating nuclear reaction

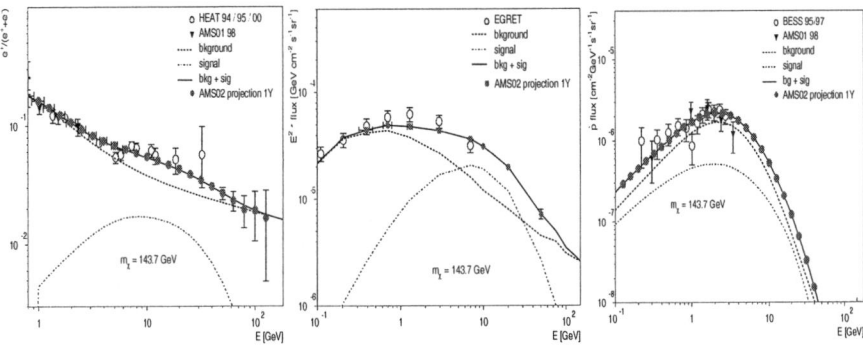

Figure 5. The expected fluxes of positrons, antiprotons and gamma rays after one year of AMS-02 data taking in comparison with existing data.

networks, cross sections for production of antiprotons, positrons, γ-rays and synchrotron radiation, energy losses, convection, diffusive reacceleration, distribution of sources, gas and radiation field etc. are needed. The most complete and publicly available code for the production and propagation of particles in our galaxy is the GALPROP code[9]. It provides a numerical solution of the transport equation including a cross section database with more than 2000 points, source functions, density distributions, etc. The cross section tables include all possible cross sections: $p+p$, $p+He$, $p+N$, $He+N$, $N+N$, where all nuclei up to the heaviest ones (Ni) are considered.

Fig. 3 shows the B/C, $^{10}Be/^9Be$ and $^3He/^4He$ ratios as expected from AMS-02 in comparison with present data. The B/C ratio shows a characteristic depletion at low energies, which is a sensitive handle on the question of diffusive reacceleration and solar modulation. This can be easily understood from the fact that Boron is a purely secondary produced spallation product of heavier nuclei, while Carbon is primarily produced and therefore dependent on the injection spectrum. Clearly better data are eagerly awaited. The same is true for the $^{10}Be/^9Be$ ratio. The lifetime of ^{10}Be is several million years, the isotope ^9Be is stable. Given the know production cross sections the low abundance of ^{10}Be is a sensitive measure of the size of the galaxy, i.e. the time it takes to diffuse from the centre of the galaxy to the solar system and the probability to escape from our galaxy or to decay beforehand. From the present poor data one can estimate that our galaxy is reasonably well described by a cylindrical symmetry with a height of about ±4 kpc and a disc radial size of 30 kpc. Fig. 4 demonstrates some of the particle identification capabilities of AMS-02: the backgrounds for antiprotons, positrons and gamma rays have a 3 to 4 orders of magnitude lower acceptance. This allows an excellent study of DMA annihilation, as shown for a WIMP mass of 147 GeV in Fig. 5.

3. Summary

The famous EGRET excess of diffuse galactic gamma rays provides an intriguing hint of DMA and with the derived WIMP mass range of 50-100 GeV[8] the AMS-02 detector in space will be the perfect instrument to study the DMA signals and background in more detail.

I want to thank the AMS collaborating institutions and individuals listed in the acknowledgements of [10] and in addition my close collaborators F. Hauler, L. Jungermann, C. Sander, M. Schmanau and V. Zhukov. This work was supported by the DLR Deutsches Zentrum für Luft- und Raumfahrt).

References

1. M. Aguilar *et al.* [AMS Collaboration], Phys. Rept. **366**, 331 (2002) [Erratum-ibid. **380**, 97 (2003)].
2. E. Cortina [AMS-02-Tracker Collaboration], *Prepared for 28th International Cosmic Ray Conferences (ICRC 2003), Tsukuba, Japan, 31 Jul - 7 Aug 2003*; W. Wallraff [AMS Collaboration], Nucl. Instrum. Meth. A **511**, 76 (2003).
3. B. Blau *et al.*, "The superconducting magnet system of the Alpha Magnetic Spectrometer AMS-02," Nucl. Instrum. Meth. A **518**, 139 (2004).
4. F. Cadoux *et al.*, Nucl. Phys. Proc. Suppl. **113**, 159 (2002).
5. D. Alvisi et al., Nucl. Instrum. and Meth. **A 437** (1999), 212.
6. D. Casadei [AMS-RICH Collaboration], Nucl. Phys. Proc. Suppl. **125**, 303 (2003) [arXiv:hep-ex/0211018].
   J. Casaus [AMS-RICH Collaboration], Nucl. Phys. Proc. Suppl. **113**, 147 (2002).
7. T. Siedenburg *et al.*, "A transition radiation detector for AMS," Nucl. Phys. Proc. Suppl. **113**, 154 (2002).
8. W. de Boer, these proceedings.
9. A. W. Strong and I. V. Moskalenko, Astrophys. J. **509**, 212 (1998); [arXiv:astro-ph/9807150].
   I. V. Moskalenko and A. W. Strong, Astrophys. Space Sci. **272** (2000) 247; [arXiv:astro-ph/9908032].
10. S. Gentile [AMS-02 Collaboration], *Prepared for ICRC 2003, Tsukuba, Japan, 2003*

SEARCH FOR DARK MATTER WITH GLAST AND PAMELA

ALDO MORSELLI, ANDREA LIONETTO, AND VLADIMIR ZDRAVKOVIĆ

INFN, Sezione di Roma II and
Dipartimento di Fisica, Università di Roma "Tor Vergata",
via della Ricerca Scientifica, Roma, Italy
E-mail: aldo.morselli@roma2.infn.it

The direct detection of annihilation products in cosmic rays offers an alternative way to search for supersymmetric dark matter particles candidates. The study of the spectrum of gamma-rays, antiprotons and positrons in space has already showed some deviation from the expected signals but with weak statistical evidence. We will review the present situation and the achievable limits with the experiments GLAST and PAMELA

The EGRET telescope has identified a gamma-ray source at the Galactic center. The spectral features of this source are compatible with the gamma-ray flux induced by pair annihilations of dark matter weakly interacting massive particles (WIMPs). We show here that the discrimination between this interpretation and other viable explanations will be possible with GLAST [1], the next major gamma-ray telescope in space, on the basis of both the spectral and the angular signature of the WIMP-induced component. If, on the other hand, the data will point to an alternative explanation, we prove that there will still be the possibility for GLAST to single out a weaker dark matter source at the Galactic center. The potential of GLAST has been explored both in the context of a generic simplified toy-model for WIMP dark matter, and in a more specific setup, the case of dark matter neutralinos in the minimal supergravity framework. In the latter, we find that even in the case of moderate dark matter densities in the Galactic center region, there are portions of the parameter space which will be probed by GLAST. Figure 1 shows the EGRET data from the Galactic center together with the diffuse gamma ray background flux expected from the standard interactions and propagation models of cosmic ray protons and electrons and an example of the flux due to neutralino annihilation in

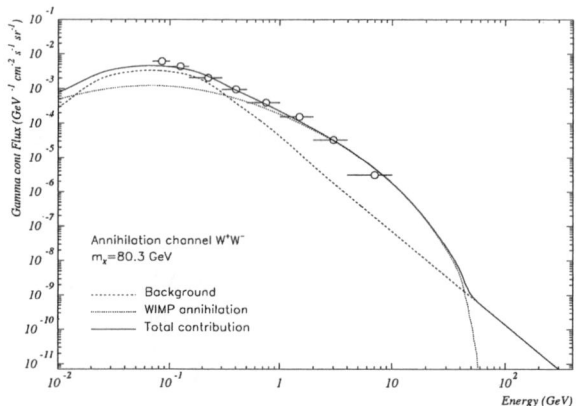

Figure 1. Fit of the EGRET Galactic Center γ-ray data for a sample WIMP models with $M_\chi = 80.3$ GeV and W^-W^+ annihilation channel.

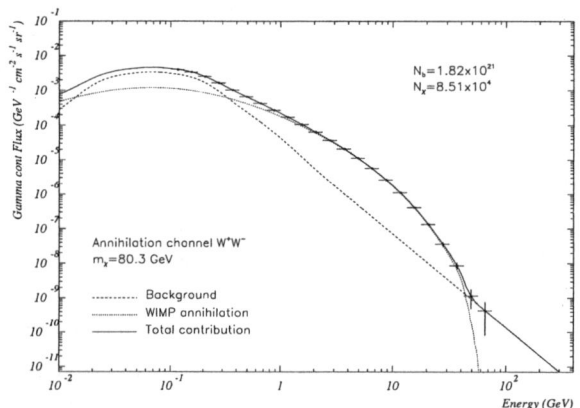

Figure 2. Same fluxes of figure 1 with the kind of statistical errors that it is expected in three years with GLAST.

the dark matter halo [2]. In this case the signal is for a ~ 80 GeV neutralino and for the W^-W^+ annihilation channel (the spectral shape of the other channels is very similar).

Figure 2 shows the same fluxes of figure 1 with the kind of statistical errors that is expected in two years with GLAST [3]. It can be seen that GLAST will have the necessary statistical, angular and energetic accuracy to distinguish the two kind of spectral shape.

We focus now on the most widely studied WIMP dark matter candidate,

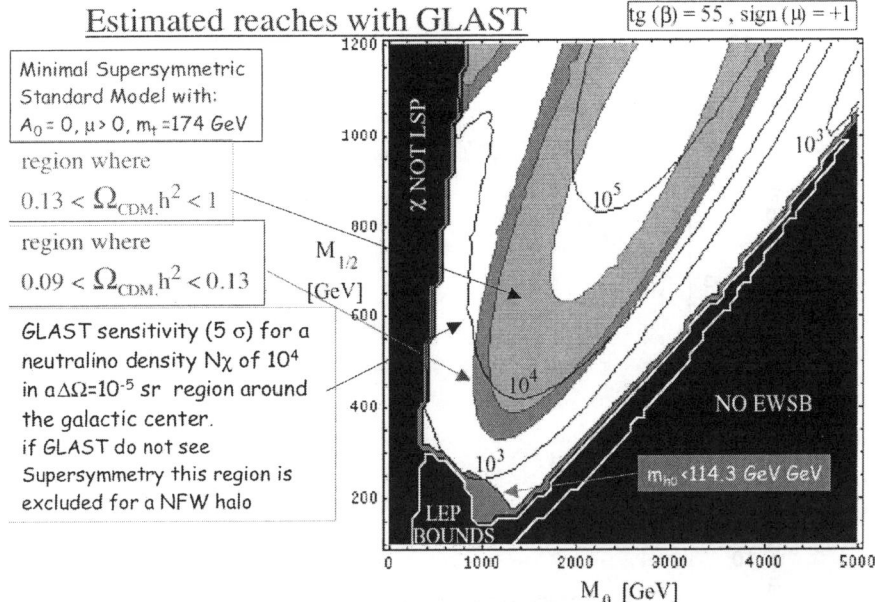

Figure 3. Contour plot in the mSUGRA $(m_0, m_{1/2})$ plane, for the value of the normalization factor N_χ, that allows the detection of the neutralino γ ray signal with GLAST. In the green region $0.13 \leq \Omega_\chi h^2 \leq 1$, while the red region corresponds to the WMAP range $0.09 \leq \Omega_\chi h^2 \leq 0.13$. The black region corresponds to models that are excluded either by incorrect EWSB, LEP bounds violations or because the neutralino is not the LSP. In the dark shaded region $m_{h_0} < 114.3$ GeV and h_0 is the lightest Higgs.

the lightest neutralino, in the most restrictive supersymmetric extension of the Standard Model, the minimal supergravity (mSUGRA) framework [5].

We fix the five mSUGRA input parameters:

$$m_{1/2}, \; m_0, \; sign(\mu), \; A_0 \text{ and } \tan\beta ,$$

where m_0 is the common scalar mass, $m_{1/2}$ is the common gaugino mass and A_0 is the proportionality factor between the supersymmetry breaking trilinear couplings and the Yukawa couplings. $\tan\beta$ denotes the ratio of the vacuum expectation values of the two neutral components of the SU(2) Higgs doublet, while the Higgs mixing μ is determined (up to a sign) by imposing the Electro-Weak Symmetry Breaking (EWSB) conditions at the weak scale. The parameters at the weak energy scale are determined by the evolution of those at the unification scale, according to the renormalization group equations (RGEs) [3]. For this purpose, we have made use of the ISASUGRA RGE package in the ISAJET 7.64 software [4]. After fixing

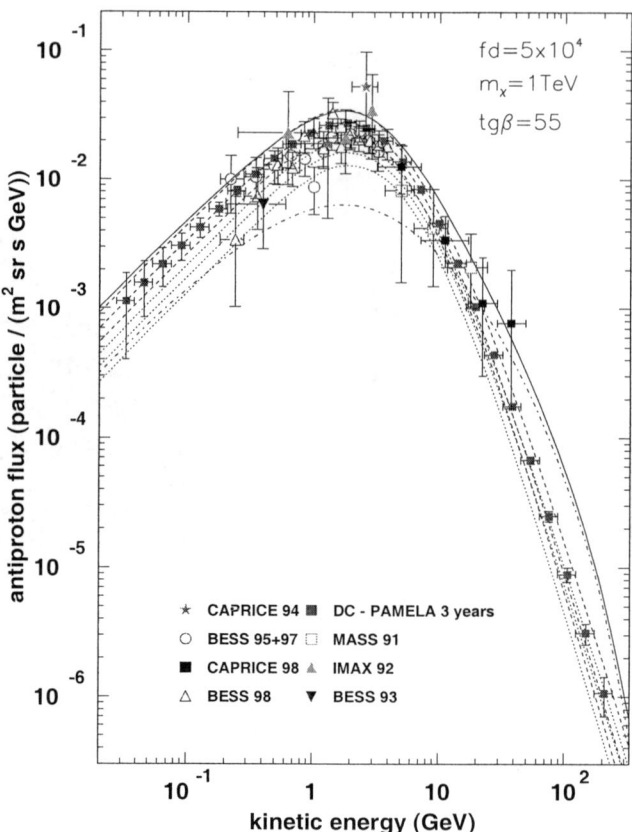

Figure 4. Total uncertainty of $\bar{p}$ fluxes and spectra that correspond to the parameters of the best B/C fit for DC (dashed line) and DRB (dotted line) model. Experimental data vs. PAMELA expectations for DC background. Neutralino induced component (dash-dotted line) and total flux (solid line) calculated with DC background.

the five mSUGRA parameters at the unification scale, we extract from the ISASUGRA output the weak-scale supersymmetric mass spectrum and the relative mixings. Cases in which the lightest neutralino is not the lightest supersymmetric particle or there is no radiative EWSB are disregarded. The ISASUGRA output is then used as an input in the DarkSUSY package[6]. The latter is exploited to:

- reject models which violate limits recommended by the Particle Data Group 2002 (PDG) [7];
- compute the neutralino relic abundance, with full numerical solution of the density evolution equation including resonances, threshold effects and all possible coannihilation processes [8];
- compute the neutralino annihilation rate at zero temperature in all kinematically allowed tree-level final states (including fermions, gauge bosons and Higgs bosons;
- estimate the induced gamma-ray yield by linking to the results of the simulations performed with the Lund Monte Carlo program Pythia [9] as implemented in the DarkSUSYpackage.

Fixing $\tan\beta$, A_0 and $sgn(\mu)$, we have performed a scan in the $(m_0, m_{1/2})$ plane searching for the minimum dark matter density, in the GC region, needed to be able to single out the neutralino annihilation signal with GLAST. First we estimate the statistical error (1σ) on GLAST data to be the square root of the number of events. To compute the latter we multiply the flux by the effective area of the detector, by the total observational time and the angular resolution $\Delta\Omega = 10^{-5}$ sr. Then for each value of the pair $(m_0, m_{1/2})$ we compute the difference between the fluxes $\phi_\gamma = \phi_b + \phi_\chi = N_b S_b + N_\chi S_\chi$ and $\phi'_\gamma = \phi_b = N_b S_b$. If $\phi_\gamma - \phi'_\gamma > 3\sigma$ we consider the SUSY model with those values of $(m_0, m_{1/2})$ to be detectable by GLAST.

Figure 3 shows the GLAST capability for $\tan\beta=55$ to probe in two years the supersymmetric dark matter hypothesis [3]. The figure shows in the $(m_0, m_{\frac{1}{2}})$ plane, the iso-contour regions for the minimum allowed value of the neutralino density in a $\Delta\Omega = 10^{-5} sr$ region around the galactic center. The density depends from the halo shape of the neutralino distribution, that is still matter of debate and can vary from a value of $N_\chi = 3 \times 10^1$ for an isotermal profile up to $N_\chi = 10^4$ for a NFW profile [11] and up to $N_\chi = 10^7$ for a Moore profile [12]. GLAST indeed can explore a good portion of the supersymmetric parameter space especially at large values of $\tan\beta$ and if the halo has a NFW (or steeper) profile. This is a very steep ($1/r$) profile but consistent with available dynamical constraints on the Galaxy.

The search for supersymmetric signal with GLAST will be complementary to the search for neutralinos looking at the distortion of the secondary positron fraction and secondary antiproton flux that will be performed with PAMELA and AMS. Figure 4 shows the PAMELA expectations for the antiproton flux for the best standard production and propagation model [13] obtained with the use of its geometrical factor and detector characteristics [10]. The primary contribution to the $\bar{p}$ flux has been computed using the public code DarkSUSY [14]. We have modified the $\bar{p}$ propagation in order to be consistent with the diffusion and convection (DC) model as implemented in Galprop. We assumed diffusion coefficient spectra used in Galprop with our best fit values for the diffusion constants D_0 and δ. In DarkSUSY the convection velocity field is constant in the upper and lower Galactic hemispheres (with opposite signs, and so it suffers unnatural discontinuity in the Galactic plane) while Galprop uses magnetohydrodynamically induced model in which component of velocity field along the Galactic latitude (the only one different from zero) increases linearly with the galactic latitude [15]. We have assumed an averaged convection velocity calculated from the Galactic plane up to the Galactic halo height z. The SUSY contribution to the $\bar{p}$ flux is shown in figure 4 for a neutralino mass of 1 TeV (obtained from a particular choice of mSUGRA parameters) and a clumpiness factor fd of $5 \cdot 10^4$. Higher neutralino masses improve high energy data fit but only with the increase of the clumpiness factor because of the dependence from the inverse neutralino mass squared m_χ in the $\bar{p}$ flux.

References

1. A.Morselli, Frascati Phys. Series, Vol.24 (2002), pp. 363-380 [astro-ph/0202340]
2. A.Morselli, A.Lionetto, A.Cesarini, F.Fucito, P.Ullio, Nuclear Physics 113B (2002) 213-220
3. A.Cesarini, F.Fucito, A.Lionetto, A.Morselli and P.Ullio, Astropart. Phys. **21** (2004) 267 [astro-ph/0305075].
4. H. Baer et. al., hep-ph/0001086.
5. L. J. Hall, J. Lykken and S. Weinberg, Phys. Rev. D **27** (1983) 2359.
6. P. Gondolo, et al., IDM 2002, York, England, [astro-ph/0211238].
7. K. Hagiwara et al., Phys. Rev. **D66** (2002) 010001.
8. J.Edsjo et. al., JCAP **0304** (2003) 001 [arXiv:hep-ph/0301106].
9. Pythia program package, T. Sjöstrand, Comp. Phys. Comm. **82** (1994) 74.
10. P.Picozza, A.Morselli, 2003. J. Phys. G: Nucl. Part. Phys., 29, 903-911
11. J.F. Navarro, C.S. Frenk and S.D.M. White, Astrophys. J. **462** (1996) 563.
12. S. Ghigna et al., Astrophys. J. **544** (2000) 616.
13. A.Lionetto, A.Morselli, V.Zdravković, astro-ph/0410409
14. P.Gondolo et. al., JCAP **0407** (2004) 008, [arXiv:astro-ph/0406204].
15. V.Zirakashvili et. al., 1996, A&A, 311, 113

DARK MATTER SEARCH BY HIGH-ENERGY GAMMA RAYS AND ELECTRONS WITH CALET*

K.YOSHIDA[†] FOR THE CALET COLLABORATION

Kanagawa University,
3-27-1 Rokkakubashi, Kanagawa-ku, Yokohama 221-8686, Japan
E-mail: yoshida@kit.ie.kanagawa-u.ac.jp

We are proposing the CALET (CALorimetric Electron Telescope) instrument for the observation of high-energy gamma rays and electrons at the Exposed Facility of the Japanese Experiment Module on the International Space Station. The CALET has a capability to observe gamma rays in 20 MeV - 10 TeV and electrons (+positrons) in 1 GeV - 10 TeV with a high energy resolution of 2 % at 100 GeV, a good angular resolution of 0.06 deg at 100 GeV, and a high proton rejection power of 10^6. The CALET has the geometrical factor of nearly $1m^2 sr$ and three-years observation is expected. The excellent energy resolution of CALET, which is much better than GLAST or air Cherenkov telescopes over 10 GeV, enables us to detect gamma-ray lines in the GeV - TeV region from WIMP dark matter annihilations. In addition, although the CALET cannot separate negative electrons and positrons, the high precise spectrum of electrons(+positrons) enables us to detect distinctive features from WIMP annihilations in the Galactic halo. Thus the CALET has a unique capability to search for WIMP dark matter by the hybrid observation of high-energy gamma rays and electrons.

1. Introduction

For the observation of high-energy gamma rays and electrons, we are proposing the CALET (CALorimetric Electron Telescope) mission at the Exposed Facility of the Japanese Experiment Module (JEM-EF) on the International Space Station (ISS)[1]. The mission goal is to investigate the high-energy phenomena in the universe by observing gamma rays in 20 MeV - several TeV and electrons in 1 GeV - 10 TeV. In addition, the CALET can observe protons up to 1000 TeV.

---

*This work is supported by a part of Ground-based Research Announcement for Space Utilization prompted by Japan Space Forum.
[†]Work partially supported by Grants in Aid for Scientific Research C (Grant No.16540268).

From 1993, we have developed an imaging calorimeter of the BETS (Balloon-borne Electron Telescope with Scintillating fibers) for the balloon experiments, and successfully observed cosmic-ray electrons of 10 GeV - 100 GeV and atmospheric gamma rays over several GeV[2,3]. In 2004, by using the Polar Patrol Balloon in Antarctica, we also carried out the observation of high-energy electrons up to 1 TeV for 13 days[4]. The CALET is designed on the basis of the BETS instrument to observe higher energy gamma rays and electrons up to 10 TeV on the ISS.

The status of the CALET mission is now under the concept study to confirm the feasibility of the detector in the performance and compatibility as a suitable mission on the JEM-EF of the ISS. The CALET will be attached on the heavy payload point of 2.5×10^3 kg at the JEM-EF. A schematic view of the CALET instrument on the JEM-EF is presented in Fig. 1. We are expecting to launch the CALET around 2010 by the Japanese H-II Transfer Vehicle (HTV), and carry out the observation more than three years.

In this paper, we present the characteristics and capability of the CALET for WIMP dark matter search by high-energy gamma ray and electron observations.

2. CALET Instrument

2.1. *Detector Structure*

The CALET detector consists of an imaging calorimeter (IMC) and a total absorption calorimeter (TASC). A schematic configuration of the CALET detector is presented in Fig. 2.

The IMC is a tracking-type calorimeter using scintillating fibers and lead plates. The scintillating fibers with a 1mm square cross section are used to observe the shower particles developing in lead plates. The IMC has 36 layers of scintillating fiber belts which are set in right angle alternately to observe the projected shower profile in x and y direction. The IMC is used for identification of the incident particle, determination of the incident angle, and energy measurement below 10 GeV. The TASC is composed of inorganic scintillators such as BGO logs with a cross section of 2.5 cm × 2.5 cm, which are aligned in x and y direction, alternately. The role of TASC is to measure the whole development of electro-magnetic showers up to 10 TeV. The TASC is used for proton rejection in the TeV region and energy measurement over 10 GeV.

The total thickness of detector is 36 radiation lengths (r.l.) for electro-

Figure 1. A schematic view of the CALET on the JEM-EF of the ISS.

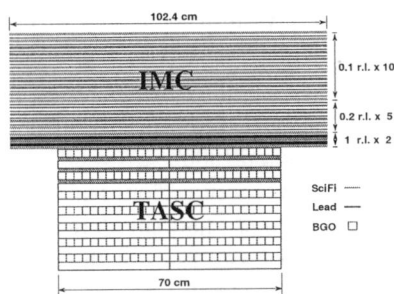

Figure 2. The configuration of the CALET detector. The anti-coincidence system covering the IMC is not presented.

magnetic particles and 1.8 mean free paths (m.f.p.) for protons. The total weight of payload is $\sim 2.5\times10^3$ kg, and the effective geometrical factor for the electrons might be ~ 1.0 m^{2}sr.

2.2. Instrument Development

In order to read out scintillating fibers, we have been developing a readout system of compact 64-multi-anode PMTs with the front-end circuit including ASIC, FPGA and 16 bit ADC. The linearity can be kept up to 3×10^3 minimum ionizing particles (MIPs), and the power consumption for the IMC can be less than 200W as required. As for the TASC, we have tested basic performance of BGO and a read-out system using photodiodes. We have carried out a beam test in 2003 by use of CERN-SPS for the proto-type detector, which consists of the IMC of 512 scintillating fibers with 4 r.l. lead and the TASC of 26 BGO logs with 23 r.l. in thickness. The performance of the proto-type has been studied on the rejection power against protons, the energy resolution, the angular resolution, and so on.

2.3. Performance

The JEM-EF on the ISS gives us a good opportunity to carry out the gamma-ray and electron observation for a long exposure. The ISS is in orbit of an inclination angle of 51.6°, changing longitudes of ascending node at the rate of −5.0° per day by the precession. The line of sight of the CALET instrument is in the opposite direction of the earth. In the ISS orbit, it is possible to cover ∼70 % of the sky for one day and all of the sky for ~20 days in a wide field of view of 2 sr without attitude control of the instrument. The CALET can survey the sky almost uniformly, observing gamma-ray point sources for 48 days on average. For diffuse gamma rays, the sensitivities for three years are 1×10^{-10} (cm^{-2}s^{-1}sr^{-1}) for the inner Galaxy region (300° < ℓ < 60°, $|b|$ < 10°) and 1×10^{-11} (cm^{-2}s^{-1}sr^{-1}) for the outer Galaxy region, respectively.

We have studied the detector performance by the balloon experiments, accelerator beam tests of the proto-type detector, and simulations. Table 1 shows the CALET performance for the gamma-ray observation compared with EGRET[5] and GLAST[6]. The CALET can observe gamma rays and electrons with the energy resolution of 2 % and angular resolution of 0.06° at 100 GeV. The CALET also has an excellent capability of the proton rejection power of $\sim 10^6$, which is necessary to select gamma rays and electrons in the TeV region. As for the separation of gamma rays and electrons, below 10 GeV, we separate by the anti-coincidence system. Over 10 GeV region, gamma rays and electrons are separated by top fibers in the IMC, since gamma rays have no tracks at the top of the IMC except for back-scattered particles. Electron rejection power for the gamma-ray observation is $> 10^5$, and gamma-ray rejection power for the electron observation is $> 5 \times 10^2$.

The event trigger will be performed in the following three modes: gamma rays in 20 MeV - 10 GeV by anti-coincidence and tracking of shower particles in the IMC, gamma rays and electrons over 10 GeV by shower trigger in the IMC, and protons over 1 TeV by shower trigger in the TASC. The electrons in 1 GeV - 10 GeV are observed only for the limited term by reducing the threshold of the shower trigger in the IMC. The trigger rates estimated by simulations are ~51 Hz for 20 MeV - 10 GeV gamma rays (~37 Hz for albedo gamma rays), ~40 Hz for gamma rays and electrons over 10 GeV, in which almost triggered events are protons and heliums as background, and <0.1 Hz for protons over 1 TeV.

Table 1. Performance for gamma rays compared with other instruments.

	EGRET	GLAST (SRD)	CALET
Energy Range (GeV)	0.02–30	0.02–300	$0.02 - 1 \times 10^4$
Effective Area (cm^2)	1500	>8000	7.9×10^3 (@10GeV) 4.6×10^3 (>100GeV)
F.O.V. (sr)	0.5	>2	0.5–1.8
Angular Res. (deg)	5.8 (@100MeV)	< 3.5 (@100MeV) < 0.15 (>10GeV)	5.0 (@100MeV) < 0.24 (> 10GeV)
Energy Res. (%)	10	< 10	$7/\sqrt{E/10\text{GeV}}$
Point Source Sensitivity (cm^{-2}s^{-1}) (> 100MeV)	5×10^{-8}	6×10^{-9}	1×10^{-8}

3. Expected Results

3.1. *Gamma-ray Observation*

In the gamma rays from WIMP dark matter annihilations, there are signatures such as distinctive substructures in the Galactic halo as seen in N-body simulations and a distinctive line feature in the gamma-ray energy spectrum. Bergström et al.[7] calculated neutralino gamma-ray signals from accreting Galactic halo dark matter. We estimated the possibility of detection of a neutralino annihilation line with their calculations. We used the gamma-ray line flux of 3×10^{-7} (cm^{-2} s^{-1} sr^{-1}) in a field of $44° \times 1°$ at the galactic latitude of $b \simeq -10°$ in the clumpy Galactic halo. Figure 3 shows the simulated energy spectrum of a gamma-ray line at 78 GeV from neutralino annihilation for the three years observation, including the background of the Galactic diffuse emission.

As shown in Fig. 3, the CALET has the excellent energy resolution of 2 % at 100 GeV that is suitable to observe line features in the gamma-ray energy spectrum. Thus, the CALET has a possibility to detect a gamma-ray line in the GeV - TeV region from dark matter annihilations.

3.2. *Electron and Positron Observation*

There are some expectations that WIMPs annihilate directly into electron-positron pairs and produce mono-energetic electrons and positrons. Although the propagation through the Galaxy would broaden the line spectrum, the observed electron and positron spectrum would still have a distinctive feature. Since there are no other known production mechanisms that would produce an electron and positron peak at energies of 10 GeV - 10 TeV, such a distinctive feature clearly indicates the existence of WIMP dark matter in the Galactic halo.

Kamionkowski & Turner[8] suggested that if the SUSY dark matter is heavier than $W^{\pm}$ boson, e.g. a Higgsino-like neutralino, a very distinctive feature in the cosmic-ray positron spectrum arises from W^+ and Z^0 decays. Cheng et al.[9] proposed that cold dark matter is made of Kaluza-Klein particles. They suggested that there is a narrow peak in the positron spectrum from direct annihilation of Kaluza-Klein gauge bosons to e^+e^-.

In the case of Kaluza-Klein dark matter for the 300 GeV mass[9], we simulated the electron + positron spectrum observed with the CALET for the three years observation. The continuum spectrum is the power-law with an index of -3.26 that well represents the observed cosmic-ray electron + positron spectra over a few 10 GeV[10]. Assuming the power-law spectrum, the expected number of electrons with the CALET for three years is 5×10^5 over 100 GeV, which is 5×10^3 times more than the present measurements of $\sim 10^2$ electrons. Although the CALET cannot separate electrons and positrons, the high precise measurements of electrons + positrons enable us to detect the distinctive features from dark matter annihilation in the Galactic halo, as shown in Fig. 4.

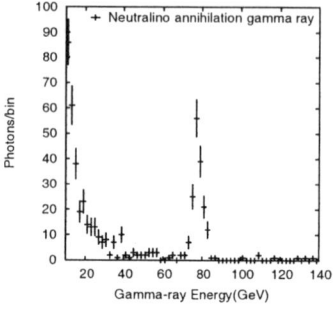

Figure 3. Simulated energy spectrum of a gamma-ray line at 78 GeV from neutralino annihilation.

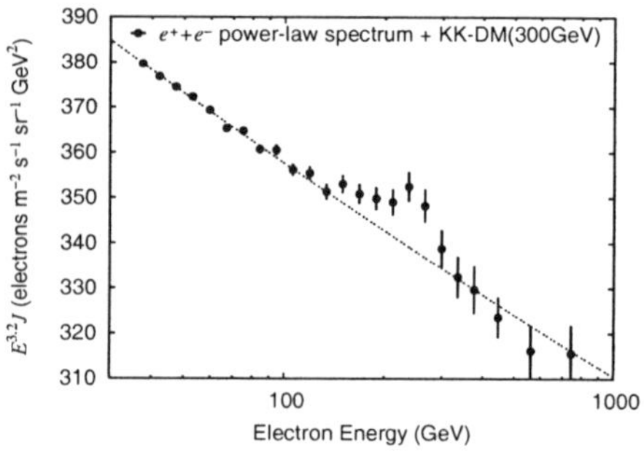

Figure 4. Simulated energy spectrum of $e^+ + e^-$ power-law spectrum with Kaluza-Klein dark matter annihilation for 300 GeV mass.

4. Summary

The CALET has a capability to observe gamma rays in 20 MeV - several TeV with an excellent energy resolution of 2 % at 100 GeV, and electrons + positrons in 1 GeV - 10 TeV with high statistical precision. This capability makes possible for us to carry out dark matter search by the detection of distinctive spectral signatures of gamma rays and electrons + positrons.

References

1. S.Torii et al., *Nucl. Phys.* **B134**, 23 (2004).
2. S.Torii et al., *Astrophys. J.* **559**, 973 (2001).
3. K.Kasahara et al., *Phys. Rev.* **D66**, 052004 (2002).
4. S.Torii et al., *Proc. of 28th International Cosmic Ray Conference*, 2085 (2003).
5. D.J.Thompson et al., *Astrophys. J. Suppl.* **86**, 629 (1993).
6. N.Gehrels, *http://glast.gsfc.nasa.gov/*, (2003).
7. L.Bergström, J.Edsjö and C.Gunnarsson, *Phys. Rev.* **D63**, 083515 (2001).
8. M.Kamionkowski and M.S.Turner, *Phys. Rev.* **D43**, 1774 (1991).
9. H.C.Cheng, J.L.Feng and K.T.Matchev, *Phys. Rev. Lett.* **89**, 211301 (2002).
10. T.Kobayashi, et al. *Proc. of 26th International Cosmic Ray Conference*, 61 (1999).

THE ANTARES NEUTRINO TELESCOPE AND ITS DARK MATTER CAPABILITIES

J. HÖßL

ON BEHALF OF THE ANTARES COLLABORATION

University of Erlangen-Nürnberg
Erwin-Rommel-Str. 1,
91058 Erlangen, Germany
E-mail: hoessl@physik.uni-erlangen.de

The ANTARES experiment is being installed in the Mediterranean Sea near Toulon, France, at a depth of 2500 m. The prospects for Dark Matter searches from the Sun, the Earth, the Galactic Centre and the status of the project are discussed.

1. Introduction

There is strong evidence for the existence of non-baryonic dark matter in the universe, for example from the fluctuations in the Cosmic Microwave Background measured by the WMAP experiment[1]. A good candidate for this dark matter is the lightest supersymmetric particle, in most models the neutralino (χ). In recent years much effort was made to search for neutralinos directly by looking for a recoil signal from neutralino scattering with detector nuclei.

A complementary indirect approach is to search for neutralino annihilation products, e. g. gammas or neutrinos. Neutralinos could be gravitationally captured by massive objects (like the Sun, Earth or Galactic Centre), and hence accumulate in the centre of those objects and produce secondary particles by annihilation, e. g. gammas or neutrinos. Large volume neutrino telescopes are ideally suited to search for high energy neutrinos produced by neutralino annihilation. In contrast to direct searches where the neutralino density can only be studied in the halo of the galaxy, indirect searches can additionally probe the neutralino density in its centre.

2. The ANTARES Neutrino Telescope

The ANTARES[2] neutrino telescope (see figure 1 for a schematic view) is being built in the Mediterranean Sea near Toulon/France at a depth of 2500 m. The main aims of the experiment are the search for high energy neutrinos from astrophysical sources, thus addressing the question of acceleration of cosmic ray particles, and the search for neutralino annihilation into neutrinos. The neutrinos will be detected by their interactions in the

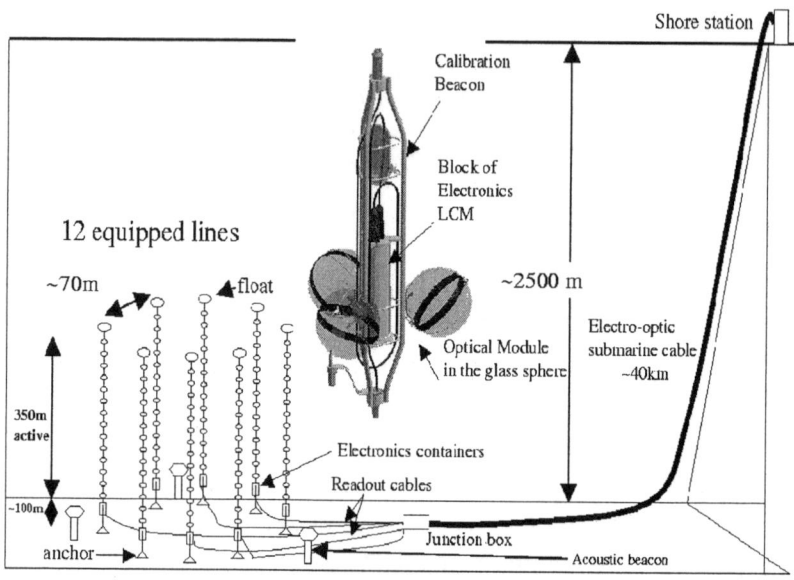

Figure 1. Schematic view of the ANTARES detector.

sea water or in the rock below, hence giving rise to charged secondary particles, which produce Cherenkov light in water. This light is detected by downward looking photomultipliers (10" Hamamatsu R7081-20) which are mounted in 17" pressure resistent glass spheres. Three of these spheres are grouped together with an electronic container and monitoring equipment to form a so called storey. A detector line comprises 25 storeys with a vertical spacing of 14.5 m between two storeys. The full detector will consist of 12 lines with an average horizontal distance of 65 m. Each line has a height of 450 m and is fixed by an anchor to the sea bed and tightened by a buoy on top of the line. The lines are connected by an electro-optical cable to a junction box, which in turn is connected to the shore station by a 40 km long cable. The lines will be operated and powered remotely from shore via the junction box, and the data from each line will be transferred from the junction box to shore.

Using the Earth as a shielding against cosmic rays the photomultipliers are mounted at an angle of 45 degrees looking downward to efficiently detect light from upward going muons or showers. They are capable of single photon counting with a time resolution of 2.7 ns FWHM dominated by the transit time spread. The analogue signals are digitised by an analogue ring sampler and sent to shore where they are processed by a PC farm.

As the lines move with the sea current (they are only fixed at the bottom), there is a monitoring system to determine the position and orientation of each storey. This system consists of hydrophones for position triangulation in combination with acoustic beacons on the sea floor (precision of a few cm) and compasses/tiltmeters to determine the orientation of the storeys (precision better than 1 degree).

The operation of prototype lines in the Mediterranean Sea in 2003 have proven the feasibility of the ANTARES concept. Useful data were obtained concerning the behaviour of the detector lines under real conditions in the sea and on the background situation at the ANTARES site. The deployment of the first complete lines will start in 2005 and the whole detector will be finished in 2007.

3. Expected Performance

Important quantities for the neutrino detection are the effective area of the telescope, which affects the event rates, and the angular resolution, which determines the size of the search cone around a point source object. Both properties influence the sensitivity of the experiment because not only event rates but also backgrounds should be considered. Monte Carlo Simulations were performed to investigate the expected performance. These take into account for example water parameters (measured at the ANTARES site) such as scattering and absorption length, transit time spread, quantum efficiency and angular acceptance of the photomultipliers to obtain a realistic description of the set-up.

The neutrino effective area, which is defined as the ratio of detected muon rate to the incoming neutrino flux, is found to be about 10^{-4} m^2 at 100 GeV neutrino energy and increases rapidly with neutrino energy (7×10^{-3} m^2 at 1 TeV, 3 m^2 at 100 TeV, saturating at about 30 m^2 for very high energies).

The expected angular resolution for different energies of the incident neutrino is shown in figure 2. At very high energies (above 10 TeV) an

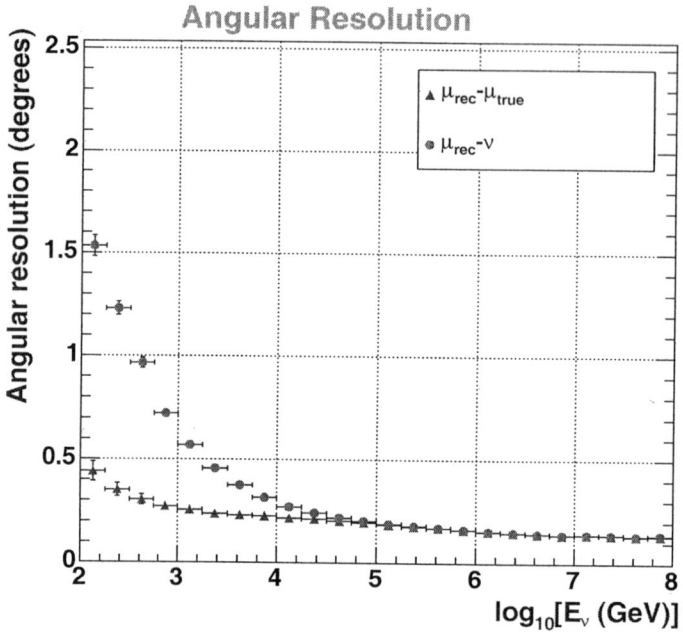

Figure 2. Median of the distribution of the angle between the direction of the reconstructed muon and of the generated one (triangles) and of the generated neutrino (circles).

excellent resolution of 0.2 degree is expected, whereas at lower energies it becomes worse mainly because the direction of the produced muon deviates from the direction of the parent neutrino due to kinematic reasons. In the energy region relevant for neutralino detection (a few 100 GeV), the resolution is of the order of 1 degree.

4. Potential for Neutralino Search

ANTARES will search for neutralinos χ by looking for neutrinos from their annihilation in the centre of massive objects. There are two main annihilation channels producing neutrinos: $\chi\chi \to W^+W^- \to \mu^+\nu_\mu\mu^-\bar{\nu}_\mu$ yielding a hard neutrino energy spectrum with typical neutrino energies about half the neutralino energy and $\chi\chi \to b\bar{b} \to c\mu^-\bar{\nu}_\mu\bar{c}\mu^+\nu_\mu$ with a softer neutrino energy spectrum.

In principle there are three promising sites which could give rise to detectable neutrino fluxes in ANTARES from neutralino annihilations: the Earth, the Sun and the Galactic Centre.

The Earth is clearly disfavoured compared to the Sun due to its smaller mass, which leads to lower capture rates[3]. Additionally, a second effect could dramatically reduce the capture rate in the Earth: The Earth does not capture neutralinos directly from the Galactic Halo but from the Solar System. Due to the depletion of the neutralino density in the Solar System resulting from the capture by the Sun, the capture rate in the Earth could be substantially suppressed for neutralino masses above 100 GeV.[4]

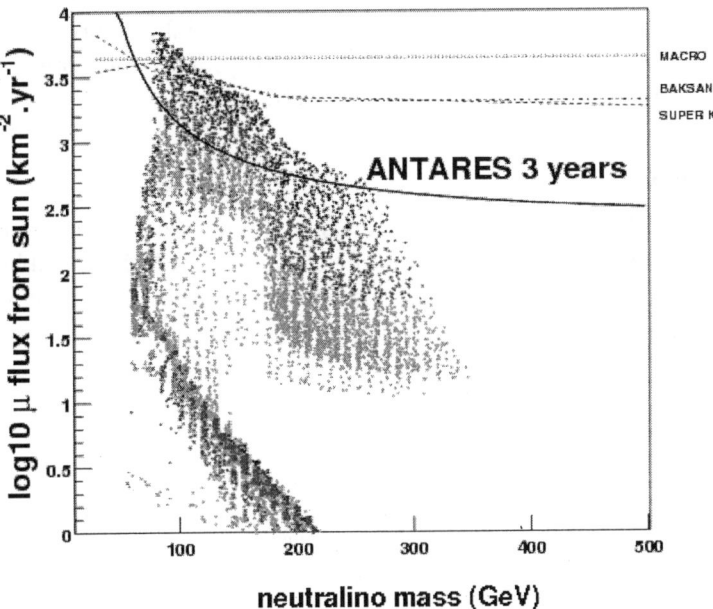

Figure 3. Expected sensitivity of ANTARES for a three year measurement period compared to existing limits and mSUGRA predictions.

The neutrino flux expected from neutralino annihilations in the Galactic Centre has large astrophysical uncertainties. The dark matter distribution in the innermost part of our galaxy is only poorly known. Usual parametrisations (e.g. Navarro-Frenk-White profile) predict only small neutrino fluxes, far below the detection threshold of ANTARES[5]. The very probable black hole at the centre of our galaxy, however, could substantially enhance the annihilation flux (by orders of magnitude) by producing a spike in the dark matter density in the vicinity of the black hole[6]. Therefore, the Galactic Centre seems to be an interesting object in terms of testing dark matter distributions, but limits on the supersymmetric parameter space have large astrophysical uncertainties.

Thus the most promising object for ANTARES to search for neutralino annihilations is the Sun. The expected sensitivity for a three year measurement period compared to existing limits and flux predictions from a mSUGRA parameter scan using SUSPECT[7] and DarkSUSY[8] is shown in figure 3. The calculation was performed using the hard neutrino energy spectrum, optimised search cones around the Sun and the Bartol atmospheric neutrino flux[9] to estimate the background within the search cone. As can be seen in figure 3, ANTARES will substantially improve existing limits from indirect neutralino annihilation in the Sun and will start to cut into the parameter space of mSUGRA models.

5. Summary

Indirect neutralino searches are a promising complementary approach to address the problem of dark matter in the Universe. ANTARES will be able to test parts of the supersymmetric parameter space in the region of the neutralino relic density favoured by the WMAP measurement ($\Omega h^2 \approx 0.1$).

References

1. D. N. Spergel et al., *arXiv:astro-ph/0302209*.
2. http://antares.in2p3.fr.
3. L. Bergström, J. Edsjö and P. Gondolo, *Physical Review* D58, 103519 (1998).
4. J. Lundberg, J. Edsjö, *arXiv:astro-ph/0401113*.
5. G. Bertone et al., *arXiv:astro-ph/0403322*.
6. P. Gondolo, J. Silk, *arXiv:astro-ph/9906391*.
7. A. Djouadi, J. L. Kneur and G. Moultaka, *arXiv:hep-ph/0211331*.
8. P. Gondolo et al., *arXiv:astro-ph/0211238*.
9. V. Agrawal et al., *Physical Review* D53, 1314 (1996).

INDIRECT SEARCHES FOR KALUZA-KLEIN DARK MATTER*

DAN HOOPER[†]

University of Oxford,
Denys Wilkinson Building, Keble Road,
Oxford, OX1-3RH, UK
E-mail: hooper@astro.ox.ac.uk

In this talk, we discuss the potential for the indirect detection of Kaluza-Klein dark matter using neutrino telescopes and cosmic positron experiments. We find that future kilometer-scale neutrino telescopes, such as IceCube, as well as future experiments capable of measuring the cosmic positron spectrum, such as PAMELA and AMS-02, will be quite sensitive to this scenario. Current data from the HEAT experiment can also be explained by the presence of Kaluza-Klein dark matter in the Galactic halo.

1. Introduction

Although there exists an enormous body of evidence for the existence of dark matter, its identity remains unknown[1]. Weakly Interacting Massive Particles (WIMPs) are, perhaps, the most well motivated class of candidates for dark matter. Among these, the lightest neutralino in models of supersymmetry is the most popular.

Models with extra spatial dimensions can provide an alternative candidate for dark matter, however. In particular, in models in which all of the Standard Models fields are free to propagate in the bulk, called universal extra dimensions, the Lightest Kaluza-Klein Particle (LKP) may be stable and a potentially viable dark matter candidate[2,3].

The most natural LKP is the first Kaluza-Klein excitation of the hypercharge gauge boson, $B^{(1)}$. We simply refer to this state as Kaluza-Klein Dark Matter (or KKDM) throughout this talk. Previous studies of KKDM have found that the relic density predicted for such a state would nat-

---

*Based on work with Graham Kribs and Joseph Silk.
[†]Supported by the Leverhulme trust

urally coincide with the measurements of WMAP for masses near about $m_{LKP} \simeq 800$ GeV if no other Kaluza-Klein states participate in the freeze-out process. If other states are light enough to significantly effect this process, however, the LKP can be substantially lighter[2].

For the purposes of indirect detection, KKDM has several interesting phenomenological features. First, approximately 60% of their annihilations are to charged lepton pairs (20% to each generation). 33% of annihilations produce pairs of up-type quarks and 3.6% produce neutrino pairs. The remaining fraction generate down type quarks and Higgs bosons. This is in stark contrast to neutralinos which do not annihilate efficiently to neutrinos, positrons, muons or other light fermions. Second, the total annihilation cross section for KKDM is given by

$$<\sigma v> = \frac{95 g_1^4}{324 \pi m_{LKP}^2} \simeq \frac{1.7 \times 10^{-26} \text{ cm}^3/\text{s}}{m_{LKP}^2(\text{TeV})}. \quad (1)$$

Notice that this consists entirely of an a-term in the expansion, $<\sigma v> = a + bv^2 + \mathcal{O}(v^4)$, thus the low velocity cross section is naturally the maximum possible for a thermal relic.

2. Indirect Detection with Neutrino Telescopes

Dark matter particles travelling through the Galactic halo can occasionally scatter and become trapped in deep gravitational wells, such as the Sun or Earth. Within these bodies, they accumulate and their annihilation rate is enhanced, potentially providing an observable flux of high-energy neutrinos[4].

The capture rate of KKDM particles in the Sun is given by[5]

$$C^\odot \simeq 3.35 \times 10^{18} \text{s}^{-1} \left(\frac{\sigma_{\text{H,SD}}}{10^{-6} \text{ pb}} \right) \left(\frac{1000 \text{ GeV}}{m_{\text{LKP}}} \right)^2, \quad (2)$$

where $\sigma_{\text{H,SD}}$ is the spin-dependent, elastic scattering cross section of KKDM off of hydrogen. This expression assumes a local dark matter density of 0.3 GeV/cm^3 and a RMS velocity of 270 km/s. The elastic scattering cross section is given by[6]

$$\sigma_{\text{H,SD}} = \frac{g'^4 m_p^2}{648 \pi m_{\text{LKP}}^4 r_{q^{(1)}}^2} \left(4 \Delta_u^p + \Delta_d^p + \Delta_s^p \right)^2, \quad (3)$$

where $r_{q^{(1)}} = (m_{q^{(1)}} - m_{\text{LKP}})/m_{\text{LKP}}$ is the fractional shift of the Kaluza-Klein quark masses over the LKP mass and the Δ_q^p's parameterize the fraction of spin carried by a constituent quark q. Inserting numerical values for the Δ_q^p's, we get

$$\sigma_{\rm H,SD} \simeq 0.9 \times 10^{-6} \, {\rm pb} \left(\frac{1000\,{\rm GeV}}{m_{\rm LKP}}\right)^4 \left(\frac{0.14}{r_{q^{(1)}}}\right)^2. \qquad (4)$$

For the annihilation and elastic scattering cross sections of KKDM, the annihilation rate in the Sun should reach (or nearly reach) equilibrium with the capture rate. These annihilations can produce neutrinos directly or in the decays of tau leptons or quarks[5].

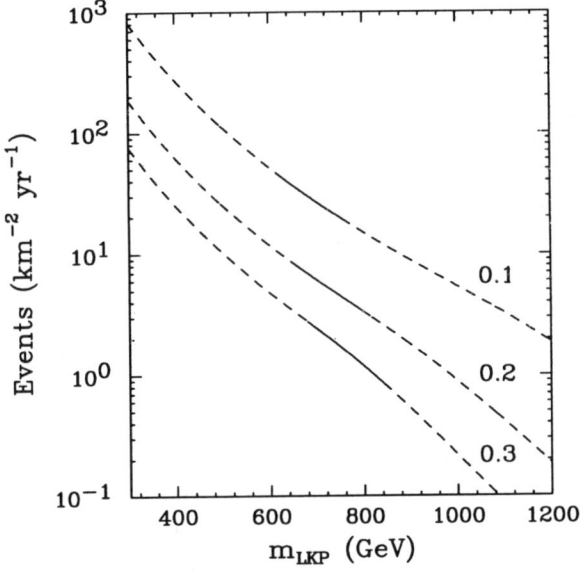

Figure 1. The rate of muon-induced neutrinos above 50 GeV predicted in a kilometer-scale neutrino telescope, such as IceCube. Curves are shown for Kaluza-Klein quarks 10%, 20% and 30% heavier than the LKP.

Muon neutrinos which reach the Earth from the Sun can scatter in charged current interactions with nucleons to produce high-energy muons. These muons produce observable "tracks" as they propagate through the medium of a neutrino telescope, such as the Antarctic Ice of the IceCube experiment. The rate of muon tracks generated in a kilometer-scale neutrino telescope from KKDM annihilations in the Sun in shown in figure 1. Notice that these results depend strongly on the LKP's mass and the mass of the Kaluza-Klein quarks. Calculations of the radiative corrections to the

Kaluza-Klein spectrum estimate values of $r_{q^{(1)}}$ roughly in the range of 0.1 to 0.2[7]. For a 800 GeV LKP, about 5 to 50 events per year are predicted in IceCube over this range. For a lighter LKP of 500-600 GeV, up to 100 events could be observed.

3. Indirect Detection with Positron Experiments

Dark matter annihilating in the Galactic halo can produce several potentially observable species of cosmic rays, including gamma-rays, anti-protons, anti-deuterons and positrons. Kaluza-Klein Dark Matter has characteristics which are favorable for detection with cosmic positrons: a large low-velocity annihilation cross section and a large fraction of annihilations which produce energetic positrons (such as the modes e^+e^-, $\mu^+\mu^-$ and $\tau^+\tau^-$).

To calculate the observed positron spectrum, the positrons injected in dark matter annihilations must be propagated through the Galactic magnetic fields, including scattering with starlight and the CMB. These effects are taken into account by solving the diffusion-loss equation:

$$\frac{\partial}{\partial t}\frac{dn_{e^+}}{dE_{e^+}} = \vec{\nabla} \cdot \left[K(E_{e^+}, \vec{x}) \vec{\nabla} \frac{dn_{e^+}}{dE_{e^+}} \right] \frac{\partial}{\partial E_{e^+}} \left[b(E_{e^+}, \vec{x}) \frac{dn_{e^+}}{dE_{e^+}} \right] + Q(E_{e^+}, \vec{x}),$$
(5)

where $K(E_{e^+}, \vec{x})$ is the diffusion constant, $b(E_{e^+}, \vec{x})$ is the energy loss rate and $Q(E_{e^+}, \vec{x})$ is the source term.

The HEAT experiment, during balloon flights in 1994-95 and 2000, observed an excess in the cosmic positron flux when compared to the electron flux. This excess peaks near 8-10 GeV and extends to above 30 GeV where the detector's sensitivity falls off[8]. It has been suggested that this excess could be the product of dark matter annihilations[9]. Neutralino dark matter, which does not annihilate to light fermions, produces positrons inefficiently, however, and therefore requires a very high annihilation rate (*i.e.* a clumpy distribution) to account for the excess observed by HEAT[10]. KKDM, on the other hand, can produce such positron fluxes more naturally[11]. Rather light LKPs (300-400 GeV) are required to generate the HEAT excess, however, which can only provide the observed relic density if other Kaluza-Klein modes play a very significant role in the freeze-out process[11].

Future experiment, such as PAMELA and AMS-02, will be capable of measuring the cosmic positron spectrum with much greater precision and to much higher energies than HEAT[12,13]. These experiments should be very sensitive to the presence of KKDM in our Galaxy[12]. In figure 3, we show the sensitivity of these experiments to KKDM.

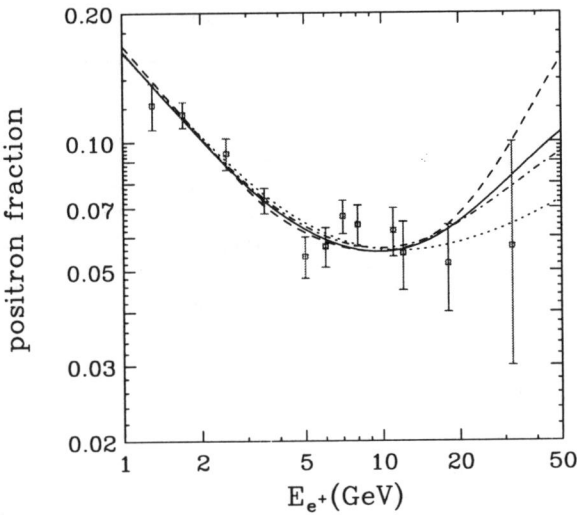

Figure 2. The ratio of positrons to positrons plus electrons with a contribution from annihilating Kaluza-Klein Dark Matter compared to the HEAT data. Results for several choices of diffusion parameters are shown. The annihilation rate was normalized to the data.

4. Summary

The prospects for the indirect detection of Kaluza-Klein Dark Matter (KKDM) are very promising for both future neutrino telescopes and cosmic positron experiments. Experiments such as IceCube, PAMELA and AMS-02 will provide powerful probes of KKDM in the coming years.

References

1. For a recent review, see G. Bertone, D. Hooper and J. Silk, Phys. Rept., in press, arXiv:hep-ph/0404175.
2. G. Servant and T. M. P. Tait, Nucl. Phys. B **650**, 391 (2003) [arXiv:hep-ph/0206071].
3. H. C. Cheng, J. L. Feng and K. T. Matchev, Phys. Rev. Lett. **89**, 211301 (2002) [arXiv:hep-ph/0207125].
4. J. Silk, K. Olive and M. Srednicki, Phys. Rev. Lett. **55**, 257 (1985); J. S. Hagelin, K. W. Ng, K. A. Olive, Phys. Lett. B **180**, 375 (1986).
5. D. Hooper and G. D. Kribs, Phys. Rev. D **67**, 055003 (2003) [arXiv:hep-ph/0208261].
6. G. Servant and T. M. P. Tait, New J. Phys. **4**, 99 (2002) [arXiv:hep-ph/0209262].
7. H. C. Cheng, K. T. Matchev and M. Schmaltz, Phys. Rev. D **66**, 036005

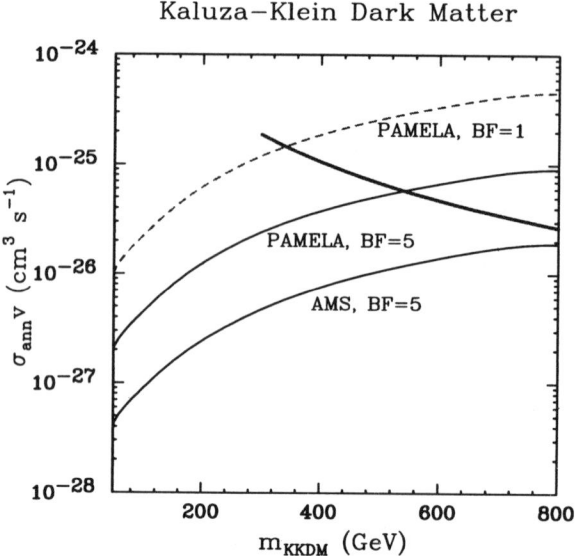

Figure 3. The sensitivity of PAMELA and AMS-02 to cosmic positrons generated in Kaluza-Klein Dark Matter (KKDM) annihilations. Contours are shown for a completely homogeneous distribution (BF=1) and a mildly clumped distribution (BF=5). The downward sloping curve represents the annihilation cross section predicted for KKDM.

(2002) [arXiv:hep-ph/0204342].
8. S. W. Barwick et al. [HEAT Collaboration], Astrophys. J. **482**, L191 (1997) [arXiv:astro-ph/9703192]; S. Coutu et al. [HEAT-pbar Collaboration], Proceedings of 27th ICRC (2001); S. Coutu et al., Astropart. Phys. **11**, 429 (1999), [arXiv:astro-ph/9902162]
9. M. Kamionkowski and M. S. Turner, Phys. Rev. D **43**, 1774 (1991); E. A. Baltz and J. Edsjo, Phys. Rev. D **59** (1999) 023511 [arXiv:astro-ph/9808243]; E. A. Baltz, J. Edsjo, K. Freese and P. Gondolo, arXiv:astro-ph/0211239; M. S. Turner and F. Wilczek, Phys. Rev. D, **42**, 1001 (1990); J. L. Feng, K. T. Matchev and F. Wilczek, Phys. Rev. D **63**, 045024 (2001).
10. D. Hooper, J. E. Taylor and J. Silk, Phys. Rev. D, **69**, 103509 (2004) [arXiv:hep-ph/0312076].
11. D. Hooper and G. D. Kribs, arXiv:hep-ph/0406026.
12. D. Hooper and J. Silk, hep-ph/0409104.
13. S. Profumo and P. Ullio, JCAP **0407**, 006 (2004) [arXiv:hep-ph/0406018]; J. Edsjo, M. Schelke and P. Ullio, arXiv:astro-ph/0405414.

SUPERHEAVY DARK MATTER

M. KACHELRIESS

Max-Planck-Institut für Physik (Werner-Heisenberg-Institut), München

Superheavy particles are produced at the end of inflation and could, for a likely set of parameters (e.g. mass $M_X \sim 10^{13}$ GeV and reheating temperature $T_R \sim 10^9$ GeV), constitute the main part of dark matter. If they are metastable, their decay products may dominate the ultra-high energy cosmic ray flux above $\sim 8 \times 10^{19}$ eV. The main signatures of this scenario, galactic anisotropy and photon dominance, should allow the Auger experiment to test this hypothesis conclusively within one year of data taking.

1. Introduction

Unitarity bounds the thermally averaged partial wave annihilation cross-section $\langle \sigma^{(l)} v \rangle$ of particles with mass M_X as

$$\langle \sigma^{(l)} v \rangle \leq \frac{4\pi}{M_X^2 v} (2l+1) \sim \frac{10^{-47} \text{cm}^2}{M_X/10^{12} \text{GeV}} \frac{100 \text{km/s}}{v} 2l. \quad (1)$$

For point-like elementary particle that were in thermal equilibrium in the early Universe the bound $M_X \lesssim 70$ TeV follows from an overclosure argument[1], $\Omega_{\text{DM}} h^2 \leq 0.2$. Superheavy dark matter (SHDM) particles should be therefore created by a non-equilibrium production mechanism. Since they are produced typically with small velocities, their relative contribution to the total density of the Universe increases as $t^{1/2}$ during the radiation-dominated phase of the Universe. The challenge is thus to create only a tiny amount of them.

Several viable non-equilibrium production mechanisms exist. The most interesting one is the gravitational production of superheavy particles X by the non-adiabatic change of the scale factor of the Universe at the end of inflation, during the transition from the de-Sitter to the radiation dominated phase[2]. In this scenario, the gravitational coupling of the X field to the background metric yields independent of any specific particle physics model the present abundance $\Omega_X h^2 \sim (M_X/10^{13} \text{GeV})^2 (10^9 \text{GeV}/T_R)$, provided that $M_X \lesssim H_*$. Here, T_R denotes the reheating temperature of the

Universe and $H_* \sim 10^{13}$ GeV the effective Hubble parameter at the end of inflation. Thus, SHDM could constitute the main component of CDM for a very interesting set of parameters. Other mechanisms proposed are thermal production during reheating, production through inflaton decay at the preheating phase, or through the decay of hybrid defects.

SHDM behaves as dissipationless dark matter regardless of whether is has electromagnetic or strong interactions with normal matter. In the latter case, direct detection experiments for DM are sensitive enough to probe SHDM. Superheavy dark matter with cross sections similar to usual strong interactions, $\sigma \sim 10^{-30} - 10^{-26}$ cm^2, were excluded[3] by an analysis of the EDELWEISS (yellow) and CDMS (red) data, cf. Fig. 1a. The figure shows also in green the region excluded by direct search experiments in space[4].

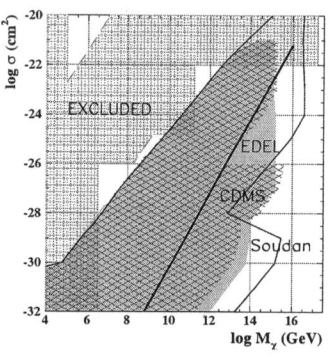 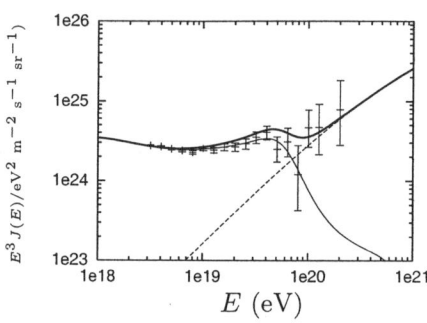

Figure 1. Left: Exclusion limits at 90% C.L. for strongly interacting SHDM from various direct detection experiments and sensitivity of ICECUBE for detection of the neutrino signal from SHDM annihilations, from [3]. Right: Comparison of the UHECR flux in the SHDM model with the AGASA data, photons from SHDM decays (dashed line), spectrum of extragalactic protons (thin solid line) and the sum of these two spectra shown by the thick curve; from [13].

Indirect detection relies normally on the search for high-energy neutrinos, $E \sim$ TeV, from DM accumulated in the Sun or Earth that annihilates. Strongly interacting SHDM can give a detectable signal in a 1 km^2 neutrino telescope: the area above the black solid line gives[5] more than 10 neutrino events from the Sun per year in such a detector. In the following, I discuss the more likely case of weakly interacting SHDM. In this case, direct detection is practically impossible, while indirect searches for annihilation products require strong clumpiness of the DM in order to have a detectable signal. However, metastable decaying SHDM can be a source of ultra-high energy cosmic rays (UHECR) and neutrinos[6,7].

2. Superheavy dark matter as UHECR source

Top–down model is a generic name for all proposals in which the observed UHECR primaries are produced as decay products of some superheavy particles X with mass $m_X \gtrsim 10^{12}$ GeV, see for a recent more complete discussion Ref. 8. These X particles can be either metastable or be emitted by topological defects at the present epoch.

Superheavy metastable relic particles constitute (part of) the CDM and, consequently, their abundance in the galactic halo is enhanced by a factor $\sim 2 \times 10^5$ above their extragalactic abundance. Therefore, the proton and photon flux is dominated by the halo component and the GZK cutoff is avoided[6]. The quotient $r_X = \Omega_X(t_0/\tau_X)$ of relic abundance Ω_X and lifetime τ_X of the X particle is fixed by the UHECR flux, $r_X \sim 10^{-11}$. The value of r_X is not predicted in the generic SHDM model, but calculable as soon as a specific particle physics and cosmological model is fixed.

The lifetime of the superheavy particle has to be in the range 10^{17} s $\lesssim \tau_X \lesssim 10^{28}$ s, i.e. longer or much longer than the age of the Universe. Therefore it is an obvious question to ask if such an extremely small decay rate can be obtained without fine-tuning. A well-known example of how metastability can be achieved is the proton: in the standard model B–L is a conserved global symmetry, and the proton can decay only via non-renormalizable operators. Similarly, the X particle could be protected by a new global symmetry which is only broken by higher-dimensional operators suppressed by M^d, where for instance $M \sim M_{\text{Pl}}$ and $d \geq 7$ is possible. The case of discrete gauged symmetries has been studied[9] also. Another possibility is that the global symmetry is broken only non-perturbatively, either by wormhole[6] or instanton[7] effects. Then an exponential suppression of the decay process is expected and lifetimes $\tau_X \gtrsim t_0$ can be naturally achieved.

An example of a SHDM particle in a semi-realistic particle physics model is the crypton[10]. Cryptons are bound-states from a strongly interacting hidden sector of string/M theory. Their mass is determined by the non-perturbative dynamics of this sector and, typically, they decay only through high-dimensional operators. For instance, flipped SU(5) motivated by string theory contains[11] bound-states with mass $\sim 10^{12}$ GeV and $\tau \sim 10^{15}$ yr. Choosing $T_R \sim 10^5$ GeV results in $r_X \sim 10^{-11}$, i.e. the required value to explain the UHECR flux above the GZK cutoff. This example shows clearly that the SHDM model has no generic "fine-tuning problem."

Superheavy dark matter has several clear signatures: 1. No GZK cutoff, instead a flat spectrum (compared to astrophysical sources) up to $m_X/2$. 2. Large neutrino and photon fluxes compared to the proton flux. 3. Galactic anisotropy. 4. If R parity is conserved, the lightest supersymmetric particle (LSP) is an additional UHE primary. 5. Small-scale clustering of the UHECR arrival directions gives possibly additional constraints.

1. *Spectral shape:* The fragmentation spectra of superheavy particles calculated by different methods and different groups[12,14,13] agree quite well. This allows one to consider the spectral shape as a signature of models with decays or annihilations of superheavy particles. The predicted spectrum in the SHDM model, $dN/dE \propto E^{-1.9}$, cannot fit the observed UHECR spectrum at energies $E \leq (6-8) \times 10^{19}$ eV. Thus only events at $E \geq (6-8) \times 10^{19}$ eV, and most notably the AGASA excess at these energies, can be explained in this model. A two-component fit[13] using protons from uniformly, continuously distributed extragalactic astrophysical sources and photons from SHDM is shown in Fig. 1 together with the experimental data from AGASA.

2. *Chemical composition*[15]: Since at the end of the QCD cascade quarks combine more easily to mesons than to baryons, the main component of the UHE flux are neutrinos and photons from pion decay. Therefore, a robust prediction of this model is photon dominance with a photon/nucleon ratio of $\gamma/N \simeq 2-3$, becoming smaller at the largest $x = 2E/M_X$.

The muon content of photon induced EAS at $E > 1 \times 10^{20}$ eV is high, but lower by a factor 5–10 than in hadronic showers[16]. It has been recently measured in a sub-array of AGASA[17]. From eleven events at $E > 1 \times 10^{20}$ eV, the muon density was measured in six. In two of them with energies about 1×10^{20} eV, the muon density is almost twice higher than predicted for gamma-induced EAS. The muon content of the remaining four EAS marginally agrees with the one predicted for gamma-induced showers. The contribution of extragalactic protons for these events is negligible, and the fraction of nucleons in the total flux can be estimated as $0.25 \leq N/\text{tot} \leq 0.33$. This fraction gives a considerable contribution to the probability of observing four showers with slightly increased muon content. Not restricting severely the SHDM model, the AGASA events give no evidence in favor of it.

The Pierre Auger Observatory[19] has great potential to distinguish between photon and proton induced EAS through the simultaneous observation of UHECR events in fluorescent light and with water Cherenkov detectors: while for a proton primary both methods should give a consistent determination of the primary energy, the ground array should systematically underestimate the energy of a photon primary. Moreover, the interaction of the photon with the geomagnetic field should induce an anisotropy in the flux.

3. *Galactic anisotropy:* The UHECR flux from SHDM should show a galactic anisotropy[20], because the Sun is not in the center of the Galaxy. The degree of this anisotropy depends on how strong the CDM is concentrated near the galactic center – a question under debate. Since experiments in the northern hemisphere do not see the Galactic center, they are not very sensitive to a possible anisotropy of arrival directions of UHECR from SHDM. In contrast, the Galactic center was visible for the old Australian SUGAR experiment. The compatibility of the SHDM hypothesis with the SUGAR data was discussed[21,22] recently: the expected arrival direction distribution for a two-component energy spectrum of UHECRs consisting of protons from uniformly distributed, astrophysical sources and the fragmentation products of SHDM calculated in SUSY-QCD was compared[21] to the data of the SUGAR experiment using a Kolmogorov-Smirnov test. Depending on the details of the dark-matter profile and of the composition of the two-components in the UHECR spectrum, the arrival directions measured by the SUGAR array have a probability of $\sim$ 5–20% to be consistent with the SHDM model. Also in the case of the galactic anisotropy, we have to wait for a definite answer for the first results of the Auger experiment.

4. *LSP as UHE primary:* An experimentally challenging but theoretically very clean signal both for supersymmetry and for top-down models would be the detection of the LSP as an UHE primary[23]. A decaying supermassive X particle initiates a particle cascade consisting mainly of gluons and light quarks but also of gluinos, squarks and even only electroweakly interacting particles for virtualities $Q^2 \gg m_W^2, M_{SUSY}^2$. When Q^2 reaches M_{SUSY}^2, the probability for further branching of the supersymmetric particles goes to zero and their decays produce eventually UHE LSPs. Signatures of UHE LPSs are a Glashow-like resonance at 10^9 GeV M_e/TeV, where M_e is the selectron mass, and up-going showers for energies where the Earth is opaque to neutrinos[23,24].

3. Conclusions

The production of superheavy particles with abundance $\Omega_X h^2 \sim (M_X/10^{13} \text{GeV})^2 (10^9 \text{GeV}/T_R)$ is a generic feature of inflation. Independent of their interactions, they behave as dissipationless dark matter and are therefore a viable dark matter candidate. Usual direct and indirect searches for WIMPs are for SHDM only relevant if they would be strongly interacting with normal matter. If SHDM is metastable, they can be also detected when they are weakly interacting. In this case, SHDM could be the source of UHECRs above the GZK-cutoff.

Acknowledgements

This work was supported by an Emmy-Noether grant of the DFG.

References

1. K. Griest and M. Kamionkowski, Phys. Rev. Lett. **64** (1990) 615.
2. D. J. Chung, E. W. Kolb and A. Riotto, Phys. Rev. **D59**, 023501 (1999); V. Kuzmin and I. Tkachev, JETP Lett. **68**, 271 (1998).
3. I. F. M. Albuquerque and L. Baudis, Phys. Rev. Lett. **90**, 221301 (2003).
4. P. C. McGuire and P. J. Steinhardt, proceedings of 27th International Cosmic Ray Conferences (ICRC 2001), Hamburg, Germany, astro-ph/0105567.
5. I. F. M. Albuquerque, L. Hui and E. W. Kolb, Phys. Rev. D **64**, 083504 (2001).
6. V. Berezinsky, M. Kachelrieß and A. Vilenkin, Phys. Rev. Lett. **79**, 4302 (1997).
7. V. A. Kuzmin and V. A. Rubakov, Phys. Atom. Nucl. **61**, 1028 (1998).
8. M. Kachelrieß, Comptes Rendus Physique **5**, 441 (2004).
9. K. Hamaguchi, Y. Nomura and T. Yanagida, Phys. Rev. **D58**, 103503 (1998); K. Hamaguchi, K. I. Izawa, Y. Nomura and T. Yanagida, Phys. Rev. **D60**, 125009.
10. J. Ellis, J.L. Lopez and D.V. Nanopoulos, Phys. Lett. **B247**, 257 (1990).
11. K. Benakli, J. Ellis and D.V. Nanopoulos, Phys. Rev. **D59**, 047301 (1999); C. Coriano, A. E. Faraggi and M. Plümacher, Nucl. Phys. B **614**, 233 (2001).
12. V. Berezinsky and M. Kachelrieß, Phys. Rev. D **63**, 034007 (2001).
13. R. Aloisio, V. Berezinsky and M. Kachelrieß, Phys. Rev. D **69**, 094023 (2004).
14. S. Sarkar and R. Toldra, Nucl. Phys. B **621**, 495 (2002); C. Barbot and M. Drees, Phys. Lett. B **533**, 107 (2002).
15. A. A. Watson, astro-ph/0312475.
16. A.V. Plyasheshnikov and F.A. Aharonian, J. Phys. **G28**, 267 (2002).
17. K. Shinozaki et al. [AGASA collaboration], Astrophys. J. **571**, L 117 (2002).
18. M. Ave et al., Phys. Rev. Lett. **85**, 2244 (2000).
19. D. Zavrtanik [AUGER Collab.], Nucl. Phys. Proc. Suppl. **85**, 324 (2000).
20. S. L. Dubovsky and P. G. Tinyakov, JETP Lett. **68**, 107 (1998).
21. M. Kachelrieß and D. V. Semikoz, Phys. Lett. B **577**, 1 (2003).
22. H. B. Kim and P. Tinyakov, Astropart. Phys. **21**, 535 (2004).
23. V. Berezinsky and M. Kachelrieß, Phys. Lett. B **422**, 163 (1998).
24. C. Barbot, M. Drees, F. Halzen and D. Hooper, Phys. Lett. B **563**, 132 (2003).

SEISMIC MOON SEARCH FOR STRANGE QUARK MATTER

W. BRUCE BANERDT AND TALSO CHUI*

*Jet Propulsion Laboratory, California Institute of Technology,
4800 Oak Grove Drive,
Pasadena, CA 91109, USA*

EUGENE T. HERRIN

*Department of Geology, Southern Methodist University
Dallas, TX 75275, USA*

DORIS ROSENBAUM

*Physics Department, Southern Methodist University
Dallas, TX 75275, USA*

VIGDOR L. TEPLITZ

*Physics Department, Southern Methodist University
Dallas, TX 75275, USA
and
NASA Goddard Space Flight Center
Greenbelt, MD 20771, USA*

We give a rough estimate on the relative advantage of attempting to detect strange quark nuggets (SQN), with seismometers, on the Moon over Earth (about 50 or more times more detections).

1. Introduction

The (US) Presidential initiative to explore Mars and the Moon offers intriguing new opportunities for searching for SQNs. The Moon seems to be particularly well suited for seismic observations as is discussed below.

---

*Work carried out at the Jet Propulsion Laboratory, California Institute of Technology, was supported by a contract with the National Aeronautics and Space Administration.

Information about Martian seismology is scarce. However overall activity is significantly less than on Earth.

1.1. Nature of Strange Quark Matter (SQM)

Matter made of up and down quarks is not stable, The quarks condense to form protons and neutrons. Witten[1] pointed out that matter made of up (charge $q_u = +2/3$), down ($q_d = -1/3$), and strange ($q_s = -1/3$) quarks, strange quark matter (SQM), may well be. It is approximately electrically neutral eliminating Coulomb repulsion energy. Three types of particles rather than two have less Pauli Exclusion Principle repulsion. Witten also suggested early universe SQN production (controversial) as well as the possibility that SQNs constitute dark matter. Under high pressure, the center of a neutron star is likely made of SQM even if early universe production does not take place. If SQN is stable under zero pressure, Collisions of neutron stars in binary systems would then give rise to a galactic SQN distribution.

Farhi and Jaffe[2] showed that being approximately neutral, SQNs would not be limited in total baryon number as is ordinary matter. Indeed, binding energy per quark would rise with the number of quarks to an asymptotic value. Thus massive, macroscopic SQNs having nuclear densities ($10^{14} gm/cm^3$) are possible. Because of the larger mass of the strange quark, the net quark charge is positive but balanced by electrons. For $M < 10^{-9}$ gram, the electron cloud would be mostly outside the nuclear (quark) part of the SQN. With high mass and low abundance, the SQN would not interact appreciably with electromagnetic energy nor affect big bang nucleosynthesis; hence its suitability as a dark matter candidate. Finally, de Rujula and Glashow[3] briefly discussed seismic detection of a massive SQN passage through the earth.

Recently, NASA's Chandra X-ray Satellite observed two neutron stars. One appeared too small[4] and one appeared too cold[5] to fit the standard model of neutron stars. These observations, barring uncertainties[6], could be consistent with the stars composed, at least in part, of SQM. It is still a matter of debate as to whether SQNs exist particularly under zero pressure. There are other quark (and gluon) models[7] that are seismically indistinguishable from the SQM model discussed here.

1.2. Previous SMU Work

Herrin and Teplitz[8] examined detection of SQN seismic signals in a Monte Carlo calculation as a guide for work with real data. Briefly, a multi-ton SQN would have dimensions of tens of microns, the size of a blood cell. Its passing through the earth would break inter- and intra-molecular bonds and produce a seismic signal. The rate of seismic energy (E) production would be given by

$$dE/dt = f\sigma\rho v^3 \quad (1)$$

where σ is the SQN cross section, ρ is earth density, v is the SQN speed (on the order of the $250 km/sec$ galactic virial velocity), and f is the fraction of SQN energy loss that results in seismic waves rather than in such dissipation as heat or breaking rock. The large density, hence small size, of an SQN passing through material, enhances coherence, depresses random motion and yields a high ratio of surface area to volume. Thus f might be significantly larger in the SQN case. The study showed that almost all (98.5%) detections of passages of SQNs of minimal detectable mass would be by 48 "Class 1" seismic stations sensitive to 1 Hz waves of power density $0.133 erg/cm^2 sec$ or better. This corresponds to a capability to detect a well coupled underground nuclear explosion of $1kT$ at $5000km$. Anderson et. al.[9] searched for seismic line events which might result from an SQN passage. Station observations of such an event would not be "associated" into Earthquakes by current methods which assume a point source for all small seismic events. Data were were collected from the United States Geological Survey (USGS) archive from 02 February 1981 through 31 December 1993. The result of of this work using the four final years of the USGS data sets a limit on the abundance of SQN in the multi-ton range with no detection.

Other SQM searches include a past one by Price[10] at Berkeley for magnetic monopole tracks in mica; a second that Price is planning with the ICE CUBE Antarctic neutrino detector now under construction; a Brookhaven heavy ion collider search for small nuggets, "strangelets", of baryon number up to twice that of gold by Armstrong et al.[11]; and the planned space station Alpha Magnetic Spectrometer (AMS) experiment of Ting[12]. All searches to date have failed to find SQNs, but not ruled them out.

1.3. Seismology on the Moon

The Apollo astronauts implanted four seismometers on the Moon. These operated over seven years finding four major sources of lunar seismic activity (see Johnson Space Center web site):

— Deep moonquakes. These most frequent events are mainly correlated with (moon) tides, strains caused by the moon rotating about the Earth.

— Shallow moonquakes. The most energetic events with frequencies which fall with energy[13] as $E^{-0.55}$. The distribution does not appear to stop at the 7-year cutoff of data from the Apollo seismometers.

— Impacts. Since there is no atmosphere to affect them, studying should provide insight into danger posed to Earth from inner Solar System loose objects.

— Other. There could be other sources such as effects of heat flow or outgassing from deep in the Moon (and SQNs).

In total, there are about 2500 seismic events each year with about one percent being classified as "major." Total seismic energy per year from the Apollo observations is about 10^{-6} that of the Earth.

2. Earth and Moon Sensitivities

We estimate relative rates of quark nugget passage detection for Earth and Moon.

1. Cross section. The ratio of the cross sectional area of the Moon to that of Earth is $\alpha_M/\alpha_E \sim 0.075$.

2. and 3. Numbers and placement of stations: Assume about 10 seismic stations for good coverage of the Moon. The number of stations needs to be 7 or more, both to fix and to confirm the 6 nugget trajectory parameters. Multiplicity of Earth stations tends to compensate particularly for absence of stations in areas of oceans. For the present we take the rough approximation that the two cancel. We believe that this approximation is conservative in the sense that it likely favors the Earth and penalizes the Moon.

4. Earthquake backgrounds. Anderson et al.[9] found it desirable to remove signals from 1/3 of the minutes in the year because of the difficulty identifying reverberations within an hour of a large quake. Though this cut is probably not necessary for the Moon with low seismic activity, there is a ringing phenomenon that may need a similar treatment.

5. Ocean and atmospheric noise. This is a very important factor for the Earth, but is almost completely absent for the Moon. Seismic detection on the Moon is only limited by instrument noise and the ringing cited just above. The relative contributions of atmospheric noise and instrument noise (ocean noise is less than atmospheric) for the Earth is unknown. We

estimate conservatively that atmospheric (amplitude) noise is the greater effect by one or possibly two orders of magnitude. 6. Attenuation with distance. Since seismic energy falls, as with other forms of energy, with distance as r^{-2}, seismic amplitude falls as $1/r$. This makes Earth seismic signals received at a station weaker, on average, than those received at lunar stations by 0.273 (the ratio of the radii) .

7. Iron core blackout. We compute for the Earth the volume of the cone segment $z \sim [r_E - r_C/2, 2r_E]$ (where the ratio of the core radius, r_C to the Earth radius r_E is about 0.5). Seismic signals from or through this volume will not reach a station at $z = 0$ in reliably predictable times because there is no reliable way of following propagation through the core. The result, weighting with the attenuation factor, is that about one third of signals are eliminated for the Earth. There is controversy with regard to a possible relatively small iron core for the Moon. We assume/approximate that none is present.

8. Nugget abundance. Based on factors 5 and 6 above, we assume that a lunar seismometer system would detect signals of amplitude at least an order of magnitude less than those on Earth. By Equation (1), signal amplitude goes as $m^{1/3}$. This means that lunar detection of masses reaches, on average, three orders of magnitude less than those detectable on Earth. We can estimate (or assume) SQN abundance as a function of mass following Dohnanyi[14]; see also Tanaka, Inaba and Nohezawa[15]. They show that a distribution in radius of collision ejecta bodies $dN/da \sim a^{-7/2}$ is self replicating. Subsequent collisions maintain the same distribution. Since $m \sim a^3$ this implies $dN/dm \sim m^{-11/6}$ and $N = \int dN/dm \sim m^{-5/6}$.

This gives, for incident detectable nuggets,

$$N_{inc}(Moon)/N_{inc}(Earth) \sim [m_{min}(Earth)/m_{min}(Moon)]^{5/6} \sim 10^{5/2} \tag{2}$$

Putting together the factors from items 1, 4, 7, and 8, we have for events expected to be detected

$$N_{det}(Moon/N_{det}(Earth) = [0.075/(2/3)^2]10^{5/2} \sim 50 \tag{3}$$

Instrument noise might be, or might be able to be made, two orders of magnitude below Earth background, since we know that it is at least a factor of five below. If that were the case, the factor estimated in Equation (3) would grow from 50 to 15,000. A seismic system on the Moon could then measure kilogram rather than ton range nuggets, and there are likely to be many more of those.

In summary, sufficient seismometers, properly placed, on the Moon should be able to detect SQNs of significantly lower mass than Earth stations, and lower mass nuggets should be significantly more plentiful than higher mass ones. The Exploration Initiative may well discover strange quark matter — if it exists.

References

1. E. Witten, *Phys. Rev* **D30**, 279 (1984).
2. E. Farhi and R.L. Jaffe, *Phys. Rev.* **D30**, 2379 (1984).
3. A. de Rujula and S. Glashow, *Nature* **312**, 734 (1984).
4. J.J. Drake, H.L. Marshall, S. Dreizler, P.E. Freeman, A. Fruscione, M. Juda, V. Kashyap, F. Nicastro, D.O. Pease, B.J. Wargelin, and K. Werner, *Astrophys. J.* **572**, 996 (2002).
5. P.O. Slane, D.J. Hefland and S.S. Murray, *Astrophys. J.* **571**, L45 (2002).
6. F.M. Walter and J. Lattimer, *astro-ph/0204199*, (2002).
7. A.R. Zhitnitsky, *hep-ph/0202161*, (2002).
8. E.T. Herrin and V.L. Teplitz, *Phys. Rev.* **D53**, 6762 (1996).
9. D.P. Anderson, E.T. Herrin, V.L. Teplitz and I.M. Tibuleac, *Bull. Seis. Soc. of Am.* **93**, 2363 (2003).
10. P.B. Price, *Phys. Rev.* **D38**, 3813 (1988).
11. T.A. Armstrong *et al. Phys. Rev.* **C63**, 054903 (2001).
12. J. Sandweiss, *J. Phys. G: Nucl. Part. Phys.* **30**, S51 (2004).
13. Y. Nakamura, *Proc. Lunar Planet. Sci. Conf. 11th*, 1847 (1980).
14. J. Dohnanyi, *J. Geophys. Res.* **74**, 2531 (1969).
15. H. Tanaka, S. Inaba and K. Nakazawa, *Icarus* **123**, 450 (1995).

DIRECT NEUTRINO MASS EXPERIMENTS

K. EITEL

Forschungszentrum Karlsruhe, Institut für Kernphysik,
P.O. Box 3640, 76021 Karlsruhe, Germany
e-mail: klaus.eitel@ik.fzk.de

The determination of the absolute neutrino mass scale is of fundamental importance in particle physics and cosmology. Apart from indirect methods (large scale structure formation in cosmology, time-of-flight measurements of Supernova neutrinos, searching for $0\nu\beta\beta$ decay) there is only one model-independent strategy to address the neutrino mass directly: investigating the kinematics of electrons from β decay. We present an overview of the experimental strategies as well as results from experiments investigating the kinematics of β decay. Activities to reach sub-eV sensitivity in the near future by analysing the electron energy of the ^{3}H β decay near its endpoint are reported.

1. INTRODUCTION

There are compelling experimental results from solar, atmospheric, reactor and accelerator neutrino experiments underlining the existence of neutrino oscillations [1]. Directly linked with the phenomenon of flavor oscillations of neutrinos is the fact that neutrinos are massive, i.e. a neutrino flavor eigenstate ν_α can be expressed as superposition of mass eigenstates $\nu_\alpha = \sum_i U_{\alpha i} \nu_i$ with non-zero mass eigenvalues $m(\nu_i)$. Therefore it is nowadays evident that the Standard Model (SM) needs to be extended to account for massive neutrinos. Since experiments investigating neutrino oscillations are only sensitive to differences $\Delta m_\nu^2 = |m^2(\nu_i) - m^2(\nu_j)|$ and not to the absolute values, the scale of neutrino masses is not determined yet. Two regimes are possible depending on the value of m_1, either $m_1 \ll m_2 \ll m_3$ ("hierarchical") or $m_1 \approx m_2 \approx m_3$ ("quasi-degenerate"). An identification of the mass scheme realized in nature would not only solve the puzzle of absolute neutrino masses but could also point to the mechanisms of mass generation in extensions of the SM.

On the other hand, in standard cosmological models, our universe is filled with primordial neutrinos arising from the freeze-out in the early universe. These neutrinos are a natural candidate for non-baryonic hot dark matter since

$$\Omega_\nu\, h^2 \approx \sum_i m(\nu_i)/93.5\,\text{eV} \qquad (1)$$

with the neutrino mass density Ω_ν expressed in terms of the critical density Ω, the Hubble constant h (in units of $100\,\text{km/s/Mpc}$, $h \approx 0.71$) and the neutrino masses in units of eV. For example, neutrinos with $m_1 \geq 1\,\text{eV}$ would make up at least $\Omega_\nu \geq 6\,\%$ of the energy content of the universe and thereby exceed the baryonic mass density.

The above arguments demonstrate the importance of the absolute neutrino mass scale for both particle physics as well as astrophysics and cosmology.

2. STRATEGIES TO DETERMINE THE ν MASS

There are 4 major approaches to determine the mass scale of neutrinos using very different sources of neutrinos as well as methods to determine the mass and which are therefore complementary. They comprise

- the deduction of the overall neutrino mass density from the formation of large scale structures as well as from anisotropies in the cosmic microwave background (CMB) using the primordial neutrino mass density relation (1)
- the measurement of the spread of arrival times of supernova neutrinos due to different time-of-flight depending on neutrino energies and masses as has been done for supernova SN1987a
- the search for neutrinoless double beta ($0\nu\beta\beta$) decay which existence would show the Majorana type of neutrinos and which decay rate would correlate with the "effective" Majorana neutrino mass $m_{ee} = |\sum_i U_{ei}^2 \cdot m(\nu_i)|$ as a coherent sum over all mass eigenstates contributing to the electron neutrino with fraction U_{ei}
- the kinematical analysis of electrons from β decay near the endpoint energy E_0. A non-vanishing neutrino mass reduces the electron endpoint energy as well as distorts the shape of the electron energy spectrum in the vicinity of $E_0 - m(\nu)$.

Combining data on CMB, large scale structure formation from surveys such as 2dF and SDSS as well as on Ly-α-forests leads to a stringent upper limit of $\sum_i m(\nu_i) < 0.69\,\text{eV}$ (95% C.L.) [2]. However, changing the data base (using data on X-ray galaxy clusters instead of Ly-α-forests) might lead to significantly different results concerning the neutrino mass, i.e. $\sum_i m(\nu_i) = 0.56^{+0.30}_{-0.26} = \text{eV}$ [3]. Both compilations assume the existence of primordial neutrinos in a cosmological concordance model, but allowing for new interactions of neutrinos, the abundance of primordial neutrinos might be strongly suppressed and the above limits on the neutrino mass considerably relaxed [4]. All this demonstrates the model dependency of this indirect measurement of $\sum_i m(\nu_i)$ which is discussed in more detail in [5].

The second approach requires the explosion of a supernova within our galaxy. Though one might expect hundreds to thousands of neutrino events in detectors such as SuperK, KamLAND and SNO, a sub-eV precision on $\sum_i m(\nu_i)$ seems unrealistic due to the limited understanding of the emission time spectrum [6].

The search for $0\nu\beta\beta$ decay is a very promising strategy to determine the neutrino mass. It requires the neutrino to be a Majorana type fermion. In addition, the decay rate depends on the complex phases of U_{ei}^2 which might lead to cancellation effects and might be influenced by other non-SM processes such as Majoron exchange. There is an evidence from $0\nu\beta\beta$ decay search with ^{76}Ge pointing to an effective mass of $m_{ee} = 0.4^{+0.5}_{-0.3}$ eV (3σ) [7] including a 50% uncertainty for the nuclear matrix element. This clearly favors a quasi-degenerate neutrino mass scheme, with neutrino masses being of cosmological relevance as hot Dark Matter contribution. For further discussion of $0\nu\beta\beta$ decay searches we refer to [8].

The direct kinematical investigation of single β decay electron spectra leads to a model-independent direct neutrino mass measurement and is therefore complementary to the approaches discussed above and has also a great potential to cross-check the $0\nu\beta\beta$ evidence reported above.

3. DIRECT NEUTRINO MASS MEASUREMENTS

In the following, we will concentrate on two specific methods, investigating the endpoint region of the electron spectrum of the Tritium β decay $^3\text{H} \to {}^3\text{He} + e^- + \bar{\nu}_e$ using an electrostatic filter, and the measurement of electron energies from $^{187}\text{Re} \to {}^{187}\text{Os} + e^- + \bar{\nu}_e$ with micro-calorimeters. The spectrum N of the electron energy E_e can be expressed as

$$N(E_e) = K \cdot F(Z, E_e) \cdot p \cdot (E_0 - E_e) \\ \cdot \sqrt{(E_0 - E_e)^2 - m^2(\nu_e)} \qquad (2)$$

with the nuclear matrix element K, the Fermi correction function F, the charge number Z of the daughter nucleus, the electron momentum p, the

β decay endpoint energy E_0 and the neutrino mass observable

$$m^2(\nu_e) = \sum_i |U_{ei}^2| \cdot m^2(\nu_i) \qquad (3)$$

3.1. ^3H decay with electrostatic filters

The scheme of a measurement of the electron energy from Tritium decay with an electrostatic filter is illustrated in figure 1. The electrons emitted

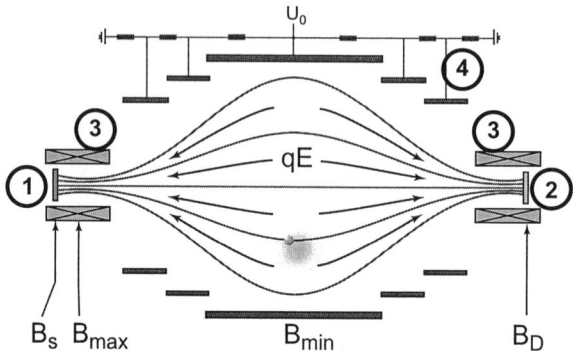

Figure 1. Scheme of a MAC-E-filter with Tritium source (1), the electron detector (2), the superconducting solenoids (3) and the HV electrode system (4).

in β decay are guided magnetically and collimated due to the magnetic field gradient to have perpendicular momentum relative to an electrostatic retarding field in which only electrons with energy greater than qE pass the electrostatic barrier and are then focused towards a detector. This magnetic adiabatic collimation followed by the electrostatic (MAC-E) filter represents an integrating high pass filter for β-electrons. This technique is used in the Mainz and Troitsk experiments with different Tritium sources, a thin film of molecular Tritium quench-condensed on a cold graphite substrate for Mainz and a windowless gaseous Tritium source (WGTS) for Troitsk.

In figure 2 we show the count rate of the Mainz experiment for the 3 measuring periods of 1994, 1998/99 and 2001. The enhancement of signal as well as the suppression of background of the 98/99 data compared with earlier runs is obvious. The final analysis [9] combines the 98/99 with 2001 data and results in

$$m^2(\nu_e) = -0.7 \pm 2.2 \pm 2.1 \, \text{eV}^2 \qquad (4)$$

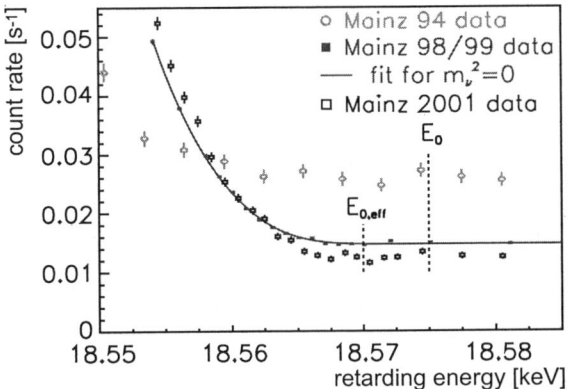

Figure 2. Mainz data points near the spectral endpoint E_0, taken within the 3 measuring periods of 1994, 1998/99 and 2001.

denoting statistical and systematical uncertainties, respectively, which corresponds to an upper limit of

$$m(\nu_e) < 2.3 \,\text{eV} (95\% \,\text{C.L.}). \tag{5}$$

During the same measurement period, the Troitsk experiment collected data showing a small anomaly near the endpoint, with fluctuating amplitude and position relative to the endpoint [10]. Accounting for this anomaly as a single additional line in the β spectrum, the analysis yields [11]

$$m^2(\nu_e) = -2.3 \pm 2.5 \pm 2.0 \,\text{eV}^2 \tag{6}$$

which corresponds to an upper limit of

$$m(\nu_e) < 2.05 \,\text{eV} (95\% \,\text{C.L.}). \tag{7}$$

With the unprecedented understanding of various systematic effects, the Mainz experiment has finally reached its sensitivity limit (as it is mainly the case for Troitsk). The two experiments yield the most stringent direct upper limits of the neutrino mass $m(\nu_e)$ so far.

3.2. ^{187}Re decay with micro-calorimeters

A very different and therefore complementary approach is the measurement of the β spectrum with micro-calorimeters where source and detector are identical. Though not yet competitive with the sensitivity of Tritium experiments, this technique also investigates potential spectral distortions far from the endpoint. Figure 3 shows a schematic view of the experimental

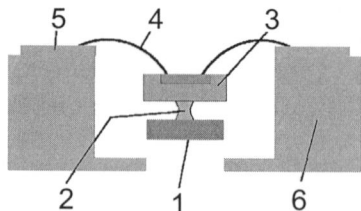

Figure 3. Scheme of the Mibeta micro-calorimeter set-up with the AgReO$_4$-crystal (1), the epoxy joint (2), silicon implanted thermistor (3), Al bonding wires as thermal link (4), the bonding pad as heat sink (5) and the OFHC Cu holder (6).

setup of the micro-calorimeter using a AgReO$_4$-crystal at operating temperature of $T = 70 - 100$ mK looking for a thermal signal from the β decay ^{187}Re $\rightarrow^{187}$ Os $+ e^- + \bar{\nu}_e$ with its endpoint energy of $E_0 = 2.46$ keV. Having taken data with 10 crystals of about 250 μg of individual mass for a total exposure time of 8751 hours×mg of AgReO$_4$, the Mibeta experiment yields the electron spectrum of $6.2 \cdot 10^6$ ^{187}Re β decays above 700 eV as shown in figure 4. Applying a fit to the data with free parameters of E_0 and $m^2(\nu_e)$

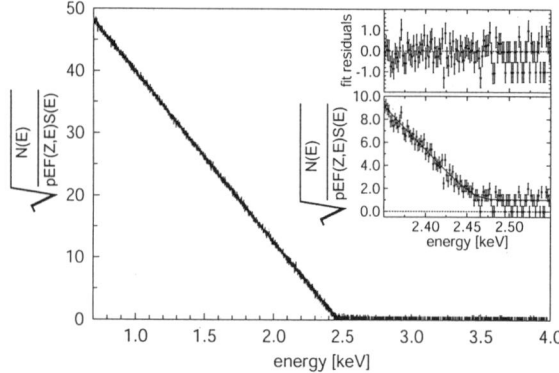

Figure 4. Mibeta Sum Kurie plot, where p and E are, respectively, the β momentum and kinetic energy, $F(Z, E)$ is the Coulomb factor and $S(E)$ is the shape factor.

yields [12]

$$E_0 = 2465.3 \pm 0.5_{\text{stat}} \pm 1.6_{\text{syst}} \text{ eV} \qquad (8)$$

$$m^2(\nu_e) = -112 \pm 207 \pm 90 \text{ eV}^2 \qquad (9)$$

leading to an upper limit of $m(\nu_e) < 15$ eV (90% C.L.). There are proposals to improve the sensitivity to values of 2-3 eV within the next years and to tackle problems of signal pile-ups and partial energy loss near surfaces as well as systematic uncertainties of calibrating the energy response of the crystals, but a sensitivity within the sub-eV range seems extremely challenging if not impossible by this technique of using micro-calorimeter arrays.

4. NEW EXPERIMENTS WITH SUB-eV SENSITIVITY

Having discussed the different methods of kinematical investigations of the β decay, the technique of a MAC-E-filter to analyse the electron energy from Tritium β decay is the only viable technique to improve the actual sensitivity by at least an order of magnitude.

The **KA**rlsruhe **TRI**tium **N**eutrino experiment will use the technique developed in the Mainz and Troitsk experiments with a strong gaseous molecular Tritium source and an electrostatic filter of unprecedented energy resolution [13]. As the experimental observable is m_ν^2, an improvement in sensitivity on $m(\nu_e)$ of an order of magnitude corresponds to an improvement of a factor ≈ 100 in accuracy on m_ν^2 which leads to numerous technical challenges for the experimental set-up as will be shown in the following.

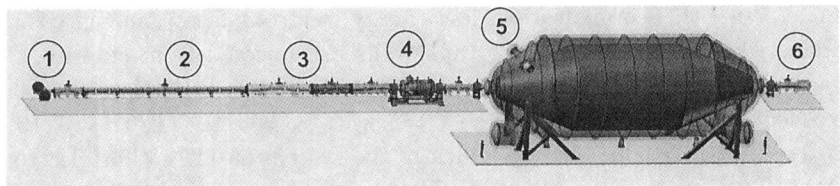

Figure 5. Scheme of the KATRIN experimental setup with the rear pump port (1), the WGTS (2), the differential- and cry-pumping section (3), the pre-spectrometer (4), the main spectrometer (5) and the electron detector array (6).

Figure 5 outlines the experimental set-up of KATRIN with the windowless gaseous Tritium source (WGTS), the differential- and cryo-pumping section within a transport section of superconducting solenoids, the tandem of electrostatic filters and the electron detector with good energy and spatial resolution. The entire set-up from the rear pump port to the detector section including the main spectrometer vessel of 10m diameter and 24m length will add up to a total length of 70m.

The WGTS will consist of a 10m long tube of 90mm diameter with Tritium injection of an injection pressure of $p_{in} \approx 3 \cdot 10^{-3}$ mbar in the center of the tube. The isotopic purity of the gas will be > 95% of Tritium. The KATRIN source strength exceeds the Mainz and Troitsk sources by a factor of ≈ 100. As an option for an alternative source, a port for a Mainz-type QCTS is foreseen in the transport section. Whereas the β-electrons are guided by a strong ($\leq 6\,\mathrm{T}$) longitudinal magnetic field into the spectrometers, the gas molecules are pumped by turbomolecular pumps and a cryogenic pumping system.

The spectrometer system consists of two ultra high vacuum ($p \leq 10^{-11}\,\mathrm{mbar}$) tanks. The pre-spectrometer with a moderate energy resolution defined by the ratio of magnetic field strengths of the maximal field in the transport section $B_{\max}$ and in the analysing plane B_A of $\Delta E/E = B_A/B_{\max} = 2 \cdot 10^{-2} T/6T \approx 60\mathrm{eV}/18575\mathrm{eV}$ will be operated with a retarding potential of $E_{\mathrm{ret.}} \approx 18.5\,\mathrm{keV}$ to reject the low energy part of the electron spectrum, thereby reducing the flux of β-electrons entering the main spectrometer from a few $10^{10}/\mathrm{s}$ by a factor of 10^7, avoiding potential background from scattering processes of low energy electrons with the rest gas in the large volume of the main spectrometer tank.

In the main spectrometer, the high resolution scanning of the electron energy will take place. The retarding potential with high voltage of the order of $18.6\,\mathrm{kV}$ is applied directly on the tank itself. Near the inner tank surface, there will be an additional wire electrode with $\Delta U = U_{\mathrm{tank}} - U_{\mathrm{wire}} \approx 100\,\mathrm{V}$. With this configuration, low energy electrons from ionizing cosmic muons which pass through the tank walls are rejected from entering the magnetic flux tube in which electrons are guided towards the detector. The excellent energy resolution of $\Delta E/E = B_A/B_{\max} = 3 \cdot 10^{-4} T/6T = 5 \cdot 10^{-5}$ corresponds to a transmission width of the system of $0.93\,\mathrm{eV}$ for electrons near the endpoint $E_0 = 18575\,\mathrm{eV}$. The electrons passing the electrostatic potential of the main spectrometer will finally be detected with a segmented Si-PIN diode detector array.

Figure 6 shows a calculation of the expected accuracy of KATRIN on m_ν^2 for a data taking of three years. Shown are the statistical uncertainties only as a function of the fit interval below the endpoint E_0 taking into account the complete transmission and response function of the experimental set-up as well as the final state distribution of the $(^3\mathrm{HeT})^+$ ionized daughter molecules. Various steps of optimization of the KATRIN configuration can be seen: the increase of source cross section corresponding to an enlargement of the main spectrometer's diameter from 7m to 10m, a higher $^3\mathrm{H}$ isotopic purity, an optimization of the measuring time spent per scanning high voltage near the endpoint as well as a reduction of the expected overall background rate from $10\,\mathrm{mHz}$ to $1\,\mathrm{mHz}$. The third line (triangles) in fig. 6 corresponds to the actual KATRIN reference set-up.

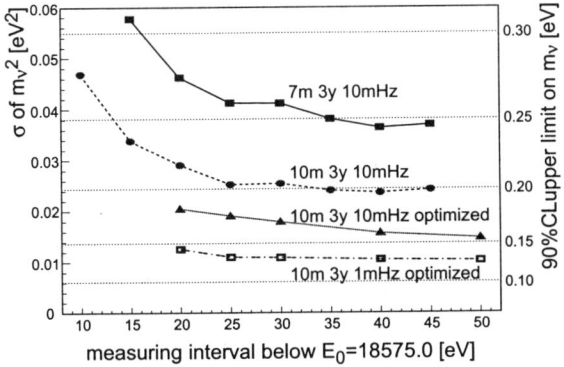

Figure 6. Statistical accuracy on the observable m_ν^2 and 90% C.L. upper limit on m_ν expected with KATRIN depending on different intervals of analysis.

Systematic uncertainties such as time variation of WGTS parameters, the determination of scattering probabilities of β-electrons within the WGTS, description of the final state distribution of $(^3\text{HeT})^+$, HV variations, limited uniformity of the magnetic and electrostatic fields in the spectrometer analysing plane are expected to amount in total to equal size compared with the shown statistical accuracy for a measuring time of 3 years and an analysing interval of 30 eV below the endpoint. This would lead to a sensitivity of the KATRIN experiment on the electron neutrino mass of

$$m(\nu_e) < 0.2\,\text{eV}\,(90\%\,\text{C.L.}) \qquad (10)$$

in case of a zero-signal measurement. The discovery potential of KATRIN is demonstrated by the separation of a neutrino mass of $m(\nu_e) = 0.35\,\text{eV}$ from zero with a statistical significance of 5σ.

4.1. Status and schedule of the KATRIN experiment

The KATRIN experiment will be built on site of the Forschungszentrum Karlsruhe, Germany. At the moment, first components such as the pre-spectrometer, magnets and a pre-spectrometer prototype detector array are available and under tests. A series of vacuum tests have been undertaken serving as tests of the pre-spectrometer itself but also as prototype studies for the main spectrometer tank. Prototype experiments of the Tritium gas circulation system as well as the cryogenic pumping section are under way. The WGTS and the main spectrometer vessel have been fully specified and will be ordered end of 2004. Commissioning of the whole set-up and the start of data taking with KATRIN is expected in 2007/2008.

5. CONCLUSIONS AND OUTLOOK

In the presence of the latest results on neutrino oscillations, the question of absolute neutrino mass scale becomes one of the most important issues, both for particle physics as well as cosmology. Although there are different approaches to address this topic, only the kinematical investigation of β decay is fully model-independent. Experiments investigating the Tritium decay with MAC-E-filters set the most stringent limits for the neutrino mass, in fact due to the mixing and the small mass differences, this kinematical limit on $m(\nu_e) < 2.3\,\text{eV}$ also holds for ν_μ and ν_τ. In the near future, the sensitivity on $m(\nu_e)$ from Tritium decay will be improved by another order of magnitude with the KATRIN experiment. Thereby, it will be possible to investigate independently from astrophysical experiments the rôle of neutrinos as candidates for the Hot Dark Matter content of the universe.

6. ACKNOWLEDGEMENTS

The author wishes to thank many colleagues from different collaborations for providing information presented here. The KATRIN experiment, in which the author is involved, is supported by the German Bundesministerium für Bildung und Forschung under contracts 05CK1VK1/7, 05CK1UM1/5 and 05CK2PD1/5.

References

1. Summary by e.g. S.T. Petcov, Neutrino2004 conference proceedings, to be published in Nuclear Physics B (Proceedings Supplement)
2. D. Spergel et al., (WMAP), Astrophys. J. Suppl. **148** (2003) 175
3. S. Allen et al., Mon. Not. Roy. Astron. Soc. 346 (2003) 593
4. J. Beacom et al., `astro-ph/0404585`
5. O. Lahav, Neutrino2004 conference proceedings, to be published in Nuclear Physics B (Proceedings Supplement)
6. P. Vogel, Prog. Part. Nucl. Phys. 48 (2002) 29
7. H.V. Klapdor-Kleingrothaus et al., Nucl. Instr. and Meth. **A522** (2004) 371
8. A. Nucciotti, H.V. Klapdor-Kleingrothaus, *these proceedings*
9. C. Kraus et al., Eur. Phys. J. **C33**, s01 (2004)
10. V. Lobashev et al., Nucl. Phys. B (Proc. Suppl.) **91**, 280 (2001)
11. V. Lobashev, Nucl. Phys. **A719** (2003) 153c
12. M. Sisti et al., Nucl. Instr. and Meth. **A520** (2004) 125

13. A. Osipowicz et al., (KATRIN collab.) hep-ex/0109033, FZKA report 6691, T. Thümmler et al., (KATRIN collab.), FZKA report 6752

THE SEARCH FOR θ_{13} WITH THE DOUBLE-CHOOZ EXPERIMENT

J.JOCHUM

Physikalisches Institut I, Universität Tübingen,
Auf der Morgenstelle 14,
72076 Tübingen, Germany
E-mail: josef.jochum@uni-tuebingen.de

The Double-Chooz experiment goal is to search for a non-vanishing value of the θ_{13} neutrino mixing angle. This is the last step to accomplish prior moving towards a new era of precision measurements in the lepton sector. The current best constraint on the third mixing angle comes from the CHOOZ reactor neutrino experiment $\sin(2\theta_{13})^2 < 0.2$ (90% C.L., $\Delta m_{atm}^2 = 2.0$ eV^2). Double-Chooz will explore the range of $\sin(2\theta_{13})^2$ from 0.2 to 0.03-0.02, within three years of data taking. The improvement of the CHOOZ result requires an increase in the statistics, a reduction of the systematic error below one percent, and a careful control of the backgrounds. Therefore, Double-Chooz will use two identical detectors, one at 150 m and another at 1.05 km distance from the Chooz nuclear cores. In addition, the near detector as a "state of the art" prototype will be used to investigate the potential of neutrinos for monitoring the civil nuclear power plants. The plan is to start operation with two detectors in 2008, and to reach a sensitivity $\sin^2(2\theta_{13})$ of 0.05 in 2009, and 0.03-0.02 in 2011.

1. Motivation

Neutrino flavor transitions have been observed in atmospheric, solar, reactor and accelerator neutrino experiments. To explain these transitions, extensions to the minimal Standard Model of particle physics are required. The simplest and most widely accepted extension is to allow neutrinos to have masses and mixing, similar to the quark sector. The flavor transitions can then be explained by neutrino oscillations.

The neutrino oscillation data can be described within a three neutrino mixing scheme, in which the flavor states ν_α ($\alpha = e, \mu, \tau$) are related to the mass states ν_i ($i = 1, 2, 3$) through the PMNS (Pontecorvo-Maki-Nakagawa-Sakata) unitary lepton mixing matrix.

It can be parameterized as $U_{\text{PMNS}} =$

$$\begin{pmatrix} 1 & & \\ & c_{23} & s_{23} \\ & -s_{23} & c_{23} \end{pmatrix} \begin{pmatrix} c_{13} & & s_{13}e^{-i\delta} \\ & 1 & \\ -s_{13}e^{i\delta} & & c_{13} \end{pmatrix} \begin{pmatrix} c_{12} & s_{12} & \\ -s_{12} & c_{12} & \\ & & 1 \end{pmatrix} \begin{pmatrix} 1 & & \\ & e^{i\alpha} & \\ & & e^{i\beta} \end{pmatrix} \quad (1)$$

where $c_{ij} = \cos\theta_{ij}$ and $s_{ij} = \sin\theta_{ij}$, δ is a Dirac CP violating phase, α and β are Majorana CP violating phases, not considered in the following. Up to now, the angles θ_{12} and θ_{23} are probed via the oscillations of solar/reactor and atmospheric neutrinos, while the angle θ_{13} is mainly constrained by the CHOOZ reactor experiment; the Dirac phase δ has not been constrained yet.

Thanks to the smallness of $\frac{\Delta m^2_{\text{sol}}}{\Delta m^2_{\text{atm}}}$ and $\sin^2\theta_{\text{CHOOZ}}$ the factorized form of this PMNS mixing matrix is often used to identify the mixing angles reported by the experiments

$$\theta_{23} \cong \theta_{\text{atm}}, \quad \theta_{12} \cong \theta_{\text{sol}}, \quad \text{and} \quad \theta_{13} \cong \theta_{\text{CHOOZ}}. \quad (2)$$

The relevant formula for the oscillation probabilities is

$$P(\nu_\alpha \to \nu_\beta) = \delta_{\alpha\beta} - 2\,\text{Re}\sum_{j>i} U_{\alpha i} U^*_{\alpha j} U^*_{\beta i} U_{\beta j} \left(1 - \exp\frac{i\Delta m^2_{ji} L}{2E}\right), \quad (3)$$

where $\Delta m^2_{ji} = m^2_j - m^2_i$.

Since the identification of the MSW-LMA mechanism as the solution of the solar neutrino anomaly [1,2,3], we now know that the mass eigenstate with the larger electron neutrino component has the smaller mass (state 1). Solar neutrino oscillations occur then mainly together with the little heavier state 2:

$$\Delta m^2_{21} = m^2_2 - m^2_1 \equiv \Delta m^2_{\text{sol}} > 0. \quad (4)$$

The large mass squared difference measured in the atmospheric sector is therefore the splitting between the mass eigenstate 3 and the more closely spaced 1 or 2. In addition, the CHOOZ reactor neutrino experiment shows that the mass eigenstate 3 has only a very small electron neutrino component. In this description, the sign of the splitting between state 3 and states 1 and 2 is unknown; this leads to two possibilities of mass ordering:

$$|\Delta m^2_{32}| = |m^2_3 - m^2_2| \equiv \Delta m^2_{\text{atm}}. \quad (5)$$

Thus, one defines the normal hierarchy (NH) scenario $m_3 > m_2 > m_1$, and the inverted hierarchy scenario (IH) $m_2 > m_1 > m_3$. The determination of the sign of Δm_{32}^2 is one of the next goals in neutrino oscillation physics.

Reactor neutrino experiments measure the disappearance of the electron antineutrinos emitted from the nuclear power plant. The low neutrino energy (a few MeV) does not allow the measurement of any appearance of another neutrino flavor. The probability for the disappearance of electron antineutrinos does not depend on the δ-CP phase. Furthermore, because of the low energy as well as the short baseline considered, matter effects are negligible [4].

According to Eq. (3) the probability for disappearance via the mixing of mass eigenstates i and k is proportional to

$$\propto \sin^2(2\theta_{ik}) \sin^2(\Delta m_{ik}^2 L/4E) \qquad (6)$$

The disappearance of reactor neutrinos at longer distances of $L > 50km$ as measured by the KAMLAND experiment is due to the mixing via θ_{12} and the small Δm_{sol}^2. With a non zero third mixing angle θ_{13}, there would be an additional small modulation of the neutrino flux. Due to the larger mass difference $\Delta m_{13}^2 \approx \Delta m_{atm}^2$ the oscillation length of this modulation would be about $1km - 2km$. The amplitude is given by $\sin^2(2\theta_{13})$, and is very small. The goal of the Double-Chooz experiment is to search for a non-vanishing value of θ_{13} by looking for such a oscillation.

The current range (90 % error intervals) of mixing parameters found in neutrino oscillation experiments are[a] [5,6]:

$$(\Delta m_{atm}^2)_{\text{SK-I}} = 2.0^{+1}_{-0.7} \cdot 10^{-3} \text{ eV}^2$$
$$(\sin^2 2\theta_{23})_{\text{SK-I}} = 1^{+0}_{-0.1}$$
$$(\Delta m_{atm}^2)_{\text{SK-L/E}} = 2.4^{+0.6}_{-0.5} \cdot 10^{-3} \text{ eV}^2$$
$$(\sin^2 2\theta_{23})_{\text{SK-L/E}} = 1^{+0}_{-0.1}$$
$$\Delta m_{sol}^2 = 7.0^{+2}_{-3} \cdot 10^{-5} \text{ eV}^2$$
$$\sin^2(2\theta_{12}) = 0.8^{+0.2}_{-0.2} \ .$$

Reactor experiments can provide a clean measurement of the mixing angle θ_{13}, free from any contamination coming from matter effects and other parameter correlations or degeneracies [4,7]. Therefore they are exclusively dominated by statistical and systematic errors.

---

[a]Two different best fit values for the atmospheric mass splitting have been released by the Super-Kamiokande collaboration, based on two different analyzes of the same data.

2. $\bar{\nu}_e$ detection principle

Reactor antineutrinos are detected through their interaction by inverse neutron decay (threshold of 1.806 MeV)

$$\bar{\nu}_e + p \to e^+ + n \ . \tag{7}$$

The cross section for inverse β-decay has approximately the form

$$\sigma(E_{e^+}) \simeq \frac{2\pi^2 \hbar^3}{m_e^5 f \tau_n} p_{e^+} E_{e^+} \ , \tag{8}$$

where p_{e^+} and E_{e^+} are the momentum and the total energy of the positron, τ_n is the lifetime of a free neutron and f is the free neutron decay phase space factor. As an approximation, we use an averaged fuel composition typical during a reactor cycle corresponding to ^{235}U (55.6 %), ^{239}Pu (32.6 %), ^{238}U (7.1 %) and ^{241}Pu (4.7 %). The mean energy release per fission W is then 203.87 MeV and the energy weighted cross section amounts to

$$<\sigma>_{\text{fission}} = 5.825 \cdot 10^{-43} \text{ cm}^2 \text{ per fission} \ . \tag{9}$$

The reactor power P_{th} is related to the number of fissions per second N_f by

$$N_f = 6.241 \cdot 10^{18} \text{sec}^{-1} \cdot (P_{th}[\text{MW}])/(W[\text{MeV}]) \ . \tag{10}$$

The event rate at a distance L from the source, assuming no oscillations, is thus

$$R_L = N_f \cdot <\sigma>_{\text{fission}} \cdot n_p \cdot 1/(4\pi L^2) \ , \tag{11}$$

where n_p is the number of protons in the target. For the purpose of simple scaling, a reactor with a power of 1 GW$_{th}$ induces a rate of $\sim$450 events per year in a detector containing 10^{29} protons, at a distance of 1 km.

Experimentally one takes advantage of the coincidence signal of the prompt positron followed in space and time by the delayed neutron capture. This very clear signature allows to strongly reject the accidental backgrounds. The energy of the incident antineutrino is then related to the energy of the positron by the relation

$$E_{\bar{\nu}_e} = E_{e^+} + (m_n - m_p) + O(E_{\bar{\nu}_e}/m_n) \ . \tag{12}$$

Experimentally, the visible energy seen in the detector is given by $E_{\text{vis}} = E_{e^+} + 511$ keV, where the additional 511 keV come from the annihilation of the positron with an electron when it stops in the matter.

3. The Chooz experimental site

The experimental site is located in the Ardennes (France), close to the Chooz nuclear power plant, operated by the French company Electricite de France (EDF). There are two N4 type PWR reactors of 4.27 GW_{th} each. We will use two almost identical detectors, containing a fiducial volume of 10 tons of liquid scintillator doped with 0.1% of Gadolinium (Gd). The laboratory of the first CHOOZ experiment, located 1.05 km (the *Chooz-far* site, overburden of 300 m.w.e.) from the cores will be used again. This is the main advantage of this site. In order to cancel the systematic errors originating from the nuclear reactors ($\overline{\nu_e}$ flux and energy spectrum), as well as to reduce the systematic errors, a second detector will be installed close to the nuclear cores (the *Chooz-near* site). Since no natural hills or underground cavity already exists at this location, an artificial overburden of about 20 meters height has to be built. At 150 m the required overburden to protect the detector from cosmic ray induced backgrounds is 60 m.w.e..

4. The new detector concept

The detector design is an evolution of the detector of the first experiment [8]. To improve the sensitivity of Double-Chooz with respect to CHOOZ it is planned to increase statistics and to reduce and better control the systematic errors and backgrounds. In order to increase the exposure to 60,000 events at Chooz-far (statistical error of 0.4%) it is planned to use a target cylinder of 120 cm radius and 280 cm height, providing a fiducial mass of 10 tons (12.7m^3), 2.3 times larger than in CHOOZ. In addition, the data taking period will be extended to at least three years, and the overall data taking efficiency will be improved. The near and far detectors will be identical inside the PMT supporting structure. This will allow a relative normalization systematic error of 0.6%. Starting from the center of the target the detector elements are as follows (Figure 1). The neutrino target: A 120 cm radius, 280 cm height, 8 mm thick acrylic cylinder, filled with 0.1% Gd loaded liquid scintillator. The baseline of the scintillator being developed for the new experiment is a mixture of 20% of PXE and 80% of dodecane, with small quantities of PPO and bis-MSB added as fluors. The γ-catcher: A 60 cm buffer of liquid scintillator not loaded with Gd, with the same light yield as the target. The role of this *new* region is to get the full positron energy, as well as most of the neutron energy released after neutron capture. It is enclosed in a 180 cm radius, 400 cm height, 10 mm thick acrylic cylinder. The non scintillating buffer: A 95 cm buffer of non scintillating oil, to decrease the level of accidental background (mainly the contribution from photomultiplier tubes radioactivity from potassium, uranium and thorium) and the PMT supporting structure. The outer veto: A 60 cm veto region filled with liquid scintillator for the far detector, and a slightly larger one (about 100 cm) for the near detector. The external shielding: A 15 cm steel shielding surrounding the far detector, and a $\sim$1 m low radioactive sand or water layer for the near detector.

We plan to build the double acrylic vessels at the manufacturer and transport it to the detector sites in a single piece. This integration procedure allow us to minimize the differences between the acrylic vessels as well as to reduce the residual mechanical stress that could favor the acrylic crazing.

5. Systematic errors and backgrounds

In the first CHOOZ experiment, the total systematic error amounted to 2.7%. Thanks to the use of the double detector concept, each error originating from the neutrino source, e.g. the reactors, cancels. Thus we can neglect the "reactor cross section error" of 1.9%, the uncertainty on the reactor power of 0.7%, as well as the lack of knowledge of the energy released per fission of 0.6%. The dominant error for Double-Chooz will thus be the relative normalization between the two detectors; it originates for instance from the detection efficiency, or a difference in the number of free protons contained in the acrylic targets. We focus our efforts towards the precise measurement of the relative volumes between the acrylic targets, and the dead time measurement. In addition, the position of the near detector with respect to the core will have to be measured with a precision better than 10 cm. In Double-Chooz, we estimate the total systematic error on the normalization between the detectors to be less than 0.6%. The main contributions come from the solid angle (0.2%), the volume (0.2%), the density and H/C ratio (0.15%), the neutron detection efficiency and energy measurement (0.2% and 0.1%), the e^+-n time delay (0.1%) and the dead time (0.25%, with a new method of fake triggers generated inside the target under development).

The signature for a neutrino event is a prompt signal with a minimal energy of about 1 MeV and a delayed 8 MeV signal after neutron capture by a Gd nucleus. This may be mimicked by background events which can be divided into two classes: accidental and correlated events. The former can be reduced by a careful selection of the materials used to build the detector.

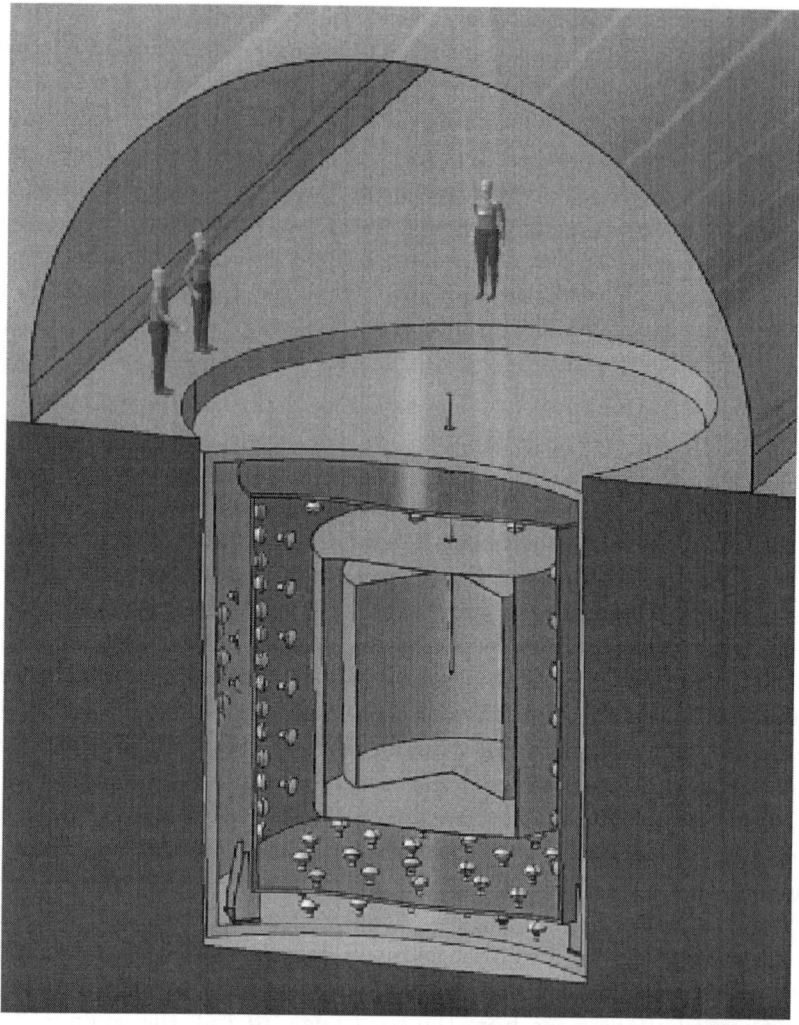

Figure 1. The detector is located in the tank used for the CHOOZ experiment (7 meters high and diameter). About 10 tons of a liquid scintillator doped with Gd is contained in a double-acrylic cylinder surrounded by the gamma-catcher region, the buffer and the muon veto. The optical coverage of the PMTs is about 15%.

In addition this background is easy to measure in-situ, and its subtraction lead to a small systemactic error. A comprehensive Monte-Carlo study shows that the correlated events are the most severe background source for the experiment. Our simulation reproduced fairly well the correlated background rate measured in the first CHOOZ experiment and is thus reliable. Two processes mainly contribute: β-neutron cascades and very fast external neutrons. Both types of events are coming from spallation processes of high energy muons. In total the background rates for the near detector will be between 9/d and 23/d, for 60 m.w.e. overburden. For the far detector a total background rate between 1/d and 2/d can be estimated. This can be compared with the signal of $\sim$ 4000/d and 80/d in the near and far detectors. The overburden of the near detector has been chosen in order to keep the signal to background ratio above 100. Under this condition, even a knowledge of the backgrounds within a factor two keeps the associated systematic error below the percent.

6. Discovery potential

The Double-Chooz experiment is searching for a deficit in the $\overline{\nu_e}$ flux at 1.05 km from the cores, while the near detector monitors the $\overline{\nu_e}$ flux and energy spectrum prior to any neutrino oscillation. This disappearance channel allows a "clean" measurement of $\sin(2\theta_{13})^2$. Assuming a relative normalization error of 0.6%, and three years of data taking with typical live time for both reactor and detector operations, the sensitivity will be $\sin(2\theta_{13})^2 < 0.025$ (at 90% C.L., for $\Delta m^2_{atm} = 2.4$ eV^2) in the case of no-oscillation. The discovery potential with a so-called "3-σ" effect will be around 0.04. For a true value $\sin(2\theta_{13})^2 = 0.1$ a rate only analysis will reject the no-oscillation scenario at 2.6σ, whereas a shape+rate analysis will reject the no-oscillation scenario at about 6σ. For a true value $\sin(2\theta_{13})^2 = 0.1$, Double-Chooz would perform a measurement of the oscillation parameter with a 67% relative error (90% C.L.); this can be compared with the potential of 100-130% of the complementary superbeam experiments.

7. Outlooks

The Double-Chooz collaboration is composed of about 16 institutes (in France, Germany, Italy, Russia and USA). A Letter of Intent has been released in may 2004 [9], and the experiment has been approved in France. The funding of the near laboratory in partnership with the EDF power company and the local authorities is currently being discussed. If fully approved in 2005, it is intended to start taking data at Chooz-far in 2007, and at Chooz-near in 2008. If the collaboration meets this goal, Double-Chooz could provide a sensitivity limit of $\sin(2\theta_{13})^2 < 0.05$ (at 90% C.L.) within the year 2009, and 0.02-0.03 in 2011.

References

1. S. Fukuda *et al.* (Super-Kamiokande Collaboration), Phys. Lett. B539, 179 (2002).
2. Q. R. Ahmad *et al.* (SNO Collaboration), Phys. Rev. Lett. 89, 011301 (2002).
3. K. Eguchi *et al.*, (KamLAND Collaboration), Phys. Rev. Lett. 90, 021802 (2003).
4. H. Minakata, H. Sugiyama, O. Yasuda, K. Inoue and F. Suekane, hep-ph/0211111 (2002)
5. M. Shiozawa (Super-Kamiokande Collaboration), Talk given at Neutrino 2002 conference, Munich, Germany. http://neutrino2002.ph.tum.de.
6. M. Ishitsuka, Ph.D. Thesis, University of Tokyo (2004).
7. P. Huber, M. Lindner, T. Schwetz and W. Winter, Nucl. Phys. B665, 487 (2003).
8. M. Apollonio, *et. al* (CHOOZ Collaboration), Eur. Phys. J. C27, 331 (2003).
9. F. Ardellier, *et. al* (Double-Chooz Collaboration), hep-exp/0405032 (2004).

NOSTOS: A NEW LOW-ENERGY NEUTRINO EXPERIMENT

S. AUNE[a], J. BUSTO[b], P. COLAS[a], J. DOLBEAU[c], G. FANOURAKIS[d],
E. FERRER RIBAS[a], H. VAN DER GRAAF[e], T. GERALIS[d],
Y. GIOMATARIS[a,*] P. GORODETZKY[c], G. J. GOUNARIS[f],
I. G. IRASTORZA[a], K. KOUSOURIS[d], V. LEPELTIER[g], T. PATZAK[c], E. A. PASCHOS[h], B. PEYAUD[a], P. SALIN[c], J. D. VERGADOS[i]

[a] DAPNIA, Centre d'Etudes de Saclay (CEA-Saclay), Gif-Sur-Yvette, France
[b] CCPM-Faculté des Sciences de Luminy, Marseille, France
[c] IN2P3/CNRS PCC-Collège de France, Paris, France
[d] National Center for Scientific Research "Demokritos" (NRCPS), Athens, Greece
[e] NIKHEF, Amsterdam, The Netherlands
[f] Aristotle University of Thessaloníki, Thessaloníki, Greece
[g] Laboratoire de l'Accélerateur Linéaire, Orsay, France
[h] Universität Dortmund, Institut für Physik, Dortmund, Germany
[i] University of Ioannina T.P.D., Ioannina, Greece

A novel low-energy (a few keV) neutrino-oscillation experiment NOSTOS, combining a strong tritium source and a high pressure spherical TPC detector of 10 m in radius has been recently proposed. The neutrino oscillation of such energies occurs within the detector itself enabling a very precise measurement of the oscillation parameters, and in particular the mixing angle θ_{13}. This detector could also be sensitive to the neutrino magnetic moment and capable of accurately measure the Weinberg angle at very low energy transfer. The same apparatus filled with high pressure Xenon, would exhibit a high sensitivity as a supernova neutrino detector. Results of a first prototype in operation will be shown in this paper.

1. Introduction

Nowadays there is a very strong evidence for neutrino oscillation from atmospheric and solar neutrino experiments. Recent results from the KamLAND experiment confirm earlier work at the Sudbury Neutrino Observatory and Super-Kamiokande that also provided strong evidence for neutrino oscillation. The fact that different neutrino types do change their identity as they propagate, suggests that they are not strictly massless as had been assumed

---

*attending speaker, e-mail: ioa@hep.saclay.cea.fr

by the standard model of weak interactions. The atmospheric neutrino oscillation data suggest that there is maximal mixing between the τ and e neutrinos and a corresponding mass squared difference of $\delta m_{23}^2 = 3 \times 10^{-3} eV^2$. On the other hand the solar neutrino and other electron neutrino disappearance experiments suggest a non-maximal mixing and a mass squared difference of $\delta m_{21}^2 = 7 \times 10^{-5} eV^2$. The other mixing angle is not known but it is constrained to be less than 15 degrees from the CHOOZ data. If one includes this small mixing angle, one can see in electron neutrino experiments the effect of the large δm_{23}^2, i.e. a transition probability associated with the small oscillation length. In other words for a detector close to the source one will see a disappearance oscillation probability of the form:

$$P(\nu_e \to \nu_{\mu,\tau}) = \sin^2 2\theta_{13} \sin^2 \pi \frac{L}{L_{23}} \qquad (1)$$

where L is the distance between source and observed interaction, E is the neutrino energy and L_{23} is related to the neutrino energy: $L_{23} = 2\pi E_\nu/(\delta m_{23}^2)$. In the case of very low energy neutrinos, such as the ones emitted by a tritium source, the oscillation length L_{23} is only 13 m. Therefore by observing neutrino interactions inside a large TPC of about 10 m in radius, surrounding a tritium source, one can contain the oscillation occurring inside the gas volume and measure the oscillation parameters by a single experiment. The idea of combining a strong tritium source and a spherical detector has been recently proposed[1]. In this proposal a large spherical drift volume filled with a suitable gas mixture at high pressure is used for detecting low energy recoils produced in neutrino interactions.

2. The spherical TPC concept

The tritium source is located in a small spherical vessel of about 25 cm radius and it is surrounded by a spherical gaseous TPC 10 m in radius as shown in figure 1. Low energy electron recoils produced in the gas by elastic scattering in the TPC volume ionise the gas. Charges drift towards the sphere center and are collected by an adequate gaseous detector. The use of a spherical Micromegas counter made out of flat detectors was advertised because of the high precision and excellent energy resolution[2,3]. The high efficiency for detecting single electrons has been demonstated[4] even at high pressures[5]. The detector is currently used for solar axion detection in the CAST experiment[6] where a great stability and ability to reject background events has been achieved.

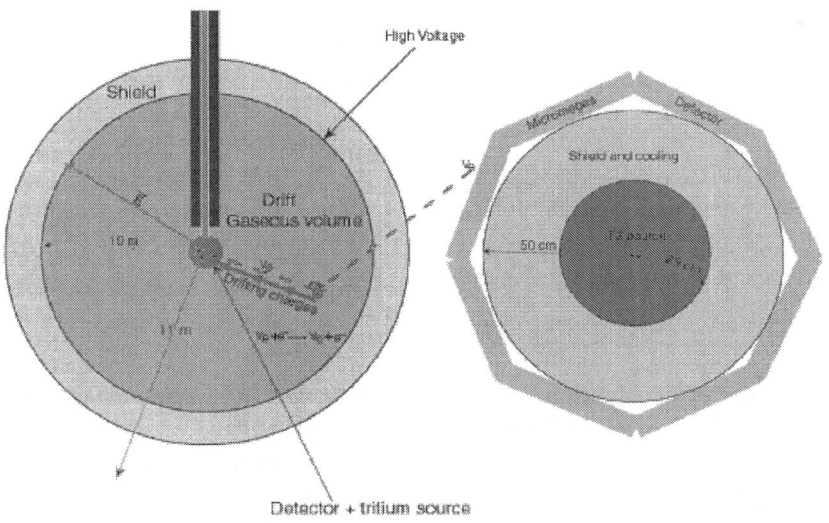

Figure 1. Scheme of the NOSTOS detector

The novel approach is radically different from all other neutrino oscillation experiments in that the neutrino source and the detector are located in the same vessel; it is then possible to measure the neutrino interactions, continuously as a function of the distance source-interaction point, with an oscillation length that is fully contained in the detector. Indeed we expect a counting rate oscillating from the centre of the sphere towards the external volume, i.e. at first a decrease, then a minimum and finally an increase. In other words we will have a full observation of the oscillation process as it has already been done in accelerator experiments with neutral strange and beauty particles. Fitting such an observed curve will provide all the relevant parameters of the oscillation by a single experiment. It is equivalent to many experiments made in the conventional way where the neutrino flux is measured in a single space point. The use of a spherical TPC detection scheme presents two main advantages. Fisrt, it is a a natural focus device requiring a small amplifying detector with only a few read-out channels. Such small size detector simplifies the construction and reduces the cost of the project. Secondly, with the neutrino source at the center of the sphere, the acceptance is greater than that provided by a cylindrical TPC with the source aside the end-cap of the TPC.

The electric field at a distance r from the center of curvature, in the case of a spherical TPC, is given by :

$$E = \frac{V_0}{r^2} \frac{R_1 R_2}{R_2 - R_1} \quad (2)$$

where R_1 and R_2 are the internal and outer sphere radii and V_0 is the applied voltage. At low electric fields (E) the drift velocity v_D is roughly proportional to E and the longitudinal diffusion coefficient is D, inversely proportional to the square of E. From the previous expressions it is easily deduced that time dispersion is :

$$\sigma_t^2 \sim \int D^2 dr \sim r^3 \quad (3)$$

From the latter equation one can conclude that in the spherical detector there is enhancement of the time dispersion of the detected signal especially at large distances. We propose to use this property for estimating the depth of the produced electron recoil during the neutrino elastic scattering. First calculations show that a precision for the depth of the interaction point of better than 10 cm will be achieved by measuring the time dispersion of drifting charges. Using a 20 kg tritium source, the total rate of emitted neutrino is $6 \times 10^{18}/s$. With the TPC filled with Xe at 1 bar the number of detected neutrinos is about 1000/year, assuming an energy threshold of 100 eV. The use of a less intense source or cheaper gases like Ar or Ne is possible at the expense of operation at higher pressures[5].

A detailed study based on Monte Carlo simulations with realistic parameters and taking into account the atomic effects[7] in neutrino interactions is currently under way to determine more precisely the NOSTOS sensitivity prospects.

3. Sensitivity for the neutrino magnetic moment

Because of the low energy of the incoming neutrinos and the low energy electron recoils detected in this experiment the sensitivity for the neutrino magnetic moment is high[8,9]. The electromagnetic contribution to the cross-section can be written as :

$$\left(\frac{d\sigma}{dT}\right)_{EM} = \sigma_0 \left(\frac{\mu_l}{10^{-12}\mu_B}\right)^2 \frac{1}{T}\left(1 - \frac{T}{E_\nu}\right) \quad (4)$$

Due to the 1/T (T being the electron recoil energy), the sensitivity for the magnetic moment is obviously higher at low energy. Indeed precise calculations show that the differential cross section due to a neutrino magnetic moment of 10^{-12} μ_B is rising at low energy and reaches 30% of the value of the weak neutrino-electron cross section which is at a first order independent of the recoil energy. Arguments from the helium ignition in globular cluster imply a limit on the neutrino magnetic moment of $< 3 \times 10^{-12}$ μ_B[10]. We would like to point out that recent measurements from the MUNU experiment set a limit of 10^{-10} μ_B for the neutrino magnetic moment. Our experiment opens the way to improve this value by two orders of magnitude.

4. The measurement of the weak charge

Another interesting quantity is the Weinberg angle which has never been measured at such low transfers. To this end atomic physics experiments, which use the neutral current, have so far been considered. Thus the atomic physics experiments suffer from the fact that the weak charges involved are extremely small[11]. Due to this smallness, complications appear in the analysis of the experiments arising from radiative corrections. This has also implications in the neutrino nucleus elastic scattering in the sense that the neutrons in the nucleus can contribute coherently. The coherence due to the protons is suppressed by the smallness of the weak charge. In the proposed experiment these cancellations do not occur and there is no need worry about such corrections.

5. Supernova sensitivity

It is generally believed that the core-collapse supernova explosion produces a large number of neutrinos and 99% of the gravitational energy is transformed to neutrinos of all types. The supernova (SN) neutrino flux consists of two main components: a very short (< 10 msec) pulse of ν_e produced in the process of neutronization of the SN matter through the reaction e + p → e + n, which is followed by a longer (< 10 sec) pulse of thermally produced ν_e, ν_μ, ν_τ, and their antiparticles. Only a small fraction, about 1%, of the neutrinos are prompt, while the rest are neutrino pairs from later cooling reactions. Therefore the neutrino signal from a supernova rises first steeply and then falls exponentially with time in a time window of about 10 seconds. It is expected that spectra of thermally produced neutrinos are characterized by the different mean energies : $\nu_e = 11$ MeV, $\bar{\nu}_e = 16$ MeV, $\nu_{e,\mu} = 25$ MeV. Our idea is to use the large cross section offered by the coherent neutrino-nucleus cross section for detecting neutrinos from supernova explosions. Coherent scattering occurs when neutrinos interact with two or more particles and the amplitudes from the various constituents of the target adding up. A consequence of the coherence is an increase of the cross section becoming proportional to the square of the number of particles in the target leading to increased counting rates :

$$\sigma = \frac{G^2 N^2 E^2}{4\pi} \quad (5)$$

where G is the weak coupling constant, N is the number of neutrons in the target nucleus and E is the neutrino energy. In order to get advantage of the coherent scattering, heavy nuclei gases are needed.

For instance, using Xenon as detector target the coherent cross sections at E=25 MeV, the energy that is relevant for Supernova detection, is quite large ($\sigma = 1.510^{-38}$ cm^2). Even at lower energy (11 MeV) where the coherent cross sections decreases quadratically with energy, the cross section is still high, in the case of gaseous Xenon detector. The recoil energy energy is quite low and it takes a maximum value of 1.5 keV for 11 MeV and 9 keV for 25 MeV neutrinos. This implies that detector thresholds must be set quite low with one advantage that backgrounds are highly suppressed given the narrow time window in which the burst takes place. The collected energy is even lower by a significant factor (quenching factor) and therefore sub-keV detector threshold is required.

A possibility to detect supernova efficiently and at low cost would be to construct an array of NOSTOS-type-spheres spread around the world. There would be no need for the nominal size NOSTOS sphere as with a 4 m radius sphere filled with Xe at 10 bar we expect 300 neutrinos to be detected per explosion. Requiring a coincidence in time between the signals of the various detectors around the world would reduce as much the background rate; therefore there is no need for deep underground laboratory. The operation and maintenance of the detector would be extremely simple as there would be only one electronic channel.

6. The first spherical prototype and results with a new proportional counter

We have built a spherical prototype 1.3 m in diameter as a first step towards the large detector needed for accomplishing the NOSTOS project. The outer diameter is made of pure Cu (6 mm thick) allowing to reach pressures as high as 5 bar. The quality of the various materials assures a good vacuum ($< 10^{-6}$ mbar) with a quite low level of outgasing that has been measured to be $< 10^{-9}$ mbar l s^{-1}. First tests were performed by filling the volume with pure gas (Argon + 10% CO$_2$ < 3 ppm impurities) and operate the detector in a seal mode. We are proposing the use of

Micromegas as amplifying structure. A particular effort is actually made to build a spherical Micromegas detector using new technologies. Another alternative will be to approximate a sphere by using several Micromegas flat elements. In view of an ideal solution for the amplification structure we decided to start the first tests using a small sphere (10 mm in diameter made of steel) as a proportional counter located at the center of curvature of the TPC. Several tests from low pressure to high pressure have shown that such a simple amplification element is able to provide high gains and stable operation as shown in figure 6. Runs using ^{109}Cd and ^{55}Fe calibration sources have been performed. Figure 6 shows an example of an energy spectrum obtained with a ^{55}Fe source

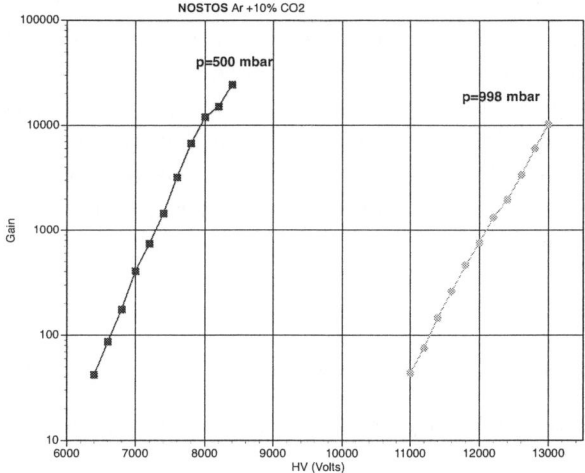

Figure 2. Measured gain as a function of the applied voltage in a Ar + 10%CO_2 for two different pressures

This simple structure will allow us to make fast progress and understand the functioning of the spherical TPC. Our plans for the short term are to optimise the detector for high gain operation with a good energy resolution and to optimise the amplification structure. We expect to measure the attenuation length of drifting electrons and the coherent neutrino scattering next to a reactor.

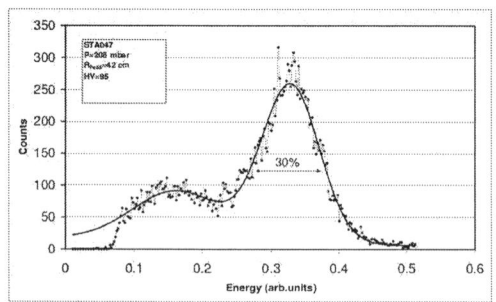

Figure 3. Pulse height distribution from a ^{55}Fe source.

7. Conclusions

NOSTOS covers an ambitious experimental program in the low energy neutrino physics sector. It includes observation of neutrino oscillations, neutrino magnetic moment and Weinberg angle at low energy measurement. The same device exhibits high sensitivity as a supernova neutrino detector. The measurement of the coherent neutrino nucleus scattering will open the way to build simple, low cost and robust telescopes dedicated for supernova neutrino detection. A small prototype is in operation in Saclay giving very encouraging results.

References

1. Y. Giomataris and J. D. Vergados, *Nucl. Instr. Meth. A* **530**, (2004) 330, [arXiv:hep-ex/0303045].
2. Y. Giomataris, P. Rebourgeard, J. P. Robert and G. Charpak, *Nucl. Instrum. Meth. A* **376** (1996) 29.
3. Y. Giomataris, *Nucl. Instrum. Meth. A* **419** (1998) 239.
4. J. Derre, Y. Giomataris, P. Rebourgeard, H. Zaccone, J. P. Perroud and G. Charpak, *Nucl. Instrum. Meth. A* **449** (2000) 314.
5. P. Gorodetzky et al., *Nucl. Instrum. Meth. A* **433** (1999) 554.
6. S. Andriamonje et al., *Nucl. Instrum. Meth. A* **518** (2004) 252.
7. G.J. Gounaris, E.A. Paschos, P.I. Porfyriadis, DO-TH-04-07, [arXiv:hep-ph/0409053].
8. G.C. MCLaughlin and C. Volpe, *Phys. Lett.* B591:229-234, 2004, [arXiv:hep-ph/0312156].
9. J.I Collar and Y. Giomataris, *Nucl. Instrum. Meth. A* **469** (2001) 249, [arXiv:hep-ex/0009063].
10. G. Raffelt, *Annu. Rev. Nucl. Part. Sci.* 49 (1999) 163-216.
11. J. Guéna et al., *Phys. Rev. Lett.* **90** 143001 (2003), [arXiv:physics/0210069].

NEUTRINO, OTHER BURSTS IN COSMOLOGY

L. STODOLSKY

Max-Planck-Institut für Physik, 80805 Munich

We recall how flight time effects for neutrinos can be used to determine the acceleration parameter q of cosmology and mention some other ideas suggested by the possibility of looking back to very early times via bursts of weakly interacting particles.

1. Introduction

There's no doubt that the most significant development in cosmology of the last decade if not the last decades is the "acceleration of the universe" and the associated notion of "dark energy". The direct evidence for the acceleration comes from the apparent dimming of distant type 1 supernovas. In view of the great importance of this question I was intrigued to realize some years ago that as a result of a non-zero neutrino mass there was a simple effect in general relativity permitting an independent direct measurement of the acceleration parameter q [1]. Since the establishment of a non-zero neutrino mass is a recent development, this may explain that the effect was overlooked in the past. While it is not at all evident that the necessary observations are easy or even possible, it is perhaps worthy of attention in view of the great importance of the question.

Furthermore, the train of thought engendered by this proposal leads one to some further novel and amusing if again not obviously practical observational ideas which Joe Silk and I have been looking into. Here I'd like to briefly recall the idea and mention some of the interesting points we have been considering.

2. Time of flight effects for neutrinos

Time of flight effects for neutrinos as far as I know, were first discussed by Zatsepin [2], back in the days when we considered only one neutrino species. Due to the non-zero mass there would be a variation of the velocity with energy, and so a spreading in arrival times for a neutrino burst traveling astronomical distances from a (core collapse) supernova.

With the discovery of different neutrino types and so presumably different masses there was an even simpler effect: the lighter neutrino will travel faster and so arrive before the heavier one [3]. In the simplest case there would be two or three closely separated bursts. Indeed, nowadays when we contemplate all kinds of new and sometimes neutral and long lived particles, we don't have to confine ourselves to neutrinos. While one particle could of course be the photon, other mass states could be an as yet undiscovered but perhaps massive particle like the WIMP, in which case the time-of-flight effects are quite large. Also note we should expect that the neutrino mass states are generally mixed flavor states, leading to a "flavor-echo", as we discussed in ref [3]. Indeed with close to maximum mixing for neutrino mass eigenstates, there should be a distinct "echo" if the pulse separation is sufficient. This is an effect where the same flavor appears twice at the detector, once in the first mass eigenstate and then again in the second one. Note further the nice point that large mixing means the mass eigenstates depart the source together.

These effects involve nothing more than the kinematics of special relativity. Two relativistic particles particles with masses m_1 and m_2 and energy or momentum p_1 and p_2 have a velocity difference

$$v_1 - v_2 \approx \frac{m_1^2}{2p_1^2} - \frac{m_2^2}{2p_2^2}. \tag{1}$$

3. General Relativistic Effect

General relativity and cosmology come in when we consider a burst from cosmological distances–well outside the galaxy, from a source with a non-negligible redshift parameter z. While the neutrino is en route to us the universe is expanding. While a photon by definition travels on the light cone, the neutrino is a bit off the light cone and becomes more so as the universe expands. When the neutrino finally reaches us, its total travel time is a record, so to speak, of the cosmological epochs through which it has passed. The difference in arrival times for particles of different masses is a record of the expansion and so in particular is sensitive to the q parameter.

To see how this comes about, take the standard FRW metric $ds^2 = dt^2 - a^2(t)(d\mathbf{x})^2$, where we define $a(t)$ to be the expansion factor of the universe normalized to its present value so that $a(now) = 1$. The calculation is simplified by the fact that in a problems where nothing depends on the coordinate "x" there is a conserved momentum. Here this is the covariant or "canonical momentum" $P_i \sim \partial_i$ in analytical mechanics.

We proceed by finding an equation for the coordinate velocity dx^i/dt, where x^i is along the particle's flight direction. First we express dx^i/dt in terms of $P^i(t)$, the spatial part of the contravariant four-momentum $m\, dx^\mu/ds$. From the definition of the metric we have $a(t)dx^i/dt = [a(t)P^i(t)]/\sqrt{m^2 + [a(t)P^i(t)]^2}$.

Expanding for the relativistic case $P \gg m$ we obtain $a(t)\, dx^i/dt \approx 1 - \frac{1}{2}m^2/[a(t)P^i(t)]^2$ Hence we can identify the constant covariant momentum as $P(now)$, the momentum at the detector. Thus from $P^i = g^{ij}P_j = 1/(a^2)P_i$, we obtain $P^i = 1/(a^2)P(now)$. Thus we finally have

$$\frac{dx}{dt} \approx \frac{1}{a(t)} - a(t)\frac{1}{2}\frac{m^2}{P^2(now)}. \qquad (2)$$

The first term $1/a(t)$ will be recognized as just the equation for motion along the light cone, $ds = 0$, and then there is the interesting correction involving the mass.

Introducing Δx for the difference in the x coordinate of two different particles of mass m_2 and m_1 emitted in the same event at the same time

$$\frac{d(\Delta x)}{dt} \approx a(t)\frac{1}{2}\left[\frac{m_1^2}{P_1^2(now)} - \frac{m_2^2}{P_2^2(now)}\right] \qquad (3)$$

At the present epoch with $a = 1$, Δx is just the spatial separation of the two particles. Integrating, we have for this separation, or in view of $v \approx c = 1$ for the time difference in arrival at a detector

$$\Delta t \approx \Delta x \approx \int a(t)\, dt \frac{1}{2}\left[\frac{m_1^2}{P_1^2(now)} - \frac{m_2^2}{P_2^2(now)}\right]. \qquad (4)$$

It is thus $\int a(t)\, dt$ which "records" the cosmological information. Observe a is small at early times so that most of the effect comes near the present time, i.e. the neutrinos are the "slowest" at present. The expression in the bracket is the factor Eq 1 for the difference in velocity for highly relativistic particles, while the cosmological effect is in $\int a dt$.

It only remains to get rid of the coordinate dependent a and t and to re-express things in terms of an observable, namely the red shift parameter z for the event. With the expansion of $a(t)$ for recent epochs $a(t) = 1 + H[t - t(now)] - \frac{1}{2}qH^2[t - t(now)]^2 + ...$, and the redshift parameter $z = 1/a - 1 = -H[t - t(now)] + (1 + q/2)H^2[t - t(now)]^2 + ...$, we find

$$\Delta t \approx \frac{z}{H}[1 - \frac{3+q}{2}z + ...]\frac{1}{2}[\frac{m_1^2}{P_1^2(now)} - \frac{m_2^2}{P_2^2(now)}] \qquad (5)$$

giving the result in terms of the directly observable z. We thus have two measured quantities for an event, Δt and z, for the present Hubble constant H and the acceleration parameter q. Thus, in principle, two good events– assuming the neutrino masses well known by the time this is done– fix these cosmological parameters.

This was for the time delay between two distinct mass states. However it may well be–unfortunately– that the mass differences aren't big enough to give cleanly separated pulses. However, even if there is effectively only one mass there is Zatsepin's original effect of a spreading of the pulse. Considering two different momenta P and P', we get a time delay between the two of

$$\Delta t \approx \frac{z}{H}[1 - \frac{3+q}{2}z + ...]\frac{1}{2}m^2[(\frac{1}{P(now)})^2 - (\frac{1}{P'(now)})^2]. \qquad (6)$$

These formulas are low z expansions, to first order z^2.

The first term in z is just what we would get without general relativity; it says that Δt is simply the velocity difference times the distance (since $z = H\ d$ is just the Hubble law, where d is the distance). Even this is not entirely trivial since it shows how, knowing the redshift of the event, one can by means of neutrinos find H without knowing the distance to the event,— without all the traditional difficulties associated with the "cosmological distance ladder".

Of course we now have the perhaps greater difficulty of detecting neutrino bursts at cosmological distances. And even if we do, the mass effects need to be observable. To get some impression of the possible size of the effects, we can evaluate the velocity difference bracket in terms of an eV (mass)^2 for the neutrinos and GeV for their energy:

$$\frac{(m/\text{eV})^2}{2(P/\text{GeV})^2} \approx 50\mu\text{sec/Mpc}. \qquad (7)$$

So at a thousand Mpc, a substantial part of the way across the visible universe, we may expect, at best, some msec delays. It is true that msec or even μsec timing doesn't seem very difficult for particle detectors where we often deal in nanoseconds. However the likely problem is the intrinsic time scale of the burst itself. That is, for an ideally clean separation we would like the spacing between bursts arriving at the detector to be greater than that of the length of the burst itself. However, if we take supernovas or gamma-ray bursts as any guide, the time scale is on the order of some seconds. It then seems that to have distinctly separated bursts we would need to have particles distinctly more massive than eV's. One interesting possibility would be not the neutrino itself, but a heavy neutral object like the WIMP. If these new objects do indeed exist we anticipate that they are stable and if they are not overly massive they could be emitted in high energy bursts.

However there is still the possibility, even if the bursts don't separate into clearly distinguishable pulses, of using the spreading of the pulse. Eq [6] involves the same cosmological factor as the separation effect Eq [5]. Our distinction of the "separation" and "spreading" effects is of course for descriptive purposes and the two would be taken into consideration together in an overall analysis.

Finally, we note the possibility of a sharp timing signal from within the burst itself. For example if neutrinos come together with the photons in gamma ray bursts, there is the phenomena of spikes or flaring in the bursts, see Fig 1, which can provide a sharper timing signal than that from the burst as a whole.

4. Deeper into the Big Bang

A fascinating aspect of neutrino or neutrino - like based observations such as these is that the particles can come to us directly from very early epochs. Observational methods using the photon, be it the optical photon of classical astronomy, the microwave photon of the background radiation or those of radio astronomy, will never allow us to look further back than "recombination", some 300,000 years after the Big Bang. On the other hand with neutrinos or other weakly interacting particles the "look-back time" is much deeper. The particle can come directly to us after its decoupling and in general the more weakly interacting the particle the further back a burst of some kind can reach us without being lost through thermalization by the intervening matter and radiation. With neutrinos we can get to a few seconds (ν_e) or milliseconds (ν_μ, ν_τ), with the yet to be discovered WIMPs to perhaps a picosecond and with gravitons to $\sim 10^{-35}s$, or the epoch of inflation.

If "Baby Universes" are created with excitations of the particle fields on a microscopic timescale, they could become "visible" via their "bursts". It is interesting that at least at the level of a *gedanken* observation there

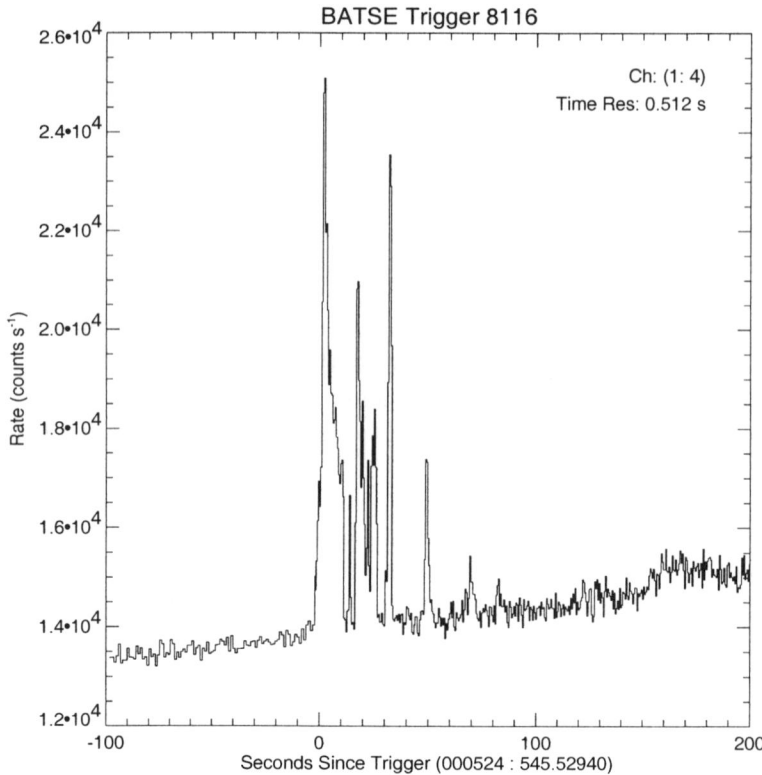

Figure 1. A gamma ray burst seen by Batse, showing sharp spikes in time.

is potentially something directly observable in connection with such happenings. Some sort of phenomenon would have to exist around the epoch in question in which powerful high energy bursts of neutrinos or our other objects are emitted. Perhaps collapse of over-dense regions or annihilation of topological defects are a possibility. Then after being strongly redshifted and diluted by the expansion of the universe, the particles must be detected at the earth.

It takes considerable fantasy to imagine that such observations might ever be possible, but even allowing for a good measure of it, at this point we seem to come to a discouraging but nevertheless novel thought: As we go back to ever- higher redhifts, there is a stretching, according to general relativity, not only of the wavelength of radiation, but of all time scales. Thus our "bursts" will be stretched out and will perhaps not look like "bursts" at all. If we go all the way back, say to neutrino decoupling, we have $z \sim (1 MeV/0.1 meV) \sim 10^{10}$, and thus for example, a millisecond event taking place around neutrino decoupling will last a year upon reaching us. This can hardly be considered a "burst" anymore. On the other hand it suggests something amusing: let us assume sometime in the future we are operating very big detector arrays and we observe long-time secular variations in the counting rate. Our first reaction would be that something's wrong with the instrumentation. However, before re-adjusting our voltages, we should consider, if perhaps we are not detecting an event out of the early universe! This poses quite a problem for future observers –they have to distinguish baseline drifts from real happenings. Furthermore the "stretching" gets worse the higher the z.

5. Preliminary considerations on very early bursts (with J. Silk, Oxford)

Even the foolhardy might be tempted to give up here. However, an interesting point arises if we think about what the natural timescale for a burst in the very early universe could be: the only natural timescale is the Hubble constant $H(t)$ for the epoch under consideration. And this gets rapidly shorter at early times and will in fact compensate for the high z "stretching". An estimate (again in the simplest FRW cosmology) for the duration τ_{now} of a burst upon reaching us yields

$$\tau_{now} \leq z\tau_{natural} = \frac{1}{a_{em}} \frac{a_{em}}{\dot{a}_{em}} = \frac{1}{\dot{a}_{em}} = 2(t_{rad} t_{em})^{1/2} \qquad t < t_{eq} \quad (8)$$

With the time unit in seconds, this gives $\tau_{now} \leq t_{em}^{1/2} \, 9 \times 10^9 s$. Here the a and t refer to the scale factor and time in the FRW metric and "em" means the emission time, t_{rad} is the time of matter-radiation equality. Now the observed burst length is actually shortening. At the QCD phase transition, for example, Eq 8 gives a burst duration of about a day.

Of course there is an enormous dilution factor for the great distance. If N is the total number of particles initially in a burst at high z, (for bursts after the inflation if there was one) the number reaching us in the simplest cosmology is

$$\text{no. crossing unit area} \approx N \frac{1}{4\pi} (\frac{1}{3 t_{now}})^2 \approx N \cdot 6 \times 10^{-59} / cm^2 \,. \quad (9)$$

This is a very great dilution, but an interesting point is that it reaches a limit once z is large and does not further increase with z.

Another interesting aspect of this problem is connected with the very large number of casually disconnected (or disconnected since inflation) regions at very early times. We might wonder if this doesn't result in a kind of "Olbers paradox" where even a small probability of a burst per region leads to an infinite amount of energy radiated to the present universe. Again making all the simplest possible assumptions, what we obtain for the present energy flux is

$$d\,(energy\ flux) = (1 \times 10^5\ eV/cm^2 s)\mathcal{P}\frac{1}{t_{em}}dt_{em} \qquad (10)$$

where $\mathcal{P}$ is the probability of a horizon-size burst at time t_{em}. Interestingly, this leads to a logarithmic divergnece if we take $\mathcal{P}$ to be constant. The dimensional quantity $(1 \times 10^5\ eV/cm^2 s)$ originates from the present intensity of the microwave background. We hope to present these and related results shortly [4].

References

1. L. Stodolsky, Physics Letters **B473** 61 (2000), and *Carolina Symposium on Neutrino Physics*, Columbia, March 2000; Eds. J. Bahcall, W. Haxton, K. Kubodera, and C. Poole; World Scientific, Singapore.
2. G. T. Zatsepin, ZhTEF Pis. Red. **8**, 333 (1968), (JETP Lett. **8** 205).
3. S. Pakvasa and K. Tennakone, *Phys. Rev. Let.* **28**, 1415 (1972); S. Pakvasa, DUMAND Symposium 1980; N. Cabibbo, *Accademia Lincei*, Meeting on Astrophysics and Elementary Particles, (1980); Tsvi Piran, *Phys. Lett.***B102**, 299 (1981); A.K. Drukier and L. Stodolsky,*Phys. Rev.* **D30**, 2295 (1984). P. Reinartz and L. Stodolsky, Z.f.Phys. **C27**, 507 (1985).
4. J. Silk and L. Stodolsky, "Bursts from the Early Universe", in preparation.

PRESENT AND FUTURE OF NEUTRINOLESS DOUBLE BETA DECAY EXPERIMENTS

A. NUCCIOTTI

Dipartimento di Fisica "G. Occhialini", Università di Milano - Bicocca
and Istituto Nazionale di Fisica Nucleare, Sezione di Milano
Piazza della Scienza, 3
I-20126, Milano, Italy
E-mail: angelo.nucciotti@mib.infn.it

With neutrino oscillations now firmly established, neutrinoless double beta decay assumes great importance since it is one of the most powerful tools to set the neutrino mass absolute scale. The present status of the experimental search for this rare decay is reported and the prospects for next generation experiments are reviewed.

1. Introduction

The double beta decay is a second order weak transition which can be energetically favoured for some even-even nuclei belonging to A even multiplets. The $(A,Z) \to (A,Z+2) + 2e^- + 2\bar{\nu}_e$ double beta ($\beta\beta$-2ν) decay process is allowed by the Standard Model and has been observed for many isotopes with lifetimes longer than 10^{19} y. A more interesting process is the so-called neutrinoless double beta ($\beta\beta$-0ν) decay given by $(A,Z) \to (A,Z+2) + 2e^-$: this process violates lepton number conservation and is therefore forbidden by the Standard Model. The lifetime for the $\beta\beta$-0ν decay is expected to be longer than 10^{25} y and so far only one evidence has been reported for ^{76}Ge (see Sec. 2.1). For recent reviews on this topic refer to Ref. 1.

1.1. *$\beta\beta$-0ν decay and neutrino physics*

Many mechanisms have been proposed for driving this decay, but the simplest one is the "mass mechanism", where a light Majorana neutrino is exchanged. Whatever is the mechanism actually causing the $\beta\beta$-0ν decay, its observation would imply that the neutrino has mass and is a Majorana particle (i.e. $\nu \equiv \bar{\nu}$). For a light Majorana neutrino mediate $\beta\beta$-0ν decay, the rate is given by $[\tau_{1/2}^{0\nu}]^{-1} = \langle m_\nu \rangle^2 F_N / m_e^2$, where the nuclear structure factor F_N contains the nuclear matrix element and the phase space. The effective neutrino Majorana mass is given by $\langle m_\nu \rangle = |\sum_k m_k \eta_k |U_{ek}|^2|$, where m_k are the mass eigenvalues of the three neutrino mass eigenstates $|\nu_k\rangle$, η_k are the CP Majorana phases ($\eta_k = \pm 1$ for CP conservation) and U_{ek} are the elements of the electron sector of the neutrino mixing matrix.

With the help of the $\Delta m_{ik}^2 = |m_i^2 - m_k^2|$ and $\sin^2 2\theta_{ik} = f(|U_{ik}|^2)$ parameters determined by neutrino flavor oscillation experiments,[2] it is possible to calculate $\langle m_\nu \rangle$ as a function of the unknown neutrino absolute mass scale and η_k phases.[3] From this analysis two possible scenarios can be devised. (1) The $\beta\beta$-0ν decay is discovered with $\langle m_\nu \rangle \geq 10\,\text{meV}$: then the neutrino is a Majorana particle and the masses are either degenerate ($m_1 \approx m_2 \approx m_3$) or follow an inverse hierarchy ($m_3 \ll m_1 \approx m_2$). If neutrinos are degenerate (for $\langle m_\nu \rangle \geq\approx 0.5\,\text{eV}$) then the absolute mass scale can be established. (2) The $\beta\beta$-0ν decay is not observed and only an upper limit $\langle m_\nu \rangle \leq 10\,\text{meV}$ is set: then either the neutrino is a Dirac particle or the masses are arranged in a normal hierarchy ($m_1 < m_2 \ll m_3$).

1.2. $\beta\beta$-0ν decay and nuclear physics

To obtain $\langle m_\nu \rangle$ from the experimental observable $\tau_{1/2}^{0\nu}$ the nuclear structure factor $F_N \equiv G^{0\nu}(Q_{\beta\beta}, Z)|M^{0\nu}|^2$ must be known. The phase space $G^{0\nu}(Q_{\beta\beta}, Z)$ can be precisely calculated, while the nuclear matrix $|M^{0\nu}|$ contains the uncertain details of the nuclear part of the process. In fact there is a large spread in the nuclear matrix elements calculated by different authors with different nuclear models.[4] Because of this spread also $\langle m_\nu \rangle$ is affected by large uncertainties (about a factor 3 on the average). Presently these uncertainties are a severe limitation to the potentialities of $\beta\beta$-0ν decay as a tool for neutrino physics: it has been recently suggested[5] that measured $\beta\beta$-2ν decay lifetimes can be used to reduce the spread in QRPA calculations. Nevertheless, it is important to search for $\beta\beta$-0ν decay of as many as possible candidate isotopes.

1.3. Experimental approaches to $\beta\beta$-0ν

There are two approaches for direct $\beta\beta$-0ν searches. In the first approach a thin $\beta\beta$-0ν decay source is sandwiched between two detectors analyzing the escaping electrons. Using tracking detectors, e.g. TPCs, a background rejection is possible using the event topology. The limits of this approach are the energy resolution and the size of the source.

It the second approach the detector contains the source (calorimeter)and only the sum energy of the 2 electrons is measured. The signature for $\beta\beta$-

0ν decay is therefore a peak at the transition energy $Q_{\beta\beta}$. Calorimeters can have large mass and high efficiency. Depending on the technique, high energy resolution and also some tracking are possible.

From statistical considerations, the sensitivity $\Sigma(\tau_{1/2}^{0\nu})$ of a $\beta\beta$-0ν decay search is given by $\Sigma(\tau_{1/2}^{0\nu}) \propto \epsilon\, i.a.(M t_{meas}/(\Delta E\, bkg))^{1/2}$, where ϵ, $i.a.$, M, t_{meas}, ΔE and bkg are the detector efficiency, the active isotope abundance, the source mass, the measuring time, the energy resolution and specific background at $Q_{\beta\beta}$, respectively. The background is a fundamental issue in $\beta\beta$-0ν searches: to reduce it, all passive (e.g. shielding, material selection) and active (e.g. Pulse Shape Discrimination, topology analysis, segmentation) measures must be taken. However the background caused by the high energy tail of the continuous $\beta\beta$-2ν spectrum cannot be avoided and must be minimized with a good enough energy resolution.

2. Present and past experiments

The following is a selection of the most sensitive experiments (see Table 1 for a more complete list of the best results to date for many isotopes).

2.1. ^{76}Ge experiments and the evidence for $\beta\beta$-0ν decay

The **Heidelberg-Moscow experiment** (hereafter HM) searched the $\beta\beta$-0ν decay of ^{76}Ge using 5 High Purity Ge semiconductor detectors enriched to 87% in ^{76}Ge. This experiment run in the Gran Sasso Underground Laboratory (Italy) from 1990 to 2003, totalling an exposure of 71.7 kg×y (i.e. 820 moles×y of ^{76}Ge). It is by far the longest running $\beta\beta$-0ν decay experiment with the largest exposure. The experiment since the end of 1995 has featured PSD on 4 crystals to reduce background by separating Single Site Events (like $\beta\beta$-0ν decay events) from Multiple Site Events (like γ interactions): the PSD is applicable to 72% of the full data set (i.e. 51.4 kg×y). The final background at $Q_{\beta\beta}$ is about 0.11 c/keV/kg/y and it is attributed mainly to U and Th contaminations in the set-up materials. The use of Ge detectors for a calorimetric $\beta\beta$-0ν decay search was first proposed in Ref. 10 and the HM experiment is the best exploitation of this technique: it represents the Status-of-the-Art for the low background techniques and has been the reference for all last generation $\beta\beta$-0ν decay experiments. After the conclusion of the experiment, part of the collaboration has reanalyzed the data[11] claiming a 4σ evidence for ^{76}Ge $\beta\beta$-0ν decay with a lifetime $\tau_{1/2}^{0\nu}$ of about 1.2×10^{25} y, corresponding to a $\langle m_\nu \rangle$ of about 0.44 eV.[12]

Igex[13] is a similar experiment which run in Homestake (USA), Canfranc (Spain) and Baksan (Russia) from 1991 to 2000 with a total exposure of

Table 1. A selection of the past and present experiments giving the best result per isotope to date.

isotope	experiment	latest result	$Q_{\beta\beta}$ [keV]	isotopic abundance natural [%]	isotopic abundance enriched [%]	exposure [kg×y]	technique	material	$\tau_{1/2}^{0\nu}$ [10^{23} y]	$\langle m_\nu \rangle$ [eV]	$\tau_{1/2}^{0\nu}$ $\langle m_\nu \rangle$ = 10 meV [10^{28} y]
^{48}Ca	Elegant VI	2004[6]	4271	0.19	–	4.2	scintillator	CaF$_2$	0.14	7.2÷44.70	8.8
^{76}Ge	Heidelberg/Moscow	2004[12]	2039	7.8	87	71.7	ionization	Ge	120.0	0.44	17.7
^{82}Se	NEMO-3	2004[16]	2995	9.2	97	0.55	tracking	Se	1.9	1.30÷3.60	5.6
^{100}Mo	NEMO-3	2004[16]	3034	9.6	95÷99	4.1	tracking	Mo	3.5	0.70÷1.20	3.9
^{116}Cd	Solotvina	2003[7]	2805	7.5	83	0.5	scintillator	CdWO$_4$	1.7	1.70	4.7
^{130}Te	Cuoricino	2004[15]	2533	34.5	–	3.8	bolometer	TeO$_2$	7.5	0.30÷1.68	5.8
^{136}Xe	DAMA	2002[17]	2476	8.9	69	4.5	scintillator	Xe	12.0	1.10÷2.90	12.1
^{150}Nd	Irvine TPC	1997[9]	3367	5.6	91	0.01	tracking	Nd$_2$O$_3$	0.012	3.00	0.1
^{160}Gd	Solotvina	2001[8]	1791	21.8	–	1.0	scintillator	Gd$_2$SiO$_5$	0.013	26.00	0.9

Note: All given $\tau_{1/2}^{0\nu}$ ($\langle m_\nu \rangle$) are lower (upper) limits with the exception of the Heidelberg-Moscow experiment where the 99.9973% CL value is given. The spread in $\langle m_\nu \rangle$ is due to the uncertainties on the nuclear factor F_N. Last column gives the expected half-life for $\langle m_\nu \rangle$ = 10 meV and the less favourable F_N.

Table 2. A selection of the proposed experiments.

experiment	isotope	$Q_{\beta\beta}$ [keV]	technique	i.a. [%]	mass [kmol]	t_{meas} [y]	σ_E [keV]	bkg [c/y]	$\tau_{1/2}^{0\nu}$ [10^{28} y]	$\langle m_\nu \rangle$ [meV]	project status
CANDLES IV+[31]	^{48}Ca	4271	scintillator	2	0.9	5	73	0.35	0.3	29÷54	R&D (III: 5 mol)
Majorana[19]	^{76}Ge	2039	ionization	90	6.6	5	2	1	0.4	21÷67	R&D (SEGA)
Genius[20]	^{76}Ge	2039	ionization	90	13.0	10	2	0.4	1	13÷42	R&D (TF: 10 kg)
MOON III[35]	^{100}Mo	3034	tracking	85	8.5	10	66	3.8	0.17	13÷48	R&D (I: small)
CAMEO III[30]	^{116}Cd	2805	scintillator	83	2.7	10	47	4	0.1	22÷69	proposed
CUORE[28]	^{130}Te	2533	bolometer	35	1.7	10	4	7.5	0.07	15÷19	approved full scale
EXO[40]	^{136}Xe	2476	tracking	65	48.0	10	49	0.55	1.3	12÷31	R&D (1.5 kmol)
DCBA-II[36]	^{150}Nd	3367	tracking	80	2.7	–	85	–	0.01	16÷22	R&D (T: small)
GSO[8]	^{160}Gd	22	scintillator	22	2.5	10	83	200	0.02	65	proposed

Note: Except for CUORE and GSO all experiments use isotopically enriched material. Background bkg is calculated on an energy interval equal to σ_E. For all tracking experiments the quoted background is due only to the $\beta\beta$-2ν tail.

only 8.87 kg×y and a background at $Q_{\beta\beta}$ of about 0.17 c/keV/kg/y: its sensitivity is not enough to check the HM claim.

2.2. *Running* ^{82}Se, ^{100}Mo, ^{130}Te *and* ^{136}Xe *experiments*

There are presently 3 running experiments (Cuoricino, NEMO-3 and DAMA), but the 2 presented in the following have the best chances to reach the sensitivity to see a $\beta\beta$-0ν signal at the level expected from the HM experiment claim. For the DAMA experiment on ^{136}Xe refer to Table 1. Cuoricino[14] is a calorimetric experiment using natural TeO$_2$ cryogenic detectors to search for ^{130}Te $\beta\beta$-0ν decay. It runs in the Gran Sasso Underground Laboratory since 2003 and consists of 62 TeO$_2$ crystals kept at a temperature of about 10 mK, arranged in a tower-like structure which is the base element of the future experiment CUORE (see Sec. 3.2). The total TeO$_2$ mass is about 41 kg. The total exposure to date is about 3.8 kg×y (i.e. 8.2 moles×y of ^{130}Te) and the measured background at $Q_{\beta\beta}$ is about 0.19 c/keV/kg/y, mainly due to U and Th contaminations on the detector and surrounding Cu surfaces. This background level is obtained after applying anti-coincidence cuts between the detectors. The present 90% CL lower limit on $\tau_{1/2}^{0\nu}$ is 7.5×10^{23} y, corresponding to an upper limit on $\langle m_\nu \rangle$ of about $0.32 \div 1.7$ eV.[15] With an exposure of about 120 kg×y (i.e. 3 years running), Cuoricino would reach a 1σ sensitivity on $\langle m_\nu \rangle$ of about $0.1 \div 0.6$ eV.

NEMO-3[16] is a tracking detector experiment running in the Frejus Underground Laboratory (France). It uses a drift chamber to analyse the electrons emitted by foils of different enriched materials. Interesting $\beta\beta$-0ν decay sensitivities are expected only for the ^{100}Mo and ^{82}Se sources. The NEMO-3 detector can reject the background by identifying γs, e^-, e^+ and αs. The present background of about 1.4 c/keV/kg/y is composed mainly by radon (72%), the $\beta\beta$-2ν tail (21%) and ^{208}Tl in the foils (7%). The data analysis using a maximum likelihood applied to 3 kinematic variables gives a 90% CL lower limit on $\tau_{1/2}^{0\nu}$ of about 3.5×10^{23} (1.9×10^{23}) y for ^{100}Mo (^{82}Se), corresponding to a limit on $\langle m_\nu \rangle$ of about $0.7 \div 1.2$ ($1.3 \div 3.6$) eV. With the expected reduction of radon, in 5 years the 90% CL sensitivity on $\tau_{1/2}^{0\nu}$ should improve to 4×10^{24} (8×10^{23}) y for ^{100}Mo (^{82}Se), corresponding to $0.2 \div 0.35$ ($0.65 \div 1.8$) eV for $\langle m_\nu \rangle$.

3. Future experiments

It is likely that presently running experiment will not be able to confirm or rule out the HM positive result: therefore this will be the task for future experiments. To have a reliable confirmation, $\beta\beta$-0ν decay must be observed for different isotopes with similar $\langle m_\nu \rangle$, while to reject the HM result much lower sensitivities on $\langle m_\nu \rangle$ are required. All proposed next generation experiments aim at a sensitivities of about 0.01 eV: whether the HM result is correct or not, they will have good chances to observe $\beta\beta$-0ν decay. The large improvement in sensitivity (a factor 10 in $\langle m_\nu \rangle$, i.e. a factor 10^2 on $\tau_{1/2}^{0\nu}$) must be obtained by going to 1 ton mass scale experiments and by further reducing the background. To be able to search for $\beta\beta$-0ν decay of as many different isotopes as possible with the required sensitivity, also the isotope enrichment techniques will become a hot issue: for many interesting isotopes large scale enrichment is still both a technical and an economical problem. A strong effort is also demanded to nuclear theory to reduce the uncertainties in the nuclear matrix calculations.

Table 2 gives some informations about the most well-defined projects. Most of the projects presented here are at a very early R&D stage, especially the ones in Sec. 3.3 and 3.4, and for all of them the predicted sensitivity heavily relies on the assumed background level.

3.1. *Calorimetric experiments with ionization detectors*

The use of Ge ionization detectors is proposed for many future experiments because this is a well established experimental technique: it is relatively easy to scale up, it guarantees high energy resolution and it provides some background rejection by PSD and segmentation. The main drawback is the high cost for the Ge enrichment and for the detectors themselves. There is also the COBRA proposal[18] for using CdTe or CdZnTe diode detectors, but the technique, though promising, is still very young.

The Ge detector proposals follow two opposite approaches, descending from the experience of HM and Igex. The first one attributes to the material surrounding the detector the main responsibility for the background observed in the HM experiment, and therefore proposes to eliminate all this material by suspending the Ge crystals bare in a highly purified cryogenic liquid. The second approach stems from the localization of the main source of the Igex background inside the Ge crystals due to cosmogenic activity.

The **Majorana** experiment[19] belongs to the second group: it would consist of 210 enriched Ge crystals with segmented electrodes giving a total mass of 500 kg. Even minimizing cosmic ray exposure, the expected background of about 17 c/keV/t/y (without cuts) is mainly due to cosmogenics, since the activity in the surrounding materials would be avoided by careful material screening and selection: the application of PSD and segmentation cuts would further reduce the background to about 0.6 c/keV/t/y, giving a sensitivity of 10^{27} y in 5 years of measuring time ($\langle m_\nu \rangle \leq 0.02 \div 0.07$ eV). The experiment is presently going through 2 preliminary R&D phases (SEGA and MEGA) and would be ready in 9 years upon starting.

The other approach is the one of Genius, GEM and Gerda.

The **Genius** experiment,[20] proposed by part of the HM collaboration, consists of 1 ton bare enriched Ge crystals suspended in a 12 m diameter liquid nitrogen tank. For a liquid purity of about 10^{-15} g/g for U and Th, the expected background is about 0.2 c/keV/t/y. A 10 year measurement would give a sensitivity of about 10^{28} y ($\langle m_\nu \rangle \leq 0.015 \div 0.05$ eV). This experiment could have also an interesting sensitivity for real time solar neutrino detection and for cold Dark Matter. Although the authors think that, given the claim of the HM experiment, it could be no longer worth to proceed with a 1 ton experiment, recently they set up the Genius Test Facility in the Gran Sasso Underground Laboratory, where 4 2.5 kg Ge crystals are presently running in liquid nitrogen.[21]

Similar to Genius is the **GEM** proposal:[22] the main difference is the reduction of the amount of liquid nitrogen by adding an external layer of pure water.

Promising and also similar to Genius, is the new ^{76}Ge $\beta\beta$-0ν decay experiment recently proposed to the Gran Sasso Underground Laboratory (also known as **Gerda**).[23] The driving idea is to scrutinize the HM evidence in a short time. The proposed set up consists of a small liquid nitrogen tank (1.5 m diameter) shielded by a layer of Pb and by 2 m of pure water. The possibility to use liquid argon instead of nitrogen is also considered: the argon scintillation would provide an active shielding, especially useful to reduce the effect of cosmogenic ^{60}Co in the detectors. The aim of a first phase of the experiment is to reduce the background to about 0.01 c/keV/kg/y (mainly detector intrinsic) and to reach an exposure of about 15 kg×y using the 20 kg of ^{76}Ge recovered from HM and Igex. If the HM evidence is correct, Gerda would detect a 5σ signal. Following phases of the experiment foresee a further background reduction to about 0.1 c/keV/t/y and an increase of the enriched mass up to 1 ton, to reach $\langle m_\nu \rangle$ sensitivities in the 10 meV range.

3.2. *Calorimetric experiments with cryogenic detectors*

The Cuoricino experiment has proved that also the cryogenic detection technique is mature for a next generation $\beta\beta$-0ν decay experiment. It is worth noting that almost all the interesting isotopes can be studied with cryogenic detectors.[24,25,35,26] Cryogenic detectors have high energy resolution, can be scaled up to a 1 ton size and their background can be reduced by segmentation. Further background rejection can be achieved by hybrid detectors where, e.g., also scintillation or ionization are detected: these techniques can provide particle identification or position information.[24,25,27] The drawbacks of this technique are the sensitivity to surface contaminations, the difficulty to reduce close materials and the still cumbersome ancillary equipments required for cooling the detectors.

To date, the **CUORE** (Cryogenic Underground Observatory for Rare Events)[28] is the only fully approved next generation $\beta\beta$-0ν decay experiment: it is being built in the Gran Sasso Underground Laboratory where it is due to start data taking in 2009. CUORE will search for $\beta\beta$-0ν decay of ^{130}Te with a detector made of about 19 towers like the one of the running Cuoricino detector. 988 natural TeO$_2$ detectors will make up a 740 kg segmented and compact calorimeter containing 200 kg of ^{130}Te. Even if it is possible to achieve a high sensitivity just with natural Te, the possibility of introducing enriched material in the core of the detector is still an open option. A background of about 1 c/keV/t/y can be reached by exploiting the segmentation and by reducing of a factor 100 the surface contaminations observed in the Cuoricino experiment. This reduction is possible with the use of specially designed cleaning processes and, if necessary, by using surface sensitive detectors.[27] Presently the refrigerator and the underground experimental building are being designed: their construction will start in 2005. The material selection and the cleaning procedure settling are also in progress. With a background of about 1 c/keV/t/y and an energy resolution FWHM of about 10 keV, a 1σ sensitivity on $\tau_{1/2}^{0\nu}$ of about 6.6×10^{26} y can be reached in 10 years ($\langle m_\nu \rangle \leq 0.011 \div 0.056$ eV). CUORE is potentially also a good detector for cold Dark Matter and Solar Axions.[29]

3.3. *Calorimetric experiments with scintillators*

Scintillators provide a relatively simple and well established instrument to search for $\beta\beta$-0ν decay of many interesting isotopes. They can be extremely large and, in order to reduce the background, they can be immersed in the ultra pure liquids of large solar neutrino experiments (e.g. Superkamiokande, SNO or Borex) and use their photomultipliers. Their background can also be reduced by PSD. The main drawback is the poor energy resolution which makes the $\beta\beta$-2ν decay tail the main component of the background at $Q_{\beta\beta}$. Moreover photomultipliers and scintillators are often not enough radiopure for low background application.

Most noticeable are the **CAMEO** proposal[30] to immerse $CdWO_4$ crystals in Borexino or CTF, the **CANDLES** project[31] to use CaF_2 crystals in a liquid scintillator active shielding to search for $\beta\beta$-0ν decay of ^{48}Ca, in spite of its exceedingly low natural isotopic abundance, and the possible use of the Dark Matter self shielding 10 ton liquid Xe **XMASS** detector to look for ^{136}Xe $\beta\beta$-0ν decay.[32] Even more challenging are the ideas to place liquefied Xe in high pressure transparent cells in SNO,[32] to dissolve Xe in Borexino,[33] and to suspend scintillating nanocrystals in SNO.[34]

3.4. Tracking experiments

Tracking detectors could potentially avoid all background sources with the exception of the $\beta\beta$-2ν decay tail. The main issue for this technique is therefore the energy resolution. In case a $\beta\beta$-0ν signal is detected, the reconstruction of the electron tracks would also provide a unique tool to distinguish the decay mechanism from the electron angular correlation.

The **MOON** proposal[35] consists of sandwiches of ^{100}Mo enriched foils, plastic scintillators, and scintillating fibers which would provide the energy and position measurements. MOON would be also a solar neutrino experiment. **DCBA** project[36] proposes a Drift Chamber Beta Ray Analyser with ^{150}Nd enriched foils. There is also an Expression of Interest[37] for a **Super-NEMO** tracking detector with about 100 kg of enriched isotopes.

The **EXO** project[38] would deserve its own section because it is actually a calorimetric experiment with moderate tracking capability. It is the evolution of the Gotthard experiment[39] on ^{136}Xe, which used a high pressure Xe TPC. The EXO proposal adds the tagging of the Ba atoms produced by the ^{136}Xe decay to completely suppress all backgrounds. A single Ba^+ ion would be detected by optical spectroscopy. Presently there are still 2 open options for the detector: a high pressure Xe TPC or a liquid Xe TPC where also scintillation would be detected to improve energy resolution. The second option is the preferred one because of its compactness (a 10 ton detector would have a $3\,m^3$ volume), but Ba tagging requires single Ba^+ ion extraction from liquid. Running for 5 years a 10 ton Xe detector with an energy resolution of 2% and with just the $\beta\beta$-2ν tail background, a sensitivity on $\langle m_\nu \rangle$ of about $16\div22\,meV$ could be reached. Presently a 200 kg liquid Xe TPC prototype without tagging is being designed.[40]

References

1. S.R. Elliott et al., *Annu. Rev. Nucl. Part. Sci.* **52**, 115 (2002); S.R. Elliott et al., hep-ph/0405078.

2. Proceedings of "21st International Conference on Neutrino Physics and Astrophysics", Paris, June 2004; to be published on *Nucl. Phys.* **B** (Proc. Suppl.).
3. F. Feruglio et al., *Nucl. Phys.* **P637**, 345 (2002); F. Feruglio et al., *Nucl. Phys.* **P659**, 359 (2003); S. Pascoli et al., *Phys. Lett.* **B544**, 239 (2002); S. Pascoli et al., hep-ph/0310003; G.L. Fogli et al., hep-ph/0408045;
4. J. Suhonen et al., *Phys. Rev.* **300**, 123 (1998); O. Civitarese et al., *Nucl. Phys.* **A729**, 867 (2003); J.N. Bahcall et al., hep-ph/0403167.
5. V.A. Rodin et al., *Phys. Rev.* **C68**, 044302 (2003).
6. I. Ogawa et al., *Nucl. Phys.* **A730**, 215 (2004).
7. F.A. Danevich et al., *Phys. Rev.* **C68**, 035501 (2003).
8. F.A. Danevich et al., *Nucl. Phys.* **A694**, 375 (2001).
9. A. De Silva et al., *Phys. Rev.* **C56**, 2451 (1997).
10. E. Fiorini et al., *Phys. Lett.* **B25**, 602 (1967).
11. H.V. Klapdor-Kleingrothaus et al., *Nucl. Instrum. Methods* **A522**, 371 (2004).
12. H.V. Klapdor-Kleingrothaus et al., *Phys. Lett.* **B586**, 198 (2004).
13. C.E. Aalseth, *Phys. Rev.* **D65**, 092007 (2002).
14. C. Arnaboldi et al., *Phys. Lett.* **B584**, 260 (2004).
15. S. Cebrian, these proceedings.
16. X. Sarazin et al., in Ref. 2.
17. R. Bernabei et al., *Phys. Lett.* **B546**, 23 (2002).
18. K. Zuber, *Phys. Lett.* **B519**, 1 (2001).
19. R. Gaitskell et al., nucl-ex/0311013.
20. H.V. Klapdor-Kleingrothaus et al., hep-ph/9910205.
21. H.V. Klapdor-Kleingrothaus et al., *Nucl. Instrum. Methods* **A511**, 341 (2003).
22. Y.G. Zdesenko et al., *J. Phys.* **G27**, 2129 (2001).
23. I. Abt et al., hep-ex/0404039.
24. A. Alessandrello et al., *Phys. Lett.* **B420**, 109 (1998).
25. G. Chardin et al., *Nucl. Instrum. Methods* **A520**, 145 (2004).
26. A. Alessandrello et al., proceedings of the "17th International Conference on Neutrino Physics and Astrophysics", 1996, World Scientific.
27. M. Pedretti in Ref. 2.
28. C. Arnaboldi et al., NIM **A518**, 775 (2004).
29. C. Arnaboldi et al., *Astropart. Phys.* **20**, 91 (2003).
30. G. Bellini et al., nucl-ex/0007012.
31. I. Ogawa, proceedings of "The 5th Workshop on Neutrino Oscillations and their Origin", 2004, World Scientific.
32. S. Moriyama these proceedings; T. Namba in Ref. 2.
33. B. Caccianiga et al., *Astropart. Phys.* **14**, 15 (2000).
34. A. McDonald, "3rd SNOLAB Workshop on Underground Science", 2004.
35. H. Ejiri et al., *Phys. Rev. Lett.* **85**, 2917 (2000); M. Nomachi in Ref. 2.
36. N. Ishihara et al., *Nucl. Instrum. Methods* **A443**, 101 (2000);
37. http://nemo.in2p3.fr/supernemo/eoi_Super-NEMO.htm
38. http://www-project.slac.stanford.edu/exo/; SLAC EPAC Letter of Intent (2001).
39. R. Luescher et al., *Phys. Lett.* **B434**, 407 (1998).
40. J-L. Vuilleumier et al., these proceedings.

FIRST EVIDENCE FOR LEPTON NUMBER VIOLATION AND THE MAJORANA CHARACTER OF NEUTRINOS

H.V. KLAPDOR-KLEINGROTHAUS*

Max-Planck-Institut für Kernphysik, PO 10 39 80,
D-69029 Heidelberg, Germany

Nuclear double beta decay provides an extraordinarily broad potential to search for beyond-standard-model physics. *The occurrence of the neutrinoless decay* ($0\nu\beta\beta$) mode has fundamental consequences: first **total lepton number is not conserved**, and second, **the neutrino is a Majorana particle**. Further the effective mass measured allows to put an absolute scale of the neutrino mass spectrum. In addition, *double beta experiments yield sharp restrictions also for other beyond standard model physics*. These include SUSY models (R-parity breaking and conserving), leptoquarks (leptoquark-Higgs coupling), compositeness, left-right symmetric models (right-handeld W boson mass), test of special relativity and of the equivalence principle in the neutrino sector and others. **First evidence for neutrinoless double beta decay was given in 2001, by the HEIDELBERG-MOSCOW experiment**. The HEIDELBERG-MOSCOW experiment is the *by far most sensitive* $0\nu\beta\beta$ experiment since more than 10 years. It is operating 11 kg of enriched ^{76}Ge in the GRAN SASSO Underground Laboratory. The analysis of the data taken from 2 August 1990 - 20 May 2003, is presented here. The collected statistics is 71.7 kg y. The background achieved in the energy region of the Q value for double beta decay is 0.11 events/ kg y keV. *The two-neutrino accompanied half-life* is determined on the *basis of more than 100 000 events* to be $(1.74^{+0.18}_{-0.16}) \times 10^{21}$ years. **The confidence level for the *neutrinoless* signal has been improved to a 4.2 σ level. The half-life is** $T^{0\nu}_{1/2} = (1.19^{+0.37}_{-0.23}) \times 10^{25}$ *years*. **The effective neutrino mass deduced is (0.2 - 0.6) eV (99.73% c.l.)**, with the consequence that neutrinos have degenerate masses. The sharp boundaries for other beyond SM physics, mentioned above, are comfortably competitive to corresponding results from high-energy accelerators like TEVATRON, HERA, etc.

1. Introduction

Since 40 years huge experimental efforts have gone into the investigation of nuclear double beta decay which probably is the most sensitive way to look for (total) lepton number violation and probably the only way to decide

---

*Spokesman of Heidelberg-Moscow (and GENIUS-TF and HDMS) Collaborations, E-mail: H.Klapdor@mpi-hd.mpg.de, Home-page: $http://www.mpi-hd.mpg.de.non\_acc/$

the Dirac or Majorana nature of the neutrino. It has further perspectives to probe also other types of beyond standard model physics. This thorny way has been documented recently in some detail [27,36,29].

The HEIDELBERG-MOSCOW experiment, proposed already in 1987[8], has been looking for double beta decay of ^{76}Ge since August 1990 until November 30, 2003 in the Gran Sasso Underground Laboratory. It was using the largest source strength of all double beta experiments at present, and has reached a record low level of background, not only for Germanium double beta decay search. It has demonstrated this during more than a decade of measurements and is since more then ten years the most sensitive double beta decay experiment worldwide. The experiment was since 2001 operated only by the Heidelberg group, which also performed the analysis of the experiment from its very beginning.

The experiment has been carried out with five high-purity p-type detectors of Ge enriched to 86% in the isotope ^{76}Ge (in total 10.96 kg of active volume). These were the first enriched high-purity Ge detectors ever produced. So, the experiment starts from the cleanest thinkable source of double beta emitter material, which at the same time is used as detector of $\beta\beta$ events.

A description of the experimental details has been given in [1,2,3,9]. This will not be repeated in this paper, instead we concentrate on the results and their consequences. But let us just mention some of the most important features of the experiment here.

1. Since the sensitivity for the $0\nu\beta\beta$ half-life is $T^{0\nu}_{1/2} \sim a \times \epsilon \sqrt{\frac{Mt}{\Delta E B}}$ (and $\frac{1}{\sqrt{T^{0\nu}}} \sim \langle m_\nu \rangle$), with a denoting the degree of enrichment, ϵ the efficiency of the detector for detection of a double beta event, M the detector (source) mass, ΔE the energy resolution, B the background and t the measuring time, the sensitivity of our 11 kg *of enriched* ^{76}Ge experiment corresponds to that of an at least 1.2 ton *natural* Ge experiment. After enrichment - the other most important parameters of a $\beta\beta$ experiment are: energy resolution, background and source strength.

2. The high energy resolution of the Ge detectors of 0.2% or better, assures that there is no background for a $0\nu\beta\beta$ line from the two-neutrino double beta decay in this experiment, in contrast to most other present experimental approaches, where limited energy resolution is a severe drawback.

3. The efficiency of Ge detectors for detection of $0\nu\beta\beta$ decay events is close to 100 % (95%, see [2]).

4. The source strength in this experiment of 11 kg is the largest source strength ever operated in a double beta decay experiment.

5. The background reached in this experiment, is 0.113±0.007 events /kg y keV (in the period 1995-2003) in the $0\nu\beta\beta$ decay region (around $Q_{\beta\beta}$). This is the lowest limit ever obtained in such type of experiment.

6. The statistics collected in this experiment during 13 years of stable running is the largest ever collected in a double beta decay experiment. The experiment took data during $\sim$ 80% of its installation time.

7. The Q value for neutrinoless double beta decay has been determined recently with high precision [31].

2. Data and Analysis

The total sum spectrum measured over the full energy range of all five detectors for the period November 1995 to May 2003 is shown in Ref. [1,2,3]. The identified lines are indicated with their source of origin (in [17]).

Fig. 1 shows the part of the spectrum around $Q_{\beta\beta}$, in the range 2000 - 2060 keV, measured in the period August 1990 to May 2003 and November 1995 to May 2003. Non-integer numbers in the sum spectra are simply a binning effect.

The spectra shown in Fig. 1 result from summing the individual 2142 runs taken with the detectors, and finally summing the sum spectra of the different detectors (in total summing 9 570 data sets).

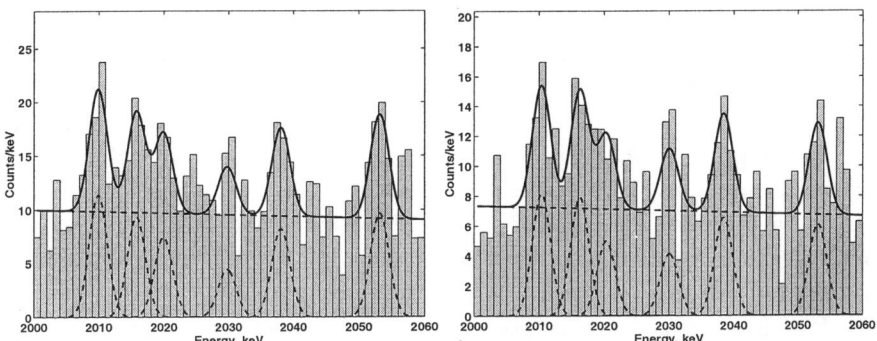

Figure 1. The total sum spectrum of all five detectors (in total 10.96 kg enriched in ^{76}Ge), for the period: left: November 1990 to May 2003 (71.7 kg y) in the range 2000 - 2060 keV right: - November 1995 to May 2003 (56.66 kg y) in the range 2000 - 2060 keV and its fit (see text and [1,2,3]).

In the measured spectra (Fig.1) we see in the range around $Q_{\beta\beta}$ the ^{214}Bi lines at 2010.7, 2016.7, 2021.8, 2052.9 keV, the line at $Q_{\beta\beta}$ and a candidate of a line at $\sim 2030\,\text{keV}$ (see also [12,18])[a]. The spectra have been analyzed by *different methods*: Least Squares Method, Maximum Likelihood Method (MLM) and Feldman-Cousins Method. The analysis is performed *without subtraction of any background*. We always process background-plus-signal data since the difference between two Poissonian variables does *not* produce a Poissonian distribution [32]. This point has been sometimes overlooked. So, e.g., in [41] a formula is developed making use of such subtraction and as a consequence the analysis given in [41] provides overestimated standard errors.

The improvement of the present analysis (for details see [1,2,3]) compared to our paper from 2001 [4,5,6], is described in detail in [1,2,3]. One reason lies in the stricter conditions for accepting data into the analysis. The second reason is a better energy calibration of the individual runs. The third reason is the refined summing procedure of the individual data sets mentioned above and the correspondingly better energy resolution of the final spectrum. (For more details see [1,2,3]). The signal strength seen in the *individual* detectors in the period 1990-2003 is shown in [3].

3. Results

3.1. *Full Spectra*

Fig.1 shows together with the measured spectra in the range around $Q_{\beta\beta}$ (2000 - 2060 keV), the result of the fit of this energy range. A linear decreasing shape of the background as function of energy was chosen corresponding to the complete simulation of the background performed in [17] by GEANT4 (see Fig.2). In the fits in Fig.1, the peak positions, widths and intensities are determined simultaneously, and also the *absolute* level of the background.

The signal at $Q_{\beta\beta}$ in the full spectrum at $\sim 2039\,\text{keV}$ **reaches a 4.2 σ confidence level** for the period 1990-2003 (28.8 ± 6.9) events, and of 4.1 σ for the period 1995-2003 (23.0 ±5.7) events. The results of the new analysis are consistent with the results given in [4,5,6]. The intensities of all other lines are given in [2,3].

We have given a detailed comparison of the spectrum measured in this experiment with other Ge experiments in [18]. It is found that the most sensitive experiment with natural Ge detectors [13], and the first experiment

---

[a]The objections raised after our first paper [4] concerning these lines and other points, by Aalseth et al. (Mod.Phys.Lett.A17:1475-1478,2002 and hep-ex/0202018 v.1), **have been shown to be wrong already** in [7] and in [6], and later in [12] and [18]. So this 'criticism' was already history, before we reached the higher statistics presented in this paper.

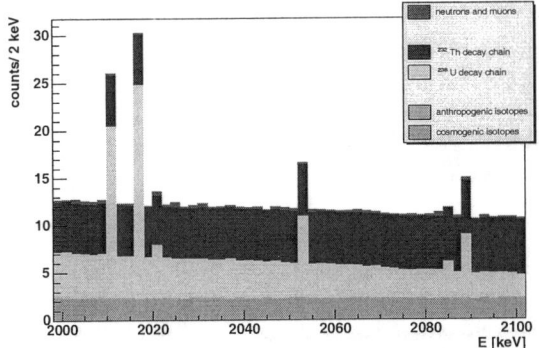

Figure 2. Monte Carlo simulation of the background in the range of $Q_{\beta\beta}$ by GEANT4, including all known sources of background in the detectors and the setup. This simulation [17] seems to be the by far most extensive and complete one ever made for any double beta experiment. The background around $Q_{\beta\beta}$ is expected to be flat, the only lines visible should be some weak ^{214}Bi lines (from [17]).

using enriched (not yet high-purity) ^{76}Ge detectors [14] find essentially the same background lines (^{214}Bi etc.), but *no* indication for the line near $Q_{\beta\beta}$. This is consistent with the rates expected from the present experiment due to their lower sensitivity: ~ 0.7 and ~ 1.1 events, respectively. It is also consistent with the result of the IGEX ^{76}Ge experiment [15], which collected only a statistics of 8.8 kg y, before finishing in 1999, and which should expect ~ 2.6 events, which they might have missed. Their published half-life limit is overestimated as result of an arithmetic mistake (see [16]).

3.2. *Time Structure of Events*

There are at present *no other* running experiments (with reasonable energy resolution) which can - not to speak about their lower sensitivity - *in principle* give *any further-going* information in the search for double beta decay than shown up to this point: namely a line at the correct energy $Q_{\beta\beta}$. Also most future projects cannot determine more. The HEIDELBERG-MOSCOW experiment developed some *additional tool* of independent verification. The method is to exploit the time structure of the events and to select $\beta\beta$ events by their pulse shape exploiting neuronal net methods. The result is shown in Fig.3 (see [1,2,3]). Except a line which sticks out sharply near $Q_{\beta\beta}$, *all* other lines are very strongly suppressed. The probability to find ~ 7 events in two neighboring channels from background fluctuations is calculated to be 0.013%. Thus, we see a line near $Q_{\beta\beta}$ at a 3.8σ level.

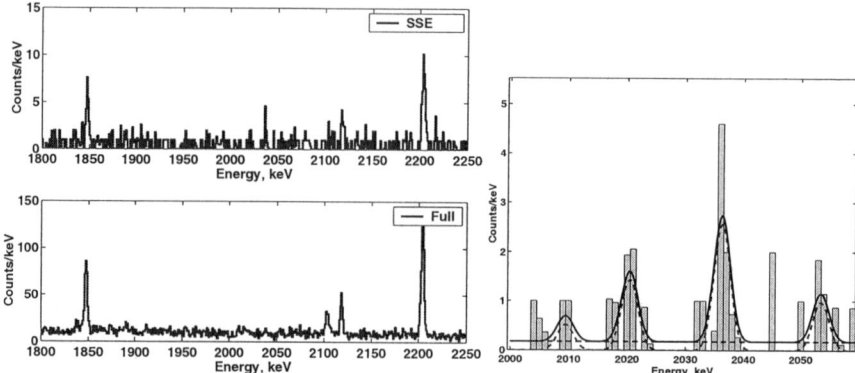

Figure 3. Left top: The pulse-shape selected spectrum of single site events measured with detectors 2,3,4,5 from 1995-2003, see text. Left below: The full spectrum measured with detectors 2,3,4,5 from 1995-2003. Right: As in top figure, but energy range 2000-2060 keV, to be compared to Fig. 1, right (see [1,2,3]).

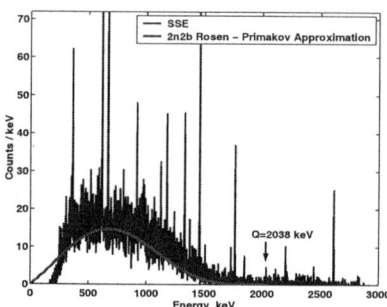

Figure 4. The pulse-shape selected spectrum measured with detectors 2,3,4,5 from 1995÷2003 in the energy range of (100÷3000) keV, see text.

The energy of this line determined by the spectroscopy ADC is slightly below $Q_{\beta\beta}$, but still within the statistical variation for a weak line (see [18]). This can be understood as result of ballistic effects (for details see[24]). Obviously the method also fulfills the criterium to select properly the *continuous* $2\nu\beta\beta$ *spectrum* (see Fig. 4).

The 2039 keV line as a single site events signal cannot be the double escape line of a γ-line whose full energy peak would be expected at 3061 keV, since no indication of a line is found there in the spectrum measured up to 8 MeV (see [2,3]).

4. Half-Life of Neutrinoless Double Beta Decay of ^{76}Ge

We have shown in chapter 4 that the signal found at $Q_{\beta\beta}$ is consisting of single site events and is not a γ line. The signal does not occur in the Ge experiments *not* enriched in the double beta emitter ^{76}Ge [13,11,18], while neighbouring background lines appear consistently in these experiments.

Table 1. Half-life for the neutrinoless decay mode and deduced effective neutrino mass from the HEIDELBERG-MOSCOW experiment (the nuclear matrix element of [22] is used). Shown are in addition to various accumulated total measuring times also the results for four *non-overlapping* data sets: the time periods 11.1995-09.1999 and 09.1999÷05.2003 for *all* detectors, and the time period 1995÷2003 for two sets of detectors: 1+2+4, and 3+5. *) denotes best value.

Significance [kg y]	Detectors	$T^{0\nu}_{1/2}$ [y] (3σ range)	$\langle m \rangle$ [eV] (3σ range)	Conf. level (σ)
Period 8.1990 ÷ 5.2003				
71.7	1,2,3,4,5	$(0.69 - 4.18) \times 10^{25}$ 1.19×10^{25} *	(0.24 - 0.58) 0.44*	4.2
Period 11.1995 ÷ 5.2003				
56.66	1,2,3,4,5	$(0.67 - 4.45) \times 10^{25}$ 1.17×10^{25} *	(0.23 - 0.59) 0.45*	4.1
51.39	2,3,4,5	$(0.68 - 7.3) \times 10^{25}$ 1.25×10^{25} *	(0.18 - 0.58) 0.43*	3.6
42.69	2,3,5	$(0.88 - 4.84) \times 10^{25}$ (2σ range) 1.5×10^{25} *	(0.22 - 0.51) (2σ range) 0.39*	2.9
28.27	1,2,4	$(0.67 - 6.56) \times 10^{25}$ (2σ range) 1.22×10^{25} *	(0.19 - 0.59) (2σ range) 0.44*	2.5
28.39	3,5	$(0.59 - 4.29) \times 10^{25}$ (2σ range) 1.03×10^{25} *	(0.23 - 0.63) (2σ range) 0.48*	2.6
Period 11.1995 ÷ 09.1999				
26.59	1,2,3,4,5	$(0.43 - 12.28) \times 10^{25}$ 0.84×10^{25} *	(0.14 - 0.73) 0.53*	3.2
Period 09.1999 ÷ 05.2003				
30.0	1,2,3,4,5	$(0.60 - 8.4) \times 10^{25}$ 1.12×10^{25} *	(0.17 - 0.63) 0.46*	3.5

On this basis we translate the observed numbers of events into half-lives for neutrinoless double beta decay. In Table 1 we give the half-lives deduced from the full data sets taken in the years 1995-2003 and in 1990-2003 and of some partial data sets. In all cases the signal is seen consistently. Also given are the deduced effective neutrino masses. The result obtained **is consistent** with the limits given earlier [10], and with the results given in [4,5,6].

Concluding **we confirm, with 4.2σ (99.9973% c.l.) probability,** our claim from 2001 [4,5,6] **of first evidence for the neutrinoless double beta decay mode.**

5. Consequences for Particle Physics, Neutrino Physics and Other Beyond Standard Model Physics

Lepton number violation: *The most important consequence* of the observation of neutrinoless double beta decay is, that **lepton number is not conserved.** This is fundamental for particle physics, and for the early Universe, e.g. for leptogenesis.

Majorana nature of neutrino: Another fundamental consequence is that **the neutrino is a Majorana particle** (see, e.g. [38,39], but also [40]). Both of these conclusions are *independent of any* discussion of nuclear matrix elements. It has been discussed that the Majorana nature of the neutrino tells us that spacetime does realize a construct that is central to construction of supersymmetric theories [33].

Effective neutrino mass: The matrix element enters when we derive a *value* for the effective neutrino mass - making the *most natural assumption* that the $0\nu\beta\beta$ decay amplitude is dominated by exchange of a massive Majorana neutrino. The half-life for the neutrinoless decay mode is under this assumption given by [22,23]

$$[T^{0\nu}_{1/2}(0^+_i \to 0^+_f)]^{-1} = C_{mm}\frac{\langle m \rangle^2}{m_e^2} + C_{\eta\eta}\langle \eta \rangle^2 + C_{\lambda\lambda}\langle \lambda \rangle^2 + C_{m\eta}\langle \eta \rangle \frac{\langle m \rangle}{m_e}$$
$$+ C_{m\lambda}\langle \lambda \rangle \frac{\langle m \rangle}{m_e} + C_{\eta\lambda}\langle \eta \rangle \langle \lambda \rangle,$$

$$\langle m \rangle = |m^{(1)}_{ee}| + e^{i\phi_2}|m^{(2)}_{ee}| + e^{i\phi_3}|m^{(3)}_{ee}|, \tag{1}$$

where $C_{mm}, C_{\eta\eta}, ...$ denote nuclear matrix elements squared, which can be calculated, (see, e.g. [27,35,34], for a review). Ignoring contributions from right-handed weak currents, on the right-hand side of eq.(1) only the first term remains.

Using the nuclear matrix element from [22,23], we conclude from the half-life given above the effective mass $\langle m \rangle$ to be $\langle m \rangle = (0.2 \div 0.6)\,\text{eV}$ (99.73% c.l.), with **best value of ∼ 0.4 eV**.

The matrix element given by [22] was the *prediction closest to* the *later* measured $2\nu\beta\beta$ decay half-life of $(1.74^{+0.18}_{-0.16}) \times 10^{25}$ y [17,9]. It underestimates the 2ν matrix elements by 32% and thus these calculations will also underestimate (to a smaller extent) the matrix element for $0\nu\beta\beta$ decay, and consequently correspondingly overestimate the (effective) neutrino mass. Allowing conservatively for an uncertainty of the nuclear matrix element of

± 50% the range for the effective mass may widen to $\langle m \rangle = (0.1 - 0.9)$ eV (99.73% c.l.).

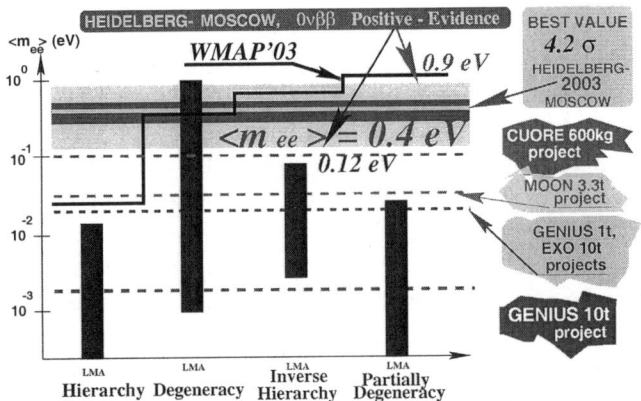

Figure 5. The impact of the evidence obtained for neutrinoless double beta decay (best value of the effective neutrino mass $\langle m \rangle = 0.4$ eV, 3σ confidence range (0.1 - 0.9) eV - allowing already for an uncertainty of the nuclear matrix element of a factor of ± 50%) on possible neutrino mass schemes. The bars denote allowed ranges of $\langle m \rangle$ in different neutrino mass scenarios, still allowed by neutrino oscillation experiments (see [26,48]). All models except the degenerate one are excluded by the new $0\nu\beta\beta$ decay result. Also shown is the exclusion line from WMAP, plotted for $\sum m_\nu < 1.0\,eV$ [49] (which is according to [62] too strict). WMAP does not rule out any of the neutrino mass schemes. Further shown are the expected sensitivities for the future potential double beta experiments CUORE, MOON, EXO and the 1 ton and 10 ton project of GENIUS [27,29,56] (from [48]).

Neutrinos degenerate in mass: With the value deduced for the effective neutrino mass, the HEIDELBERG-MOSCOW experiment excludes several of the neutrino mass scenarios allowed from present neutrino oscillation experiments (see Fig. 5, and Fig.1 in [48]), - allowing only for degenerate mass scenarios [26,48,25].

Other beyond Standard Model Physics: Assuming *other* mechanisms to dominate the $0\nu\beta\beta$ decay amplitude, which have been studied extensively in our group, and other groups, in recent years, the result allows to set stringent limits on parameters of SUSY models, leptoquarks, compositeness, masses of heavy neutrinos, the right-handed W boson and possible violation of Lorentz invariance and equivalence principle in the neutrino sector. For a further discussion and for references we refer to [27,28,29,30].

6. Conclusion - Perspectives

Recent information from many *independent* sides seems to condense now to a nonvanishing neutrino mass of the order of the value found by the

HEIDELBERG-MOSCOW experiment. This is the case for the results from CMB, LSS, neutrino oscillations, particle theory and cosmology (for a detailed discussion see [1,2,3]). To mention a few examples: Neutrino oscillations require in the case of degenerate neutrinos common mass eigenvalues of m > 0.04 eV. An analysis of CMB, large scale structure and X-ray from clusters of galaxies yields a 'preferred' value for $\sum m_\nu$ of 0.6 eV [50]. WMAP yields $\sum m_\nu < 1.0$ eV [49], SDSS yields $\sum m_\nu < 1.7$ eV [62]. Theoretical papers require degenerate neutrinos with m > 0.1, or 0.2 eV or 0.3 eV [45,43,51,52,44], and the recent alternative cosmological concordance model requires relic neutrinos with mass of order of eV [53]. As mentioned already earlier [37,2] the results of double beta decay and CMB measurements together indicate that the neutrino mass eigenvalues have the same CP parity, as required by the model of [45]. Also the approach of [61] comes to the conclusion of a Majorana neutrino. The Z-burst scenario for ultra-high energy cosmic rays requires $m_\nu \sim 0.4$ eV [46,47], and also a non-standard model (g-2) has been connected with degenerate neutrino masses >0.2 eV [42]. The neutrino mass determined from $0\nu\beta\beta$ decay is consistent also with present models of leptogenesis in the early Universe [57].

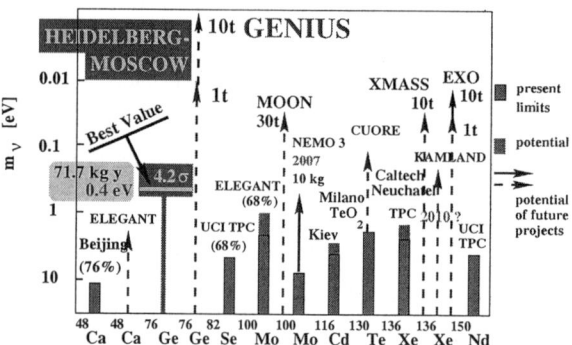

Figure 6. Present sensitivity, and expectation for the future, of the most promising $\beta\beta$ experiments. Given are limits for $\langle m \rangle$, except for the HEIDELBERG-MOSCOW experiment where the measured *value* is given (3σ c.l. range and best value). Framed parts of the bars: present status; not framed parts: future expectation for running experiments; solid and dashed lines: experiments under construction or proposed, respectively. For references see [27,5,6].

Future: With the HEIDELBERG-MOSCOW experiment, *the era of small smart experiments is over.* Fig. 6 shows the present result and a comparison to the potential of the most sensitive other double beta decay experiments and the possible potential of some future projects. It is visible that the presently running experiments have hardly a chance, to reach the sensitivity of the HEIDELBERG-MOSCOW experiment. New approaches and considerably *enlarged* experiments would be required to fix the $0\nu\beta\beta$ half life with higher accuracy. **This will, however, only marginally improve the precision of the deduced neutrino mass,** because of the uncertainties in the nuclear matrix elements, which probably hardly can be reduced to less than 50%.

One has to keep in mind further, that **no more can be learnt** on *other* beyond standard model physics parameters from future more sensitive experiments. The reason is that there is **a half-life** now, and **no more a limit** on the half-life, which could be further reduced.

From future projects one has to require that they should be able to differentiate between a β and a γ signal, or that the tracks of the emitted electrons should be measured. At the same time, as is visible from the present information, the energy resolution should be *at least* in the order of that of Ge semiconductor detectors, or better. These requirements exclude at present calorimeter experiments like CUORE, CUORICINO, which *cannot* differentiate between a β and γ signal, etc, but also experiments like EXO [60], *if* the latter will not be able to reconstruct the tracks of the electrons, as it seems at present. The NEMO project *can* see tracks, but unfortunately has at present only a small efficiency, and a low energy resolution of more than 200 keV. The most sensitive future project, is probably the GENIUS project, proposed already in 1997 [54,58,59,55,56,28,27].

A GENIUS Test Facility, (which could already be used to search for cold dark matter by the annual modulation effect) has started operation with 10 kg of natural Germanium detectors in liquid nitrogen in Gran-Sasso on May 5, 2003 [21,20,19].

However, if one wants to get *independent* evidence for the neutrinoless double beta decay mode, one would probably, wish to see the effect in *another* isotope, which would then simultaneously give additional information also on the nuclear matrix elements. In view of these considerations, future efforts to obtain *deeper* information on the process of neutrinoless double beta decay, would require *a new experimental approach, different from all, what is at present persued*.

Acknowledgement:
The author would like to thank all colleagues, who have contributed to the experiment over the last 15 years. He thanks in particular Irina Krivosheina, for her important contribution to the analysis of this experiment.

Our thanks extend also to the technical staff of the Max-Planck Institut für Kernphysik and of the Gran Sasso Underground Laboratory. We acknowledge the invaluable support from BMBF and DFG, and LNGS of this project. We are grateful to the former State Committee of Atomic Energy of the USSR for providing the enriched material used in this experiment.

The author thanks Prof. Neil Spooner for the kind invitation to IDM 2004.

References

1. H.V. Klapdor-Kleingrothaus, I.V. Krivosheina, A. Dietz et al., Phys. Lett. B 586 (2004) 198-212.
2. H.V. Klapdor-Kleingrothaus, A. Dietz, I.V. Krivosheina et al., NIM A 522 (2004) 371-406.
3. H.V. Klapdor-Kleingrothaus, in Proc. of BEYOND03, Castle Ringberg, Germany, 9-14 June 2003, Springer (2004), ed. H.V. Klapdor-Kleingrothaus, 307.
4. H.V. Klapdor-Kleingrothaus et al. Mod. Phys. Lett. A 16 (2001) 2409-2420.
5. H.V. Klapdor-Kleingrothaus, A. Dietz, I.V. Krivosheina, Part. and Nucl. 110 (2002) 57-79.
6. H.V. Klapdor-Kleingrothaus, A. Dietz, I.V. Krivosheina, Foundations of Physics 31 (2002) 1181-1223 and Corrigenda, 2003.
7. H.V. Klapdor-Kleingrothaus et al., hep-ph/0205228
8. H.V. Klapdor-Kleingrothaus, Proposal, MPI-1987-V17, September 1987.
9. HEIDELBERG-MOSCOW Coll., Phys. Rev. D 55 (1997) 54.
10. H.V. Klapdor-Kleingrothaus et al., (HEIDELBERG-MOSCOW Coll.), Eur. Phys. J. A 12, 147 (2001) and in Proc. of 3-rd Int. Conf. DARK2000, H.V. Klapdor-Kleingrothaus, ed., (Springer, Heidelberg, 2001) pp. 520-533.
11. H.V. Klapdor-Kleingrothaus et al., Nucl. Instr. Meth. 510A (2003) 281.
12. H.V. Klapdor-Kleingrothaus et al., Nucl. Instr. Meth. 511 A (2003) 335.
13. D. Caldwell, J. Phys. G 17 (1991) S137-S144.
14. A.A. Vasenko et al., Mod. Phys. Lett. A5 (1990) 1299, and I. Kirpichnikov, Preprint ITEP (1991).
15. C.E. Aalseth et al., Phys. Rev. D65 (2002) 092007.
16. H.V. Klapdor-Kleingrothaus et al., Phys. Rev. D70 (2004) 078301.
17. Ch. Dörr, H.V. Klapdor-Kleingrothaus, Nucl. Instr. Meth. A513 (2003) 596.
18. H.V. Klapdor-Kleingrothaus et al., Phys. Lett. 578 B (2004) 54 and NIM A510 (2003) 281.
19. H.V. Klapdor-Kleingrothaus, CERN Courier 43 N6 (2003) 9 and hep-ph/0307329, "'Naked' Crystals go Underground".
20. H.V. Klapdor-Kleingrothaus et al., Nucl. Instr. Meth. A 511 (2003) 341-346.
21. H.V. Klapdor-Kleingrothaus et al., Nucl. Instr. Meth. A 530 (2004) 410-418.
22. A. Staudt, K. Muto, H.V. Klapdor-Kleingrothaus, Eur. Lett. 13 (1990) 31.
23. K. Muto, E. Bender, H.V. Klapdor, Z. Phys. A 334 (1989) 187.
24. H.V. Klapdor-Kleingrothaus et al., in preparation.
25. H.V. Klapdor-Kleingrothaus, H. Päs, A.Yu. Smirnov, Phys. Rev. D 63 (2001) 073005; in Proc. of DARK'2000, Heidelberg, 10-15 July, 2000, Germany, ed. H.V. Klapdor-Kleingrothaus, Springer, Heidelberg (2001) 420-434.

26. H.V. Klapdor-Kleingrothaus, U. Sarkar, Mod. Phys. Lett. A 16 (2001) 2469.
27. H.V. Klapdor-Kleingrothaus, "60 Years of Double Beta Decay - From Nuclear Physics to Beyond the Standard Model", World Scientific, Singapore (2001) 1281 pages.
28. H.V. Klapdor-Kleingrothaus, Int. J. Mod. Phys. A 13 (1998) 3953.
29. H.V. Klapdor-Kleingrothaus, Springer Tracts in Modern Physics, 163 (2000) 69-104, Springer-Verlag, Heidelberg, Germany (2000).
30. H.V. Klapdor-Kleingrothaus, U. Sarkar, hep-ph/0302237.
31. G. Douysset et al., Phys. Rev. Lett. 86 (2001) 4259;J.G. Hykawy et al., Phys. Rev. Lett. 67 (1991) 1708; G. Audi, A.H. Wapstra, Nucl. Phys. A 595 (1995) 409-480; R.J. Ellis et al., Nucl. Phys. A 435 (1985) 34-42.
32. M.D. Hannam, W.J. Thompson, Nucl. Instr. Meth. A 431 (1999) 239-251.
33. D.V. Ahluwalia, in Proc. of BEYOND'02, Oulu, Finland, 2-7 Juni, 2002, IOP, Bristol, 2003, ed. H.V. Klapdor-Kleingrothaus, 143-160; D.V. Ahluwalia, M. Kirchbach, Phys. Lett. B529 (2002) 124.
34. K. Muto, H.V. Klapdor, in "Neutrinos", Graduate Texts in Contemporary Physics", ed. H.V. Klapdor, Berlin, Germany: Springer (1988) 183-238.
35. K. Grotz, H.V. Klapdor, "Die Schwache Wechselwirkung in Kern-, Teilchen- und Astrophysik", B.G. Teubner, Stuttgart (1989), "The Weak Interaction in Nuclear, Particle and Astrophysics", IOP Bristol (1990), Moscow, MIR (1992) and China (1998).
36. H.V. Klapdor-Kleingrothaus, A. Staudt, "Teilchenphysik ohne Beschleuniger", B.G. Teubner, Stuttgart (1995), "Non–Accelerator Particle Physics", IOP Pub., Bristol and Philadelphia (1995) and 2. ed. (1998), Moscow, Nauka, Fizmalit (1998).
37. H.V. Klapdor-Kleingrothaus, in Proc. of Int. Conf. BEYOND'02, Oulu, Finland, 2-7 Jun. 2002, IOP, Bristol, 2003 ed. H.V. Klapdor-Kleingrothaus, 215.
38. J. Schechter, J.W.F. Valle, Phys. Rev. D 25 (1982) 2951-2954.
39. M. Hirsch, H.V. Klapdor-Kleingrothaus, Phys. Lett. B 398 (1997) 311; Phys. Rev. D 57 (1998) 1947; M. Hirsch, H.V. Klapdor-Kleingrothaus, St. Kolb, Phys. Rev. D 57 (1998) 2020.
40. G. Bhattacharyya, H.V. Klapdor-Kleingrothaus, H. Päs, A. Pilaftsis, Phys. Rev. D 67 (2003) 113001 and hep-ph/0402071 in Proc. of Int. Worksh. on Astr. and HE Phys. (AHEP-2003), Valencia, Spain, 14-18 Oct 2003.
41. Yu. Zdesenko et al., Phys. Lett. B 546 (2002) 206-215.
42. E. Ma, M. Raidal, Phys. Rev. Lett. 87 (2001) 011802; Erratum-ibid. 87 (2001) 159901 and hep-ph/0102255.
43. K.S. Babu et al., Phys. Lett. B552 (2003) 207-213.
44. E. Ma in Proc. of Intern. Conf. BEYOND'02, Oulu, Finland, 2-7 Jun. 2002, IOP, Bristol, 2003, and BEYOND 2003, Ringberg Castle, Tegernsee, Germany, 9-14 Juni 2003, Springer, Heidelberg, Germany, 2004, ed. H.V. Klapdor-Kleingrothaus.
45. R.N. Mohapatra, M.K. Parida, G. Rajasekaran, (2003) hep-ph/0301234.
46. D. Fargion et al., in Proc. of DARK2000, Heidelberg, Germany, July 10-15, 2000, Ed. H.V. Klapdor-Kleingrothaus, *Springer*, (2001) 455-468 and in Proc. of Beyond the Desert 2002, BEYOND02, Oulu, Finland, June 2002,

IOP 2003, and BEYOND03, Ringberg Castle, Tegernsee, Germany, 9-14 Juni 2003, Springer, Heidelberg, Germany, 2003, ed. H.V. Klapdor-Kleingrothaus.
47. Z. Fodor, S.D. Katz, A. Ringwald, Phys. Rev. Lett. 88 (2002) 171101 and Z. Fodor et al., *JHEP* (2002) 0206:046, and in Proc. of Intern. Conf. Beyond the Desert 02, BEYOND'02, Oulu, Finland, 2-7 Jun 2002, IOP, Bristol, 2003, ed. H V Klapdor-Kleingrothaus and hep-ph/0210123.
48. H. V. Klapdor-Kleingrothaus, U. Sarkar, Mod. Phys. Letter. A18 (2003) 2243.
49. S. Hannestad, CAP 0305 (2003) 920030 004, in Proc. of 4th Int. Conf. BEYOND03, Ringberg Castle, Germany, 9-14 Juni 2003, Springer, Heidelberg, Germany, 2003, ed. H.V. Klapdor-Kleingrothaus.
50. S.W. Allen, R.W. Schmidt, S.L. Bridle, astro-ph/0306386.
51. K.S. Babu, E. Ma and J.W.F. Valle, Phys. Lett. B 552 (2003) 207-213.
52. M. Hirsch et al., Phys. Rev. D69 (2004) 093006.
53. A. Blanchard, M. Douspis, M. Rowan-Robinson, S. Sarkar, astro-ph/0304237.
54. H.V. Klapdor-Kleingrothaus in Proc. of BEYOND'97, Castle Ringberg, Germany, 8-14 June 1997, ed. by H.V. Klapdor-Kleingrothaus et al.,IOP Bristol (1998) 485-531, and Int. J. Mod. Phys. A13 (1998) 3953.
55. H.V. Klapdor-Kleingrothaus, J. Hellmig et al.,J. Phys. G24 (1998) 483-516.
56. H.V. Klapdor-Kleingrothaus et al. MPI-Report MPI-H-V26-1999, hep-ph/9910205, in Proc. of the 2nd Int. Conf. BEYOND'99, Castle Ringberg, Germany, 6-12 June 1999, eds. H.V. Klapdor-Kleingrothaus and I.V. Krivosheina, IOP Bristol (2000) 915-1014.
57. M.N. Rebelo, Proc. of BEYOND'2003, Castle Ringberg, Germany, July 2003, ed. H.V. Klapdor-Kleingrothaus, Springer, Heidelberg (2004) 267.
58. H.V. Klapdor-Kleingrothaus, M. Hirsch, Z. Phys. A 359 (1997) 361-372.
59. J. Hellmig, H.V. Klapdor-Kleingrothaus, Z. Phys. A 359 (1997) 351-359.
60. G. Gratta, ApPEC Paris, France 22.01.2002, and in Proc. of LowNu2, Dec. 4-5 (2000) Tokyo, Japan, ed: Y. Suzuki, World Scientific (2001) p.98.
61. R. Hofmann, hep-ph/0401017 v.1.
62. M. Tegmark et al., astro-ph/0310723, subm. Phys. Rev. D.

RESULTS FROM CUORICINO AND PROSPECTS FOR CUORE

S. CEBRIAN[a], R. ARDITO[b], C. ARNABOLDI[c], D.R. ARTUSA[d],
F.T. AVIGNONE III[d], M. BALATA[e], I. BANDAC[d], M. BARUCCI[f],
J.W. BEEMAN[g], F. BELLINI[h], C. BROFFERIO[c], C. BUCCI[e], S. CAPELLI[c],
F. CAPOZZI[c], L. CARBONE[c], C. COSMELLI[h], O. CREMONESI[c],
R.J.CRESWICK[d], A. DE WAARD[i], M. DOLINSKI[j], H.A.FARACH[d],
F. FERRONI[h], E. FIORINI[c], G.FROSSATI[i], C. GARGIULO[h],
A. GIULIANI[k], P. GORLA[a], E. GUARDINCERRI[l], T. GUTIERREZ[g],
E.E. HALLER[j], I. G. IRASTORZA[a], E. LONGO[h], G. MAIER[b], R.
MARUYAMA[j], R.J. MCDONALD[g], S. MORGANTI[h], S. NISI[e],
E.B. NORMAN[g], A. NUCCIOTTI[c], E. OLIVIERI[f], P. OTTONELLO[l],
M. PALLAVICINI[l], E. PALMIERI[m], E. PASCA[f], M. PAVAN[c],
M. PEDRETTI[k], G. PESSINA[c], S. PIRRO[c], E. PREVITALI[c], B. QUITER[j],
L. RISEGARI[f], C. ROSENFELD[d], S. SANGIORGIO[k], M. SISTI[c],
A.R. SMITH[g], S. TOFFANIN[m], L. TORRES[c], G. VENTURA[f], N. XU[g]

[a] Laboratorio de Fisica Nuclear y Alta Energias, Universidad de Zaragoza, 50009 Zaragoza, Spain

[b] Dipartimento di Ingegneria Strutturale del Politecnico di Milano, Milano I-20133, Italy

[c] Dipartimento di Fisica dell'Università di Milano-Bicocca e Sezione di Milano dell'INFN, Milan I-2016, Italy

[d] Department of Physics and Astronomy, University of South Carolina, Columbia, South Carolina, 29208 USA

[e] Laboratori Nazionali del Gran Sasso, I-67010, Assergi (L'Aquila), Italy

[f] Dipartimento di Fisica dell'Università di Firenze e Sezione di Firenze dell'INFN, Firenze I-50125, Italy

[g] Lawrence Berkeley National Laboratory, Berkeley, California 94720, USA

[h] Dipartimento Fisica dell'Università di Roma e Sezione di Roma 1 dell'INFN, Roma I-00185, Italy

[i] Kamerling Onnes Laboratory, Leiden University, 2300 RAQ, Leiden, The Netherlands

[j] University of California, Berkeley, California 94720, USA

[k] Dipartimento di Fisica e Matemtica dell'Università dell'Insubria e Sezione di Milano dell'INFN, Como I-22100, Italy

[l] Dipartimento di Fisica dell'Università di Genova e Sezione di Genova dell'INFN, Genova I-16146, Italy

^{m} *Laboratori Nazionali di Legnaro, Via Romea 4, I-35020 Legnaro (Padova), Italy*

CUORE will be an observatory for rare events consisting of a tightly packed array of TeO$_2$ bolometers, with a total mass of ~740 kg, operating in the underground Gran Sasso laboratory. A first step towards CUORE is CUORICINO, a running experiment with 40.7 kg of TeO$_2$. Present results from CUORICINO for the neutrinoless double beta decay of ^{130}Te will be presented here ($T^{0\nu}_{1/2} \geq 1.0 \times 10^{24}$ y, $\langle m_\nu \rangle \leq 0.26\text{-}1.4$ eV at 90% C.L.). The status of CUORE preparation and its physics potential, including dark matter searches, will be shown.

1. Introduction

Experiments searching for neutrinoless Double Beta Decay (DBD) can prove the Majorana character of neutrinos and determine the scale of the neutrino mass. CUORE (Cryogenic Underground Observatory for Rare Events)[1,2] is a proposed next generation experiment with almost 1000 bolometers of TeO$_2$ and a total mass of ~740 kg to operate at the Laboratori Nazionali del Gran Sasso (LNGS) in Italy. It is intended to investigate mainly double beta decay but searches for other rare events like the direct detection of WIMPs are also envisaged[3]. The first step towards CUORE is CUORICINO[4], running in Gran Sasso since the beginning of 2003 with 40.7 kg of TeO$_2$. In this paper, status and present results for CUORICINO will be presented and preparation and physics potential for CUORE will be described.

2. CUORICINO results

The CUORICINO detector consists of a tower of TeO$_2$ crystals with 13 planes (see Fig. 1)[4]. Eleven of these planes are 4-crystal modules like those foreseen for CUORE. Two additional planes are made of 9-crystal modules. CUORICINO contains therefore 44 5×5×5 cm^3 crystals (average mass 790 g) and 18 3×3×6 cm^3 crystals (average mass 330 g, previously used in the MiDBD experiment[5]). The same dilution refrigerator used in MiDBD at Hall A of LNGS is being used to operate at ~10 mK.

CUORICINO was cooled for the first time at the beginning of 2003. Unfortunately during the cooling procedure some of the signal wires disconnected so that only 32 of the large crystals and 16 of the small ones could be read. Nevertheless, in April 2003 the first background measurement started and went on up to October 2003, when data taking was stopped to solve

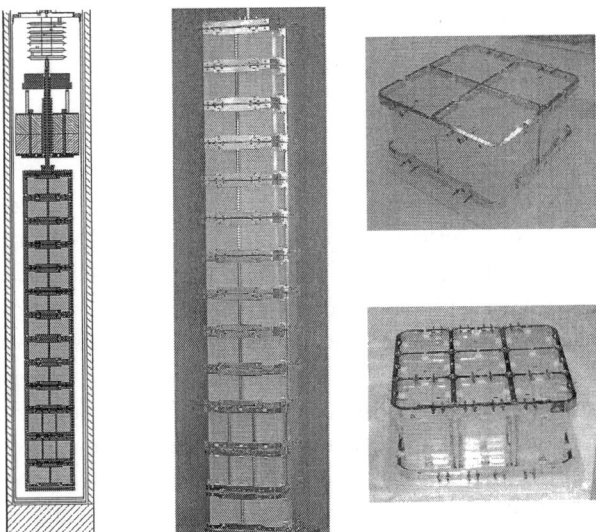

Figure 1. The CUORICINO detector: scheme of the tower and internal roman lead shields (left), the 13 planes tower (center), the 4-crystal module (top right) and the 9-crystal module (bottom right).

the wiring problem. CUORICINO was cooled again in March 2004, losing only two of the electrical contacts this time. Data taking was resumed and is progressing.

Up to June 2004, the total collected statistics are of 52785 h×crystal for the 5×5×5 cm^3 detectors and 14115 h×crystal for the 3×3×6 cm^3 ones. Preliminary analysis indicates that ∼70% of the events in the neutrinoles DBD region can be due to surface contaminations on crystals and copper detector holders. By comparing Monte Carlo simulations with measured data in CUORICINO it has been deduced that the shapes and rates of the observed degraded alpha peaks are well reproduced assuming surface contaminations on the crystals and copper holders having a density profile decaying exponentially with constants between 0.1 and 10 μm and being from 2 to 3 orders of magnitude greater than the bulk contaminations[6].

The present background in the neutrinoless DBD region is 0.17±0.04 c/keV/kg/y. Since no peak appears in the anticoincidence background spectrum at the DBD transition energy (2529 keV) a 90% C.L. lower limit on the ^{130}Te $\beta\beta(0\nu)$ half-life of 1.0×10^{24} years has been derived. The corresponding constraint on the effective neutrino mass ranges from 0.26 to 1.4 eV according to the various evaluations of nuclear matrix elements.

3. Prospects for CUORE

CUORE[1,2,3] will be a high granularity detector consisting of an array of 19 towers, containing each 13 CUORICINO-like modules (with 4 5x5x5 cm^3 TeO$_2$ crystals). The total mass will be of ∼740 kg. A new dilution refrigerator is being built to operate at ∼10 mK.

The surface contaminations on crystals and copper holders detected in CUORICINO are a serious problem for the CUORE sensitivity; different R&D works are being devoted to solve it. A direct measurement of the ^{238}U and ^{232}Th surface contaminations on copper attacking samples with acid during different times and analyzing the obtained solutions with ICPMS (Inductive Coupled Plasma Mass Spectroscopy) is being attempted. A preliminary result for a sample of the CUORICINO copper shows that the contamination quickly decreases with depth, in good agreement with our background model. If the elimination of the surface contaminations was not possible, an active rejection of surface events could be the solution. This technique has been proved using composite bolometers made covering the TeO$_2$ crystal with very thin Ge/Si detectors; pulse amplitudes from the Ge/Si thermistor are much higher for surface events allowing their identification[7].

The sensitivity of CUORE will strongly depend on the final background levels achieved. CUORE background predictions are based on Monte Carlo simulations and data from CUORICINO (see the CUORE Proposal[1]). Assuming for surface contaminations a reduction of a factor of 10 with respect to CUORICINO, the CUORE background could be of ∼0.007 c/keV/kg/y in the neutrinoless DBD region and of ∼0.05 c/keV/kg/d near the threshold.

The physics potential of CUORE has been studied in detail in Ref.[3]. Regarding the search for neutrinoless DBD of ^{130}Te, assuming five years of data, in a conservative (optimistic) situation with background b=0.01 (0.001) c/keV/kg/y and energy resolution Γ=10 (5) keV, the expected sensitivity is of 1.5 (6.5) $\times 10^{26}$ y for the half-life. The corresponding regions explored for the neutrino effective mass are 23-118 (11-57) meV, depending on the nuclear matrix elements considered.

Figure 2 shows the expected exclusion plots for the direct detection of WIMPs calculated assuming an energy threshold of 10 keV, an exposure of one year and two different background levels in the low energy region, 0.05 and 0.01 c/keV/kg/d[3]. As a reference, the limit obtained in the previous MiDBD experiment is also shown[8].

Since CUORE will be a very massive experiment, a search for the annual modulation in the WIMP signal might be possible provided a good stability was achieved. Following the method from Ref.[9], the sensitivity of CUORE has been evaluated in Fig. 3 assuming an energy threshold of 10 keV, an exposure of two years and background levels of 0.05 and 0.01 c/keV/kg/d. The more optimistic case of a 5 keV threshold with the lowest background has been also plotted. The regions of WIMPs above the curves would give a positive annual modulation signal at 90% C.L. with a 50% probability.

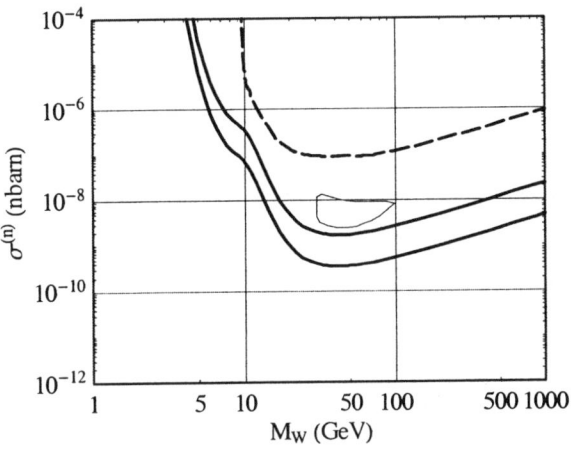

Figure 2. Expected exclusion plots for WIMP direct detection in CUORE (solid lines) and MiDBD result (dashed line). See text for details.

Acknowledgments

One of us (S. Cebrián) greatly acknowledges Spanish foundation "Fundación Areces" for the support of a fellowship. Grants from the Commission of European Communities under contract HPRN-CT-2002-00322 are also deeply acknowledged.

References

1. CUORE Proposal, January 2004,
   http://crio.mib.infn.it/wig/Cuorepage/CUORE.php.
2. C. Arnaboldi et al., Nucl. Instrum. Meth. **A518**, 775 (2004).
3. C. Arnaboldi et al., Astrop. Phys. **20**, 91 (2003).
4. C. Arnaboldi et al., Phys. Lett. **B584**, 260 (2004).

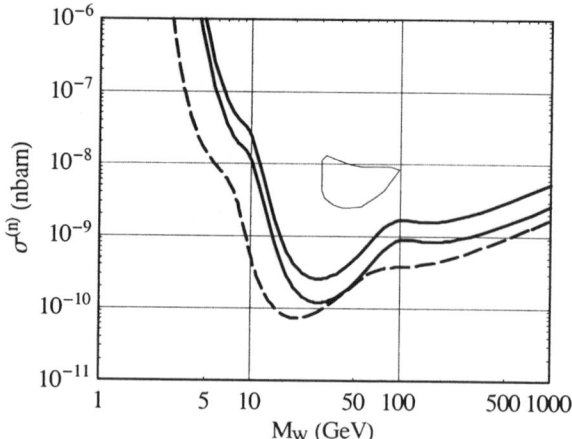

Figure 3. Sensitivity plots for the annual modulation effect in the WIMP signal for CUORE with threshold of 10 keV (thick solid lines) and 5 keV (dashed line). See text for details.

5. C. Arnaboldi et al., *Phys.Lett.* **B557**, 167 (2003).
6. S. Capelli, Poster Session at Neutrino 2004, Paris, June 2004. To appear in *Nucl. Phys. B (Proc. Suppl.)*.
7. M. Pedretti, Poster Session at Neutrino 2004, Paris, June 2004. To appear in *Nucl. Phys. B (Proc. Suppl.)*.
8. A. Giuliani et al., *Nucl. Phys. (Proc. Suppl.)* **B110**, 64 (2002).
9. S. Cebrián et al., *Astrop. Phys.* **14**, 339 (2001).

THE EXO DOUBLE BETA DECAY EXPERIMENT

J.-L. VUILLEUMIER, on behalf of the EXO collaboration
Institut de physique
A.-L. Breguet 1,
CH-2000 Neuchâtel, Switzerland
E-mail: jean-luc.vuilleumier@unine.ch

Studying double beta decay is possibly the best way to determine the absolute mass scale of neutrinos. The EXO experimental program is devoted to the search of neutrinoless double beta decay in ^{136}Xe. A sensitivity of order 0.01 eV to the effective neutrino mass is eventually aimed for. This will be achieved by combining a large source mass and state of the art background reduction. This includes an improved event signature, by identifying the daughter ion by laser tagging. A first prototype liquid xenon TPC is being designed, and will be described.

1. Double beta decay

As discussed by A. Nuccioti[1], studies of solar and atmospheric neutrinos, and also reactor neutrinos, convincingly show that neutrinos oscillate. All data together determine two mass squared differences. The oscillation experiments however cannot determine the absolute masses. Two other types of experiments provide informations on these, the study of the kinematics in weak decays, and the search for neutrinoless double beta decay.

Neutrinoless double beta decay $(\beta\beta0\nu)$ $(A,Z) \longrightarrow (A,Z+2) + e^- + e^-$ can only occur if neutrinos have Majorana masses. The electrons carry away the entire energy released in the decay. The half life $T_{1/2}^{0\nu}$ is inversely proportional to the square of the effective mass

$$|\langle m_\nu \rangle| = |\sum_{i=1}^{N} U_{ei} U_{ei} m_i|. \tag{1}$$

The effective mass, an average of the neutrino masses m_i weighed by the mixings U_{ei}, is a direct measure of the absolute mass scale. $T_{1/2}^{0\nu}$ also depends on matrix elements which must be calculated. The QRPA technique and the shell model are used. The authors of ref. [2] realized that QRPA matrix elements reproduce well the half lives of allowed two neutrino double beta decay $(\beta\beta 2\nu)$ $(A,Z) \longrightarrow (A,Z+2)+e^-+e^-+\bar{\nu}_e+\bar{\nu}_e$ in various nuclei. From that they determine more precisely the particle-particle coupling, the parameter poorest known in QRPA. With that they calculate the matrix elements for the neutrinoless mode. This procedure, called renormalized QRPA (RQRPA), may provide the safest matrix elements, which we shall use systematically in the following.

The most sensitive experiment at present is that of the Heidelberg-Moscow collaboration in the Gran Sasso lab [3], with an array of germanium crystals (11 kg total mass) enriched to 87 % in ^{76}Ge. The final data analysis provides a hint of a peak at the decay energy, at the 4.2 σ level, giving a half-life of $T^{0\nu}_{1/2} = (0.69 - 4.18) \cdot 10^{25}$ yr (95 % CL). It implies a mass $\langle m_\nu \rangle$ of order 0.37 to 0.91 eV in RQRPA. Clearly this result calls for confirmation in a different nucleus.

In addition, should the Heidelberg-Moscow signal turn out to be real, the determination of $T^{0\nu}_{1/2}$ in a second nucleus will provide a direct test of the matrix elements, more precisely their ratios, since $T^{0\nu}_{1/2}$ factorizes with the matrix elements and the effective mass.

Beyond that there is a strong motivation to push the sensitivity to double beta decay as far down as 0.01 eV. As shown by various authors (see ref. [4]), this would make it possible, assuming masses below 0.1 eV, to find out whether neutrino masses follow the normal hierarchy, or the inversed one. This requires large source mass, as well as much improved low background techniques.

2. EXO

EXO [5] will search for the decay ^{136}Xe $\longrightarrow e^- + e^- + ^{136}$Ba^{++}. Russia has impressive facilities to enrich xenon to 80 % in ^{136}Xe so that a large source mass can be contemplated. As in other experiments, great care will be taken to minimize the background from natural and cosmogenic activities. But the unique feature to defeat the background is the improved event signature. The detector will be made so as to observe not only the two double beta decay electrons, as usual, but to identify also, in coincidence, the positive ^{136}Ba^{++} ion left behind [6]. It is first brought to the singly charged state ^{136}Ba$^+$, the most favorable for laser tagging. A 493 nm laser excites the ion from the $6^2S_{1/2}$ ground state to the $6^2P_{1/2}$ state (fig. 1). From there it decays back to the ground state, or to the $5^4D_{3/2}$ metastable state. If this happens, a second 650 nm laser will bring it again to the $6^2P_{1/2}$ state. As long as the laser irradiation continues, photons are emitted, at a

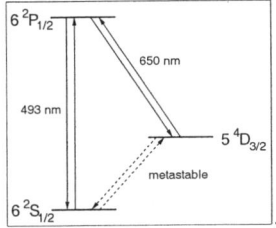 

Figure 1. Left: the level scheme of ^{136}Ba$^+$. Right: a single Ba$^+$ ion in the Stanford ion trap, under vacuum, seen by a CCD camera

rate up to 10^7 per second, which allows to identify the ion.

This extra feature should reduce all backgrounds to a negligible level, except that from $\beta\beta 2\nu$ decay when searching for $\beta\beta 0\nu$. Good energy resolution on the two electrons is required to distinguish the $\beta\beta 0\nu$ peak from the high tail of the $\beta\beta 2\nu$ energy distribution. In ^{136}Xe $\beta\beta 2\nu$ is comparatively long, and has never been observed. Calculations indicate that the half life should be around 10^{21} yr. From that we estimate that we need an energy resolution of order $\sigma(E)/E = 2\%$ at 2.48 MeV, the transition energy in ^{136}Xe.

The detector

Two options are being considered for the detector, a gas and a liquid time projection chamber (TPC). The gas TPC has the advantage that the electron tracks are long enough to be visualized, which provides an additional selection criterion. The disadvantage is that it is large, and difficult to shield against outside activities. The liquid TPC is much more compact, given the high density of liquid xenon (2.95 g/l), and much easier to shield. But the double beta decay electrons only appear as a point energy deposition. A complication is the need to work at temperatures of order 163 K. Presently EXO favors the liquid version. But R&D continues on the gas TPC, to keep the option open.

The liquid TPC

The ionization electrons produced in an event are collected by x and y wire planes at the end of the drift volume. The primary scintillation light is detected as well. There is no amplification in the liquid, which puts strong constraints on the read-out electronics. The tracks are too short to be visualized. Nevertheless a fiducial volume can be defined in x, y and also z, the drift direction, since the scintillation light provides a reference for the absolute drift time measurement. A good event appears as a single energy deposition in that volume. This way α and β activities from the detector walls can be eliminated, as well as a large fraction of Compton events, which deposit energy in more than one site.

The ion tagging must be done outside. The ion must first be grabbed with a tip at negative potential dipped into the liquid xenon after detection of the electrons, and then deposited, in the singly ionized state, in an ion trap illuminated by the 493 and 650 nm lasers. R&D is in progress in Stanford. The tagging presently works fairly well in a quadrupole RF trap, operated in vacuum, and loaded with a barium oven. The number of trapped ions can be determined from the amount of reemitted light. Figure 1 shows the light from a single Ba$^+$ ion. Work is continuing to improve the resolution, limited by Doppler broadening, simplify the operations, and above all to make the system work in a rest atmosphere of xenon.

Energy resolution is more delicate in liquid xenon. But the EXO collaboration has demonstrated that there is a clear anticorrelation between the primary scintillation light and the ionization [7]. Therefore, by measuring both, in a small test device, it was possible to significantly improve the energy resolution. A value of $\sigma(E)/E = 1.6$ % at 2.48 MeV was extrapolated. Work is continuing at Stanford/SLAC to improve both the ionization and light collection to get a better resolution.

The 200 kg prototype

While the R&D work on the ion tagging goes on, a prototype liquid TPC (fig. 2) is being built to use the 200 kg of xenon enriched to 80 % in ^{136}Xe already available. This prototype without laser tagging, will

- test all technical aspects of EXO (except Ba identification)
- measure for the first time $\beta\beta 2\nu$ in ^{136}Xe,
- have a sensitivity sufficient to explore the $\beta\beta 0\nu$ range in $|\langle m_\nu \rangle|$ suggested by the Heidelberg Moscow experiment.

The xenon of the inner TPC will be contained in a vessel made from a highly radiopure material, which can be welded without introducing any foreign material. Various teflons are being investigated. A central cathode separates two drift volumes, with x and y readout wires (0.9 mm spacing), and 16 mm diameter APD's (356) at each end. The APD response peaks

657

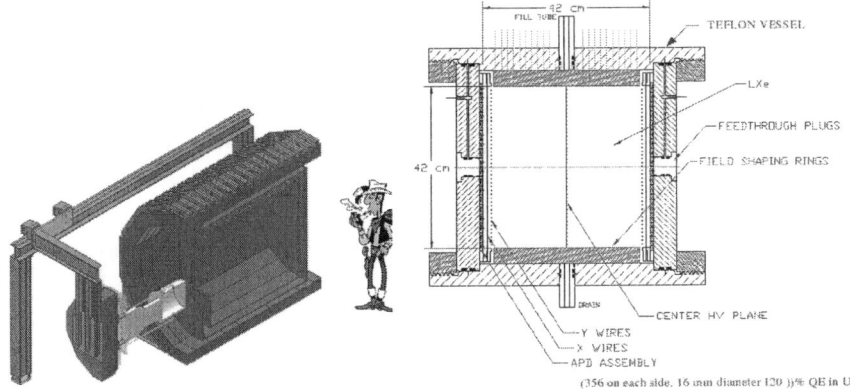

Figure 2. Left: the 200 kg prototype, with the inner liquid xenon TPC moved out for servicing, the copper cryostat and the lead shielding. Right: the inner TPC.

in the UV (120 % quantum efficiency) and is well matched to the xenon scintillation light.

The inner TPC is immersed in a high density (1.8 g/cm^3) cooling liquid acting as passive shielding, itself contained in a double wall cryostat made from high purity copper. Simulations (see below) indicate that a thickness of 50 cm of coolant is necessary between the teflon vessel and the cryostat. The copper cryostat is surrounded by 25 cm of high purity lead to shield against local activities.

A major effort is made to select clean materials for the detector components. All materials are tested. Several techniques are used: neutron activation whenever possible, γ activity, α counting, high sensitivity mass spectrometry. Finally radon outgassing is being investigated.

The measured activities are incorporated into a full simulation program, which serves to estimate the background. Conversely it can be used to set tolerances on activities in function of the mass of a component and the position in the detector.

Work is progressing fast on the construction of the detector. It will be located in a clean room, to avoid pollutions by radioactive dust. It will be assembled in Stanford, in a building offering some protection against the cosmics, and thoroughly tested, before being shipped to the final location, the WIPP (New Mexico) underground laboratory. Data taking should then last for two years to exploit the full potential.

Various parameters are summarized in table 1. The energy resolution is assumed to be the same as in the small test device. The detection efficiency, 70 %, is obtained from the simulation of the experiment. The background is dominated by activities from the detector components themselves. It was estimated from our materials radioactivity tests. The background from $\beta\beta 2\nu$ decay is negligible.

Table 1. Projected sensitivity for the EXO 200 kg prototype, and for two possible configurations of the final version (see text).

Xenon mass [kg]	$\frac{\sigma(E)}{E}$ [%]	Time [yr]	background [events]	$T^{0\nu}_{1/2}$ [yr]	$\langle m_\nu \rangle$ RQRPA [eV]
200 (proto)	1.6	2	40	$6.4 \cdot 10^{25}$	0.28
1000	1.6	5	0.5 (use 1)	$2.0 \cdot 10^{27}$	0.05
10000	1.0	10	0.7 (use 1)	$4.1 \cdot 10^{28}$	0.01

The final configuration

The exact configuration of the final detector will depend on the experience gained with the prototype, and the development of the ion tagging. If the background goals without tagging can be met in the prototype, then using the same techniques with Ba tagging in addition will indeed practically eliminate the background from activities. The background from $\beta\beta 2\nu$ decay only will be left. Estimated rates and projected sensitivities for two possible configurations are given in table 1: a 1 t device with the energy resolution achieved in the test device and operated for 5 years, and a bolder set-up with 10 t, slightly improved energy resolution, and operated for 10 years. In the latter case the sensitivity definitely covers all the allowed range in $|\langle m_\nu \rangle|$ in the case of inverse hierarchy.

References

1. A. Nuccioti, *Present and future of double beta decay*, these proceedings
2. V. A. Rodin et al., *Phys.Rev.* **C68** 044302 (2003)
3. H.V. Klapdor-Kleingrothaus et al., *Phys.Lett.* B **586**, 198 (2004) 198
4. R.D. McKeown and P. Vogel, *Phys.Rept.* **394**, 315 (2004)
5. M. Danilov et al., *Phys.Lett.* B **480** (2000)
6. M. Moe, *Phys. Rev.* **C44**, 931 (1991)
7. E. Conti et al., *Phys. Rev.* B **68**, 054201 (2003)

LIST OF PARTICIPANTS

Delegate	Institution Affiliation	e-mail
R. Ruiz De Austri	U. of Sheffield	r.ruizdeaustri@shef.ac.uk
M. Axenides	NCSR Demokritos Greece	axenides@inp.demokritos.gr
E. Baltz	U. of Stanford	eabaltz@slac.stanford.edu
M. Bauer	U. of Tuebingen	bauer@pit.physik.uni-tuebingen.de
P. Belli	U. of Rome 2 and INFN	belli@roma2.infn.it
A. Blanchard	LATT, France	ablancha@ast.obs-mip.fr
W. de Boer	U. of Karlsruhe	wim.de.boer@cern.ch
J. Bourjaily	U. of Michigan	j.bourj@umich.edu
Y. Bunkov	CRTBT-CNRS Grenoble, France	yuriy.bunkov@grenoble.cnrs.fr
B. Cabrera	U. of Stanford	cabrera@stanford.edu
J. Carmoná	Canfranc Lab.	jcarmona@unizar.es
G. Cantatore	U. of Trieste	cantatore@trieste.infn.it
M. Carson	U. of Sheffield	m.j.carson@sheffield.ac.uk
S. Cartwright	U. of Sheffield	susan.cartwright@cern.ch
S. Cebrián	U. of Milano-Bicocca	susana.cebrian@mib.infn.it
D. Cerdeño	U. of Hamburg	cerdeno@mail.desy.de
J. Collar	U. of Chicago	collar@uchicago.edu
A. Cooray	Caltech	asante@caltech.edu
C. Cozzini	U. of Oxford	c.cozzini@physics.ox.ac.uk
T. Dafni	TU Darmstadt	theopisti.dafni@cern.ch
J. Davies	U. of Sheffield	j.c.davies@sheffield.ac.uk
E. Daw	U. of Sheffield	e.daw@sheffield.ac.uk
V. Dokuchaev	INR, Moscow	dokuchaev@inr.npd.ac.ru
M. Dragowsky	CWRU, USA	dragowsky@case.edu
J. Drexler	NJ Inst. Techn.	drexlerastro@aol.com
E. Drobyshevski	Ioffe Institute, Russia	emdrob@mail.ioffe.ru
C. Duffy	U. of Sheffield	pha01cd@sheffield.ac.uk
A. Eckart	U. of Cologne	eckart@ph1.uni-koeln.de
J. Edsjö	Stockholm U.	edsjo@physto.se
K. Eitel	FZ Kernphysik	klaus.eitel@ik.fzk.de
W. Evans	IOA, Cambridge	new@ast.cam.ac.uk
N. Ferrari	Gran Sasso Lab.	nicola.ferrari@lngs.infn.it
D. Finkbeiner	U. of Princeton	dfink@astro.princeton.edu
L. Fu-Sin	U. of Libre de Bruxelles	fu-sin.ling@ulb.ac.be

R. Gaitskell	Brown University	gaitskell@brown.edu
G. Gerbier	LSM-CEA/CNRS	ggerbier@cea.fr
C. Ghag	U. of Edinburgh	c.ghag@ed.ac.uk
I. Giomataris	DAPNIA, CEA/Saclay	ioa@hep.saclay.cea.fr
F. Giuliani	U. of Lisboa	criodets@cii.fc.ul.pt
M. Gómez	U. of Huelva	mario.gomez@dfa.uhu.es
A. Goobar	U. of Stockholm	ariel@physto.se
L. Grandi	INFN Pavia	luca.grandi@pv.infn.it
A. Green	U. of Sheffield	a.m.green@sheffield.ac.uk
B. Heinemann	U. of Liverpool	beate@hep.ph.liv.ac.uk
A. Helmi	Kapteyn Astron Inst.	ahelmi@astro.rug.nl
A. Hitachi	Kochi Medical School, Japan	hitachia@med.kochi-ms.ac.jp
J. Hoeßl	U. of Erlangen	hoessl@physik.uni-erlangen.de
S. Hofmann	U. of Frankfurt	stehof@th.physik.uni-frankfurt.de
R. Hollingworth	U. of Sheffield	r.j.hollingworth@shef.ac.uk
D. Hooper	U. of Oxford	hooper@astro.ox.ac.uk
I. Irastorza	CERN	igor.irastorza@cern.ch
P. Jetzer	U. of Zurich	jetzer@physik.unizh.ch
J. Jochum	U. of Tuebingen	josef.jochum@uni-tuebingen.de
M. Kachelriess	Max-Planck Institute	michael.kachelriess@mppmu.mpg.de
D. Kinion	LLNL, U. of Berkeley	kinion1@llnl.gov
J. Kirkpatrick	U. of Sheffield	j.kirkpatrick@sheffield.ac.uk
H. Klapdor-Kleingrothaus	MPI, Heidelberg	klapdor@gustav.mpi-hd.mpg.de
I. Krivosheina	MPI, Heidelberg	irina@gustav.mpi-hd.mpg.de
V. Kudryavtsev	U. of Sheffield	v.kudryavtsev@shef.ac.uk
J. Kwak	Seoul National U.	jwkwak@hep1.snu.ac.kr
H. Lee	Seoul National U.	tgsh@hep1.snu.ac.kr
R. Lieu	U. of Alabama	lieur@cspar.uah.edu
P. Lightfoot	U. of Sheffield	p.k.lightfoot@shef.ac.uk
P. Loaiza	LSM, France	ploaiza@lsm.in2p3.fr
G. Luzón	Canfranc Lab.	luzon@unizar.es
A. Magnon	Blaise Pascal	matiman@wanadoo.fr
B. Majorovits	U. of Oxford	majorovits@physics.ox.ac.uk
M. Martínez	Canfranc Lab.	mariam@unizar.es
R. McAlpine	Electron Tubes Ltd	info@electron-tubes.co.uk
A. McConnachie	IOA, Cambridge	alan@ast.cam.ac.uk
J. McMillan	U. of Sheffield	j.e.mcmillan@shef.ac.uk

H. Menghetti	U. of Bologna and INFN	helenia.menghetti@bo.infn.it
M. Merrifield	U. of Nottingham	michael.merrifield@nottingham.ac.uk
D. Miller	U. of Glasgow	d.miller@physics.gla.ac.uk
K. Miuchi	Kyoto U.	miuchi@cr.scphys.kyoto-u.ac.jp
J. Morales	Canfranc Lab.	j.morales@unizar.es
B. Morgan	U. of Sheffield	b.morgan@sheffield.ac.uk
S. Moriyama	ICRR, U. of Tokyo	moriyama@icrr.u-tokyo.ac.jp
A. Morselli	U. of Rome 2 and INFN	aldo.morselli@roma2.infn.it
D. Muna	U. of Sheffield	d.muna@sheffield.ac.uk
A. Murphy	U. of Edinburgh	a.s.murphy@ed.ac.uk
A. Nucciotti	U. of Milano-Bicocca	angelo.nucciotti@mib.infn.it
K. Olive	U. of Minnesota	olive@physics.umn.edu
J. Peacock	Royal Observatory, U. of Edinburgh	jap@roe.ac.uk
W. Piechocki	Soltan Inst. for Nucl. Studies	piech@fuw.edu.pl
J. Quenby	Imperial College London	j.quenby@imperial.ac.uk
Y. Ramachers	U. of Warwick	yorck@ings.infn.it
W. Rau	TUM, Munich	wolfgang.rau@ph.tum.de
M. Robinson	U. of Sheffield	matthew.robinson@shef.ac.uk
M. Roos	U. of Helsinki	matts.roos@helsinki.fi
D. Rosenbaum	SMU, USA	drteplitz@aol.com
L. Roszkowski	U. of Sheffield	l.roszkowski@shef.ac.uk
K. Rottler	U. of Tuebingen	rottler@pit.physik.uni-tuebingen.de
G. Ruoso	INFN, Legnaro	ruoso@legnaro.infn.it
V. Sanglard	IPNL, France	sanglard@ipnl.in2p3.fr
D. Santos	ISN, Grenoble	daniel.santos@isn.in2p3.fr
C. Savage	U. of Michigan	cmsavage@umich.edu
R. Schild	Smithsonian Astrophysical Observatory	rschild@cfa.harvard.edu
S. Scholl	U. of Tuebingen	scholl@pit.physik.uni-tuebingen.de
H. Sekiya	U. of Kyoto	sekiya@cr.scphys.kyoto-u.ac.jp
P. Sikivie	U. of Florida	sikivie@phys.ufl.edu
D. Sinclair	Carleton U.	sinclair@physics.carleton.ca
N. Spooner	U. of Sheffield	n.spooner@sheffield.ac.uk
L. Stodolsky	Max-Planck Institute	les@mppmu.mpg.de
C. Suneel	ETH Zurich	rico.chandrasekharan@cern.ch
A. Taylor	U. of Edinburgh	ant@roe.ac.uk

A. Teixeira	U. of Madrid	teixeira@delta.ft.uam.es
V. Teplitz	NASA Goddard	vigdor.l.teplitz@nasa.gov
L. Thompson	U. of Sheffield	l.thompson@sheffield.ac.uk
C. Tomei	MPI, Heidelberg	claudia.tomei@mpi-hd.mpg.de
L. Torres	Canfranc Lab.	ltorres@unizar.es
C. Van De Bruck	U. of Sheffield	c.vandebruck@shef.ac.uk
J. Vergados	U. of Ioannina	vergados@cc.uoi.gr
J. Vuilleumier	Institut de Physique, Neuchatel	jean-luc.vuilleumier@ ⌐nine.ch
H. Wang	U. of California, LA	wangh@physics.ucla.edu
G. Weiglein	U. of Durham	georg.weiglein@durham.ac.uk
N. Weiner	U. of New York	nealw@phys.washington.edu
M. Weston	QMUL	m.j.weston@qmul.ac.uk
L. Widrow	Queen's U., Kingston	widrow@astro.queensu.ca
C. Winkelmann	CNRS-CRTBT, Grenoble	clemens.winkelmann@grenoble.cnrs.fr
H. Wulandari	Technical U. of Munich	hesti.wulandari@ph.tum.de
K. Yoshida	U. of Kanagawa	yoshida@kit.ie.kanagawa-u.ac.jp
G. Zavattini	U. of Ferrara	zavattini@fe.infn.it
A. Zentner	U. of Chicago	zentner@kicp.uchicago.edu